Satellite Altimetry
and Earth Sciences

Satellite Altimetry and Earth Sciences

A Handbook of Techniques and Applications

Edited by

Lee-Lueng Fu

Jet Propulsion Laboratory
California Institute of Technology
Pasadena, California

Anny Cazenave

Laboratoire d'Etudes en Geophysique et Oceanographie Spatiales
Centre National d'Etudes Spatiales
Toulouse, France

ACADEMIC PRESS

A Harcourt Science and Technology Company

San Diego   San Francisco   New York   Boston   London   Sydney   Tokyo

Copyright © 2001 by ACADEMIC PRESS

All Rights Reserved.
No part of this publication may be reproduced or transmitted in any form or by any means, electronic or mechanical, including photocopy, recording, or any information storage and retrieval system, without permission in writing from the publisher.

Requests for permission to make copies of any part of the work should be mailed to: Permissions Department, Harcourt, Inc., 6277 Sea Harbor Drive, Orlando, Florida 32887-6777

Explicit permission from Academic Press is not required to reproduce a maximum of two figures or tables from an Academic Press chapter in another scientific or research publication provided that the material has not been credited to another source and that full credit to the Academic Press chapter is given.

Academic Press
A Harcourt Science and Technology Company
525 B Street, Suite 1900, San Diego, California 92101-4495, USA
http://www.academicpress.com

Academic Press
Harcourt Place, 32 Jamestown Road, London NW1 7BY, UK
http://www.academicpress.com

Library of Congress Catalog Card Number: 00-102572

International Standard Book Number: 0-12-269545-3

PRINTED IN THE UNITED STATES OF AMERICA
00 01 02 03 04 05   EB   9   8   7   6   5   4   3   2   1

Contents

CHAPTER

7

Ocean Surface Waves

J.-M. LEFÈVRE AND P. D. COTTON

CHAPTER

8

Sea Level Change

R. S. NEREM AND G. T. MITCHUM

CHAPTER

9

Ice Sheet Dynamics and Mass Balance

H. JAY ZWALLY AND ANITA C. BRENNER

Contributors

The numbers in parentheses indicate the pages on which the authors' contributions begin.

Antonio J. Busalacchi (217) Laboratory for Hydrospheric Processes, NASA/Goddard Space Flight Center, Greenbelt, Maryland 20771.

Anita C. Brenner (351) Raytheon Technical Services Company at NASA Goddard Space Flight Center, Oceans and Ice Branch, Greenbelt, Maryland 20771.

Philip S. Callahan (1) Jet Propulsion Laboratory, California Institute of Technology, Pasadena, California 91109.

Anny Cazenave (407) Laboratoire d'Etudes en Geophysique et Oceanographie Spatiales, Centre National d'Etudes Spatiales, 31400 Toulouse, France.

Dudley B. Chelton (1, 133) College of Oceanic and Atmospheric Sciences, Oregon State University, Corvallis, Oregon 97331.

P. David Cotton (305) Satellite Observing Systems, Godalming, Surrey GU7 1EL, United Kingdom.

Lee-Lueng Fu (1, 133) Jet Propulsion Laboratory, California Institute of Technology, Pasadena, California 91109.

Ichiro Fukumori (237) Jet Propulsion Laboratory, California Institute of Technology, Pasadena, California 91109.

Bruce J. Haines (1) Jet Propulsion Laboratory, California Institute of Technology, Pasadena, California 91109.

Myung-Chan Kim (371) Center for Space Research, The University of Texas at Austin, Austin, Texas 78759.

Jean-Michel Lefèvre (305) Météo–France, Coriolis, 31057 Toulouse, France.

Christian Le Provost (267) Laboratoire d'Etudes en Geophysique et Oceanographie Spatiales, Groupe de Recherche en Geodesie Spatiale, UMR CNES/CNRS/UPS, 31401 Toulouse, France.

Pierre-Yves Le Traon (171) CLS Space Oceanography Division, 31526 Ramonville St. Agne, France.

Gary T. Mitchum (329) Department of Marine Science, University of South Florida, St. Petersburg, Florida 33701.

Rosemary Morrow (171) Laboratoire d'Etudes en Geophysique et Oceanographie Spatiales, 31055 Toulouse, France.

R. Steve Nerem (329) Center for Space Research, The University of Texas at Austin, Austin, Texas 78759.

Joel Picaut (217) Institute de Recherche pour le Developpement, LEGOS/GRGS, Toulouse, France and Laboratory for Hydrospheric Processes, NASA Goddard Space Flight Center, Greenbelt, Maryland 20771.

John C. Ries (1) Center for Space Research, The University of Texas at Austin, Austin, Texas 78712.

Jean-Yves Royer (407) UBO–IUEM Domaines Océaniques, 29280 Plouzane, France.

David T. Sandwell (441) University of California, San Diego, Scripps Institution of Oceanography, La Jolla, California 92093.

Walter H. F. Smith (441) NOAA Laboratory for Satellite Altimetry, Silver Spring, Maryland 20910.

Byron D. Tapley (371) Center for Space Research, The University of Texas at Austin, Austin, Texas 78759.

H. Jay Zwally (351) Oceans and Ice Branch, NASA Goddard Space Flight Center, Greenbelt, Maryland 20771.

Preface

Satellite altimetry was developed in the 1960s soon after the flight of artificial satellites became a reality. From the vantage point in space, a radar altimeter is able to measure the shape of the sea surface globally and frequently. Such measurements have a wide range of applications to oceanography, geodesy, and geophysics. The results are often revolutionary. For example, in oceanography; it takes a ship weeks or months to cross the ocean making measurements while the ocean is constantly changing its circulation, temperature, and salinity. Therefore, it is unfeasible to make synoptic observations of the global ocean using *in situ* instrumentation. The advent of satellite altimetry has given oceanographers a unique tool for mapping the global ocean topography for studying the ocean circulation and its changes with time.

The permanent undulations (ranging from a few meters to more than 100 m) of the sea surface reflect the geoid determined by the geographical variations of the earth's gravity field. These variations are linked to the heterogeneous mass distribution inside or at the surface of the planet. Altimetric measurements of the shape of the sea surface are thus useful to the study of marine geodesy and geophysics. Deviations (on the order of 1 m) of the sea surface from the geoid are associated with the ocean topography, which is a measure of the dynamic pressure at the sea surface relating to the current velocity of the ocean's surface. To be useful for the study of ocean circulation, ocean topography must be measured with an accuracy of a few centimeters. Such a requirement presents the most demanding challenge to satellite altimetry as a remote-sensing tool.

Technological and scientific developments have gone a long way towards meeting the stringent measurement requirement since the first flight of a satellite altimeter in the early 1970s. Owing to two decades of effort involving close international collaborations among radar engineers, geodsists, geophysicists, oceanographers, as well as program managers, satellite altimetry has benefited from a series of missions, leading to an improvement in measurement accuracy by three orders of magnitudes, from tens of meters to a few centimeters. The evolution from Seasat (1978), Geosat (1985–1989), ERS (1991 to the present), to TOPEX/POSEIDON (1992 to the present) has created a wealth of data of progressively improved quality. A host of advances in various branches of the earth sciences have taken place as a result of the data collected.

This book provides a comprehensive description of the techniques of satellite altimetry and a summary of the scientific applications to a variety of topics in the earth sciences. The style of each chapter allows the reader a broad exposure to the subject from the basics to the state of the art. This book can be considered a handbook for students and researchers who are interested in using altimeter data for the study of the earth as well as a reference for a graduate-level course on satellite remote sensing.

Chapter 1 is a treatise of the techniques of satellite altimetry. It provides comprehensive descriptions of a wide range of aspects of the techniques: the fundamentals of radar measurement and precision orbit determination; effects of the atmosphere (including the ionosphere) on the propagation of radar signals and the methodology of the corrections for these effects; various geophysical effects on the sea surface height measurements and correction algorithms; radar measurement of wind speed and wave height; and mission design and evaluation.

Chapters 2 through 9 address the applications to the study of the hydrosphere—including the oceans and ice. Chapters 2 to 4 discuss ocean circulation, covering large-scale circulation, mesoscale eddies and boundary currents, and tropical circulation. Specific topics include: the determination of the ocean general circulation in relation to geoid errors; large-scale variability on time scales from intraseasonal to interannual in relation to forcing mechanisms; El Niño and La Niña; planetary wave dynamics; eddy dynamics; evaluations of numerical models; and descriptions of the world's major ocean currents. Satellite altimetry provides the first global synoptic datasets for the study of these topics. Chapter 5 provides a discussion of the assimilation of altimeter data by ocean circulation models. The development of oceanographic data assimilation is primarily motivated by the advent of satellite altimeter data. Constrained by the global data, ocean models allow the estimation of the ocean's subsurface fields (temperature, salinity, and current velocity) from altimeter data.

Chapter 6 presents results on the ocean tides, including numerical models, tidal dissipation, and internal tides. Satellite altimetry has provided a new look on the old topic of tides, leading to the best tidal charts of the world's oceans and a new understanding of the tidal energetics. Chapter 7 deals with ocean surface waves: their global distributions and low-frequency variabilities as well as numerical models of ocean waves. Sea level changes on a global scale are addressed in Chapter 8. The state of the art performance of satellite altimetry has provided a new approach to the determination of global mean sea level change. Chapter 9 provides an overview of the use of satellite altimetry over Greenland and Antarctica to study the dynamics of ice sheets and their mass balance.

Chapters 10 to 12 address the applications for the study of the solid earth. Chapter 10 talks about marine geodesy, concentrating on the shape of the mean sea surface, the earth's gravity field and the geoid, and the time-varying geocenter and the earth's rotational parameters. Chapter 11 examines marine geophysics focusing on the small-scale features of the mean sea surface and their relations to geophysical processes: isostatic compensation, thermal and mechanical properties of the oceanic plates, hotspot swells, and mapping of the seafloor tectonic fabric. Chapter 12 presents the detection of bathymetric features from the gravity anomalies determined from altimetry. Many previously unknown features of the seafloor topography have been discovered by the study, leading to a new bathymetric chart of the world's oceans.

Satellite altimetry has proven a powerful new tool for the study of the earth, but the useful data record is short when compared to the slow changes of the oceans, the ice sheets, and the solid earth. Continuation and maintenance of long-term data records from satellite altimetry for the benefit of society and future generations is a pressing issue facing the international scientific community as well as governmental organizations. This book, by introducing the techniques and the wide range of applications of satellite altimetry, provides a basis for the need of establishing a long-term data record for the study and monitoring of the various key processes of the dynamic earth.

A volume of such a diverse set of topics cannot be accomplished without the effort of a group of very dedicated people. We would like to thank all the authors who made contributions to this book. They have not only made a great deal of effort in writing and revising as required by the peer-review process, some of them also suffered from lengthy delays in the publication of the book resulting from the varying degree of effort involved in the various chapters. We are also indebted to the outstanding staff at Academic Press, especially Frank Cynar and Paul Gottehrer, who exhibited a great deal of professionalism and patience in dealing with the numerous rounds of revisions and schedule delays. Another important group of people who have made tremendous efforts in elevating the quality of the book are the reviewers. Due to the agreement of anonymity with most of them, we cannot reveal their names, but we would like to express our appreciation to them for their efforts. The effort of producing the book was sponsored by the National Aeronautics and Space Administration of the United States and the Centre National d'Etudes Spatiales of France.

Lee-Lueng Fu
Pasadena, California
Anny Cazenave
Toulouse, France

1

Satellite Altimetry

DUDLEY B. CHELTON,* JOHN C. RIES,† BRUCE J. HAINES,‡ LEE-LUENG FU‡
and PHILIP S. CALLAHAN‡

*College of Oceanic and Atmospheric Sciences
Oregon State University
Corvallis, Oregon
†Center for Space Research
University of Texas
Austin, Texas
‡Jet Propulsion Laboratory
California Institute of Technology
Pasadena, California

1. INTRODUCTION

At distances of more than 100 km or so from the equator, ocean currents on time scales longer than a few days and space scales longer than a few tens of kilometers are very nearly in geostrophic balance. The current velocity can therefore be accurately calculated from the pressure gradient on a gravitational equipotential surface. In accord with the hydrostatic approximation, the pressure gradient at the sea surface is proportional to the horizontal slope of the sea surface on these time and space scales. Satellite altimeter observations of sea level, coupled with knowledge of the marine geoid (the gravitational equipotential closest to the time-averaged sea-surface height), therefore provide global information on the ocean-surface velocity. Moreover, because surface currents are generally coupled to subsurface variability through relatively simple vertical modal structures, sea-level variations contain information about fluid motion deep in the ocean interior. In combination with vertical profiles of ocean density and current velocity and dynamical modeling with appropriate data assimilation schemes, the global coverage and frequent sampling of sea-level variability by satellite altimetry is significantly advancing the understanding of large- and mesoscale ocean circulation throughout the water column.

The basic concept of satellite altimetry is deceptively straightforward. The principal objective is to measure the range R from the satellite to the sea surface (see Figure 1). The altimeter transmits a short pulse of microwave radiation with known power toward the sea surface. The pulse interacts with the rough sea surface and part of the incident radiation reflects back to the altimeter. The techniques for radar determination of the time t for the pulse to travel round trip between the satellite and the sea surface are described in Section 2.4. The range R from the satellite to mean sea level is estimated from the round-trip travel time by

$$R = \hat{R} - \sum_j \Delta R_j, \qquad (1)$$

where $\hat{R} = ct/2$ is the range computed neglecting refraction based on the free-space speed of light c and ΔR_j, $j = 1, \ldots$ are corrections for the various components of atmospheric refraction and for biases between the mean electromagnetic scattering surface and mean sea level at the air-sea interface. As described in Section 3, all of the corrections ΔR_j are positive quantities that would lead to overestimates of the range if not accounted for in the processing of altimeter data.[1]

---

[1]The convention adopted here differs from that used, for example, in the *TOPEX/POSEIDON Geophysical Data Record (GDR) Users Handbook* (Callahan, 1993) in which the corrections are defined to be negative quantities to be added to $\hat{R}$. For convenience of discussion, the corrections ΔR_j are defined here to be positive quantities to be subtracted from $\hat{R}$ to reduce the altimeter overestimate of the range from the satellite to mean sea level.

TABLE 1.   Summary of Past and Present Satellite Altimeter Measurement Precisions
and Orbit Accuracies

Satellite	Mission period	Measurement precision (cm)	Orbit accuracy (cm)
GEOS-3	April 1975–December 1978	25	~500
Seasat[a]	July 1978–October 1978	5	~100
Geosat[a]	March 1985–December 1989	4	30–50
ERS-1[b]	July 1991–May 1996	3	8–15
TOPEX/POSEIDON	October 1992–present	2	2–3
ERS-2[c]	August 1995–present	3	7–8

[a]The orbit accuracies listed for Seasat and Geosat refer to the orbits used to produce the sea surface height fields analyzed in most of the published studies from these altimeter datasets. Subsequent improvements in precision orbit determination applied retroactively have reduced the orbit errors to about 20 cm for Seasat and 10–20 cm for Geosat.

[b]An orbit accuracy of about 5 cm has recently been obtained for ERS-1.

[c]An orbit accuracy of better than 5 cm has recently been obtained for ERS-2 with PRARE tracking (see Section 4.2.4).

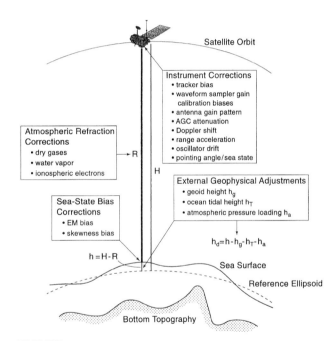

FIGURE 1 A schematic summary of the corrections that must be applied to the altimeter range measurement R and the relations between R, the orbit height H and the height h of the sea surface relative to an ellipsoidal approximation of the equipotential of the sea surface from the combined effects of the earth's gravity and centrifugal forces (the geoid). The dynamic sea surface elevation h_d that is of interest for ocean circulation studies is obtained from h by subtracting the height h_g of geoid undulations relative to the reference ellipsoid and the height variations h_T and h_a from tides and atmospheric pressure loading, respectively.

The range estimate Eq. (1) varies along the satellite orbit from along-track variations of both the sea-surface topography and the orbit height relative to the center of the earth. To be useful for oceanography, the range estimate must be transformed to a fixed coordinate system. As described in Section 4, this is achieved by precision orbit determination of the height H of the satellite relative to a specified reference ellipsoid approximation of the geoid.[2] The range measurement is then converted to the height h of the sea surface relative to the reference ellipsoid by

$$h = H - R$$
$$= H - \hat{R} + \sum_j \Delta R_j. \qquad (2)$$

Note that the signs of the corrections are reversed in the computation of the sea-surface height from the measurements of range and orbit height.

State-of-the-art satellite altimetry is based upon more than 2 decades of heritage (see Table 1). As documented in previous reviews of satellite altimetry by Brown and Cheney (1983), Fu (1983), Cheney *et al.* (1984; 1986), Douglas *et al.* (1987), Boissier *et al.* (1989), Fu *et al.* (1990), Fu and Cheney (1995), and Wunsch and Stammer (1998), progressive technological developments have improved the point-to-point measurement precision in 1-sec averages of range estimates R from 25 cm for GEOS-3 launched by the National Aeronautics and Space Administration (NASA) in April 1975 (Stanley, 1979) to 1.7 cm for the NASA dual-frequency altimeter onboard the TOPEX/POSEIDON dedicated altimetric satellite launched jointly by NASA and the French space agency Centre National d'Etudes Spatiales (CNES) in October 1992 (Fu *et al.*, 1994). The decorrelation time scale of the measurement noise of the dual-frequency altimeter is a few seconds (Rodríguez and Martin, 1994a). Since the TOPEX/POSEIDON orbital velocity projected onto the sea surface is about 6 km sec^{-1}, the

---

[2]The orbit height H is referred to as the orbit altitude in the *TOPEX/POSEIDON GDR Users Handbook.*

measurement imprecision can be reduced to less than 1 cm by averaging the measurements over along-track distances of $\sim$100 km. Other sources of measurement error such as corrections for atmospheric refraction and sea-state biases in the range estimates are of much greater concern since they typically have length scales on the order of 100 km or longer. Through a concerted effort devoted to algorithm improvement, the root-sum-of-squares of these various measurement errors has been reduced to about 2.7 cm (see Table 11 in Section 8.3). Equally important, there has been a parallel improvement in precision orbit determination that has reduced root-mean-square errors of estimated orbit height H from $\sim$10 m for GEOS-3 to 2.5 cm for TOPEX/POSEIDON (see Table 1). The overall root-sum-of-squares measurement accuracy for the TOPEX/POSEIDON dual-frequency altimeter estimates of sea-surface height is therefore about 4 cm (see Section 8.3).

One of the purposes of this chapter is to pay tribute to the success of the intensive effort that has been devoted to the technical details of altimetry over the past 2 decades. The present $\sim$4 cm state-of-the-art overall accuracy of the sea-surface height estimates h has been achieved through major technological advancements in precision orbit determination and a dedicated effort to improve each of more than 40 sensor and geophysical algorithms. This attention to algorithm improvements has transformed altimetry from a semiquantitative measurement of the sea-surface height for which the distinction between measurement errors and geophysical signals was sometimes difficult to discern, to a highly quantitative measure of sea-surface height variability that is providing insight into the wide range of dynamical processes summarized in later chapters. Moreover, a major benefit of the high degree of accuracy that has been achieved with the TOPEX/POSEIDON dual-frequency altimeter is that it is no longer essential for users to be deeply versed in all of the idiosyncrasies of satellite altimetry. Altimetry has thus become a standard tool for oceanographic research.

The interest in ever more subtle signals with smaller-amplitude sea-surface height signatures in the altimeter data demands a thorough understanding of the errors of each of the corrections applied to altimeter data. As various algorithms have improved, errors in other algorithms that were once considered of secondary concern have become the focus of attention. Each of the corrections that must be applied for accurate determination of the sea-surface height by satellite altimetry is described from basic principles in this chapter. Although full wavenumber-frequency descriptions of the various measurement errors are not yet available, the magnitudes of the errors of each of the corrections can be estimated as described herein. If not taken into consideration, some of these errors might be mistakenly interpreted as variations of the sea surface height.

As summarized above and shown schematically in Figure 1, accurate determination of the orbit height H is an integral part of altimetric determination of the sea-surface height h. Indeed, the improvement in orbit accuracy by more than two orders of magnitude over the past decade is the primary reason for the improved accuracy and utility of altimeter data. A detailed discussion of the evolution of precision orbit determination is therefore included as Section 4 in this chapter.

Accurate estimates of R and H are not sufficient for oceanographic applications of altimeter range measurements. The sea-surface height given by Eq. (2) relative to the reference ellipsoid is the superposition of a number of geophysical effects. In addition to the dynamic effects of geostrophic ocean currents that are of primary interest for oceanographic applications (see Chapters 2, 3, and 4), h is affected by undulations of the geoid h_g about the ellipsoidal approximation (Chapter 10), tidal height variations h_T (Chapter 6), and the ocean surface response h_a to atmospheric pressure loading (Chapter 2). These effects on the sea-surface height must be modeled and removed from h in order to investigate the effects of geostrophic currents on the sea surface height field. The geoid undulations, tidal height and atmospheric pressure loading contributions to h are briefly described here in Section 5; more detailed discussions are given in later chapters. The dynamic sea-surface height is thus estimated as

$$
\begin{aligned}
h_d &= h - h_g - h_T - h_a \\
&= H - R + \sum_j \Delta R_j - h_g - h_T - h_a. \quad (3)
\end{aligned}
$$

While complicating altimetric estimation of the range R, the alteration of the incident radar pulse by the rough sea surface can be utilized to extract other geophysical information from the radar returns. In particular, the significant wave height[3] can be estimated from the shape of the returned signal as described in Sections 2.4.2 and 6. In addition, since the sea-surface roughness is highly correlated with near-surface winds, the wind speed can be estimated from the power of the returned signal as described in Section 7.

The emphasis throughout this chapter is on the correction algorithms applied to the dual-frequency altimeter onboard the TOPEX/POSEIDON satellite (referred to hereafter as T/P). This state-of-the-art altimeter sets the standard for future altimeter missions since it is significantly more accurate than any of the other altimeters that have been launched to date. After discussions of the various elements of satellite estimation of the sea surface height as outlined above, the chapter concludes with a summary of the T/P mission de-

---

[3]As discussed in Sections 2.4.1 and 6, the significant wave height is a traditional characterization of the wave height field that is approximately equal to four times the standard deviation of the wave heights within the altimeter footprint.

sign and an assessment of the performance of the T/P dual-frequency altimeter in Section 8. An overview of future altimeter missions is given in Section 9.

2. RADAR MEASUREMENT PRINCIPLES

Because of a fortuitous combination of physical properties of the atmosphere and the sea surface, the frequencies most well suited to satellite altimetry fall within the microwave frequency range of 2–18 GHz. According to the frequency band allocations defined in Section 1–3.2 of Ulaby *et al.* (1981), this encompasses S-band (1.55–4.20 GHz), C-band (4.20–5.75 GHz), X-band (5.75–10.9 GHz), and K_u-band (10.9–22.0 GHz) radar frequencies. Graybody emission of electromagnetic radiation from the sea surface is very weak and the reflectivity of water is high in this frequency band, thus allowing easy distinction between radar return and natural emission. At frequencies higher than 18 GHz, atmospheric attenuation rapidly increases, thus decreasing the power of the transmitted signal that reaches the sea surface and the reflected signal that is received by the altimeter. At lower frequencies, Faraday rotation and refraction of electromagnetic radiation by the ionosphere increase and interference increases from ground-based civilian and military sources of electromagnetic radiation related to communications, navigation, and radar. In addition, practical design constraints on the size of spaceborne antennas set a lower limit on the frequencies useful for satellite altimetry. The antenna footprint size on the sea surface is proportional to the wavelength of the electromagnetic radiation and inversely proportional to the antenna size. A beam-limited footprint diameter of 5 km, for example, would require an impractically large antenna diameter of about 7.7 m for the K_u-band microwave frequency of 13.6 GHz at the T/P orbit height of 1336 km [see Eq. (22) in Section 2.4]. Spatial resolution of altimetric measurements from smaller antennas is enhanced by a technique known as pulse compression. These and other principles of radar altimetry are summarized in this section.

2.1. Normalized Radar Cross Section

The relationship between the power of the signal transmitted by a radar altimeter and the backscattered power that is received by the altimeter is fundamentally important to altimetry. The electromagnetic radiation that is transmitted toward the sea surface is attenuated by the intervening atmosphere. The signal that reaches the sea surface is partly absorbed by seawater and partly scattered from the rough sea surface over a wide range of directions. The power that is reflected back to the altimeter is then attenuated by the intervening atmosphere. The power of the returned signal

measured by the radar thus depends upon the scattering character of the sea surface, the parameters of the radar system, and two-way attenuation by the intervening atmosphere.

The physical factors that affect the signal received by the altimeter can be understood by considering a radar antenna illuminating a footprint area A_f on the sea surface from a height R above the sea surface at a pointing angle θ with corresponding slant range R_θ, as shown in Figure 2. Because of the curvature of the sea surface owing to the ellipsoidal shape of the earth and the presence of local geoid undulations, there is a distinction between the pointing angle and the antenna incidence angle (see Figure 2). The antenna pointing angle θ is the angle between the boresight of the antenna (i.e., the peak of the antenna gain) and the line from the satellite center of mass to the point on the earth's surface that is vertically beneath the satellite (referred to as the nadir point). The incidence angle θ' is the angle between the line of propagation and the normal to the sea surface at the point of incidence. For the small incidence angles less than $\sim 1°$ relevant to altimetry and the small geoid slopes that only very rarely exceed 10^{-4} (Brenner *et al.*, 1990; Minster *et al.*, 1993), the two angles are very nearly identical. Likewise, R and R_θ are very nearly identical for altimetry.

The parameters of the radar system are the wavelength λ of the transmitted and received electromagnetic radiation, the transmitted power P_t in watts, and the transmitting and receiving antenna gains G_t and G_r. Since altimeter systems use the same antenna to transmit and receive the radar pulses, G_t and G_r are essentially the same and both can therefore be denoted simply by G. The backscattered power from a

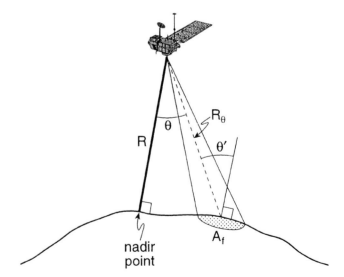

FIGURE 2 Schematic representation of satellite measurements of the radar return from the approximately ellipsoidal sea surface with large-scale geoid undulations. The angles θ and θ' are the antenna pointing angle and incidence angle, respectively; R is the satellite altitude above the nadir point; R_θ is the slant range of the radar measurement at pointing angle θ; and A_f is the antenna footprint area.

differential area dA within the antenna footprint on the sea surface is related to λ, P_t and the range R from the satellite to the target area by

$$dP_r = t_\lambda^2 \frac{G^2\lambda^2 P_t}{(4\pi)^3 R^4}\sigma, \qquad (4)$$

where the proportionality factor σ has units of area and $t_\lambda(R,\theta)$ is the atmospheric transmittance, defined to be the fraction of electromagnetic radiation at wavelength λ that is transmitted through the atmosphere from an altitude R at off-nadir angle θ. The two-way transmittance is $t_\lambda^2(R,\theta)$. A detailed derivation of Eq. (4) can be found in Section 7.1 of Ulaby *et al.* (1986a). This so-called radar equation is equivalent to characterizing the backscattered power from the location dA as having been isotropically scattered by a fictitious sphere with cross-sectional area σ. The proportionality factor σ is thus referred to as the radar cross section of the differential target area dA.

For oceanographic applications, the total returned power received by the radar antenna is backscattered from a distributed target (the rough sea surface) over the area A_f that is illuminated by the antenna, rather than from a point-source target within the antenna footprint. In this case, dP_r and σ are meaningful only when they are considered to be the average quantities over all of the differential target areas dA within the field of view. The scattering properties of the sea surface within the antenna footprint are characterized by the differential radar cross section per unit area, which is denoted as σ_0 and is related to the radar cross section by $\sigma = \sigma_0 dA$. In general, the dimensionless quantity σ_0 (referred to as the normalized radar cross section and expressed in decibels) varies spatially over the antenna footprint. The total returned power is obtained by integrating over the returns from each differential area illuminated by the antenna,

$$P_r = t_\lambda^2 \frac{\lambda^2 P_t}{(4\pi)^3 R^4}\int_{A_f} G^2\sigma_0\, dA. \qquad (5)$$

This expression assumes that the antenna footprint area is small enough that R and t_λ can be considered constant over the footprint. The antenna gain is retained inside of the integral since G depends on off-axis angle and, to a lesser extent, azimuthal angle and therefore varies across the antenna footprint area.

With the exception of σ_0, each quantity on the right side of Eq. (5) is either a known parameter of the radar system (G, λ, and P_t) or a physical parameter of the medium between the altimeter and the target (R and t_λ). If the normalized radar cross section σ_0 is spatially homogeneous over the antenna footprint or is considered to be the average differential cross section per unit area over the antenna footprint, then it can be passed through the integral on the right side of Eq. (5). Then Eq. (5) can be rearranged to express σ_0 in terms of the measured returned power P_r and the parameters

of the radar system. The radar equation can then be written in the more useful form

$$\sigma_0 = \frac{(4\pi)^3 R^4}{t_\lambda^2 G_0^2\lambda^2 A_{\text{eff}} P_t}P_r, \qquad (6)$$

where G_0 is the boresight antenna gain and A_{eff} is the "effective footprint area" defined to be the area integral of $G^2(\theta)/G_0^2$ over the angles θ subtended by the antenna beamwidth. In Section 2.4.2, a distinction is made between the footprint area within the beamwidth of the antenna and the area within this footprint that is actually illuminated by the radar.

Since all of the quantities in the multiplicative factor on the right side of Eq. (6) are known parameters of the radar system or can be determined from the measurement geometry, the returned power P_r and hence σ_0 depend only on the radar scattering characteristics (the "roughness") of the target area. The sea-surface roughness increases with increasing wind speed. At the small incidence angles relevant to satellite altimetry, P_r and therefore σ_0 decrease monotonically with increasing wind speed. Near-surface wind speed can therefore be inferred from altimetric measurements of the radar return. Expressed in decibels, a typical altimetric measurement of σ_0 is about 11 dB. It is shown in Section 7 that σ_0 decreases from 20 dB at very low wind speed to about 5 dB at a wind speed of 30 m sec^{-1} (see Figure 56).

An important property of σ_0 can be deduced from Eq. (6). If the scattering characteristics of the sea surface are statistically homogeneous over the antenna footprint, then σ_0 is insensitive to the distance between the radar antenna and the sea surface. This is easily seen by noting that the radiant flux density of electromagnetic radiation emanating from a point source falls off as the square of the distance from the source. The transmitted power that reaches the sea surface therefore decreases as R^{-2}. Likewise, the backscattered power that reaches the satellite also falls off as R^{-2}. The power P_r received by the radar therefore decreases as R^{-4} from spreading loss, in accord with Eq. (5). The resulting R^{-4} dependence of P_r is exactly offset by the R^{-4} dependence in the numerator of Eq. (6). Thus, assuming that proper corrections for atmospheric transmittance $t_\lambda(R,\theta)$ are applied, measurements of σ_0 are independent of the altitude of the measurement. Aircraft measurements of σ_0 are therefore the same as those made from a satellite at any altitude,[4] as long as the scattering properties of the sea surface are equivalent over the footprints of the different radars.

[4] As noted by Rodríguez (1988) and Chelton *et al.* (1989), σ_0 measured by a satellite altimeter must be adjusted to account for the approximate spherical shape of the earth. Depending on the orbit height R, this results in a 10–20% reduction of σ_0 compared with the flat-earth approximation (see Section 2.4.1).

2.2. Ocean Surface Reflectivity

Thorough discussions of the theory of radar backscatter from the sea surface can be found in Ulaby *et al.* (1986a, 1986b). A brief overview is given here. The phenomena responsible for radar return from a rough sea surface are reasonably well understood theoretically. The solution for the radar backscatter can be obtained based on the principles of physical optics, for which the scattered electromagnetic fields are determined from the electromagnetic current distribution induced on the sea surface by the incident radar signal. Alternatively, the solution can be obtained based on geometrical optics, for which the wavelength of the radar signal is considered to vanish and the electromagnetic radiation is treated as a bundle of rays. Both approaches lead to essentially the same result. At the small incidence angles relevant to altimetry, Barrick (1968) and Barrick and Peake (1968) showed that the radar return measured at the satellite consists of the total specular return from all of the mirror-like facets oriented perpendicular to the incident radiation within the antenna footprint. The normalized radar cross section for small pointing angle θ is given by

$$\sigma_0(\theta \approx 0) = \pi \rho^2(0°)\sec^4\theta p(\zeta_x, \zeta_y)\big|_{\zeta_x=\zeta_y=0}, \qquad (7)$$

where ζ is the sea-surface height relative to mean sea level, $\rho(0°)$ is the Fresnel reflectivity for normal incidence angle (defined to be the fraction of incident radiation that is reflected) and $p(\zeta_x, \zeta_y)$ is the joint probability density function of long-wave sea surface slopes ζ_x and ζ_y in two orthogonal directions. Here "long wave" means the portion of the ocean wave spectrum with wavelengths longer than that of the radar signal (approximately 2.2 cm for the 13.6 GHz primary frequency of the T/P dual-frequency altimeter). The joint probability density function $p(\zeta_x, \zeta_y)$ in Eq. (7) is evaluated at $\zeta_x = \zeta_y = 0$ for the specular scatterers.

The radar return at small incidence angles thus depends explicitly on the spectral characteristics of long waves on the sea surface and on the reflectivity of sea water. It also depends on the radar frequency through the implicit dependence of $\rho(0°)$ on frequency. The reflectivity for normal incidence angle is shown in Figure 3 as a function of frequency for fresh water. (The presence of dissolved salts introduces minor changes to the reflectivity only at frequencies lower than about 1 GHz.) It is apparent that water is about an order of magnitude more reflective at microwave frequencies than at visible or infrared frequencies. For a given antenna beamwidth, the much lower power requirement for a radar transmitter at microwave frequencies than at infrared or visible frequencies is a distinct advantage for radar measurements at small incidence angles.

To a first degree of approximation, the long-wave sea-surface height distribution can be modeled as isotropic with a Gaussian probability density function. Then Eq. (7) becomes

$$\sigma_0(\theta \approx 0) = \frac{\rho^2(0°)}{2s^2}\sec^4\theta \exp\left(-\frac{\tan^2\theta}{2s^2}\right), \qquad (8)$$

where s^2 is the mean square of the sea surface slopes. The radar return at small incidence angles thus decreases with increasing roughness of the sea surface (i.e., increasing s^2). Physically, a greater fraction of the incident radiation reflects specularly in directions away from the radar receiver. Since the sea surface roughness increases with increasing wind speed, σ_0 at small incidence angles is inversely related to wind speed. For a specified spectral density of the sea-surface elevation, σ_0 can be computed accurately from theory. What is required in order to determine wind speed from measurements of σ_0 is a theoretical relation between the wind field and the ocean wave spectrum.

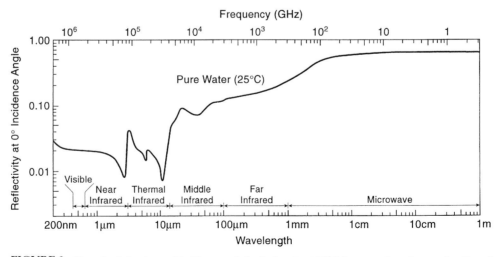

FIGURE 3   The reflectivity at normal incidence angle for fresh water at 25°C for a smooth surface as a function of frequency and the corresponding wavelength of the electromagnetic radiation. (From Maul, 1985. With permission.)

At present, the relation between wind speed and the ocean wave spectrum is not well understood. Indeed, the mechanism for transfer of energy from the wind to ocean waves is one of the important unsolved problems in physical oceanography. The input of wind energy to short capillary and gravity-capillary waves depends on the velocity profile near the sea surface and on wind-induced turbulence in the air. The surface-wave amplitude is thus presumably related in some way to the friction velocity or the wind stress. Although significant progress has been made in relating the wave spectrum to the wind (e.g., Plant, 1986; Donelan and Pierson, 1987), the most accurate models relating σ_0 to the near-surface wind are still purely empirical. These models express σ_0 as a function of the wind speed at a reference height of 10 m above the sea surface (see Section 7).

2.3. Atmospheric Attenuation

2.3.1. Clear-Sky Attenuation

The transmittance t_λ for cloud-free subpolar, midlatitude, and tropical atmospheres is shown in Figure 4 as a func-

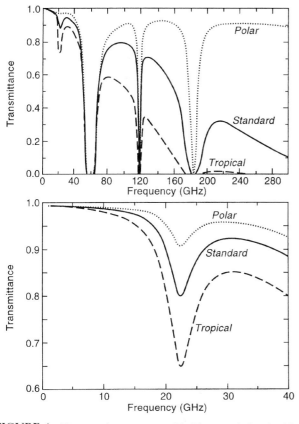

FIGURE 4 The transmittance at normal incidence angle for cloud-free subpolar (dotted line), midlatitude (solid line) and tropical (dashed line) atmospheres as a function of frequency. The frequency range 0 to 300 GHz (after Grody, 1976) is shown in the upper panel and an enlargement for the frequency range 0 to 40 GHz (data courtesy of F. Wentz, based on the atmospheric model of Liebe, 1985) is shown in the lower panel.

tion of frequency at incidence angle $\theta' = 0°$ (generally referred to as normal incidence) for microwave frequencies between 1 and 300 GHz. The one-way attenuation is correspondingly defined as $(1 - t_\lambda)$. The salient features of Figure 4 are a moderately strong water vapor absorption line centered at 22.235 GHz, a strong oxygen absorption band between 50 and 70 GHz, a strong oxygen absorption line at 118.75 GHz, and a strong water vapor absorption band centered at 183.31 GHz. The oxygen and water vapor molecules that contribute to the attenuation are almost entirely confined to the troposphere, which extends to altitudes of less than 10 km at midlatitudes and about 18 km in the tropics (see, for example, Figure 1.10 of Wallace and Hobbs, 1977). From the differences between the curves in Figure 4 for a dry subpolar atmosphere and a moist tropical atmosphere, it is apparent that atmospheric transmittance generally decreases with increasing frequency, owing primarily to the presence of water vapor.

At the K_u-band frequency of 13.6 GHz that is the primary frequency of the dual-frequency T/P altimeter, the clear-sky one-way transmittance at normal incidence angle is seldom less than 0.96 (see Figure 4), even in a moist, tropical atmosphere. From the radar Eq. (5), the power of the received signal is determined by the two-way transmittance t_λ^2. The corresponding two-way attenuation $(1 - t_\lambda^2)$ at 13.6 GHz is therefore generally less than 8%. At the secondary C-band frequency of 5.3 GHz used to correct for ionospheric refraction of the T/P range estimates (see Section 3.1.3), the clear-sky one-way transmittance exceeds 0.98 (see Figure 4). The clear-sky atmospheric attenuation at 5.3 GHz is thus less than half that at 13.6 GHz.

The total attenuation of electromagnetic radiation with wavelength λ is characterized by the opacity (also referred to as the optical thickness) of the atmosphere along the propagation path. The opacity τ_λ is related to the transmittance t_λ by

$$t_\lambda = e^{-\tau_\lambda}. \tag{9}$$

Since the transmittance is the fraction of electromagnetic radiation that is transmitted through the atmosphere, it is apparent from Eq. (9) that the transmittance decreases exponentially with increasing opacity.

The water vapor and dry-air (primarily oxygen) attenuations of the returned power from which σ_0 is calculated by the radar Eq. (6) are easily corrected. Since atmospheric gas molecules are much smaller than the radar wavelength, attenuation of σ_0 through a cloud-free atmosphere is governed by Rayleigh scattering. The attenuation at any point z along the path of propagation is therefore proportional to the air density at that location. The total opacity of the cloud-free atmosphere is thus proportional to the integrals of dry-air gas and water vapor densities ρ_{dry} and ρ_{vap} along the propagation path between the altimeter and the sea surface,

$$\tau_\lambda = \tau_\lambda^{\text{dry}} + \tau_\lambda^{\text{vap}}, \tag{10}$$

where λ is the radar wavelength and

$$\tau_\lambda^{\text{dry}} = \alpha_\lambda^{\text{dry}} \int_0^R \rho_{\text{dry}}(z)\,dz, \tag{11}$$

$$\tau_\lambda^{\text{vap}} = \alpha_\lambda^{\text{vap}} \int_0^R \rho_{\text{vap}}(z)\,dz. \tag{12}$$

The coefficients in Eqs. (11) and (12) have been determined from radiosonde observations[5] to be approximately $\alpha_\lambda^{\text{dry}} = 0.013 \; (\text{g cm}^{-2})^{-1}$ and $\alpha_\lambda^{\text{vap}} = 0.0035 \; (\text{g cm}^{-2})^{-1}$ for K_u-band frequencies near 13.6 GHz.

The measured normalized radar cross section σ_0^{meas} is corrected for two-way atmospheric attenuation by the radar Eq. (6), which can be written as

$$\sigma_0 = \frac{\sigma_0^{\text{meas}}}{t_\lambda^2}. \tag{13}$$

Expressed in decibels, the two-way correction Eq. (13) becomes

$$\Delta\sigma_0(\text{dB}) = \sigma_0(\text{dB}) - \sigma_0^{\text{meas}}(\text{dB}) = -10\log_{10} t_\lambda^2. \tag{14}$$

Substitution of Eqs. (9) and (10) into Eq. (14) yields the cloud-free two-way attenuation of σ_0 in the form

$$\Delta\sigma_0(\text{dB}) = \frac{20}{\log_e 10}(\tau_\lambda^{\text{dry}} + \tau_\lambda^{\text{vap}})$$

$$= 8.686(\tau_\lambda^{\text{dry}} + \tau_\lambda^{\text{vap}}). \tag{15}$$

Air density decreases approximately exponentially with altitude with an e-folding height of about 7 km (see, for example, Figure 1.4 of Wallace and Hobbs, 1977). Virtually all of the mass of the atmosphere is therefore at altitudes below the approximate 800–1300 km altitudes of altimeter satellites and the integral Eq. (11) can be replaced with an integral from the sea surface to the top of the atmosphere with negligible loss of accuracy,

$$\int_0^R \rho_{\text{dry}}(z)\,dz \approx \int_0^\infty \rho_{\text{dry}}(z)\,dz. \tag{16}$$

The hydrostatic equation (Gill, 1982, Section 3.5) relates the vertical integral of total air density ρ_a to the sea level pressure P_0 by

$$P_0 = \int_0^\infty g(z)\rho_a(z)\,dz. \tag{17}$$

Because the gravitational acceleration g decreases by less than 0.3% over the 7-km scale height of dry gases (see, for example, Table 2.1 of Wallace and Hobbs, 1977), $g(z)$ can

be considered approximately constant with height z and can therefore be passed through the integral in Eq. (17) with negligible loss of accuracy. Since $\rho_a \approx \rho_{\text{dry}}$, Eqs. (16) and (17) can be combined to obtain

$$\int_0^R \rho_{\text{dry}}(z)\,dz \approx P_0/g_0, \tag{18}$$

where P_0 is in units of millibars (mbar) and the average gravitational acceleration $g_0 = 980.6 \; \text{cm sec}^{-2}$ at the sea surface varies latitudinally by less than 0.3% (see Section 3.1.1).

Sea-level pressure seldom falls outside of the range 980–1035 mbar. Except in infrequent extreme events, P_0 thus seldom differs by more than $\pm 3\%$ from its global average value of 1013 mbar (see Figure 24 in Section 3.1.1). Attenuation of the radar signal by oxygen is therefore approximately constant over the global ocean. Dry-air attenuation can thus be effectively accounted for by replacing the first term in parentheses on the right side of Eq. (15) with a constant, $\tau_\lambda^{\text{dry}} \approx \alpha_\lambda^{\text{dry}} P_0/g_0 \approx 0.014$. The contribution of dry-air to the attenuation of σ_0 in Eq. (15) is therefore only about 0.12 dB.

Water vapor is much more variable spatially and temporally than oxygen (see Figure 26 in Section 3.1.2). As summarized in Section 3.1.2, the columnar integral of water vapor in Eq. (12) can be estimated to an accuracy of better than $0.15 \; \text{g cm}^{-2}$ from a three-frequency microwave radiometer. The T/P satellite carried a nadir-looking microwave radiometer with frequencies of 18.0, 21.0, and 37.0 GHz (Ruf *et al.*, 1995). Because the primary purpose of this radiometer was to correct for the effects of wet tropospheric refraction on the range measurements as described in Section 3.1.2, the algorithm for water vapor retrievals was cast in terms of wet tropospheric path delay (see Section 3.1.2) rather than columnar water vapor (Keihm *et al.*, 1995). Consequently, the integral of the water vapor density in Eq. (12) was replaced with the wet tropospheric path delay and the coefficient $\alpha_\lambda^{\text{vap}}$ in Eq. (12) was adjusted accordingly for the correction Eq. (15) for water vapor attenuation, (see Callahan, 1991).

Since the integrated columnar water vapor can reach values as high as $7 \; \text{g cm}^{-2}$ in the tropics (see Figure 26 in Section 3.1.2), the water vapor contribution to the attenuation of σ_0 in Eq. (15) can be as large as about 0.2 dB. As noted above, the dry-air contribution is about 0.12 dB. The total cloud-free attenuation of σ_0 from dry air and water vapor can therefore be as large as a few tenths of a decibel. Although this is only a few percent of the signal level, even in high wind-speed conditions where σ_0 can be as small as 5 dB, corrections for clear-sky attenuation must be applied for accurate estimation of wind speed. For a wind speed of $10 \; \text{m sec}^{-1}$, for example, failure to correct for a 0.3 dB attenuation of σ_0 would lead to about a 20% overestimate of wind speed (see Section 7).

---

[5]These values were deduced from the T/P algorithm for σ_0 attenuation described by Callahan (1991).

Although clear-sky attenuation of σ_0 is relatively small, it should be noted that oxygen and water vapor have a very large effect on the two-way propagation speed of the radar signal. Corrections for the oxygen and water vapor two-way path delays are therefore critical for accurate altimetric estimation of the range from the satellite to the sea surface by satellite altimetry. These refractive effects of the troposphere are discussed in Sections 3.1.1 and 3.1.2.

2.3.2. Cloud Attenuation

While clouds are completely opaque to infrared and visible radiation, they are relatively transparent to microwave radiation. Microwave remote sensing is therefore not restricted to cloud-free conditions. This a major advantage since about 60% of the tropical ocean and more than 75% of the mid-latitude ocean are typically cloud covered at any given time (Rossow and Schiffer, 1991; Hahn *et al.*, 1995). Like the attenuation by oxygen and water vapor molecules, attenuation of altimeter radar signals by cloud liquid water droplets is governed by Rayleigh scattering since the droplets are much smaller than the K_u-band and C-band radar wavelengths of $\sim$2 and $\sim$6 cm, respectively. Cloud attenuation at any point z along the path of propagation is therefore proportional to the cloud liquid water droplet density at that location. Attenuation by clouds also depends on the temperature of the droplets.

The two-way attenuation of microwave radiation from scattering and absorption by cloud liquid water droplets is shown as a function of frequency in Figure 5 for three different droplet temperatures. Attenuation increases with increasing frequency of the radar signal and with decreasing tem-

perature (i.e., increasing altitude). Cloud attenuation of the radar signals thus depends on the vertical profile of the cloud droplets. At the K_u-band frequency of 13.6 GHz, cloud attenuation is generally less than a few tenths of a dB per km of cloud thickness. Cloud attenuation is a factor of 2–3 smaller at the secondary C-band altimetric frequency of 5.3 GHz (see Figure 5).

Since the dependence of attenuation on cloud-droplet temperature is relatively small (see Figure 5) and vertical profiles of cloud liquid water content cannot be obtained from a nadir-looking radiometer, a practical solution (Goldstein, 1951) is to approximate the total opacity from cloud droplets in analogy to the formulations Eqs. (11) and (12) for dry air and water vapor attenuations of σ_0. The opacity from cloud droplets is thus expressed in terms of the integrated liquid water droplet density $\rho_{\text{liq}}(z)$ along the path of propagation between the satellite and the sea surface,

$$\tau_\lambda^{\text{liq}} = \alpha_\lambda^{\text{liq}} L_z, \qquad (19)$$

where

$$L_z = \int_0^R \rho_{\text{liq}}(z)\,dz. \qquad (20)$$

The coefficient in Eq. (19) has been determined from radiosonde observations to be approximately $\alpha_\lambda^{\text{liq}} = 0.33$ dB $(\text{g cm}^{-2})^{-1}$ (Callahan, 1991). The cloud liquid water attenuation of σ_0 is then estimated analogous to Eq. (15) by

$$\Delta\sigma_0^{\text{liq}} = 8.686\tau_\lambda^{\text{liq}}. \qquad (21)$$

The vertical integral of cloud liquid water density Eq. (20) is estimated from a multi-frequency passive microwave radiometer simultaneously with the columnar water vapor (e.g., Alishouse *et al.*, 1990a; Keihm *et al.*, 1995; Wentz, 1997). The accuracies of these columnar cloud liquid water estimates are believed to be better than 0.003 g cm^{-2} in rain-free conditions.

For nonraining clouds, the liquid water density ranges from $1 - 4 \times 10^{-6}$ g cm^{-3}, but rarely exceeds 2.5×10^{-6} g cm^{-3} (Maul, 1985). For a thick cloud with a vertical extent of 1 km, this corresponds to a vertically integrated liquid water density of 0.25 g cm^{-2}. The corresponding cloud contribution Eq. (21) to the attenuation of σ_0 is about 0.7 dB, which cannot be neglected. For a 10 m sec^{-1} wind speed, for example, failure to correct for a 0.7 dB attenuation of σ_0 would result in a 40% overestimate of the wind speed (see Section 7). The attenuation by deep convective clouds found in the tropics can be much larger. It is noteworthy, however, that these deep convective clouds are usually associated with heavy rainfall that is easily identified from passive microwave brightness temperatures, as well as from the radar returns themselves (see Section 2.3.3). Rain-contaminated observations are routinely eliminated in the geophysical processing of altimeter data.

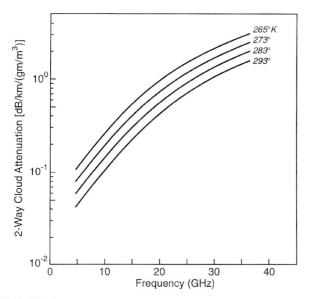

FIGURE 5 Two-way cloud attenuation of the power transmitted and received by a microwave radar as a function of frequency for four different cloud temperatures. (From Chelton *et al.*, 1989. With permission.)

2.3.3. Rain Attenuation

Rain has a much greater effect on the radar signal than clouds, water vapor or dry gases. Radar signals are attenuated by raindrops from both absorption and scattering. In addition to reducing the measured value of σ_0, rain cells that are smaller than the illuminated area of the antenna footprint distort the shape of the radar signal that is returned from the sea surface. Attenuation of selected portions of the returned waveform measured by the altimeter can corrupt altimeter estimates of two-way travel time as well as altimeter estimates of significant wave height (see Section 2.4.4). The effects of rain contamination are often apparent from erratic variation of σ_0, two-way travel time and significant wave height (see Goldhirsh and Rowland, 1982, for an example). In some cases, however, the effects of rain contamination can lead to more subtle but significant errors in altimetric estimates of these three quantities. It is therefore important to identify altimeter observations for which rain contamination is highly probable.

At frequencies below 10 GHz, rain attenuation is dominated by absorption. The scattering contribution becomes increasingly important with increasing frequency and increasing rain rate. As shown in Figure 6, the total attenuation of a radar signal increases with increasing frequency, increasing rain rate and increasing columnar thickness of the rain. Precise correction for two-way attenuation by raindrops thus requires knowledge of the vertical distribution of raindrops, and not just the rain rate. Such information is difficult to obtain from satellite measurements. Goldhirsh (1988) investigated three methods of estimating vertical profiles of rain-rate from radar measurements of backscatter at frequencies of 13.6 and 35 GHz from a hypothetical dual-frequency altimeter. He concluded that difficulties exist with each method because of the complexity of the combined effects of backscatter and attenuation on the signals received

by the radar. Additional research is needed before the methods could be applied operationally to correct for rain effects on multi-frequency altimetric estimates of σ_0.

Because of the difficulties in obtaining rain-rate profiles from satellite data, no attempt is made to correct radar measurements for rain attenuation of σ_0. Rather, rain-contaminated altimeter observations are flagged and excluded from further geophysical analysis. Ideally, the flagging would be based on a threshold rain rate of a few millimeters per hour. For smaller rain rates, the two-way attenuation of the radar signal at 13.6 GHz is generally less than a few tenths of a dB, even for a very thick rain column (see Figure 6). Errors of this magnitude are comparable to the attenuation from dry air, water vapor and cloud liquid water droplets discussed in Sections 2.3.1 and 2.3.2.

As reviewed by Wilheit *et al.* (1994), Petty (1997), and Smith *et al.* (1998) (see also Wentz and Spencer, 1998), rain rate can be estimated from measurements of the microwave radiance emitted by the ocean and the intervening atmosphere at an appropriate combination of frequencies. The algorithms that have been developed to date all require measurements of brightness temperatures at both horizontal and vertical polarization at an oblique incidence angle. Because of the highly transient nature of rainfall, the ability to identify rain-contaminated altimeter data requires coincident measurements from a nadir-looking passive microwave radiometer onboard the altimeter satellite. Since there is no distinction between horizontal and vertical polarization at normal incidence angle, presently available rain-rate algorithms cannot be used with a nadir-looking radiometer.

In lieu of a nadir rain-rate algorithm, the probability that an altimetric range estimate is contaminated by rain is flagged based on a threshold in the integrated columnar liquid water content L_z defined by Eq. (20). Selecting a threshold has proven problematic. Initially, T/P range measurements were flagged if L_z exceeded 0.1 $\mathrm{g\,cm^{-2}}$ (Callahan, 1991). The threshold was later reduced to 0.06 $\mathrm{g\,cm^{-2}}$ when it was determined that the higher threshold failed to flag many rain events. Subsequent studies of the effects of rain on altimeter observations have shown that a simple threshold on L_z is inadequate for flagging rain-contaminated altimeter observations. Since rain attenuation is an order of magnitude larger at K_u-band than at C-band (see Figure 6), rain-contaminated observations from the T/P dual-frequency altimeter can usually be identified as an abrupt decrease in K_u-band σ_0 relative to C-band σ_0. Quartly *et al.* (1996), Chen *et al.* (1997), Tournadre and Morland (1997), Quartly *et al.* (1999), and Cailliau and Zlotnicki (2000) have shown that the flagging of rain-contaminated altimeter data can be greatly improved by using a flag that is based on a combination of a threshold columnar liquid water content and a threshold reduction of K_u-band σ_0 relative to C-band σ_0. In the most recent of these studies, Cailliau and Zlot-

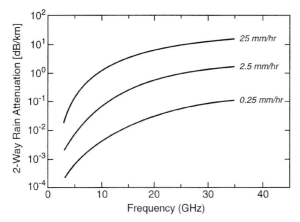

FIGURE 6 Two-way rain attenuation of the power transmitted and received by a microwave radar as a function of frequency for three different rain rates at a temperature of 291°K. (From Chelton *et al.*, 1989. With permission.)

nicki (2000) suggest using a columnar liquid water threshold of $0.01 \, \mathrm{g \, cm^{-2}}$ and a $\mathrm{K_u}$-band σ_0 reduction in excess of 1.5 standard deviations from a statistically derived relation between $\mathrm{K_u}$-band and C-band σ_0. The rain flag studies have recently been extended by Quartly (1998) and Quartly *et al.* (1999) to obtain quantitative estimates or rain rate from the T/P dual-frequency altimeter.

2.4. Two-Way Travel Time

2.4.1. Pulse-Limited Altimetry

A primary consideration in the design of an altimeter system is the area on the sea-surface over which the range from the altimeter to the mean sea surface height is measured. The footprint size should be large enough to filter out the effects of gravity waves on the sea surface, yet small enough to resolve the first internal Rossby radius of deformation that characterizes baroclinic mesoscale variability (see Chapter 3). The wavelength equivalent of the Rossby radius of deformation decreases from about 1500 km near the equator to about 60 km at 60° latitude (Chelton *et al.*, 1998). To simplify interpretation of the integrand in the radar Eq. (5), the footprint should also be small enough that the wave field and wind-induced roughness of the sea surface (i.e., the normalized radar cross section σ_0) are approximately homogeneous over the footprint. A footprint diameter of 1–10 km satisfies all of these criteria.

The footprint of an antenna is traditionally described in terms of the beam-limited footprint, defined to be the area on the sea surface within the field of view subtended by the beamwidth (full width at half power) of the antenna gain pattern. For a narrow-beam antenna, the antenna beamwidth γ, orbit height R, and footprint radius r are related by $\gamma = 2 \tan^{-1}(r/R) \approx 2r/R$. From the T/P orbit height of $R = 1336$ km, a footprint radius of 2.5 km, for example, corresponds to an antenna beamwidth of $\gamma \approx 3.74 \times 10^{-3} \, \mathrm{rad} = 0.21°$. The antenna beamwidth for a circularly symmetric antenna gain pattern is related to the antenna diameter d by

$$\gamma = k \frac{\lambda}{d}, \qquad (22)$$

where λ is the radar wavelength and k is a constant particular to the details of the illumination pattern across the antenna aperture (see Ulaby *et al.*, 1981, p. 141). For an antenna constant of about 1.3 that characterizes the T/P antenna,[6] the antenna diameter for a 0.21° beamwidth at the T/P primary frequency of 13.6 GHz (a radar wavelength of $\lambda = 2.21$ cm) is

$d = 7.7$ m. In addition to the impracticality of building and deploying an antenna this large, the accuracy of the range measurement from a beam-limited altimeter design is highly sensitive to antenna pointing errors. If not corrected for mispointing, a pointing error of only $\theta = 0.02°$, for example, introduces a range error of $\Delta R = 8$ cm from a very narrow-beam antenna at the T/P orbit height of $R = 1336$ km (see Figure 7a).

The limitations of the beam-limited altimeter design can be overcome by transmitting a very short pulse with a duration of a few nanoseconds from an antenna with a smaller diameter and correspondingly wider beamwidth. As shown below, the footprint size over which the range to nadir mean sea level is estimated by the altimeter is effectively defined by the pulse duration. The antenna diameters for altimeters launched to date have been $d = 1.6$ m for Seasat, 2.1 m for Geosat, 1.2 m for ERS-1 and ERS-2, and 1.5 m for T/P. At the T/P orbit height of $R = 1336$ km, the 1.5-m antenna diameter corresponds to a beamwidth-limited footprint diameter of about 25 km on a flat sea surface. Since the transmitted pulse expands spherically as it propagates away from the antenna, the time required for the leading edge of the short pulse to reach the nearest point on the sea surface is independent of the antenna pointing angle θ (see Figure 7b). As long as the pointing angle does not exceed the half beamwidth of the antenna, the range measurement from a pulse altimeter is insensitive to antenna pointing angle.[7]

The pulse-limited footprint associated with a pulse of duration τ can be derived analytically for the case of a sea surface consisting of a monochromatic, unidirectional wavetrain with crest-to-trough wave height H_w. This simple wave field is therefore given special consideration here. The results provide important insight into later discussion of the pulse-limited footprint for the more realistic case of a random wave field with a Gaussian sea surface height distribution. Although not essential to the analysis that follows, it will be further assumed that the phase of the monochromatic wavetrain is such that a wave crest is located at the nadir point. The pulse width is $c\tau$, where c is the speed of light. The spherically expanding altimeter pulse for an altimeter orbiting at a height R above mean sea level begins to illuminate the monochromatic sea surface at nadir range $R_0 = R - H_w/2$ when the leading edge of the pulse strikes the wave crest at the height $H_w/2$ above mean sea level (see Figure 8a). Thereafter, the areal extent of the region on the sea surface that contributes to the radar return measured by the altimeter is an expanding circle defined by the intersection of the leading edge of the pulse with the planar surface of wave crests at off-nadir angle θ_{out} (Figure 8b).

---

[6]From on-orbit measurement of the antenna pattern, Callahan and Haub (1995) determined that the beamwidth of the T/P antenna is $\gamma = 1.11° = 1.937 \times 10^{-2}$ rad for the primary frequency of 13.6 GHz. For the T/P antenna diameter of $d = 1.5$ m, this corresponds to an antenna constant of $k = 1.31$.

[7]As described in Section 2.4.6, corrections must be applied for off-nadir pointing angle to account for the combined effects of antenna gain rolloff and the method used onboard the satellite to estimate the two-way travel time of the pulse.

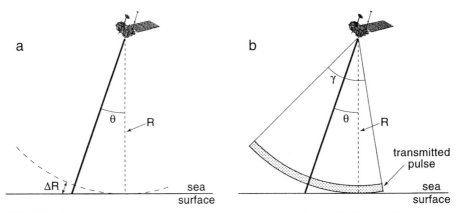

FIGURE 7 The measurement geometry for (a) a very narrow beamwidth-limited altimeter; and (b) a pulse-limited altimeter with a relatively large antenna beamwidth γ. In both cases, the boresight of the antenna views the sea surface at off-nadir angle θ from a height R above the sea surface.

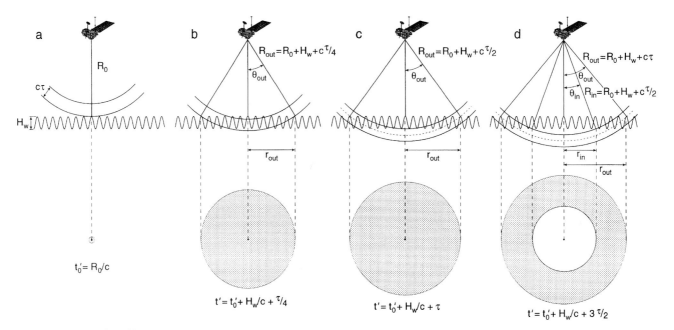

FIGURE 8 A schematic geometrical description of a pulse of duration τ and width $c\tau$ incident on a monochromatic wave surface with crest-to-trough wave height H_w. One-way travel times t' are labelled relative to the time $t'_0 = R_0/c$ when the leading edge of the pulse intersects the planar surface of wave crests at nadir. The angle θ_{out} corresponds to the angle from which radar returns arrive at the satellite at twice the one-way travel times displayed on each panel. As described in the text, the reflection of the midpoint of the pulse from wave crests at one-way travel time t' and angle θ_{out} arrives at the satellite simultaneously with the trailing edge of the pulse reflected from wave troughs at the same one-way travel time t'. The bounds of the pulse-limited footprint area contributing to the radar return at two-way travel time $2t'$ are shown in the lower panels.

A more precise description of the detailed illumination pattern within this expanding circle is given in Section 2.4.2.

The path between the antenna and the sea surface is traversed twice by the radar return measured by the altimeter. The reflection of the leading edge of the pulse from the wave crest at nadir therefore arrives at the altimeter at two-way travel time $t_0(H_w) = 2R_0/c$, where the speed of light c must be corrected for atmospheric refraction as described in

Section 3.1. At a two-way travel time $t > t_0$, the altimeter measures the reflection of the leading edge of the pulse from wave crests along the outer perimeter of the illuminated region at off-nadir angle θ_{out} and corresponding radial distance r_{out} from the nadir point. Define $R_{out} = R_0 + \Delta R_{out}$ to be the slant range to the planar surface of wave crests at off-nadir angle θ_{out}. Note that θ_{out}, r_{out}, R_{out} and ΔR_{out} all depend on t and H_w. It can be seen geometrically from Figure 8b that the radius of the circular footprint on the sea surface that contributes to the signal received by the altimeter at two-way travel time t is

$$r_{\text{out}}(t, H_w) = \left[2R_0 \Delta R_{\text{out}} + \Delta R_{\text{out}}^2\right]^{1/2}. \quad (23)$$

Since the incremental slant range ΔR_{out} is much smaller than the nadir range R_0 at all off-nadir angles within the small 1–2° beamwidths characteristic of altimeter antennas, Eq. (23) for the outer perimeter of the expanding footprint can be accurately approximated as

$$r_{\text{out}}(t, H_w) \approx [2R_0 \Delta R_{\text{out}}]^{1/2}. \quad (24)$$

The measurement geometry shown schematically in Figure 8 neglects the curvature of the earth's surface. The off-nadir angle θ_{out} for a given two-way travel time t is slightly smaller than the flat-earth approximation shown in the figures (see Figure 19 of Chelton et al., 1989). The arc length r_{out} defining the footprint on the curved surface of the earth is therefore somewhat smaller than the flat-earth approximation Eq. (24). It is shown in Appendix A1 of Chelton et al. (1989) that the corrected footprint radius on a spherical earth is

$$r_{\text{out}}(t, H_w) \approx \left[\frac{2R_0 \Delta R_{\text{out}}}{1 + R_0/R_e}\right]^{1/2}, \quad (25)$$

where $R_e \approx 6371$ km is the radius of the earth. The flat-earth approximation Eq. (24) is equivalent to considering the earth radius to be infinite. For the Seasat, Geosat, ERS-1, and ERS-2 orbit heights of about 785 km, the normalization by $(1 + R_0/R_e)$ in Eq. (25) represents a 5.8% reduction of the flat-earth footprint radius Eq. (24). The reduction is 9.1% for the higher T/P orbit height of 1336 km.

The elapsed two-way travel time between the arrival time t_0 of the first radar return from the wave crest at nadir and the arrival time t of the leading edge of the pulse reflected from wave crests at off-nadir angle θ_{out} is related to the incremental slant range ΔR_{out} by

$$\Delta t(H_w) \equiv t - t_0 = \frac{2\Delta R_{\text{out}}}{c}. \quad (26)$$

Substituting Eq. (26) into Eq. (25) expresses the radius of the outer perimeter of the circular footprint in terms of elapsed time relative to two-way travel time t_0,

$$r_{\text{out}}(\Delta t, H_w) = \left[\frac{c\Delta t R_0}{1 + R_0/R_e}\right]^{1/2}. \quad (27)$$

The corresponding area of the footprint contributing to the measured radar return is

$$A_{\text{out}}(\Delta t, H_w) = \pi r_{\text{out}}^2 = \frac{\pi c \Delta t R_0}{1 + R_0/R_e}. \quad (28)$$

The outer perimeter of the area on the sea surface that contributes to the radar return measured by a pulse-limited altimeter is thus a circle with area that increases linearly with

elapsed two-way travel time Δt relative to the arrival time t_0 of the leading edge of the pulse reflected from the wave crest at nadir. It should be noted that the flat-earth approximation Eq. (24) for r_{out} results in a footprint area that is overestimated by 11.2% for the Seasat, Geosat, ERS-1, and ERS-2 orbit heights of about 785 km and 17.3% for the T/P orbit height of 1336 km.

The pulse-limited footprint area on the monochromatic wave surface continues to grow linearly with time according to Eq. (28) until the trailing edge of the pulse intersects the planar surface of wave troughs at nadir (see Figure 8c). Thereafter, the footprint becomes an expanding annulus. The two-way travel time for the trailing edge of the pulse to return from the wave troughs near nadir is $2(R_0 + H_w)/c$. Accounting for the additional lag τ between the leading and trailing edges of the pulse, this reflection of the trailing edge of the pulse arrives at the altimeter at two-way travel time $t_1(H_w) = \tau + 2(R_0 + H_w)/c$. The corresponding total elapsed time since the arrival time $t_0 = 2R_0/c$ of the leading edge of the pulse reflected from the planar surface of wave crests at nadir is

$$\Delta t_1(H_w) \equiv t_1 - t_0 = \tau + 2H_w/c. \quad (29)$$

At the two-way travel time t_1 when the pulse-limited footprint contributing to the measured radar return first becomes an annulus, the radius Eq. (27) and area Eq. (28) of the circular footprint are

$$r_1(H_w) = \left[\frac{c\Delta t_1 R_0}{1 + R_0/R_e}\right]^{1/2} = \left[\frac{(c\tau + 2H_w)R_0}{1 + R_0/R_e}\right]^{1/2}, \quad (30)$$

$$A_1(H_w) = \pi r_1^2 = \frac{\pi c \Delta t_1 R_0}{1 + R_0/R_e} = \frac{\pi(c\tau + 2H_w)R_0}{1 + R_0/R_e}. \quad (31)$$

The "illumination hole" in the annular footprint that forms behind the trailing edge of the pulse is a circle with radius and area that expand at the same rates as the radius Eq. (27) and area Eq. (28) defined by the outer perimeter of the footprint, but lagged in time by Δt_1. At a total elapsed time $\Delta t > \Delta t_1$, the inner and outer perimeters of the annulus are defined by off-nadir angles θ_{in} and θ_{out} (see Figure 8d). The radius and area within the outer perimeter of the annulus continue to expand as Eqs. (27) and (28). The radius and area of the illumination hole behind the annulus expand as

$$r_{\text{in}}(\Delta t, H_w) = \left[\frac{c(\Delta t - \Delta t_1)R_0}{1 + R_0/R_e}\right]^{1/2}$$
$$= \left[r_{\text{out}}^2(\Delta t, H_w) - r_1^2(H_w)\right]^{1/2}, \quad (32)$$

$$A_{\text{in}}(\Delta t, H_w) = \frac{\pi c(\Delta t - \Delta t_1)R_0}{1 + R_0/R_e}$$
$$= A_{\text{out}}(\Delta t, H_w) - A_1(H_w). \quad (33)$$

The total area of the expanding annulus contributing to the measured radar return after elapsed time $\Delta t > \Delta t_1(H_w)$ is thus

$$A_{ann}(\Delta t, H_w) = A_{out}(\Delta t, H_w) - A_{in}(\Delta t, H_w)$$
$$= A_1(H_w). \quad (34)$$

While the radii r_{out} and r_{in} defining the outer and inner perimeters of the annulus for wave height H_w continue to grow with time according to Eqs. (27) and (32), the pulse-limited footprint area Eq. (34) contributing to the radar return remains constant after the time t_1 when the trailing edge of the pulse returns from the planar surface of wave troughs at nadir.

In the preceding description, it should be noted that simultaneous illumination of different locations on the sea surface by different portions of the pulse of duration τ arrive back at the altimeter at different two-way travel times. Consider the one-way travel time $t' = t_1/2$ when the trailing edge of the pulse intersects the planar surface of wave troughs at nadir as shown in Figure 8c. The maximum radial extent of the off-nadir return that arrives at the altimeter at two-way travel time t_1 is not the perimeter that was illuminated by the leading edge of the pulse at the same one-way travel time $t_1/2$. The leading edge of the pulse has propagated a distance $c\tau$ farther from the altimeter than has the trailing edge of the pulse. The off-nadir angle θ_{out} and associated radial distance r_{out} at two-way travel time t_1 are therefore defined by the perimeter on the planar surface of wave crests that was illuminated by the leading edge of the pulse at the earlier one-way travel time $(t_1 - \tau)/2$. This $\tau/2$ time lag coincides with the intersection of the midpoint of the radar pulse with the planar surface of wave crests at the one-way travel time $t_1/2$ when the trailing edge of the pulse intersects the planar surface of wave troughs at nadir (see Figure 8c).

At two-way travel times $t > t_1$, the off-nadir angle θ_{in} of the inner boundary of the expanding annulus that contributes to the radar return is defined by the intersection of the trailing edge of the pulse with the planar surface of wave troughs at one-way travel time $t/2$ (see Figure 8d). The angle θ_{out} defining the outer perimeter of the annulus contributing to the radar return at the same two-way travel time t is defined by the intersection of the midpoint of the pulse with the planar surface of wave crests at one-way travel time $t/2$.

The evolution of the pulse-limited footprint area and the radii that define the associated outer and inner perimeters of the expanding footprint annulus are shown in Figure 9 for monochromatic crest-to-trough wave heights of $H_w = 1, 5,$ and 10 m. After the leading edge of the pulse reflected from the wave crest at nadir arrives at the altimeter at two-way travel time $t_0 = 2R_0/c$, the area contributing to the measured radar return increases linearly until the arrival time $t_1 = \tau + 2(R_0 + H_w)/c$ of the trailing edge of the pulse reflected from wave troughs at nadir. Thereafter, the footprint becomes an annulus with constant area $A_1(H_w)$ given by Eq. (31). Note, however, that the radii of the outer and inner perimeters continue to expand as the square root of the elapsed time after t_0 and t_1, respectively (see Figure 9b). The time axis in Figure 9 is displayed relative to the arrival time $t_{1/2}$ defined as

$$t_{1/2} \equiv \frac{t_0 + t_1}{2} = \frac{2R_0 + H_w}{c} + \frac{\tau}{2} = \frac{2R}{c} + \frac{\tau}{2}, \quad (35)$$

where $R = R_0 + H_w/2$ is the range from the altimeter to the mean sea-surface height at nadir. The time $t_{1/2}$ thus represents the two-way travel time for the midpoint of the pulse to return from mean sea level at nadir.

It is apparent from Figure 9 and the associated Eqs. (34) and (31) that the area A_1 of the expanding annulus depends only on the wave height H_w, the pulse duration τ and the orbit height R (approximately equal to the range R_0 from the altimeter to the planar surface of monochromatic wave crests). The factor $R_0/(1 - R_0/R_e)$ in the expressions for footprint radius and area increases by 56% from the Seasat, Geosat, and ERS orbit heights of about 785 km to the T/P orbit height of 1336 km. This corresponds to only a 25% increase in the effective footprint radius r_1 of a circle with area equal to the annulus area A_1. Orbit height is therefore a relatively minor consideration in determining pulse-limited footprint size.

The only other adjustable parameter controlling the annulus area is the pulse duration τ. The effective pulse duration that has been used in the Seasat, Geosat, and T/P altimeters[8] is $\tau = 3.125$ nsec. At an orbit height of 785 km, this yields annulus areas A_1 with equivalent footprint diameters $2r_1$ that increase from 1.6 km for a flat sea surface to 7.7 km for a monochromatic crest-to-trough wave height of 10 m. At the higher T/P orbit height of 1336 km, the corresponding equivalent footprint diameters increase from 2.0–9.6 km.

The monochromatic wave field considered above provides a framework for interpretation of the footprint characteristics for a more realistic wave field. Consider a specular reflector on a wavy sea surface where the height above mean sea level is ζ. The coordinates of this location can be defined by the radial distance from the altimeter to the point on the sea surface, the off-nadir angle θ and the azimuth angle χ about the axis defined by the line between the altimeter and the nadir point. Because of the approximate spherical geometry of the earth, it is convenient to replace the angle θ with the colatitude angle ϕ subtended by rays from the center of the earth to the nadir point and the off-nadir point $\zeta(\chi, \theta, t) = \zeta(\chi, \phi, t)$ (see Figure 10). Define $R'_\phi(\chi, t)$ to be the radial distance from the altimeter to $\zeta(\chi, \phi, t)$. Reflected power is received from this specular reflector at two-way travel time t if $\zeta(\chi, \phi, t)$ falls within the spherically expanding pulse of duration τ, i.e., if $2R'_\phi(\chi, t)/c \leq t \leq \tau + 2R'_\phi(\chi, t)/c$. To lowest order for the

---

[8] A slightly shorter pulse duration of 3.03 nsec is used in the ERS-1 and ERS-2 altimeters.

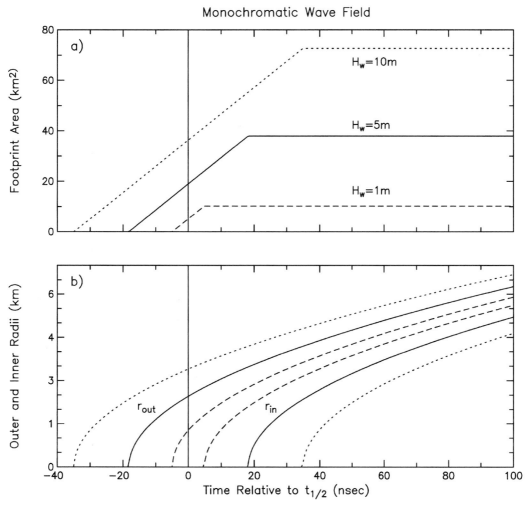

FIGURE 9 The evolution of (a) the footprint area; and (b) the associated radii that define the outer and inner perimeters of the footprint annulus contributing to the radar return as a function of two-way travel time for monochromatic crest-to-trough wave heights of $H_w = 1$, 5, and 10 m (dashed, solid, and dotted lines, respectively). Time is displayed in nanoseconds relative to the two-way arrival time $t_{1/2}$ of the midpoint of the pulse reflected from mean sea level at nadir for a pulse duration of $\tau = 3.125$ nsec.

small off-nadir angles θ relevant to altimetry, the distance $R'_\phi(\chi, t)$ is related to the distance R_ϕ from the altimeter to mean sea level at angle ϕ by $R'_\phi(\chi, t) \approx R_\phi - \zeta(\chi, \phi, t)$. A more precise relation is easily obtained from the geometry in Figure 10, but this approximate solution is adequate for present purposes. The condition for receipt of power at two-way travel time t can therefore be expressed in terms of an "indicator function" defined as

$$I(\chi, \phi, t) = \begin{cases} 1 & \text{if } \zeta_1 \leq \zeta(\chi, \phi, t) \leq \zeta_2 \\ 0 & \text{otherwise,} \end{cases} \quad (36)$$

where

$$\zeta_1(\phi, t) = R_\phi - ct/2, \quad (37)$$
$$\zeta_2(\phi, t) = R_\phi - c(t - \tau)/2. \quad (38)$$

The instantaneous illumination pattern depends on the specific wave height field at time t. Because of the random nature of the sea surface height distribution, it is more instructive to consider the probabilistic illumination pattern. The probability that the sea surface at a location defined by the colatitude angle ϕ and azimuth angle χ is illuminated at time t is given by the expected value of the indicator function defined by Eq. (36). This can be expressed as

$$\langle I(\chi, \phi, t) \rangle = \int_{-\infty}^{\infty} q_s(\zeta) I(\chi, \phi, t) \, d\zeta$$
$$= \int_{\zeta_1}^{\zeta_2} q_s(\zeta) \, d\zeta, \quad (39)$$

where angle brackets are used to denote expected value and $q_s(\zeta)$ is the probability density function for the sea surface

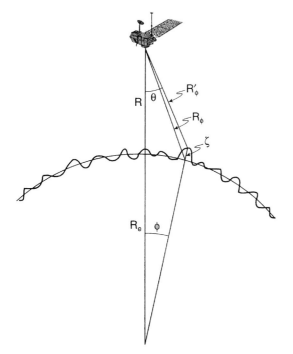

FIGURE 10   A schematic geometrical description of the radar return from a point on the sea surface at off-nadir angle θ and associated colatitude angle ϕ where the sea surface elevation is ζ. The distances from the altimeter to the sea surface and to mean sea level at angle ϕ are R'_ϕ and R_ϕ, respectively.

elevation.[9] For a spatially homogeneous wave height distribution, the right side of Eq. (39) is independent of azimuth angle χ, which implies that $\langle I(\chi, \phi, t)\rangle$ is a function of angle ϕ only. This probability of illumination can be thought of as the fractional illumination of the sea surface at radial distance $r(\phi)$.

Equation (39) is equivalent to a convolution of the probability density function of ζ with an ideal pulse consisting of a rectangle function of width $c\tau$. In reality, altimeter pulses have a shape that is given by $p_\tau(t) \approx [\sin(\pi t/\tau)/(\pi t/\tau)]^2$, referred to as the point target response (see, for example, Figure 8 of Lipa and Barrick, 1981, and Figure 6 of Rodríguez, 1988). The center lobe of this functional form for the point target response has a half-power width of 0.886τ. The point-target response is discussed further in the derivation and discussion of Eq. (52) in Section 2.4.3. The details of the pulse shape are not a major concern here. The representation of $p_\tau(t)$ as a rectangle of duration τ is adequate for present purposes.

To first order, the distribution of the sea surface height is Gaussian. The scale of the wave-height distribution is tradi-

tionally characterized by the significant wave height, which is defined to be the average crest-to-trough height of the 1/3 highest waves. The significant wave height is therefore denoted as $H_{1/3}$ and is usually considered to be equivalent to four times the standard deviation of the wave height distribution.[10] The Gaussian wave height distribution can therefore be expressed as

$$q_s(\zeta) = \frac{4}{\sqrt{2\pi}\,H_{1/3}} \exp\left[-\frac{1}{2}\left(\frac{4\zeta}{H_{1/3}}\right)^2\right]. \qquad (40)$$

The character of the probabilistic illumination pattern is summarized in Figure 11, which shows the time evolution of $\langle I(\phi, t)\rangle$ as a function of the radial distance $r(\phi)$ corresponding to the arc length on the sea surface associated with the angle ϕ. For the Gaussian wave-height distribution with $H_{1/3} = 5$ m shown in the figure, the fractional illumination is negligibly small until about 20 nsec before the time $t_{1/2}$ when the midpoint of the pulse returns from mean sea level at nadir. Thereafter, the fractional illumination increases at nadir, and the region of significant illumination expands radially outward from the nadir point. After time $t_{1/2}$, the fractional illumination decreases at nadir as the region of largest fractional illumination spreads radially away from the nadir point.

The outer and inner perimeters of the radially expanding illumination pattern can be defined somewhat arbitrarily by the radii where less than 1% of the sea surface is illuminated.[11] These points are shown by the short vertical tic marks in Figure 11. Plan views of the spatial patterns of fractional illumination at selected two-way travel times are shown in Figure 12 for significant wave heights of $H_{1/3} = 1$, 5, and 10 m. Circles with radii $r_{out}(t)$ and $r_{in}(t)$ defined by the 1% threshold illumination criterion are overlaid as thin lines on these fractional illumination patterns. The time evo-

[9]More precisely, $q_s(\zeta)$ is the distribution of specular scatterers on the sea-surface, which is the same as the distribution of the sea-surface elevation to lowest order. As discussed in Section 3.2, however, corrections must be applied to account for differences between the distributions of the sea surface elevation and the specular scatterers.

[10]As summarized in Section 6, there is confusion in the literature about the relationship between $H_{1/3}$ and the standard deviation σ_ζ of the sea surface elevation ζ relative to mean sea level. The reason for the confusion is that this relationship depends on the precise nature of the wave field. From Figure 4 of Cartwright and Longuet-Higgins (1956), the ratio $H_{1/3}/\sigma_\zeta$ decreases from 4 for a very narrow-banded wave spectrum to about 3 for a very broad-banded wave spectrum. A ratio of 4 is a reasonably good approximation over a wide range of bandwidths. For satellite altimetry, it is traditional to assume that $H_{1/3} = 4.0\,\sigma_\zeta$ (e.g., Fedor et al., 1979) and that is the value adopted for the pulse-limited footprints determined probabilistically as described in this section. The footprint characteristics for two-way travel times longer than the time t_1 when the footprint becomes an expanding annulus are not strongly dependent on this ratio; a value of $H_{1/3} = 3.0\,\sigma_\zeta$ yields radii that are larger by only a few percent than the radii obtained for the value of $H_{1/3} = 4.0\,\sigma_\zeta$ used here.

[11]Although a direct analogy to the method used here is difficult, Parke and Walsh (1995) effectively define the altimeter footprint by a much higher threshold illumination area. The corresponding outer and inner radii of the expanding footprint are much smaller than the radii for the 1% threshold illumination area used here to define the footprint.

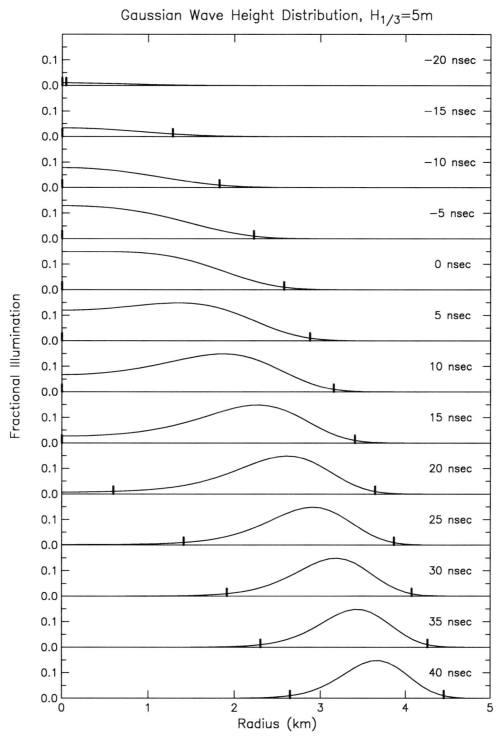

FIGURE 11 The fractional illumination $\langle I(\phi, t) \rangle$ as a function of radial distance from nadir at selected two-way travel times t for a Gaussian sea surface height distribution with a significant wave height of $H_{1/3} = 5$ m. The short vertical tic marks on the tails of the fractional illumination patterns indicate the 1% thresholds of fractional illumination.

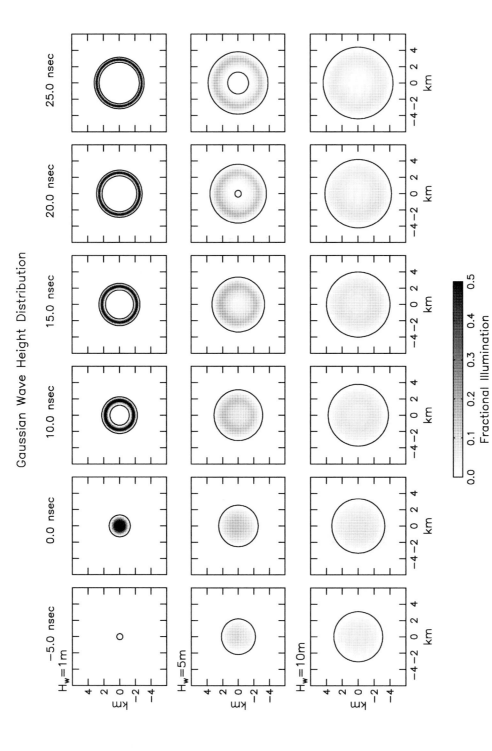

FIGURE 12 Plan views of the fractional illumination patterns at two-way travel time intervals of 5 nsec for a pulse of duration 3.125 nsec and significant wave heights of $H_{1/3} = 1$, 5, and 10 m. The circles in each panel represent the perimeters where the fractional illumination is 1%.

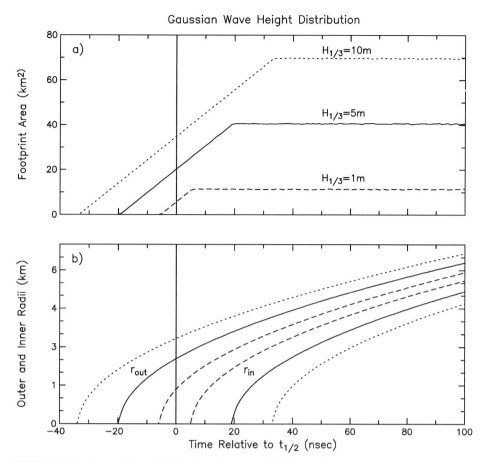

FIGURE 13 The evolution of (a) the footprint area; and (b) the associated radii that define the outer and inner perimeters of the footprint annulus contributing to the radar return as a function of two-way travel time for Gaussian sea surface height distributions with $H_{1/3} = 1$, 5, and 10 m (dashed, solid, and dotted lines, respectively). The radii and areas are defined probabilistically by the 1% thresholds of fractional illumination as described in the text.

lutions of $r_{out}(t)$ and $r_{in}(t)$ for the three values of $H_{1/3}$ are shown in the bottom panel of Figure 13.

Comparing Figure 13 with Figure 9, it is apparent that the probabilistic time series of $r_{out}(t)$ and $r_{in}(t)$ defined by the 1% threshold illumination criterion for Gaussian wave-height distributions with $H_{1/3} = 1$, 5, and 10 m are almost identical to the time series derived analytically for the simple case of monochromatic wave fields with crest-to-trough wave heights of $H_w = 1$, 5, and 10 m. With simple substitution of $H_{1/3}$ for H_w, the analytical formulas derived previously for $r_{out}(t)$ and $r_{in}(t)$ for the case of a monochromatic wave field are therefore good approximations for the outer and inner radii of the pulse-limited footprint for a realistic wave field.

The pulse-limited footprint areas $A(t)$ within the perimeters defined by the probabilistic outer and inner radii $r_{out}(t)$ and $r_{in}(t)$ are shown in the top panel of Figure 13. The footprint area increases linearly from the time t_0 when more than 1% of the wave crests near nadir are illuminated to the time t_1 when less than 1% of the wave troughs near nadir are illu-

minated. The rise time of the footprint area is approximately equal to that given by Eq. (29) with H_w replaced by $H_{1/3}$. After time t_1, the pulse-limited footprint area becomes the annulus with constant area given approximately by the analytical expression Eq. (31) with the monochromatic crest-to-trough wave height H_w replaced with $H_{1/3}$.

The time evolutions of the probabilistic radii and the corresponding footprint area for a Gaussian wave height distribution characterized by $H_{1/3}$ are essentially the same as the values given by Chelton *et al.* (1989). It should be noted, however, that these values are significantly larger than the radii and area suggested by Fedor *et al.* (1979) and Parke and Walsh (1995). It should also be noted that characterization of the footprint over which the altimeter estimates the range to mean sea level requires specification of a particular two-way travel time in the evolution of the outer footprint radius $r_{out}(t)$. Rational specification of this two-way travel time is deferred to Section 2.4.7 after the instrumental details of altimeter estimation two-way travel time have been described in Sections 2.4.2–2.4.6.

2.4.2. Average Returned Waveform

The pulse-limited footprint area described in Section 2.4.1 represents the region on the sea surface within which specularly reflecting wave facets contribute to the radar return measured by the altimeter. The radius $r_{out}(t)$ of the outer perimeter of the expanding footprint defines the maximum radial extent of the region from which specularly reflecting wave facets can contribute significantly to the signal measured by the altimeter at two-way travel time t. However, the full area within this expanding perimeter is not the reflecting area that determines the power of the received signal in the radar Eq. (5) during the early period of the radar return when the pulse-limited footprint is an expanding circle. Nor does the full area between $r_{out}(t)$ and $r_{in}(t)$ determine the power of the received signal after the pulse-limited footprint becomes an expanding annulus. At any instant in time, the sea surface within the pulse-limited footprint is only sparsely illuminated by the short pulse. The area that determines the returned power measured by the altimeter at two-way travel time t consists of the spatially integrated area of the specular scatterers where the short pulse intersected the instantaneous sea surface at one-way travel time $t/2$.

For the monochromatic wave considered in Section 2.4.1, the area illuminated within the pulse-limited footprint can be determined analytically by simple geometry for any specified two-way travel time. The sparse illumination of the monochromatic wave surface at any instant in time is apparent from the discontinuous intersections of the pulse with the wavy sea surface in the top panels of Figure 8. Plan views of the instantaneous illumination patterns contributing to the radar return at two selected two-way travel times are shown in Figure 14 for the cases of a flat sea surface and a monochromatic wavetrain with crest-to-trough wave height $H_w = 2$ m and a wavelength of 200 m. It is apparent that the monochromatic waves distribute the specular reflectors along closed loci of points that are stretched parallel to wave crests and troughs over a larger region on the sea surface.

The pulse-limited footprint area is similarly sparsely illuminated for a more realistic Gaussian sea-surface height distribution, as evidenced by the fact that the fractional illumination in Figure 12 is much less than 1 everywhere. The total area determining the power of the signal received by the altimeter is the spatial integral of the instantaneous illuminated area over the pulse-limited footprint. For two-way travel time t, this is given by the area integral of the indicator function Eq. (36),

$$A_I(t) = \int_0^{2\pi} \int_0^\pi I(\chi, \phi, t) R_e^2 \sin\phi \, d\phi \, d\chi. \qquad (41)$$

This total area $A_I(t)$ fluctuates with time as the pulse spreads radially across the wavy sea surface. Any individual time series $A_I(t)$ is thus very noisy. The characteristics of the illuminated area can be described probabilistically by considering the average illuminated area over a hypothetical infinite

ensemble of realizations. In practice, this infinite ensemble is approximated by averaging the radar returns from a large number of independent pulses as described below. Because the probability density function of the sea-surface height distribution is assumed to be spatially homogeneous over the pulse-limited footprint, the right side of Eq. (39) is independent of azimuth angle χ. The average of the illuminated area Eq. (41) within the pulse-limited footprint that contributes to the radar return at two-way travel time t is therefore

$$\langle A_I(t) \rangle = 2\pi R_e^2 \int_0^\pi \int_{\zeta_1}^{\zeta_2} q_s(\zeta) \sin\phi \, d\zeta \, d\phi, \qquad (42)$$

where $\zeta_1(\phi, t)$ and $\zeta_2(\phi, t)$ are defined by Eqs. (37) and (38).

The time evolution of the average illuminated area given by Eq. (42) is shown in Figure 15 for Gaussian sea surface height distributions with $H_{1/3} = 1$, 5, 10, and 15 m. The rise time increases with increasing $H_{1/3}$ and the slope of the average area time series at the midpoint $t_{1/2}$ is inversely proportional to $H_{1/3}$. Although the locations of the specular reflectors are spread over a larger pulse-limited footprint area on a wavy sea surface, it is apparent from the figure that the average illuminated area contributing to the radar return is independent of wave height after the two-way travel time $t_1(H_{1/3})$ when the footprint area becomes an expanding annulus. The illuminated area in this "plateau region" is therefore given by the pulse-limited footprint area Eq. (31) of the annulus for a flat sea surface,

$$A_\tau = \frac{\pi c \tau R_0}{1 + R_0/R_e} \approx \frac{\pi c \tau R}{1 + R/R_e}, \qquad (43)$$

which depends only on the pulse duration τ and the orbit height R.

The distance from the altimeter to mean sea level at off-nadir angle ϕ and the sea surface elevation ζ can both be expressed in terms of two-way travel time. As first shown by Moore and Williams (1957) (see also Brown, 1977; Hayne, 1980; Barrick and Lipa, 1985; and Rodríguez, 1988), the average illuminated area Eq. (42) can therefore be written as the convolution of three terms,

$$\langle A_I(t) \rangle = A_\tau U(t - t_{1/2}) * q_s(t) * p_\tau(t), \qquad (44)$$

where the asterisk denotes convolution, $q_s(t)$ is the probability density function for the sea surface height distribution ζ expressed in terms of two-way travel time $t = -2\zeta/c$ relative to the mean sea surface height, $p_\tau(t) \approx [\sin(\pi t/\tau)/(\pi t/\tau)]^2$ is the point target response introduced following Eq. (39) that describes the shape of the altimeter pulse [see the derivation and discussion of Eq. (52) in Section 2.4.3] and $U(t - t_{1/2})$ is the unit step function,

$$U(t - t_{1/2}) \equiv \begin{cases} 1 & \text{if } t \geq t_{1/2} \\ 0 & \text{otherwise.} \end{cases} \qquad (45)$$

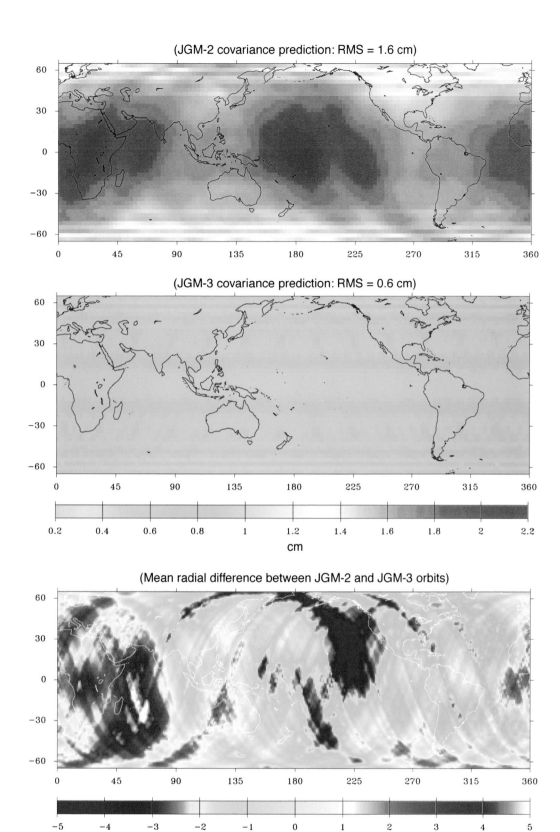

CHAPTER 1, FIGURE 40 Example of the mean geographically correlated orbit error for T/P predicted by the JGM-2 (top) and JGM-3 (middle) covariances. The averaged differences between orbits computed with JGM-2 and JGM-3 (bottom) show good correlation with the orbit errors predicted for JGM-2.

Color plates for Chapters 3 through 9 are located after page 308.

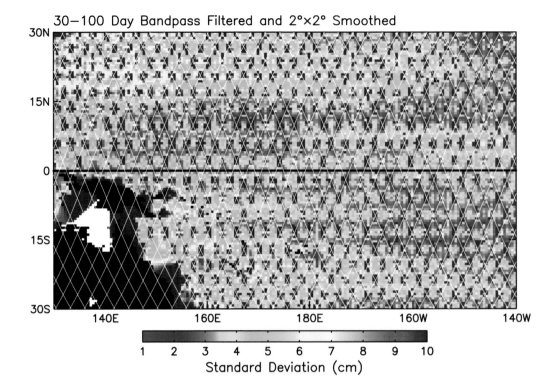

30–100 Day Bandpass Filtered and 2°×2° Smoothed

Standard Deviation (cm)

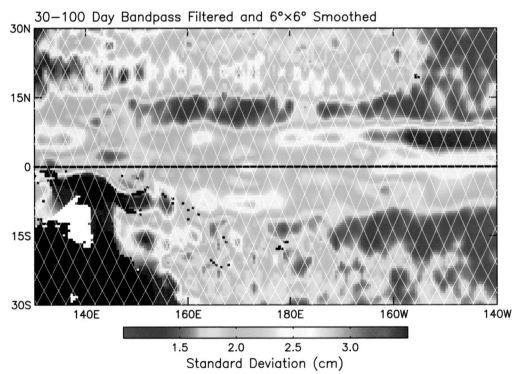

30–100 Day Bandpass Filtered and 6°×6° Smoothed

Standard Deviation (cm)

CHAPTER 1, FIGURE 66 The standard deviation computed from SSH fields constructed from 7 years of T/P data with 30–100-day band-pass filtering and two different degrees of spatial smoothing applied. For the top panel, the data were smoothed on a 0.5° grid with a loess smoother (Cleveland and Devlin, 1988) that retained wavelengths longer than 2° of latitude by 2° of longitude and periods longer than 30 days (Schlax and Chelton, 1992; Greenslade *et al.*, 1997). For the bottom panel, the data were loess smoothed 6° x 6° x 30 days on a 1° grid. For both panels, the smoothed SSH fields were then high-pass filtered with a loess smoother to attenuate periods longer than 100 days, thus isolating band-pass filtered estimates of mesoscale SSH variability with periods between 30 and 100 days. Note the different dynamic ranges of the color bars for the two panels.

LARGE SCALE 20-100 DAY SEA SURFACE HEIGHT VARIABILITY

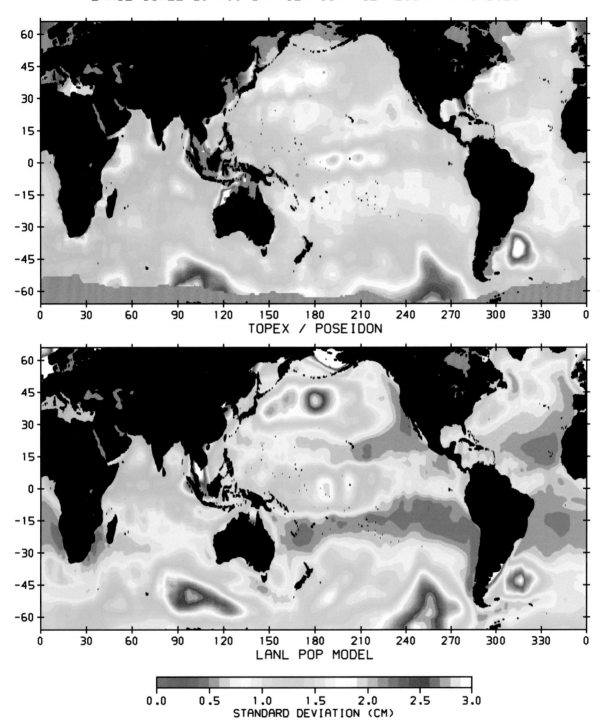

CHAPTER 2, FIGURE 7 Root-mean-square variability of sea surface height measured by T/P (top) and simulated by the ocean general circulation model developed by the Parallel Ocean Program of the Los Alamos National Laboratory (bottom). Both were filtered to retain energy at spatial scales larger than 1000 km and temporal scales shorter than 100 days. (From Fu, L.-L., and Smith, R.D., 1996. With permission.)

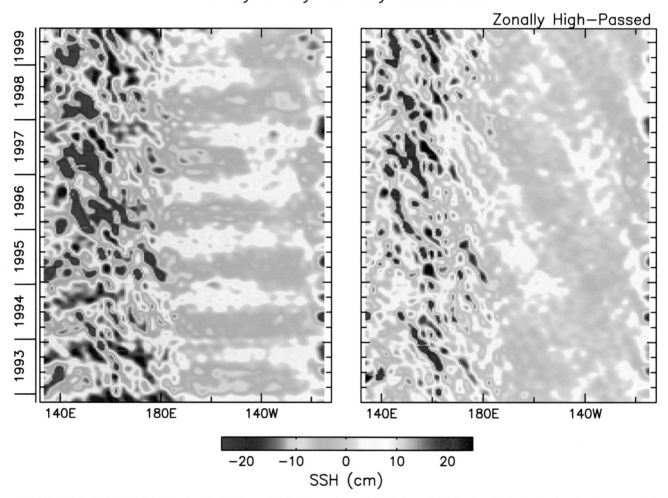

North Pacific, 32°N
6° by 6° by 60 day Smoothed

Zonally High—Passed

SSH (cm)

CHAPTER 2, FIGURE 11 Time-longitude plots of T/P observations of SSH along 32°N in the North Pacific for the 7-year period 1993–1999. In both panels, the raw data were smoothed to retain periods longer than 60 days and wavelengths longer than 6° of longitude by 6° of latitude. This filter transfer function is analogous to block averaging the data over 40 days by 3.5° of longitude by 3.5° of latitude. In the right panel, the steric sea level signal was removed by further zonally high-pass filtering the data to attenuate wavelengths longer than about 6000 km.

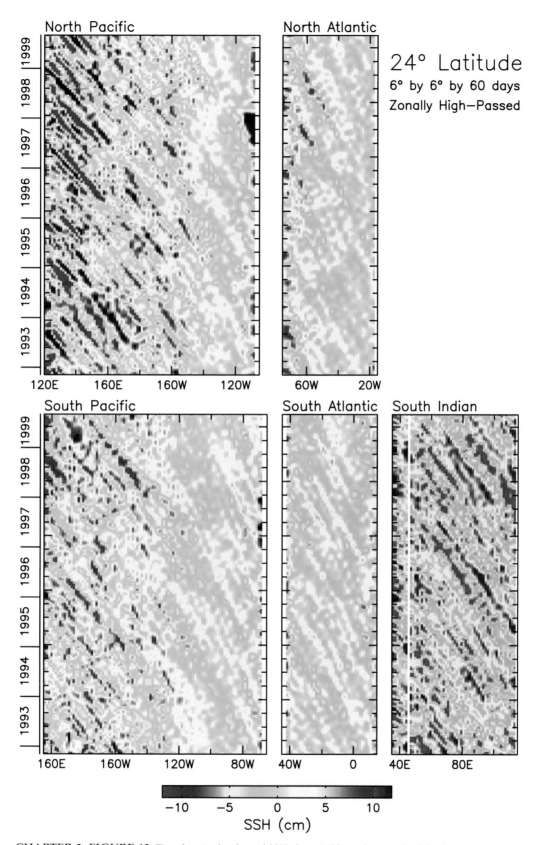

CHAPTER 2, FIGURE 12 Time-longitude plots of SSH along 24° latitude in each of the five ocean basins. The raw SSH data were smoothed as in Figure 11 and zonally high-pass filtered to attenuate wavelengths longer than about 8000 km.

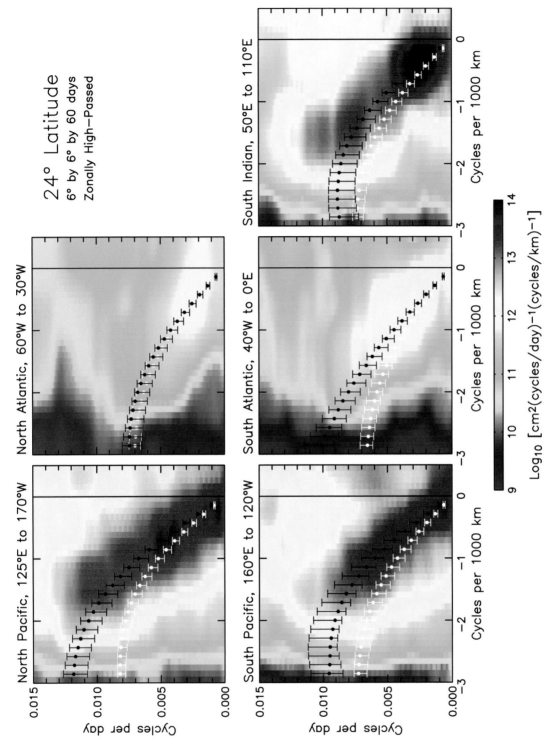

CHAPTER 2, FIGURE 15 Frequency-wavenumber spectra computed from 7 years of T/P data over subranges of the time-longitude plots shown in Figure 12. The SSH data were truncated to the longitude range labeled on each panel to restrict the spectral estimates to regions over which the phase speed characteristics were visually homogeneous longitudinally. Each spectral estimate was band averaged in space and time to obtain 18° of freedom. The dispersion relation computed from the eigenvalue problems for the standard theory with zero mean background flow and the extended theory that includes the baroclinic background mean flow are shown by the white and black circles, respectively. These circles and the associated confidence intervals at each wavenumber correspond to the median and central 75% of the distribution of individual eigenvalues at 1°-intervals along the longitudinal section over which the spectra were computed.

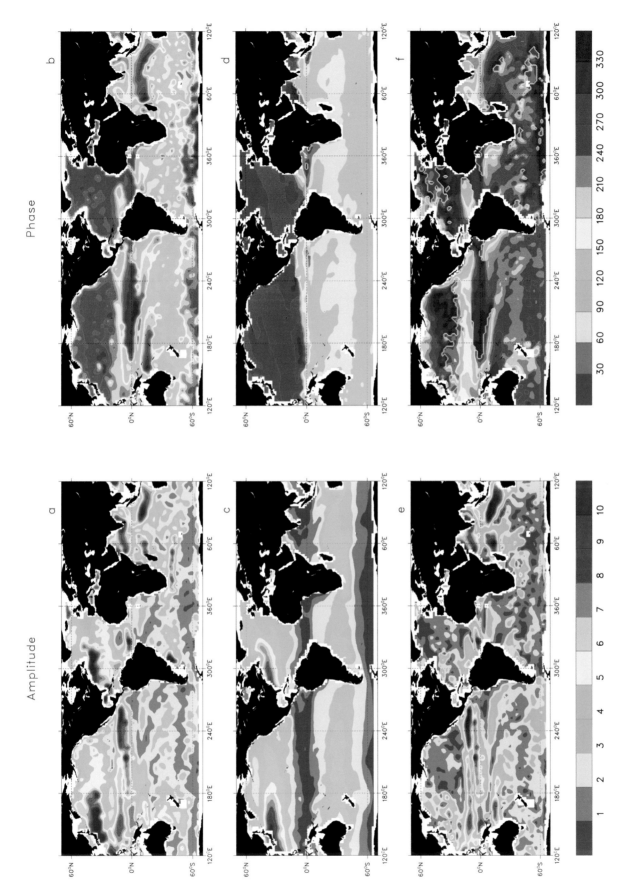

CHAPTER 2, FIGURE 17 Panels a and b: Amplitude (cm) and phase (degree) of the annual cycle of sea level variation determined from 3 years of T/P data (1993–1995); panels c and d: the steric component estimated from Eq. (7) based on the ECMWF heat flux; panels e and f: the residual signal after subtracting the steric component from the T/P observations. The phase is expressed in a way that its value is roughly the year-day number when the sea level is maximum. (From Stammer, D., 1997. With permission.)

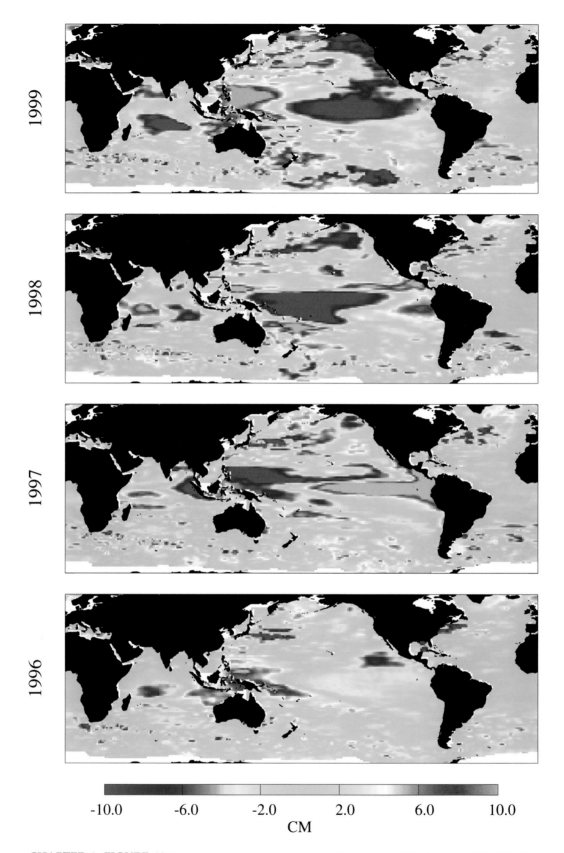

CHAPTER 2, FIGURE 18 Yearly averages of sea level anomalies from the T/P data for 1996–1999. The anomalies were computed relative to the 1993–1996 mean.

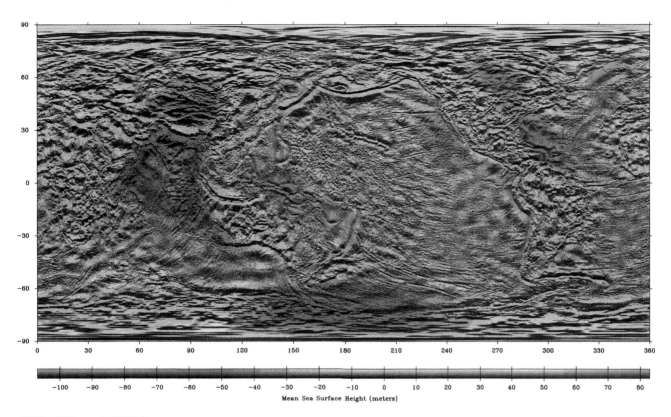

CHAPTER 10, FIGURE 11 The CSR98 global inverted barometric mean sea-surface model determined using satellite altimeter measurements collected from the Geosat, ERS 1, ERS 2, and TOPEX/POSEIDON missions. The map is color contoured according to the mean sea-surface height above the earth reference ellipsoid with the semi-major axis of 6378136.3 m and the reciprocal flattening of 298.257. The mean sea-surface height has variations from −107 m (south of India in purple) to 84 m (northeast of Australia in orange). Areas without mean sea-surface heights are filled with the EGM96 geoid undulation complete to the spherical harmonic degree 360. In order to emphasize the short-wavelength features, the map is illuminated (from the north). Note that most of the major sea-floor topography features produce a signature in the mean sea surface. The combination of coloring and shading demonstrates the value of image visualization techniques.

CHAPTER 10, FIGURE 14 Spectral power comparison of high degree geoid models with respect to the Kaula's rule. The OSU86F model (Rapp and Cruz, 1986) incorporates the GEOS 3 and Seasat data, the OSU91A model (Rapp *et al.*, 1991b) adds the Geosat ERM data, the EGM96 model adds the TOPEX/POSEIDON and Geosat GM data, and the CSR98 mean sea-surface model adds the ERS 1 and ERS 2 data. The sudden reduction of the CSR98 mean sea-surface curve at degree 360 is due to the absence of higher degree information over the land areas. The EGM96 geoid closely approximates the spectral power of the mean sea-surface field, indicating the enhancement of data density from old models. As Jekeli (1999) states, the EGM96 model may be under-powered in the high-degree spectrum, i.e., beyond degree 200.

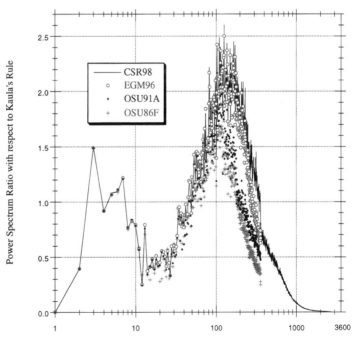

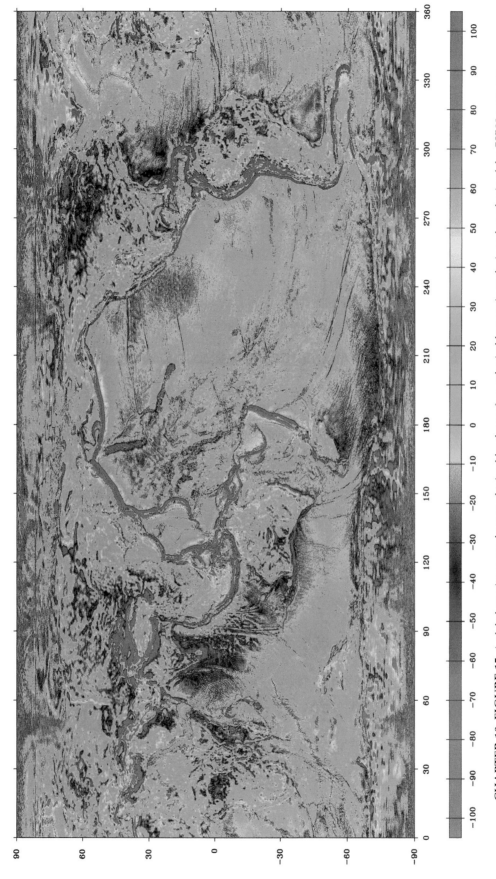

CHAPTER 10, FIGURE 15 A global gravity anomaly map obtained by the surface spherical harmonic analysis and synthesis of the CSR98 mean sea surface. No shading is applied. The magnitude of gravity anomaly can reach more than 400 mGal at some island and trench locations.

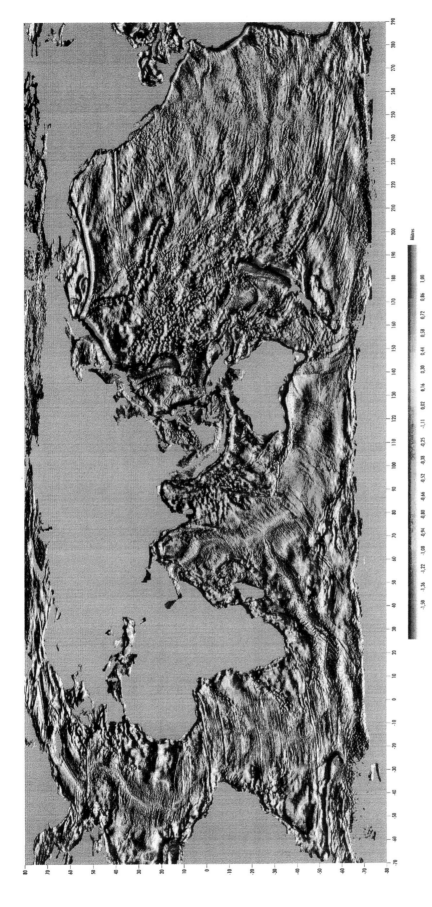

CHAPTER 11, FIGURE 1 Map of the marine geoid computed from high-resolution altimetry data of the ERS-1 and Geosat geodetic mission (from Cazenave *et al.*, 1996). A high-pass filter with a cut off at 2000 km has been applied.

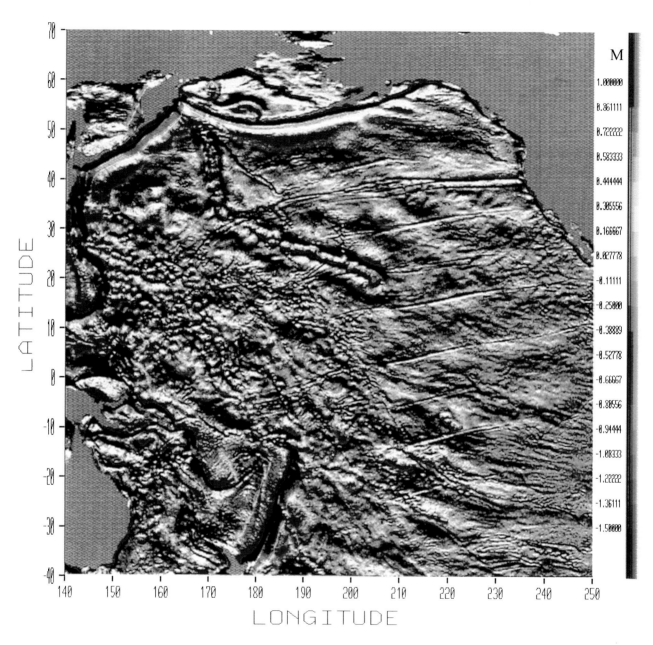

CHAPTER 11, FIGURE 8 High-pass filtered geoid map in the Pacific showing lineations at short and medium wavelengths elongated in the direction of absolute plate motion. These lineations are mostly visible between 20 °N and 30 °S latitude and 200 °E and 250 °E longitude.

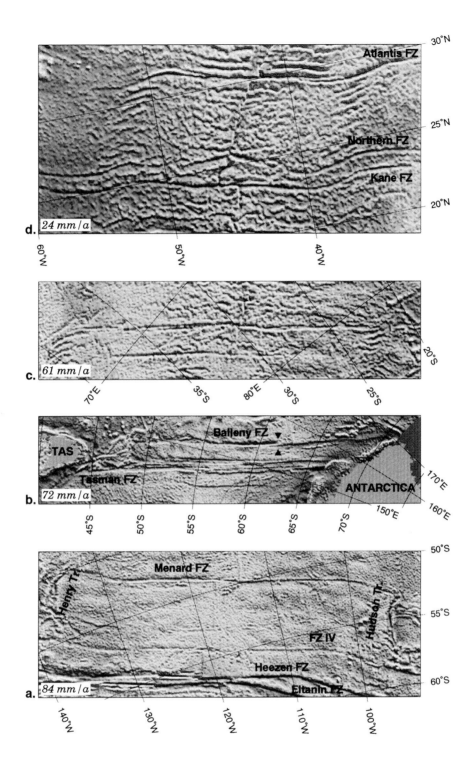

CHAPTER 11, FIGURE 10 Satellite-derived gravity anomalies along oceanic fracture zones in different spreading contexts: (a) Pacific-Antarctic Ridge; (b) easternmost termination of Southeast Indian Ridge between Tasmania (TAS) and Australia; (c) westernmost termination of Southeast Indian Ridge just south of the Rodrigues Triple Junction; (d) Central mid-Atlantic Ridge. Spreading rates correspond to present-day full rates at the center of each plot (after NUVEL-1A model : DeMets *et al.*, 1994). All plots are in oblique Mercator projections about chron 5 (11 Ma) Euler poles for the relevant plates. In such projection, the young parts (last 11 Ma) of fracture zones follow small circles (horizontal lines) about the Euler pole. Fracture zones along fast spreading ridges (a, b) are marked by a step in the gravity. This signature is asymmetric relative to the ridge axis; arrows in (b) show the change in polarity at the exact mid-point of the active transform fault along the Tasman FZ, where the age offset is equal to zero. Fracture zones in slower regimes (c, d) are outlined by a continuous gravity trough. Note the contrast in the gravity roughness of the oceanic crust between (a, b) and (c, d), and particularly the differernt gravity signatures of the spreading axes: axial rises in (a, b) vs. deep valley in (c, d). Same color scale as in Fig. 12 (Chapter 11).

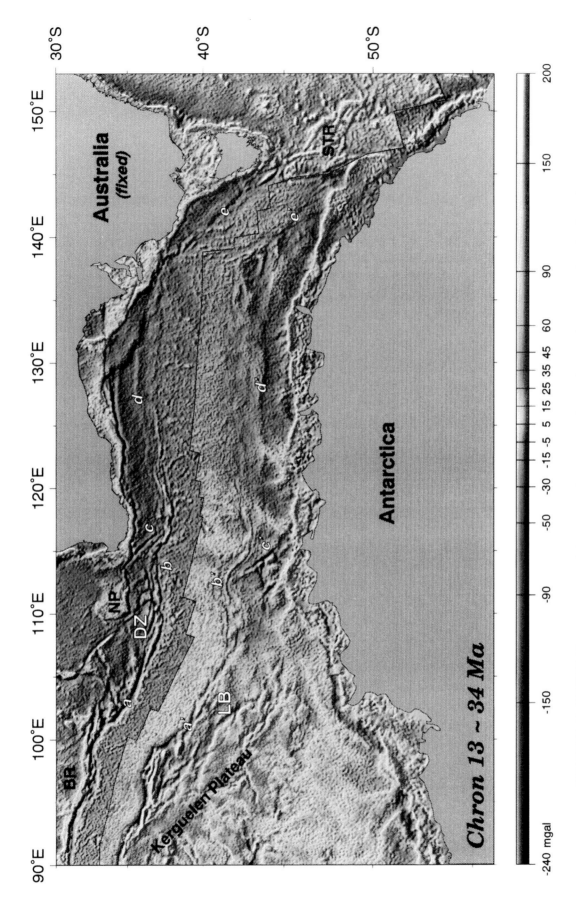

CHAPTER 11, FIGURE 12 Reconstruction of the eastern Indian Ocean at chron 13 (34 Ma; after Royer and Sandwell, 1989) using present-day satellite-derived gravity (Sandwell and Smith, 1997). All the oceanic crust younger than 34 Ma is removed; the continuous stair-stepped line shows the location and geometry of the Southeast Indian Ridge at this time. Note the close correspondence of gravity anomalies along the conjugate Australian and Antarctic margins: (a-a′) the rifted edges of Kerguelen Plateau and Broken Ridge (BR) and limits of the Labuan Basin (LB) and Diamantina Zone (DZ) that split apart in the Middle Eocene (~45 Ma); (b-b′) the limits of fast spreading oceanic crust; (c-c′) the conjugate limbs of the Leeuwin Fracture Zone, southeast of the Naturaliste Plateau (NP); (d-d′) and (e-e′) lineated gravity anomalies marking the landward limits of the oceanic crust; and further east, the conjugate transform margins of the South Tasman Rise (STR) and Antarctica.

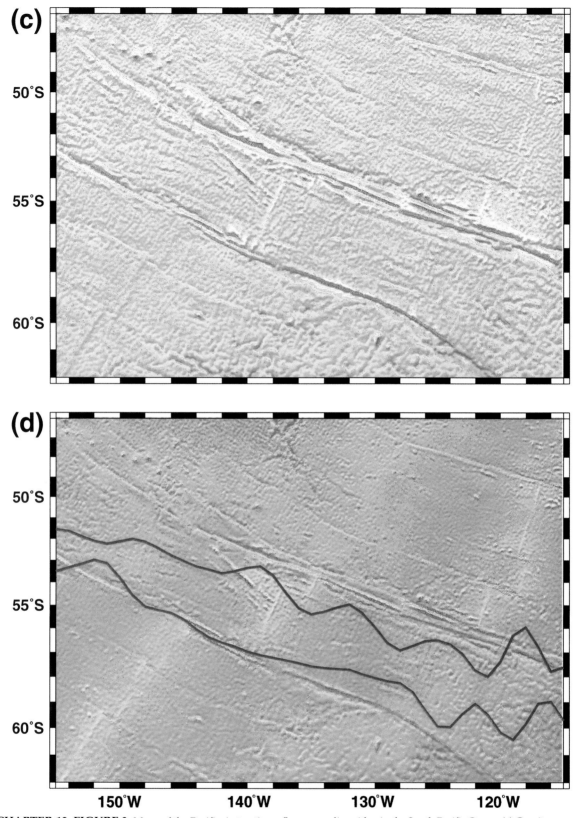

CHAPTER 12, FIGURE 3 Maps of the Pacific-Antarctic seafloor spreading ridge in the South Pacific Ocean. (c) Gravity anomaly (mGal) derived from all four altimeter data sets. (d) Bathymetry (m) estimated from ship soundings and gravity inversion. Red curves mark the Sub-Antarctic and polar fronts of the Antarctic Circumpolar Current (Gille, 1994). The Sub-Antarctic Front (SAF-red) passes directly over a NW-trending ridge having a minimum ocean depth of 135 m. The Polar Front (PF) is centered on the 6000-m deep valley of the Udintsev transform fault.

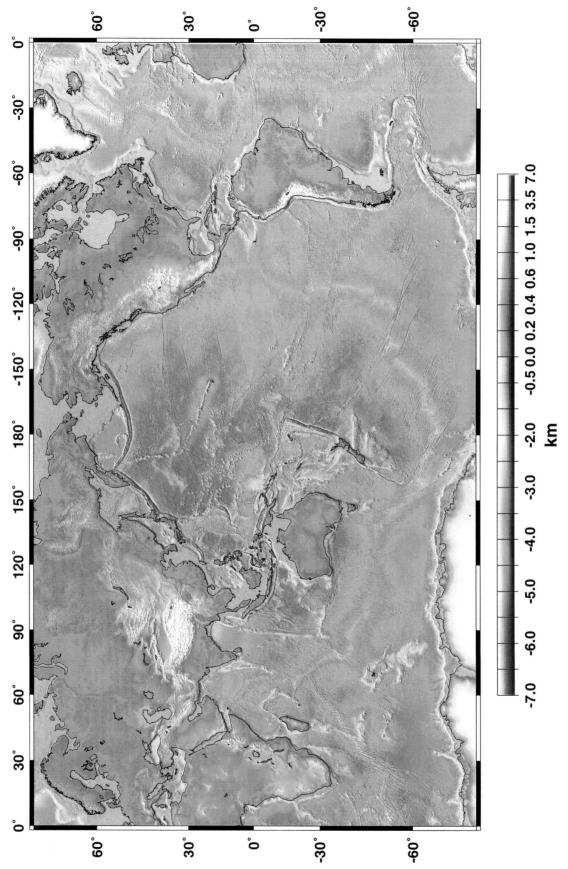

CHAPTER 12, FIGURE 4 Global map of predicted seafloor depth (Smith and Sandwell, 1997) and elevation from GTOPO-30.

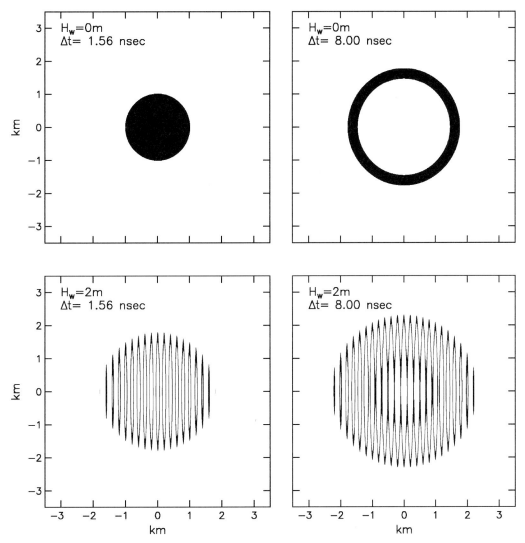

FIGURE 14 Plan views of the instantaneous illumination patterns for a pulse of duration $\tau = 3.125$ nsec incident on a flat sea surface (top panels) and a monochromatic wave surface with crest-to-trough wave height $H_w = 2$ m and 200 m wavelength (lower panels). The left panels correspond to the two-way travel time when the pulse-limited footprint area on the flat sea surface first becomes an annulus. The right panels correspond to the two-way travel time when the footprint on the wavy surface first becomes an annulus.

The time $t_{1/2}$ of the discontinuity of the step function is defined by Eq. (35), which is the two-way travel time for the midpoint of the pulse to return from mean sea level at nadir.

The preceding discussion clarifies the distinction between the pulse-limited footprint area and the illuminated area within the pulse-limited footprint that actually contributes to the radar return measured by the altimeter. The footprint area $A_{out}(t)$ within the outer perimeter defined by the radius $r_{out}(t)$ of the expanding projection of the radar pulse onto the sea surface represents the region within which specular reflectors contribute to the radar return received by the altimeter at any time and is therefore the area over which the range to mean sea level is estimated by the altimeter. This is discussed further in Section 2.4.7. The footprint area $A_I(t)$

is the integral of the sparsely illuminated area within the circular area with radius $r_{out}(t)$.

It should be emphasized that each of the probabilistic time series $\langle A_I(t) \rangle$ shown in Figure 15 represents the time evolution of the illuminated area averaged over a hypothetical infinite ensemble of realizations. Any particular realization $A_I(t)$ will be very noisy owing to the random nature of the phases of the various components of the wave field over the antenna footprint that contribute to the radar return measured at any particular two-way travel time. As the altimeter moves along the satellite orbit, the path lengths to the specular reflectors on the various wave facets change, resulting in pulse-to-pulse fluctuations in the $A_I(t)$ time series. The noisiness of each individual time series can be re-

Gaussian Wave Height Distribution

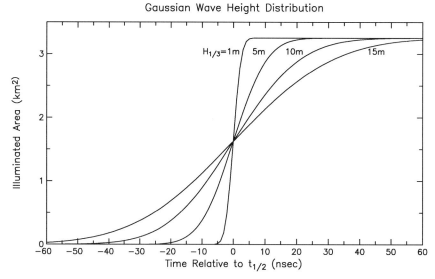

FIGURE 15 The time evolution of the average illuminated area for a pulse of duration $\tau = 3.125$ nsec and Gaussian sea surface height distributions with $H_{1/3} = 1$, 5, 10, and 15 m. Time is displayed in nanoseconds relative to the two-way arrival time $t_{1/2}$ of the midpoint of the pulse reflected from mean sea level at nadir.

duced by averaging as illustrated in Figure 16 from averages of 1, 25, and 1000 simulated $A_I(t)$ time series for a Gaussian wave-height distribution. If the individual time series in the average are statistically independent (which depends on the pulse repetition rate and the satellite ground-track velocity), the noise decreases as the square root of the number of time series in the average. Walsh (1982) has shown that the maximum pulse repetition rate for independent samples is proportional to the square root of $H_{1/3}$, increasing from about 1000 pulses per second for $H_{1/3} = 2$ m to about 3000 pulses per second for $H_{1/3} = 10$ m. The pulse repetition rate was 1000 pulses per second for the Seasat and Geosat altimeters and approximately 4000 pulses per second for the T/P dual-frequency altimeter. Successive altimeter pulses were therefore statistically independent for the Seasat and Geosat altimeters but are somewhat correlated for the T/P altimeter.

In addition to a dependence on the illuminated area within the pulse-limited footprint, the returned power measured by an altimeter depends on the antenna gain pattern and the normalized radar cross section of the sea surface, denoted as σ_0 in the radar Eq. (5). It is shown in Section 7 that σ_0 decreases with increasing wind-induced roughness of the sea surface. The implementation of an automatic gain control (AGC) loop in the electronics package effectively eliminates the σ_0 dependence of the returned power processed in the altimeter electronics. The power of the returned signal analyzed by the altimeter is therefore proportional to the illuminated area scaled by the off-nadir rolloff of the antenna gain pattern. The time evolution of the average returned power is given by the convolution Eq. (44) with an additional factor

to account for the antenna gain,

$$W(t) = W_{max}\, P_{FS}(t) * q_s(t) * p_\tau(t), \qquad (46)$$

where

$$P_{FS}(t) = G(t)U(t - t_{1/2}), \qquad (47)$$

is called the radar impulse response for a flat sea surface on a spherical earth, W_{max} is the power of the average signal output by the AGC at the time t_1 when the illuminated area becomes constant and $G(t)$ is the two-way antenna gain pattern $G(\theta)$ with off-nadir angle θ expressed in terms of two-way travel time t. The precise form of $G(t)$ is rather complicated (see, for example, Hayne, 1980; Barrick and Lipa, 1985; and Rodríguez, 1988). For present purposes, it suffices to say that $G(t)$ decreases approximately exponentially with increasing two-way travel time t.

The time series of returned power measured by an altimeter as described by the convolution Eq. (46) is referred to as the returned waveform. Simulated waveforms based on pre-launch determination of the radar impulse response Eq. (47) for the T/P dual-frequency altimeter are shown in Figure 17 for Gaussian wave-height distributions with $H_{1/3} = 2$ and 8 m. After the time t_1 when the average illuminated area within the pulse-limited footprint becomes constant, the power decreases approximately exponentially with increasing time owing to the rolloff of the antenna gain pattern. This is sometimes referred to as "plateau droop" of the waveform.

As discussed previously from the time evolution of the average illuminated area $\langle A_I(t) \rangle$, the time $t_{1/2}$ represents the two-way travel time Eq. (35) for the midpoint of the pulse to

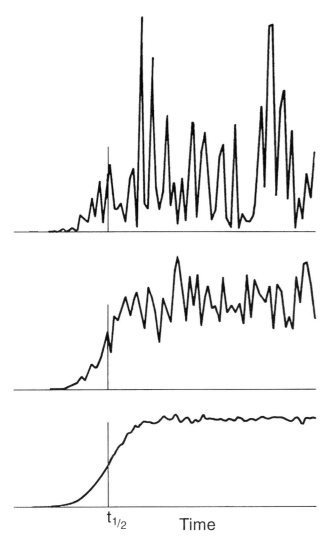

FIGURE 16   Averages of 1, 25, and 1000 (top to bottom) simulated independent time series of illuminated area for a Gaussian wave height distribution. (After Townsend *et al.*, 1981. With permission.)

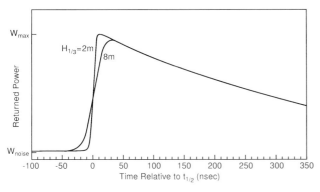

FIGURE 17   Simulated returned waveforms for the T/P dual-frequency altimeter for significant wave heights of $H_{1/3} = 2$ and 8 m. (Data provided courtesy of G. Hayne.)

reflect from mean sea level at nadir. Since the returned power is proportional to illuminated area, this two-way travel time corresponds to the midpoint on the leading edge of the returned waveform where the power $W_{1/2}$ is half the difference between W_{max} and the background power W_{noise} from electronic noise (see Figure 17). The altimeter thus determines the two-way travel time by identifying the half-power point $W_{1/2}$ on the leading edge of the returned waveform. Detailed descriptions of the methods used to track this half-power point are given in Sections 2.4.3 and 2.4.4. The range from the satellite to mean sea level at nadir is determined from this two-way travel time with adjustments for atmospheric refraction (see Section 3.1) and biases introduced by the non-Gaussian distribution of the specular scattering surfaces that contribute to the radar return (see Section 3.2). In addition to two-way travel time, the altimeter estimates $H_{1/3}$ from the slope of the leading edge of the waveform at the half-power point $W_{1/2}$ (see Section 6).

2.4.3. Pulse Compression

The description of pulse-limited altimetry in Sections 2.4.1 and 2.4.2 provides a very useful framework for understanding how the two-way travel time to mean sea level at nadir is measured by an altimeter. As previously discussed, a pulse duration of a few nanoseconds satisfies the requirements for a pulse-limited footprint diameter of 1–10 km, which is large enough to average out the effects of surface gravity waves yet small enough to resolve the Rossby radius of deformation that characterizes the spatial scales of mesoscale variability in the ocean. An adequate signal-to-noise ratio for a pulse duration this short requires a very high transmit power that places unacceptably high demands on the satellite power system and limits the lifetime of the transmitter. In practice, these limitations are overcome by a radar-ranging technique called pulse compression (see, for example, Ulaby *et al.*, 1981, Section 1–6; Ulaby *et al.*, 1986a, Section 7–5). A complete description of the technical details of the implementation of pulse compression in satellite altimetry is given by Chelton *et al.* (1989). Only a brief overview is given here.

All radar transmitters are plagued by internally generated white noise. Because the power of this white noise is uniformly distributed with frequency, the total noise energy is proportional to the frequency bandwidth of the receiver. For a given bandwidth, the signal-to-noise ratio of the radar system is therefore determined by the total energy of the pulse, i.e., the received power integrated over the pulse duration. A high signal-to-noise ratio can thus be achieved by transmitting a long pulse. As described in Section 2.4.1, however, the pulse-limited footprint diameter increases with increasing pulse duration. The small footprint size of a very short pulse can be preserved as summarized below by transmitting a frequency-modulated long pulse and analyzing the returned signal in a way that is equivalent to having transmitted a short pulse.

The frequency modulation that has traditionally been used in altimeter radars consists of a transmitted signal that is swept linearly from a frequency f_1 to a lower frequency f_2 over the pulse duration τ'. Such a linearly frequency-modulated pulse is called a chirp. The chirp duration τ' can be chosen to satisfy the energy requirements of the system within the limitations of the time-bandwidth product of the chirp generator. The relatively long chirp can be passed through a dispersive filter that produces a time delay that increases linearly with frequency, thus compressing the chirp to a pulse of duration $\tau \ll \tau'$ by delaying the early portion of the chirp more than the later portion. The power of the compressed pulse is thereby greatly increased. This method of pulse compression was used on the GEOS-3 altimeter. A disadvantage of this technique is that it is sensitive to variable filter gains and other problems with dispersive filters that complicate interpretation of the returned signal.

Because the dispersive filtering is a linear operation, it can be applied at any stage of the processing. In particular, the uncompressed chirp can be transmitted and the compression can be applied to the radar return within the altimeter electronics onboard the satellite. This is achieved as summarized below by comparing the radar return with a delayed replica of the transmitted chirp and analyzing the difference signal in the frequency domain rather than in the time domain. This equivalent technique of pulse compression was first used on the Seasat altimeter and has been used on all subsequent satellite altimeters.

The basic technique can be understood by describing the processing of an echo from a single specular scatterer on the sea surface. Consider first the radar return from nadir mean sea level at a distance R from the altimeter. Neglecting the effects of a Doppler shift from the relative velocity between the satellite and the sea surface (see Section 2.4.6), the returned signal is a replica of the transmitted chirp, except delayed by two-way travel time $t_{1/2} = 2R/c$ as shown by the short-dashed line in Figure 18a. The reflected chirp is passed through a low-noise amplifier and mixed with an internally generated "deramping chirp" that is identical to the transmitted and reflected chirps, except offset for electronic convenience by an "intermediate frequency" f_{IF}, as shown by the long-dashed line in Figure 18a. The deramping chirp is generated at a time lag t_d that is determined by the onboard adaptive tracking unit as described in Section 2.4.4. For the present discussion, it will be assumed that this timing of the deramping chirp is correct. The difference signal obtained by subtracting the deramping chirp from the chirp reflected from nadir mean sea level consists of a harmonic signal with a frequency f_0 (see short-dashed line in Figure 18b) that is specified prior to launch as part of the overall design of the radar system as described below. From the geometry of the

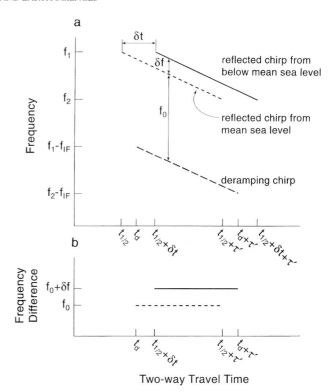

FIGURE 18 A schematic description of the chirp reflection from a single specular scatterer at nadir mean sea level $\zeta = 0$ received at two-way travel time $t_{1/2}$ (short-dashed line) and the chirp reflection from a single specular scatterer received at a two-way travel time δt later from a point $\zeta = -c\delta t/2$ on the sea surface below mean sea level (solid line). The long-dashed line is the deramping chirp that is offset by frequency f_{IF} from the transmitted chirp and generated at a time delay $t_d = t_{1/2} + (f_{IF} - f_0)/Q$, where $Q = (f_1 - f_2)/\tau'$ is the sweep rate of the chirp (i.e., the inverse of the slope of the reflected and deramping chirps shown in the figure.) As shown in the bottom panel, the difference signals obtained by subtracting the deramping chirp from the reflected chirps received from mean sea level and $\zeta = -c\delta t/2$ are harmonic signals with frequencies f_0 and $f_0 + \delta f$, respectively, where $\delta f = Q\delta t$.

figure, f_0 is related to the intermediate frequency by

$$f_0 = f_{IF} - Q(t_d - t_{1/2}), \qquad (48)$$

where

$$Q = \frac{\Delta F}{\tau'} \qquad (49)$$

is the sweep rate of the chirp for a swept frequency range of $\Delta F = f_1 - f_2$.

Now consider a specular scatterer at nadir that is displaced by a height ζ relative to mean sea level. The reflected chirp arrives at the altimeter with a two-way travel time that differs from $t_{1/2}$ by $\delta t = -2\zeta/c$. Positive and negative ζ thus correspond to arrival times that are, respectively, earlier and later than the two-way travel time $t_{1/2}$ for a chirp reflected from nadir mean sea level. The difference signal obtained by subtracting the deramping chirp from a chirp reflected from a point above or below nadir mean sea level is therefore a harmonic signal with frequency lower or higher, respectively, than the frequency f_0 of the deramped chirp from mean sea level. The case of the echo from a scatterer below mean sea level is shown by the solid line in Figure 18a. From the geometry of the figure, the frequency shift of the deramped signal is seen to be

$$\delta f = Q \delta t. \qquad (50)$$

Deramping thus uniquely maps the two-way travel time difference δt relative to $t_{1/2}$ into a frequency shift δf relative to f_0 that is linearly proportional to δt. In other words, there is a one-to-one mapping between the frequency of the output signal and the two-way travel time between the altimeter and the specular scatterer on the sea surface.

Because of the finite duration of the chirp, the frequency power spectral density of the deramped signal is smeared by a convolution of the harmonic signal with the "spectral window" associated with the transmitted signal. For an ideal chirp of duration τ' that is turned on and off instantaneously at the beginning and end of the chirp, the spectral window of the chirp is given by

$$p_\tau(f) = \left[\frac{\sin(\pi f \tau')}{\pi f \tau'} \right]^2. \qquad (51)$$

From Eqs. (49) and (50), this frequency-domain representation of the chirp spectral window can be expressed equivalently in terms of two-way travel time $t = f/Q$ as

$$p_\tau(t) = \left[\frac{\sin(\pi \Delta F t)}{\pi \Delta F t} \right]^2. \qquad (52)$$

This time-domain description of the idealized chirp spectral window is the point target response introduced in Sections 2.4.1 and 2.4.2. The two-way travel time resolution of this point target response is defined by the half-power width of the center lobe, which is $0.886 \Delta F^{-1}$. Actual chirps cannot be turned on and off instantaneously. The tapering of the chirp at the beginning and end results in a chirp spectral window with a center lobe that has a slightly wider half-power width and smaller side lobes. The point target response is usually described as having a characteristic width of $\tau = \Delta F^{-1}$.

It is thus seen that the deramping applied to the reflected frequency-modulated chirp of duration τ' yields a frequency-domain description of the deramped signal with a two-way travel time resolution that is exactly equivalent to having transmitted a much shorter pulse of duration $\tau = \Delta F^{-1}$. This effective pulse duration τ depends only on the swept frequency range of the chirp and is thus independent of the actual chirp duration τ'. The frequency bandwidth ΔF is chosen to satisfy the desired effective pulse duration τ. For the $\Delta F = 320$ MHz bandwidth that has been used with the Seasat, Geosat, and T/P altimeters,[12] the two-way travel time resolution of the chirp is $\tau = 3.125$ nsec. The relevance of the discussion of an altimeter with a pulse duration of 3.125 nsec in Sections 2.4.1 and 2.4.2 thus becomes apparent. The point target response $p_\tau(t)$ in Eqs. (44) and (46) is given approximately by the time-domain description Eq. (52) of the chirp spectral window. The actual point target responses for the Seasat and Geosat altimeters are shown in Figure 6 of Rodríguez (1988) (see also Figure 8 of Lipa and Barrick (1981) for the Seasat point target response).

For the Seasat altimeter, the dispersive wave devices that were available for chirp generation were limited to a time-bandwidth product of a few hundred. The sweep time of the signal was $\tau' = 3.2$ μsec with the frequency range $\Delta F = 320$ MHz centered on 13.5 GHz. Beginning with Geosat, a digital chirp generation technique was used which allowed a longer sweep time of $\tau' = 102.4$ μsec with the same frequency range of $\Delta F = 320$ MHz. The Geosat chirp was centered on 13.5 GHz. The K_u-band and C-band frequencies of the T/P dual-frequency altimeter are centered on 13.6 and 5.3 GHz, respectively.

The 32-fold increase of the chirp duration for the Geosat and T/P altimeters could result in an improved signal-to-noise ratio. Alternatively, the increased signal energy from the longer pulse could be offset by transmitting with a lower power, achieving approximately the same signal-to-noise ratio. The power requirement of the travelling wave tube amplifier used in the Seasat altimeter was 2 kW. The longer signal duration reduced the power requirement to only 20 W for the Geosat and K_u-band T/P altimeters, thus significantly extending the lifetimes of the amplifiers, as well as reducing the overall power requirements of the satellite. With the lower power requirements of the longer pulse duration, the conventional travelling wave tube amplifier can be replaced with more reliable solid-state amplifiers. Solid-state amplifiers were implemented for the C-band frequency of the T/P dual-frequency altimeter and the CNES K_u-band single-frequency altimeter also onboard the T/P satellite. The peak powers of these solid-state amplifiers are 20 W and 5 W, respectively. Future altimeters will all use solid-state amplifiers transmitting with peak power of 6–8 W.

The above description of the returned chirp from a single specular scatterer on the sea surface is easily extended to the case of multiple scatterers. For a finite number of scatterers, the deramped signal would consist of a superposition of discrete harmonic signals with frequencies linearly related to two-way travel time by Eq. (50). In the frequency domain, the power spectral density would consist of a series of line spectra that are each broadened somewhat by convolution

---

[12]The frequency bandwidth of the ERS-1 and ERS-2 altimeters was chosen to be $\Delta F = 330$ MHz, which corresponds to a two-way travel time resolution of $\tau = \Delta F^{-1} = 3.03$ nsec.

with the chirp spectral window Eq. (51) because of the finite record length of the chirp.

For the actual case of a continuous distribution of specular scatterers on the sea surface, the power spectral density of the deramped radar return is continuous with a shape that depends on the probability density function of the specular scatterers and on the antenna gain characteristics. To first order, the distribution of specular scatterers is the same as the distribution of the sea surface elevation ζ relative to mean sea level, which is approximately Gaussian as described by Eq. (40) with a scale that is directly related to $H_{1/3}$ and the associated standard deviation of ζ from the presence of waves. This probability density function can be expressed in terms of two-way travel time $\delta t = -2\zeta/c$, which can be transformed into frequency by Eq. (50) to obtain a frequency-domain description $q_s(f)$ of the probability density function of specular scatterers.

From any small area on the sea surface, the power spectral density of the deramped radar return from the transmitted chirp is thus approximately Gaussian with a small amount of smoothing from convolution with the chirp spectral window Eq. (51). If the deramping chirp is timed correctly to coincide with the radar return from nadir mean sea level, the Gaussian spectrum from any particular small area is centered at the frequency corresponding to the two-way travel time between the altimeter and that point on the sea surface. The total power spectral density of the deramped radar return consists of the superposition of the Gaussian spectra contributed from points on the sea surface at different two-way travel times corresponding to different off-nadir angles across the antenna beamwidth. This superposition can be described by the convolution of the Gaussian probability density function of specular scatterers with the frequency-domain equivalent of the unit step function Eq. (45). The power from off-nadir angles is attenuated by the rolloff of the antenna gain pattern so that the total power spectral density of the radar return, including the effects of the point target response $p_\tau(f)$, is given by the convolution

$$W(f) = W_{\max} P_{\mathrm{FS}}(f) * q_s(f) * p_\tau(f), \qquad (53)$$

where

$$P_{\mathrm{FS}}(f) = G(f) U(f - f_0) \qquad (54)$$

and $W_{\max}$ is the maximum of the power spectral density over all frequencies. The frequency-domain description $G(f)$ of the angular dependence of the two-way antenna gain pattern is an exponentially decaying function of frequency (see Appendix A3 of Chelton *et al.*, 1989).

With the transformation Eq. (50), the frequency-domain description Eq. (53) of the power spectral density of the deramped radar return from the transmitted chirp of duration τ' is identical to the time-domain description Eq. (46) of the power of the radar return from a much shorter pulse of duration τ. The shape of the spectral waveform Eq. (53) is thus

the same as the time-domain returned waveform shown in Figure 17, with the distinction that the abscissa is frequency rather than two-way travel time. If the deramping chirp is timed correctly, the half-power point on the leading edge of the spectral waveform Eq. (53) corresponds to the frequency f_0 of the deramped chirp from nadir mean sea level (see Figure 19). As discussed in Section 2.4.2 for the case of the time-domain returned waveform, the power spectral density Eq. (53) is achieved from an ensemble average of noisy spectra constructed from the deramped radar returns from a large number of successive chirps.

Analysis of the radar return from the transmitted chirp is carried out in the frequency domain by spectral analysis of the deramped signal. Because of the smearing effects of the point target response, the frequency resolution of the sample power spectral density obtained from the record length τ' of the deramped signal is $\delta f' = 1/\tau'$. For the T/P chirp duration of $\tau' = 102.4 \ \mu\mathrm{sec}$, this frequency resolution is $\delta f' = 9.766$ kHz. There is no advantage to computing the sample power spectral density at a frequency interval smaller than $\delta f'$. From Eqs. (49) and (50), this is equivalent to computing the spectrum with an effective two-way travel time resolution of $\delta t_{\mathrm{eff}} = \Delta F^{-1}$. For the T/P swept frequency range of $\Delta F = 320$ MHz, sampling of the spectral waveform at discrete frequencies $\delta f'$ is thus equivalent to "range gating" the time-domain description of the returned waveform at two-way travel time intervals δt_{eff} equal to the effective pulse duration $\tau = 3.125$ nsec.

It can be noted that the Fourier frequency interval $\delta f'$ depends only on the chirp duration τ'. The sample spectrum can therefore be computed by discretely sampling the deramped signal at an arbitrary sample interval $\delta t'$. Care must be taken, however, to avoid aliasing of unresolved variabil-

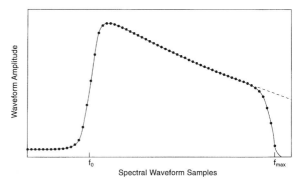

FIGURE 19   A schematic representation of discrete sampling of the T/P spectral waveform. The waveform rolloff is based on pre-launch simulations (courtesy of G. Hayne). The actual T/P waveforms are sampled at 128 frequencies. For clarity of presentation, only 64 discrete samples are shown on the figure. The frequency f_0 corresponds to the frequency of the deramped chirp received from nadir mean sea level. Only the portion of the waveform with frequencies below $f_{\max}$ is used for waveform processing. The steep dropoff of the spectral waveform at frequencies near $f_{\max}$ is caused by the effects of the anti-aliasing filter applied to attenuate frequencies higher than $f_{\max}$.

ity with frequencies higher than the associated Nyquist frequency $f_N = (2\delta t')^{-1}$. For a specified sample interval $\delta t'$, the analog deramped signal must therefore be low-pass filtered with an "anti-aliasing" filter designed to attenuate frequencies higher than f_N. This filtering must be applied prior to discretely sampling the deramped signal.

The Nyquist frequency of the anti-aliasing filter (and hence the sample interval $\delta t'$) can be selected to retain any desired range of frequencies in the spectral waveform Eq. (53). As described in Section 2.4.2, the two-way travel time $t_{1/2}$ and the significant wave height $H_{1/3}$ are estimated from the timing and slope at the half-power point on the leading edge of the time-domain description of the returned waveform (see Figure 17). In the frequency domain, this is equivalent to the frequency f_0 and slope of the leading edge of the spectral waveform at frequencies near f_0. Some portion of the spectral waveform prior to f_0 is also needed in order to estimate the noise floor of the waveform. In addition, the trailing edge of the waveform provides important information about antenna pointing errors (see Sections 2.4.4 and 2.4.6). The portion of the spectral waveform that is needed for altimetry is thus restricted to the frequency range below a specified frequency f_{max} as shown in Figure 19. By taking advantage of this reduced frequency range requirement, the deramped signal can be sampled at a coarse sample interval after applying the anti-aliasing filter.

In the final stages of the T/P onboard digital processing, the deramped signal is mixed with a harmonic signal to remove the f_{IF} frequency offset of the deramped signal. An anti-aliasing filter with a maximum frequency of $f_{max} = 625$ kHz is then applied. This low-pass filtered signal is then discretely sampled with the corresponding Nyquist sample interval of $\delta t' = 0.8$ μsec. For the record length $\tau' = 102.4$ μsec, this corresponds to a total of 128 samples. Samples of the spectral waveform at 128 discrete frequencies uniformly spaced by $\delta f' = 9.766$ kHz are obtained from the discretely sampled signal by fast Fourier transforming onboard the satellite.

A final consideration in the design of the digital processing system of the altimeter is specification of the frequency f_0 corresponding to the two-way travel time of the chirp reflected from nadir mean sea level. As discussed previously, this corresponds to the half-power point on the leading edge of the spectral waveform if the timing t_d of the deramping chirp is correct. For T/P, the time delay t_d between generation of the transmitted chirp and the deramping chirp was chosen so that this point on the waveform is aligned half way between frequency samples 32 and 33 (see Figure 19). This placement at a relatively low frequency provides a long record (large frequency range) of the trailing edge of the returned waveform for estimation of antenna pointing errors as described in Sections 2.4.4 and 2.4.6. The frequency f_0 of the track point is maintained by the onboard adaptive tracking unit that sets the timing of the deramping chirp based on

analysis of previous waveforms. It is evident from Eq. (50) and Figure 18 that the frequency of the half-power point on the leading edge of the spectrum of the deramped signal is higher than f_0 if the deramping chirp is generated too early. Likewise, late generation of the deramping chirp results in a half-power point at a frequency lower than f_0. The details of the adaptive tracking unit adjustments of the timing of the deramping chirp are described in Section 2.4.4.

The two-way travel time $t_{1/2}$ to nadir mean sea level is thus obtained from the timing t_d of the deramping chirp that is continuously updated by the adaptive tracking unit. It should be noted that the accuracy of this estimate of the two-way travel time to nadir mean sea level is sensitive to the assumption that the probability density function of the specular scatterers is symmetric. The actual distribution of specular scatterers is skewed from Gaussian. The non-Gaussian distribution of scatterers results in a half-power point at a frequency that differs from the frequency f_0 for two-way travel time to nadir mean sea level. As discussed in Section 3.2, empirical corrections must be applied to remove this sea-state bias in the estimate of the two-way travel time.

2.4.4. Onboard Adaptive Tracking

The factors affecting the shape of altimeter returned waveforms were described in Section 2.4.2. The practical needs for a space-borne altimeter are to extract information from the waveform about the two way travel time, the significant wave height $H_{1/3}$ and the normalized radar cross section σ_0 in an efficient manner and to maintain the leading edge of the returned waveform within a tracking or processing window under varying sea-state conditions and varying range from the altimeter to nadir mean sea level. These functions are performed by the adaptive tracking unit (ATU). The ATU consists of software in a microprocessor onboard the altimeter that performs high-level processing on signals provided by the hardware and generates information for telemetry to ground receiving stations. The ATU also controls the hardware via the "synchronizer" which generates the timing of the transmitted and deramping chirps and other signals that control the altimeter hardware on a pulse-by-pulse basis.

Prior to analysis by the ATU, the returned signal from each transmitted chirp is deramped, low-pass filtered to prevent aliasing, discretely sampled and fast Fourier transformed to obtain spectral estimates at 128 discrete frequencies (referred to here as waveform samples[13]) as described in Section 2.4.3. A very important function of the ATU is to analyze the 128 waveform samples to provide information to the synchronizer for the timing of the deramping chirp.

---

[13]The discrete frequencies of the waveform samples are sometimes referred to as "gates," but this term is reserved here for combinations of waveform samples used in tracking as described below.

For the K_u-band frequency of the T/P altimeter, the leading edge of the returned waveform falls outside of the frequency range passed by the narrow anti-aliasing filter if the deramping chirp is generated more than about 200 nsec early or about 100 nsec late. As described in Section 2.4.3, this low-pass filter is intentionally narrow in order to reduce the required sampling rate of the returned signal and the amount of onboard signal processing. The narrow frequency range translates into a small span of two-way travel times available to the ATU; the time span for 128 waveform samples at the two-way travel time resolution of 3.125 nsec is 400 nsec. The corresponding one-way range interval sampled by the waveform is 60 m.

The ATU software is programmed to time the deramping chirp to align the half power point of the leading edge of the waveform at a specific point in the set of waveform samples. This point with frequency f_0 is called the track point. As discussed in Section 2.4.3, this half-power point corresponds to the radar return from nadir mean sea level at two-way travel time $t_{1/2}$. For the T/P dual-frequency altimeter, the deramping chirp is timed for the track point to be midway between waveform samples 32 and 33 for the K_u-band altimeter and midway between waveform samples 35 and 36 for the C-band altimeter. The offset between the two track points allows for the larger ionospheric delay of the C-band pulses (see Section 3.1.3). Different choices for the track point have been adopted for other altimeters based on the number of waveform samples in the signal processor (see Chelton *et al.*, 1989, for descriptions of the track points for the Seasat and Geosat altimeters). The ATU determines the offset of the half-power point on the leading edge of the waveform from the track point and provides information for the synchronizer to adjust the timing of the deramping chirp to realign subsequent waveforms with the half-power point at the frequency f_0.

To reduce the inherent noisiness of individual returned waveforms owing to the random nature of the wave height field (see Figure 16) that determines the distribution of scattering surfaces, the waveform samples from successive pulses are averaged over a "track interval" spanning about 50 msec. The specific number of pulses in a track interval depends on the pulse repetition rate of the altimeter. For the high pulse repetition rate of about 4000 pulses per second for the K_u-band frequency of the T/P dual-frequency altimeter, there are 240 pulses in a track interval. The range can change by as much as 30 m sec^{-1} from along-track variations of the orbit height and geoid variations. This corresponds to a 150-cm change in the range to mean sea level over a 50-msec track interval. The range rate must therefore be taken into consideration by the ATU in determining the adjustment of the timing of the deramping chirp for each pulse. In practice, the deramping chirp is timed in coarse steps of 12.5 nsec. Finer timing adjustments are effectively achieved by applying a "linear phase rotation" to the der-

amped signal in the time domain before further signal processing. The discrete sample at two-way travel time t_k in the deramped time series is multiplied by $\exp(i2\pi \delta f_d t_k)$, where δf_d is the frequency offset of the waveform determined by the ATU as described below. By the frequency-shift theorem (Bracewell, 1986), this is equivalent to shifting the spectral waveform in frequency by δf_d, thus aligning the waveform with the track point. By this technique, a two-way time resolution of 0.0488 nsec is obtained.

For the T/P dual-frequency altimeter, the 128 waveform samples averaged over each 240-pulse track interval are analyzed by the ATU by forming "gates" consisting of averages of neighboring waveform samples as shown in Figure 20. These gates are used in the onboard tracking. The noise level in the altimeter waveforms is estimated by a "noise gate" that consists of the average over waveform samples 5–8, denoted here as S_{noise}. The average signal level across the leading edge of the waveform is represented by the "AGC gate," denoted as S_{agc}, which consists of the average of 16 waveform samples on each side of the track point (i.e., waveform samples 17–48). Over the full range of significant wave-height conditions, the AGC gate includes some waveform samples from the "noise floor" preceding the leading edge of the waveform and some waveform samples from the "plateau region" following the leading edge of the waveform. Because it is the average of a large number of waveform samples, S_{agc} is relatively insensitive to noise in the individual waveform samples and to misalignment of the waveform. Assuming that the misalignment is not large, the difference $(S_{\text{agc}} - S_{\text{noise}})$ is therefore very nearly equal to the power at the midpoint of the leading edge of the waveform, regardless of the precise alignment of the waveform.

The heart of the onboard tracking lies in the computation of five combinations of "early," "middle," and "late" gates of varying widths centered on the track point as shown in Figure 20. As discussed in Section 2.4.2, the slope of the leading edge of the waveform is inversely related to $H_{1/3}$ (see Figures 15 and 17). A large gate width can therefore be used for large wave heights because of the spread of the leading edge of the waveform. For small wave heights, however, the gate widths must be narrow in order to resolve the steeper leading edge of the waveform. The different gate widths thus accommodate the wide range of wave conditions sampled globally by an altimeter. The averaging of neighboring waveform samples wherever possible mitigates the effects of waveform noise on the onboard estimates of the power in the early, middle, and late portions of the leading edge of the waveform.

The five sets of early, middle, and late gates, denoted here by the gate index m, are formed simultaneously by the ATU. A sixth set of very wide early and late gates is formed in order to select the proper set from among the other five as described below. For each set of tracking gates, the middle gate $S_{\text{mid}}(m)$ is centered on the track point and consists of

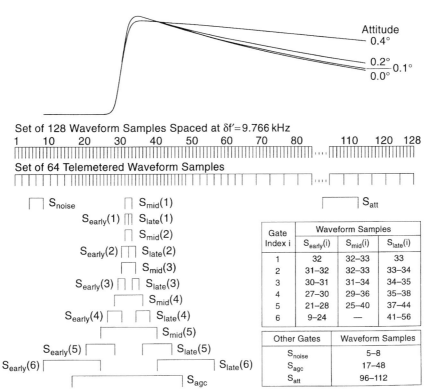

FIGURE 20 The expected shapes of the waveforms for the K_u-band frequency of the T/P dual-frequency altimeter for various off-nadir antenna pointing angles. The 128 waveform samples that are computed and processed onboard the satellite and the compressed 64 samples that are telemetered to ground receiving stations are shown below the waveforms. The combinations of the waveform samples that form the various gates used in the adaptive tracking unit and for ground-based attitude determination are indicated on the figure. (Adapted from Zieger *et al.*, 1991. With permission.)

averages over two or more waveform samples as shown in Figure 20. The associated early gate $S_{early}(m)$ and late gate $S_{late}(m)$ are averages of waveforms preceding and following the track point, respectively. The number of waveform samples in the early and late gates increases from one to eight for sets $m = 1$–5 and sixteen for set $m = 6$ (see Figure 20). The slope of the leading edge of the waveform is determined from the difference between the late and early gates. The "voltage" related to significant wave height (swh) is determined for each of the five ratios

$$V_{swh}(m) = \frac{S_{late}(m) - S_{early}(m)}{S_{late}(6) - S_{early}(6)}. \qquad (55)$$

Normalizing by the difference for gate 6 serves to reduce the sensitivity of $V_{swh}(m)$ to misalignment of the leading edge of the waveform and to errors in the AGC that are applied to the returned signal by the hardware. The value of $V_{swh}(m)$ that is closest to a reference value in an onboard look-up table determines the gate index m to be used in subsequent processing (see Townsend, 1980, and Zieger *et al.*, 1991, for detailed discussions). The ATU is programmed for the transitions from one gate index to the next to occur at $H_{1/3}$ values of 1, 3, 6.2, and 13 m. The ATU is referred to as "adap-

tive" because of this ability to adjust the width of the gates according to $H_{1/3}$.

The middle gate for the chosen index m is used to determine the misalignment δf_d of the waveform relative to the track point f_0. Assuming that tracking of the half-power point is not off by a large amount, the frequency offset of the waveform is given by

$$\delta f_d = a_{agc}(S_{agc} - S_{noise}) - S_{mid}(m), \qquad (56)$$

where $a_{agc} \approx 1$ is a scaling factor that is determined empirically from pre-launch testing. As noted previously, $(S_{agc} - S_{noise})$ is very nearly equal the power at the midpoint on the leading edge of the waveform. If the deramping chirp is properly timed so that this midpoint is aligned at frequency f_0, then $S_{mid}(m)$ is also equal to the power at the midpoint. The scaling factor a_{agc} is intended to make $a_{agc}(S_{agc} - S_{noise})$ equal to $S_{mid}(m)$ for proper waveform alignment. As described below, the precise value of a_{agc} depends on the antenna pointing angle. The value adopted for the ATU corresponds to the anticipated nominal pointing of the altimeter (0.1° for the T/P altimeter).

Assuming that the scaling factor a_{agc} is correct and that the deramping chirp has been timed for proper alignment of the waveform, the terms on the right side of Eq. (56) sum to zero and there is no frequency offset of the waveform. If these terms sum to a positive value, then $S_{\mathrm{mid}}(m)$ is too small which implies that the half-power point of the leading edge of the waveform was misplaced in a gate later than the track point. Similarly, a negative value for the sum of these terms implies that the half-power point of the leading edge of the waveform was misplaced in a gate earlier than the track point. Note that waveform misalignment is not actually found by comparing the middle gate to half of the maximum value of the waveform. As the individual waveform samples are quite noisy, the error or fluctuation of δf_d from such an approach would be unacceptably large. Matching the middle gate to the average over the wider AGC gate provides much greater stability in the estimates of δf_d.

From Eqs. (49) and (50), the mistiming of the deramping chirp is linearly related to the frequency offset Eq. (56) by

$$\delta t_d = \frac{\tau' \delta f_d}{\Delta F}, \tag{57}$$

where ΔF is the swept frequency range over the chirp duration τ' (see Section 2.4.3). Estimates of δt_d obtained by Eq. (57) from waveform averages over a single track interval of about 50 msec are still relatively noisy while the expected change of range varies relatively smoothly along the satellite ground track. To mitigate the effects of noise, the values of δt_d are smoothed by an "$\alpha - \beta$ tracker" that consists of a set of recursion relations for estimating the range rate for the current track interval and the range for the next track interval based on previous estimates of range and range rate (see Chelton et al., 1989, for a detailed description). The recursion relations include parameters α and β that determine the weighting of the previous range estimates (i.e., the two-way travel time measurements). These parameters are specified prior to launch as part of the overall design of the altimeter with values that allow the tracker to be reasonably responsive to two-way travel time variations along the satellite ground track without being overly sensitive to noise. For the values of α and β used for the T/P dual-frequency altimeter and most other altimeters, approximately 60 update intervals are required for the two-way travel time to change by the full amount of δt_d determined by Eq. (57). Thus, although the two-way travel time estimate is updated approximately 20 times per second, it contains information spanning about three seconds along the satellite ground track.

The recursion relations of the $\alpha - \beta$ tracker do not take into account the relative acceleration between the altimeter and the sea surface (the second derivative of the range). For a constant acceleration, this results in a constant error of the two-way travel time estimate (see Section 2.4.6). While the recursion relations could easily be extended to account for acceleration in the onboard estimation of two-way travel time, this would likely introduce additional noise with little improvement in accuracy. As described in Section 2.4.6,

a correction for acceleration is applied in ground processing based on acceleration estimates obtained by along-track smoothing of the two-way travel times estimated by the $\alpha - \beta$ tracker and telemetered to ground receiving stations. This correction is not needed in the ground-based waveform retracking described in Section 2.4.5.

The performance of the ATU can also be affected by atmospheric attenuation. Selective attenuation of altimeter waveform samples near the leading edge of the waveform can corrupt estimates of $S_{\mathrm{mid}}(m)$ and S_{agc}, thereby affecting the ATU estimates of waveform misalignment by Eq. (56). As summarized in Section 2.3, the radar return is attenuated by atmospheric dry gases, water vapor, clouds, and rain. Since the horizontal scales of dry gases and water vapor are larger than the altimeter footprint, $S_{\mathrm{mid}}(m)$ and S_{agc} are attenuated equally, which has little effect on Eq. (56). These sources of attenuation can therefore be neglected with negligible loss of accuracy in the two-way travel time estimates. Likewise, homogeneous clouds across the altimeter footprint have little effect on the tracker performance. Attenuation by isolated patchy clouds is usually not a serious problem because cloud attenuation is so small. However, Walsh et al. (1984) have shown that the regularity of parallel cloud streets oriented orthogonal to the satellite ground track can lead to substantial tracking errors. These effects are generally small, however, in the 1-sec averaging that is usually applied to reduce noise in the ATU estimates of two-way travel time for geophysical applications of altimeter data.

Rainfall is a more serious source of noise in ATU estimates of two-way travel time. Attenuation of the radar return by rain is much larger than attenuation by dry gases, water vapor and clouds (see Section 2.3). The spatial scales of rain cells are often smaller than the altimeter antenna footprint size. The sizes of rain cells generally decrease with increasing rain rate (Goldhirsh and Rowland, 1982). Because attenuation by heavy rainfall is large (see Figure 6), distortion of portions of the leading edge of the returned waveform by non-uniform attenuation from patchy rain cells can lead to errors in the sea level tracking algorithm (Monaldo et al., 1986). An attempt is made to eliminate rain-contaminated altimeter measurements by estimating the rain rate from coincident passive microwave radiometer measurements of brightness temperature. As summarized in Section 2.3.3, these rain rate estimates are imperfect. Moreover, the footprint size of the radiometer is much larger than the altimeter footprint size. For small rain cells, the average rain rate over the radiometer footprint can be small while the rain rate is large within the individual rain cells that attenuate selected waveform samples. Depending on the location of the rain cell within the altimeter footprint, rain contamination is often identifiable as spurious altimeter estimates of range, wave height, or σ_0 (an example is shown in Goldhirsh and Rowland, 1982). In such cases, rain-contaminated data can be detected and flagged on the basis of the standard deviation of the range, wave height or σ_0 estimates within the 1-sec averages analyzed for geophysical applications. Since not all rain contamination is detectable by this method, rain is sometimes an unavoidable source of noise in altimeter data.

Antenna pointing errors are another source of error in the onboard estimates of two-way travel time. The frequency offset δf_d determined from Eq. (56) assumes that the scale factor a_{agc} is correct and that there is no bias in the AGC gate S_{agc}. The "plateau droop" of the tail of the waveform is sensitive to the pointing accuracy of the altimeter. The power in the tail of the waveform decreases with increasing two-way travel time (i.e., increasing waveform sample number) because of the rolloff of the antenna gain with increasing angle away from antenna boresight. If the antenna is pointed off nadir, the peak gain is directed to a point on the sea surface that is displaced from nadir. The longer two-way travel time to this off-nadir location corresponds to a higher frequency in the spectral waveform. This increases the power in the later waveform samples as shown at the top of Figure 20. In this case, the value of the AGC gate is too large since S_{agc} includes some waveform samples from the tail of the waveform. This results in a small error in the onboard estimate of two-way travel time. This error could be mitigated by adjusting the value of a_{agc} to account for off-nadir pointing. Although the tail of the waveform contains information about the antenna pointing angle, the onboard tracker does not make use of this information. A fixed value of a_{agc} appropriate for the nominal pointing angle is used by the ATU for all antenna pointing angles. As described in Section 2.4.6, this instrumental error is corrected in ground-based processing by estimating the off-nadir antenna pointing angle from the power in the tail of the waveform.

It is important to bear in mind that the convolution expression Eq. (46), or equivalently the frequency-domain expression Eq. (53), for the shape of the returned waveform depends on the probability density function $q_s(t)$ of the specular reflectors on the sea surface. The two-way travel time to nadir mean sea level estimated by the $\alpha - \beta$ tracker assumes that the specular reflectors are symmetrically distributed, as described by Eq. (40). As noted previously, the actual distribution of specular reflectors is skewed toward wave troughs. The onboard estimate of two-way travel time is therefore biased toward wave troughs. This sea-state bias must be corrected in ground-based processing as described in Section 3.2. The skewness-bias component of the total sea-state bias (see Section 3.2.2) owing to the non-Gaussian character of $q_s(t)$ can be explicitly accounted for in ground-based waveform processing by estimating the two-way travel time simultaneously with the statistical moments of the sea surface height distribution as summarized in Section 2.4.5.

In addition to onboard tracking of the two-way travel time to nadir mean sea level, the ATU inserts the waveform samples into the telemetry transmitted to ground receiving stations. To reduce the data rate, the telemetry samples for the

T/P dual-frequency altimeter are averaged over groups of waveform samples and over two successive track intervals (about 100 msec) for the K_u-band altimeter and four successive track intervals (about 200 msec) for the C-band altimeter. As shown in Figure 20, the telemetry samples preserve individual waveform samples in the leading edge of the waveform but condense the waveform samples into groups of two or four in the portion of the returned waveform preceding and following the leading edge. The 128 waveform samples are reduced to a total of 64 compressed samples for telemetry.

While not strictly part of the tracking described above, the altimeter software also controls signal acquisition through an automated series of stages. Acquisition begins with the altimeter searching for a return from the surface over a very wide range of about 60 km. In accord with Eqs. (49) and (50) for the one-to-one correspondence between two-way travel time and frequency in the spectral waveform, this is achieved by reducing the bandwidth ΔF of the chirp to a very narrow range of frequencies (5 MHz for the T/P dual-frequency altimeter), thereby sampling a coarse two-way travel time and hence a large time span in the 128 waveform samples. The two-way travel time with the strongest reflected signal that is found during this initial sweep is then selected to attempt tracking in a "coarse-tracking" mode with $\Delta F = 5$ MHz using a simplified tracking algorithm. This allows the ATU to locate the leading edge of the returned waveform over a relatively broad range of two-way travel times. If the strongest signal in the initial sweep is not near the nadir point, coarse tracking will fail to produce an adequate signal, and the ATU will revert to the acquisition mode with 60-km range sweep. If the signal strength and the alignment of the leading edge of the waveform in coarse-track mode meet specific criteria programmed into the ATU software, the ATU switches to the full frequency sweep of $\Delta F = 320$ MHz to initiate full-precision "fine tracking" as described previously. The capability for coarse tracking is essential in order for the altimeter to acquire "lock" on the range to nadir mean sea level after the satellite ground track passes from land or ice to open water. Fine tracking is typically initiated within 1–3 sec. This acquisition time results in a loss of data within 6–20 km of land or ice.

For a reasonably flat land or ice surface, the altimeter is able to maintain tracking, although usually with reduced two-way travel time resolution in coarse-track mode. The T/P dual-frequency altimeter is designed to switch autonomously to fine-track mode after acquiring lock. Some altimeters, such as those on the ERS-1 and ERS-2 satellites, are designed to operate in coarse-track mode (called "ice mode"), specifically for tracking over ice. Range measurements over irregular terrain for which the altimeter is able to maintain tracking can be difficult to interpret because the leading edge of the waveform may represent the return from a high-elevation topographic feature away from the nadir point. In this case, the waveform often deviates from the shape expected by the ATU. To obtain accurate range estimates over such surfaces, it is generally necessary to perform ground-based waveform processing, often with special models for the waveform shape. This has been done by a number of investigators for studies of high-latitude ice topography (e.g., Martin *et al.*, 1983; Partington *et al.*, 1989, Bamber *et al.*, 1998, Bindschadler, 1998; see also Chapter 9).

2.4.5. Ground-Based Waveform Retracking

The set of 64 waveform samples that are telemetered to ground receiving stations are used in regular ground-based processing to determine the correction for antenna pointing errors as described in Section 2.4.6 below. These errors arise from the simplifications inherent in the onboard tracking algorithm summarized in Section 2.4.4. The 64 waveform samples can also be retracked in ground-based analysis using sophisticated algorithms to retrieve more accurate estimates of range and $H_{1/3}$, as well as additional information such as off-nadir antenna pointing angle and skewness of the surface height distribution.

Waveform retracking is based on fitting a convolution model of the form given by Eq. (46) to the observed waveforms. The effects of skewness can be included in the model for the probability density function $q_s(t)$ of the specular reflectors on the sea surface. In addition, the effects of off-nadir pointing on the shape of the tail of the waveform can be explicitly represented in the impulse response $P_{FS}(t)$. These other quantities can then be estimated along with range and $H_{1/3}$. The effects of skewness are most noticeable in the early part of the rise of the leading edge of the waveform. Off-nadir pointing is detectable as a flattening of the tail of the waveform (see top of Figure 20).

The range, off-nadir pointing angle, and the sea-surface height distribution parameters ($H_{1/3}$ and skewness) are estimated from the waveforms by performing iterative nonlinear fits of waveform models to the observed returned waveforms. The range and surface height distribution parameters can be estimated by simpler deconvolution methods as discussed by Lipa and Barrick (1981), Rodríguez (1988), and Rodríguez and Chapman (1989). These methods concentrate on the leading edge of the waveform to extract the surface height distribution parameters from noisy waveforms with constraints imposed on $q_s(t)$. Deconvolution methods require independent estimates of the noise floor in the waveform and the off-nadir pointing of the altimeter. Inaccurate values of these variables add error to the estimates of the surface height distribution parameters. Rodríguez and Chapman (1989) investigated the use of two different deconvolution methods and several mathematical techniques. Each approach had strengths and weaknesses. They recommended a technique that is based on singular value decomposition with spectral filtering.

For simultaneous estimation of the range, off-nadir pointing angle, $H_{1/3}$ and skewness by full waveform processing, the nonlinear fitting requires some simplification in order to be carried out efficiently with available computing resources. The first simplification is to use least squares rather than the more accurate maximum likelihood estimation. In practice, maximum likelihood estimates are probably not significantly better because of unreliability of the statistical weights owing to unmodeled errors in the waveform samples. In addition, the portions of the waveform with low signal power tend to be weighted more heavily in maximum likelihood estimation.

Rodríguez and Martin (1994a) present the results of retracking from full waveform processing of 0.1-sec averages of T/P waveforms. The nonlinear fit for each 0.1-sec average waveform was initialized based on least-squares fitting of equations linearized about parameter values obtained from the previous nonlinear least-squares fit. After data gaps from land, for example, the nonlinear estimation is initialized based on estimates of range, off-nadir pointing, and $H_{1/3}$ obtained from the standard ground-based processing. The computational efficiency was greatly improved by expanding the altimeter point target response $p_\tau(t)$ in the convolution model as a series of Gaussian functions. With this approach, the model waveform and its derivatives can be calculated analytically for efficient generation of the terms in the matrix equation for the least-squares solution.

In practice, nonlinear least-squares fitting of the T/P dual-frequency altimeter waveforms is made difficult by the presence of "anomalies" in the waveforms. These anomalies result from spurious energy entering the altimeter receiver, probably from switching harmonics in the digital chirp generator within the altimeter, and from finite arithmetic effects in the digital signal processor (see Section 2.4.3). The anomalous characteristics of the waveforms from the T/P dual-frequency altimeter are summarized by Hayne *et al.* (1994). The most significant feature (see Figure 21) appears near waveform sample 64, which corresponds to the zero frequency in the fast Fourier transform of the in-phase and quadrature signals analyzed by the digital processor (see Section 2.4.3). This feature probably arises from a bias in the voltage output by the filter or the analog-to-digital converter. Since this portion of the waveform is far from the leading edge, it does not interfere with the timing of the deramping chirp determined by the ATU as described in Section 2.4.4. Nor does this portion of the waveform affect ground-based estimates of off-nadir pointing angle described in Section 2.4.6 since they are determined from the "attitude gate" S_{att}, which consists of the average of waveform samples 96–112 (see Figure 20). The anomalies near waveform sample 64 are masked out in ground-based waveform retracking so that they will have no significant effect on the fitted parameters.

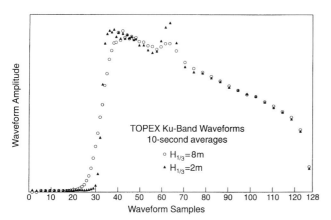

FIGURE 21 Examples of 10-sec averages of actual waveforms for the K_u-band frequency of the T/P dual-frequency altimeter for significant wave heights of $H_{1/3} = 2$ m (triangles) and 8 m (open circles). A large waveform anomaly is centered at waveform sample 64. It is evident from the figure that this anomaly does not corrupt the leading edge of the waveform, which is the portion of the waveform analyzed by the onboard adaptive tracking unit. Nor does it corrupt the antenna pointing angle estimated from waveform samples 96–112 (compressed telemetry waveform samples 57–60) in the tail of the waveform (see Figure 20). Smaller waveform anomalies are evident in the noise floor and especially near the peak power of the waveform. (Data provided courtesy of G. Hayne.)

There are several other waveform anomalies with amplitudes of about 0.5–1% of the peak power that are more problematic because they occur in the earlier portion of the waveform and near the leading edge of the waveform that is used to estimate two-way travel time. The anomalies in the early waveform samples result in less reliable estimates of the noise gate S_{noise}, which introduces small errors in the estimate of waveform misalignment by Eq. (56). The small amplitudes of the waveform anomalies closer to the track point have a negligible effect on range estimates in small $H_{1/3}$ conditions for which the leading edge of the waveform has a steep slope. However, because of the flattening of the leading edge of the waveform for large wave heights, waveform anomalies in this region of the waveform can affect the accuracy of the two-way travel time estimates obtained by the ATU. If not accounted for in the waveform model, these anomalies also affect the accuracy of the range estimates obtained by ground-based waveform retracking. These anomalies near the leading edge of the T/P waveforms may render the estimates of skewness only marginally useful oceanographically. However, estimation of the apparent skewness is useful numerically in order to mitigate the effects of the waveform anomalies on the ground-based waveform retracking estimates of range and $H_{1/3}$.

It is noteworthy that the exact waveform samples in which the anomalies occur near the leading edge of the waveform vary because of a number of factors, most notably the sign of the range rate along the satellite ground track. For the very nearly circular orbits of altimeter satellites, the primary source of range rate along the satellite ground track is the oblateness of the earth that results in about a 20-km decrease of the range measurements between the equator and the turning latitude of the satellite. The sign of the range rate thus depends primarily on whether the altimeter is approaching or receding from the equator. Hayne *et al.* (1994) give average corrections to remove the anomalies from the waveforms.

The errors of the ATU estimates of two-way travel time that are associated with the sign of the range rate were discovered by Rodríguez and Martin (1994a) from waveform retracking. Although systematic depending on whether the satellite is approaching or receding from the equator, they showed that, after applying the empirical attitude and sea-state corrections as described in Section 2.4.6, the errors are only 1–2 cm for range estimates determined from onboard measurements of two-way travel time to nadir mean sea level and only 0.1 m for $H_{1/3}$ obtained by the standard ground-based processing.

2.4.6. Instrument Corrections

To achieve the goal of an overall accuracy of 2 cm or better for range estimates, the onboard tracker estimates of two-way travel time must be carefully corrected for a number of instrumental errors. While the requirements for altimeter estimates of σ_0 and $H_{1/3}$ are not as stringent as for the range estimates, corrections must be applied to them as well. Most of the instrumental corrections have been described in detail by Chelton *et al.* (1989). Only brief summaries are given here. Corrections for basic frequency and signal processing effects are described first, followed by corrections for pointing-angle and sea-state effects on the performance of the ATU.

Doppler-shift error   A short pulse with a duration of a few nanoseconds is synthesized as discussed in Section 2.4.3 by transmitting a chirp with linearly decreasing frequency over a relatively long duration. The returned signal is deramped and analyzed in the frequency domain where there is a one-to-one correspondence between frequency and two-way travel time. Any change in the frequency of the returned signal thus appears as an error in the estimated two-way travel time. One source of frequency change is the Doppler shift from the relative velocity between the altimeter and the sea surface. The two-way Doppler shift is given by

$$\delta f_{\mathrm{dopp}} = \frac{2v}{c} F, \qquad (58)$$

where $v = dR/dt$ is the rate of change of the range R (the "range rate"), c is the speed of light and F is the center frequency of the transmitted chirp (13.6 and 5.3 GHz for the K_u-band and C-band frequencies of the T/P dual-frequency altimeter). From Eqs. (49) and (50), the equivalent error of the two-way travel time estimate is

$$\delta t_{\mathrm{dopp}} = \frac{\tau' \delta f_{\mathrm{dopp}}}{\Delta F} = \frac{2\tau' v F}{c \Delta F}, \qquad (59)$$

where ΔF is the swept frequency range over the chirp duration τ'. The frequency bandwidth for the T/P dual-frequency altimeter is 320 MHz (see Section 2.4.3) and the range rates typically observed over the ocean are 3–30 m sec^{-1}. The Doppler shift error Eq. (59) in the estimated two-way travel time thus corresponds to range errors $\delta R_{\rm dopp} = 0.5c\delta t_{\rm dopp}$ of 1–13 cm for the K$_{\rm u}$-band altimeter and 0.5–5 cm for the C-band altimeter.[14]

The range rate v used to estimate the Doppler correction by Eq. (59) merits careful consideration. As described in Section 2.4.4, the range rate is estimated and telemetered to ground receiving stations by the ATU. These estimates of range rate are noisy. More accurate estimates are obtained from ground-based processing by least-squares fitting to the range estimates. For T/P, the range rate and acceleration (see below) are estimated from least-squares fits of a quadratic function to successive range measurements over approximately 3 sec (60 track intervals). The first derivative of the quadratic function is the range rate used in the Doppler shift correction Eq. (59).

Acceleration error   The $\alpha - \beta$ tracker described in Section 2.4.4 does not account for range accelerations $dv/dt = d^2R/dt^2$ of the altimeter relative to the sea surface. The effects of acceleration errors were analyzed for the Geosat altimeter by Hancock et al. (1990). Acceleration errors for the T/P dual-frequency altimeter have been extensively analyzed by Rodríguez and Martin (1994a) who showed that the correction for two-way travel time can be expressed as

$$\delta t_{\rm acc} = \frac{2\Delta t_{\rm track}^2}{c\beta}\frac{dv}{dt}, \qquad (60)$$

where $\Delta t_{\rm track} \approx 50$ msec is the track interval and $\beta = 1/64$ is one of the two parameters of the $\alpha - \beta$ tracker (see Section 4 of Schelton et al., 1989). The two-way travel time error, and therefore the range error $\delta R_{\rm acc} = 0.5c\delta t_{\rm acc}$, are thus proportional to the acceleration. For a constant acceleration, the tracker response lags the true range. Rodríguez and Martin (1994a) find that the acceleration correction suppresses noise in the data for times shorter than about 10 sec (approximately 60 km along-track).

The acceleration used in the correction Eq. (60) is estimated for the T/P dual-frequency altimeter from the second derivative of the above-noted least-squares fits of a quadratic function to successive range estimates over time intervals of about 3 sec. The largest range accelerations over the ocean occur at deep-ocean trenches where an extreme acceleration might reach 10 m sec^{-2}. The corresponding two-way travel time correction Eq. (60) is equivalent to a range correction

of 160 cm. Over most of the ocean, however, the corrections are much smaller, typically only a few centimeters.

Oscillator drift error   The fundamental altimeter measurement is the two-way travel time of the radar pulse. The altimeter measures time by counting cycles of an oscillator. Any error in knowledge of the oscillator frequency therefore results in an error in the estimated two-way travel time that is proportional to the number of cycles counted. The frequency and stability of the oscillator are known prior to launch. Because of slow drifts of the oscillator from aging and the effects of radiation on the crystal that controls the oscillator frequency, it is necessary to calibrate the oscillator on a weekly or more-frequent basis. This calibration is based on the timing of the reception of telemetry signals at the ground receiving stations.

For T/P, the oscillator frequency has changed by several parts in 10^9 per year over the duration of the mission. The estimated two-way travel time $t_{1/2}$ is corrected for this oscillator drift based on a look-up table derived from the ground-based updates of the measured frequency. The oscillator drift correction is formulated as

$$\delta t_{\rm osc} = \left[\frac{C_{\rm meas}}{C_{\rm nom}} - 1\right]t_{1/2}, \qquad (61)$$

where $C_{\rm nom}$ and $C_{\rm meas}$ are, respectively, the nominal and measured seconds per count (the inverse of oscillator frequency).

The distinction between "frequency" and "seconds per count" was the source of a significant error in the original two-way travel time estimates from the T/P dual-frequency altimeter. The ground-based oscillator measurement was changed late in the mission development from frequency to seconds per count, but this change was inadvertently not implemented in the processing software. The sign of the oscillator correction was therefore effectively reversed. This software error resulted in a bias of approximately 13 cm in the T/P dual-frequency range measurements. The drift of the oscillator resulted in a systematic decrease of the two-way travel time that corresponded to an apparent rise of the global mean sea level of approximately 8 mm year^{-1} (Nerem et al., 1997; see also Chapter 8). Comparisons of the T/P dual-frequency altimeter estimates of sea surface height with tide-gauge estimates (Mitchum, 1998) and skepticism about the 13-cm bias relative to the T/P single-frequency altimeter and ground calibration site measurements (Christensen et al., 1994a) and the large magnitude of the drift eventually led to the discovery of the error in the processing software (P. Escudier, P. Vincent and O. Zanife, 1996, personal communication). When the software was corrected, the relative bias between the NASA dual-frequency altimeter and the CNES single-frequency altimeter decreased to about 1 cm.

---

[14]Even for the maximum observed range rate of 30 m sec^{-1}, the dispersion of the Doppler shift over the 320-MHz bandwidths of the K$_{\rm u}$-band and C-band chirps results in negligible error of the range estimate.

Pointing-angle and sea-state corrections   The largest source of instrumental error arises from mispointing of the altimeter antenna. To varying degrees, off-nadir pointing affects the ATU estimates of two-way travel time (from which the range is estimated), $H_{1/3}$ and the AGC (from which σ_0 is determined). These errors occur because of the approximations made in the onboard tracker as summarized in Section 2.4.4 (see Chelton *et al.*, 1989, for a more detailed discussion). These approximations include errors in the calibration of the scaling factor a_{agc} in Eq. (56), anomalies in the waveform samples (see Section 2.4.5 and Figure 21), the effects of skewness of the sea surface height distribution, and small variations with $H_{1/3}$ and off-nadir pointing angle in the expected position of the leading edge of the waveform relative to the track point. The latter errors were observed in pre-launch altimeter testing and from numerical simulations. These variations arise because several onboard constants used in the ATU are calculated for nominal conditions of surface skewness (a value of 0.1), off-nadir angle (a value greater than 0 based on the expected satellite attitude control, approximately 0.1° for T/P), and average $H_{1/3}$ within each gate index m (see Section 2.4.4).

Empirical corrections for the effects of pointing angle and sea state are based on the outputs of two-way travel time (equivalent to range), AGC, and $H_{1/3}$ from simulations of the altimeter signal processing and adaptive tracking compared with the values used in the convolution model that generated the simulated data (Hayne and Hancock, 1990; Hayne *et al.*, 1994). The corrections are implemented as polynomials using estimates of attitude and $H_{1/3}$ with coefficients that depend on the $H_{1/3}$-selected gate index m. The attitude and $H_{1/3}$ in the empirical corrections are characterized by "voltages" derived directly from the waveform samples. The effects of $H_{1/3}$ are represented by the quantity V_{swh} defined by Eq. (55) that is included in the telemetry. An "attitude voltage" (V_{att}) is derived in ground-based processing by forming the ratio

$$V_{\text{att}} = \frac{S_{\text{att}} - S_{\text{noise}}}{S_{\text{agc}} - S_{\text{noise}}}, \qquad (62)$$

where S_{att} is the attitude gate shown in Figure 20. The polynomial for the range and $H_{1/3}$ corrections includes cubic functions of V_{swh} and V_{att} (the cubic terms are small compared with the quadratic terms). The AGC correction is limited to a quadratic function of V_{swh} and V_{att}. The corrections are typically a few centimeters in range, about 0.1 m in $H_{1/3}$ and less than 0.5 dB in σ_0 for $H_{1/3}$ smaller than 6 m and the small off-nadir pointing angles of about 0.1° for the T/P altimeter. For satellites with poorer attitude control where off-nadir pointing angles exceed 0.5° (e.g., Geosat) the corrections, especially to AGC, are much larger.

For T/P, the range and $H_{1/3}$ estimates obtained after applying these empirical corrections for pointing errors and sea-state effects were compared with the estimates obtained from ground-based waveform retracking by Rodríguez and Martin (1994a). Their analysis accounted for the effects of the waveform anomalies discussed in Section 2.4.5 (see Figure 21). Despite the fact that adjustments were not explicitly made to account for the effects of these waveform anomalies in the standard ground-based processing, the differences from the retracked estimates were generally small. Based on these comparisons, the coefficients of the empirical polynomial correction for pointing-angle and sea-state effects for the range estimates from gate index $m = 3$ ($H_{1/3}$ in the range of 3–6.2 m) were adjusted by 1 cm to bring them into closer agreement with the retracking results.

2.4.7. Footprint Size and Location

As described in Section 2.4.1, the projection of the radar pulse onto the sea surface consists of a circular area with radius that grows as the square root of time after the leading edge of the pulse begins to illuminate wave crests at nadir. Although the sea surface is not uniformly illuminated within this pulse-limited footprint area (see Section 2.4.2), radar returns are received by the altimeter from specular reflectors distributed over the full area of this expanding footprint. For a realistic wave field with a Gaussian wave height distribution, the probabilistic time series $r_{\text{out}}(t)$ for a specific significant wave height $H_{1/3}$ is almost identical to the time series Eq. (27) derived analytically for the simple case of a monochromatic wave field with crest-to-trough wave height H_w equal to $H_{1/3}$ (see Figures 9 and 13). At any particular two-way travel time t during the evolution of the expanding pulse-limited footprint, the footprint radius increases with increasing wave height.

Ideally, the characteristic footprint over which the sea-surface height is averaged in radar altimeter measurements of two-way travel time should be described in terms of the length scales of features in the sea-surface height field that can be resolved. This footprint diameter increases with increasing $H_{1/3}$. Lacking such a filter transfer function description, the footprint size must be estimated in some other manner. Because of the rather complicated details of the pulse compression and onboard adaptive tracking techniques by which the two-way travel time to nadir mean sea level is estimated as described in Sections 2.4.3 and 2.4.4, characterization of the footprint size is not entirely straightforward. The central issue is specification of a particular two-way travel time t that represents the footprint over which the range to mean sea level is estimated from the returned waveform. As described by Eq. (56), the onboard adaptive tracking unit estimates the two-way travel time to nadir mean sea level from the difference between the AGC gate S_{agc} and the middle gate $S_{\text{mid}}(m)$, which consist of averages of waveform samples centered on the midpoint of the leading edge of the returned waveform. The AGC gate is the average of 32 waveform samples centered on the track point. The middle gate is also the average of waveform samples centered

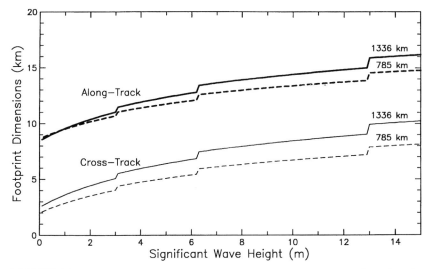

FIGURE 22   The cross-track (thin lines) and along-track (thick lines) footprint dimensions defined by the width of the middle gate $S_{\mathrm{mid}}(m)$ as a function of significant wave height for 1-sec averages of the range measurements from altimeters at orbit heights of 1336 km (solid lines) and 785 km (dashed lines). Discontinuities of the slopes of each curve occur at the transitions from one gate index to the next.

on the track point. The width of $S_{\mathrm{mid}}(m)$ varies from two waveform samples to 16 waveform samples, depending on which of the five gate indexes m is selected by the ATU (see Figure 20).

Regardless of the gate index m, S_{agc} is always wider than $S_{\mathrm{mid}}(m)$. Since the tracking algorithm utilizes information from the full width of S_{agc}, it could be argued that the footprint resolution should be defined as the radius $r_{\mathrm{out}}(t)$ at the time t corresponding to the last waveform sample in S_{agc}, i.e., the two-way travel time that is 15.5 waveform samples after the track point. For the effective 3.125 nsec separation between T/P waveform samples (see Section 2.4.3), this corresponds to 48.4375 nsec after the two-way travel time to nadir mean sea level. To distinguish it from other definitions of altimeter footprints, this is sometimes called the "AGC footprint." The probabilistic diameter of the circular AGC footprint at this two-way travel time is only weakly dependent on wave height. For the T/P orbit height of 1336 km, the AGC footprint diameter increases from about 8 km for a flat sea surface to about 11 km for $H_{1/3} = 15$ m.

In actuality, the tracking of the midpoint on the leading edge of the returned waveform is much more sensitive to $S_{\mathrm{mid}}(m)$ than to S_{agc}. Indeed, the broad width of the AGC gate is intended to make S_{agc} relatively insensitive to $H_{1/3}$. The AGC footprint is probably therefore an overly conservative estimate of the footprint size. It seems more appropriate to characterize the altimeter footprint in terms of the "middle-gate footprint" defined by the half-width of $S_{\mathrm{mid}}(m)$. As the width of $S_{\mathrm{mid}}(m)$ increases with increasing $H_{1/3}$, the radius $r_{\mathrm{out}}(t)$ at the time t corresponding to the last waveform sample in $S_{\mathrm{mid}}(m)$ increases discontinuously with increasing $H_{1/3}$. As noted in Section 2.4.4,

the adaptive tracking unit is programmed for the transitions from one gate index to the next to occur at $H_{1/3}$ values of 1, 3, 6.2, and 13 m. The probabilistic diameter of the circular middle-gate footprint at the $H_{1/3}$-dependent two-way travel times of the last waveform sample in $S_{\mathrm{mid}}(m)$ is shown as a function of $H_{1/3}$ by the thin lines in Figure 22. For the T/P orbit height of 1336 km, the middle-gate footprint diameter (thin solid line) increases from about 2.5 km for a flat sea surface to about 10 km for $H_{1/3} = 15$ m. At the lower Seasat, Geosat, and ERS orbit heights of about 785 km, the footprint diameter (thin dashed line) increases from about 2–8 km over the same range of $H_{1/3}$.

An additional factor that must be taken into consideration in characterizing the altimeter footprint is the along-track motion of the satellite. The circular middle-gate footprint described above is for an instantaneous measurement. As noted previously, waveforms must be averaged to reduce the noise of individual waveforms before processing to estimate the two-way travel time to nadir mean sea level. Typical onboard averaging times are about 50 msec. For oceanographic applications, altimeter data are generally further averaged over time periods of about 1 sec. The along-track motion of the satellite over a period of 1 sec is 5.9 and 6.6 km for orbit heights of 1336 and 785 km, respectively (see Section 8.1.2). The footprints in 1-sec averages are thus stretched by these amounts along the satellite ground tracks to form ovals as shown in Figure 23. The along-track dimensions of these ovals are shown for the two orbit heights by the solid and dashed lines in Figure 22. At both orbit heights, the along-track dimension of the footprint is about 8.5 km for a flat sea surface. For $H_{1/3} = 15$ m, the along-track dimension in-

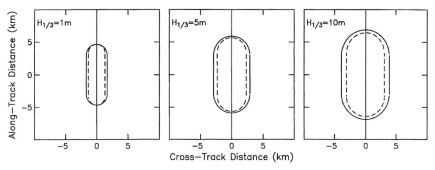

FIGURE 23   The oval footprint characteristics defined by the width of the middle gate $S_{mid}(m)$ for significant wave heights of $H_{1/3} = 1$, 5, and 10 m for 1-sec averages of altimeter measurements of nadir mean sea level from orbit heights of 1336 km (solid lines) and 785 km (dashed lines).

creases to about 15 and 16 km for orbit heights of 1336 and 785 km, respectively.

The center of the oval footprint described above is at the subsatellite point along the ground track. There is some ambiguity in the definition of the subsatellite point. One definition is the point where the line connecting the earth and satellite centers of mass intersects the earth's surface (more precisely referred to as the geocentric subsatellite point). The more common definition is the geodetic subsatellite point, which is the point on the earth's surface defined by the line from the satellite center of mass to the local normal to an ellipsoidal approximation of the earth's surface. Because of the oblateness of the earth, the geodetic and geocentric subsatellite points are coincident at the poles and the equator but differ at intermediate latitudes. The flattening of the earth (defined to be the difference between the semimajor and semiminor axes of an ellipsoidal approximation of the earth's surface, normalized by the semimajor axis) is 3.35×10^{-3}. The maximum latitudinal separation of the geodetic and geocentric subsatellite points for this flattening is about 21 km at 45° latitude.

The geodetic subsatellite point more nearly represents the altimetric measurement location. Even this is not precisely correct, however. Altimeters measure the two-way travel time to the nearest point on the earth's surface within the antenna beamwidth. Steep geoid topography can have a slope as large as 10^{-4} in extreme cases (Brenner *et al.*, 1990; Minster *et al.*, 1993). The effects of geoid slopes are therefore relatively minor; the nearest point on the sea surface differs from the geodetic subsatellite point by less than 1 km horizontally for measurements from the T/P orbit height of 1336 km.

3. RANGE ESTIMATION

Implementation of satellite altimetry with the 1-cm accuracy desired for ocean circulation studies is an extreme challenge. The technique by which the two-way travel time be-

tween the altimeter and nadir mean sea level is measured by a radar altimeter was described in Section 2.4. If the atmosphere were a vacuum and the probability density function of the wave field were Gaussian, the range from the altimeter to nadir mean sea level would be easily determined from the two-way travel time and the free-space speed of light. As shown schematically in Figure 1, the presence of water vapor, dry gases, and free electrons in the atmosphere reduces the propagation speed of the radar pulse. Additional biases in the range estimates are introduced by the non-Gaussian distribution of the wave field. Failure to account for the effects of atmospheric refraction and sea-state bias introduces errors in the range measurements that are more than an order of magnitude larger than the 1-cm accuracy goal, even when considering only the time-variable component of each correction (i.e., the correction after removing the time-average range correction at each location along the satellite ground track). Corrections for refraction and sea-state bias must therefore be applied to transform the round-trip travel time into a range estimate with the high degree of accuracy required for oceanographic applications of altimeter data.

A sense of the relative magnitudes of the various range corrections is conveyed in Table 2 from a tabulation of the typical values for the total and time-variable component of each correction for latitude ranges 30°N to 30°S and 30° to 60° (both hemispheres). An approximate formulation for each correction is included in the footnotes of the table. More detailed formulations along with a derivation of the physical basis for each correction are presented in this section. It should be noted that these range corrections vary considerably, both geographically and temporally. A rough sense of the geographical variability can be inferred from the differences in the magnitudes of the corrections for the latitude bands listed in the table.

3.1. Atmospheric Refraction

For altimetric purposes, the effects of atmospheric refraction are generally expressed in terms of range correction

TABLE 2.   Typical Values for the Total and Time-variable Component of Each of the
Corrections Applied to the T/P Range Estimates at Low Latitudes (30°N to 30°S) and
Midlatitudes (30° to 60° in Both Hemispheres)

Range correction	Total (cm)	Variable (cm)	Uncertainty (cm)
Dry tropospheric refraction[a]			~1
Latitudes 30°N to 30°S	226	0.5	
Latitudes 30° to 60°	226	2	
Wet tropospheric refraction[b]			~1
Latitudes 30°N to 30°S	24	6	
Latitudes 30° to 60°	10	5	
Ionospheric refraction[c]			<1
Latitudes 30°N to 30°S	12	5	
Latitudes 30° to 60°	6	2	
Sea-state bias[d]			~2
Latitudes 30°N to 30°S	4	1	
Latitudes 30° to 60°	6	3	

[a]The dry tropospheric range correction is ΔR_{dry} (cm) $\approx 0.223 P_0$, where P_0 is sea-level pressure in mbar. The estimated uncertainty assumes an uncertainty of less than 5 mbar in operational weather analyses of P_0.

[b]The wet tropospheric range correction is ΔR_{vap}(cm) $\approx 6.4 V$, where V is the vertically integrated water vapor in g cm^{-2}. The estimated uncertainty is based on an uncertainty of 0.15 g cm^{-2} in estimates of V from a three-frequency microwave radiometer.

[c]The ionospheric range correction for a K_u-band radar frequency of 13.6 GHz is ΔR_{ion}(cm) $\approx 0.22 E$, where E is the vertically integrated electron content in electrons cm^{-2}. The listed values correspond to afternoon values of E near the peak of the diurnal cycle during the period 1998–1999 of moderately high solar activity. The estimated uncertainty is based on 50-km along-track smoothed estimates of E from a dual-frequency altimeter.

[d]The sea-state bias correction is ΔR_{ssb}(cm) $\approx 0.02 H_{1/3}$, where $H_{1/3}$ is the significant wave height in centimeters. The estimated uncertainty assumes an uncertainty of 1% of $H_{1/3}$ in the sea-state bias correction and a typical wave height of $H_{1/3} = 2$ m. It is apparent from the table that the uncertainty of the sea-state bias correction is smaller for the time-variable component of the range estimates because variations of $H_{1/3}$ are much smaller than the mean $H_{1/3}$ at any particular location.

(also referred to as path delay), defined to be the path length that must be subtracted from the range that is estimated from the measured two-way travel time assuming the free-space value for the speed of light. Failure to correct for atmospheric refraction results in a range estimate that is longer than the true range.

Formulation of the range correction is straightforward for a nondispersive medium such as the troposphere for which the refraction of electromagnetic radiation from atmospheric gases is independent of frequency. In this case, the range correction to account for atmospheric refraction for a measured two-way travel time $t_{1/2}$ between the satellite at height R and nadir mean sea level is

$$\Delta R = \hat{R} - R = \frac{c}{2} \int_0^{t_{1/2}} \frac{\eta - 1}{\eta} \, dt, \qquad (63)$$

where c is the speed of light in free space, $\hat{R} = ct_{1/2}/2$ is the range estimate neglecting refraction and η is the real part of the index of refraction, which is 1 for free space and larger than 1 for a nondispersive refractive medium. For a dispersive medium such as the ionosphere, η can be less than 1,

which implies that the phase of electromagnetic radiation advances faster than the group velocity of energy propagation. The index of refraction η in Eq. (63) must therefore be replaced with the group index of refraction $\eta' = d(f\eta)/df$ (see Section 3.1.3), which is equal to η for a nondispersive medium in which η is independent of frequency.

Because the index of refraction of the atmosphere is very nearly 1 at microwave frequencies, it is convenient to express atmospheric refraction in terms of the refractivity, defined as $N = 10^6(\eta' - 1)$. Travel time and path length are related by $dz = c/\eta' dt$ where $\eta' = \eta$ for a nondispersive medium as noted above. The range correction Eq. (63) can then be expressed in terms of distance along the path of the radar pulse by

$$\Delta R = \frac{10^{-6}}{2} \int_0^{2R} N(z) \, dz = 10^{-6} \int_0^R N(z) \, dz. \qquad (64)$$

The magnitude of atmospheric refractivity is governed by the temperature, pressure, water vapor density, cloud liquid water droplet density, and ionospheric electron density, all of which vary temporally and spatially (both vertically and horizontally). It is convenient to express N in terms of its constituents: the dry tropospheric refractivity N_{dry} associated with dry gases, the water vapor refractivity N_{vap}, the cloud liquid water droplet refractivity N_{liq}, and the ionospheric refractivity N_{ion}. The refractive range correction Eq. (64) relies upon independent estimates of the vertical integrals of each of these four components. The physical basis for each of these path delays is summarized in this section.

The presence of raindrops also introduces a range delay. Refraction of radar signals by raindrops increases with increasing rain rate and varies somewhat with the frequency of the radar signal (Goldhirsh and Rowland, 1982). Moderate rain rates of less than 10 mm hr^{-1} result in less than a 0.5-cm range error, but a heavy thunderstorm with a rain rate of 100 mm hr^{-1} can introduce a range error of 2.5 cm or more. The refractive effects of rain are not nearly as serious a problem as the attenuation effects discussed in Section 2.3.3. Isolated heavy rain cells attenuate portions of the altimeter-returned waveforms, thus distorting the waveforms and introducing errors in the range estimates by the onboard tracking algorithm described in Section 2.4.4 (Monaldo *et al.*, 1986). Widespread moderate rain rates could also result in large errors in the range estimates because the presence of rain decreases the accuracy of passive microwave estimates of columnar water vapor that are used in the water vapor range correction described below in Section 3.1.2. Altimeter observations for which contamination by rain is deemed to be highly probable are thus flagged and eliminated from further analysis as summarized in Section 2.3.3. The refractive effects of raindrops are therefore not discussed here.

3.1.1. Dry Tropospheric Refraction

The correction for the dry gas component of atmospheric refraction is by far the largest adjustment that must be applied to altimeter measurements of two-way travel time to estimate the range from the satellite to nadir mean sea level. Smith and Weintraub (1953) expressed the dry tropospheric refractivity in the form

$$N_{dry}(z) = \beta_{dry} P(z)/T(z), \qquad (65)$$

where P is atmospheric pressure in units of mbar, T is temperature in units of $^\circ$K and the parameter in Eq. (65) has been empirically estimated to be $\beta_{dry} = 77.6\,^\circ$K mbar^{-1}. More precise expressions for N_{dry} have been given by Thayer (1974) and Liebe (1985) in the same form as Eq. (65), but with total atmospheric pressure P replaced by the partial pressure of dry air P_{dry} and accordingly slightly different values for the parameter β_{dry}. The formulation Eq. (65) in terms of the total atmospheric pressure P thus includes a small contribution from water vapor. Since the dry component generally makes up more than 99% of the total pressure, the difference between the expressions in terms of P and P_{dry} is small. For practical applications, the form given by Eq. (65) is more convenient since observations of P are

more routinely available. In any case, Thayer (1974) finds that the Smith and Weintraub (1953) expression Eq. (65), when combined with their expression for the water vapor refractivity N_{vap} given in Section 3.1.2, is accurate to better than 0.2%.

Using the ideal gas law $P(z) = R_a \rho_a(z) T(z)$, Eq. (65) can be expressed as

$$N_{dry}(z) = \beta_{dry} R_a \rho_a(z), \qquad (66)$$

where $R_a = 2.8704 \times 10^6$ ergs (g $^\circ$K)$^{-1}$ is the gas constant for 1 g of air and ρ_a is the total air density in g cm^{-3}. The correction for non-ideal gas behavior of the atmosphere is small (Thayer, 1974). The dry tropospheric component of the range correction Eq. (64) can therefore be expressed as

$$\Delta R_{dry} = 10^{-6} \int_0^R N_{dry}(z)\,dz = \beta'_{dry} \int_0^R \rho_a(z)\,dz, \quad (67)$$

where $\beta'_{dry} = 10^{-6} R_a \beta_{dry} = 222.74$ cm^3 g^{-1}. The 0.2% uncertainty in the refractivity as given by Eq. (66) introduces negligible error in the range correction. It was shown in Section 2.3.1 that the vertical integral of air density in Eq. (67) can be closely approximated by

$$\int_0^R \rho_a(z)\,dz \approx P_0/g_0(\varphi), \qquad (68)$$

where $g_0(\varphi)$ is the gravitational acceleration in cm sec^{-2} on the earth's surface at latitude φ and P_0 is the sea level pressure in millibars. The dry tropospheric range correction in units of cm can then be approximated by

$$\Delta R_{dry} \approx 222.74\, P_0/g_0(\varphi). \qquad (69)$$

The dry tropospheric range delay is thus proportional to sea level pressure (SLP).

The form Eq. (69) for ΔR_{dry} has long been used routinely in ground-based radio ranging of satellites and in radio astronomy (e.g., Hopfield, 1971; Saastamoinen, 1972; Moran and Rosen, 1981; Dodson, 1986). In practice, a latitudinal dependence of $g_0(\varphi)$ must be used in Eq. (69) to account for the oblateness of the earth; the variation of $g_0(\varphi)$ from 978.04 cm sec^{-2} at the equator to 983.21 cm sec^{-2} at the poles results in more than a 1-cm change in range delay. The dependence of $g_0(\varphi)$ on latitude is given approximately by $g_0(\varphi) = \overline{g}_0(1 - 0.0026 \cos 2\varphi)$, where $\overline{g}_0 = 980.6$ cm sec^{-2} is the standard reference value for the gravitational acceleration. Replacing the latitudinal dependence of $g_0(\varphi)$ in the denominator of Eq. (69) with a Taylor expansion and keeping only the dominant latitude-dependent term, the dry tropospheric range correction in units of cm for SLP in units of millibars becomes

$$\Delta R_{dry} \approx 0.2277 P_0 (1 + 0.0026 \cos 2\varphi). \qquad (70)$$

There is evidence suggesting that it may be possible to estimate SLP from satellites to an accuracy of a few millibars or better by laser altimetry (lidar) (Singer, 1968; Garner *et al.*, 1983) or microwave radar measurements near the wide oxygen absorption band centered near 60 MHz (Smith *et al.*, 1972; Peckham *et al.*, 1983). Since both of these spaceborne techniques for estimation of SLP are still in the developmental stages, estimates of SLP along the satellite ground track that are required for the dry tropospheric correction Eq. (70) must be obtained from some other source.

Direct observations of SLP are sparsely distributed over the world ocean. The only viable source of SLP estimates for the dry tropospheric range correction is therefore the analyzed SLP fields produced operationally from numerical weather prediction models that assimilate all available in situ observations. These analyses have errors that vary geographically and seasonally in ways that are difficult to quantify. The source of SLP for present altimeters is the European Centre for Medium-range Weather Forecasts (ECMWF), which produces operational weather analyses four times daily. The ECMWF analyzed meteorological fields are widely considered to be the best global analyses available (e.g., Trenberth and Olson, 1988a; 1988b).

The geographical distributions of summertime and wintertime means and standard deviations of SLP computed from 5 years of ECMWF analyses are shown in Figure 24. Except in rare cases of extreme atmospheric conditions, SLP ranges globally from about 980–1035 mbar with a global average value of about 1013 mbar. The dry tropospheric range correction is therefore large (225–235 cm) but is only moderately sensitive to errors in SLP. An SLP error of 5 mbar, for example, corresponds to a range error of about 1 cm.

A rough measure of the uncertainties of meteorological analyses of SLP can be obtained from Figure 25, which shows the standard deviation of the differences between P_0 from operational weather analyses from ECMWF and from the U.S. National Centers for Environmental Prediction (NCEP, formerly the National Meteorological Center, NMC). Typical discrepancies are 2–3 mbar at low and middle latitudes, about 4 mbar north of 40°N and 4–7 mbar south of about 40°S (see also Trenberth and Olson, 1988a; 1988b). In each hemisphere, the discrepancies are somewhat larger during the corresponding winter season than during summer. To the extent that the differences between ECMWF and NCEP can be considered a rough measure of the uncertainties of the analyzed SLP fields, Figure 25 suggests that the root-mean-square errors of the dry tropospheric range correction Eq. (70) are only about 0.5 cm equatorward of about 40°. At higher latitudes, the rms errors of ΔR_{dry} may exceed 1 cm.

It should be noted that differences between ECMWF and NCEP analyses of SLP in excess of 10 mbar are not unusual at the high southern latitudes and in the northern-hemisphere westerlies. These large errors correspond to range errors of more than 2 cm, which is large enough to be a significant

concern. Trenberth and Olson (1988a) have documented an absolute error of 40 mbar in a case study of ECMWF and NCEP analyses of a rapidly developing low-pressure system over the South Pacific during 1985. The dry tropospheric range error for such extreme cases of poorly resolved storms (probably mostly at high southern latitudes) is about 9 cm. Fortunately, such extreme events are highly transient and infrequent.

Because of large uncertainties in present knowledge of the geoid height (see Section 5.1), most applications of altimeter data consider the time-variable sea-surface height obtained by removing the time average at each location along the satellite ground track. The large dry tropospheric correction contributed by the mean SLP is eliminated along with all other time-averaged quantities in altimetric studies of sea-surface height variability. It is seen from the standard deviations of SLP in Figure 24 that the time-variable component of the dry tropospheric correction seldom exceeds 10 cm since the maximum standard deviation of SLP is only about 15 mbar. The uncertainties of the time-variable component of SLP are the same as the overall uncertainties of SLP, which Figure 25 suggests are about 4 mbar, which corresponds to an uncertainty of less than 1 cm for the time-variable dry tropospheric range correction.

3.1.2. Wet Tropospheric Refraction

The wet tropospheric correction includes both the water vapor and the cloud liquid water droplet contributions to atmospheric refraction. From measurements of cloud liquid drop size distribution over land, the effective refractivity from cloud liquid water droplets has been found to be very nearly a linear function of the liquid droplet density $\rho_{liq}(z)$. Then N_{liq} can be parameterized as

$$N_{liq}(z) = \beta_{liq}\rho_{liq}(z), \qquad (71)$$

where the parameter in Eq. (71) has been empirically estimated by Resch (1984) to be $\beta_{liq} = 1.6 \times 10^6 \text{ cm}^3 \text{ g}^{-1}$. This parameter may be uncertain by as much as a factor of two. With this form for the liquid water refractivity, the component of the range correction Eq. (64) owing to liquid water droplets along the path of propagation becomes

$$\Delta R_{liq} = 10^{-6} \int_0^R N_{liq}(z)\,dz = 1.6L_z, \qquad (72)$$

where L_z is the integrated columnar liquid water defined by Eq. (20).

As noted in Section 2.3.2, for nonraining conditions and a cloud thickness of 1 km, the vertically integrated liquid water density is about 0.25 g cm^{-2}. The corresponding cloud liquid water range correction is only 0.38 cm. The vertically integrated liquid water density can be estimated from passive microwave measurements as summarized in Section 2.3.2. Even if the multiplicative factor of 1.6 in Eq. (72) is in

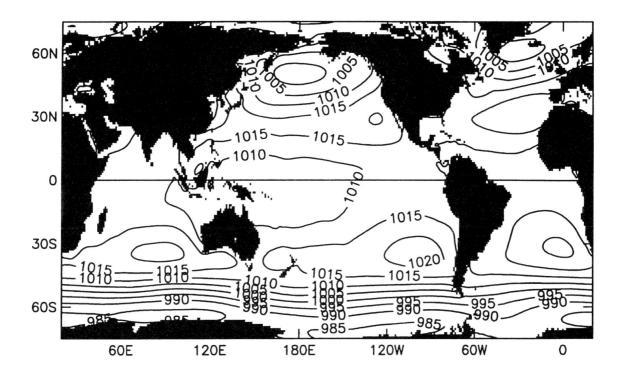

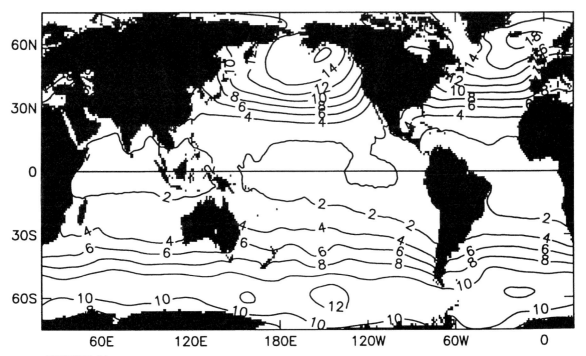

FIGURE 24a The December-January-February mean (top) and standard deviation (bottom) of sea level pressure over the global ocean in mbar computed from analyses by the European Centre for Medium-Range Weather Forecasts for the 5-year period 1993–1997. (Data provided courtesy of D. Stammer and C. King.)

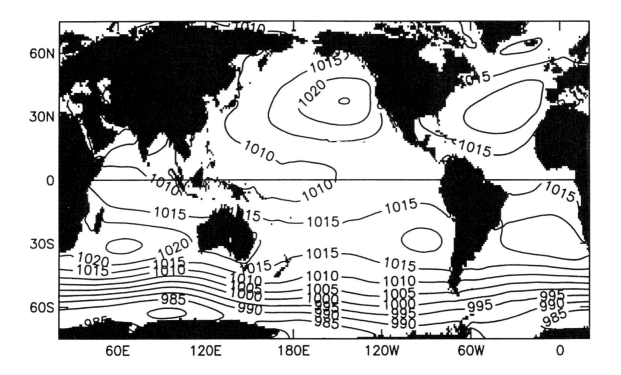

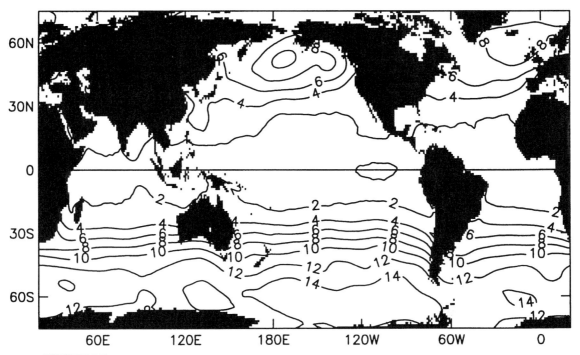

FIGURE 24b   The same as Figure 24a, except for the June-July-August period. (Data provided courtesy of D. Stammer and C. King.)

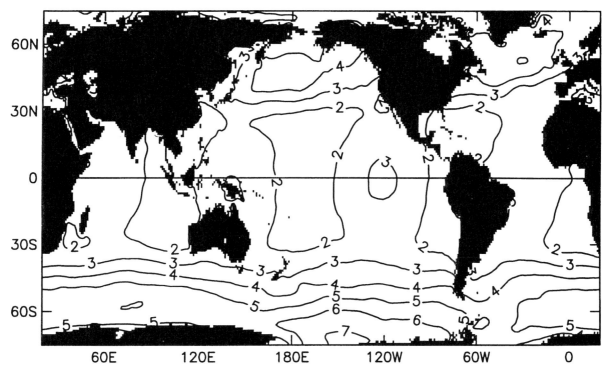

FIGURE 25 The standard deviation of the differences between 6-hr synoptic maps of sea level pressure in mbar from meteorological model analyses by the European Centre for Medium-Range Weather Forecasts (ECMWF) and the U.S. National Centers for Environmental Prediction (NCEP) over the 5-year period 1993–1997. (Data provided courtesy of D. Stammer and C. King.)

error by a factor of two, the cloud liquid water range delay for small-to-moderate cloud thicknesses is usually less than 1 cm. For convective cumulus clouds, however, Goldhirsh and Rowland (1982) have noted from aircraft measurements of N_{liq} that refractive range delays can sometimes be as large as a few centimeters.

The effects of water vapor on atmospheric refraction are 1–2 orders of magnitude larger than the effects of cloud liquid water droplets. Smith and Weintraub (1953) expressed the refractivity of water vapor in the form

$$N_{\mathrm{vap}}(z) = \beta_{\mathrm{vap}} \frac{P_{\mathrm{vap}}(z)}{T^2(z)}, \tag{73}$$

where T is temperature in units of $^\circ$K, P_{vap} is the partial pressure of water vapor in units of mbar and the parameter in Eq. (73) is empirically estimated as $\beta_{\mathrm{vap}} = 3.74^\circ\mathrm{K}^2\,\mathrm{mbar}^{-1}$ (Liebe, 1985). A more precise expression for the total water vapor refractivity includes a term proportional to P_{vap}/T (Thayer, 1974; Liebe, 1985). In the formulation Eq. (73), this term is absorbed into Eq. (65) for the dry refractivity, as discussed in Section 3.1.1. Thayer (1974) finds that the Smith and Weintraub (1953) expression Eq. (73), when combined with their expression Eq. (65) for N_{dry}, is accurate to better than 0.2%.

The ideal gas law expresses the partial pressure of water vapor as $P_{\mathrm{vap}} = R_{\mathrm{vap}} \rho_{\mathrm{vap}}(z) T(z)$, where $R_{\mathrm{vap}} = 4.613 \times$

10^6 ergs $(\mathrm{g}\,^\circ\mathrm{K})^{-1}$ is the gas constant for 1 g of water vapor. Then Eq. (73) can be written as

$$N_{\mathrm{vap}}(z) = \beta_{\mathrm{vap}} R_{\mathrm{vap}} \frac{\rho_{\mathrm{vap}}(z)}{T(z)}. \tag{74}$$

For temperature in units of $^\circ$K and water vapor density in units of g cm^{-3}, the water vapor component of the range correction Eq. (64) in units of cm then becomes

$$\Delta R_{\mathrm{vap}} = 10^{-6} \int_0^R N_{\mathrm{vap}}(z)\, dz$$

$$= \beta'_{\mathrm{vap}} \int_0^R \frac{\rho_{\mathrm{vap}}(z)}{T(z)}\, dz, \tag{75}$$

where $\beta'_{\mathrm{vap}} = 10^{-6} R_{\mathrm{vap}} \beta_{\mathrm{vap}} = 1720.6^\circ\mathrm{K}\,\mathrm{cm}^3\,\mathrm{g}^{-1}$. The 0.2% uncertainty of β_{vap} in Eq. (74) introduces less than 1-cm error in the range correction.

For ground-based satellite ranging measurements and for radio astronomy applications, attempts have been made to derive an approximate expression for ΔR_{vap} in terms of surface meteorological data because of the ease with which these observations can be made. As a first approximation, the water vapor density decreases exponentially with height in the atmosphere with an e-folding scale between 2.0 and 2.5 km (Ulaby *et al.*, 1981, Section 5–2.2). For this form of water vapor profile, ΔR_{vap} can be expressed in terms of the surface temperature and the surface water vapor density or partial pressure of water vapor (e.g., Hopfield, 1971; Saastamoinen, 1972; Black and Eisner, 1986). Bisagni (1989) obtained a root-mean-square accuracy of better than 2 cm for ΔR_{vap} estimated from surface meteorological data in the western North Atlantic during the period January to June 1986. However, as pointed out by Goldfinger (1980), Moran and Rosen (1981), Liu *et al.* (1991), Esbensen *et al.* (1993), and others, the vertical structure of the water vapor density profile is quite variable, often deviating significantly from exponential. This can result in errors of 5 cm or more in ΔR_{vap} estimated from surface meteorological data. Some information about the vertical distribution of water vapor is therefore necessary for accurate estimates of ΔR_{vap}.

It is possible in principle to obtain vertical profiles of the water vapor and temperature in the integrand of Eq. (75) from atmospheric sounding by passive remote sensing (see, for example, Houghton *et al.*, 1984; Ulaby *et al.*, 1986b). In practice, however, infrared sounding techniques are restricted to clear sky conditions, and microwave sounding techniques are limited to very coarse vertical resolution.

As a practical solution to the problem of evaluating Eq. (75) for routine application to altimeter data, the denominator in the integrand can be replaced with a constant "effective temperature," T_{eff}. This procedure could be justified by modeling the temperature profile as having a constant lapse rate κ in the lower $\sim$10 km of the atmosphere where $\rho_{vap}(z)$ is significant, i.e., $T(z) = T_0 - \kappa z$, where $T_0 = T(z = 0)$ is about 300°K. Since κ is approximately 7°K km^{-1} (see, for example, Section 1.5 of Wallace and Hobbs, 1977), the second term in this expression is much smaller than the first. Then the denominator of Eq. (75) can be approximated by a binomial expansion, resulting in

$$\Delta R_{vap} \approx \frac{\beta'_{vap}}{T_0} \left[\int_0^R \rho_{vap}(z)\, dz + \frac{\kappa}{T_0} \int_0^R z\rho_{vap}(z)\, dz \right]. \quad (76)$$

For a realistic model temperature and water vapor profile, Goldfinger (1980) showed that the magnitude of the second term in Eq. (76) is only 4% that of the first term.

Replacing T_0 with an effective temperature T_{eff} to absorb some of the error from neglecting the second term in Eq. (76), the water vapor range correction can be approximated by the simpler expression

$$\Delta R_{vap} \approx \hat{\beta}'_{vap} \int_0^R \rho_{vap}(z)\, dz, \quad (77)$$

where $\hat{\beta}'_{vap} = \beta'_{vap}/T_{eff}$ has units of cm^3 g^{-1}. This form conveniently expresses the water vapor range delay as a linear function of the columnar water vapor.

The vertical integral of water vapor is easily estimated from a satellite passive microwave radiometer. As discussed in detail by Wilheit (1978), Wilheit *et al.* (1980), and Swift (1980), the microwave radiation received by a satellite-borne

passive radiometer at frequencies between 10 and 50 GHz is determined almost entirely by columnar water vapor, columnar cloud liquid water droplets, raindrops, and wind-induced changes in the emissivity of the sea surface. For rain-free conditions, an estimate of columnar water vapor can be obtained from a dual-frequency microwave radiometer that measures the brightness temperatures within the water vapor absorption band centered at 22.235 GHz and at a nearby frequency to provide a reference for the background microwave radiation from the other factors affecting microwave brightness temperatures. For ground-based upward-looking radiometers, measurements at these two frequencies are generally sufficient (see, for example, Moran and Rosen, 1981). For satellite-based downward-looking radiometers, however, the accuracy of columnar water vapor estimates is significantly improved by including measurements at a third frequency to correct for the effects of wind-induced changes in sea-surface emissivity on the brightness temperatures measured from the satellite (see, for example, Keihm *et al.*, 1995).

Algorithms for satellite retrievals of columnar water vapor from three-frequency microwave radiometers have been developed using frequencies of 18.0, 21.0, and 37.0 GHz from the scanning multichannel microwave radiometer (SMMR) on the NASA satellites Seasat and NIMBUS-7 (Staelin *et al.*, 1976; Grody *et al.*, 1980; Wilheit and Chang, 1980; Alishouse, 1983) and frequencies of 19.35, 22.235, and 37.0 GHz from the special sensor microwave/imager (SSM/I) (Alishouse *et al.*, 1990b; Wentz, 1997) on the series of Defense Meteorological Satellite Program satellites that have been in continuous operation since June 1987. These studies have shown that the columnar integral of the water vapor can be estimated to an accuracy of better than 0.15 g cm^{-2} from space with three-frequency passive microwave radiometers. This accuracy is degraded when rain is present in the radiometer footprint.

It is noteworthy that the 22.235 water vapor absorption line is broadened by molecular collisions (e.g., Ulaby *et al.*, 1981). Collisions of water molecules in the atmosphere depend on pressure, temperature and the water vapor density itself. The detailed shape of the absorption line therefore depends on the vertical distribution of these three quantities. The high atmospheric pressure at low altitudes where the water vapor is most concentrated broadens the absorption spectrum by reducing the absorption at the center of the absorption band and increasing the absorption in the wings of the band. Based on a simple model atmosphere, Barrett and Chung (1962) showed that the region of the absorption band near the center frequency is most sensitive to profiles of water vapor, pressure, and temperature.

Radiometer measurements at the 22.235 GHz peak of the water vapor absorption band are thus sensitive to more than just the columnar integrated water vapor [see also the discussion and references in Wu (1979) and Keihm *et al.*,

1995]. By comparison, the frequencies of about 21.0 GHz and 23.8 GHz are "hinge points" that are relatively insensitive to variations in the atmospheric profiles of pressure, temperature, and water vapor (see, for example, Figure 5 of Keihm et al., 1995). For passive microwave estimates of columnar water vapor, radiometer measurements at 21.0 or 23.8 GHz are therefore somewhat preferable to measurements at the center frequency 22.235 GHz of the absorption band. It should be noted, however, that water vapor estimates at the hinge-point frequencies are somewhat more sensitive to instrument calibration since the absolute sensitivity to water vapor content is lower than at the center frequency.

Globally, the columnar water vapor generally ranges from 0.5–7 g cm^{-2} and varies geographically and temporally over a broad range of space and time scales. The mean and standard deviation of the columnar water vapor estimated from 11 years of SSM/I data are shown in Figure 26. The largest columnar water vapor values are found in the western tropical Pacific and eastern tropical Indian Ocean and in association with the inter-tropical convergence zones in the eastern tropical Pacific and Atlantic. These are also the regions where the columnar water vapor is most variable, with typical standard deviations of about 1 g cm^{-2}. The columnar water vapor values are lowest and least variable at latitudes higher than about 55° in both hemispheres.

The choice of the effective temperature T_{eff} to be used in the coefficient $\hat{\beta}'_{vap}$ in Eq. (77) presents some difficulty in the development of a global algorithm for the wet tropospheric range correction. Estimates of T_{eff} can be obtained based on assumptions about the water vapor density and temperature profiles. Because of questions about the validity of these assumptions, an alternative approach based on analysis of a limited number of radiosonde profiles of water vapor density was used for the Seasat altimeter (Tapley et al., 1982). The exact water vapor range correction Eq. (75) was computed from the vertical integral of the ratio $\rho_{vap}(z)/T(z)$ obtained from the radiosondes. The coefficient $\hat{\beta}'_{vap}$ in the approximate expression Eq. (77) was evaluated by regression of the exact range correction onto the columnar water vapor. The resulting least-squares estimate was $\hat{\beta}'_{vap} = 6.36$ cm^3 g^{-1}, which corresponds to a value of $T_{eff} = 270.6°$. For this value of $\hat{\beta}'_{vap}$ and the 0.5–7 g cm^{-2} global dynamic range of columnar water vapor, the wet tropospheric path delay ranges from about 3–45 cm with standard deviations ranging from about 3–6 cm. The corresponding accuracy of the range correction for an uncertainty of 0.15 g cm^{-2} in estimates of columnar water vapor from a three-frequency passive microwave radiometer is about 1 cm.

Based on an analysis of a large collection of radiosondes at a wide variety of locations, Liu and Mock (1990) questioned the use of a constant value for T_{eff} in the coefficient $\hat{\beta}'_{vap}$ in Eq. (77). They found that the mean value of T_{eff} varied significantly latitudinally from a value of about 288°K in the tropics to a value of about 268°K at high latitudes,

with an overall mean of 278°K. The use of the constant value $T_{eff} = 270.6°$K thus introduces a latitudinally dependent bias. The standard deviation of T_{eff} over the radiosonde dataset was found to range from about 2–5°K. These latitudinal and temporal variations in the value of T_{eff} correspond to a mean error in the wet tropospheric range correction that varies from about 1.8 cm in the tropics to near zero at high latitudes with a standard deviation of less than 0.6 cm everywhere. A random variability of 0.6 cm is a relatively minor concern, but the latitudinally varying systematic error is significant for studies of the mean ocean circulation from altimetric estimates of absolute sea surface height (see Chapter 2).

For T/P, the question of what value to use for T_{eff} was side-stepped by estimating the wet tropospheric range delay directly from the brightness temperatures at 18.0, 21.0, and 37.0 GHz measured by the T/P microwave radiometer. The coefficients of the three brightness temperatures were derived by least-squares analysis of a large global database of radiosonde profiles for four different ranges of columnar water vapor conditions as described in Keihm et al. (1995). While such an approach may seem less physical than estimating the range delay from the columnar water vapor, the two methods are essentially equivalent since passive microwave estimates of columnar water vapor are based on linear combinations of the brightness temperatures with coefficients derived by least squares techniques. A two-step procedure is used to estimate ΔR_{vap}. In the first step, a rough estimate is obtained based on a single global set of coefficients. Then a more accurate estimate is obtained using the coefficients for the range of water vapor conditions corresponding to the initial estimate. The intent of this two-step procedure is to model different water vapor mixing conditions.

Based on post-launch validation from radiosondes at a wide range of geographical locations and weather conditions, Ruf et al. (1994) have estimated that the accuracy of the T/P wet tropospheric range correction is 1.1 cm in rain-free conditions, even when heavy clouds and winds are present in the radiometer footprint. This accuracy estimate takes into account assumptions regarding the inherent variability of the water vapor field owing to space-time separation between the satellite and radiosonde observations. Keihm and Ruf (1995) compared the T/P estimates of ΔR_{vap} with values obtained from highly calibrated upward-looking radiometers synchronized with T/P overflights at four locations within 30 km of the satellite ground track over the first year of the T/P mission. The rms differences were less than 1 cm at all four locations, lending support to the Ruf et al. (1994) estimate of 1.1 cm uncertainty in the T/P microwave radiometer estimates of ΔR_{vap}.

It should be emphasized that the availability of three frequencies on the T/P microwave radiometer is crucial to the high accuracy of the estimates of ΔR_{vap}. The accuracies of dual-frequency estimates of ΔR_{vap} have been thoroughly

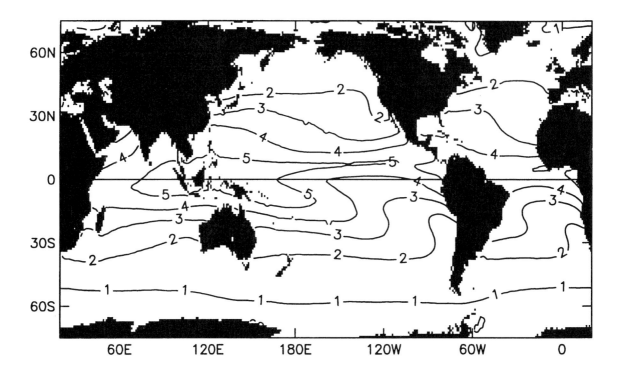

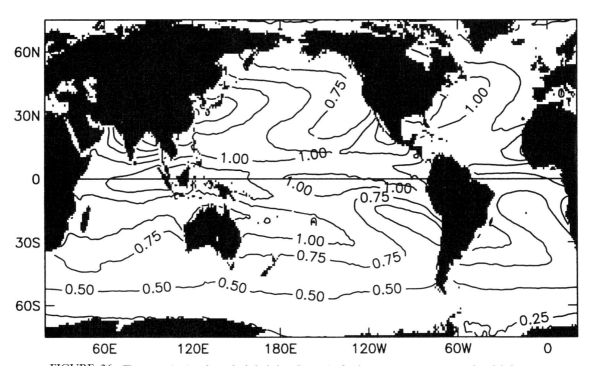

FIGURE 26 The mean (top) and standard deviation (bottom) of columnar water vapor over the global ocean in $g\,cm^{-2}$ from measurements by the series of Special Sensor Microwave Imagers (SSM/Is) operated by the Defense Meteorological Satellite Program over the 11-year period 1988–1998. (Data provided courtesy of C. Gentemann.)

investigated by Keihm *et al.* (1995). Water vapor effects on radiometer measurements of brightness temperature are mostly restricted to the frequency range 20–25 GHz. Cloud effects on brightness temperature increase with increasing frequency (Figure 5). Brightness temperature is also affected by the surface wind speed from changes in the surface emissivity owing to wind-generated roughness of the sea surface. The sensitivity of brightness temperature to variations of wind speed is relatively constant over the range 10–40 GHz (Webster *et al.*, 1976). The combination of brightness temperature measurements at the three frequencies 18.0, 21.0, and 37.0 of the T/P radiometer allows the simultaneous estimation of water vapor, cloud liquid water, and wind speed. Keihm *et al.* (1995) estimate that the lack of the 37.0-GHz channel would increase uncertainties in ΔR_{vap} to 1.5–2 cm owing to the inability to correct for the effects of clouds. Likewise, the lack of the 18.0 GHz channel would lead to errors in excess of 1.5 cm when the wind speed exceeds 12 m sec^{-1}. This conclusion is consistent with the empirical results obtained by Eymard *et al.* (1996), who showed that estimates of ΔR_{vap} with rms accuracies of about 2 cm can be derived from the dual-frequency passive microwave radiometer (frequencies of 23.8 and 36.5 GHz) that is onboard the ERS satellites. This uncertainty is about a factor of two higher than the uncertainty of the T/P three-frequency radiometer estimates of ΔR_{vap}.

An important point to note is that, because the columnar water vapor at any particular location varies temporally, the ability to correct satellite radar data for water vapor attenuation requires coincident measurements from a passive microwave radiometer onboard the satellite. The need for coincident estimates of the wet tropospheric range correction has been clearly demonstrated from numerous analyses of the Geosat altimeter data. Apparently because of the lower accuracy requirements of the defense-related geodetic origins of the mission, there was no microwave radiometer onboard Geosat. Estimates of water vapor were derived from the Fleet Numerical Oceanography Center (FNOC) operational analyses of surface humidity. This significantly degraded the quality of the Geosat range estimates, especially in tropical regions where columnar water vapor densities are high with large variability over a wide range of space and time scales.

The problem of estimating the wet tropospheric range correction for Geosat received considerable attention. Emery *et al.* (1990) showed that the FNOC-based correction systematically underestimated the columnar water vapor in the tropics. Zimbelman and Busalacchi (1990) and Didden and Stammer (1994) showed that even major features such as the intertropical convergence zones in the Pacific and Atlantic were virtually absent from the FNOC-based estimates of ΔR_{vap}. Even outside of the tropics, Phoebus and Hawkins (1990), Emery *et al.* (1990), Monaldo (1990) and Jourdan *et al.* (1990) all showed that the FNOC-based correction failed to resolve significant variability in the wet tropospheric range delay on the spatial and temporal scales of oceanographic mesoscale variability. This resulted in significant errors in the Geosat range estimates on the mesoscale space and time scales that received the greatest attention in the literature (see Chapter 3).

Because of the temporal variability of columnar water vapor at any given location, it is also not possible to obtain accurate estimates of ΔR_{vap} from microwave radiometers on other satellites orbiting simultaneously with an altimeter satellite. An important legacy of the Geosat mission is thus that any well-designed altimeter mission must include an onboard microwave radiometer for estimation of the wet tropospheric range delay.

The errors in analyzed water vapor fields deduced from the Geostat analyses summarized above are not unique to the FNOC-based estimates of ΔR_{vap}. It has been shown that the ECMWF estimates of columnar water vapor underestimate the variability at all spatial scales shorter than about 800 km in the northern hemisphere and all scales south of $20°S$ (see, for example, Plate 5 of Stum, 1994). In addition, Stum (1994) showed an example in which the ECMWF analyses of columnar water vapor misplaced and underestimated the meridional scale of the ITCZ in the Pacific. Globally, the rms difference between T/P microwave radiometer and ECMWF estimates of the wet tropospheric range correction was 3 cm. Morris and Gill (1994) found that T/P microwave radiometer estimates of the wet tropospheric range correction agreed with estimates obtained from the French Meteorological Office (FMO) analyzed columnar water vapor fields (which are essentially equivalent to the ECMWF analyses) during the fall and winter. During the spring and summer, however, the rms errors of the FMO estimates were a factor of two larger than the T/P microwave radiometer estimates. The use of analyzed columnar water vapor fields from operational weather centers is clearly inadequate for highly precise altimetry.

An important limitation of microwave radiometer estimates of ΔR_{vap} is that radiometers are prone to calibration drifts. This was especially a problem for the Nimbus-7 SMMR (Francis, 1987; see also the discussion in Ruf *et al.*, 1995). Despite considerable pre-launch attention that was devoted to minimizing calibration drifts in the T/P microwave radiometer (Ruf *et al.*, 1995), a slow drift of the 18.0-GHz channel became apparent over the long duration of the T/P mission. In terms of the wet tropospheric range correction, this calibration drift resulted in a downward drift of about 1.5 mm year^{-1} in ΔR_{vap} over the first four years of the T/P mission. The drift rate appears to have decreased during 1997 but there is evidence that it may have increased again in 1998 (Keihm *et al.*, 2000).

The drift of the T/P microwave radiometer was first noted from comparisons of T/P measurements of sea-surface height with the global network of tide gauges (Mitchum,

1998). As summarized by Keihm *et al.* (2000), this calibration drift has been confirmed from a variety of other sources, including radiosondes, upward-looking water vapor radiometers, the ERS-1 and ERS-2 radiometers, the SSM/I radiometers, and estimates of water vapor from the Global Positioning System satellite constellation. The small drift of the T/P microwave radiometer is inconsequential for many applications of altimeter data. For accurate monitoring of global sea-level change, however, corrections for a 1.5 mm year^{-1} calibration drift in the onboard microwave radiometer are essential since the predicted rise of global sea level is only 1.8 mm year^{-1} (Douglas, 1995; see also Chapter 8). A continuous and sustained calibration and validation program is thus a necessary element of any altimeter mission (see Section 8.2).

3.1.3. Ionospheric Refraction

Ionospheric refraction of altimetric radar signals is determined by the dielectric properties of the upper atmosphere associated with the presence of free electrons. At frequencies higher than $\sim$2 GHz, the magnetic effects of the ionosphere that give rise to Faraday rotation can be neglected and the real part of the ionospheric index of refraction can be shown to be related to the electromagnetic radiation frequency f by

$$\eta_{\text{ion}} = \left(1 - \frac{f_p^2}{f^2}\right)^{1/2}, \tag{78}$$

where f_p is the natural frequency of oscillation of electrons in a plasma (e.g., Ginzburg, 1964; Lawrence *et al.*, 1964). This plasma frequency is related to the electron density n_e by $f_p^2 = 80.6 \times 10^6 n_e$, where n_e has units of electrons cm^{-3}. A typical daytime maximum electron density at the approximate 250–400-km altitudes of maximum ionospheric electron content at midlatitudes is $n_e \approx 10^6$ electrons cm^{-3} (see, for example, Sojka, 1989), which corresponds to a plasma frequency of $f_p \approx 9$ MHz. The ionospheric index of refraction becomes imaginary for electromagnetic frequencies lower than f_p, implying that the electromagnetic signals are reflected by the ionosphere since they cannot propagate with a purely imaginary index of refraction.

The K$_u$-band and C-band frequencies of 13.6 and 5.3 GHz on the T/P dual-frequency altimeter are about three orders of magnitude higher than the plasma frequency f_p. For altimeter applications, Eq. (78) for the ionospheric index of refraction can thus be approximated by the binomial expansion

$$\eta_{\text{ion}} \approx 1 - \frac{f_p^2}{2f^2} = 1 - \frac{40.3 \times 10^6 n_e}{f^2}. \tag{79}$$

Unlike the nondispersive nature of tropospheric gases, it is apparent from the frequency dependence of η_{ion} that the ionosphere is a dispersive medium. The fact that η_{ion} is less than 1 implies that the phase speed $c_p = c/\eta_{\text{ion}}$ exceeds the free-space speed of light c. The phases of electromagnetic

radiation thus advance faster than the electromagnetic energy, which propagates through a dispersive medium at the group velocity. Group velocity can be determined from the dispersion relation $c_p = c/\eta_{\text{ion}}$, where $c_p = \omega/k$ is the phase speed for wavenumber k and the angular frequency $\omega = 2\pi f$ associated with electromagnetic frequency f. The group velocity $c_g = d\omega/dk$ is thus

$$c_g = \frac{c}{\eta_{\text{ion}}} - \frac{\omega}{\eta_{\text{ion}}} \frac{d\eta_{\text{ion}}}{dk}. \tag{80}$$

The chain rule for differentiation transforms Eq. (80) into

$$\begin{aligned} c_g &= \frac{c}{\eta_{\text{ion}}} - \frac{\omega}{\eta_{\text{ion}}} \frac{d\eta_{\text{ion}}}{d\omega} \frac{d\omega}{dk} \\ &= \frac{c}{\eta_{\text{ion}}} - c_g \frac{\omega}{\eta_{\text{ion}}} \frac{d\eta_{\text{ion}}}{d\omega}, \end{aligned} \tag{81}$$

which can be rearranged and solved for c_g to obtain

$$c_g = \frac{c}{\eta'_{\text{ion}}}, \tag{82}$$

where, from Eq. (79),

$$\begin{aligned} \eta'_{\text{ion}} &= \eta_{\text{ion}} + \omega \frac{d\eta_{\text{ion}}}{d\omega} = \frac{d}{d\omega}\left(\omega\eta_{\text{ion}}\right) \\ &= \frac{d}{df}\left(f\eta_{\text{ion}}\right) \\ &= 1 + \frac{40.3 \times 10^6 n_e}{f^2} \end{aligned} \tag{83}$$

is the group index of refraction. It can be noted that $\eta'_{\text{ion}} > 1$, thus assuring that the group velocity is slower than the free-space speed of light. The group ionospheric refractivity is

$$N_{\text{ion}}(z) = 10^6(\eta'_{\text{ion}} - 1) = \frac{40.3 \times 10^{12}}{f^2} n_e(z). \tag{84}$$

The ionospheric component of the range correction Eq. (64) in units of cm then becomes

$$\begin{aligned} \Delta R_{\text{ion}}(f) &= 10^{-6} \int_0^R N_{\text{ion}}(z)\,dz \\ &= \frac{40.3 \times 10^6}{f^2} \int_0^R n_e(z)\,dz. \end{aligned} \tag{85}$$

For a K$_u$-band frequency of 13.6 GHz, the sensitivity of the ionospheric range correction to the vertically integrated electron content is 0.22 cm per 10^{12} electrons cm^{-2}. Expressed in terms of the standard unit for vertically integrated electron content, the ionospheric range correction is 0.22 cm per TECU, where 1 TECU $\equiv 10^{16}$ electrons m^{-2} is called a "total electron content unit."

The ionospheric range correction Eq. (85) for a given electromagnetic frequency f is fully determined from knowledge of the integrated columnar electron content along the path of propagation. The vertically integrated electron content ranges from about 1 TECU during the nighttime to as high as 180 TECU in the daytime during peak periods of

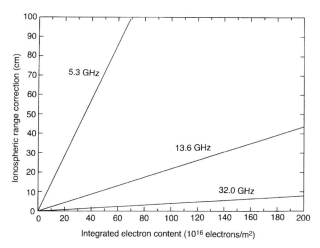

FIGURE 27 Ionospheric range correction as a function of vertically integrated electron content at frequencies of 5.3, 13.6, and 32.0 GHz.

ionospheric activity (Davies *et al.*, 1977; Soicher, 1986). The ionospheric range delays for three different radar altimeter frequencies are shown in Figure 27 as a function of columnar electron content. The rapid decrease in the magnitude of the ionospheric range correction Eq. (85) with increasing frequency favors the use of a high microwave radar frequency. As discussed in Section 2.3, however, higher frequencies are more attenuated by water vapor, clouds, and oxygen. These atmospheric effects dictate an upper limit of around 15 GHz for altimeter systems. At the K_u-band frequency of 13.6 GHz that is the primary frequency of the T/P altimeter, the ionospheric range correction Eq. (85) ranges from 2–40 cm (see Figure 27).

Most of the charged particles that interfere with the propagation of electromagnetic radiation are at altitudes ranging from 50–2000 km, with the highest concentration between about 250 and 400 km (e.g., Rush, 1986; Sojka, 1989). Ionization of this region of the atmosphere is attributed mostly to extreme-ultraviolet radiation from the sun. As reviewed by Callahan (1984), the vertically integrated electron density varies diurnally by as much as an order of magnitude with maxima occurring near 1500 local time (Figure 28). The total electron content also varies latitudinally by a factor of two with maxima occurring near ± 15° geomagnetic latitude (Figures 28 and 29) and decreased total electron content at the geomagnetic equator. This latitudinal structure (called the "equatorial anomaly") is most clearly developed near the peak of the diurnal cycle and in the early evening hours. Meridional gradients as large as 2 cm/100 km can occur during mid to late afternoon at latitudes of 20–30° (Goldhirsh and Rowland, 1982). The geographical structure of the region of highest electron content is tipped with respect to latitude (Figure 29) because of the influence of the geomagnetic field, which is oriented with an axis that differs from the earth's rotational axis by about 11°.

On longer time scales (Figure 30), there are significant seasonal variations with maxima near the solar equinoxes and large variations (about a factor of 5) associated with the approximate 11-year solar cycle (Gorney, 1990). As shown in Figure 31, the most energetic solar cycle on record peaked in 1957 with an annual average of about 190 sunspots. The two most recent solar cycles were both well above average with annual averages of about 160 sunspots at their peaks in 1979 and 1989. Solar activity is expected to peak again in late 2000 with an amplitude comparable to the last two solar cycles (Joselyn *et al.*, 1997; Sofia *et al.*, 1998).

In addition to the periodic diurnal, seasonal, and 11-year variations, the ionosphere is also affected by aperiodic variations in the number of sunspots (Figure 32), and by episodic X rays and solar particle fluxes associated with solar flares. Significant ionospheric disturbances also occur from coupled interactions between the ionosphere, thermosphere, and magnetosphere that are most dramatic during geomagnetic storms caused by the interaction between the solar wind and the earth's magnetic field. Along-track gradients of the range delay associated with these unpredictable ionospheric disturbances can be as high as 4 cm/100 km (Goldhirsh and Rowland, 1982).

Prior to the launch of the T/P, the ionospheric range correction was based upon semi-empirical model estimates of the vertically integrated electron density. The global model used for Seasat was based upon Faraday rotation of signals transmitted from geosynchronous satellite beacons measured twice daily at two locations, one in the western United States and the other in eastern Australia (Goldhirsh and Rowland, 1982; Lorell *et al.*, 1982). These measured columnar electron densities were extrapolated to the Seasat subsatellite points using a simple functional dependence on latitude, sun angle and local time of day. For Geosat, the ionospheric correction was obtained from a simple model that was scaled by ground-based measurements of solar energy flux at a wavelength of 10.7 cm (Klobuchar, 1987). From Eq. (85), an accuracy of 1-cm ionospheric range correction for a 13.6-GHz altimeter is equivalent to an accuracy of 4.5 TECU in the total electron content. The uncertainties of the empirical ionospheric range corrections applied to Seasat and Geosat data were not known quantitatively. Klobuchar (1987) estimated that the total electron content had an rms accuracy of 50%. For a typical daytime total electron content of 50 TECU during moderate solar activity, this corresponds to an rms ionospheric range error of about 5 cm. Errors could be much higher during periods of high solar activity such as the last year of the Geosat mission as it approached the 1989 maximum of the solar cycle (see Figure 31).

Much of the error in the modeled ionospheric corrections applied to Seasat and Geosat altimeter data had large spatial scale and was therefore mitigated in the empirical orbit error corrections that had to be applied to these datasets. As

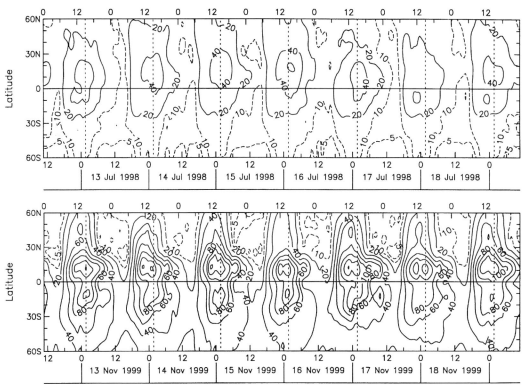

FIGURE 28   Vertically integrated ionospheric electron content in units of TECU (1 TECU $\equiv 10^{16}$ electrons m^{-2}) estimated from GPS data along 160°W in the central Pacific at hourly intervals over two 7-day periods: (top) a period of low electron content in July 1998, and (bottom) a period of high electron content in November 1999. The data are contoured as a function of latitude and time, which is displayed as UTC along the bottom of each panel and as local time along the top of each panel. Contour interval is 20 TECU with additional contours shown as dashed lines for 5 and 10 TECU. (Data are from the Global Ionospheric Maps produced at the Jet Propulsion Laboratory, with sponsorship from the U.S. Naval Oceanographic Office and the National Aeronautics and Space Administration.)

discussed in Section 4, these empirical orbit error corrections corrupted the altimeter data in other ways that complicated interpretation of altimetric estimates of sea-surface height. Indeed, the dramatic improvement in the accuracy of precision orbit determination reviewed in that section is the primary reason for the vastly expanded utility of satellite altimetry for ocean circulation studies. However, the benefits of this improved orbit accuracy could not be fully realized without concomitant improvements in the accuracies of the environmental corrections that are applied to the altimeter range measurements. Of these various corrections, model-based estimates of the ionospheric range delay posed one of the greatest challenges.

Progress in physical modeling of the ionosphere has been reviewed by Rush (1986) and Sojka (1989). The most widely used models are empirically based. These include the Bent model (Bent *et al.*, 1976), the International Reference Ionosphere (IRI) model (Bilitza, 1997) and the Parameterized Real-time Ionospheric Specification Model (PRISM) (Daniell *et al.*, 1995). While the accuracies of these model estimates of ionospheric electron content have improved over the past decade, the errors remain unacceptably large for highly accurate altimeter systems. For the Bent model, for example, Ho *et al.* (1997) and Mannucci *et al.* (1998) showed that errors of the ionospheric range corrections can be more than 8 cm at low latitudes. Similarly, Schreiner *et al.* (1997) showed examples of more than 8-cm errors in the ionospheric range correction deduced from the PRISM model. They also showed that the errors of the IRI model were comparable to those of PRISM. The overall rms errors in the case studies conducted to date from all three of these models were equivalent to more than 2-cm errors of the ionospheric range correction deduced from the model estimates of total electron content.

Extreme ionospheric range errors of 8 cm and even rms errors of 2 cm far exceed the 1-cm accuracy goal of the T/P altimeter system. As first described in detail by Goldhirsh and Rowland (1982), the inverse-square dependence of the ionospheric range correction Eq. (85) on frequency allows direct simultaneous estimation of the total electron content and the range from altimeter measurements at two frequencies. T/P is the first satellite to carry a dual-frequency altimeter. The dual-frequency range estimation technique and the associated measurement errors can be elucidated by ex-

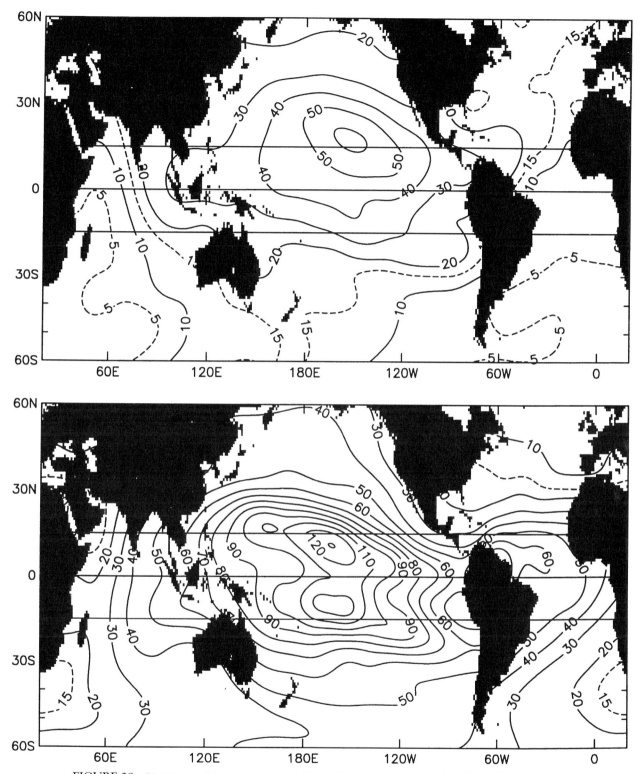

FIGURE 29 Global maps of the vertically integrated ionospheric electron content estimated from GPS data at 1500 local time along 160°W (0140 UTC) on July 16, 1998 (top) during a period of low electron content and on November 16, 1999 (bottom) during a period of high electron content. Contour interval is 10 TECU with additional contours shown as dashed lines for 5 and 15 TECU. (Data are from the Global Ionospheric Maps produced at the Jet Propulsion Laboratory, with sponsorship from the U.S. Naval Oceanographic Office and the National Aeronautics and Space Administration.)

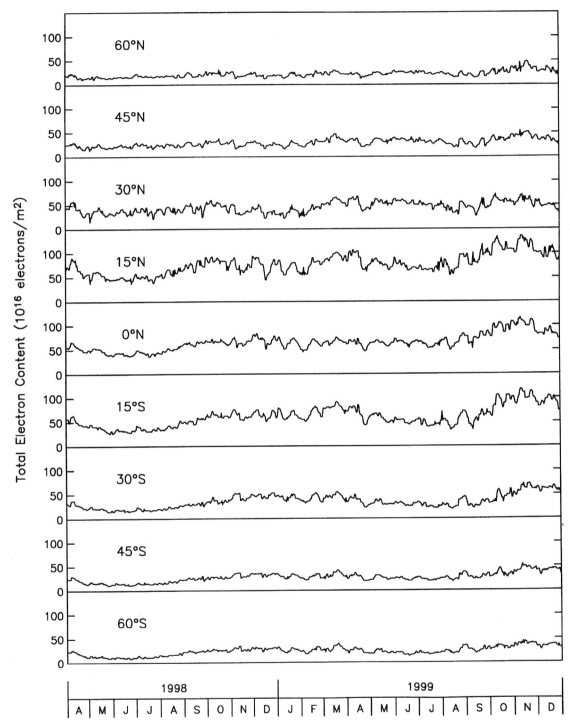

FIGURE 30 Daily vertically integrated ionospheric electron content in units of TECU estimated from GPS data at selected latitudes along 160°W at 1500 local time (0140 UTC) over the 21-month period April 1998 through December 1999. The data values have been smoothed with a three-point median filter. (Data are from the Global Ionospheric Maps produced at the Jet Propulsion Laboratory, with sponsorship from the U.S. Naval Oceanographic Office and the National Aeronautics and Space Administration.)

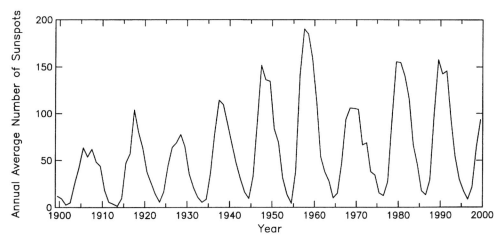

FIGURE 31 Annual average number of sunspots over the time period 1900–1999.

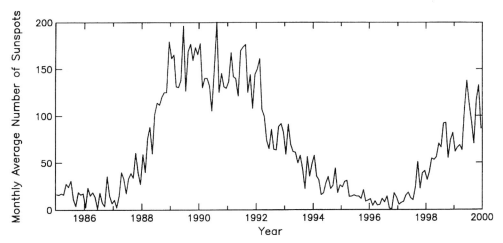

FIGURE 32 Monthly average number of sunspots over the time period 1985–1999.

pressing the true range R from the satellite to the sea surface in terms of the range measurement $\hat{R}(f_j)$ computed from radar measurements at frequency f_j neglecting the effects of ionospheric refraction on the two-way travel time,

$$R = \hat{R}(f_j) - \Delta R_{\mathrm{ion}}(f_j) + \varepsilon(f_j), \qquad (86)$$

where $\Delta R_{\mathrm{ion}}(f_j)$ is the ionospheric range correction and $\varepsilon(f_j)$ is the total measurement error from other sources. Substituting Eq. (85) for $\Delta R_{\mathrm{ion}}(f_j)$ and multiplying both sides of Eq. (86) by f_j^2 yields

$$f_j^2 R = f_j^2 \hat{R}(f_j) - 40.3 \times 10^6 \int_0^R n_e(z)\,dz + f_j^2 \varepsilon(f_j). \quad (87)$$

For the T/P dual-frequency altimeter, the subscript j will be denoted as k for the K_u-band frequency $f_k = 13.6$ GHz and c for the C-band frequency $f_c = 5.3$ GHz. The separate expressions Eq. (87) for the two frequencies can be differenced and normalized by $(f_k^2 - f_c^2)$ to obtain

$$R = \tilde{R} + \tilde{\varepsilon}, \qquad (88)$$

where

$$\tilde{R} = a_k \hat{R}(f_k) - a_c \hat{R}(f_c) \qquad (89)$$

is the so-called combined range estimate from the dual-frequency altimeter and

$$\tilde{\varepsilon} = a_k \varepsilon(f_k) - a_c \varepsilon(f_c) \qquad (90)$$

is the combined measurement error of the dual-frequency range estimate. The nondimensional factors a_k and a_c in Eqs. (89) and (90) are

$$a_k = \frac{1}{1 - \delta_f^2}, \qquad (91)$$

$$a_c = \frac{\delta_f^2}{1 - \delta_f^2}, \qquad (92)$$

where $\delta_f = f_c/f_k$ is the ratio of the two frequencies.

For the T/P dual-frequency altimeter, the squared frequency ratio is $\delta_f^2 = (5.3/13.6)^2 = 0.152$. The corresponding coefficients defined by Eqs. (91) and (92) are $a_k = 1.18$ and $a_c = 0.18$. It is thus apparent why K_u band is referred to as the primary frequency of the T/P dual-frequency altimeter; the ratio a_k/a_c is about 6.6, which implies that the combined height estimate $\tilde{R}$ weights the K_u-band range estimate $\hat{R}(f_k)$ about 6.6 times greater than the C-band range estimate $\hat{R}(f_c)$. Furthermore, the contribution of the K_u-band measurement errors to the combined height uncertainty is amplified by the multiplicative factor $a_k = 1.18$. The rms instrumental errors of the K_u-band and C-band altimeters are estimated to be about 2 cm and 3–3.5 cm, respectively. The higher C-band measurement errors are of secondary concern since they are reduced by the multiplicative factor $a_c = 0.18$ in Eq. (90). It is apparent from Eq. (90), however, that measurement errors that are the same for K_u-band and C-band (e.g., residual errors in the nondispersive dry and wet tropospheric range corrections) are unaltered in magnitude in the combined range estimate since, from Eqs. (91) and (92), $a_k - a_c = 1$.

The individual measurement errors contributing to the combined measurement error Eq. (90) have been discussed by Callahan (1987) and Monaldo (1993). The variance of the combined measurement error can be shown to be

$$\tilde{\sigma}^2 = a_k^2 \sigma_k^2 + a_c^2 \sigma_c^2 - 2 a_c a_k \sigma_c \sigma_k \rho_{ck}, \quad (93)$$

where ρ_{ck} is the cross-correlation between the C-band and K_u-band measurement errors with variances of σ_c^2 and σ_k^2, respectively. From this formulation, it becomes apparent that the variance of the overall measurement error in the dual-frequency altimeter system depends not only on the separate measurement error variances at the two frequencies but also on the cross correlations between the various components of the measurement errors at the two frequencies. Thus, in addition to magnifying the K_u-band measurement errors by the factor a_k, the combined measurement error is increased or decreased by the correlated measurement errors, depending on the sign of the correlation. The magnitudes of the coupled measurement errors are proportional to the correlation coefficient and are moderated by the multiplicative factor $(2 a_c a_k)^{1/2}$, which is about 0.65 for the T/P dual-frequency altimeter.

In practice, the T/P range estimates are obtained by a 3-step procedure. First, the individual T/P range estimates $\hat{R}(f_k)$ and $\hat{R}(f_c)$ are computed from the two-way travel times of the K_u-band and C-band radar signals neglecting the effects of atmospheric refraction. Then $\hat{R}(f_k)$ and $\hat{R}(f_c)$ are separately corrected for sea-state bias effects (see Section 3.2), which are the only non-random frequency-dependent range correction other than the ionospheric correction that is applied to altimeter data. Finally, the K_u-band ionospheric range correction is computed and removed from $\hat{R}(f_k)$ along with the other refractive effects on the two-way

travel time (the dry and wet tropospheric range corrections described in Sections 3.1.1 and 3.1.2). The K_u-band ionospheric range correction obtained from Eqs. (86) and (88)–(90) after correcting the individual range estimates $\hat{R}(f_k)$ and $\hat{R}(f_c)$ for sea-state bias effects is

$$\Delta R_{\text{ion}}(f_k) = a_c \hat{R}(f_c) - (a_k - 1) \hat{R}(f_k) \\ + a_c \varepsilon(f_c) - (a_k - 1) \varepsilon(f_k). \quad (94)$$

This expression can be simplified by noting from Eqs. (91) and (92) that $(a_k - 1) = a_c$. The K_u-band ionospheric range correction is therefore

$$\Delta R_{\text{ion}}(f_k) = \Delta \hat{R}_{\text{ion}}(f_k) + \varepsilon_{\text{ion}}(f_k), \quad (95)$$

where

$$\Delta \hat{R}_{\text{ion}}(f_k) = a_c \big[\hat{R}(f_c) - \hat{R}(f_k) \big] \quad (96)$$

is the dual-frequency estimate of the K_u-band range correction and

$$\varepsilon_{\text{ion}}(f_k) = a_c [\varepsilon(f_c) - \varepsilon(f_k)] \quad (97)$$

is the error of the estimated range correction.

The range measurement determined from the two-way travel time of the K_u-band radar signal and corrected for the dual-frequency estimate of ionospheric effects is thus

$$\hat{R}(f_k) - \Delta \hat{R}_{\text{ion}}(f_k) = (1 + a_c) \hat{R}(f_k) - a_c \hat{R}(f_c). \quad (98)$$

Since $(1 + a_c) = a_k$, this reduces to

$$\hat{R}(f_k) - \Delta \hat{R}_{\text{ion}}(f_k) = \tilde{R}, \quad (99)$$

where $\tilde{R}$ is the combined range estimate defined by Eq. (89). Likewise, Eqs. (86), (95), and (97) can be combined to show that the total error of the corrected K_u-band range measurement is

$$\varepsilon(f_k) - \varepsilon_{\text{ion}}(f_k) = (1 + a_c) \varepsilon(f_k) - a_c \varepsilon(f_c) = \tilde{\varepsilon}, \quad (100)$$

where $\tilde{\varepsilon}$ is the combined error estimate Eq. (90).

It is evident from Eq. (96) that any systematic errors in the range estimates $\hat{R}(f_k)$ and $\hat{R}(f_c)$ will contaminate the dual-frequency estimate of the ionospheric range correction $\Delta \hat{R}_{\text{ion}}(f_k)$. The instrumental measurement errors in the K_u-band and C-band altimeter range measurements are essentially random and therefore do not contribute any systematic error to $\Delta \hat{R}_{\text{ion}}(f_k)$. The dry and wet tropospheric range corrections are the same for the two altimeter frequencies and therefore also do not contaminate $\Delta \hat{R}_{\text{ion}}(f_k)$. The sea-state bias, however, is different for the two altimeter frequencies (see Section 3.2). Any systematic errors in the K_u-band or C-band sea-state bias corrections will therefore be misinterpreted as an ionospheric range error. By examining errors of T/P estimates of the total electron content as a function of wave height, Imel (1994) presented evidence that the error in the difference between the K_u-band and C-band sea-state bias correction has an upper bound of about 0.5 cm, suggesting that the wave-height dependencies of the sea-state bias correction have been well modeled at both frequencies.

The measurement precision for the T/P dual-frequency estimate of the K_u-band ionospheric range correction is believed to be about 1.1 cm in 1-sec averages of T/P data (Zlotnicki, 1994; Imel, 1994). Most of this error arises because of the cumulative effects of the 2–3 cm instrumental noise at the two radar frequencies and therefore has short along-track scale. Imel (1994) and Zlotnicki (1994) recommend low-pass filtering the ionospheric correction over 15–25 successive measurements (90–150 km along the satellite ground track), thereby reducing the effects of instrumental noise by a factor of 4–5 (i.e., to less than 0.5 cm).

Single-frequency altimeters must rely on some independent estimate of the total electron content for the ionospheric range correction. As noted previously, models of the ionosphere do not meet the 1-cm accuracy requirements of the T/P mission. One source of information about ionospheric electron content is the Doppler Orbitography and Radiopositioning Integrated by Satellite (DORIS) system used for precise orbit determination for T/P (see Section 4.2.3). DORIS consists of a ground network of about 50 beacons that transmit at microwave frequencies of 0.4 and 2.0 GHz specifically to correct for ionospheric delays of the two-way propagation between the satellite and the ground station. The Doppler data can thus also be used to estimate the integrated electron content along the line of sight between the satellite and the ground-based Doppler transmitter. The primary limitation of this method is that the Doppler radar only measures the rate of change of range between the transmitter and the onboard receiver and thus provides information only on along-track changes of electron content. The Doppler data must therefore be combined with a model to estimate the total electron content. An additional limitation is that the electron content along the slant line of sight differs from the vertical profile of the ionosphere that is sampled by the altimeter.

The principles for using measurements of slant electron content from ground-based Doppler transmitters to estimate the vertical electron content at the subsatellite point are described by Leitinger et al. (1975; 1984) and the feasibility of using these slant measurements of electron content to estimate the ionospheric correction of the altimeter range measurements is discussed by Fleury et al. (1991). CNES uses the dense set of Doppler measurements in combination with the Bent model of the ionosphere (Bent et al., 1976) to derive an estimate of the ionospheric range correction at every point along the T/P ground track. The method essentially consists of least-squares estimation of the vertical electron content from slant measurements of range rate at multiple ground stations for a large number of successive points along the satellite ground track. The method assumes a smooth latitudinal variation of the vertical electron content, which restricts the along-track span of the fit and the latitudinal range of the ground stations that can be used in the least-squares estimation.

The few comparisons of the DORIS-based estimate of the ionospheric range correction with the T/P dual-frequency estimate that have been published to date suggest that the DORIS estimate does not meet the 1-cm rms accuracy goal. Imel (1994) found that the mean and standard deviation of the differences between the DORIS and 20-sec averages of the T/P dual-frequency estimates were 1.0 cm (DORIS smaller than T/P) and 1.9 cm, respectively. The power spectral density of the DORIS estimates were less energetic than the T/P estimates at all wavelengths. Morris and Gill (1994) found that the rms error of the DORIS estimate was about 1 cm larger than the T/P dual-frequency estimate at a midlatitude location where the total electron content is only moderately large. Zlotnicki (1994) showed an example where DORIS estimates of the ionospheric range correction differed from the T/P dual-frequency estimates by more than 2 cm over an along-track distance of more than 6000 km. Moreover, he showed examples of 2–3 cm discrepancies between DORIS and T/P ionospheric range corrections over large geographical regions spanning entire ocean basins. There is a need for further quantitative assessment of the accuracy of the DORIS estimates of the ionospheric range correction.

Another promising source of information about global ionospheric variability is the range delay computed from ground-based measurements of the dual-frequency microwave signals transmitted from the constellation of Global Positioning System (GPS) satellites (Mannucci et al., 1999). The network of about two dozen GPS satellites is intended for navigation purposes (see Section 4.2.6). To correct for ionospheric effects on the estimate of range between a GPS satellite and a ground-based receiver, the GPS satellites transmit at microwave frequencies of 1.228 and 1.575 GHz.

As summarized by Lanyi and Roth (1988), Monaldo (1991) and others, GPS-based estimates of the vertically integrated ionospheric electron density can be obtained from the differential travel times at the two frequencies, analogous to the methods used with the T/P dual-frequency altimeter as described above. The accuracy of GPS-based estimates is limited somewhat by an unknown bias associated with differences between the propagation times of the signal paths for the two frequencies through the hardware of the GPS transmitters and receivers. These biases can be estimated to an accuracy of better than 1 TECU (X. Pi, personal communication, 2000).

For application of ionospheric range corrections at each point along a satellite ground track, the discrete GPS observations must be interpolated to obtain global estimates of the total electron content. The interpolation methods that have been used are all based on the technique presented by Lanyi and Roth (1988) for converting GPS slant line-of-sight estimates of total electron content to the vertical electron content. This conversion of slant to vertical total electron content is a significant source of error in GPS-based estimates of total electron content. In essence, the ionosphere is assumed to be regionally uniform so that slant-range total electron content can be approximately converted to vertical total electron content by a simple geometrical mapping. This is achieved by approximating the ionosphere as a thin shell near the mean altitude of maximum electron density (about 350 km). At the point where the line of sight between a GPS satellite and a ground receiver intersects the shell (called the ionospheric pierce point), the vertical electron content is assumed to be approximately equal to the slant measurement of total electron content multiplied by a geometric mapping function (Coster *et al.*, 1992). Lanyi and Roth (1988) showed that the accuracies of total electron content estimated from GPS data by their method are comparable to those estimated from the traditional Faraday rotation measurements of signals transmitted from geostationary satellites. The much larger number of high-quality GPS stations distributed worldwide yields global maps of total electron content that are much more accurate than can be obtained by extrapolation of the small number of Faraday rotation measurements available worldwide.

In addition to errors from the transformation from slant measurements to the vertical total electron content, the fixed altitude of the shell in the pierce-point method introduces errors in the GPS-based estimates of total electron content. This results in systematic errors because of latitudinal variations and day-night differences in the altitude of maximum electron density. Lanyi and Roth (1988) estimate that the errors introduced by fixing the altitude of the shell are less than 10%, which corresponds to about 5 TECU for a typical daytime total electron content of 50 TECU. This represents an ionospheric range correction error of about 1 cm. Another limitation is that the GPS-based estimates of total electron content are for the full ionosphere, whereas altimetry samples only the portion of the ionosphere between the satellite and the sea surface. Electron content above the 1336-km altitude of T/P is usually only a few TECU (Mannucci *et al.*, 1998). This corresponds to an ionospheric range correction error less than 1 cm. Moreover the altitude of the shell can be chosen to best fit the T/P dual-frequency range estimates of total electron content in a least squares sense, thus mitigating the relative bias between GPS full-ionosphere and T/P partial-ionosphere estimates of total electron content.

The accuracies of GPS-based global maps of ionospheric electron content have steadily improved over the past decade as the network of GPS tracking stations has expanded. Various techniques have been used to estimate the ionospheric range correction along the T/P ground track from the GPS pierce points based on an ionospheric shell model similar to that introduced by Lanyi and Roth (1988). Christensen *et al.* (1994a) found that GPS and T/P estimates of K_u-band ionospheric range correction agreed to better than 1 cm when

T/P flew directly over a GPS receiver. Globally, Imel (1994) showed that the GPS-based estimates of the ionospheric range correction during the early period of the T/P mission were limited to wavelengths longer than about 2000 km because of the coarse coverage of the 39 ground-based GPS receivers that were available at that time. The global distributions of pierce-point estimates of vertical electron content over 24-hr periods were fit to spherical harmonics up to degree and order 20. The rms difference between these GPS and T/P dual-frequency estimates of the ionospheric range correction was about 8 cm. The GPS-based estimates were especially degraded south of 35°S where the distribution of GPS receivers was very poor.

Schreiner *et al.* (1997) estimated the ionospheric range correction along the T/P ground track by assimilation of GPS-based estimates of vertical ionospheric content at the pierce points into the PRISM model (Daniell *et al.*, 1995) based on a global network of 33 GPS receivers. They found that the PRISM adjustment procedure yielded estimates of ionospheric range correction that were almost identical to the T/P dual-frequency estimates when the T/P ground track passed near a GPS station. When a point along the ground track was more than a few hundred kilometers away from the ionospheric pierce point of a GPS measurement, errors of the adjusted PRISM estimates of ionospheric range correction typically exceeded 2 cm and were sometimes as large as 8 cm.

Similar conclusions have been reached by Ho *et al.* (1997) and Mannucci *et al.* (1998) based on global networks of 60 and 100 GPS receivers, respectively. They describe a Kalman filtering technique for interpolation of the GPS estimates of vertical electron content at the ionospheric pierce points to construct global maps of the total electron content at time intervals of an hour or less. As in the other GPS studies of ionospheric range delay, they found that the accuracy of the GPS-based total electron content degraded with increasing distance from the GPS station. During a period when the ionosphere was actively changing, the overall rms error of the GPS-based estimates of total electron content ranged from 5.8 TECU within 100 km of a GPS station to 12.5 TECU at a distance of 4000 km from a GPS station. These correspond to rms errors of 1.3 and 2.75 cm, respectively, for the ionospheric range correction. Errors were larger in the equatorial region and southern hemisphere than in the middle- and high-latitude northern hemisphere.

It can be concluded that GPS-based estimates of total electron content cannot meet a 1-cm accuracy requirement for the ionospheric range correction with the present network of GPS stations. The accuracy will improve as the terrestrial network of GPS receivers continues to expand. There are logistical problems with deploying GPS receivers over the vast expanses of ocean where there are no islands. The possibility of deploying geodetic-quality GPS receivers on buoys looks promising (e.g., Born *et al.*, 1994). With the development of improved dynamical models of the iono-sphere and sophisticated data assimilation techniques, it may be possible in the future to obtain GPS-based estimates of the ionospheric correction with the global 1-cm accuracy required for highly accurate altimeter missions. At present, however, a dual-frequency altimeter is essential to achieve this degree accuracy for the ionospheric range correction.

While the precision of 1-sec averages of T/P dual-frequency estimates of the K_u-band ionospheric range correction has been shown to be about 1 cm, it should be noted that the overall accuracy is more difficult to assess. Several studies have found the T/P ionospheric range corrections to be about 1 cm higher than the range corrections computed from DORIS- and GPS-based estimates of total electron content (Imel, 1994; Morris and Gill, 1994; Schreiner *et al.*, 1997). The possibility that this relative bias is indicative of a systematic error in the T/P ionospheric range correction is under investigation.

3.2. Sea-State Effects

As described previously in Section 2.4.2, the returned signal measured by an altimeter is the pulse reflected from the small wave facets within the antenna footprint that are oriented perpendicular to the incident radiation. The shape of the returned waveform is thus determined by the distribution of these specular scatterers rather than by the actual sea surface height distribution within the footprint. The altimeter range measurements must be corrected for biases in the estimate of mean sea level that arise because of differences between the distributions of the scatterers and the sea surface height. The range from the satellite to the sea surface is estimated from the time interval between the time that the pulse is transmitted and the time that the midpoint of the leading edge of the returned waveform is received. This half-power point corresponds to the return from the median height of the specular scatterers (Barrick and Lipa, 1985; Rodríguez, 1988), referred to as the electromagnetic (EM) sea level.

The EM sea level estimated by the onboard tracking algorithm from averaged returned waveforms as described in Section 2.4.4 differs from the true mean sea level because of two effects. As shown schematically in Figure 33, there is

an EM bias that arises because of the height difference between mean sea level and the mean scattering surface. There is also a skewness bias that arises because of the height difference between the mean scattering surface and the median scattering surface that is actually measured by the onboard tracker as the 2-way travel time corresponding to the half-power point on the leading edge of the returned waveforms (see Section 2). The sum of the EM and skewness biases is referred to as the total sea-state bias. The two components of the sea-state bias are described separately in this section. As discussed below, the skewness bias can be estimated from the ground-based waveform retracking techniques summarized in Section 2.4.5. However, waveform retracking cannot provide insight into the EM bias.

3.2.1. Electromagnetic Bias

The physical basis for the EM bias is well established. It arises because of a greater backscattered power per unit surface area from wave troughs than from wave crests. In part, this is because the power backscattered from a small wave facet is proportional to the local radius of curvature of the long-wavelength portion of the wave spectrum. Because of the non-Gaussian distribution of the sea surface height, ocean waves are generally skewed such that wave troughs have a larger radius of curvature than wave crests. The result is a bias in backscattered power toward wave troughs that was first noted experimentally by Yaplee *et al.* (1971) from tower-based radar measurements of a wavy sea surface. This bias is further enhanced by greater wind-generated small-scale roughness near wave crests which scatters the altimeter pulse in directions away from that of the incident radiation (see Section 7), even if there is no skewness in the height distribution. The backscattered power measured by the altimeter is therefore greater from wave troughs than from wave crests, thus biasing the EM sea level toward wave troughs.

Formulations of the EM bias have historically been expressed in terms of the significant wave height, which is defined to be the height of the 1/3 highest waves in the field of view (hence the notation $H_{1/3}$) and is usually considered to be equivalent to four times the standard deviation of the sea surface height (see discussion in Section 6). All previous studies have indicated that the EM bias increases monotoni-

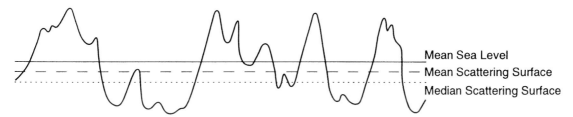

FIGURE 33 Schematic representation of the distinctions between mean sea level (thin horizontal line), the mean scattering surface (dashed line) and the median of the distribution of specular scatterers (dotted line) for a rough sea surface. (From Chelton *et al.*, 1989. With permission.)

cally with increasing wave height. Empirical estimates of the EM bias toward wave troughs are thus formulated as

$$\Delta R_{\text{EM}} = -bH_{1/3}, \tag{101}$$

where the positive nondimensional coefficient b is equivalent to the EM bias normalized by $H_{1/3}$. The EM bias surely depends on additional characteristics of the wave field but $H_{1/3}$ is the only characteristic that can be unambiguously extracted directly from altimeter waveforms. It is important to bear in mind that the simple model Eq. (101) for the EM bias absorbs any other biases that have the same dependence on the $H_{1/3}$ characterization of the sea state. Empirical estimates of the coefficient b in the model Eq. (101) are therefore more appropriately referred to as the total sea-state bias coefficient.

Theoretical studies of electromagnetic bias   A theoretical explanation of the EM bias was first attempted by Jackson (1979) who used geometric optics to describe the microwave backscatter at nadir incidence. To the extent that geometric optics is a valid approximation, the radar cross section is proportional to the distribution of the specular scatterers on the sea surface. The EM bias is given by the difference between the mean height of the specular scatterers and the mean height of the sea surface. Determination of the EM bias thus requires knowledge of the joint probability density function of the sea-surface height and slope. Based on the joint probability density function derived by Longuet-Higgins (1963) using a weakly nonlinear approximation of the wave dynamics, Jackson (1979) showed that the EM bias coefficient b for a unidirectional wave field is

$$b = \frac{1}{8}\lambda_{12}, \tag{102}$$

where λ_{12} is the cross skewness coefficient which is defined in terms of the sea surface elevation ζ relative to mean sea level as

$$\lambda_{12} = \frac{\langle \zeta \zeta_x^2 \rangle}{\langle \zeta^2 \rangle^{1/2} \langle \zeta_x^2 \rangle}. \tag{103}$$

The angle brackets in Eq. (103) denote expected value and the subscript x denotes the spatial derivative in the direction of wave propagation.

Barrick and Lipa (1985) and Srokosz (1986) extended the Jackson (1979) analysis to a two-dimensional wave field and derived a more accurate theoretical representation of the EM bias coefficient that can be expressed as

$$b = \frac{1}{8}\left(\frac{\lambda_{120} + \lambda_{102} - 2\lambda_{011}\lambda_{111}}{1 - \lambda_{011}^2}\right), \tag{104}$$

where the λ_{ijk} are the cross skewness coefficients defined analogous to Eq. (103),

$$\lambda_{120} = \frac{\langle \zeta \zeta_x^2 \rangle}{\langle \zeta^2 \rangle^{1/2} \langle \zeta_x^2 \rangle} \tag{105a}$$

$$\lambda_{102} = \frac{\langle \zeta \zeta_y^2 \rangle}{\langle \zeta^2 \rangle^{1/2} \langle \zeta_y^2 \rangle} \tag{105b}$$

$$\lambda_{011} = \frac{\langle \zeta_x \zeta_y \rangle}{\langle \zeta_x^2 \rangle^{1/2} \langle \zeta_y^2 \rangle^{1/2}} \tag{105c}$$

$$\lambda_{111} = \frac{\langle \zeta \zeta_x \zeta_y \rangle}{\langle \zeta^2 \rangle^{1/2} \langle \zeta_x^2 \rangle^{1/2} \langle \zeta_y^2 \rangle^{1/2}}. \tag{105d}$$

The subscripts i, j, and k on the λ_{ijk} correspond to the exponents of ζ, ζ_x, and ζ_y, respectively, in the numerators of Eqs. (105a–d). The coefficient λ_{011} represents the correlation between the two wave slope components ζ_x and ζ_y in an arbitrary Cartesian coordinate system. As noted by Srokosz (1986) and Lagerloef (1987), a principal-axis coordinate system can always be chosen for which $\langle \zeta_x \zeta_y \rangle = 0$. By so doing, Eq. (104) reduces to

$$b = \frac{1}{8}(\lambda_{120} + \lambda_{102}). \tag{106}$$

Evaluation of the skewness coefficients and hence the EM bias coefficient b requires knowledge of the two-dimensional wavenumber spectrum of the wave field. Glazman *et al.* (1996) computed the EM bias from Eq. (104) using a wavenumber spectrum that is a combination of a one-dimensional spectrum (Glazman and Srokosz, 1991) with a directional spread function (Apel, 1994). In addition to the spectrum for wind-driven waves, they included a Gaussian spectrum for the background swell. Their analysis considered the combined effects of EM bias and skewness bias (see Section 3.2.2 below), although the skewness bias accounts for only about 10–20% of the total sea-state bias. The results of the study reveal the effect of wind fetch on sea-state bias in various parts of the world's oceans. For example, the sea-state bias coefficient b in the western North Atlantic, where the wind fetch is relatively short, is much larger than that in the North Pacific where the wind fetch is very long. The effects of swell are to reduce the sea-state bias coefficient, which is consistent with the observations reported by Minster *et al.* (1992). It should be noted, however, that the theoretical values of 2–5% of $H_{1/3}$ for the sea-state bias predicted by Glazman *et al.* (1996) are significantly larger than the values of 1–3% of $H_{1/3}$ obtained empirically from observations as described below.

The theoretical analyses of Srokosz (1986) and Glazman *et al.* (1996) are based on the assumption of weak nonlinearity underlying the development of Longuet-Higgins (1963) on ocean wave dynamics. Elfouhaily *et al.* (1999) pointed out that the weakly nonlinear theory should apply only to long gravity waves, and that the inclusion of short waves in the studies by Srokosz (1986) and Glazman *et al.* (1996) violates this assumption. The effects of short waves on EM bias were studied by Rodríguez *et al.* (1992) using numerical simulations. When the wavelengths of the short ocean waves become close to that of the radar wavelength, the assumption of geometric optics breaks down

(Arnold *et al.*, 1991). Rodríguez *et al.* (1992) performed numerical scattering experiments in which the radar cross section of the short waves was computed without the assumption of geometric optics. The linear wave field was simulated based on a prescribed wavenumber spectrum. Superimposed on the linear waves were perturbations caused by nonlinear interactions among the long waves as well as perturbations of the short waves owing to modulations by the long waves. The results indicate that, for typical wind speeds, the contribution to EM bias from the short wave modulations is comparable to that from the modulations by the nonlinearity of the long waves. It is therefore important to revise the treatment of the short waves in the EM bias theories.

The statistics of the short waves are treated separately from those of the long waves in a theory developed by Elfouhaily *et al.* (2000). The EM scattering in this theory is based on geometric optics. The radar cross section is thus proportional to the probability density function for the slopes of the short waves modulated by the long waves. Under the same conditions specified by Lagerloef (1987), the EM bias predicted by this theory can be written as

$$b = \frac{1}{8}\left(\lambda_{120}\frac{\langle \zeta_x^2\rangle_L}{\langle \zeta_x^2\rangle_L + \langle \zeta_x^2\rangle_S} + \lambda_{102}\frac{\langle \zeta_y^2\rangle_L}{\langle \zeta_y^2\rangle_L + \langle \zeta_y^2\rangle_S}\right), \quad (107)$$

where the subscripts L and S on the angle brackets denote the expected values for the long and short wave fields, respectively. Compared with Eq. (106), the two cross skewness coefficients in Eq. (107) are weighted by the ratios of the slope variances of the long waves to the slope variances of the total wave field (long waves plus short waves). When the slope variance of the short waves is zero, Eq. (107) becomes identical to Eq. (106). The Srokosz (1986) theory is thus a special case of the Elfouhaily *et al.* (2000) theory in which the effects of the short waves are neglected.

The presence of short waves also leads to a sensitivity of the EM bias to the radar frequency that has been observed as described below. This is because the radar wavelength determines the cutoff wavenumber of the short waves that affect the radar backscatter. The bias tends to be higher for lower radar frequencies (longer wavelengths). Figure 34 shows the total sea-state bias coefficient (EM bias plus skewness bias) predicted by Elfouhaily *et al.* (2000) as a function of wind speed, together with the theoretical prediction of Glazman *et al.* (1996) and the empirical estimates of Gaspar *et al.* (1994) and Chelton (1994) obtained from T/P data as summarized below. The new theory is fairly close to the empirical estimates. The large values of Glazman *et al.* (1996) probably arise from the mistreatment of the statistics of the short waves.

Although the theory of Elfouhaily *et al.* (2000) is the most comprehensive that has been developed to date, there is still much room for improvement. For example, the calculation underlying Figure 34 is based on the assumption that the

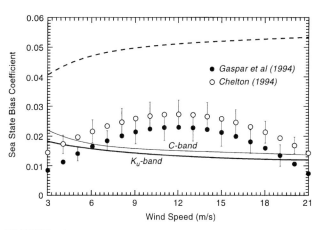

FIGURE 34 The sea-state bias coefficient b estimated by the theoretical model of Elfouhaily *et al.* (2000) for the K_u-band frequency of 13.6 GHz (heavy solid line) and the C-band frequency of 5.3 GHz (thin solid line). Note that positive values of b in the formulation Eq. (101) for the sea-state bias correspond to negative bias, i.e., a bias toward wave troughs. The dashed line shows the estimate obtained from the theoretical model of Glazman *et al.* (1996). Also shown are empirical estimates from the three-parameter model Eq. (110) obtained from T/P data by Gaspar *et al.* (1994) (solid circles) and Chelton (1994) (open circles). (After Elfouhaily *et al.*, 2000. With permission.)

short waves and long waves are propagating in the same direction. Furthermore, only the tilt modulations of the short waves by long waves are accounted for, while the hydrodynamic interaction between the short and long waves has been neglected. These are just two examples of aspects of the new theory in need of additional work. The current state-of-the-art of the theoretical understanding of the EM bias is thus not sufficiently well developed to predict the bias from a purely theoretical basis. Moreover, the theoretical formulations express the bias in terms of variables (skewness coefficients, statistical moments and the wave spectrum) that cannot be readily recovered directly from altimeter data or from other remotely sensed data and are therefore not practical for global applications. It has therefore been necessary to estimate the EM bias empirically from analyses of satellite, aircraft and tower-based altimeter measurements as summarized below. These observations provide datasets that are critical for testing the theoretical predictions.

Empirical estimation of electromagnetic bias Empirical estimation of EM bias from Seasat altimeter data was complicated by the existence of a large tracker bias that apparently had a linear dependence on $H_{1/3}$ similar to that of the EM bias but with more than twice the magnitude. Hayne and Hancock (1982) estimated the Seasat tracker bias to be 4–5% of $H_{1/3}$. This large tracker bias was never satisfactorily explained, but it is thought to have arisen because of incomplete corrections for the other sources of tracking errors summarized in Section 2.4.6. Since the tracker bias has a linear dependence on $H_{1/3}$, it is absorbed into the formulation Eq. (101) for the total sea-state bias.

Because of large residual orbit errors, it was necessary to remove empirical estimates of orbit error in order to investigate sea level variability from all altimeters launched prior to T/P. Statistical methods of reducing the orbit error and the errors of the sea surface height estimates that are introduced by these empirical corrections have been discussed by Tai (1989; 1991), Le Traon *et al.* (1991) and Wagner and Tai (1994) (see also Section 4). As pointed out by Zlotnicki *et al.* (1989) and Witter and Chelton (1991b), whatever orbit error approximation is removed from the range data also must be removed from the $H_{1/3}$ data to assure compatibility of the range and $H_{1/3}$ datasets for the purposes of empirical estimation of sea-state bias. After mitigating the effects of orbit errors, empirical determinations of sea-state bias must further address the large uncertainties of the geoid height (see Section 5.1) by removing the mean sea-surface elevation at each altimeter measurement location. Empirical estimates of EM bias from altimeters must therefore be based upon residual sea surface height relative to an unknown reference height. If the relation between height bias and $H_{1/3}$ is truly linear, as parameterized by the simple model Eq. (101), then the sea-state bias coefficient b deduced from the relative height and relative $H_{1/3}$ data should be the same as the value that would be deduced from absolute height (if it were obtainable) and $H_{1/3}$.

The sea-state bias for the Geosat altimeter was first estimated by Cheney *et al.* (1989). Residual orbit errors were approximated by a linear trend over orbit arcs from 45–72°S. Assumed values for the sea-state bias coefficient b in Eq. (101) ranging from 0.002–0.018 (i.e., a bias of 0.2–1.8% of $H_{1/3}$) were arbitrarily applied and the rms variability of residual sea level was calculated along collinear ground tracks. The value of the sea-state bias coefficient that gave the smallest residual sea level variability was found to be 0.01. They concluded that this offered the best estimate of the sea-state bias coefficient. When the linear trend approximation of the orbit error was replaced by a quadratic trend, the sea-state bias coefficient that gave the smallest residual sea level variability increased by about a factor of three (B. Douglas, personal communication, 1990). This sensitivity to the form of the orbit error approximation is very disconcerting.

Many subsequent studies attempted more direct estimates of the sea-state bias coefficient from analysis of Geosat range and $H_{1/3}$ data. All of these studies obtained larger sea-state bias coefficients than the value of 0.01 suggested by Cheney *et al.* (1989). Using a quadratic approximation for residual orbit error and regression analysis of residual range and $H_{1/3}$ data, Zlotnicki *et al.* (1989) estimated the sea-state bias for six geographical regions from 17 months of Geosat data. They obtained estimates for b ranging from 0.023–0.029, with an rms scatter of about 0.003 for each

region, and concluded that no single value of the sea-state bias coefficient could be recommended. They also found that their estimate of sea-state bias was sensitive to the form of the orbit error approximation; the sea-state bias coefficient increased with decreasing arc length used in the quadratic fit.

A regression model of the sea-state bias as a function of $H_{1/3}$ was also considered by Fu and Glazman (1991). They analyzed 57 repeat cycles (2.7 years) of Geosat data along 16 ground tracks between 60°S and 50°N. Orbit error was modeled as a sinusoid along each ground track with wavelength equal to the circumference of the earth. The sea-state bias coefficient b estimated from residual height and $H_{1/3}$ data was 0.014 ± 0.006, which is significantly smaller than the earlier estimates obtained by Zlotnicki *et al.* (1989).

Yet a different approach was used by Ray and Koblinsky (1991) to estimate the sea-state bias. The orbit error was modeled as a sinusoid with wavelength equal to the circumference of the earth. However, rather than estimate the orbit error and sea-state bias separately, the two were estimated simultaneously by a least squares procedure. This approach was adopted because of a noted sensitivity of the orbit error amplitudes to the magnitude of sea-state bias coefficient in preliminary investigations similar to those conducted by Cheney *et al.* (1989). Based upon 20 repeat cycles of global Geosat data, Ray and Koblinsky (1991) obtained an estimate of 0.026 ± 0.002 for the sea-state bias coefficient b, which is comparable to the estimates obtained by Zlotnicki *et al.* (1989).

The sensitivity of empirical estimates of b to the large Seasat and Geosat orbit errors complicated the interpretation of the on-orbit altimeter estimates of the sea-state bias coefficient. As an alternative to the use of satellite data, the sea-state bias can be investigated from aircraft or tower-based radar measurements. From simultaneous measurements of range and the normalized radar cross section σ_0 from a low-altitude radar with beamwidth narrow enough to resolve the dominant ocean wavelength, the EM bias component of sea-state bias can be determined directly. This approach eliminates the orbit error and tracking error problems associated with bias estimates based on satellite data. However, it should be noted that the EM bias estimated from aircraft and tower-based observations may not be directly applicable to satellite data since the nature of the radar return from a distributed target (i.e., a large footprint) can be much different than that from a small footprint. The results of the aircraft and tower-based measurements have nonetheless provided valuable guidance for satellite-based analyses.

Aircraft and tower radar observations isolated two important characteristics of the EM bias. The most fundamental is a dependence on frequency that was demonstrated by Walsh *et al.* (1989) from aircraft measurements at 10 and 36 GHz and by Walsh *et al.* (1991) from aircraft measurements at the 13.6 and 5.3 GHz frequencies of the T/P dual-frequency altimeter. The EM bias increases approximately linearly with decreasing frequency. For the two frequencies of the T/P dual-frequency altimeter, the EM bias differs by about 30%. A difference of about the same magnitude was found from tower-based measurements at 14 GHz and 5 GHz (Melville $et\ al.$, 1990). This frequency dependence of the EM bias must be quantified for accurate determination of the ionospheric range correction from the dual-frequency altimeter as discussed in Section 3.1.3.

Tower-based radar observations found that the EM bias also depends on wind speed, increasing approximately linearly by a factor of about two as wind speed increases from $4\ \mathrm{m\,sec}^{-1}$ to $12\ \mathrm{m\,sec}^{-1}$ (Melville $et\ al.$, 1990, 1991). This dependence on wind speed was observed from aircraft observations as well (Walsh $et\ al.$, 1991). The possibility of a wind-speed dependence in the sea-state bias in satellite observations was investigated from Geosat data by Ray and Koblinsky (1991). They were able to identify a dependence of the sea-state bias coefficient b on wind speed, but the increase with increasing wind speed was smaller than that observed in the aircraft and tower data. The addition of a wind-speed dependence of the bias coefficient b in Eq. (101) provided a small but statistically significant improvement over the simple expression as a linear function of $H_{1/3}$ only.

Fu and Glazman (1991) proposed an alternative formulation for the sea-state bias that attempts to account for the effects of wave type (e.g., fully-developed swell versus wind waves) rather than just $H_{1/3}$. The sea-state bias was expressed in terms of a quantity that they called the "pseudo wave age," which was computed from $H_{1/3}$ and wind-speed U based on the model described in detail by Glazman $et\ al.$ (1988) and Glazman and Pilorz (1990). The pseudo wave age was defined as

$$\Psi = \left(\frac{g H_{1/3}}{U^2} \right)^{\psi}, \qquad (108)$$

where ψ is a nondimensional constant close to 0.6 and g is the gravitational acceleration. The EM bias model was then expressed as

$$b = B \left(\frac{\Psi}{\Psi_m} \right)^{\beta}, \qquad (109)$$

where Ψ_m is a globally averaged pseudo wave age, $B = 0.013 \pm 0.005$ and $\beta = -0.88 \pm 0.37$. The inclusion of pseudo wave age effects was found to give a slightly improved estimate of sea-state bias over the model based only on a linear function of $H_{1/3}$. Numerical experiments conducted by Glazman and Srokosz (1991) produced results similar to the empirical results of Fu and Glazman (1991).

The physical basis for the dependence on pseudo wave age is essentially that the young waves are more nonlinear than old waves and hence cause larger EM bias as well as skewness bias. According to the Fu and Glazman model, the sea-state bias increases with increasing wind speed and/or decreasing $H_{1/3}$. The Fu and Glazman (1991) model may thus offer insight into a possible physical explanation for the wind speed and wave height dependence found by other studies as summarized above.

An analysis of Geosat data by Witter and Chelton (1991b) raised new questions about the sea-state bias. The simple linear dependence on $H_{1/3}$ was investigated from a detailed analysis of Geosat data for the region 10–65°S. This latitude band was selected because it represents the region in the world ocean with the largest dynamic range in $H_{1/3}$. Orbit errors were modeled as sinusoids with wavelength equal to the circumference of the earth. The sea-state bias coefficient b in Eq. (101) was found to vary geographically with a value of 0.022 for the region south of 40°S and a value of 0.032 for the region between 10 and 40°S. A more detailed investigation suggested that this geographical dependence was actually due to a nonlinear sea-state bias dependence on wave height. For $H_{1/3}$ smaller than about 4 m, b was approximately constant and equal to 0.035 ± 0.003. For larger $H_{1/3}$, the bias coefficient b decreased approximately linearly to a value of 0.015 ± 0.007 for $H_{1/3} = 6$ m. Witter and Chelton (1991b) speculated that the apparent $H_{1/3}$ dependence of the sea-state bias coefficient b may be due to the effects of errors in the pointing-angle and sea-state correction discussed in Section 2.4.6. However, the dependence on the pseudo wave age discovered by Fu and Glazman (1991) offers another explanation for the smaller sea-state bias in large $H_{1/3}$ conditions.

The preceding overview summarizes the state of confusion about the nature of the EM bias deduced from the large number of empirical analyses of altimeter, tower and aircraft data prior to the launch of T/P. The satellite-based estimates were clouded by the fact that the Seasat and Geosat results were compromised by the presence of large orbit errors and other factors (the tracker bias for Seasat and large attitude errors for Geosat). The correction algorithm adopted by the T/P Project before launch was based primarily on aircraft measurements (Hevizi $et\ al.$, 1993). The sea-state bias coefficient b in Eq. (101) was formulated parametrically as a quadratic function in wind speed,

$$b = a_0 + a_1 U + a_2 U^2. \qquad (110)$$

The parameters a_0, a_1 and a_2 were different for the K_u-band frequency of 13.6 GHz and the C-band frequency of 5.3 GHz of the T/P dual-frequency altimeter. As described in Section 3.1.3, the EM bias correction must be applied separately to the K_u-band and C-band range measurements prior to applying the ionospheric range correction in order to remove any contamination owing to the different EM bias at the two frequencies.

The performance of the pre-launch algorithm for the K_u-band altimeter was evaluated from T/P data by Rodríguez and Martin (1994b), Gaspar $et\ al.$ (1994), and Chelton (1994). Within estimated uncertainties, the pre-launch model parameters for the K_u-band frequency of the T/P dual-frequency altimeter were found to be in agreement with the empirical estimates from on-orbit data. These studies all indicated that the single most effective quantity in describing the variability of the sea-state bias coefficient b is wind speed. However, the inclusion of a linear dependence on $H_{1/3}$ in the bias model further improved the accuracy of the sea-state bias correction (Gaspar et al., 1994; Chelton, 1994). The functional form of the four parameter model for the sea-state bias coefficient in Eq. (101) is

$$b = a_0 + a_1 U + a_2 U^2 + a_3 H_{1/3}. \qquad (111)$$

Although there was no discernible improvement at low latitudes, the increase in the sea surface height variance accounted for by the four parameter model Eq. (111) over the three parameter model Eq. (110) was about 1 cm^2 at middle and high latitudes where the variability of $H_{1/3}$ is largest.

Ideally, the wind-speed and wave-height dependence of the sea-state bias coefficient b should be determined nonparametrically, rather than by presupposing a functional form in terms of the parameters a_0, a_1, a_2 and a_3. Nonparametric approaches to estimating the sea-state bias coefficient were applied by Witter and Chelton (1991a, b), Chelton (1994), and Rodríguez and Martin (1994b). The most comprehensive nonparametric investigation of the sea-state bias to date was presented by Gaspar and Florens (1998). They demonstrated that the parametric approach is not optimal in the sense of least-squares estimation. By not imposing a functional form for the dependence on wind speed and wave height a priori, they obtained an optimal nonparametric solution using a statistical technique called kernel smoothing. This new model improved upon the parametric model by reducing the sea surface height variance by 1 cm^2 at middle and high latitudes.

The Gaspar and Florens (1998) nonparametric estimate of the sea-state bias coefficient b is contoured in the upper panel of Figure 35 as a function of wind speed and $H_{1/3}$. The model captures the well-known features of the wind-speed dependence of the sea-state bias coefficient. For a given $H_{1/3}$ larger than about 3 m, b increases with wind speed up to about 12 m sec^{-1} and then decreases with higher wind speeds. For wind speeds less than about 9 m sec^{-1}, b is only weakly dependent on $H_{1/3}$. At higher wind speeds, b decreases with increasing $H_{1/3}$. This wave-height dependence, first discovered by Witter and Chelton (1991b), seemingly contradicts the positive correlation between wind speed and $H_{1/3}$. The apparent inconsistency might reflect the presence of swell, which would degrade the correlation between wind speed and $H_{1/3}$. There are conditions in which $H_{1/3}$ is large because of the presence of swell and not because of high local wind speed. The EM bias caused by swell is less than that caused by wind waves (Glazman et al., 1996; Minster et al., 1992). The increasing dominance of swell in high sea-state

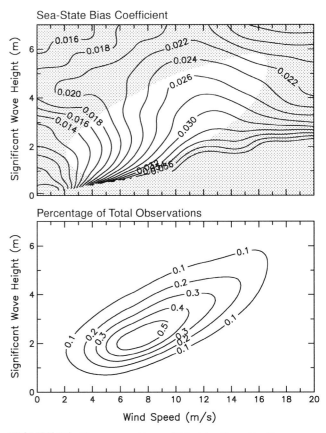

FIGURE 35 The sea-state bias coefficient b estimated by the nonparametric model of Gaspar and Florens (1998) as a function of significant wave height and wind speed (upper panel). Lower panel shows binned values of the percentage of total observations for a bin size of 0.25 m in $H_{1/3}$ by 0.5 m sec^{-1} in wind speed. The shading in the upper panel indicates regions where the bins contain less than 0.1% of the observations (see lower panel), which is too small to obtain reliable estimates of b. (Data provided courtesy of P. Gaspar.)

would cause the sea-state bias coefficient to decrease with increasing $H_{1/3}$. This is also consistent with the dependence on pseudo wave age discussed above.

The standard deviations of the nonparametric estimates of sea-state bias obtained by Gaspar and Florens (1998) are a negligibly small fraction of $H_{1/3}$ in the regime of $H_{1/3}$ and wind speed where there is a large amount of data (the unshaded region in the upper panel of Figure 35). While these standard deviations provide a measure of the reliability with which b can be estimated nonparametrically from the full dataset, they do not constitute a measure of the confidence with which the bias coefficient b is known. Chelton (1994) found that parametric estimates of b obtained from individual 10-day cycles of T/P data varied with a standard deviation of about 1% of $H_{1/3}$ (see, for example, Figure 5 of Chelton, 1994). This is probably a good estimate for the uncertainty of the bias coefficient b estimated either parametrically or nonparametrically.

The validity of the parametric model for the C-band sea-state bias coefficient has not received as much attention as the K_u-band sea-state bias. Imel (1994) compared the T/P dual-frequency estimates of the ionospheric range correction with GPS-based estimates and found less than 0.5 cm difference for $H_{1/3}$ values ranging from 0–10 m. This suggests that the relative difference between K_u-band and C-band sea-state bias corrections is very small. Stewart and Devalla (1994) reached a similar conclusion. They examined the differences between the range measurements made by the K_u-band and C-band altimeters at nighttime over $H_{1/3}$ conditions ranging from 5–15 m. At nighttime, the ionospheric effect on the range difference is only about 2 cm, making the differences between nighttime range estimates at the two frequencies a good indicator of possible problems in the differential sea-state bias correction applied at the two frequencies. They reported no discernible relationship between the range difference and $H_{1/3}$, indicating that there are no significant relative errors between the sea-state bias corrections applied at the two frequencies. It should be noted, however, that the Imel (1994) and Stewart and Devalla (1994) studies cannot detect any errors that are common to the sea-state bias corrections at the two frequencies.

3.2.2. Skewness Bias

The physical basis for the skewness bias is also reasonably well understood. The skewness bias is the height difference between the mean scattering surface and the median scattering surface (see Figure 33) that corresponds to the 2-way travel time for the half-power point on the leading edge of the returned waveform (see Section 2). Unfortunately, the skewness bias is not related in any simple way to significant wave height or any other geophysical quantity that can easily be inferred from altimeter data. Srokosz (1986), Lagerloef (1987), and Rodríguez (1988) have shown that the skewness bias is given approximately by

$$\Delta R_{skew} = -\lambda_\zeta \frac{H_{1/3}}{24}, \tag{112}$$

where λ_ζ is the height skewness. This corresponds to an error of 0.8 cm for a skewness of 0.1 and a wave height of 2 m. This error increases to about 4 cm for extreme wave heights of 10 m. With the weakly nonlinear approximation of Longuet-Higgins (1963), the skewness bias has been computed by Glazman et al. (1996) and Elfouhaily et al. (2000) from assumed forms for the wave spectrum. The skewness bias was actually included in the total sea-state bias in these calculations. The characteristics of the skewness bias that depend linearly on $H_{1/3}$ are also included in most empirical estimates of the sea-state bias because there is no easy way to separate it from the EM bias in the altimeter data. An exception is the work of Rodríguez and Martin (1994b) who used retracked data in which the skewness bias had been removed before the EM bias analysis.

Because of its effect on the shape of the leading edge of the returned waveform, numerous studies have suggested that the ground-based waveform retracking described in Section 2.4.5 could be used to estimate the skewness (e.g., Lipa and Barrick, 1981; Hayne and Hancock, 1982; Barrick and Lipa, 1985; Srokosz, 1986; 1987; Lagerloef, 1987; Rodríguez 1988). The traditional onboard tracking algorithm (see Sections 2.4.3 and 2.4.4) is based on the assumption of a Gaussian distribution of specular reflectors. As described in detail by Brown (1977), Barrick and Lipa (1985), Rodríguez (1988), and others, the time history of the returned waveform can be expressed as the convolution Eq. (46). The term $q_s(t)$ in this convolution equation represents the probability density function of the specular reflectors within the antenna footprint. It is thus apparent that a skewness in the height distribution, through its effect on the probability density function of the specular scatterers q_s, affects the ability of the onboard tracking algorithm to identify the return from nadir mean sea level. These skewness effects were apparently first noted by Hayne (1980).

The probability density function $q_s(t)$ of the specular reflectors that is obtained by deconvolution of the returned waveform can be fitted to model probability density functions to estimate simultaneously the skewness parameter and the skewness bias in the range estimate. Rodríguez and Chapman (1989) presented a numerical deconvolution technique that iteratively varies the waveform model parameters to achieve the best fit in a least-squares sense. The method is reportedly more efficient than the Fourier technique described by Hayne and Hancock (1990). The technique was applied to simulated and actual Seasat waveforms. The skewness of the sea-surface height distribution is not presently well known over the open ocean. The deconvolution of Seasat data yielded suspiciously high skewness values of about 0.3. However, this analysis was compromised by poor calibration of the point-target response p_τ in Eqs. (46) and (53); when this term is properly calibrated, the skewness estimates from deconvolution of Seasat data decrease to a mean of about 0.15 (E. Rodríguez, personal communication, 1991).

Rodríguez and Martin (1994b) applied the waveform deconvolution method of Rodríguez and Chapman (1989) to the T/P data. In the course of estimating the skewness bias, they discovered a subtle problem in the waveform data caused by a power spike migrating in and out of the leading edge of the returned waveforms. As discussed in Section 2.4.5, the location of the contamination of the waveform depends on the sign of the relative velocity between the satellite and sea surface. Because of the very nearly circular T/P orbit and the equatorial bulge from the oblateness of the earth, the relative velocity changes sign when the satellite passes over the equator. The net effect of the spike is to create an erroneous jump in the estimated skewness when crossing the equator.

Despite the spike problem when crossing the equator, skewness estimates were found to be qualitatively consistent with expectations. The relationship between skewness and the nature of the wave field was investigated by characterizing the wave maturity in terms of wind speed U and significant wave height $H_{1/3}$ by

$$\mu = \frac{U}{\sqrt{g H_{1/3}}}, \qquad (113)$$

which is seen to be related to the inverse of the pseudo wave age Eq. (108). The skewness increased monotonically with μ. Because the equatorial jump does not alter the spatial mean of the skewness, the measured skewnesses between 0.03–0.12 may be valid estimates of the wave skewness.

3.2.3. Accuracy of the Estimated Total Sea-State Bias

Although significant advances have been made toward developing a theoretical understanding of the sea-state bias, empirical estimates continue to offer the most accurate estimates of the sea-state bias. The nonparametric approach has proven more effective than the more traditional parametric approaches. However, the uncertainty of empirical estimates of the sea-state bias coefficient b remains at the level of about 1% of $H_{1/3}$. From the geographical distributions of the mean and standard deviation of $H_{1/3}$ computed from 7 years of T/P data shown in Figure 36, it is apparent that this uncertainty corresponds to errors as large as 5 cm in absolute sea-surface height and more than 1.5 cm in estimates of the time-variable sea-surface height in the subpolar oceans. These uncertainties of the sea-state bias have become one of the leading sources of error in satellite altimetry.

An important limitation of empirical estimates of sea-state bias is that it is difficult to account for the variability of the sea-state bias owing to the wide range of spatially- and temporally-varying sea-state conditions that are effectively ensemble averaged into the bias estimates. Accounting for the sea-state variability will require improvements in the theoretical understanding of the physics of the EM bias that address such problems as the hydrodynamic interactions between short and long waves, as well as the effects of large directional spread of the waves of varying scales. Testing these theories will require observations and estimation of the global ocean wave spectrum simultaneous with the altimeter observations. Improvements of the sea-state bias correction thus pose difficult challenges, both theoretically and observationally.

4. PRECISION ORBIT DETERMINATION

The procedure for determining the three-dimensional location of a satellite's center-of-mass (called the orbit ephemeris) at regularly spaced time intervals in a specified reference frame with high accuracy is known as precision orbit determination (POD). Knowing the location of the radar altimeter antenna relative to the center-of-mass of the satellite, the altimeter range measurement can be used to determine the height of the ocean with respect to a reference surface. Variations in the ocean height of a few cm over length scales ranging from tens of kilometers to ocean basins must be accurately observed to utilize the altimeter data fully. To obtain the satellite height to this accuracy, POD combines accurate and complex mathematical models for the dynamics of a satellite's motion with high precision physical observations of the satellite's position or velocity (see, for example, Seeber, 1993 and Tapley et al., 2000 for more detailed reviews of orbit determination).

Advancements in the determination of precise orbits received a considerable stimulus with the initiation of the T/P altimeter mission. At the time of the launch of Seasat in 1978, the rms error in the radial component of the orbit ephemerides (hereinafter referred to simply as the orbit error since it is the radial component that is of primary interest to altimetry) was about 5 m (Lerch et al., 1982). The orbit errors for altimeter satellites were still at the one-meter level when the T/P mission was conceived in the early 1980s (Schutz et al., 1985). While a 1-cm orbit was desired, a goal of 13 cm for T/P was considered attainable if a major effort were initiated towards improving the model for the earth's gravity field, the accuracy and spatial coverage of the tracking data, and all other aspects of the models and orbit determination procedures. As a result of the improvements supported and stimulated by the T/P mission, the orbit accuracy for all altimeter satellites has increased dramatically. The reduction in orbit errors achieved for typical altimeter satellites, excluding T/P, is illustrated in Figure 37. As discussed in this chapter, orbit determination for T/P has been particularly successful, with orbit errors reduced to the 2-cm rms level.

Because orbit errors have historically been a major limitation in the utility of the altimeter data, it was necessary to develop techniques to remove the orbit errors for the high precision required for oceanographic applications. As discussed below, most of the orbit error power is centered around the frequency of one cycle per orbit revolution (1-cpr) and is thus very long wavelength (Francis and Bergé, 1993). It was traditional to reduce the error by removing a bias and slope in the altimeter data track over an ocean basin. Later, recognizing the 1-cpr nature of the orbit error, methods to empirically remove error power at that frequency were employed (Tai, 1989; Wakker et al., 1991; Chelton and Schlax, 1993a). Considerable analysis has been devoted to characterize the level of ocean signal removed in such procedures. It is generally conceded that at least some important sea-surface variations at the longer temporal and spatial wavelengths can be contaminated through orbit-error removal techniques (Tai, 1989, 1991; Wagner, 1990; Le Traon et al., 1991; Wagner

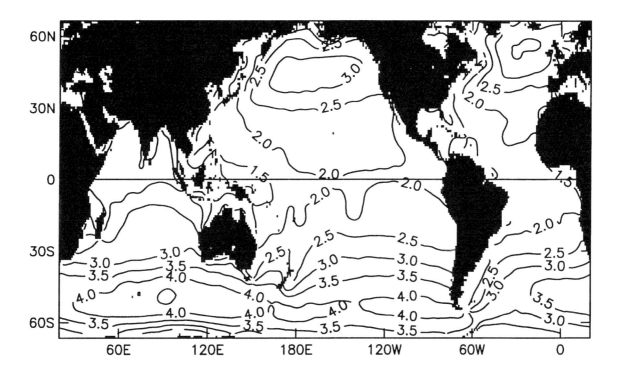

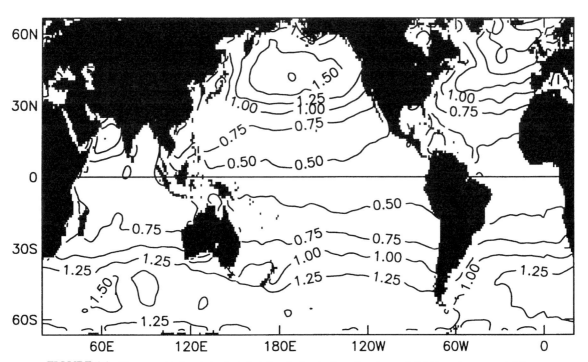

FIGURE 36 The mean (top) and standard deviation (bottom) of significant wave height in meters computed from 7 years of T/P data.

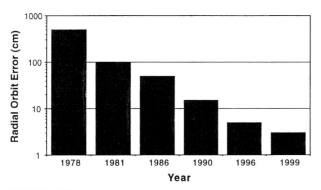

FIGURE 37 Improvement in radial orbit accuracy for altimeter satellites at the 800 km altitude typical of most altimeter satellites.

and Tai, 1994; Tai and Kuhn, 1995). It is a particular success of the T/P mission that the orbit errors have been reduced to the level where such techniques are no longer required. Moreover, the ERS-1 and ERS-2 satellite altimeters have also benefited from the highly accurate T/P orbits through the use of crossover techniques in which the simultaneous T/P observations of the sea surface are used to correct the larger ERS-1 and ERS-2 orbit errors (Moore and Ehlers, 1993; Smith and Visser, 1995; Kim, 1995; Scharroo and Visser, 1998).

It is not practical to attempt a comprehensive review of POD here for all past, present, and planned altimeter missions, but they all have many aspects in common. Given the dramatic impact the T/P mission has had on POD for altimeter satellites, this mission will be used to illustrate the principal issues associated with very precise orbit determination and the qualitative characteristics of the orbit error that can be expected.

4.1. The Orbit Determination Concept

The basic procedure for orbit determination starts with an initial model for the trajectory of a satellite over some interval of time. This initial orbit will be incorrect owing to errors in the estimate for the starting point, deficiencies in the mathematical model for the forces acting on the satellite, and errors in the parameters used in the model. To correct the model, independent observations of the satellite's motion must be obtained. These observations generally measure only some component of the motion, such as the height above the ocean, or the line-of-sight range or the change in the line-of-sight range over some short interval of time. Measurements of the full three-dimensional position or velocity are usually not available, but as long as the observations depend on the satellite's motion in some way, they contain information that helps determine the orbit. The evolution of the satellite's position and velocity must be consistent with both the physics of the mathematical model and the sequence of

observations that are used to constrain the orbit estimate to a specific solution (within some envelope of uncertainty).

The observations must also have a corresponding mathematical model in order to be usable in the orbit estimation problem. The observation model depends not only on the satellite's motion, but also on the orientation of the spacecraft and the motion of the observing station. The measurement model must relate the location of the tracking instrument to the spacecraft's center-of-mass, which may change with time as onboard fuel is consumed. This model also requires accurate knowledge of the spacecraft's orientation in space, which may be based on a model of the satellite's attitude control program or a time series of measurements from an onboard attitude sensor. At the same time, the tracking station is on a rotating Earth with surface deformations from tidal distortions and plate tectonics. The observing "station" may even be another orbiting satellite. Finally, the observational model must account for atmospheric refraction and other biases or instrument effects.

Assuming that the measurements are reliable, the discrepancies (called the residuals) between the computed observables and the real observations represent errors in the initial conditions as well as deficiencies in the dynamical and observational models. Through a linearized least-squares solution process discussed in Section 4.3.1, the initial conditions and selected model parameters are adjusted to minimize the residuals. Considerable experience is reflected in choosing the model parameters that are best suited for adjustment. The mathematical models will always be imperfect in some respects, and the adjusted parameters are chosen on the basis of their ability to compensate for the deficiencies. The orbit is then recalculated based on improved initial conditions and parameters. The observations are again compared with their computed counterparts and the initial conditions and parameters are readjusted. During this iterative process, unreliable observations can be identified and removed. Given a set of observations that contain sufficient information, the adjustments become smaller and smaller with each iteration, and the process is judged to have converged when a satisfactory and stable orbit solution is obtained.

The advent of the GPS and space-qualified GPS receivers have allowed continuous kinematic (i.e., purely geometric) positioning of satellites (see Section 4.2.6). However, the dynamical techniques (including the "reduced-dynamics" variations discussed in Section 4.3.2) still provide, and are likely to continue to provide, the most accurate orbits.

4.1.1. Dynamics of Satellite Motion

For precise applications, the trajectory of a satellite has generally been obtained by integrating the dynamical equations of motion using numerical methods (Shampine and Gordon, 1975; Montenbruck, 1992). The mathematical representation of the motion of the center of mass of a spacecraft is given as a function of time t by

$$\mathbf{r}(t) = \int_{t_0}^{t} \mathbf{v}(t')\, dt' \qquad (114)$$

$$\mathbf{v}(t) = \int_{t_0}^{t} \mathbf{a}[\mathbf{r}(t'), \mathbf{v}(t'), t', p]\, dt', \qquad (115)$$

where $\mathbf{r}$ and $\mathbf{v}$ are the position and velocity vectors of the spacecraft center-of-mass with initial conditions

$$\mathbf{r}(t_0) = \mathbf{r}_0 \qquad (116)$$
$$\mathbf{v}(t_0) = \mathbf{v}_0. \qquad (117)$$

The acceleration $\mathbf{a}$ (force per unit mass) of the spacecraft is a function of the spacecraft instantaneous position, velocity and all of the parameters p that are employed in the models for the reference frame, the forces and the observations. The interval of time from the initial point to some chosen final time is called the arc length. This may be several hours, a day, several days, a month or even years. In the case of altimeter satellites, the repeat period of the ground track is sometimes chosen as a convenient arc length.

The forces acting on the satellite can be broadly classified as either gravitational or nongravitational, as discussed separately in the following sections.

4.1.2. Gravitational Forces

Uncertainties in models for the earth's gravity field have long been the leading source of orbit error in POD. Considerable discussion is therefore warranted concerning the nature of the gravity-induced orbit perturbations and the efforts that have been so effective in reducing this source of error. The gravitational forces acting on a satellite are independent of the size or shape of the satellite. Temporal variations in the earth's gravity field tend to be fairly regular at the spatial wavelengths of concern for POD, occurring at predictable frequencies. Consequently, many of the satellites that have flown to date have contributed to a dramatic improvement in the knowledge of the gravitational field of the earth. In contrast, each satellite experiences nongravitational surface forces peculiar to it alone, and modeling the surface forces has emerged as a much greater limitation to POD accuracy than in the past (see Section 4.1.3).

Among the gravitational forces, the two-body term (where the central body is assumed to be perfectly spherical) dominates by far the orbital motion. As a consequence, an orbit is well characterized by the Keplerian elements of an elliptical orbit (Danby, 1962; Geyling and Westerman, 1971). The geometric properties of the set of orbit elements that are generally used to describe the motion of a satellite in orbit about the earth are shown in Figure 38. The principal concern for an altimeter mission is the distance to the center of mass of the central body. This distance is characterized by the orbit's semimajor axis a, the variation in the radial distance associated with the ellipticity of the orbit (the eccen-

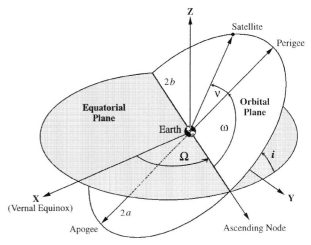

FIGURE 38 The elements of an elliptical orbit: semimajor axis a, inclination i, longitude (or right ascension) of the ascending node Ω, argument of perigee ω, and true anomaly ν. The eccentricity e relates the semimajor axis a to the semiminor axis b through the expression $e^2 = (a^2 - b^2)/a^2$.

tricity e), and the angular distance ν (the true anomaly) from the point of closest approach in the orbit (called the perigee). In addition to ν, other angular measures of the position along the orbit that are used to relate time and the motion of the satellite along the orbit include the eccentric anomaly E and the mean anomaly M. The three anomalies are very similar for the near-circular orbits relevant to altimeter satellites. The tilt and orientation of the orbital plane are given by the inclination i and the longitude of the ascending node Ω. These two angles are related to the out-of-plane components of the orbit and thus do not directly affect the radial distance, as do the in-plane elements. The argument of perigee ω is the angular distance along the orbit from the equatorial plane to the perigee, which determines the orientation of the long axis of the elliptical orbit within the orbital plane. The motion of a satellite in earth orbit is principally characterized by these six orbit elements, with the satellite moving along an elliptical orbit within a plane that tends to slowly precess in space (see Section 8.1.4).

The earth's gravity field (the geopotential) is, however, not perfectly spherical, and undulations in the gravity field, corresponding to the variations in the earth's shape and density, cause perturbations from perfectly elliptical motion. The deviations in the geopotential are typically represented using a spherical harmonic expansion (see, for example, Kaula, 1966; Seeber, 1993). The effects of the gravity variations on the orbit decrease rapidly with altitude, which is one of the principal reasons for the selection of 1336 km for the altitude of the T/P mission. Most other altimeter satellites have been flown at altitudes around 800 km, where the gravity perturbations are larger. The earth's oblateness is by far the largest deviation from sphericity and is the result of the centrifugal force from rotation of the elastic earth. Beyond the oblateness, the earth's gravity field has large varia-

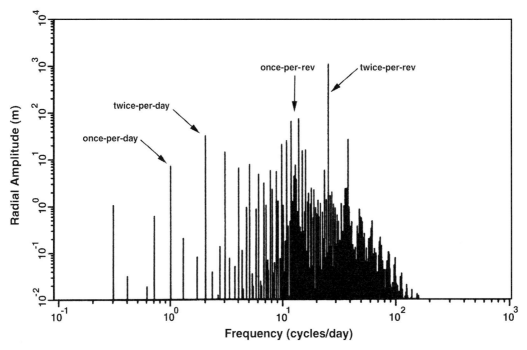

FIGURE 39   Spectrum of the radial orbit perturbations cause by the earth's gravity field for the T/P satellite.

tions with length scales ranging from 10,000 to 100 km (and shorter), and the spectrum of the orbit perturbations due to the gravity field is consequently very complex. The undulations in the geopotential cause perturbations in the orbit at all wavelengths, and the rotation of the earth underneath the satellite orbit generates additional modulations of these perturbations. In addition, resonance occurs if the ground track has a near-repeat pattern after an integer number of orbit revolutions, resulting in large-amplitude, long-period effects on the orbit. The closer the ground track repeats, the deeper the resonance.

The complex structure in the orbit perturbations owing to the gravity field is evident from Figure 39. While there is a great deal of power in the vicinity of the orbital period (12.8 cycles per day for T/P), much of the power lies in the high-frequency portion of the spectrum. One of the largest deviations from a perfectly elliptical orbit occurs at a frequency of twice-per-revolution, due to the oblateness term. The magnitude and complexity of the gravity perturbations have motivated efforts, starting with the first launch of an artificial satellite, to improve the accuracy of the gravity field model (King-Hele, 1992).

There are a large number of periodic variations in the geopotential due to the solid-earth and ocean tides (Lambeck *et al.*, 1974; Cazenave *et al.*, 1977; Wahr, 1981). There are also secular variations in the earth's gravity field caused by deformation of the solid earth (e.g., post-glacial rebound) and quasi-periodic variations due to atmosphere and watermass motion within the earth system. The effect of solid-earth and ocean tides on an orbit are the same as those of the

constant part of the geopotential model, except that the perturbation spectrum is even more complex due to additional modulations by the tidal frequencies. The solid-earth tides can be described almost entirely by the elastic response of the solid earth to the tidal potential of the sun and moon. Both the luni-solar tidal potential and the elastic properties of the solid earth are well known (Wahr, 1981). Consequently, the solid-earth tide can be modeled quite well. The ocean tides are considerably more complex because of the disturbances introduced by the effects of varying ocean depth and continental land masses. This creates a much broader spectrum of frequencies and requires a much higher spatial resolution in the model. Because of the rich spectrum of the ocean-tide perturbations, resonances with particular tides are more likely to occur, which magnifies their effect. In addition, the shorter periods associated with the diurnal and semidiurnal tides can, through the sampling by the orbital motion, alias into undesirable longer periods. This can distort the altimetric measurements of long-period tides and certain seasonal sea-surface variations (Bettadpur and Eanes, 1994; Marshall *et al.*, 1995a).

In addition to the perturbations by the geopotential, other gravitational effects include the direct attraction of the sun, moon and planets on the satellite. For high-precision applications, the effects of General Relativity, principally the precession of perigee and the relativistic effects on the observations, must also be considered. The current models for the various gravitational effects are well documented in the International Earth Rotation Service (IERS) Conventions documents (McCarthy, 1996).

Gravitational force modeling improvement Error analysis of the best general gravity models available in 1980, e.g., the Goddard Earth Model (GEM)-10B (Lerch *et al.*, 1981), predicted radial orbit errors well above 50 cm for the 1336-km T/P orbit height and meter-level orbit errors for lower satellites. With the observation that the gravity model was the primary error source, an intense effort aimed at improving the gravity model was initiated by the T/P project. This effort spanned almost a decade and led to a dramatic improvement in the modeling of the earth's geopotential.

Based on the best geodetic constants, improved models, improved models and reference frame definitions, and software capabilities which were available in the mid-1980's, the GEM-T1, GEM-T2, Texas Earth Gravity (TEG)-2B and GEM-T3 solutions were developed (Marsh *et al.*, 1988; Marsh *et al.*, 1990, Shum *et al.*, 1990a; Lerch *et al.*, 1993). Improved methods for obtaining realistic weights for the data in the solution and a reliable error estimate were also developed (Lerch *et al.*, 1991; Yuan, 1991). An accurately calibrated covariance matrix, which represents the uncertainties and cross-correlations of the coefficients estimated in the solution, is critical for assessing the quality of the solution and estimating the errors in the orbit due to the gravity model. Upon the completion of GEM-T3, which incorporated the satellite and surface gravity datasets available at the time, a reiteration of the solution process was undertaken. The resulting gravity model was called the Joint Gravity Model (JGM-1) (Nerem *et al.*, 1994).

Geopotential models have been traditionally derived from satellite tracking data for a large number of satellites at a variety of orbit inclinations and altitudes, in combination with surface gravity and ocean altimetry information. As a consequence, the magnitude of the orbit error from the gravity model error depends upon whether the inclination and altitude of the satellite are similar to those of past satellites. Because none of the satellites used in deriving the JGM-1 model had an orbit similar to T/P, a post-launch model improvement effort was conducted to refine the gravity and surface force models. This effort, using the SLR and DORIS (see Sections 4.2.2 and 4.2.3) tracking data newly available from T/P, resulted in the JGM-2 model (Nerem *et al.*, 1994). The addition of even more T/P tracking data, most notably from the GPS receiver onboard the spacecraft, culminated in the production of the JGM-3 gravity model (Tapley *et al.*, 1996).

Building on these results, additional gravity model improvement was achieved for the European Remote Sensing (ERS) altimeter satellites (Scharroo *et al.*, 1998; Tapley *et al.*, 1997). Significant reductions in the orbit errors were obtained by incorporating data from the ERS satellites themselves or from satellites in similar orbits, such as SPOT-2 (a CNES remote sensing satellite) and Stella (a geodetic satellite which serves as a target for ground-based lasers). Tracking information from a new

German microwave system (PRARE, see Section 4.2.4) aboard the ERS-2 spacecraft has enabled further reduction in the gravity-induced orbit errors for the ERS satellites (Bordi *et al.*, 2000).

The improvement in the radial component of the T/P and ERS-2 orbits as manifested in the repeatability of global sea-surface height measurements taken at the locations of ground-track crossovers is apparent from Table 3. To reduce the effects of ocean variability, crossovers were considered only if the elapsed time between the ascending and descending ground tracks was less than 10 days (6 days for ERS-2). Other editing criteria were also applied to reduce the influence of non-orbit signals on the crossover rms. The surface force models were identical within each of the two series of comparisons in order to isolate the gravity model improvement. Also given are the statistics of the fit to the range measurements from ground-based lasers that tracked the satellites (see Section 4.2.2). Note that, while the crossover statistics provide a more global measure of the radial orbit accuracy than the fit to the laser ranges, they also include residual ocean and altimeter error signals at the 5–6 cm level (see Section 4.2.5). For very low levels of orbit error, the crossover residual rms will be limited by non-orbit error signals. The orbit accuracy for both satellites improved by more than an order of magnitude from GEM-10B to current geopotential models. Further improvements are becoming increasingly difficult as other (mainly nongravitational) orbit errors are becoming relatively more important as the geopotential errors decrease.

Progress in gravity modeling continues today with the recent release of the EGM96 geopotential, based on additional satellite tracking data and a more modern and extensive set of surface gravity measurements (Lemoine *et al.*, 1998). This model has much higher spatial resolution than models such as JGM-3 and represents a significant improvement in modeling the earth's geoid (see Chapter 10). Work is in progress on the European gravity model GRIM-5 (Schwintzer *et al.*, 1998), and the upcoming CHAllenging Mini-Satellite Payload mission (CHAMP) will help to resolve long-term temporal variations in the earth's gravity field (Reigber *et al.*, 2000). Following that, the Gravity Recovery and Climate Experiment (GRACE) mission, a joint project between NASA and the Deutsches Zentrum für Luft und Raumfahrt (DLR), has been approved for launch in mid-2001 (Davis *et al.*, 1999). This dedicated gravity mission is expected to provide, with unprecedented accuracy, a determination of the mean earth gravity field and its seasonal variations, which will eliminate most of the gravity-model contributions to the orbit error.

Geographically correlated orbit errors Much of the radial orbit error caused by gravity model error has a time-invariant geographic pattern (Tapley and Rosborough, 1985; Rosborough and Tapley, 1987; Schrama, 1992). Some of this

TABLE 3. Typical Laser Range and Altimeter Crossover Residual rms Using Various Gravity Models for T/P and ERS-2

Gravity model	Reference	Crossover rms (cm)		Laser range rms (cm)	
		T/P	ERS-2	T/P	ERS-2
GEM-10B	Lerch *et al.* (1981)	66	267	45	55
GEM-T1	Marsh *et al.* (1988)	40	205	20	56
GEM-T2	Marsh *et al.* (1990)	18	148	10	58
TEG-2B	Shum *et al.* (1990a)	12	12	7	12
GEM-T3	Lerch *et al.* (1993)	9	58	6	48
JGM-1	Nerem *et al.* (1994)	7	9.9	4	14
JGM-2	Nerem *et al.* (1994)	6.5	9.6	2.7	13
GRIM4-C4	Schwintzer *et al.* (1997)	6.5	9.2	2.6	6.9
JGM-3	Tapley *et al.* (1996)	6.1	8.4	2.3	6.8
EGM96	Lemoine *et al.* (1998)	6.2	8.0	2.4	6.8
TEG-3	Tapley *et al.* (1997)	6.1	7.5	2.3	3.3
DGM-E04	Scharroo and Visser (1998)	6.1	6.7	2.3	4.3
JGM-3P/TEG-3P	Bordi *et al.* (2000)	6.1	6.7	2.3	3.1

Note: The larger errors for ERS-2 reflect the increased gravity errors and higher atmospheric drag associated with the lower altitude, as well as a lower level of tracking coverage compared with T/P.

error is common to both the ascending and descending tracks at a location of interest. This component is known as the mean geographically correlated error (GCE); it cannot be reduced through averaging of the altimeter data. The magnitude and distribution of this error can be predicted using the gravity solution covariances. In the case of T/P, a direct observation of the error was obtained early in the mission with the help of GPS-based orbits (Christensen *et al.*, 1994b). A large part of the mean correlated error resembles a miscentering of the orbit in the equatorial plane, with orbit errors that are high on one side of the earth opposite a low on the other side [Figure 40 (see color insert)].

The other component of the GCE is observed in crossover differences and represents the variation of the ascending and descending track errors about the mean GCE. By definition, this variable component of the GCE is equal in magnitude but opposite in sign for ascending and descending tracks at the location of interest. It is important to note that variable does not imply temporal variability; rather, it simply refers to the portion of the GCE that depends on the manner of the satellite approach (ascending or descending). For this reason, it is sometimes referred to as the geographically anticorrelated error (Scharroo and Visser, 1998; Moore *et al.*, 1998). A geopotential coefficient error generates both mean and variable error components; the difference is that the mean component is in phase and the variable component is out of phase. Both the mean and variable errors are important to satellite altimetry. The mean GCE will distort the mean sea-surface determination, whereas the variable GCE can affect estimates of sea-surface height variability derived from altimeter crossovers (Rosborough and Marshall, 1990). Estimates of sea-surface height variations derived from collinear repeat-track differences are immune to both types of GCE (Schrama, 1992; Chelton and Schlax, 1993a). In this special case, the entire GCE is common to the along-track profiles and cancels in the difference. (Note that other contributions to the orbit error do not generally have this same geographic dependence and thus will not difference out. Because of the dramatic improvement in the gravity models, orbit errors are no longer necessarily dominated by the gravity model errors.)

For T/P, the total rms radial orbit error (mean and variable) attributable to errors in the JGM-2 gravity model is estimated to be about 2 cm; the rms of the mean part only is about 1.6 cm. For JGM-3, the corresponding errors are 0.9 and 0.6 cm respectively. The geographic distribution of the mean GCE for both models, as predicted by the respective calibrated covariances, is shown for T/P in Figure 40. Note the larger orbit errors in regions where there is little history of ground-based satellite tracking (e.g., the Pacific Ocean, the South Atlantic Ocean, and Africa). In contrast, the orbit errors are smaller in the regions where the tracking has generally been more abundant (e.g., North America and Europe). The overall pattern is similar for most satellites, with the principal difference being the amplitude of the errors. In the same figure, the radial differences between actual T/P orbits calculated with JGM-2 and JGM-3 are averaged geographically. As the GCE for JGM-2 is much larger than its JGM-3 counterpart, the geographic patterns predominantly reflect the orbit errors in the JGM-2 model. The regions of large orbit differences tend to be located in the same geographic areas predicted by the covariances, and the magnitudes are also in good agreement. The near elimination of the GCE in JGM-3 is directly attributable to the nearly continuous, highly accurate, three-dimensional information provided by the GPS tracking (see Section 4.2.6). T/P is the first satellite to benefit so dramatically from the GPS tracking concept. The separation of the gravity signals from drag effects was made considerably easier by the choice of a high altitude for the satellite orbit, which reduced the atmospheric drag modeling errors to negligible levels (see Section 4.1.3).

The redistribution of the water mass associated with the ocean tides causes errors in the orbit in the same manner as the static gravity field, except that there are modulations at the tidal frequencies. In a regional sense, there is generally strong coherence between the orbit error due to the tide model and the ocean-tide height itself. Although the tide-induced orbit perturbation is much smaller than the actual tide height, the recovery of ocean tides from the altimeter data can be significantly affected by the ocean-tide model used for the orbit computations (Bettadpur and Eanes, 1994). In addition, some of the tides with short-period effects on the orbit, when sampled by the altimeter repeat track at a fixed geographic point, can alias into long-period, long-wavelength height errors (Colombo, 1984; Marshall et al., 1995a). Mismodeling the semidiurnal S2 and M2 tides, for example, induces spurious sea-surface height variations at a fixed point with a period close to 60 days for T/P (see Table 8 in Section 8.1.5; see also Chapter 6). This is true of the tide-induced error on the radial orbit, as well as the direct influence of the tide extraction on the sea surface height.

Because much of the orbit error is short-period in the time domain, traditional POD methods tend to be ineffective at removing tide-induced orbit errors (see Section 4.3.1). However, tide-induced errors have been observed directly using orbits derived from GPS tracking data (Haines et al., 1996). As with the static component of the gravity field, the most effective recourse has been to improve the ocean tide model itself. The altimeter data from T/P have made it possible to improve the ocean-tide models (Chapter 6), which then enabled a further increase in the orbit accuracy during subsequent reprocessing of the T/P orbits. This in turn is enabling a more accurate recovery of the ocean tides from the altimeter data (Bettadpur and Eanes, 1994), an iterative process that continues even now.

Reference frame effects It is easily verified that a small offset or miscentering of the model for the terrestrial reference frame (specifically, the tracking station coordinates) in the equatorial plane does not induce a corresponding miscentering of a dynamically computed orbit for arcs longer than a day or so. Only the mean GCE caused by the gravity model is able to generate the error patterns observed, for example, in Figure 40. Orbits computed from very short arcs or kinematic techniques, on the other hand, will be directly biased by any errors in the terrestrial frame, and short-arc analyses need to consider the likely contributions from this source of error. Because the error in the gravity model can have a geographically fixed component, station coordinates estimated from satellite tracking data would reflect that error. Conversely, when estimating the gravity field from satellite tracking data, biases in the reference frame defined by the network of tracking stations can affect the gravity solution. The gravity field will be distorted as the least-squares process (see Section 4.3.1) attempts to accommodate the biased reference frame. Simultaneously estimating both quantities is helpful, but the correlation cannot be entirely eliminated. This is one of many reasons why the gravity solutions have required a number of iterations to achieve the current level of accuracy. The gravity models have been determined more accurately as the station coordinates have improved, and the station coordinates have improved with the gravity model improvements.

As orbit accuracies have approached the level of a few centimeters, another kind of correlated orbit error, unrelated to the gravity field, has emerged as a concern. While a long-arc orbit solution does not readily accommodate a miscentering of the reference frame in the equatorial $(X - Y)$ plane, it easily follows small shifts along the earth's spin axis (Z). While the true satellite orbit is naturally centered at the earth's mass center, the calculated orbit can be shifted in Z if the coordinates for the station network have a net bias along the Z direction. This translates into a North-South hemispherical bias in the sea surface determined by the altimeter data, which, due to the asymmetric distribution of the oceans, also affects determination of the global mean sea level (Nerem et al., 1998). A Z-shift in the orbit may also result from weak centering of the orbit by the tracking data, either because of the sparseness of the data, the poor geographical distribution of the tracking stations, or the nature of the tracking measurement itself.

4.1.3. Nongravitational Forces

The nongravitational perturbing forces are also known as surface forces, since they describe the transfer of momentum from photons or atmospheric molecules as they interact with the surfaces of the spacecraft (Milani et al., 1987; Ries et al., 1993; Ries, 1997). The surface forces can be broadly classified into the categories of atmospheric drag and radiation pressure. Up to altitudes of 600–700 km, depending on the level of solar activity, drag is the dominant nongravitational force. Above that, as shown in Figure 41, solar and terrestrial radiation pressure become the largest forces. Even at altitudes where radiation pressure is larger, atmospheric drag acts in a dissipative fashion, constantly removing energy from the orbit. Ultimately, this results in a much larger

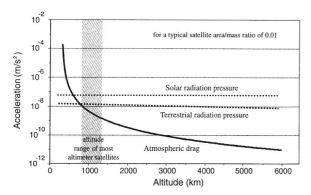

FIGURE 41 Comparison of the magnitude of the acceleration caused by various surface forces as a function of altitude for a typical altimeter satellite.

cumulative effect on the orbit. In addition, there are large, sometimes rapid, variations in the atmospheric density at the lower altitudes associated with geomagnetic activity and solar storms, which are generally not well modeled. As shown in Figure 41, atmospheric drag drops off rapidly with altitude, decreasing by about an order of magnitude from an altitude of 800 to an altitude of 1300 km. The 1336-km altitude of T/P was chosen to reduce the drag perturbations of the satellite orbits. Radiation pressure includes the effects of the direct solar light pressure on a satellite, the indirect reflected light and re-emitted heat from the earth, and radiation emitted by the satellite itself, mainly in the form of heat. All of these phenomena are complicated by the uncertain and variable nature of the incident momentum flux as well as the complex interaction with the satellite. These various surface forces are further discussed below.

Atmospheric drag   Over the last 3 decades, extensive effort has been devoted to developing models for the composition, density and temperature of the earth's atmosphere at high altitudes. Marcos *et al.* (1993), King-Hele (1992), Biancale *et al.* (1993), Hedin (1991), and Killeen *et al.* (1993) discuss some efforts at density model improvement and reviews of the various models. Recent progress appears to have been modest, however, with some atmosphere density models 2 decades old performing nearly as well for orbit determination as more recent models (Ries *et al.*, 1993). Uncertainties of the air density at high altitudes remain at a globally averaged value of about 15% (Marcos *et al.*, 1993). Significant improvements in the density models will be difficult without dedicated missions to measure the density, composition and temperature of the various constituents directly at the altitudes of interest over a variety of solar activity levels.

Along with the uncertainties in the atmospheric density, there is the problem of determining the aerodynamic drag coefficient (C_d) for a spacecraft (Herrero, 1988; Moe *et al.*, 1993). The uncertainty in C_d is probably still at the 20–30% level, and many of the questions regarding the basic

atom-surface interactions remain open (Harrison and Swinerd, 1994). In the presence of these uncertainties, the only alternative is to compensate for the various modeling deficiencies by estimating C_d using the tracking data as described in Section 4.3.1. The estimated C_d will not represent the true drag coefficient, but rather the lumped effect of all of the drag and other force modeling errors that resemble drag, averaged over the data span used in the estimation. It is thus possible to overcome the drag modeling limitations through parameterization if a sufficient amount of high-accuracy tracking data are available. In addition, other empirical parameters can be included in the estimation problem to further minimize the model error effects (see Section 4.3.1).

Radiation pressure   Even for satellites with simple spherical shapes, the precise calculation of the light pressure from the sun is not trivial (Milani *et al.*, 1987; Ries *et al.*, 1993; Marshall *et al.*, 1995b). The net reflectivity of the spacecraft is generally not known and must be estimated as part of the orbit determination process. Modeling the transition between sunlight and the shadows of the earth or moon is also essential. Numerical orbit integrators typically employ time steps that are large relative to the time spent in the penumbra, resulting in systematic orbit errors. The effects of reflected sunlight from the earth are considerably more complicated because of the highly variable nature of the earth's specular and diffuse reflectivity (Knocke *et al.*, 1988).

The problems of radiation pressure modeling are further exacerbated for satellites with complex shapes and large, rotating solar panels, as is typical of altimeter satellites. Complete modeling must take into account the orientation, reflectivity and emissivity of each surface, as well as shadowing and multiple-reflections among the various spacecraft surfaces. In addition, thermal radiation emitted by the spacecraft from heating by external sources or generated internally must be considered. Precise knowledge of all of these factors is unattainable, so even the most complex models employing very detailed and time-consuming calculations fall short of exactly modeling the radiation pressure forces. Consequently, some level of approximation must be employed, with model parameters usually adjusted to best fit the observed orbital perturbations. Various versions of these strategies have been applied for precision orbit determination for T/P (Marshall and Luthcke, 1994), the GPS satellites (Fliegel *et al.*, 1992; Beutler *et al.*, 1994), and the ERS satellites (Klinkrad *et al.*, 1991).

Nongravitational force modeling improvements for TOPEX/POSEIDON   Because of the significant improvements in the gravity model that resulted from the T/P effort (see Section 4.1.2), the nongravitational force models have become a comparable error source. To address the modeling limitations, intensive efforts were devoted to improving the satellite force models for T/P. Although atmospheric drag is considerably reduced at the 1336-km altitude (see Figure 41), radiation pressure is a very important source of error since the T/P satellite is large with a 25.5 m^2 solar panel and a mass of 2400 kg.

To meet the T/P surface-force error requirement, a detailed model was developed to account for atmospheric drag and radiation pressures for the complex geometry and attitude variations of the satellite and its thermal and radiative surface properties. The accelerations produced by a computationally intensive "micro-model" were treated as the truth for evaluation of a comparatively simple "macro-model" approximation (Antreasian and Rosborough, 1992). To overcome the excessive computation requirements associated with the micro-model, the complex spacecraft is represented in the macro model as an eight-surface box-wing composite (a combination of flat plates arranged in the shape of a box and the front and back surfaces of the solar panel). The parameters of the macro-model were determined by making it match, in a least squares adjustment, the acceleration histories generated by the micro-model for various orbital configurations (Marshall and Luthcke, 1994). The macro-model clearly cannot duplicate the micro-model precisely. Moreover, the micro-model itself was based on imperfect knowledge of the surface properties, a nominal mission profile and theoretical spacecraft performance. An observable subset of the macro-model parameters was therefore adjusted using post-launch tracking data to obtain a better representation of the actual, on-orbit satellite accelerations (Nerem et al., 1994; Marshall et al., 1995a).

Even after the improvements in the surface force models, residual errors remain which would compromise the orbit accuracy if not accommodated. A strategy is typically adopted which consists of estimating a set of empirical accelerations to absorb the errors in the surface force model that still remain (see Section 4.3.1).

4.2. Precision Satellite Tracking Systems

As discussed in Section 4.1, accurate observations of the satellite's motion are required to support the precision or-

bit determination process. Various types of tracking systems have been employed to obtain these observations, each with different measurement characteristics, temporal coverage, geographic coverage and accuracy levels. The principal tracking systems that have supported altimeter missions are summarized in Table 4. For most satellites, a laser retroreflector array onboard the spacecraft supports tracking by the satellite laser ranging (SLR) system. For Geosat, tracking was provided by only the U.S. Defense Mapping Agency's Transit Network (TRANET) and the smaller U.S. Navy Operational Tracking Network (OPNET). T/P was the first altimeter mission to carry a Doppler Orbitography and Radiopositioning Integrated by Satellite (DORIS) receiver as well as an experimental GPS receiver. In cases where the tracking is inadequate, the altimeter data, used in the form of crossovers, can be used to support the orbit determination process. Finally, when the Tracking and Data Relay Satellite System (TDRSS) data transmission link is used, the range-rate information can be used for orbit determination purposes (Teles et al., 1980). Although the TDRSS data can support orbit accuracies at the meter level in some cases (Rowlands et al., 1997; Visser and Ambrosius, 1997), they are not regularly used for precise orbit determination for altimeter satellites and will not be considered further here.

Of the five systems available for T/P, only the SLR and DORIS data are used to determine the orbits appearing on the final T/P Geophysical Data Records. Although of high accuracy, the SLR tracking is limited to non-overcast weather conditions and operator availability. The all-weather coverage provided by the DORIS system provides the requisite temporal and geographical coverage to complement the absolute accuracy of the SLR tracking. The combination of the SLR and DORIS tracking datasets for T/P provides nearly continuous geographical and temporal coverage from high-precision tracking systems.

The six-channel Motorola Global Positioning System Demonstration Receiver (GPSDR) carried onboard T/P provides independent high-accuracy tracking data with continuous coverage in three dimensions (Melbourne et al., 1994). This experiment confirmed that GPS data collected by a satellite could support very accurate orbits, and it afforded a

TABLE 4. Present Altimeter Satellite Tracking Systems and Approximate Precision

Technique	Measurement	Precision	Missions
SLR	Range	0.5–5 cm	All but Geosat
DORIS	Range rate	0.5 mm sec^{-1}	T/P
PRARE	Range, range-rate	2.5 cm, 0.25 mm sec^{-1}	ERS-2
GPS	Phase	0.2–0.5 cm	T/P
TRANET/OPNET	Range-rate	2–10 mm sec^{-1}	Seasat, Geosat
TDRSS	Range-rate	0.3 mm sec^{-1}	T/P
Altimeter	Height crossovers	5 cm	All

valuable means for model improvement and an independent orbit accuracy evaluation. However, after the precise ranging codes from the GPS constellation were encrypted (see Section 4.2.6) starting in January 1994, the orbit accuracy achievable with the GPS data was degraded. The GPSDR is an older design, and as such did not employ the "codeless" tracking techniques commonly used in modern GPS receivers. Consequently, the GPS data were used primarily to validate the concept and as an independent orbit error assessment, rather than for POD production. Recent advances, however, have achieved orbit accuracies that begin to approach that of the T/P production orbits, even with the GPS encryption function activated (Muellerschoen et al., 1994; Haines et al., 1999).

4.2.1. TRANET/OPNET

A primary means of tracking the Seasat satellite in 1978 was the TRANET Doppler system. This global network of over 40 ground receivers was designed to support geodetic surveys as enabled by the determination of precise orbits for the U.S. Navy Navigation Satellite System (NAVSAT, an antecedent to GPS). However, the receivers could track other satellites carrying transmitters operating near the NAVSAT frequencies of 150 and 400 MHz. The two frequencies allow for correction of the ionosphere delay to first order (see Section 3.1.3). The TRANET system also provided the primary tracking dataset for the U.S. Navy's Geodetic Satellite (Geosat) from 1985–1989. Geosat could also be tracked by OPNET. The OPNET network is comprised of only four sites in the continental United States and Hawaii and is used to support operational NAVSAT orbit determination. Representing the state-of-the art in satellite tracking at radio wavelengths prior to the advent of GPS, DORIS and PRARE, the TRANET and OPNET systems provided one-way dual-frequency range-rate with a precision at the few mm sec^{-1} level. Anderle (1986) gives a detailed discussion of NAVSAT and the supporting TRANET and OPNET Doppler tracking systems (see also Seeber, 1993). It should be noted that, while these data types are often referred to as range-rate measurements, Doppler systems actually observe the change in range over a finite count interval, rather than the instantaneous velocity along the line of sight.

After considerable improvement in the force models developed for Geosat, particularly the gravity model, the TRANET tracking system supports a retroactive orbit accuracy that decreased from approximately 10 cm early in the mission to perhaps 30–40 cm near the end of the mission when increased solar activity (see Figure 31) reduced the quality of the Geosat altimeter data and increased the atmospheric drag (Shum et al., 1990b; Chelton and Schlax, 1993a). One of the principal problems for Geosat was uncertainty in the reference frame determined by the TRANET tracking stations. Geosat did not carry a laser reflector array, so it was not possible to relate the Geosat orbit and

the TRANET stations to the same reference frame as the SLR stations, nor were there adequate surveys between the two systems to provide the tie. In the EGM-96 gravity solution (Lemoine et al., 1998), an attempt was made to tie the Geosat tracking stations to the other tracking networks via common tracking of Seasat, but this was considered to be a tenuous link at best. As a consequence, it is expected that there might be significant miscentering of the TRANET reference frame, especially in the Z direction (along the earth's spin axis), which would be reflected in the orbits and in the sea surfaces derived from Geosat altimeter data (Klokocnic et al., 1999).

4.2.2. SLR

The satellite laser ranging (SLR) system has been one of the primary geodetic tracking systems for more than two decades (Degnan, 1985; Smith et al., 1991; Tapley et al., 1993). SLR serves as the baseline tracking system for the T/P mission, although the DORIS system described in Section 4.2.3 plays a crucial role in attaining the highly accurate orbits required for altimetry. The laser range measurement is the time for an optical pulse to travel from the tracking system transmitter to a reflector on the satellite and back to the tracking system. Many modern systems operate at the few- or even single-photon level. This measurement represents the state-of-the-art in satellite tracking accuracy, with a precision of a few millimeters and an absolute accuracy better than 1 cm for the best instruments. As with all terrestrial tracking data, the neutral-atmosphere delay (~2 m at zenith) is calculated using a model for the zenith delay along with a mapping function to relate the zenith value to the line of sight to the spacecraft. In contrast to radio wavelengths, the optical wavelengths used by lasers are not influenced by ionosphere refraction. Moreover, the effect of water vapor is much smaller than for radiometric tracking systems. As with all tracking observations, a correction to the center of mass of the satellite is required, since this is the point whose position is represented by the numerically integrated orbit ephemeris.

Because SLR observations provide an accurate and unambiguous measurement of the range from the tracking station to a satellite, they provide good resolution of all three components of the spacecraft position with respect to the tracking network when assimilated into a model of the orbital motion over an extended period. This is particularly important for purposes of orbit error assessment. In addition, the positions and velocities of the laser stations are well determined as the result of tracking high precision targets such as LAGEOS (Ray et al., 1991). The SLR data thus provide an accurate determination of the terrestrial reference frame (Watkins et al., 1994), and they provide a strong constraint on the orbit in the Z direction. The geographic distribution of the SLR stations during much of the T/P mission is shown in Figure 42. The principal weaknesses in the SLR system are

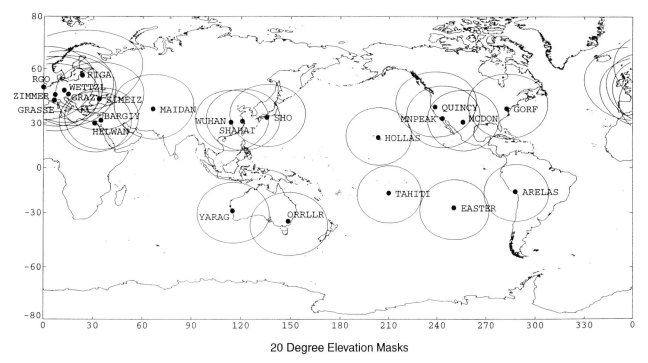

FIGURE 42   The geographic distribution of the SLR tracking stations during the T/P, ERS-1, and ERS-2 missions.

the sparse geographic distribution of tracking stations and the fact that optical ranging cannot occur without relatively clear skies.

4.2.3. DORIS

Doppler Orbitography and Radiopositioning Integrated by Satellite (DORIS) is a one-way ground-to-satellite Doppler system which uses a set of ground beacons that broadcast continuously and omnidirectionally at frequencies of 2036.25 and 401.25 MHz (Nöuel *et al.*, 1988; Kuijper *et al.*, 1995; Agnieray, 1997). The DORIS system was developed by CNES, the Institut Geographic National (IGN) and the Groupe de Recherche de Geodesie Spatiale (GRGS) to support 10-cm-level orbit determination for low-altitude earth satellites, particularly T/P. A receiver onboard the satellite receives this signal and measures the Doppler shift, from which the average range-rate of the satellite with respect to the beacon can be inferred. Average range-rate is here defined as the range change over a short count interval, usually 7–10 sec for DORIS. By receiving the tracking data onboard the satellite, there are no difficulties with timely collection of data from a network of stations scattered around the world. Moreover, all of the data are time tagged in a consistent manner. With ground-based tracking, time tagging problems can occur at individual stations, although the routine use of time transfer with GPS is steadily eliminating this as a source of error. The use of two frequencies allows for the removal of the effects of ionosphere refraction (Section 3.1.3). With the excellent short-period stability of the

beacon oscillators and the in situ meteorological data provided by beacon sensors, the DORIS system produces range-rate observations with an average precision over the network of 0.5 mm sec^{-1}.

Since Doppler observations are a differenced range measurement, they are inherently less capable of resolving the spacecraft position, particularly the component normal to the satellite orbital plane. This is, however, largely compensated by the dense coverage and precision of the DORIS data, which provide the frequent, accurate observations required for accurate orbit determination. The change in position of the satellite is compared to the predicted change over the count interval, and, given enough high precision measurements of this change, the estimation procedure arrives at the only orbit (within some error) that is consistent with the observed series of range changes. The extensive geographic distribution of the approximately 50 DORIS stations that are currently tracking T/P and will track the upcoming ENVISAT-1 and Jason-1 missions is shown in Figure 43.

Because of the near-continuous data coverage, the DORIS system is an excellent complement to the high precision SLR system and plays an essential role in meeting the rigorous tracking demands of the T/P mission (Nöuel *et al.*, 1994). Analyses of the contributions of SLR and DORIS to orbit determination for T/P indicate that the DORIS data provide the dominant contribution to the orbit accuracy. The SLR data provide a slight improvement to the overall orbit accuracy, but their main contribution is to align the center of the orbit with the geocenter.

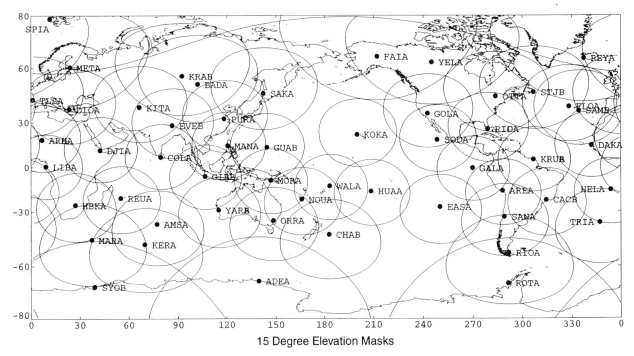

FIGURE 43   The geographic distribution of the DORIS tracking stations during the T/P mission.

As with all radiometric-tracking systems operating in this frequency range, the correction for the wet component of the troposphere refraction (Section 3.1.2) is difficult to model accurately. However, given the robust dataset provided by these systems, a zenith scaling parameter to account for troposphere refraction model error is estimated as part of the orbit determination process (see Section 4.3.1). In addition, because the DORIS measurement is one-way (ground-to-satellite), offsets between the frequency of the transmitting and receiving oscillators must also be estimated for each pass. (A "pass" is defined to be a transit across the sky over an individual ground-based tracking station.)

4.2.4. PRARE

The Precise Range and Range-rate Equipment (PRARE) was developed in the late 1980s on behalf of the German Space Agency. Although the PRARE system failed very early in the ERS-1 mission, it has been functioning without significant problems on ERS-2. The PRARE system provides two-way dual-frequency range and range-rate data (Wilmes *et al.*, 1987; Massmann *et al.*, 1997). The data are collected onboard the satellite and processed at a central location by GeoForschungs Zentrum, Potsdam (GFZ) (Reigber *et al.*, 1997).

The PRARE space segment transmits two microwave signals toward the ground in the X band (8.489 GHz) and S band (2.048 GHz). Both signals are modulated with the same Pseudo-random Noise (PN) codes (see Section 4.2.6 below). Upon receipt, the ground station demodulates the two signals and correlates the two reconstructed PN-codes to determine

the time delay of the S band versus the X band. This time delay is a measure of the one-way ionosphere delay (Section 3.1.3). In addition to the PN-codes, the carrier signals are modulated with a navigation message. These low-rate data include time information that is used to synchronize the tracking station's clock and predicted ephemeris information that is used by the station to acquire and track the satellite for future passes.

Before acquisition of the satellite's signal for each pass, the ground receiver performs an initial internal delay measurement using a built-in test transponder. The internal delay is the time required for the station to process the incoming signal and retransmit it. During the pass, the ground station makes corrections to this initial internal delay measurement every few seconds based on monitoring of the voltage and noise levels in the receivers, the Doppler frequency, and the temperatures of different parts of the station. These calculated internal delays are then used to correct the PRARE range measurements. The errors in the calculated internal delay can be significant and variable over time. The result of these errors is a bias in the range measurements, which varies from station to station over time with typical values between ±30 cm that must be estimated as part of the orbit determination procedure. Depending on the apparent stability of the bias for an individual station, it may be assumed to be constant for several days at a time (i.e., arc-dependent), or a separate bias can be estimated for each pass (i.e., pass-dependent).

Because the PRARE system is two-way, the ground stations are considerably more complex than the DORIS bea-

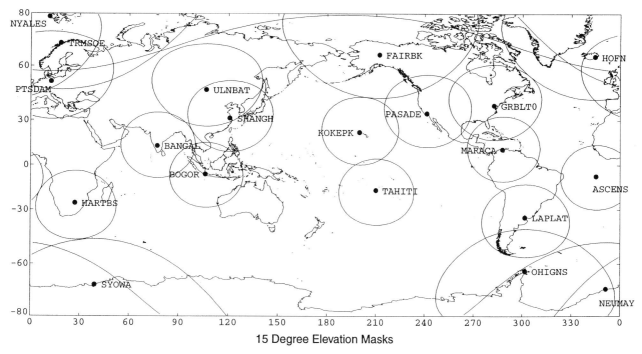

FIGURE 44 The geographic distribution of the PRARE tracking stations during the ERS-2 mission.

cons, resulting in fewer stations and less extensive global coverage (see Figure 44). However, in full operation, the system collects four to ten times the number of passes for ERS-2 than are obtained by the SLR network, since the radiometric measurements are not limited by the weather or operator scheduling. Using a gravity model improved with the PRARE data, orbit accuracies better than 4 cm can be achieved for ERS-2 (Bordi *et al.*, 2000).

4.2.5. Altimeter

The altimeter provides height measurements with an accuracy of a few centimeters over the ocean. By including the sea-surface height differences at the ground track crossover points (where the contribution to the altimeter range from the mean sea surface cancels), some additional information about the radial orbit error can be obtained to augment sparse tracking (Born *et al.*, 1986; Shum *et al.*, 1990c). In the case of ERS-1, for example, the PRARE space segment failed, and the orbit determination had to rely on SLR tracking alone. Initially, the SLR data were sparse, and the orbit quality suffered. By incorporating the single-satellite crossovers (ERS-1 with ERS-1) and dual-satellite crossovers (ERS-1 with T/P) into the POD process, the orbit errors were significantly reduced. This proved to be beneficial for the ERS-1 mission. With PRARE tracking available for ERS-2, the altimeter crossovers are not especially helpful (Anderson *et al.*, 1998).

The altimeter data can be used for orbit accuracy assessment in those cases where the orbit error is still a significant

contribution to the altimeter rms crossover differences. For T/P, the rms crossover difference is about 6 cm. This value includes a number of atmospheric and ocean surface signals whose exact magnitude can only be estimated. In the case of T/P, the contribution of the orbit error is so small (in a root-sum-squared sense) that the rms crossover difference is almost entirely from other errors and is thus a less valuable measure of the orbit error. Consequently, the 6-cm rms crossover difference represents an estimate of the contributions of the ocean variability, altimeter measurement errors and residual tide model errors. Examination of the crossover residuals over the Great Lakes, where the surface height and tide variability are much smaller, indicates that the orbit error is less than a few cm (Morris and Gill, 1994), as do direct comparisons of altimeter and *in situ* sea-surface height measurements at calibration sites (e.g., Christensen *et al.*, 1994a).

4.2.6. GPS

The Global Positioning System (GPS) is a space-based radionavigation system developed under the auspices of U.S. Department of Defense (Hoffman-Wellenhof *et al.*, 1993; Parkinson, 1996). The GPS space segment consists of a minimum of 24 satellites in circular orbits 20,200 km above the earth's surface. A group of 4 satellites is assigned to each of 6 orbital planes evenly spaced in longitude and inclined at 55° with respect to the equatorial plane. The design of the constellation implies that a GPS user anywhere on the surface of the earth will nearly always observe 6–11 GPS satellites simultaneously.

The foundation of classical GPS positioning is one-way ranging from GPS spacecraft that carry atomic clocks and continuously broadcast accurate predictions of their geocentric orbital locations. Ranges from a minimum of four GPS spacecraft are used to determine simultaneously the three Cartesian coordinates of a user's terrestrial position, and a correction to the potentially aberrant clock in the user's GPS receiver. Since the ranges are one-way, the clock correction is fundamental to the positioning: The user's receiver determines the range to any single GPS spacecraft by correlating an incoming ranging code with an internal replica generated off the receiver clock. The "clock error" of the receiver introduces a potentially large bias in the one-way ranges, which are therefore called "pseudoranges." Traveling at the speed of light, the GPS transmission will appear to be biased by 300 km for every 1-msec error in the receiver clock. The bias is common to the ranges from all satellites being tracked and, as such, can be determined explicitly along with the three-dimensional user position. The accuracies with which a typical GPS receiver can determine its position and clock correction in real time are limited primarily by the imperfect knowledge of the orbits and clocks of the GPS spacecraft. Under typical conditions, the system can support 10–30 m positioning in real time.[15]

Understanding how POD can be performed to the few-cm level with a system that is designed to deliver 10–30 m accuracies requires some basic understanding of the structure of the GPS signal and the manner in which various GPS measurements can be combined. A vast majority of common GPS applications use only the specialized ranging codes to determine position and time. These codes are composed of a repeating sequence of pseudo-random bits (ones or zeros) unique to each GPS spacecraft. The ranging codes are designed so that they can be carried on the same broadcast frequency, regardless of the spacecraft from which they originate. (This implies that signals coming from different GPS spacecraft are essentially orthogonal and cannot be confused with one another even though they are at the same frequency.) A precise (P) code is modulated on two L-band carriers: L1 at 1575.42 MHz and L2 at 1227.60 MHz. The presence of two microwave frequencies enables correction for the effects of the ionosphere (Section 3.1.3). The L1 transmission also carries a coarse acquisition (C/A) code, also known as the "civilian code." Owing to its lower bandwidth and the lack of a second frequency for the ionospheric

correction, the C/A pseudorange is less accurate than its P-code counterpart. However, the C/A code is always available, whereas the P code is generally encrypted to prevent jamming (spoofing) of the code by hostile forces. When the encryption function is activated, the GPS constellation is said to be in "Anti-Spoofing" (AS) mode. Excepting a few short periods, AS has been activated continuously since January 1994. When AS is active, the encrypted P code (known as the Y code) can be directly tracked only by users authorized to carry receivers with decryption keys. However, advanced "codeless" civilian receivers are capable of obtaining ranging information from the Y code without decryption keys, albeit with slightly degraded accuracy. Moreover, in what promises to be a significant benefit to many civilian users of GPS, C/A code will be added to the L2 carrier by 2003 (Divis, 1999). This second frequency will enable direct correction of the C/A ionosphere delay to first order.

For the most demanding scientific applications, the key to precise positioning comes not from the ranging codes but from the L-band carriers. By beating against a reference frequency generated internally, state-of-the-art GPS receivers can track the advance and retreat of the Doppler-shifted carrier phase with a precision approaching 1% of the wavelength (~19 cm for L1), or about 2 mm (e.g., Spilker and Parkinson, 1996). However, the carrier is simply a continuous flow of sinusoids and, as such, there is ambiguity in the integer number of cycles when tracking is initiated. Thus, while very precise measurement of range changes can be made from monitoring of the carrier phase alone, the absolute range is not directly observed. Despite this, the carrier phase observation is extremely valuable owing to its high precision and continuity. This allows measurement of range change over long intervals punctuated by occasional losses of lock, as opposed to a shorter accumulation interval that is used in most traditional Doppler systems.

When dual-frequency carrier phase and pseudorange are processed together, the full potential of the GPS measurements can be realized. While much less precise than the carrier phase, the pseudorange supplies complementary absolute range information. The precision of the pseudorange data is generally well under 1 m for most advanced codeless GPS receivers; the pseudorange can thus be used to provide some constraint on estimating the integer number of cycles in the carrier. The carrier phase in turn can be used to smooth and condition the much noisier pseudoranges.

The final impediments to precise centimeter-level positioning are the errors, both real and intentional, in the clock and orbit information broadcast by the GPS spacecraft. Lacking a means of mitigating the GPS spacecraft errors, the sub-centimeter precision of the carrier phase is essentially wasted. To circumvent this problem, an additional GPS receiver can be placed at a precisely known location. Differencing data collected simultaneously by the user receiver and the reference GPS receiver removes the spacecraft clock

---

[15]Until recently, civilian users of GPS data have been limited to accuracies of 50–100 m because of an intentional degradation of the GPS transmissions. Known as "Selective Availability" (SA), this degradation was realized by dithering the spacecraft clocks and by introducing errors into the broadcast orbit information. SA was deactivated completely on May 2, 2000, thus significantly improving civilian GPS positioning capabilities. Uncertainties in the broadcast orbit and clock information, however, remain important sources of positioning error at the 5–10 m level.

errors, including the effects of SA. Differencing also substantially reduces the effects of the error in the GPS spacecraft orbits, with increasing success as the baseline length decreases. An attractive alternative to explicit differencing is to take advantage of one of the several sources offering precise estimates of the GPS spacecraft positions and clocks ex post facto. Especially noteworthy is the International GPS Service (IGS), which provides GPS ephemeris and clock products with accuracies better than 10 cm and 1 nsec respectively (Beutler *et al.*, 1999). The IGS products are based on weighted combinations of solutions from various participating agencies, and are well suited to many of the most demanding scientific applications of GPS including POD.

Concepts capitalizing on the obvious potential of GPS to determine satellite orbits began to emerge with the GPS system itself (e.g., Van Leeuween *et al.*, 1979; Yunck, 1996). The beams from the GPS signals extend approximately 3000 km above the earth's limb. A GPS receiver in low-earth orbit therefore receives nearly the same level of GPS spacecraft observability as a ground-based user (see Figure 45). This rich observing geometry enables continuous three-dimensional positioning of a low-earth orbiter without any underpinning from dynamic models, a powerful capability that distinguishes GPS from other spacecraft tracking systems.

For the vast majority of the commercial and government spacecraft now carrying GPS receivers, the 10–30 m accuracies of the satellite positions determined solely from the onboard receiver (the so-called navigation solution) are adequate. Altimeter satellites clearly demand higher accuracies of GPS. As a means of delivering sub-decimeter accuracy for such missions, Ondrasik and Wu (1982; see also Yunck *et al.*, 1985) proposed a GPS-based POD technique based on post-processing of the differential carrier phase measurements formed between the satellite receiver and a global "fiducial" network of ground receivers. Although this technique has evolved considerably in detail, it is fundamentally the same procedure whose application culminated in the achievement of rms radial accuracies near 2 cm for the GPS-based estimates of the T/P orbit (Bertiger *et al.*, 1994; Yunck *et al.*, 1990, 1994; Schutz *et al.*, 1994).

4.3. Orbit Estimation

As summarized in Section 4.1, knowledge of the forces acting on the satellite is imperfect, and the initial conditions required to start the integration cannot be known exactly. Observations of the position or velocity of the satellite must be obtained and incorporated into an orbit determination system which estimates these initial conditions as well as corrections to the parameters in the force and measurement models. Starting with estimates for the initial satellite position and velocity, the orbit is predicted to the times of the observations. Using a model for each measurement based on a priori estimates for the tracking station position, earth orientation, atmospheric refraction effects and measurement biases, a computed range or range rate C is formed and compared to the observed range or range rate O. The measurement residual $O - C$ is an indication of the mismatch between the mathematical model and the actual orbit, as illustrated schematically in Figure 46. With an appropriate estimation procedure, the residuals from the data fits can be used to improve the estimate of the initial satellite state, as well as estimates of various parameters in the force or measurement model.

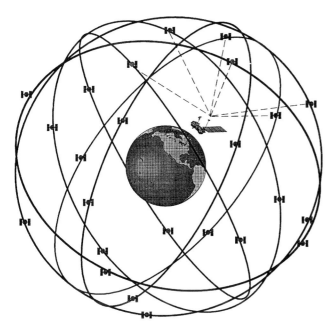

FIGURE 45 The Global Positioning System as a satellite tracking system.

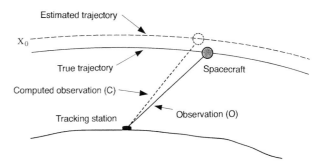

FIGURE 46 A schematic overview of the orbit estimation problem. An observed value O for the range or range-rate of the satellite is compared with the range or range-rate C computed based on the numerically integrated satellite position and velocity to compute an observation residual $O - C$. With many such observations, the initial conditions r_0 and v_0 and other dynamical and measurement model parameters are iteratively adjusted to bring the estimated orbit closer to the true orbit by minimizing the residuals in a least-squares sense. Clearly, the more observations available and the more accurate they are, the better the orbit can be determined.

This is a complex and nonlinear estimation problem that is usually linearized and solved numerically through an iterative method (Tapley, 1973; Bierman, 1977; Tapley, 1989). This may be a "batch" process in which all of the measurements within the arc are combined to form the estimates of the initial conditions and the other force and measurement model parameters. Various "sequential" methods are also available in which the orbit and model parameter estimates are updated as soon as new data are available. The fidelity of the models for the dynamics and the measurements, the precision of the tracking data, and the quality of the orbit determination technique all determine the accuracy of the resulting satellite ephemerides. There are a variety of estimation strategies, and it is not possible to describe them all here. The following discussion addresses the principal aspects of the orbit estimation methods generally employed.

4.3.1. Dynamical Methods

The most general method of POD typically involves an iterative least-squares fit of selected model parameters and model-predicted satellite position and velocity to tracking data along an orbit arc. This least-squares adjustment produces a new initial state vector, which corrects for much of the error in the previous initial conditions and errors in the force model (Tapley, 1973). The atmospheric drag and radiation pressure forces in the orbit models are generally approximated by relatively simple formulations with model parameters that can be adjusted for each arc as part of the POD estimation procedure. More sophisticated modeling of these forces would be of limited value because of the uncertainties in the drag and reflectivity coefficients, the air density at the altitude of the satellite and the cross-sectional area of the satellite. Some of the parameters for these force models may be updated more frequently than once per arc (typically every day or even every few hours). The magnitude of the orbit errors arising from uncertainties in atmospheric drag and radiation pressure are thus dependent on the density and quality of the tracking data and the frequency of the updates of these coupling coefficients.

The arc lengths for altimeter satellites generally range from a day to a few weeks, depending on the density of tracking data. Longer arcs are necessary if the tracking data are sparse, poorly distributed geographically or of low quality. In these cases, the fidelity of the dynamic model becomes critical. Shorter arcs are generally preferred if the tracking data will support them, but there are additional considerations. Short arcs have a larger number of discontinuities, since the orbit solutions at the end points of successive fit intervals do not match exactly. These discontinuities would appear as artifacts in the altimetric estimates of the sea surface height. Smoothing or blending procedures, overlapping the orbit fit intervals, and placing the arc boundaries over land are all methods that reduce the effects of these discontinuities. A subtler problem with short arc lengths is a reduction in the accuracy of the "centering" of the orbit relative to the mass center of the earth. A long arc has the benefit of averaging the modeling and tracking errors over many orbital revolutions. Short arcs tend to display more variation from arc to arc in terms of the centering. Finally, short arcs can be more sensitive to tracking data problems. The trade-offs between long and short arcs depend on a number of factors, with the choice for the balance of global altimetric studies often being the longest arc possible that still fulfills the accuracy requirements.

If only the initial conditions of an arc are estimated, the mean orbit elements for the arc interval will be corrected by the tracking data.[16] The secular drifts in the orbit elements (the first term in a Taylor series expansion of the errors in the orbit elements) will remain, resulting in a so-called "bow-tie effect" in the residual orbit error when projected into the radial, transverse (along-track) and normal (cross-track) directions (e.g., Colombo, 1989; Chelton and Schlax, 1993a). Since the tracking data are not always uniformly distributed over the orbit arc, the minimum residual orbit error (the "knot" of the bow tie) can occur at points other than the middle of the arc, as shown in Figure 47a. Left alone, the secular errors grow with the arc length, leading to unacceptably large orbit errors when long arc lengths are used. However, one can choose to estimate special empirical accelerations which vary sinusoidally with a period that matches the orbital period (i.e., one cycle per revolution or 1-cpr). These parameters are very effective at removing the secular component of the orbit-element errors and dramatically reducing the bow-tie effect (see Figure 47b). This parameterization has become a common practice for precise orbit determination of altimeter satellites (e.g., Tapley et al., 1994; Scharroo and Visser, 1998; Moore et al., 1998).

In the case of T/P, the nominal set of parameters estimated for the 10-day arcs are the initial conditions, a constant along-track acceleration for each 8-hr period (i.e., 8-hr subarcs), and daily sinusoidal (1-cpr) along-track and cross-track accelerations. The effectiveness of this parameterization is demonstrated in Table 5, where all models are identical except for the increasing level of parameterization for each entry in the table. It may be surprising to see the magnitude of the modeling errors without the empirical parameters. This indicates the difficulty in completely and accurately modeling the surface forces, particularly for arcs as

---

[16]The adjustment of the initial conditions is equivalent to an adjustment of the mean orbit elements over the arc. Because of the fitting of an imperfect model to imperfect tracking data, the adjusted mean orbit will not match the true mean orbit, resulting in errors that are largely 1 cpr. Errors in the mean eccentricity or perigee, for example, lead to 1-cpr radial and along-track errors, while errors in the inclination or node lead to 1-cpr cross-track errors. If the orbit elements evolve slowly with time, the adjusted orbit will match the true orbit best at some point within the fit interval, depending on the distribution of the tracking data. At this point (the "knot" in the bow-tie pattern of orbit errors), the 1-cpr error is the smallest.

TABLE 5.   The Effects of Various Levels of Empirical Parameterization on T/P Orbit Accuracy for a Typical 10-Day Arc

Case	Number of parameters	SLR rms (cm)	Radial rms (cm)
Estimate initial position and velocity only	6	870	115
Add 8-hr constant transverse acceleration	36	80	100
Add daily 1-cpr transverse acceleration	56	22	3.5
Add daily 1-cpr normal acceleration	76	<3	<3

Note: As discussed in the text, transverse and normal refer to along-track and cross-track, respectively.

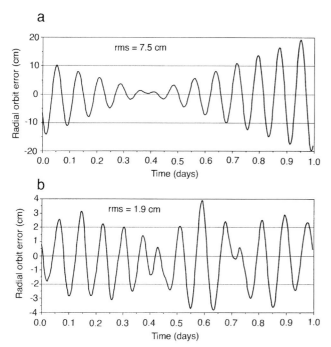

FIGURE 47   An example of the radial orbit error for a one-day T/P arc where (a) only initial conditions and empirical constant along-track accelerations are estimated, and (b) an empirical 1-cpr along-track acceleration is also estimated. The typical "bow-tie" error is clearly evident in the top panel but eliminated in the bottom panel (note the change of scale in the two panels). The remaining error in the bottom panel, while still exhibiting the expected 1-cpr variation, is much smaller in amplitude and more random.

long as 10 days. However, it also demonstrates that the orbit error tends to be dominated by slow variations in the orbit elements caused mainly by the surface-force modeling errors, which are very effectively accommodated by estimating the 1-cpr empirical accelerations using well distributed and accurate tracking data.

The reason for the effectiveness of this particular acceleration parameterization is apparent upon examination of Gauss' form of the Lagrange planetary equations, which express the time derivative of the orbit elements in terms of radial, transverse, and normal (RTN) perturbations (e.g.,

Danby, 1962; Geyling and Westerman, 1971). If the perturbing RTN acceleration is resonant with the period of the orbit, then the variation in the affected orbit element (or elements) has infinite period, i.e., it is secular. The in-plane (radial and transverse) 1-cpr accelerations are effective in removing the errors in eccentricity e and argument of perigee ω. This is particularly important for an altimeter mission, since these errors affect the satellite orbit-height accuracy. It should be noted that estimating either the radial or the transverse empirical 1-cpr acceleration is sufficient to remove the in-plane errors; estimating both results in a near-singular solution. The out-of-plane (normal) 1-cpr acceleration removes the secular errors in the node Ω and inclination i. This is less critical for the height component, but the best height determination is generally achieved when all three components (RTN) of the orbit are accurately determined. If the tracking data residuals are large because of errors in the normal component of the orbit, the radial and transverse components will be affected as the least-squares process attempts to minimize the overall fit to the data.

It is thus necessary to estimate four acceleration parameters to mitigate the slowly varying orbit element errors: a sine and cosine term for the transverse (or radial) direction and the same for the normal direction. Errors in the angular position along the satellite orbit that build up quadratically owing to a secular change in the semimajor axis a are easily accommodated by estimating the drag coefficient C_d or a constant empirical transverse acceleration C_t. These five components are highly effective at reducing the slowly varying orbit element errors caused mainly by the surface forces, but also by some terms in the geopotential model (notably the zonal harmonics of a spherical harmonic expansion of the geopotential and certain resonance coefficients). The secular drift in the mean anomaly is automatically accommodated in the least-squares fitting process by a tiny adjustment of the semimajor axis a, thus requiring only five parameters in the empirical model rather than six.

The orbit element errors will also exhibit slow oscillations superimposed on the secular drifts. Provided that the tracking data are dense enough, these variations can be accommodated to some degree by estimating empirical parameters on a more frequent basis than once per arc. It should be noted that these parameters will not be effective in removing the higher frequency errors due to the geopotential (twice-per-rev and above). These errors are most effectively reduced through improvements in the gravity model itself. In addition, as the sub-arc length of these estimated empirical acceleration parameters is reduced (i.e., more parameters are estimated per arc), the dynamical method begins to approach the reduced-dynamics method discussed in Section 4.3.2.

4.3.2. Reduced-Dynamic Methods

Using a Kalman-filter formulation (e.g., Lichten, 1990) in the estimation step enables some interesting departures from the traditional dynamical POD strategy. In particular, it provides a unique ability to liberate, in varying degrees, the POD process from the force models underlying the satellite motion. This approach, referred to as reduced-dynamic tracking (Wu et al., 1991; Yunck et al., 1994), is realized by treating unmodeled or mismodeled spacecraft accelerations as stochastic processes which explain departures of the integrated orbit from the observed orbit. The concept of a satellite ephemeris being precisely observed at all times and in three dimensions is unique to GPS; however, other data combinations such as DORIS/SLR can support certain variations of reduced-dynamic tracking (Barotto, 1995). In general terms, reduced-dynamic tracking seeks to exploit the strong tracking observability by making local geometric (or nearly geometric) corrections to a converged, dynamically determined orbit solution.

Reduced-dynamic POD actually refers to a continuum of strategies bounded by the extreme cases of "fully dynamic" and "kinematic" POD. The placement of any particular reduced-dynamic strategy on this scale is controlled by setting the parameters that govern the behavior of the stochastic spacecraft accelerations. Generally, the accelerations are treated as a first-order Gauss-Markov process characterized by a correlation time scale and a steady-state variance. With the correlation time scale set to a large number and the variance set to zero, the strategy will yield a fully dynamic orbit solution. Conversely, with the variance unbounded and the correlation time scale set to zero (white noise), the strategy will yield a purely kinematic orbit solution. In the latter case, relevant to GPS tracking observations, the filter estimates a three-dimensional acceleration correction freely and independently at each measurement time. It should be kept in mind that the kinematic solutions are very sensitive to aberrations in the observing geometry and to anomalies in the tracking data. If the GPS receiver loses lock, the kinematic adjustments spanning the lapse in tracking can become unbounded. In contrast, such an outage would not compromise a purely dynamic solution.

In reduced-dynamic POD, the optimal balance of dynamics and kinematics is not straightforward to determine. The considerations that are significant factors in choosing the weights assigned to the observations versus the dynamics include the satellite orbit characteristics (altitude, in particular), the quality of the force modeling and the tracking capability of the onboard GPS receiver. For T/P, early solutions (Yunck et al., 1994) departed from the dynamic extreme more than later solutions. This is because the intensive post-launch tuning of the T/P force models yielded significant improvement in the ability to model the satellite orbit dynamically (e.g., Nerem et al., 1994, Marshall et al., 1995a; Tapley et al., 1996). Prior to these tuning efforts, geographically correlated orbit errors (see Section 4.1.2) stemming from errors in both the static and time-varying components of the gravity field were revealed by the GPS reduced-dynamic orbits (Christensen et al., 1994b; Haines et al., 1996), testifying to the strength of the GPS observations when processed in this manner.

In view of the unprecedented accuracy of the current T/P force models, the benefit of reduced-dynamics is now tangible only when the weighting of the observations versus the dynamics is very conservative. For example, Lough et al. (1998) allowed stochastic variations in the reduced-dynamic accelerations only at the 1-cpr frequency. This reduced-dynamic approach resulted in a modest improvement in the POD accuracy. It should be noted, however, that it is not dissimilar from the dynamical POD practice of estimating piece-wise (sub-arc) empirical acceleration terms at the same 1-cpr frequency. For satellites at lower altitudes, where the errors in the dynamic models are larger, the reduced-dynamic approach offers the potential to achieve orbit accuracies comparable to T/P (Yunck, 1996).

4.4. Performance Assessment

The radial orbit accuracy assessment is an estimate of the rms radial differences between the estimated orbit and the true orbit. This estimate is essential to the error assessment for the overall altimeter error budget, especially since the orbit error is correlated over long distances owing to the 1-cpr nature of orbit error. The error is generally on the order of three to five times larger in the transverse and normal directions than in the radial direction. The use of an rms metric should not be taken to imply that orbit error is a random process. Much of the orbit-error spectrum tends to congregate around the 1-cpr frequency, with the power dropping rapidly at higher and lower frequencies (see Figure 48). Since the orbit error cannot be measured directly, the power of the orbit error is estimated using covariance analysis and simulations of the effect of the expected model errors on the orbit, as well as indirect observational measures of the orbit error (e.g., Tapley et al., 1994; Marshall et al., 1995a; Smith et al., 1996; Klokocnik et al., 1999). At the few-centemeters level of orbit error, it is difficult to find a single test that confidently quantifies the remaining error. The assessment must

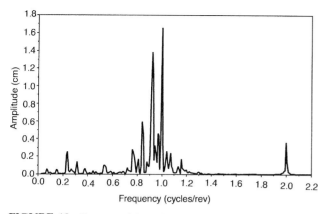

FIGURE 48 Spectrum of the radial orbit differences between a T/P dynamic orbit determined with SLR/DORIS tracking and a reduced-dynamic orbit determined with GPS tracking.

be gathered from a number of independent tests which measure various aspects of the orbit error, and these must then be combined in a manner that is not overly optimistic. Various methods of performance assessment are summarized in this section.

4.4.1. Internal Orbit Quality Tests

An obvious but important measure of the quality of an orbit is how well it fits the tracking data. The overall residual rms is indicative of both the model accuracy and the data quality since large residuals, relative to the theoretical noise, indicate either model deficiencies or systematic errors in the tracking observations. The most recent models produce fits for 10-day T/P arcs of approximately 2 cm for the SLR data and 0.55 mm sec^{-1} for the DORIS data. Whereas the DORIS data are fitting close to the measurement noise level (see Section 4.2.3), the SLR data do not fit at the centimeter-level accuracy of the SLR measurements, hinting at the possibility of additional orbit accuracy improvement. This test must be interpreted with caution, however, since the residual rms can be reduced to the noise limit if enough empirical parameters are adjusted in the orbit solution.

In the analysis of the post-fit SLR residuals, a timing bias and a range bias can be estimated for each pass, effectively removing all orbit-error signals from the residuals. Given the accurate time tagging of the SLR data (generally better than a microsecond), and assuming accurate knowledge of the station location, the inferred timing bias can be used as an indicator of the level of along-track orbit error. The range bias indicates the combined level of radial and cross-track orbit error in the pass at the point of closest approach to the station. For high-elevation passes, however, the cross-track orbit errors contribute very little to the range biases. In these cases, the range bias provides an absolute and unambiguous measure of the radial orbit error, albeit with limited geographic and temporal coverage. The rms of the range biases for the high-elevation passes is typically on the order

of 2 cm for T/P, even when the high-elevation passes are excluded from the orbit solution (Ries and Tapley, 1999). For ERS-2, using the PRARE tracking data and the TEG-3P gravity model, the rms of the excluded SLR data is less than 4 cm (Bordi *et al.*, 2000). This test represents one of the most accurate and reliable indicators of the radial orbit accuracy, especially when the data are excluded from the fit, and illustrates one of the unique contributions of the SLR tracking to satellite altimetry.

In the absence of more reliable external tests, a useful internal test is the mismatch of the endpoints of adjacent arcs. Because of the adjustment of the initial conditions of each arc, the final point from the previous arc will generally not match the initial point of the next arc exactly, even though both are supposed to represent the satellite position at the same point in time. The rms of these overlaps has been demonstrated to be a useful indicator of the level of orbit error, assuming that the errors common to both arcs are not large relative to the other errors. In conjunction with the other internal tests described above, this test can provide some additional confidence in the orbit error assessment, but it must be interpreted with caution. In cases where the two orbits actually overlap by hours or days, some of the tracking data are used for both orbit fits, and there is likely to be more common-mode orbit error, which will not sensed by the overlap test. Nevertheless, it is still a useful test of the orbit consistency, and it can help to identify problems that are not obvious from the tracking fits alone.

4.4.2. External Orbit Quality Tests

A unique and valuable measure of the orbit quality is obtained by comparing orbits determined independently with different tracking systems and different orbit-determination strategies. In the case of T/P, the orbits produced independently from GPS tracking data using the reduced-dynamic technique could be compared to the production orbits based on the dynamical approach. Given the different modeling, tracking data, and estimation techniques, the level of agreement provides a good indication of the combined accuracies of both orbits. The complementary nature of the tracking systems and the two orbit determination strategies (SLR/DORIS dynamic and GPS-based reduced-dynamic) has proven to be valuable in several other regards. The intercomparison provided direct observations of the magnitude and spatial distribution of the gravity-induced errors in the SLR/DORIS orbits, corroborating the predictions from the JGM-2 covariance (Christensen *et al.*, 1994b; Haines *et al.*, 1996). This helped to assure the reliability of the gravity covariance calibration procedure. The GPS data also offered the opportunity for additional gravity-model improvement, resulting in the JGM-3 model (see Section 4.1.2).

The intercomparisons also yielded important insight on an approximate 6-cm radial bias in early GPS-based reduced-dynamic orbit solutions (e.g., Bertiger *et al.*, 1994).

TABLE 6. Estimates of Radial Orbit Error Budgets for T/P in Millimeters

Error Source	Mission specifications	JGM-2	JGM-3	Goal
Static gravity	100	22	9	4
Earth and ocean tides	30	13	7	4
Temporal gravity[a]	not considered	8	8	4
Surface forces[b]	70	15–25	10–15	5
Data errors[c]	10	5–10	5–10	4
Station location[d]	20	10	5	4
RSS	130	35–40	20–25	~10

[a] Seasonal and other temporal variations apart from tides.

[b] Solar, terrestrial and thermal radiation, atmospheric drag, bias forces.

[c] Data noise, biases, troposphere, center-of-gravity offset errors, attitude errors.

[d] Random and systematic station position and velocity errors, geocenter motion.

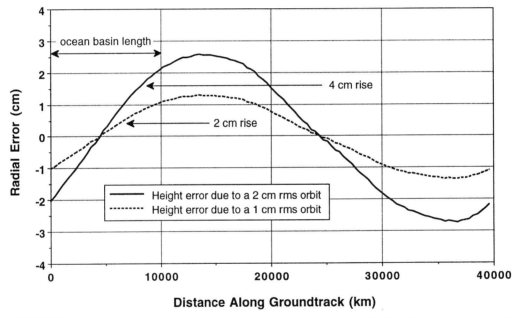

FIGURE 49   Effect on sea surface height determination at basin-scale wavelengths for orbits with rms errors of 1 and 2 cm.

Although the exact origin of the bias remains unknown, comparisons with SLR/DORIS-based orbits corroborated that an anomaly exists somewhere in the T/P GPS observation model. The reason for this anomaly is not yet understood. An additional example is provided by the 1–2 cm centering problems commonly experienced by T/P orbits that are computed solely from radiometric data. As noted in Section 4.1.2, the centering error is manifest as a spurious shift of the orbit solution along the earth's spin (Z) axis. SLR data are particularly well suited to diagnosing this class of errors, since the unambiguous range information provides a more accurate tie to the earth's center of mass than is currently possible with radiometric data alone. These are a few

of the myriad examples underscoring the value of external orbit quality tests in detecting anomalies not revealed by the internal consistency checks.

4.4.3. TOPEX/POSEIDON Orbit Error Budget

The confidence in the assessment of the orbit accuracy for a satellite depends on the quality and independence of the tests available. Not all of the tests used to evaluate the T/P orbit are available for other altimeter satellites. The multiple tracking systems on T/P have provided an unprecedented opportunity to test orbit-determination strategies and to accurately assess the level of orbit error. In the case of T/P, the tests described previously indicate that the radial accuracy is close to 2 cm rms (e.g., Marshall *et al.*, 1995a; Tapley *et al.*, 1996; Ries and Tapley, 1999).

The estimated orbit error budget given in Table 6 indicates that the primary improvement beyond the original mission requirements has been in the gravity model. The effects of the surface forces, and to a lesser degree, the ocean tides, have been reduced by improvements in the models and by the estimation of the empirical accelerations as summarized in Section 4.3.1. The last column in Table 6 illustrates the level of improvement necessary to achieve the 1-cm goal for the orbit accuracy. It should be noted that, at this level, the time variation of the gravity field associated with the seasonal migration of mass between the atmosphere, oceans, and solid earth must be considered.

The distinction between a 1-cm orbit versus a 2-cm orbit may not seem important compared to the estimated noise of a few cm for the altimeter measurements. However, the characteristics of the orbit error require every effort to reduce it as far as possible. A typical radial orbit error as a function of time along the satellite ground track is illustrated in Figure 49 for orbits with 1- and 2-cm rms errors. As discussed previously, the two-body problem of a satellite in orbit tends to map a part of every perturbation into the 1-cpr region of the spectrum. Various parameterizations have been shown to be effective in removing much of this error, but the remaining error still retains the long-wavelength 1-cpr signal. Over an ocean basin, for example, the orbit error can appear as a several-cm slope in the observed sea level. Most applications of altimeter data analyze spatially and temporally smoothed fields of sea surface height (see Section 8.1.7), and the random errors in these smoothed estimates have probably been reduced to 1 cm or less. Consequently, any systematic errors are important and must be reduced as much as possible to avoid contaminating the large-scale ocean signals of interest.

4.5. Future Prospects

Future prospects for altimeter satellite POD are probably best understood in the context of advances in the tracking systems. As amply demonstrated by T/P, deficiencies in the force models underlying the satellite motion can be largely overcome with dense, continuous and precise tracking data. In recognition of this, the T/P follow-on mission (Jason-1) will carry advanced DORIS and GPS receivers and laser retroreflector arrays. The new DORIS system, also slated for inclusion on the European ENVISAT-1 mission, will feature lower noise and the ability of the DORIS receiver onboard the satellite to track two ground stations simultaneously. The latter capability will enable further relaxation of the dynamic models in DORIS-based POD strategies. The satellite will also carry an advanced codeless GPS TurboRogue space receiver (TRSR) capable of simultaneously tracking all GPS spacecraft in view (up to 16) on two frequencies, independent of the encryption status of the GPS constellation. It is

anticipated that the orbits for Jason-1 will thus be significantly improved over their T/P counterparts. The fidelity of the observations is also expected to support the computation of precise kinematic orbit solutions. Missions beyond Jason-1 may benefit from augmentation of the GPS tracking with observations from other satellite radio-navigation constellations, such as the Russian GLobal NAvigation Satellite System (GLONASS) and the proposed European system Galileo.

The SLR system also continues to be improved with ongoing efforts to increase the accuracy of the measurement to the few-millimeters level. Additional stations are expected soon in South Africa and South America, which will significantly improve the geographical distribution of the network. Recognizing that the SLR network is sparse and expensive to operate owing to the equipment cost and manpower requirements of the current station designs, NASA and CNES have both embarked on programs to develop less-expensive, more compact SLR stations. The French Transportable Laser Ranging Station (FTLRS) is being developed in cooperation with Institut Géographique National (IGN), CNES and Observatoire de la Côte d'Azur/Centre d'Etudes et de Recherches en Géodynamique et Astronomie (OCA/CERGA). The NASA SLR 2000 system (Degnan, 1998) will feature full automation and is expected to be online by 2003 with systems installed at the nine NASA stations and partner sites. These advances will make possible a much broader distribution of SLR stations, essential for the high-accuracy requirements of the many missions planned for the future. Together, advances in the DORIS, GPS and SLR systems are expected to help challenge the 1-cm rms threshold for radial orbit error that is a stated goal for the Jason-1 mission.

Advances in the force and measurement models used for POD continue to be made. As indicated in Section 4.1.2, the GRACE mission will dramatically improve the model for the static and seasonal gravity variations. Analyses of the altimeter data will continue to improve the models for the tidal variations. The models for the surface forces have proven difficult to improve, but this limitation can be overcome with the improved tracking systems and appropriate orbit error parameterizations. The accuracy of the terrestrial reference system is also improving, as long time series and multiple geodetic methods are combined to provide better estimates of the tracking station positions and velocities, as well as the millimeter-level motions of the geocenter.

There is potential for significant improvements in POD turnaround time to support emerging operational oceanography programs. There are already systems producing 3-cm (rms) T/P orbits within 1–2 days of real time (Muellerschoen *et al.*, 1995; Haines *et al.*, 1999; Berthias and Houry, 1999). The near-real-time orbits for T/P support a growing number of diverse operational oceanographic applications, and figured positively in the early forecast of the 1997 to 1998 El Niño event by the National Center for Environmental Prediction (Cheney *et al.*, 1997; Ji *et al.*, 2000). Motivated in large measure by POD requirements and associated scientific goals, the International GPS Service (Beutler *et al.*, 1999) is preparing to make available precise GPS clock and orbit estimates within hours of real time. Also on the horizon are global GPS augmentation systems (Yunck *et al.*, 1996; Bertiger *et al.*, 1998), which disseminate clock and orbit corrections in real time. For the DORIS tracking system, a new onboard orbit determination capability DIODE (Détermination Immédiate d'Orbite par DORIS Embarqué) was recently demonstrated on SPOT-4 (Berthias and Houry, 1998) and will fly on Jason-1. These systems hold great promise for enabling routine real-time precise orbits for altimetry missions.

5. GEOPHYSICAL EFFECTS ON THE SEA SURFACE TOPOGRAPHY

The end-to-end altimeter system has been described in the preceding sections. The two-way travel time $t_{1/2}$ between the altimeter and nadir mean sea level is determined from the radar return as described in detail in Section 2.4. The range R from the altimeter to nadir mean sea level is then determined from $t_{1/2}$ with corrections for atmospheric refraction and sea-state biases as described in Sections 3.1 and 3.2. For typical wave conditions of $H_{1/3}$ ranging from 2–4 m, it was shown in Section 2.4.7 that measurements of R in the 1-s averages that are generally analyzed in applications of altimeter data are based on the radar return from an oval footprint with cross-track and along-track dimensions of about 5 and 11 km, respectively (see Figures 22 and 23). The range measurements are transformed into the sea-surface topography relative to a reference ellipsoid by independently estimating the orbit height H relative to a reference ellipsoid by precision orbit determination as described in Section 4. With accurate estimates of R and H, the sea-surface topography relative to the reference ellipsoid is $h = H - R$. The reference ellipsoid that is used corresponds to the ellipse that is closest to the equipotential of the sea surface topography, which is determined primarily by the combined effects of the earth's gravity and centrifugal forces (see Section 5.1).

As summarized by Eq. (3) in the introduction and shown schematically in Figure 50, the sea surface topography can be decomposed as $h = h_g + h_d + h_T + h_a$, where h_g is the height of geoid undulations relative to the reference ellipsoid, h_d is the dynamic sea-surface height associated with geostrophic surface currents, h_T is the tidal height and h_a is the sea-surface response to atmospheric pressure loading. Applications of altimeter data to investigate each of these four components of the sea-surface topography are summarized in detail in later chapters. A brief overview is given

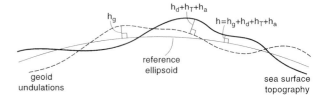

FIGURE 50 Schematic summary of the various contributions to the sea surface height h relative to a reference ellipsoid. The components of h include the geoid undulations h_g, tidal variations h_T, atmospheric pressure loading h_a and the dynamic sea surface height h_d associated with geostrophic surface currents.

here with the primary objective of providing a sense of the relative magnitudes of h_g, h_d, h_T, and h_a.

5.1. Geoid Undulations

In the absence of all forces other than gravity and the centrifugal force from the earth's rotation, the sea surface would coincide with a surface of equipotential (the marine geoid). As shown in Figure 51, the geoid undulations h_g relative to an ellipsoidal approximation of the geoid range from -105 m south of India to $+85$ m off the northeast coast of New Guinea. The geographic variability of the geoid undulations is determined by perturbations in the gravity field owing to variations in ocean bottom topography and spatial inhomogeneities in the density of the earth's interior. As summarized in the following sections, the variations of the sea-surface topography h arising from other forces at the sea surface (h_d associated with pressure gradient forces and wind and buoyancy forces, h_T associated with tidal forces and h_a associated with atmospheric pressure force) are about two orders of magnitude smaller than the dynamic range of h_g. Moreover, the spatial variability of h_g is more energetic than h_T, h_a, and h_d on all length scales (see, for example, Figure 8 of Wagner, 1979). It is therefore essential to remove h_g from h in order to investigate these smaller-amplitude signals from altimeter data.

The issue of the accuracy of the geopotential that is critically important to precision orbit determination as discussed in Section 4.1.2 is thus also fundamentally important to determination of the geoid. As summarized in Chapter 10, the global geoid undulations are estimated from terrestrial gravimetric data and by inference from ground-based tracking of the perturbations of satellite orbits. The satellite data provide information on the large-scale geoid undulations while the terrestrial gravity measurements provide information on short-scale variations of geoid undulations. The most accurate global geoid models that do not also utilize altimeter data are obtained by combining satellite tracking data and terrestrial gravity measurements. Because of the nonuniform geographical distribution of terrestrial gravity observations (see Figures 2 and 3 of Rapp and Pavlis, 1990), the accuracies of these geoid estimates vary geographically. The

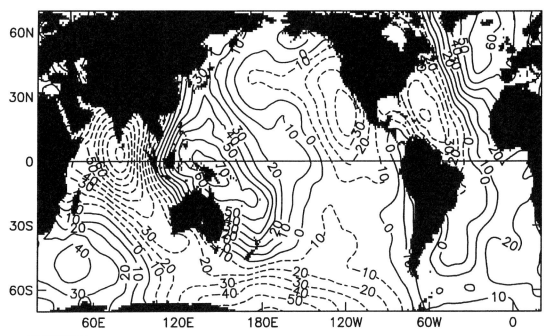

FIGURE 51   Geoid undulations about a reference ellipsoid in meters from the mean sea surface estimated by Bašić and Rapp (1992) that is included on the TOPEX/POSEIDON Geophysical Data Records (see also Rapp *et al.*, 1994). (Data provided courtesy of V. Zlotnicki.)

rms uncertainties of the most accurate global geoid models that are presently available are about 37 cm for the ocean areas of the earth's surface (see Figure 16 of Chapter 10). This accuracy will improve considerably with the planned launch of two gravity satellite missions in the near future (see Section 9).

5.2. Dynamic Sea-Surface Height

For altimetric studies of ocean circulation, the interest is in the dynamic sea-surface height h_d. Because the ocean circulation is very nearly in geostrophic balance on time scales longer than a few days and spatial scales longer than a few tens of kilometers, the surface velocity can be accurately determined from the horizontal gradient of h_d. Contours of h_d therefore correspond approximately to streamlines of the surface flow. The long-term mean dynamic height of the sea surface computed from a compilation of 100 years of ship observations of temperature and salinity profiles is shown in Figure 52. Although heavily smoothed (see, for example, Lozier *et al.*, 1994, 1995), a general sense of the magnitudes of h_d can be inferred from this figure. Over the entire world, the dynamic height values span a range of about 250 cm. The lowest dynamic heights are found near the Antarctic continent and the highest values are found in the western subtropical North Pacific.

The mean dynamic height field shown in Figure 52 was computed in the traditional manner by integrating the spe-

cific volume anomaly (which is directly related to the reciprocal of the water density) from a reference surface of 3000 db (see, for example, Gill, 1982). This dynamic height field is therefore not the absolute dynamic sea-surface height. The absolute sea-surface height can be determined with accurate knowledge of the horizontal velocity (or, equivalently through geostrophy, the horizontal pressure gradient) at the 3000-db reference level. The absolute dynamic sea-surface height could alternatively be determined from the dynamic height relative to 3000 db with adjustments for the absolute velocity or pressure gradient specified at any arbitrary level (e.g., the sea surface).

Since the launch of the first satellite altimeters in the 1970s, a long-term hope has been that altimetry will some day provide an absolute sea-surface height h_d. This could then be used as a reference surface for dynamic-height calculations from temperature and salinity profiles. In this manner, altimetry and hydrography could be combined to determine vertical profiles of the absolute velocity throughout the water column. Accurate determination of h_d from altimetry requires accurate independent estimates of the other contributions to h on the right side of Eq. (3). The most limiting factor in determination of h_d is uncertainty in the height h_g of the geoid undulations. As noted in Section 5.1, errors of h_g are estimated to be 37 cm. Uncertainties this large in estimates of h_d obtained by subtracting h_g from h render the data only marginally useful for studies of the absolute dynamic topography of the sea surface.

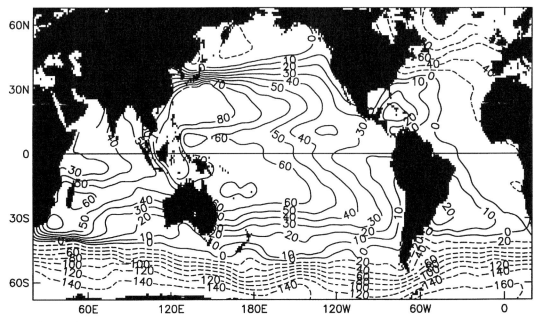

FIGURE 52   The mean dynamic height of the sea surface relative to 3000 db in centimeters computed from 100 years of ship observations of temperature and salinity profiles compiled on a 1° latitude by 1° longitude grid and spatially smoothed by Levitus and Boyer (1994) and Levitus *et al.* (1994). The dynamic height at each grid point was computed in the traditional manner as described by Gill (1982), for example. The global average dynamic height was then removed at each grid point for display purposes. The contour interval is 10 cm with negative contours shown as dashed lines.

Until accurate high-resolution estimates of the geoid undulations become available, studies of the mean ocean circulation will be restricted to only the large spatial scales for which h_g is known accurately. At present, the errors of h_g are larger than the signal levels of h_d for length scales shorter than degree and order 14 in an orthonormal expansion over the ocean (see Figure 2 of Chapter 2). This is equivalent to a wavelength of approximately 3000 km. Accurate estimates of the geoid undulations on the length scales of 100 km or shorter that are of interest for ocean circulation studies will require the launch of one or more gravity satellites. Two such gravity satellite missions are planned for the near future (see Section 9). As accurate gravity fields become available from these satellites, historical altimeter data can be analyzed retrospectively to study the mean ocean circulation.

The limitations imposed by the large uncertainties in the geoid undulations are addressed in oceanographic applications of altimeter data by considering the time-variable component of h_d. The earth's gravity field varies somewhat temporally from a variety of effects, including lunar and solar tides, atmospheric and oceanic mass redistribution, variations in ground water storage and snow and ice cover, earthquakes, post-glacial rebound in the earth's mantle and long-term variations of mantle convection. Since these variations of the gravity field are small, the geoid undulations can be considered essentially time invariant for oceanographic applications. Altimetric studies of the time variability of large-scale and mesoscale sea surface height are therefore not lim-

ited by inaccuracies in the global geoid. The geoid undulations are removed (along with all other time-invariant contributions to h, unfortunately including the time average of the dynamic sea-surface height h_d) by gridding the altimeter data along the satellite ground track and removing the time-averaged value of h at each grid location. The standard deviations of the residual sea surface height variations (after correcting for tidal variations and atmospheric pressure loading as described in Sections 5.3 and 5.4) are shown in Figure 53. Typical variations of h_d range from about 5 cm in the quiescent eastern ocean basins to more than 30 cm in the western boundary current extensions, the Agulhas Return Current and the Brazil-Malvinas Confluence.

5.3. Ocean Tides

To a very close degree of approximation, tides on the earth are controlled by the moon and the sun. While other heavenly bodies contribute tidal-generating forces, their relative strengths are very small by comparison. As the motions of the moon and the sun relative to the earth are known very precisely, it is possible to compute the tidal-generating potential to great accuracy at any point on the earth. Doodson (1922) decomposed the tidal-generating potential into 389 harmonic constituents. In most regions of the world oceans, the total tidal-generating potential can be closely approximated by only the six constituents with the largest amplitude, all of which are semidiurnal or diurnal (periods near

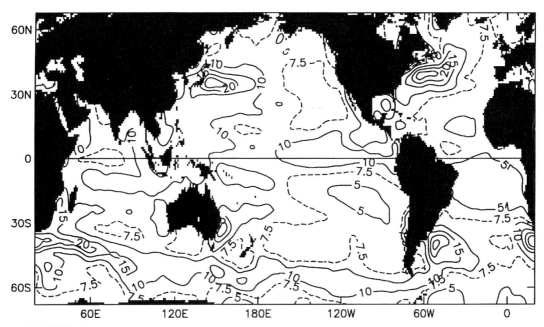

FIGURE 53 The standard deviation of the sea surface height in centimeters computed from 7 years of T/P data. The contour interval is 5 cm with an additional contour for 7.5 cm shown as a dashed line.

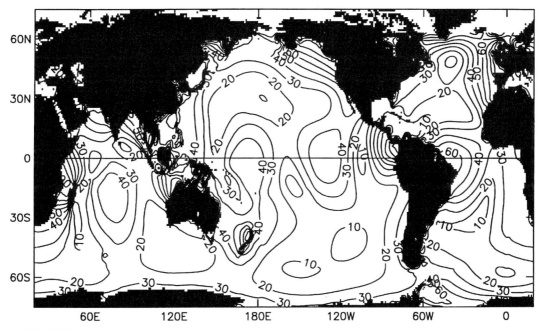

FIGURE 54 The standard deviation of tidal height variations in centimeters computed from the 8 dominant tidal constituents by the model described by Egbert (1997) that was derived from 116 cycles (3.2 years) of T/P data. (Data provided courtesy of G. Egbert.)

0.5 or 1 day). These periods are much shorter than the repeat period of an altimeter satellite orbit. The unresolved tidal variations therefore alias into the lower frequencies that are of interest for ocean circulation studies (see Section 8.1.5 and Chapter 6).

As shown in Figure 54, the standard deviations of tidal variations h_T in the open ocean are 10–60 cm with larger values near coastal regions and in marginal seas. Tidal variations are thus larger in magnitude than the 5–30 cm standard deviations of the dynamic sea surface height h_d (Figure 53) that is of interest to altimetric studies of large-scale and mesoscale ocean circulation. In this respect, ocean tides can be regarded as noise and must therefore be removed to estimate h_d. It is imperative that the tide model used to correct the altimeter data be accurate. If errors in the tide model are large compared with oceanographic variability at the alias frequencies, then oceanographic applications will be compromised at these frequencies.

Prior to the launch of T/P, knowledge of tides over the global ocean was based primarily on hydrodynamical models. The accuracies of these models were limited by uncertainties in the parameterization of friction. The models were therefore constrained by empirically determined ocean tides from a worldwide network of coastal and island tide gauges and bottom pressure gauges. Because these in situ observations of tides were not uniformly distributed globally, the quality of the global tide models varied geographically in ways that were difficult to quantify. Tide errors exceeded 10 cm in many areas of the world ocean. One of the important accomplishments of the T/P mission has been the global estimation of tides to an accuracy of 2–3 cm (see Chapter 6). As summarized below in Section 8.1.5, this was achieved by carefully selecting the satellite orbit configuration with the specific objective of aliasing the energetic tidal constituents to periods that can be well resolved from an altimeter data record shorter than a few years and are easily distinguished from other narrow-band ocean signals such as the annual and semiannual cycles. The altimeter data at the aliased tidal periods have been analyzed by numerous techniques to extract estimates of tidal amplitudes and phases at the T/P observation locations. These empirical estimates have been assimilated into sophisticated hydrodynamical and statistical models to estimate the tides globally with high spatial resolution (see Chapter 6).

A noteworthy distinction between altimeter and tide gauge data is that altimeter data include the geocentric tide (solid earth tide plus ocean tide), whereas tide gauges include only the ocean tide component since the gauges move with the solid earth. Care must therefore be taken when comparing altimeter and tide gauge data. Both the solid earth and ocean tides must be removed from altimeter data for studies of the dynamic sea surface height variations associated with large-scale and mesoscale ocean circulation. Solid earth tides include the effects of direct astronomical forcing (the body tide) and the effects of crustal loading by ocean tides (see Chapter 6).

5.4. Atmospheric Pressure Loading

Atmospheric pressure exerts a downward force on the sea surface. Spatial and temporal variations of this force are compensated at least partially by variations of the sea surface elevation. These variations of sea surface elevation are unrelated to sea surface topographic features associated with geostrophic currents and therefore must be removed to obtain accurate estimates of the dynamic sea surface height h_d. The effects of atmospheric pressure loading have been comprehensively reviewed by Wunsch and Stammer (1997). A brief overview is given here.

The simplest model for the effects of atmospheric pressure loading is the hydrostatic equation

$$\frac{dp}{dz} = \rho_w g, \qquad (118)$$

where p is pressure, z is depth, ρ_w is the water density and g is the magnitude of the gravitational acceleration. Defining $z = 0$ to be the mean free sea surface in the absence of pressure forcing, Eq. (118) can be integrated from a depth z_0 to the actual sea surface height h_a where the atmospheric pressure is p_a to obtain the total pressure at z_0,

$$p(z_0) = p_a + \int_{z_0}^{0} \rho_w g \, dz + \int_{0}^{h_a} \rho_w g \, dz. \qquad (119)$$

Because the compressibility of water is exceedingly small, any variations of $p(z_0)$ associated with changes in atmospheric pressure result from a horizontal redistribution of water mass in response to horizontal variations of atmospheric pressure. If atmospheric pressure changed uniformly over the entire ocean, there would be no change of $p(z_0)$. As noted by Ponte *et al.* (1991), the time-varying average atmospheric pressure over the global ocean should therefore be removed to estimate the sea surface response to atmospheric pressure loading (see also the discussions by Wunsch and Stammer, 1997, Dorandeu and Le Traon, 1999, and others). The atmospheric pressure can thus be decomposed as

$$p_a(x, y, t) = \overline{p}_a(t) + \delta p_a(x, y, t) \qquad (120)$$

where

$$\overline{p}_a(t) = \iint_{oceans} p_a(x, y, t) \, dx \, dy \qquad (121)$$

is the instantaneous average atmospheric pressure over the global ocean and δp_a is the local departure from this global average. Then Eq. (119) can be rewritten as

$$\delta p(z_0) = \delta p_a + \int_{z_0}^{0} \rho_w g \, dz + \int_{0}^{h_a} \rho_w g \, dz, \qquad (122)$$

where $\delta p(z_0) = p(z_0) - \overline{p}_a$ is the perturbation pressure associated with variations δp_a of the local atmospheric pressure.

The middle term on the right side of Eq. (122) is essentially time invariant, except for the very small variations of water density ρ_w associated with temperature and salinity variations. If the ocean response to variations of atmospheric pressure is isostatic (i.e., there is no net pressure change at depth associated with atmospheric pressure changes) then $\delta p(z_0)$ on the left side of Eq. (122) also vanishes. In this case, the first and last terms on the right side of Eq. (122) balance so that the isostatic response to atmospheric pressure variations is

$$\delta p_a = - \int_{0}^{h_a} \rho_w g \, dz \approx -\rho_w g h_a. \qquad (123)$$

The approximation in Eq. (123) comes from the fact that ρ_w and g are very nearly constant over the shallow depth range h_a near the sea surface. For atmospheric pressure p_a in mbar, ρ_w in g cm^{-3}, and g in cm sec^{-2}, the isostatic response of the sea surface in centimeters is

$$h_a = -\frac{\delta p_a}{\rho_w g}. \qquad (124)$$

Using characteristic sea-surface values of $\rho_w = 1.025$ g cm^{-3} and $g = 980.6$ cm sec^{-2}, this so-called "inverted-barometer response" is -0.995 cm mbar^{-1}.

Typical magnitudes of the inverted barometer response Eq. (124) can be inferred from the standard deviations of sea level pressure in Figure 24. In the tropics, h_a is only about 2 cm. Over the subpolar gyres of the northern hemisphere, h_a ranges from 6–10 cm in the boreal summertime to 10–16 cm in the boreal wintertime. Over the Southern Ocean, h_a is 8–10 cm in the austral summertime and 10–14 cm in the austral wintertime. To the extent that a static response to atmospheric pressure loading is valid, the magnitudes of the sea surface height variations h_a in middle and high latitudes are comparable to the standard deviations of the dynamic surface height h_d shown in Figure 53. It is therefore essential that corrections be applied for altimetric studies of ocean circulation.

In reality, the sea surface response to atmospheric pressure loading depends both statically and dynamically on the spatial and temporal scales of the forcing. Wunsch and Stammer (1997) showed that an inverted barometer response is anticipated at all frequencies and wavenumbers, except for resonances near the dispersion curves of Rossby waves and internal gravity waves. A complete description of the resonant dynamic response computed for realistic ocean basin geometry is not available. The modeling studies of Ponte *et al.* (1991), Ponte (1992; 1993) and Ponte and Gaspar (1999) suggest that the resonant response at high frequencies (periods shorter than about 10 days) is concentrated in the Southern Ocean and in the tropics. These dynamic responses to high-frequency pressure variations are summarized in Chapter 2. The modeling predictions have been confirmed from empirical analyses of altimeter data. Fu and Pihos (1994) and Gaspar and Ponte (1997; 1998) found that the inverted barometer response is generally accurate at middle and high latitudes on time-scales longer than a few days. In the tropics, however, the sea-surface response to atmospheric pressure loading was found to be systematically lower than the 1 cm mbar^{-1} inverted barometer response.

Lacking a well-established procedure for applying a correction for the dynamic sea-surface response to atmospheric pressure loading, the first approximation is to remove the static inverted barometer response Eq. (124). After correcting for the tidal signals summarized in Section 5.3, Wunsch and Stammer (1997) showed that this static correction sig-

nificantly reduces the global rms sea level variability from 11.8 to 9.7 cm. Based on the results of the modeling study by Ponte and Gaspar (1999), the errors in this inverted barometer correction owing to dynamic response to atmospheric pressure forcing are probably on the order of 10% of the correction.

The accuracy of the inverted-barometer correction is also limited by uncertainty in the actual sea surface atmospheric pressure. The only sources of atmospheric pressure estimates that are available globally for corrections of altimeter data are the analyzed sea-level pressure fields from operational numerical weather prediction models. As noted in Section 3.1.1 (see Figure 25), the uncertainties of these sea-level pressures may be as large as 2–3 mbar in the tropics and subtropics, 4 mbar north of 40°N and 4–7 mbar south of 40°S. The corresponding uncertainties of the inverted barometer correction of about 1 cm mbar^{-1} are thus a major source of error in altimetric estimates of dynamic sea-surface height.

5.5. Aliased Barotropic Motion

In addition to the tidal variations discussed in Section 5.3, there are other high-frequency ocean signals that are not resolved in altimeter observations. The nature of sea-level variability over a wide range of the frequency-wavenumber spectrum was investigated by Fukumori *et al.* (1998) from an ocean general circulation model forced by realistic daily winds and climatological heat fluxes. They focused on analysis of variability with periods shorter than a year. Over a large part of the ocean at middle and high latitudes, half of the energy in this frequency band consisted of barotropic variability with periods shorter than 20 days. This short-period variability cannot be resolved by the 10-day repeat period of the T/P orbit. The amplitude of this high-frequency variability was typically a few centimeters but exceeded 5 cm in some regions of the Southern Ocean (see Plate 2b of Fukumori *et al.*, 1998).

If the altimeter samples at grid points along the satellite ground track are considered pointwise, classical considerations of aliasing conclude that periods shorter than 20 days are aliased in the time series constructed at each grid point for the T/P 10-day orbit repeat pattern. Because of the complex space-time sampling of an altimeter (see Section 8.1.4), the traditional concept of aliasing and an associated Nyquist sampling period are not strictly applicable to spatially and temporally smoothed and interpolated sea-surface height fields constructed from altimeter data. It is nonetheless clear that energetic variability at periods shorter than 20 days could significantly contaminate altimeter observations of sea-surface height variations. To the extent that the barotropic variability has large spatial scales, the detrimental effects of this high-frequency variability can be mitigated to some degree by sufficient space-time smoothing of the altimeter data that capitalizes on the 3-day subcycles in the sampling patterns (see Section 8.1.4). For the T/P orbit, for example, neighboring ground tracks separated by about 2.8° of longitude in the 10-day repeat orbit are sampled at 3-day time intervals. The contamination of raw T/P observations can thus be reduced by smoothing over spatial scales of a few degrees or more in latitude and longitude (see Section 8.1.7).

Fukumori *et al.* (1998) suggested that models could be used to "correct" the T/P data by removing the simulated high-frequency variability in a manner similar to the tidal and inverted barometer corrections that are routinely applied to altimeter data. The possibility of model-based corrections for high-frequency barotropic variability is an area of active research (see Chapter 2). Analyses of the sea-surface height variations associated with fast barotropic motions in other models finds results similar to those reported by Fukumori *et al.* (1998). When model output is subsampled along T/P ground tracks to simulate altimeter data and then averaged over small areas of about 2° (e.g., Stammer *et al.*, 2000), the aliased barotropic variability shows up as "trackiness" in sea-surface height fields, similar to the trackiness from uncorrected tidal errors and orbit errors. Stammer *et al.* (2000) report a correlation of 0.63 between their model simulation and T/P observations of sea surface height variability. They showed that the variance of T/P data was reduced by 10–20% over large areas of the high-latitude oceans in both hemispheres after removing the model simulation of high-frequency sea level variability. Tierney *et al.* (2000) report similarly encouraging results from a more sophisticated ocean general circulation model.

It should be noted that the skill of the model used to remove the barotropic contribution to sea-surface height variations in altimeter data is dependent upon the quality of the wind and pressure fields that force the model. These models are forced by surface wind fields from operational weather analyses. Quantitative estimates of the errors of these wind and pressure fields are not presently available. The quality of the model simulations is also sensitive to the model representation of the bottom topography and basin geometry. Although the recent analyses of the relationship between barotropic model simulations and T/P observations are encouraging, further studies are needed to understand the quality of model-based corrections for the barotropic component of sea surface height variability. Notwithstanding the fact that model-based corrections for high-frequency barotropic variability reduce the variance of T/P estimates of the sea surface height field, it should be borne in mind that the present understanding of these model-based corrections is not yet sufficiently mature to implement reliable corrections for barotropic variability in routine altimeter data processing.

6. SIGNIFICANT WAVE HEIGHT ESTIMATION

The presence of waves on the sea surface alters the shape of the returned waveform measured by an altimeter as described in Section 2.4.2. The leading edge of the returned waveform is stretched as a result of earlier returns from wave crests and later returns from wave troughs. This stretching increases with increasing wave height (see Figures 15, 17, and 21). Wave conditions can therefore be estimated from the slope of the leading edge of the returned waveform. The slope decreases approximately linearly with increasing wave height. The algorithms for altimeter estimates of wave height are described in this section and comparisons between altimeter and buoy estimates of wave height are briefly summarized. More extensive overviews of validation studies and applications of altimeter data to investigate geographical and temporal variability of the global ocean wave field and to improve wave forecasts are presented in Chapter 7.

Ocean wave conditions are generally characterized by the significant wave height, originally defined by Sverdrup and Munk (1947) and Bretschneider (1958) to be the crest-to-trough height of the 1/3 highest waves in the field of view and therefore denoted as $H_{1/3}$. There is a surprising amount of confusion in the literature about the relationship between $H_{1/3}$ and statistical characterization of the wave field in terms of the standard deviation σ_ζ of the sea surface elevation ζ the latter of which is equal to half of the wave height. This confusion is perhaps best illustrated by the different mathematical relationships reported in successive editions of *The Dynamics of the Upper Ocean*, by O. M. Phillips. On page 188 of Phillips (1977), the relationship is said to be $H_{1/3} = 3.13\,\sigma_\zeta$. On the same page of Phillips (1980), the relationship is said to be $H_{1/3} = 4.0\,\sigma_\zeta$. Both editions attribute these relations to Longuet-Higgins (1952), who gives the altogether different relationship $H_{1/3} = 2.83\,\sigma_\zeta$ for the case of a narrow-banded wave spectrum. For a similar characterization of the wave field, Neumann and Pierson (1966) give the relationship $H_{1/3} \approx 4\,\sigma_\zeta$ and report that this agrees well with observations. The reason for the confusion is that the ratio $H_{1/3}/\sigma_\zeta$ depends on the width of the wave spectrum. Cartwright and Longuet-Higgins (1956) examined the ratio as a function of the spectral bandwidth of the waves and obtained values ranging from about 4 for a very narrow-banded wave spectrum to about 3 for a very broad-banded wave spectrum (see Figure 4 of Cartwright and Longuet-Higgins, 1956). For a wide range of bandwidths, a ratio of $H_{1/3}/\sigma_\zeta$ near 4 is a reasonably good approximation.

6.1. Significant Wave-Height Algorithms

As described briefly by Townsend (1980) and in greater detail by MacArthur (1978), $H_{1/3}$ was estimated onboard the

Seasat and Geosat satellites as part of the sea-level tracking algorithm by comparing averages of the early and late portions of the leading edge of the returned waveform. Analogous to the description in Section 2.4.4 for the T/P dual-frequency altimeter, the widths of these gated averages were selected by the adaptive tracking unit according to the rise time of the leading edge of the waveform. In order to minimize the effects of measurement noise and maximize the sensitivity to $H_{1/3}$, the gate widths are increased for the longer rise time associated with larger waves. For the Seasat and Geosat altimeters, $H_{1/3}$ was determined onboard the satellite from the difference between the selected pair of late and early gates based on comparisons with an onboard look-up table. There was a total of 80 discrete entries spanning a range of $H_{1/3}$ from 0–19.92 m. The increments between tabulated values of $H_{1/3}$ were nearly constant percentage-wise, never exceeding 8% of $H_{1/3}$. Small wave heights were therefore finely resolved, with half of the table entries for $H_{1/3}$ between 0 and 5.0 m. Large wave heights were less well resolved, with only 1/4 of the table entries for $H_{1/3}$ greater than 7.0 m.

As with the range measurements, it is necessary to correct the estimates of $H_{1/3}$ in ground-based post processing to compensate for deficiencies in the idealized returned waveform used in the onboard tracking algorithm as described in Section 2.4.4. The Geosat onboard estimates of $H_{1/3}$ that were telemetered to ground receiving stations were corrected with a third-order polynomial function of antenna pointing angle and $H_{1/3}$. The coefficients of this pointing-angle and sea-state correction algorithm were based upon pre-launch modeling and simulation studies. When actual Geosat returned waveforms become available after launch, a waveform processing deconvolution procedure (see Section 2.4.5) was used to examine the accuracy of the pre-launch coefficients. The estimates of $H_{1/3}$ were found to be systematically smaller than the waveform processing estimates by about 10%. Based on the results of the waveform processing, Hayne and Hancock (1990) derived a second-level correction expressed in exactly the same form as the third-order polynomial function of antenna pointing angle and $H_{1/3}$ used in the original pre-launch algorithm.

An improved method of wave height estimation was implemented for T/P. Rather than estimating the wave height onboard the satellite, $H_{1/3}$ was estimated in ground-based processing. The quantities $V_{swh}(m)$ defined by Eq. (55) for each of five gate indices m are used by the adaptive tracking unit as described in Section 2.4.4 to select the best gate index for onboard estimates of the two-way travel time to nadir mean sea level. The value of $V_{swh}(m)$ for the selected gate index m provides a measure of the slope of the leading edge of the returned waveform. This selected value of V_{swh} is telemetered to ground-receiving stations for estimation of $H_{1/3}$ by an empirical algorithm derived from prelaunch testing and simulations. The use of ground-based processing al-

lows finer resolution of $H_{1/3}$ than the onboard look-up tables used to compute $H_{1/3}$ on the Seasat and Geosat altimeters. As described in Section 2.4.6, the value of $H_{1/3}$ estimated by the ground-based algorithm is corrected for pointing-angle and sea-state effects with a cubic polynomial in V_{swh} and the quantity V_{att} defined by Eq. (62). The corrected values of $H_{1/3}$ agree with values estimated by waveform processing to within about 0.1 m (Hayne et al., 1994).

6.2. Significant Wave Height Measurement Accuracy

The overall accuracy of altimeter measurements of $H_{1/3}$ has been investigated by numerous comparisons with buoy observations. Seasat and Geosat studies indicated that the altimeter measurements were systematically low by somewhere between a few percent and 10% (Webb, 1981; Fedor and Brown, 1982; Shuhy et al., 1987; Dobson et al., 1987). It is noteworthy, however, that systematic errors in the buoy data may also be a problem in these studies because of the fact that buoys tend to be partly submerged at wave crests, possibly leading to systematic underestimates of $H_{1/3}$ in the buoy data as well. The altimeter measurements of $H_{1/3}$ could therefore be systematically low by an even greater amount than the buoy comparisons suggest. Since the accuracies of $H_{1/3}$ estimates are sensitive to antenna mispointing, the large attitude variations of the Geosat altimeter owing to the lack of an active attitude control system on the satellite may have contributed to the above-noted large 10% error in $H_{1/3}$ estimated from the Geosat data.

The bias in altimeter estimates of $H_{1/3}$ has been found to vary somewhat from mission to mission. Careful calibration of altimeter estimates against in-situ measurements is thus very important for accurate global estimation of $H_{1/3}$. A long-standing difficulty in validation of altimeter estimates of $H_{1/3}$ is acquisition of collocated buoy and altimeter observations over a wide range of wave conditions. This is especially problematic for large $H_{1/3}$ since large wave conditions occur infrequently at any particular location. The probability of simultaneous observations of large $H_{1/3}$ by a buoy and an altimeter is therefore very small. In addition, as discussed by Monaldo (1988), buoy and altimeter estimates of $H_{1/3}$ can differ because of differences in the spatial and temporal scales that characterize the measurements. Altimeter measurements are essentially an instantaneous spatial average of $H_{1/3}$ over the altimeter footprint area, which increases from about 3 km for small wave conditions to about 10 km for large wave conditions (see Figure 22). Buoy measurements are time-averaged measurements of $H_{1/3}$ at a point location.

To avoid problems with incompatibility between spatially averaged altimeter measurements and temporally averaged buoy measurements at a point location, Cotton and Carter (1994) compared monthly averages of the buoy time series

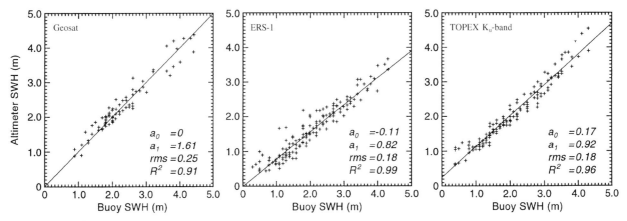

FIGURE 55   Scatter plot comparisons between monthly average estimates of significant wave height from buoys and Geosat, ERS-1 and the K_u-band frequency of the T/P dual-frequency altimeter. (After Cotton and Carter, 1994. With permission.)

with monthly averages of altimeter data over $2° \times 2°$ areas surrounding the buoys. The results for Geosat, ERS-1 and the T/P dual-frequency altimeter are shown in Figure 55. The Geosat data were corrected for a 13% bias (low relative to the buoy data) documented in an earlier study by Carter *et al.* (1992) and consistent with other early studies as noted above. The buoy data used for the comparisons with T/P and ERS-1 were from 24 National Data Buoy Center (NDBC) buoys; for an independent assessment, the buoy data used for the Geosat comparison were restricted to 6 NDBC buoys that had not been used in the earlier Geosat calibration study by Carter *et al.* (1992).

It is evident from the figure that the relationships between monthly averaged buoy and altimeter measurements of $H_{1/3}$ are well described by straight-line least-squares fits through the data. Although the y-axis intercepts and slopes of the least-squares fits of the data in Figure 55 differ somewhat for each altimeter, the rms errors of the various straight-line fits are 0.14–0.18 m, which can be considered an estimate of the measurement uncertainty of monthly averages after a calibration against buoy data is performed. Estimates of $H_{1/3}$ from the ERS-1 altimeter and the T/P dual-frequency altimeter suggest a possible overestimation at low $H_{1/3}$. This "flattening" of the scatter plot at small wave heights is characteristic of the expected relationship between two measurements of a non-negative quantity such as $H_{1/3}$ when there are random errors in both measurements (Freilich, 1997; Freilich and Vanhoff, 2000). The flattening is therefore not necessarily indicative of systematic errors in T/P estimates of small wave heights.

A limitation of the comparisons in Figure 55 is that the dynamic range of $H_{1/3}$ is restricted to about 5 m by the monthly averaging. Gower (1996) compared pointwise buoy measurements of $H_{1/3}$ off the west coast of Canada with measurements by the T/P dual-frequency altimeter during the first 15 months of the T/P mission. This pointwise validation study is unusual in the wide range of wave conditions

that was observed over the duration of the data record. For three open-ocean buoys and six nearshore buoys with good exposure to the open ocean, wave heights generally ranged from 1–8 m. The buoy and T/P measurements were in very good agreement over this full range of wave heights. In one collocated measurement during extreme wave conditions, the buoy and T/P estimates of $H_{1/3}$ were 12.8 and 12.5 m, respectively. For all of the buoys, the slopes of straight-line fits through the scatter plots were slightly less than one. On average, the T/P estimates were found to be smaller than the buoy estimates by about 5%.

It should be noted that a slope of less than one for a least-squares fit straight line is characteristic of measurement errors in the independent variable (in this case, the buoy estimates of $H_{1/3}$), as well as in the dependent variable (the T/P estimates of $H_{1/3}$). The scatter plots presented by Gower (1996) are therefore not necessarily indicative of a systematic error in the altimeter data. The methods developed by Freilich (1997) and Freilich and Vanhoff (2000) should be applied to take into account the measurement errors in both the independent and dependent variables. Despite the fact that buoy measurement errors were not taken into consideration, the results obtained by Gower (1996) are very encouraging. The overall rms difference between the buoy and T/P estimates of $H_{1/3}$ was 0.3 m. This rms difference decreased to 0.15 m for buoys located within 10 km of a T/P ground track.

The relative bias and rms difference between T/P and buoy estimates of $H_{1/3}$ has been investigated from a number of other studies in various regions of the world ocean. The statistics of the comparisons vary somewhat from one study to another, which is likely an indication of differences in the calibrations of the buoys rather than geographical variations in the quality of the T/P data. As another example of pointwise comparisons, Ebuchi and Kawamura (1994) compared T/P estimates of $H_{1/3}$ with estimates from three buoys off the coast of Japan during the first 10 months of the T/P mission. The significant wave height ranged from 0 to about 4 m over the duration of the data record. The T/P estimates were found to be biased about 0.3 m high relative to the buoy estimates and the rms difference was about 0.45 m. Other $H_{1/3}$ validation studies are summarized in Chapter 7. The rms differences typically fall in the range 0.25–0.45 m, which can be considered an upper bound on the uncertainty of the T/P estimates since some of this rms difference can be attributed to errors in the buoy data. If the measurement errors are equally partitioned between the buoys and the altimeter, the rms accuracy of the T/P estimates of $H_{1/3}$ is 0.2–0.3 m.

7. WIND-SPEED ESTIMATION

In addition to measuring the shape of the returned pulse for estimating the range and $H_{1/3}$, the altimeter also measures the power of the returned signal, which is related to the wind-induced roughness of the sea surface. The power measurement is not direct, however. In order to operate the altimeter electronics within the linear response region of all receiver stages, an automatic gain control (AGC) loop is implemented in the electronics package (Townsend, 1980). The AGC determines the attenuation that must be applied to the returned signal power to keep constant the total power of the signal processed by the altimeter. From the radar Eq. (5), the power of the returned signal is directly related to the normalized radar cross section σ_0 of the sea surface. In altimetry, σ_0 is computed from the AGC value with corrections for losses from variations in satellite altitude and attitude.

In accord with the radar Eq. (6), σ_0 must be corrected for the effects of atmospheric attenuation. As summarized in Sections 2.3.1 and 2.3.2, the two-way attenuation by a non-raining atmosphere at the T/P primary frequency of 13.6 GHz is seldom greater than a few percent at nadir, even for high water-vapor content or thick, dense clouds. Although small, errors of this magnitude must be corrected for accurate altimeter estimates of wind speed. This is especially true for high wind-speed conditions (see below). Observations contaminated by rainfall (estimated from satellite observations as described in Section 2.3.3) exceeding a threshold of a few mm hr^{-1} are flagged and excluded from further analysis.

7.1. Wind-Speed Model Functions

The physics of the relation between wind speed and altimeter measurements of σ_0 are different from the Bragg scattering mechanism responsible for the backscattered power at the 20–60° incidence angles measured by a scatterometer (Naderi et al., 1991). At the small incidence angles relevant to altimetry (less than 1° from satellite nadir), the backscattered power results almost totally from spec-

ular reflection from favorably oriented facets with wavelengths longer than about three times that of the incident radiation (Brown, 1990). As the wind speed increases, the sea-surface roughness increases and a greater fraction of the incident radiation is reflected away from the satellite. Altimeter measurements of σ_0 are therefore inversely related to wind speed. There is no significant dependence of σ_0 on wind direction at small incidence angles so only the wind speed can be inferred from altimetry.

Historically, altimeter wind-speed model functions relate σ_0 to the neutral-stability wind speed at a height of 10 m above the sea surface. Although a theory-based model function for altimeter estimates of wind speed has recently been developed by Elfouhaily et al. (1998) (see discussion below), the model functions used most widely to date have been purely empirical. The wind-speed model function initially adopted for Seasat consisted of a three-branch logarithmic model function with a fifth-order polynomial correction (Brown et al., 1981). The complicated form of this model function arose as a result of extensive numerical experimentation. The Brown model function was actually derived from least-squares analysis of comparatively noisy GEOS-3 altimeter measurements of σ_0 and 184 coincident buoy observations. Based on only 19 nearly coincident Seasat and GEOS-3 observations of σ_0, a bias adjustment of 1.6 dB was incorporated to apply the Brown model function to Seasat altimeter measurements of σ_0.

A thorough analysis by Chelton and McCabe (1985) of the statistical properties of wind speeds computed from Seasat altimeter measurements of σ_0 using the Brown model function identified some peculiar features of histograms of σ_0 and wind speed. These were eventually traced to two sources. The most fundamental problem was an error in the Seasat σ_0 algorithm which resulted in a discontinuous dependence of σ_0 on AGC. Even after correcting the σ_0 algorithm, the discontinuous derivatives at the two branch points in the three-branch Brown model function lead to undesirable properties in the distribution of wind-speed estimates. Chelton and McCabe (1985) proposed a much simpler form for the wind speed model function which consisted of the same power-law relation used in scatterometry,

$$\sigma_0(dB) = 10[A + B \log_{10} U], \qquad (125)$$

where U is the wind speed and the parameters A and B were determined empirically from comparisons of σ_0 with wind-speed observations. At the small incidence angles relevant to altimetry, the parameter B is negative (i.e., σ_0 decreases with increasing wind speed). At very low wind speeds, the sea surface is highly reflective and the relationship between nadir measurements of σ_0 and wind speed is not well defined (see Figure 56). It is also evident from Figure 56 that, with increasing wind speed, altimeter estimates of wind speed become increasingly sensitive to errors in σ_0 because of the flattening of the wind-speed model function at high wind

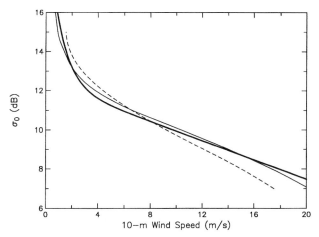

FIGURE 56   A comparison of empirical model functions for altimeter estimates of wind speed at a height of 10 m above the sea surface. The three curves correspond to the Smoothed Brown (SB) model function (dashed line), the Modified Chelton and Wentz (MCW) model function (thin solid line) and the Freilich and Challenor (FC) geometric model function (heavy solid line). Note that the SB model function was derived only for the 7–15 dB σ_0 range shown for the dashed line.

speeds. The importance of correcting σ_0 for atmospheric attenuation in high wind speed conditions is thus apparent. If not accounted for, an attenuation of 0.5 dB results in more than a 30% overestimate of a 10 m sec^{-1} wind speed.

An alternative method of eliminating the discontinuous derivatives in the Brown model function was proposed by Goldhirsh and Dobson (1985) and Dobson et al. (1987). They smoothed the Brown model function by a least-squares fit to a fifth-order polynomial over the range 7 dB $\leq \sigma_0 \leq$ 15 dB. This so-called Smoothed Brown (SB) polynomial model function has the undesirable characteristics of double valued solutions outside of the region of fit. Also undesirable is the fact that the SB model function was developed for application to Geosat altimeter data but is based on coincident buoy observations of wind speed and measurements of σ_0 by the comparatively noisy past-generation GEOS-3 altimeter with the 1.6-dB bias adjustment based on the previously described validation against a small number of Seasat observations of σ_0. No attempt was made by Goldhirsh and Dobson (1985) or Dobson et al. (1987) to estimate calibration differences between the Seasat and Geosat measurements of σ_0.

Over the range of wind speeds well sampled by the 184 buoy observations originally used to derive the Brown model function, the Chelton and McCabe (1985) (CM) and SB model functions fit the data equally well. For wind speeds greater than about 8 m sec^{-1}, however, the two model functions diverge dramatically. Assessment of the relative accuracies of the CM and SB model functions requires an extensive set of coincident buoy observations at the high wind speeds for which the two model functions differ the most.

Collection of an adequate in situ database from which to develop an accurate empirical wind speed model function

will always be a problem. This is especially true for altimetry which measures σ_0 for only a small area (a footprint diameter less than 10 km, see Section 2.4.7) at satellite nadir. Chelton and Wentz (1986) utilized the simultaneous measurements of σ_0 by the Seasat altimeter and the Seasat-A Satellite Scatterometer (SASS) to obtain a greatly expanded calibration dataset for deriving a wind speed model function (denoted as the CW model function hereafter) for the Seasat altimeter. The primary interest for scatterometry is in measurements of σ_0 at incidence angles greater than about 20° where both wind speed and direction can be inferred from the radar backscatter. The CW model function was derived in tabular form by comparison of altimeter measurements of σ_0 with 241,000 SASS observations of wind speed at 24° incidence angle (the nearest off-nadir incidence angle at which estimates of vector winds were derived from the SASS measurements of σ_0). Although the 200-km spatial separation between the points on the sea surface at nadir and 24° incidence angle is a concern (see Monaldo, 1988), Chelton and Wentz (1986) showed that spatial variability in the wind field averages to zero in the large number of comparison observations considered. The clear advantage of a SASS-based altimeter wind-speed model function is the much larger calibration database spanning a broader range of wind speeds than could ever practically be obtained from buoy observations over the lifetime of a single altimeter mission. The SASS-2 wind speeds used to derive the CW model function were extensively compared with buoy observations and found to have an rms accuracy of better than 1.6 m sec^{-1} with negligible bias over the range from 0 to about 17 m sec^{-1} (Wentz et al., 1986).

The tabular CW model function is almost identical to the earlier CM power-law model function for wind speeds less than about 14 m sec^{-1}. For higher wind speeds, the CM power-law relation overestimates the wind speed. Chelton and Wentz (1986) cautioned, however, that the tabular CW model function is only appropriate for the Seasat altimeter. Although of no interest for scatterometry because of the lack of sensitivity to wind direction, σ_0 was measured by SASS at 0° incidence angle as well as the 20–60° range of off-nadir incidence angles where σ_0 is sensitive to both wind speed and direction. From a cross-calibration of σ_0 measured by the Seasat altimeter and the nadir SASS measurements of σ_0, Chelton and Wentz (1986) presented evidence for two systematic errors in Seasat altimeter estimates of σ_0. The first of these consisted of an overall negative bias of about −0.5 dB that was later shown by Chelton et al. (1989) to be attributable to the use of a flat-earth approximation of the altimeter footprint area used to compute σ_0 from the returned power measured by the altimeter. The second apparent error in the Seasat altimeter measurements of σ_0 consisted of an approximately linearly increasing bias with decreasing σ_0 (i.e., increasing wind speed) for σ_0 less than about 10.5 dB. Chelton and Wentz (1986) suggested that at least part of this error might be the result of the tracker bias known to exist in the Seasat altimeter onboard sea level tracking algorithm (see Section 3.2.1). However, they estimated that such a tracking error would account for less than half of the observed bias. The reason for the apparent linearly increasing bias at small σ_0 remains unexplained.

Dobson *et al.* (1987) criticized the CW model function because it was developed without the use of any direct comparisons with in situ wind observations. They applied the polynomial SB and the tabular CW model functions to estimate wind speeds from the Geosat altimeter. Based on a comparison with 119 buoy observations, they concluded that the SB model function marginally outperformed the CW model function. However, their database included very few buoy observations of wind speed greater than 10 m sec^{-1} and they made no attempt to correct for possible differences in σ_0 calibration between the Geosat altimeter and the Seasat altimeter for which the CW model function was derived. Furthermore, there is clear evidence in their comparisons that the SB model function systematically underestimates wind speeds greater than 6 m sec^{-1} and overestimates wind speeds lower than 6 m sec^{-1} (see their Figure 8), despite the fact that the overall rms agreement with the buoy observations was slightly better than for the CW model function.

The question of calibration differences between Seasat and Geosat estimates of σ_0 was investigated by Witter and Chelton (1991a). Comparisons of histograms of globally distributed observations of σ_0 from the two altimeters for the July to October season suggested errors in the Seasat estimates of σ_0 that were consistent with the earlier calibration errors surmised by Chelton and Wentz (1986). Differences between the Seasat and Geosat σ_0 histograms were much greater than the differences between σ_0 histograms from two separate years of Geosat data alone, ruling out interannual variability in the surface wind field as an explanation for the discrepancy. The largest discrepancies between the Seasat and Geosat data were for σ_0 less than about 11 dB. Because of the calibration differences between Seasat and Geosat estimates of σ_0, the CW wind-speed model function derived for the Seasat altimeter could not be directly applied to Geosat data.

When the SASS nadir estimates of σ_0 are assumed correct and the Seasat altimeter estimates of σ_0 are adjusted by the calibration difference deduced by Chelton and Wentz (1986), the histogram of adjusted Seasat σ_0 is very similar to histograms computed from each year of Geosat data. Witter and Chelton (1991a) showed that essentially the same calibration adjustment could be deduced from the σ_0 histograms alone, without the use of SASS data. The histogram-matching method can therefore be used to cross-calibrate σ_0 measurements from any future altimeter. A wind-speed model function derived for a particular altimeter can then be applied to later altimeters with an appropriate σ_0 calibration adjustment.

The Witter and Chelton (1991a) analysis did not prove that the Seasat altimeter estimates of σ_0 were in error. However, the fact that SASS and Geosat estimates of σ_0 were consistent and that the Seasat altimeter estimates differed in a very similar manner from both SASS and Geosat is highly suggestive of a calibration error in the Seasat altimeter data. Witter and Chelton (1991a) therefore modified the tabular CW wind speed model function to correct for the apparent calibration errors in the Seasat altimeter estimates of σ_0. The resulting modified CW (MCW) model function (see Figure 56) should then be applicable to the Geosat data. The MCW model function was shown to be more accurate than the CW model function in a comparison with the same 119 buoy observations used by Dobson *et al.* (1987).

The MCW wind speed model function was adopted for the processing of T/P measurements of σ_0. Since the MCW model function had been tuned to estimate wind speed from the Geosat and (corrected) Seasat σ_0 values as summarized above, it was necessary to calibrate the T/P measurements of σ_0 by a method similar to that used by Witter and Chelton (1991a) for calibration of Seasat measurements of σ_0. Callahan *et al.* (1994) compared global histograms of T/P and Geosat σ_0 values by month and by latitude to investigate the possibility of seasonal and latitudinal variations of σ_0. Overall, T/P values of σ_0 were found to be 0.7 dB higher than the Geosat values with seasonal and latitudinal variations of less than 0.1 dB. The relative offset was later refined to be 0.63 dB. This offset was explained by Callahan *et al.* (1994) in terms of the differences in the Geosat and T/P retrieval algorithms for σ_0; the Geosat algorithm was based on a flat-earth approximation (see appendix A2 of Chelton *et al.*, 1989) and was not corrected for atmospheric attenuation (see Section 2.3). The T/P measurements of σ_0 are therefore more accurate. The σ_0 values included in the T/P Geophysical Data Records are therefore not adjusted to match the Geosat values. In order to estimate wind speed from the MCW model function, however, the 0.63 dB offset between T/P and Geosat values of σ_0 must be subtracted from the T/P data.

Because of the sensitivity of wind speed estimation to variations of σ_0, it is essential that the calibration of σ_0 be continuously monitored over the duration of an altimeter mission. The T/P altimeter includes circuitry to monitor the transmit power of the altimeter and to calibrate the receiver gain. The transmitted power has remained remarkably stable over the first 7 years of the mission but the receiver gain has been found to drift. The drift was small during the first 2 years of the mission and the internal calibration could be used to maintain consistency of the global histograms of σ_0. As the mission continued, however, the internal calibration mode systematically diverged from the σ_0 calibration inferred from the global histogram analyses. The reasons for this discrepancy are not yet understood. To provide long-term consistency of the σ_0 values and consistent accuracy of the wind speed effects in the sea-state bias correction (see Section 3.2.1), calibration adjustments are applied to the T/P data based on a low-order polynomial fit to global histograms of σ_0 for each 10-day cycle of the T/P exact-repeat orbit. Over the first 6 years of the mission, the drift of T/P measurements of σ_0 resulted in a decrease of about 2.5 dB. The empirical corrections applied to the data have maintained the σ_0 calibration to within 0.1 dB, which corresponds to an accuracy of better than 1 m sec^{-1} at moderate wind speeds (see Figure 56).

It is noteworthy that the differences between the SB, CW, and MCW model functions are not statistically significant in terms of the rms differences from the 119 buoy wind speeds in the Dobson *et al.* (1987) dataset. Quantitative assessment of the model functions is limited by the small *in situ* database and the restricted range of wind speeds in the buoy data (very few observations greater than 10 m sec^{-1}). If a larger number of coincident buoy observations of wind speeds in excess of 10 m sec^{-1} were available, a clear distinction between the relative accuracies of the SB and MCW model functions would emerge since the two model functions diverge at these higher wind speeds. It should be noted, however, that the dependence of σ_0 on wind speed is very flat in this regime (see Figure 56). Altimeter estimates of wind speed are therefore most sensitive to errors in σ_0 at high wind speeds. It will never be possible to measure high wind speeds as accurately from an altimeter as from a scatterometer.

To circumvent the limited availability of coincident buoy and altimeter estimates of high wind speeds, Freilich and Dunbar (1993) proposed the use of wind speeds estimated by numerical weather prediction (NWP) models to create a large database from which a model function for altimeter estimates of wind speed could be derived. The NWP estimates were interpolated to the locations and times of the Geosat measurements of σ_0. The correspondence between wind speed and σ_0 was determined for a set of wind-speed bins. Weighted averages of wind speed and σ_0 were formed within each bin with the weights determined by the degree of consistency between NWP wind speed estimates obtained from the operational surface analyses produced by ECMWF and NCEP. The mismatch between altimeter measurements and NWP estimates owing to underestimation of mesoscale wind variability in the NWP estimates at scales shorter than a few hundred kilometers was addressed by performing spatial averages of the altimeter measurements. To mitigate the effects of NWP errors on synoptic scales longer than about 2000 km, data segments where the differences between NWP estimates and altimeter measurements were large were eliminated from the analysis. The resultant model function was found to be very similar to the MCW model for wind speeds from 5–15 m sec^{-1} (see Figure 56). For wind speeds higher than 15 m sec^{-1}, however, the MCW model showed a negative bias relative to the NWP estimates. The bias of the SB model function at these high wind speeds was

more than twice as large as that of the MCW model function (see Figure 56).

Although the primary factor affecting σ_0 is wind speed, the distribution of the specular facets reflecting the incident radar energy likely depends on additional properties of the sea state. The effects of wave age and wave height on altimetric estimates of wind speed were investigated by Glazman and Greysukh (1993). They considered model functions that related wind speed to both σ_0 and $H_{1/3}$ using Chebyshev polynomials. Such model functions were derived by fitting to buoy data collocated with the Geosat altimeter data. Comparing the resulting altimeter wind speeds to the collocated buoy data, they found marginal improvement in the new model function over previous model functions in terms of bias, rms difference, and the dependence of the difference on wave age. They also considered a second approach in which the wind speed was expressed as a combination of polynomials and power laws in σ_0 only. The exact form and coefficients of the model function are different in two overlapping regimes of wave age estimated from the buoy data. This complicated model function yielded improved accuracy but the statistical significance of the improvement was not assessed by Glazman and Greysukh (1993).

The dependence of wind speed on wave height was also investigated by Lefebvre *et al.* (1994) using T/P-measurements of σ_0 and $H_{1/3}$ in conjunction with NWP estimates of wind speed. Wind speed was expressed as a quadratic polynomial in σ_0 and $H_{1/3}$. The performance of the model function was assessed by examining the mean and rms difference from NWP estimates of wind speed. The rms difference was 1.67 m sec^{-1} compared with an rms difference of 1.75 m sec^{-1} for a fifth-degree polynomial model based on σ_0 only (the same functional form as that of the SB model). The improvement is marginal, but reportedly statistically significant. The model function derived for the K_u-band frequency of the T/P dual-frequency altimeter was applied by Lefebvre *et al.* (1994) to σ_0 measured by the single-frequency CNES altimeter onboard the T/P satellite. The resulting wind speeds were compared with the wind speed estimates from an independent NWP database. They again found that the improvement over the σ_0-only models was only marginally significant. The dependence of wind speed on wave age was not investigated by Lefebvre *et al.* (1994).

Freilich and Challenor (1994) proposed a fundamentally different approach for deriving a wind speed model function that obviates the need for collocated altimeter measurements of σ_0 and buoy or NWP estimates of wind speed. They showed that the relationship between wind speed and σ_0 can be determined solely from the separate, rather than joint, probability density functions of wind speed and σ_0. The only requirement is that σ_0 vary monotonically and uniquely with wind speed. This requirement is apparently consistent with all available observations. By eliminating the need for collocated altimeter and buoy observations, this probabilistic approach addresses the most fundamental limitation of model function development. Denoting the cumulative probability density functions for wind speed and σ_0 as Γ_U and Γ_{σ_0}, respectively, Freilich and Challenor (1994) showed that the wind speed U is related to the probability distribution functions by

$$U = \Gamma_U^{-1}[1 - \Gamma_{\sigma_0}(\sigma_0)], \qquad (126)$$

where Γ_U^{-1} is the inverse function of Γ_U, i.e., the value of the cumulative probability density function corresponding to the particular value of the wind speed U. Because Γ_U is monotonic, such an inverse function is assured to exist. They applied the formulation to one year of Geosat data. The wind-speed distribution function was determined from both buoy data and NWP wind-speed estimates. Three different model functions were constructed from the different distributions.

The Freilich and Challenor (1994) NWP-based "geometric" model function (referred to here as the FC model function) can be accurately expressed analytically[17] by

$$\sigma_0 = 12.40 - 0.2459U + 8.956 \exp(-0.9593U), \quad (127)$$

where U is the wind speed at 10 m. As shown in Figure 56, the FC and MCW model functions agree to within 1 m sec^{-1} for wind speeds over the full range of σ_0 values shown in the figure. At wind speeds higher than about 15 m sec^{-1}, the two model functions diverge with the MCW estimates becoming systematically lower than the FC estimates. It can be noted that there is significant discrepancy between the SB model and both of the other model functions at all wind speeds except near the intersection of the three model functions at about 7 m sec^{-1}. A surprising result of the Freilich and Challenor analysis is that, of all of the wind speed model functions considered, their model functions exhibited the weakest dependence of wind speed on wave age. Equally surprising, the Glazman and Greysukh (1993) model that was specifically derived to minimize the dependence on wave age exhibited the strongest dependence on wave age (see Table 2 of Freilich and Challenor, 1994).

Elfouhaily et al. (1998) have recently presented a theoretically based method for estimating surface wind speed from linear combinations of K$_u$-band and C-band measurements of σ_0 by the T/P dual-frequency altimeter. The basis for the approach is that the difference between σ_0 at the two frequencies is related to the spectrum of short gravity waves with wavelengths in the range responsible for the difference in the backscatter at the two frequencies. As noted previously, Brown (1990) showed that the backscattered power from radar measurements at nadir results almost totally from specular reflection from wavelengths longer than about three

times that of the incident radiation (i.e., wavelengths longer than about 6 and 16 cm for the K$_u$-band and C-band frequencies, respectively). The short gravity waves with wavelengths of 6–16 cm are closely coupled to the wind stress at the surface (and hence the friction velocity). Elfouhaily et al. (1998) derived a theoretical relationship between the surface friction velocity and the two σ_0 measurements based on a prescribed wave spectrum. The estimates of friction velocity were then transformed into the neutral-stability wind speed at 10 m height using a sea-state dependent drag law with the sea-state characterized by the pseudo wave age Eq. (108) computed from wind speed and $H_{1/3}$. Because the drag law is dependent on wind speed itself, the wind speed must be estimated iteratively.

From comparisons with collocated buoy data, Elfouhaily et al. (1998) showed that the accuracy of their theory-based wind speed estimates (referred to here as the E98 model function) was somewhat better than the empirical wind speed model functions (see Table 7). At wind speeds above about 15 m sec^{-1}, where there are few observations for comparison, the E98 model function diverges from the MCW model function (see their Figure 6) in a manner that is consistent with the divergence of the FC and MCW model functions noted above (see Figure 56). Elfouhaily et al. (1998) found that the E98 wind-speed estimates were more consistent with a model for hurricane winds than were the MCW wind speed estimates.

7.2. Wind Speed Measurement Accuracy

In summary, there is remarkable agreement between empirical and theoretical model functions for altimeter estimates of wind speed. The MCW, FC and E98 model functions are all in close agreement for wind speeds up to 15 m sec^{-1}. The rms errors of the three model functions are 1.5–1.75 m sec^{-1} with biases of less than 0.5 m sec^{-1} (see Table 7). At wind speeds higher than 15 m sec^{-1}, the MCW model function yields lower wind speed estimates than the FC and E98 model functions. An expanded database of wind speeds in excess of 15 m sec^{-1} will be necessary to determine which model function is most accurate at high wind speeds.

8. TOPEX/POSEIDON MISSION DESIGN AND PERFORMANCE

8.1. Orbit Considerations and Altimeter Sampling Patterns

Many factors must be considered when selecting the orbit for an altimeter satellite. A comprehensive treatise on the subject has been given by Parke et al. (1987). The issues

---

[17]The analytical formulation in Freilich and Challenor (1994) is in terms of the wind speed at 19.5 m. The coefficients have been adjusted here to express σ_0 in terms of the wind speed at 10 m.

TABLE 7.   Comparisons of Collocated Buoy Wind Speeds with Wind Speeds Estimated
from the Smoothed Brown (SB) Model Function, Modified Chelton and Wentz (MCW)
Model Function, the Freilich and Challenor (1994) (FC) Geometric Model Function and
the Dual-Frequency Elfouhaily *et al.* (1998) (E98) Model Function

Model function	Mean error (m sec^{-1})	RMS error (m sec^{-1})	Error standard deviation (m sec^{-1})
SB	-0.02	1.68	1.68
MCW	0.48	1.75	1.68
FC	-0.12	1.72	1.72
E98	~ 0	1.67	1.45

Note: The statistics for the SB, MCW and FC model functions are based on a comparison with more than 800 collocated observations from NOAA NDBC buoys spanning the 3-year period 1987–1989 (see Freilich and Challenor, 1994). The statistics for the E98 model function are based on a comparison with 1282 collocated observations from NOAA NDBC buoys spanning the 3-year period 1993–1995 (see Elfouhaily *et al.*, 1998). (The bias of the E98 model function was not reported, but appears to be very small from their Figure 4.)

that are pertinent to the altimeter sampling pattern are summarized here. The orbit selection criteria are discussed in a general manner for all altimeter satellites with a goal of motivating the particular orbital parameters selected for the T/P satellite.

8.1.1. Altitude

A primary consideration when selecting the orbit for a satellite altimeter is the accuracy of precision orbit determination (see Section 4). At the time of the launch of T/P, GPS tracking of satellites was not available operationally. The tracking data available for POD were therefore restricted to ground-based tracking systems. Because the ground-tracking data are of varying quality and are not available continuously along the satellite orbit (see Section 4.2), orbit determination restricted to ground-based tracking requires accurate dynamical modeling of the satellite orbit, with constraints imposed by the available tracking data to adjust for modeling errors. The dominant source of error in orbit modeling is uncertainty in the short scales of the geopotential. Gravity-induced perturbations of the satellite orbit are strongly dependent on the orbit height. For wavelengths of about 1200 km in the gravity field, for example, the gravity-induced perturbations of a satellite orbit are about an order of magnitude larger at an orbit height of 500 km than at 1000 km (see Figure 2 of Pisacane, 1986). For wavelengths of about 250 km, the gravity-induced orbit perturbations are five orders of magnitude larger at the 500 km orbit height.

Another altitude-dependent source of error in orbit modeling is uncertainty in the air drag on the satellite. Air drag is proportional to the air density, which decreases approximately exponentially with increasing altitude (see, for example, Figure 1.4 of Wallace and Hobbs, 1977). The air density at high altitudes varies over a broad range of space and time scales owing primarily to solar effects and variations of the

geomagnetic field. These variations of the high-altitude air density are not completely predictable, which introduces uncertainties in the air drag term in the dynamical model. As shown previously in Figure 41, air drag decreases by about an order of magnitude between the approximate 785 km orbit heights of the Seasat, Geosat and ERS satellites and the 1336 km orbit height of T/P.

Because of the decreased reliance of orbit modeling on the accuracies of the gravity field and air drag with increasing orbit height, a high-altitude orbit is highly desirable. Indeed, the 1336 km orbit height of T/P is a major factor contributing to the dramatic improvement in the orbit accuracy compared with the orbit accuracies achieved for the approximate 785-km orbits of other altimeter satellites.

There are, however, limitations on the orbit height that are imposed by other considerations. As described by the radar Eq. (5), the power of the radar return decreases as the fourth power of the orbit height. The signal-to-noise ratio for a given radar system is thus nearly an order of magnitude lower at 1336 than at 785 km. Within limitations of the physical size of the antenna (see Section 2.4.1) and power availability imposed by the overall design of the satellite, this can be compensated by increasing the antenna transmit power or by increasing the antenna diameter, thereby decreasing the beamwidth and focusing the radar signal. With the T/P antenna diameter of 1.5 m and the transmit power of 20 W for the K$_u$-band frequency of the dual-frequency altimeter, an adequate signal-to-noise ratio can be obtained for altitudes up to about 1500 km (Parke *et al.*, 1987).

The orbit height is also limited by the harshness of the radiation environment at high altitudes. The Van Allen belts of intense radiation extend from an altitude of about 1000 km to altitudes of about 20,000 km. Exposure to radiation shortens the lifetime of the satellite electronics. For the T/P orbit inclination, the radiation exposure increases by about

40% from an orbit height of 1300 km to an orbit height of 1500 km (Parke *et al.*, 1987).

Consideration of the tradeoffs between the advantages and disadvantages of high-altitude orbits ultimately led to the selection of an orbit height of about 1300 km for T/P. The precise value of 1336 km was chosen to satisfy the other constraints on the orbit eccentricity, inclination and exact-repeat period as discussed in the following sections.

It can be noted that the availability of continuous GPS tracking of satellites greatly reduces the reliance of POD on the accuracy of the orbit model (see Section 4.3.2). With GPS tracking, an orbit accuracy comparable to that achieved for T/P can now be achieved for a lower orbit height, thus obviating the requirements that originally dictated the choice of 1336 km for the orbit of T/P. The 1-cm accuracy goal of ongoing precision orbit determination efforts is more likely to be attainable at the higher orbit height. Moreover, having established a long record length of sea-surface height variations along the T/P ground tracks, it is highly desirable for studies of short-term climate variability to maintain continuity of the data record along the same ground tracks with future high-accuracy satellite altimeters. This ground-track pattern can only be achieved with orbit parameters identical to those of T/P. The Jason-1 altimeter to be launched in early 2001 and Jason-2 with a planned launch 3 years later will both operate in the present T/P orbit configuration.

The selected orbit height of a satellite specifies the orbital period. For the small eccentricities of altimeter satellite orbits (see Section 8.1.2), the time between successive ascending crossings of the equatorial plane is

$$T_{\text{orb}} = 2\pi \left(\frac{a^3}{\mu_e}\right)^{1/2} \left[1 - \frac{3J_2}{2}\left(\frac{R_e}{a}\right)^2 (4\cos^2 i - 1)\right], \quad (128)$$

where a is the semimajor axis of the orbit, $\mu_e = 3.98600 \times 10^5$ km^3 sec^{-2} is the product of the Universal Gravitational Constant and the mass of the earth, $R_e = 6378$ km is the equatorial radius of the earth, i is the orbit inclination (see Section 8.1.3) and $J_2 = 1.0826 \times 10^{-3}$ is the second zonal harmonic of the earth's gravity field, which accounts for most of the equatorial bulge of the earth. The second term in square brackets on the right side of Eq. (128) is a small correction to the orbital period owing to the oblateness of the earth. Neglecting this term, the orbital period is fully determined by the semimajor axis of the orbit. For the nearly circular orbits of altimeter satellites, the semimajor axis for an orbit height R is $a \approx R_e + R$. The orbital period for the approximate 785 km orbit heights of Seasat, Geosat and ERS is about 101 min. At the higher 1336 orbit height of T/P, the orbital period increases somewhat to about 112 min.

Because the earth rotates 0.25° min^{-1} relative to the orbital plane of a satellite, the ground tracks of successive orbits migrate westward longitudinally, as shown in Figures 57 and 58 for the T/P and ERS orbits. The separation between successive ground tracks depends on the orbital period of the satellite. For the 112-min period of the T/P orbit, successive ground tracks are separated by 28° of longitude. For the somewhat shorter 101-min period of the Seasat, Geosat and ERS satellites, the ground tracks migrate westward by 25.25° per orbit.

8.1.2. Eccentricity

An important consideration in the design of an altimeter mission is the ability of the onboard adaptive tracking unit to track the radar return from mean sea level. As described in Section 2.4.4, the onboard sea-level tracking algorithm for the T/P dual-frequency altimeter is limited to a total range of about 60 km in acquisition mode. In addition, the tracking circuitry imposes limitations on the ability of the tracker to maintain lock with the radar return from mean sea level in the presence of variations of the range from the altimeter to the sea surface between successive radar pulses. Because of the oblateness of the earth, the range rate for a circular orbit can be as high as 30 m sec^{-1}. The range rate would be higher for an elliptical orbit. The limitations imposed by the electronic circuitry of the altimeter require that the orbit eccentricity be less than 0.001 for altimeter satellites.

A secondary consideration for the orbit eccentricity is that an approximately constant ground speed is desirable so that the range measurements are approximately evenly spaced along the satellite ground track. The variations of the ground speed during the course of each orbit increase with increasing orbit eccentricity. Except for small variations owing to the oblateness of the earth, the ground speed for a circular orbit is constant. From Eq. (128), this ground speed is approximately

$$v_s = \frac{2\pi R_e}{T_{\text{orb}}} \approx \frac{R_e}{a}\left(\frac{\mu_e}{a}\right)^{1/2}. \quad (129)$$

For orbit heights of 1336 km and 785 km, the ground track speeds are about 5.9 and 6.6 km sec^{-1}, respectively.

8.1.3. Inclination

The primary consideration in selection of the orbit inclination is the latitudinal range over which the sea surface is to be measured. By convention, the orbit inclination i is less than 90° for satellites that orbit in the direction of the earth's rotation, referred to as prograde orbits. The orbit inclination is greater than 90° for satellites that orbit in the direction opposite that of the earth's rotation, referred to as retrograde orbits. To within a fraction of a degree (owing to the oblateness of the earth), the turning latitude of the satellite is $|\varphi_{\max}| = i$ for prograde orbits and $|\varphi_{\max}| = 180° - i$ for retrograde orbits. In order to sample the full latitudinal extent of the Antarctic Circumpolar Current, for example, the orbit inclination must be between 65° and 115°.

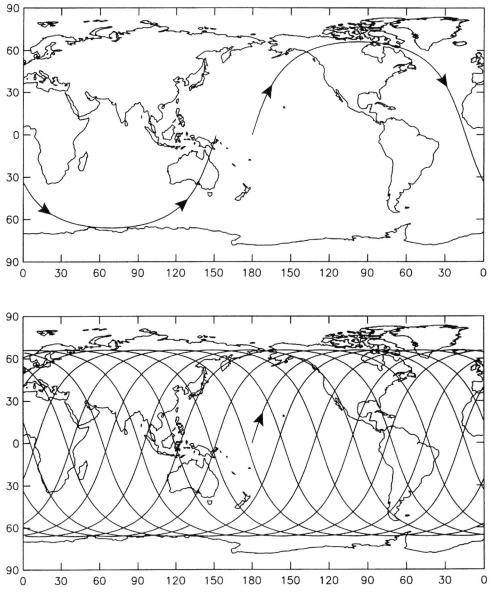

FIGURE 57   The ground track for a single orbit (top) and the ground track pattern traced out in one day (bottom) for the T/P prograde orbit with 66° inclination.

The Seasat and Geosat orbits were both retrograde with an inclination of 108°, thus sampling the world ocean over a latitudinal range from 72°N to 72°S. The ERS orbits are also retrograde with a 98° inclination that samples more of the polar ocean (to a latitude of 82°, see Figure 58). This high inclination was driven by the constraint that the ERS orbit be sun synchronous, which requires an orbit inclination of 98° at the ERS orbit height of 787 km (see, for example, Figure 15.4 of Stewart, 1985). The T/P orbit is prograde with an inclination of 66° (see Figure 57). The prograde nature and much lower inclination of the T/P orbit compared with other altimeter satellites were dictated by the tidal aliasing considerations summarized in Section 8.1.5.

A secondary consideration in the selection of an orbit inclination is the angle between ascending and descending ground tracks at the crossover points. Because the earth rotates relative to the satellite orbital plane, the ground tracks deviate from great circles. The angle of a satellite ground track relative to north is given by

$$\xi = \tan^{-1}\left[\frac{v_s \sin\xi_0 \pm v_e \cos\varphi}{v_s \cos\xi_0}\right], \qquad (130)$$

where v_s is the satellite ground speed Eq. (129), $v_e \cos\varphi$ is the rotational velocity of the earth's surface at latitude φ and

$$\xi_0 = \sin^{-1}\left[\frac{|\cos i|}{\cos\varphi}\right] \qquad (131)$$

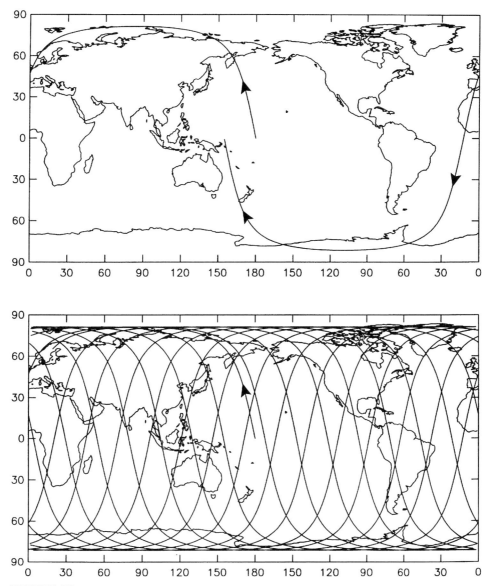

FIGURE 58 The ground track for a single orbit (top) and the ground track pattern traced out in one day (bottom) for the ERS retrograde orbit with 98° inclination.

is the angle relative to north of the instantaneous great circle of the ground track at the crossover point. The plus and minus signs in Eq. (130) pertain to retrograde and prograde orbit inclinations, respectively. Because of the symmetry of the ascending and descending ground tracks with respect to the meridian at the crossover point, the crossover angle is 2ξ. It can be noted that ξ_0, and hence ξ, increase with increasing latitude φ for any given orbit inclination i. At the turning latitudes where $|\varphi| = i$ for prograde orbits and $|\varphi| = 180° - i$ for retrograde orbits, $\xi_0 = 90°$. The denominator of Eq. (130) therefore vanishes, which implies that the crossover angle is $2\xi = 180°$ at the turning latitudes. (In reality, a crossover is not possible at the precise turning latitude because of the finite longitudinal separa-

tion of individual ground tracks.) The crossover angle for a wide range of orbit inclinations is contoured as a function of latitude in Figure 6 of Parke *et al.* (1987). The latitudinal variations of the crossover angles for the T/P, ERS and Geosat orbits are shown in Figure 59. The different effects of the earth's rotation on the crossover angles for prograde and retrograde orbits are evident from the intersection of the curves for the T/P and Geosat crossover angles at a latitude of about 27°.

For estimation of surface geostrophic velocity at the crossover points (e.g., Morrow *et al.*, 1994), it is desirable to have the crossover angles as near to 90° as possible, thus minimizing the errors of the geometrical transformations of along-track sea level slopes to estimate two orthogonal com-

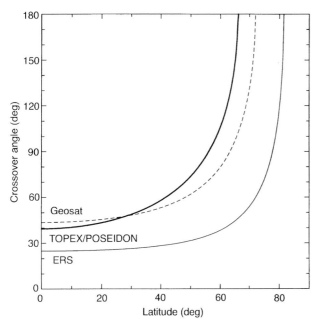

FIGURE 59   The crossover angle between ascending and descending ground tracks as a function of latitude for the T/P, ERS, and Geosat orbit configurations.

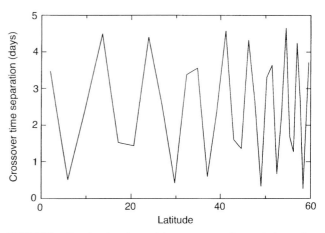

FIGURE 60   The time interval between ascending and descending ground tracks at crossover points as a function of latitude for the T/P orbit configuration. Note that a time lag between 5 and 10 days within a single 10-day repeat period is equivalent to a time lag between −5 and 0 days in successive repeat periods. The time intervals shown in the figure correspond to the magnitudes of the minimum crossover separation time.

ponents of velocity. For any particular orbit inclination, it is evident from Figure 59 that this can be achieved only over a very restricted range of latitudes. For the T/P orbit, for example, the crossover angle is $90° \pm 20°$ within the latitude range from about $48°$ to $61°$. For the ERS orbit, crossover angles of $90° \pm 20°$ are obtained only within the narrow latitudinal range from about $75°$ to $80°$. It is clear that the desire for orthogonal crossovers cannot be a major constraint on selection of the orbit configuration.

Another consideration for velocity estimation at the crossover points is the time separation between ascending and descending ground tracks. Ideally, the ground tracks will intersect within a very short time interval, thus sampling two projections of the slope of the sea surface at essentially the same time to eliminate time evolution of the surface velocity field as a source of noise in the velocity estimates. The magnitude of the time separation between ascending and descending ground tracks at the crossover points for the T/P orbit is shown in Figure 60. The crossover time separation varies from a minimum of about 0.25 days to a maximum of about 4.7 days. The latitudinal variations of crossover time separations are similarly erratic for other orbit configurations (see, for example, Figure A.1 of Morrow *et al.*, 1994, for the case of the Geosat orbit, for which the crossover time separation ranges from a fraction of a day to about 8 days). Clearly, crossover time separation also cannot be a constraint on selection of the orbit configuration.

The orbit inclination also affects the orbit precession and is therefore an important consideration when selecting the exact-repeat period of the satellite orbit (see Section 8.1.4).

8.1.4. Exact-Repeat Period

For a variety of reasons, it is desirable to have an exactly repeating orbit for satellite altimetry. The most important of these is the need to compute an accurate mean sea surface along the satellite ground track in order to remove the unknown geoid contribution to the sea surface topography as summarized in Section 5.2. The earth's rotation rate in radians per solar day (also referred to as a calendar day) is

$$\omega_e = 2\pi(1 + 1/365.2422). \tag{132}$$

The first term in this expression is the rotation of the earth in one sidereal day and the second term accounts for the revolution of the earth with respect to the sun each solar day. The repeat period of a satellite is specified in terms of the nodal day, defined to be the time required for the earth to rotate once with respect to the orbital plane of the satellite. Because of orbital precession, a nodal day differs slightly from a solar day. Exact-repeat orbits consist of a satellite ground track that repeats an integer number of revolutions in an integer number of nodal days. For a repeat period of D nodal days (referred to simply as a "D-day repeat orbit"), the exact-repeat period in solar days is

$$P_{\text{orb}} = \frac{2\pi D}{\omega_e - \dot{\Omega}}, \tag{133}$$

where

$$\dot{\Omega} = -\frac{3J_2}{2}\left(\frac{\mu_e}{a^3}\right)\left[\frac{R_e}{a(1-e^2)}\right]^2 \cos i \tag{134}$$

is the orbital precession rate induced by the equatorial bulge of the earth.

For the orbit heights and inclinations of altimeter satellites, $\dot{\Omega}$ falls within the range $-3° \leq \dot{\Omega} \leq +3°$ per solar day (see Figure 3 of Parke *et al.*, 1987). The precession rate is more sensitive to the orbit inclination than to the orbit height. Retrograde orbits for which $i > 90°$ and $\dot{\Omega} > 0$ precess eastward. From Eq. (133), the corresponding repeat periods are slightly longer in solar days than in nodal days. Thus, for example, the repeat period of the "17-day Geosat orbit" is 17.0505 solar days. Similarly, prograde orbits for which $i < 90°$ and $\dot{\Omega} < 0$ precess westward and the repeat periods are slightly shorter in solar days than in nodal days. The repeat period of the "10-day T/P orbit" is 9.9156 solar days. For sun-synchronous orbits such as that of the ERS satellites, the precession rate is $\dot{\Omega} = 2\pi/365.2422$ so that the orbital plane precesses eastward at exactly the same rate that the earth revolves about the sun. For sun-synchronous orbits, the repeat period is the same in solar and nodal days since $\omega_e - \dot{\Omega} = 2\pi$.

The pattern of ground tracks on the sea surface is determined by the orbit inclination i, the nodal period D of the exact repeat and the number of revolutions N per repeat period. For all latitudes sampled by the satellite, the longitudinal separation of ground tracks in radians is $2\pi/N$. In kilometers, this ground track spacing as a function of latitude φ is

$$\Delta x = \frac{222.4\pi}{N} \cos \varphi. \tag{135}$$

It is not possible to specify N arbitrarily for any particular repeat period in nodal days. Only certain combinations of D and N yield valid orbits. Furthermore, any specific combination of D and N dictates the orbit height R. The complicated interplay between D, N, and R is shown in Figure 4 of Parke *et al.* (1987). The strategy for orbit selection is thus to choose i, D, and an approximate orbit height R. The number of orbits N is then selected to achieve a specific orbit height near R. This procedure can be illustrated for the T/P orbit. With an orbit inclination of 66° and a repeat period of 10 nodal days, odd integer values of N between 143 and 129 (excluding 135) yield possible orbits with orbit heights in the range from 700–1400 km (see Figure 4 of Parke *et al.*, 1987). Near the desired ~1300 km orbit height, values of $N = 127$ and 129 yield orbit heights of the 1336 km and 1254 km, respectively. The value for the 1336-km orbit height was adopted for the T/P 10-day repeat orbit.

Some general statements can be made about the character of satellite ground track patterns. Over the range of orbit heights suitable for altimeter satellites, N is much more dependent on the repeat period than on the orbit height. For orbits in the vicinity of any particular orbit height, N therefore increases with increasing orbit repeat period. There is thus a tradeoff between spatial and temporal resolution. From Eq. (135), the ground track spacing therefore decreases with increasing orbit repeat period. This is evident from Figure 61, which shows the longitudinal spacing of ground tracks as a function of latitude for the 10-day T/P orbit with

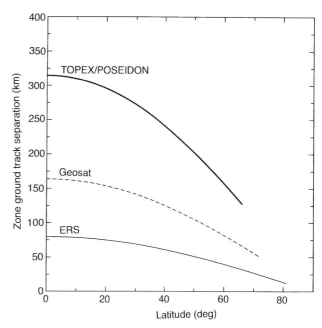

FIGURE 61 The zonal separation of ground tracks in kilometers as a function of latitude for the T/P, ERS and Geosat orbit configurations.

$N = 127$, the 35-day ERS orbit with $N = 501$ and the 17-day Geosat orbit with $N = 244$. The spatial patterns of the T/P, ERS, and Geosat ground tracks over the respective repeat periods are shown in the upper panels of Figures 62–64.

Spatially adjacent ground tracks in the grid mapped out during a satellite repeat period are not sequential in time. As shown in Figures 57 and 58, successive ground tracks are separated by about 28° and 25° of longitude for orbit heights of 1336 and 785 km, respectively. Over the repeat period, these wide gaps between successive orbits are filled in by later orbits. There are numerous "subcycles" in the ground track patterns. For example, a 1-day subcycle is evident by the interleaved pattern of ground tracks during three successive days shown in the middle panels of Figures 62–64. The coarse ground-track pattern that is mapped out during one day is essentially shifted longitudinally during each successive day. The longitudinal shift of this 1-day subcycle is westward for the T/P orbit and eastward for the ERS and Geosat orbits.

Three-day subcycles are also evident for all three orbits (see the lower panels of Figures 62–64). The coarse grid of ground tracks that is mapped out over a 3-day period is essentially shifted longitudinally during each of the next 3-day periods. This 3-day pattern continues shifting longitudinally over the course of the repeat period until the ground-track pattern from the first 3-day period is repeated. For the T/P and Geosat orbits, successive 3-day subcycles shift eastward. The 3-day subcycles shift westward for the ERS orbit.

For the ERS orbit, there is also a 17.5-day subcycle in the ground track pattern. The ground-track pattern mapped out

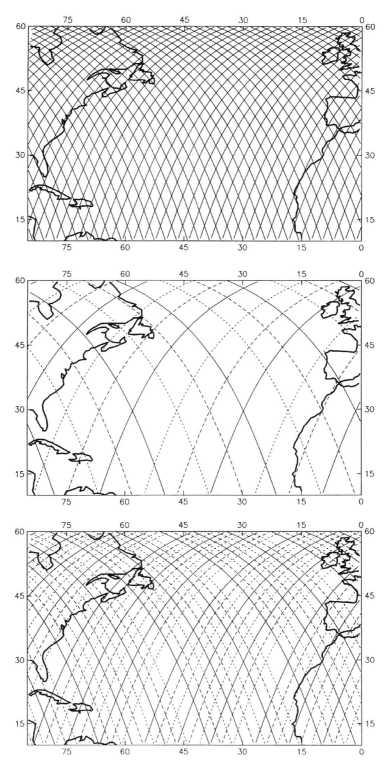

FIGURE 62 The ground tracks over the North Atlantic Ocean for the T/P 10-day exact repeat orbit configuration. The top panel shows the ground track pattern for the full 10-day repeat period. The middle panel shows the ground track pattern during a 3-day period, with solid, dashed, and dotted lines corresponding, respectively, to the first, second, and third day of the orbit repeat period. Note the westward shift of a 1-day subcycle. The bottom panel shows the ground-track pattern during a 9-day period, with solid, dashed, and dotted lines corresponding, respectively, to days 1–3, days 4–6 and days 7–9 of the orbit repeat period. Note the eastward shift of a 3-day subcycle.

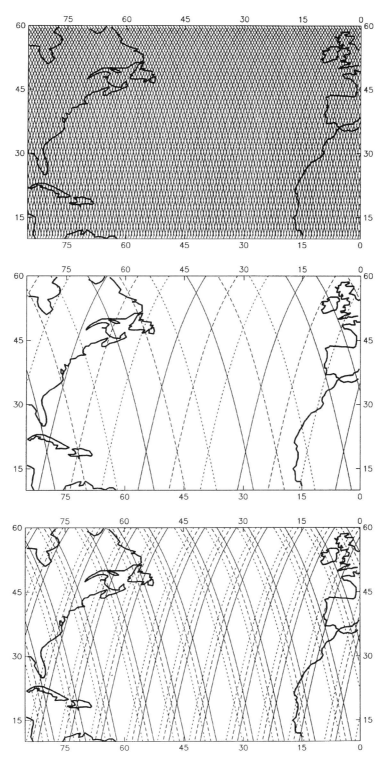

FIGURE 63 The same as Figure 62, except for the ERS 35-day exact-repeat orbit configuration. Note the eastward shift of a 1-day subcycle in the middle panel and the westward shift of a 3-day subcycle in the bottom panel.

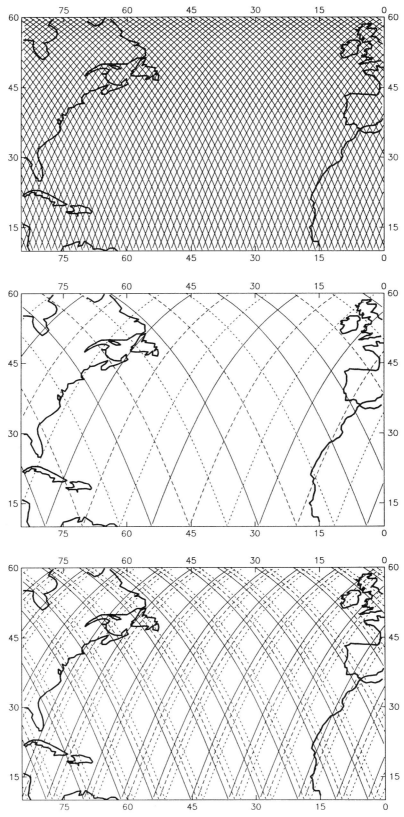

FIGURE 64   The same as Figure 62, except for the Geosat 17-day exact-repeat orbit configuration. Note the eastward shift of a 1-day subcycle in the middle panel and the eastward shift
of a 3-day subcycle in the bottom panel.

during the first half of the 35-day orbit is repeated during the second half of the repeat period, except shifted longitudinally so as to interleave the ground tracks sampled during the first half of the repeat period.

8.1.5. Tidal Aliasing Considerations

The most energetic tides are the semidiurnal and diurnal constituents with frequencies near 2 and 1 cpd. Such short-period variability clearly cannot be resolved in altimeter observations and consequently aliases to the lower frequencies that are of interest for oceanographic studies. As noted previously in Section 5.3, the rms errors in the global tide models that were available prior to the launch of T/P exceeded 10 cm in many regions of the world oceans. The T/P orbit was specifically designed to avoid aliasing of major tidal constituents to undesirable frequencies. In fact, as discussed extensively by Parke *et al.* (1987), consideration of tidal aliasing was one of the two primary constraints on selection of the T/P orbit (the other being precision orbit determination as discussed in Section 4). Because of the careful selection of the orbit configuration, it has been possible to estimate the tides globally from T/P data to an accuracy of 2–3 cm (see Chapter 6).

Estimation of tidal amplitudes and phases from simple time series analyses of altimeter data with a sample interval equal to the orbit repeat period imposes three constraints on the orbit. First, since an altimeter mission is typically designed for a duration of a few years, the orbit configuration should be conservatively chosen so that the alias frequencies are high enough to be resolved by a record length of about a year or two. This requires that the alias frequencies be higher than 1 cpy (i.e., periods shorter than a year). Second, the orbit should be chosen so that the alias frequencies of the major tidal constituents are separable by the same record length of a year or two. Ideally, then, alias frequencies should be separated by at least 1 cpy. Third, orbit configurations for which the alias frequencies of any of the major tidal constituents coincide with energetic variability from other causes (e.g., annual or semiannual variability or the zero frequency that defines the mean sea-surface height) should be avoided.

The formalism for determining the tidal alias frequencies is straightforward. For a particular tidal constituent j with period $T_{\text{tide}}(j)$, the phase change at a particular location between successive observations separated by the orbit repeat period P_{orb} in solar days is

$$\Delta\phi_{\text{tide}}(j) = \frac{2\pi P_{\text{orb}}}{T_{\text{tide}}(j)}. \qquad (136)$$

The phase change $\Delta\phi_{\text{tide}}(j)$ is cyclical with increasing P_{orb} and can be restricted to the range from $-\pi$ to π. The principal alias is the period required for one complete cycle of the

aliased tidal phase, which is given by

$$T_{\text{alias}}(j) = \frac{2\pi P_{\text{orb}}}{|\Delta\phi_{\text{tide}}(j)|}. \qquad (137)$$

The corresponding alias frequency is $f_{\text{alias}}(j) = T_{\text{alias}}^{-1}(j)$. Both $T_{\text{tide}}(j)$ and P_{orb} must be specified very precisely since small errors accumulate to very large phase differences $\Delta\phi_{\text{tide}}(j)$ over the duration of a repeat period. For example, the 317-day alias period of the dominant M_2 tidal constituent for the Geosat exact-repeat period of 17.0505 solar days (see Table 10 below) becomes 112 days if the repeat period is approximated as 17.0000 solar days.

From Eq. (133), the change in tidal phase given by Eq. (136) over the orbit repeat period P_{orb} can be written in terms of the satellite orbital parameters as

$$\Delta\phi_{\text{tide}}(j) = \frac{2\pi D}{T_{\text{tide}}(j)}\left(\frac{2\pi}{\omega_e - \dot{\Omega}}\right). \qquad (138)$$

It is thus apparent that the alias period Eq. (137) for a particular tidal constituent depends on the orbit repeat period D in nodal days and the orbital precession rate $\dot{\Omega}$. Because D and $\dot{\Omega}$ are a function of orbit inclination, altitude, eccentricity and the number N of orbits per repeat period at the particular orbit height (see Section 8.1.4), a wide range of orbit repeat periods is possible near any particular orbit altitude and inclination (see, for example, Figure 4 of Parke *et al.*, 1987). For the range of altitudes relevant to satellite altimetry, tidal alias periods are only a weak function of orbit altitude since the dependence of $\dot{\Omega}$ on altitude is relatively weak, as noted previously. The alias period is thus most sensitive to D, N, and the orbit inclination. The details for determining the aliasing period from Eq. (138) are described by Parke *et al.* (1987). An alternative formalism is presented by Schlax and Chelton (1994b).

Because of the generally large scales of the tides in the open ocean and the 3-day interval between measurements along adjacent ground tracks (see bottom panels of Figures 62–64), aliased tidal variations are phase shifted on neighboring ground tracks. This in effect introduces spatial aliasing of tides in an altimeter sampling pattern. The combined effects of the spatial and temporal aliasing result in systematic propagation of the aliased tidal variability. The formalism for determining the zonal wavelength of the aliased variability is described by Schlax and Chelton (1994b; 1996; see also Okkonen and Jacobs, 1996). For any particular alias period $T_{\text{alias}}(j)$, there is a multiplicity of alias wavelengths $\lambda_k(j)$, where the subscript k is used to denote the different aliasing wavelengths for a given aliasing period. Parke *et al.* (1998) extended the analysis of Schlax and Chelton (1994b; 1996) to describe the tidal aliasing along the vector direction of propagation perpendicular to satellite ground tracks.

The details of application of the formalism to determine the aliasing periods and wavelengths are somewhat tedious and the dependencies of alias frequency and wavelengths on orbit altitude, inclination and repeat period are complex. Some general statements can nonetheless be made about the aliasing periods (see, for example, Parke *et al.*, 1987). Tidal aliases are constrained to lower frequencies with increasing orbit repeat period. For all retrograde orbits, at least one of the dominant tidal constituents aliases to a frequency near zero (i.e., the mean). Such orbits are to be avoided if at all possible since they will compromise future estimates of the mean ocean circulation when accurate geoid information becomes available (see discussion in Section 5.2). Prograde orbits with a westward precession rate $\dot{\Omega}$ of more than $-2°$ per solar day are most desirable for purposes of tidal estimation. For an orbit altitude near 1300 km, this corresponds to a maximum inclination of about 66° (see Figure 3 of Parke *et al.*, 1987). A 10-day repeat is one of the few choices of repeat periods longer than a week for which inclinations exist so that the sampling pattern aliases all of the major tidal constituents into frequencies higher than 2 cpy. These various considerations ultimately led to the choice of a 10-day repeat period with a 66° inclination and 1336 km orbit height for the T/P satellite.

To illustrate the importance of the orbit configuration on the aliasing characteristics of the tides, the aliasing periods and associated three longest aliasing wavelengths are listed for the six dominant tidal constituents in Tables 8–10 for the T/P, ERS and Geosat orbits, respectively. The superiority of the T/P orbit for tidal estimation is readily apparent; the K_1 constituent aliases to a frequency near 2 cpy (a period of 6 months) but all of the other dominant tidal constituents alias to shorter periods that are resolvable and separable by a record length of about a year. In contrast, the ERS and Geosat orbits alias at least half of the six dominant constituents to a frequency near 0, 1, or 2 cpy. The sun-synchronous ERS orbit is particularly undesirable since the S_2 constituent (and all other solar tidal constituents not listed in Table 9) alias to exactly the zero frequency. Estimates of the mean ocean circulation from ERS altimeter data are therefore contaminated by errors in the model solar tide corrections. The K_1 and P_1 constituents alias to exactly the annual cycle for a sun-synchronous orbit. Errors in the model tides for these constituents thus compromise studies of the seasonal cycle of sea-level variations. For altimetry, it is clear that sun-synchronous orbits are to be avoided, if at all possible. The choice of a sun-synchronous orbit for the ERS satellites was dictated by requirements for other instruments onboard the spacecraft.

As noted above, the coupling between the aliasing period and an aliasing wavelength results in systematic propagation of aliased tidal errors. For some orbit configurations, the wavelength and period of the propagating tidal errors resembles westward propagating baroclinic Rossby waves over certain latitude ranges of the ocean (Jacobs *et al.*, 1992; Schlax and Chelton, 1994a; 1994b; 1996; Parke *et al.*, 1998).

This was especially a problem for the Geosat orbit for which the M_2 tidal constituent, which is the dominant contribution to the tides in many areas of the world ocean, propagated westward with a period of 317 days and a wavelength of about 800 km. Over the approximate 3-year duration of the Geosat exact-repeat mission, this aliasing period was indistinguishable from the annual period. There are examples in the published literature where westward propagating M_2 tidal aliases were misinterpreted as near-annual Rossby waves in sea surface height fields derived from Geosat data with tidal corrections based on pre-T/P tide models that had errors of order 10 cm.

The accuracies of global tide models have improved dramatically as a result of optimization of the T/P orbit parameters for the sampling of tidal variations. As summarized in Chapter 6, the rms errors of these models have been reduced to 2–3 cm over the open ocean. Tidal aliasing is therefore a less stringent constraint on orbit configuration now than it was prior to the launch of T/P. Owing to the sea-surface height signatures of the effects of incoherent tidal variations from oceanic internal tides (Ray and Mitchum, 1996; 1997), it may not be possible to reduce tidal errors much below this level. Errors of this magnitude are large enough to be a concern in the overall error budget for altimetric estimates of the dynamic sea surface topography for ocean circulation studies. In particular, altimeter satellite orbit configurations that alias any of the dominant tidal constituents into 0, 1, or 2 cpy continue to be undesirable for accurate determination of the mean and seasonal cycle of the sea-surface height field.

8.1.6. Ground Track Repeatability

As noted in Sections 5.1 and 5.2, the rms uncertainties of presently available estimates of the geoid undulations are about 37 cm (see Chapter 10). Except at the very largest scales that are accurately resolved in these geoid models, analyses of altimeter data are therefore restricted to the time-variable component of the sea-surface height. For exact-repeat orbits, this is achieved by subtracting the time average sea-surface height at each grid location along the satellite ground track. Because of cross-track gradients in the geoid, even small cross-track variations of the ground track location can corrupt the time average. More importantly, if not accounted for, the different geoid undulations sampled along laterally shifted ground tracks are indistinguishable from time variability of the sea surface height along the nominal ground track, thus introducing fictitious sea level variability in the altimeter data. To avoid this source of noise in estimates of sea surface height, the cross-track variations of altimeter satellite ground tracks are maintained within ± 1 km of the nominal ground track.

From analyses of Geosat data, Brenner *et al.* (1990) and Minster *et al.* (1993) found that geoid slopes can be as large as 2×10^{-4} near abrupt features in the geoid (ocean trenches,

TABLE 8.   Tidal Periods for the Six Major Tidal Constituents, Along with the Alias Period and the Three Longest Zonal Alias Wavelengths for Altimeter Measurements in the 10-Day Repeat T/P Orbit Configuration

Tide	Period (hr)	T_{alias} (days)	λ_{-1} (degrees)	λ_0 (degrees)	λ_1 (degrees)
M_2	12.420601	62.11	2.16 E	9.01 E	4.14 W
S_2	12.000000	58.74	2.79 W	183.01 W	2.88 E
N_2	12.658348	49.53	2.16 W	9.01 W	4.14 E
K_1	23.93447	173.19	2.81 W	366.03 W	2.86 E
O_1	25.819342	45.71	2.17 E	9.24 E	4.09 W
P_1	24.06589	88.89	2.81 W	366.03 W	2.86 E

Note: Generally, the longest wavelength is the one that is most apparent in time-longitude plots of sea surface height variability (see Schlax and Chelton, 1994b; 1996). The direction of propagation for each alias is denoted as E for eastward and W for westward.

TABLE 9.   The Same as Table 8, Except for the 35-Day Repeat ERS Orbit Configuration

Tide	T_{alias} (days)	λ_{-1} (degrees)	λ_0 (degrees)	λ_1 (degrees)
M_2	94.49	0.78 W	8.79 E	0.67 E
S_2	∞	0.72	179.76	0.72
N_2	97.39	0.86 E	4.29 W	0.62 W
K_1	365.25	0.72 E	359.70 E	0.72 W
O_1	75.07	0.79 W	8.58 E	0.66 E
P_1	365.25	0.72 W	359.52 W	0.72 E

TABLE 10.   The Same as Table 8, Except for the 17-Day Repeat Geosat Orbit Configuration

Tide	T_{alias} (days)	λ_{-1} (degrees)	λ_0 (degrees)	λ_1 (degrees)
M_2	317.13	1.25 W	8.00 W	1.81 E
S_2	168.81	1.46 E	179.89 E	1.49 W
N_2	52.07	1.08 E	4.09 E	2.31 W
K_1	175.45	1.47 E	359.79 E	1.48 W
O_1	112.95	1.25 W	8.18 W	1.80 E
P_1	4465.22	1.47 E	359.78 E	1.48 W

continental margins, islands and seamounts). In such extreme cases, a cross-track drift of the satellite by 1 km would result in a 20-cm change of the sea-surface height. More typical geoid slopes are an order of magnitude smaller, corresponding to less than 2 cm of geoid change over a cross-track distance of 1 km. With enough repeat cycles, the cross-track geoid variations can be estimated from the altimeter data and a correction for cross-track drift within the ±1 km ground track window can be applied to the sea-level estimates (Brenner *et al.*, 1990; Minster *et al.*, 1993). Alternatively, the effects of cross-track gradients of the geoid can

be mitigated by correcting altimeter data based on a high-resolution geoid constructed as described in Chapter 10. Such corrections are routinely applied by most investigators based on the high-resolution mean sea surface that is interpolated to each observation location and included in the T/P Geophysical Data Records (Callahan, 1993).

Depending on the orbit configuration, maintaining the ground track within a window of ±1 km along a nominal ground track can be a challenge. Air drag and radiation pressure cause the orbit height to decrease slowly, thus causing the orbit period to decrease in accord with Eq. (128). A de-

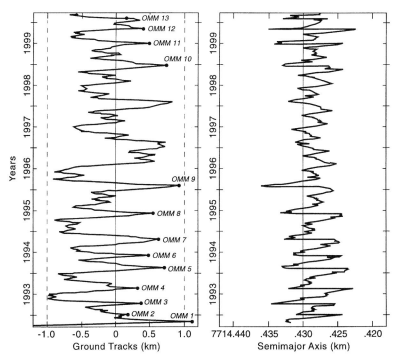

FIGURE 65   Time series of the cross-track drift of ground tracks (left) and the semimajor axis of the orbit (right) for the T/P satellite during the first 7.5 years of the mission. The times of orbit maintenance maneuvers are indicated by the dots in the left panel.

crease of the orbit period causes the satellite ground track to drift slowly to the west. The satellite ground track also changes because of the effects of higher order terms in the gravity field on the argument of perigee (see Section 4.1.2) and eccentricity of the orbit that cause the locations of the ground tracks to vary in a periodic fashion. When the ground track drifts out of the ±1 km window, maintenance maneuvers are required to adjust the satellite orbit height, thereby maintaining the nominal ground track. This is achieved by a sequence of thruster firings that can be conducted over continental land masses to minimize the corruption of the sea level data record during the period of an orbit maintenance maneuver.

A variety of techniques are utilized to minimize the frequency of orbit maintenance maneuvers. The higher orbit height of T/P is advantageous because of the previously noted order-of-magnitude lower air drag at 1336 km than at the approximate 785-km orbit heights of other altimeter satellites (see Figure 41). In addition, the T/P orbit is a "frozen orbit" for which the forces on the satellite from the higher order terms of the gravity field balance so that the eccentricity and argument of perigee remain within a very small region of parameter space and vary along a closed path within this parameter space. The periodicity of the eccentricity and argument of perigee for the T/P frozen orbit is about 26 months. The T/P Navigation Team has developed a new method of minimizing the need for orbit maintenance ma-

neuvers. By small changes in the orientation of the solar array panel, the radiation pressure forces (direct solar radiation and re-radiation from the back of the solar array) have been used to "sail" the satellite to assist in maintaining the orbit within the required ±1 km window (Shapiro *et al.*, 1994).

As a result of the T/P orbit configuration and the efforts of the T/P Navigation Team, remarkably few orbit maintenance maneuvers have been necessary. As shown in the left panel of Figure 65, only 13 orbit maneuvers were required over the first 7.5 years of the mission. Most remarkably, the utilization of radiation pressure to sail the satellite eliminated the need for orbit maneuvers during the 3-year period of low solar activity from January 1996 through December 1998 (see Figures 31 and 32). Because of the increased solar activity during the present solar cycle, it has become necessary to increase the frequency of orbit maneuvers.

It is noteworthy that maintaining the ground track within a ±1 km window requires that the semimajor axis of the orbit be maintained within a very narrow window. As shown in the right panel of Figure 65, the semimajor axis of the T/P orbit has been maintained within a window of 15 m over the first 7.5 years of the T/P mission.

8.1.7. Interpolation and Smoothing

For most analyses of altimeter data, it is necessary to interpolate the irregularly sampled measurements to map the sea surface height (SSH) field on a regular space-time grid. A wide variety of mapping algorithms have been applied to altimeter data. The details of whatever mapping formalism is adopted merit careful consideration. Every mapping algorithm has an associated filter transfer function and is therefore equivalent to smoothing the data. As altimeter data are exploited to investigate ever more subtle signals in the SSH field, it becomes increasingly important to understand the filtering properties of the interpolation and smoothing algorithm and the limitations imposed by the space-time sampling pattern of the altimeter data. The appropriate degree of smoothing depends on the specific orbit configuration of the altimeter satellite and on the particular application of interest. The issues to be considered in the selection of a mapping algorithm and specification of the degree of smoothing to be applied to the data are summarized in this section.

Before discussing the effects of sampling errors on SSH fields constructed from altimeter data, some general comments about interpolation and smoothing are in order. Objective mapping[18] based on the Gauss-Markov theorem that minimizes the mean squared error of the estimate (Gandin, 1965; Alaka and Elvander, 1972; Bretherton et al., 1976) is often considered the best method of mapping irregularly spaced observations. This is more precisely referred to as "optimal estimation" to better distinguish the method from other mapping formalisms that abound in the literature. The primary attraction of optimal estimation is that it is the minimum mean squared error estimate, presuming that the variance and autocorrelation functions of SSH have been specified accurately. Optimal estimation also allows a straightforward explicit treatment of measurement errors. The effects of measurement errors can be explicitly included in all non-optimal mapping formalisms as well (e.g., Press et al., 1992, Section 13.3), but this is seldom done in practice. The signal-to-noise ratio of the measurements significantly affects the filter transfer function of an estimate derived by any mapping formalism. With increasing measurement error variance, the amplitude of the filter transfer function decreases at the long wavelengths and periods that are passed by the filter (Press et al., 1992). The effects of measurement errors on the filter transfer function are shown for the case of optimal estimation in Figure 20 of Chelton and Schlax (1993b).

For SSH fields constructed from altimeter data, measurement errors are generally of less concern than sampling errors from the irregular and sparse space-time distribution of the observations owing to the coarse spacing of satellite ground tracks (Wunsch, 1989). The effects of measurement errors become increasingly important in SSH fields constructed from multiple simultaneous altimeter datasets because of the improved sampling provided by the dense network of intersecting ground tracks for the various altimeters (see further discussion below).

It is noteworthy that the Gauss-Markov formulation of optimal estimation is much more computationally intensive than other mapping formalisms. As a consequence, it is often necessary in practice to reduce the computational demands by subsampling the data and restricting the spatial and temporal radii from which data are considered for any particular optimal estimation time and location (e.g., Le Traon and Hernandez, 1992; Le Traon et al., 1998; Le Traon and Dibarboure, 1999). An additional limitation of the Gauss-Markov formalism is that the optimality of the estimate is dependent upon the accuracy of the time and length scales adopted for the autocorrelation function; if the variance and autocorrelation function are not quantitatively correct, then the Gauss-Markov estimate is not truly the minimum mean squared error estimate and the attraction of optimal estimation is lost. In practice, optimal estimates seldom differ significantly from estimates obtained by other less computationally intensive mapping techniques for which the smoothing parameters are suitably chosen to match the filter transfer function properties of the optimal estimate. The mathematical details of the formalism used to map altimeter data are therefore of secondary concern.

The length and time scales of mesoscale variability are on the order of 50–100 km and 10–100 days at midlatitudes (Shen et al., 1986; Le Traon et al., 1990; Stammer, 1997). Because of the tradeoff between the ground-track spacing and orbit repeat period (see Section 8.1.4), it is not possible with any mapping algorithm to resolve the structure of mesoscale variability in maps constructed from the measurements obtained from any single satellite altimeter. Spatial resolution is especially a problem since a ground track spacing capable of resolving 50-km variability at midlatitudes would require an unacceptably long orbit repeat period of about two months that would severely alias the mesoscale variability temporally. Spatial aliasing is thus unavoidable from a single altimeter. It is highly desirable to select an orbit configuration that does not also alias the temporal characteristics of mesoscale variability. In this regard, the 10-day repeat of the T/P sampling pattern is preferable to the 17- and 35-day repeats of the Geosat and ERS sampling patterns. A 3-day repeat would be even better from the perspective of minimizing the effects of temporal aliasing of mesoscale variability. However, the corresponding longitudinal spacing of about 8.5° between neighboring ground tracks is so coarse that only the very largest spatial scales of variability could be resolved. While mesoscale variability exists on time scales shorter than the approximate 20-day periods that can be resolved by the T/P sampling pattern, it is often less energetic than the mesoscale variability on time scales from a few weeks to a few months. There is thus at least some

---

[18]The terminology "objective mapping" is usually considered to be synonymous with "optimal interpolation." This is clearly a misnomer since any space-time interpolation algorithm that can be formulated mathematically is "objective," and is thus distinguished from subjective hand contouring of data that cannot be automated on a computer. The synonymy of "objective analysis" and "optimal interpolation" is so steeped in the literature that it would be hopeless to try to change this unfortunate but accepted jargon.

solace in the fact that the T/P data are less contaminated by temporal aliasing than are the data from altimeter satellites with longer repeat periods. The tradeoff, however, is that the ground tracks are more coarsely spaced than are the ground tracks of altimeter satellites with longer orbit repeat periods (see Figures 61–64).

Most interpolation and smoothing algorithms can be expressed as a linear combination of the raw data values from which the smoothed estimate is constructed. In this case, it is straightforward to determine the filter transfer function for the particular parameters of the specific algorithm. Regardless of the details of the formulation, every linear estimate can be characterized by the half-power filter cutoff frequency and wavenumber that are effectively prescribed by the parameters of the mathematical formalism for the linear estimate. Even the optimal estimates obtained by the Gauss-Markov theorem based on the variance and autocorrelation structure of the SSH field can be characterized in terms of filtering properties (see, for example, Appendix B of Chelton and Schlax, 1993b); the filter cutoff wavenumber and frequency are directly related to the specified space-time autocorrelation function.

A formalism for determining the filter transfer function of an arbitrary linear estimate has been developed by Schlax and Chelton (1992). With presumed variance and space-time autocorrelation structure of the SSH field, but without the need for actual observations, estimates of the mapping errors can be derived based on the filter transfer function for any specific linear interpolation and smoothing scheme and an arbitrary space-time sampling pattern of the dataset. The methodology is therefore easily applied to investigate the predicted mapping errors for various actual or hypothetical altimeter satellite orbit configurations (Chelton and Schlax, 1994; Greenslade et al., 1997). For the specific case of optimal estimation, the analogous formalism for determining the expected mapping errors has been applied by Le Traon and Dibarboure (1999) to investigate the errors in SSH fields constructed from various altimeter orbit configurations.

As long as the filtering properties of the mapping algorithm are taken into consideration in the selection of the smoothing parameters of the algorithm, the SSH fields constructed from different mapping algorithms are generally very similar. The key issue is selection of an appropriate degree of smoothing. Examples can be found in the literature in which it is argued that mesoscale variability can be adequately resolved from maps constructed from altimeter data. Using different (but related) statistical formalisms, Greenslade et al. (1997) and Le Traon and Dibarboure (1999) have shown that this cannot be done with acceptable accuracy from SSH measurements from any single altimeter. For the T/P sampling pattern, the mapping errors are spatially inhomogeneous owing to the coarse longitudinal spacing of the ground tracks (see Figure 66 below). Because

of the short 10-day repeat period, the T/P mapping errors are relatively homogeneous temporally. The characteristics of the mapping errors for the ERS and Geosat sampling patterns are much more complex. The errors are smaller at some locations and times because of the closer spacing of ground tracks but are inhomogeneous temporally because of the relatively long repeat periods of the orbits (35 days for ERS and 17 days for Geosat).

For some applications such as assimilation of altimetric SSH fields into numerical ocean circulation models, spatial and temporal inhomogeneities of the mapping errors are not a major concern. As long as quantitative estimates of the mapping errors are available, the uncertainties can be formally accounted for in the data assimilation procedure. For descriptive analyses of SSH fields, however, inhomogeneous mapping errors pose difficulties. If the data have not been smoothed sufficiently, it can be difficult to distinguish real SSH variability from spurious variability owing to mapping errors.

The limitations in the resolution capability of SSH fields constructed from altimeter data are often obscured by the fact that SSH varies over a continuum of space and time scales. The large-scale, low-frequency signals are often the most energetic. The detrimental effects of unresolved mesoscale variability can therefore be difficult to discern. The desire to map the SSH field with high spatial resolution is presumably driven by an interest in this mesoscale variability since mapping the larger-scale variability does not require high spatial resolution. As the spatial and temporal scales of SSH variability are generally coupled, the mesoscale variability can be isolated by suitable high-pass filtering in time, thereby removing the background large-scale, low-frequency signals that are easily resolved by the altimeter sampling pattern.[19]

The mesoscale mapping limitations of altimeter data are illustrated in Figure 66 (see color insert) from SSH fields constructed from T/P data for a region of the western and central Pacific that encompasses a variety of mesoscale variability regimes. Perhaps most notable is a band of very energetic mesoscale variability centered near 5°N between about 100°W and the dateline that is associated with tropical instability waves generated by shear instabilities of the equatorial current system (Weidman et al., 1999; Chelton et al., 2000).

---

[19]For the discussion in this section, mesoscale variability refers to variability with short space and time scales that is associated with baroclinic eddies. The unresolved barotropic variability discussed in Section 5.5 has short time scales but large spatial scales. Since the T/P data analyzed for Figure 66 were not corrected for barotropic variability, the results presented here may be contaminated to some extent by large-scale barotropic signals that are passed by the 30–100-day band-pass filter. The pattern of the spurious SSH standard deviation in the top panel of Figure 66 and the similarity to the baroclinic mapping errors presented by Le Traon and Dibarboure (1999) suggests that the checkerboard pattern in the standard deviation field is primarily from unresolved baroclinic variability.

There is also a zonal band of energetic mesoscale variability near 20°N in the western Pacific in association with baroclinic instability of the North Pacific Subtropical Countercurrent (Qiu, 1999). A third band of energetic variability exists to the west of the Hawaiian Islands, extending westward at least to Wake Island. This mesoscale variability evidently forms from the impingement of the North Equatorial Current on the Hawaiian Islands (Mitchum, 1995).

To illustrate the effects of undersampling of mesoscale variability, T/P data were mapped with a high-resolution smoothing algorithm[20] that retained wavelengths longer than 2° of latitude by 2° of longitude and periods longer than 30 days. Estimates of mesoscale variability were isolated in the $2° \times 2° \times 30$-day smoothed SSH fields by high-pass filtering to eliminate variability with periods longer than 100 days. The standard deviation of this $2° \times 2°$ smoothed and 30–100 day bandpass filtered SSH is shown in the top panel of Figure 66. Behind the checkerboard pattern of large standard deviation at the crossovers of ascending and descending ground tracks, the three regions of energetic mesoscale variability are apparent as zonal bands of variability with 5–7 cm amplitude. These banded structures are punctuated by patches with more than 10-cm variability at the crossovers. This systematic geographical variation of the SSH variability is a clear indication of spurious variability owing to mapping errors from the inadequacy of the $2° \times 2°$ smoothing applied to the data. The checkerboard pattern of spurious SSH standard deviation is almost identical to the pattern of mapping errors predicted theoretically by Le Traon and Dibarboure (1999) from the optimal estimation formalism (see their Figure 4 and compare with the ground track pattern in their Figure 2). This similarity is because the filter transfer function of the $2° \times 2°$ smoothing applied here is very similar to the filter transfer function of the autocorrelation function adopted for the optimal estimation. The checkerboard pattern in the standard deviation map in the upper panel of Figure 66 can thus be interpreted as mapping errors.

It is perhaps surprising that the largest mapping errors in SSH fields constructed from T/P data occur at the crossover points. This is because smoothed estimates of SSH at the diamond centers with a 2° spatial span of the smoothing utilize observations from all of the surrounding ground tracks.[21] At the T/P crossover points, however, a 2° span is not large enough to include neighboring crossover points, either meridionally or zonally. As a consequence, the optimal estimate and the $2° \times 2°$ smoothed estimate of SSH at a crossover point are constrained by observations only along the satellite ground tracks that pass through the crossover point. The mean squared error of the SSH estimates are therefore large at the crossover points and small at the diamond centers.

An important distinction of optimal estimation from other mapping algorithms is that the filtering properties of the optimal estimation formalism are fixed by the true autocorrelation function of the SSH field. The inhomogeneous error structure (larger at the crossover points) is clearly an undesirable feature of SSH fields constructed by optimal estimation or by high-resolution mapping using any other mapping algorithm. The inhomogeneous mapping errors in the fields obtained by the Gauss-Markov formalism of optimal estimation can be addressed by artificially increasing the length and time scales of the autocorrelation function, but then the Gauss-Markov estimates are no longer the minimum mean squared error estimates. In this case, the Gauss-Markov formalism becomes equivalent (except for its much greater computational burden) to smoothing the data by other mapping techniques.

The effects of increasing the spatial smoothing of the T/P data are evident from the lower panel of Figure 66, which shows the standard deviation of SSH fields computed with the same 30–100 day bandpass filtering in time but with $6° \times 6°$ spatial smoothing. The checkerboard pattern of spurious large SSH variability is no longer present and the bands of energetic mesoscale variability are more clearly apparent than in the upper panel of Figure 66, albeit with somewhat smaller standard deviation because of the increased smoothing applied to the data. Even with this large degree of smoothing, however, there is evidence of sampling errors from unresolved mesoscale variability. Poleward of about 15°, the variability is locally largest at the crossover points. This is especially noticeable at the crossover points within the band of energetic mesoscale variability near 20°N. Spurious regions of locally high SSH variability with smaller amplitude are also evident at the crossover points near 20°S.

The dependence of the patterns of mapping errors on the smoothing applied to the data has been investigated for other

---

[20]Although not a crucial factor in the results presented here, the mapping algorithm used to produce these fields is the loess smoother introduced by Cleveland and Devlin (1988), which consists of a weighted least squares fit to a quadratic surface in space and time. The filtering characteristics of the loess smoother are described by Schlax and Chelton (1992) (see also Chelton and Schlax, 1994; Greenslade et al., 1997). A convenient property of the loess smoother is that the half-power point of the filter transfer function is approximately equal to the half span over which the data are fit to a quadratic surface. It is thus very straightforward to control the filtering properties of the loess smoother by a priori specification of the span of the smoother.

[21]Altimeter data are often mapped by simple block averaging of the data. The filter transfer function of the block-average smoother has been discussed by Schlax and Chelton (1992). For the $2° \times 2° \times 30$-day filter cutoffs in the SSH fields shown in the top panel of Figure 66, the equivalent block average smoother has dimensions of about $1.2° \times 1.2° \times 20$ days. Because of the coarse 2.8° longitudinal spacing of T/P crossover points, SSH estimates cannot be computed at the diamond centers for a block average with dimensions this small. The spatial dimension of the block-average smoother must be increased to approximately 2.5°, which corresponds to a filter transfer function cutoff of approximately 4°.

single satellite altimeters. The mapping errors for optimal estimation based on the Geosat sampling pattern are shown in Figure 1 of Le Traon and Dibarboure (1999) and Figure 8 of Chelton and Schlax (1994). An interesting feature of the mapping errors for Geosat is that the errors are smallest at the crossover points and largest at the diamond centers. This difference between the patterns of the Geosat and T/P mapping errors is because of the much closer spacing of the ground tracks in the Geosat 17-day repeat orbit; unlike the case for T/P, neighboring crossover points lie within the radius of influence of the optimal estimates at a given crossover point. The SSH estimates at Geosat crossovers are therefore constrained in all directions, thus explaining the smaller mapping errors. The mapping errors at the diamond centers of the Geosat sampling pattern are very similar in magnitude to the mapping errors at the T/P diamond centers (compare Figures 1 and 4 of Le Traon and Dibarboure, 1999, noting the different color bars in the two figures). Although the mean mapping error is smaller for the Geosat sampling pattern, the mapping errors are nonetheless very inhomogeneous spatially. Because of the longer Geosat orbit repeat period, the regions of maximum mapping errors migrate to different geographical locations, depending on the time during each 17-day repeat period (see Figure 1 of Le Traon and Dibarboure, 1999; see also Chelton and Schlax, 1994). This temporal inhomogeneity is also undesirable.

It is clear from the Figure 66 and the above discussion that mesoscale variability cannot be resolved by the sampling pattern of any single satellite altimeter. As shown in the lower panel of Figure 66, the effects of unresolved mesoscale variability can be greatly reduced by suitable smoothing of the data. Quantifying the degree of smoothing that should be applied is more problematic. Evaluation of the adequacy of the smoothing inevitably involves the establishment of subjective criteria for what constitutes "adequate." Greenslade et al. (1997) proposed two criteria: the average and the variability of the expected mapping errors over the spatial and temporal grid points at which SSH is estimated. A large mean mapping error obviously implies that the altimeter observations have not been smoothed sufficiently. Likewise, gross spatial or temporal inhomogeneity of the expected mapping errors is also a clear indication of insufficient smoothing. The mapping errors implied by the checkerboard pattern of spurious variability in the top panel of Figure 66 are obviously unacceptable. It is less clear, however, whether the inhomogeneities of the mapping errors that are apparent from the locally high SSH variability at the crossover points in the lower panel of Figure 66 are tolerable.

By the criteria proposed by Greenslade et al. (1997), assessment of the adequacy of the smoothing applied to construct SSH fields from altimeter data thus requires subjective specification of threshold tolerances on the magnitudes of the mean mapping error and the degree of spatial and temporal inhomogeneity of the mapping errors over the interpolation grid. With the thresholds proposed by Greenslade et al. (1997), mesoscale variability cannot be resolved even in SSH fields constructed from tandem altimeter datasets. It is noteworthy, however, that the threshold criteria proposed by Greenslade et al. (1997) suggest that the $6° × 6° × 30$ day smoothing applied for the SSH standard deviation shown in the bottom panel of Figure 66 is adequate to resolve large scales of variability. The effects of sampling errors are nonetheless evident from the spurious high standard deviations near the crossover points.

Different conclusions about the resolution capability can be reached, depending on the thresholds that are adopted for the Greenslade et al. (1997) criteria. Le Traon and Dibarboure (1999) argued that the thresholds proposed by Greenslade et al. (1997) are overly restrictive. By relaxing the threshold for the mapping error inhomogeneity criterion, they suggest that mesoscale variability can be adequately resolved by tandem altimeter missions. Quantitative specification of the most appropriate thresholds for the mean and inhomogeneity of expected mapping errors is thus a controversial issue.

As high-quality data have become available from simultaneous operation of multiple satellite altimeters (e.g., T/P, ERS and the upcoming Jason-1 and ENVISAT altimeters, see Section 9), altimeter measurement errors have become increasingly important for high-resolution mapping of the SSH field. The higher spatial and temporal density of the measurements available from the multiple satellite ground tracks reduces the effects of sampling errors that are so limiting in SSH fields constructed from a single altimeter satellite, thus increasing the relative importance of measurement errors. If not explicitly accounted for in the mapping algorithm, measurement errors in the dense network of intersecting ground tracks of multiple altimeter satellites can result in "trackiness" in the SSH fields. Of particular concern are the measurement errors that are introduced when orbit errors are larger for one of the satellites (e.g., the factor of 2–3 times larger orbit errors for the ERS satellites than for T/P, see Table 1). Le Traon et al. (1998) have presented an optimal estimation mapping technique that takes into account the effects of orbit errors (or any other errors with large along-track coherence). Similar modifications could be made to other mapping algorithms to mitigate the effects of measurement errors.

The preceding discussion has outlined the issues that must be taken into consideration when constructing SSH fields from altimeter data. The two criteria proposed by Greenslade et al. (1997) for assessing the quality of the interpolated and smoothed SSH fields (i.e., the mean mapping error and the spatial and temporal inhomogeneity of the mapping errors over the interpolation grid points) provide quantitative guidelines that can be used to select the smoothing parameters for any particular mapping formalism. Other criteria could be developed to assess the adequacy of the degree of smoothing applied to altimeter data. Determination of the most appropriate degree of smoothing is presently an active and debatable topic of research. Although it is easy to discern when the data have not been sufficiently smoothed in extreme cases (e.g., the top panel of Figure 66), the best choice of smoothing parameters will depend, to some extent, on the application of interest.

8.2. Calibration and Validation

To provide assurance that the performance requirements for altimeter measurement accuracy are met or exceeded, extensive calibration and validation (cal/val) are important elements of altimeter missions. Cal/val embraces a wide variety of activities, ranging from the interpretation of information from internal-calibration modes of the sensors to validation of the fully corrected sea-level estimates using in situ data (see Born, 1995). The activities begin well before launch, with rigorous laboratory calibration of the satellite sensors and development of a functional science-data processing system. During the initial months after launch, engineering assessments and statistical evaluations of the geophysical data products are given high priority. This initial "verification" phase also features intensive comparisons of the geophysical data against data from *in situ* sources. The results of the comparisons performed during the verification phase lead invariably to a tuning of the algorithms that are at the heart of the science-data processing system. Establishing that the performance requirements are met is a prerequisite to releasing the geophysical data records to the science community.

The end of the verification phase does not signal a termination of cal/val activities. Ongoing verification, albeit at a reduced level, is essential throughout the life of the altimeter mission. A particular challenge is the monitoring of subtle bias and drift errors at levels that are necessary to support studies of basin- and global-scale sea-level changes (Chapter 8). Dedicated and continuous *in situ* observations from a variety of sources are indispensable for meeting this challenge.

While the outputs of all cal/val activities are crucial to assuring the integrity of the measurements, we focus herein principally on the *in situ* comparisons. They provide a unique perspective on the accuracy of sea surface height, wind speed, significant wave height, and other higher-level geophysical measurements. Such a perspective cannot be gained from individual analyses of the diversity of component measurements (e.g., altimeter range and backscatter measurements, orbit height estimates and corrections for atmospheric refraction and attenuation).

8.2.1. Point Calibration

The traditional "overhead" concept of altimeter *in situ* calibration invokes direct satellite overflights of a thoroughly

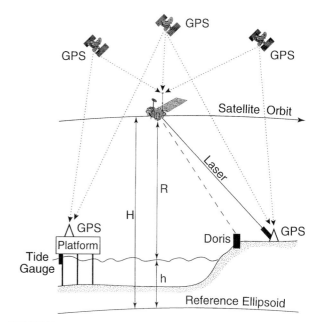

FIGURE 67 Schematic diagram showing the geometry of altimeter point calibration. At the instant the satellite passes overhead, geocentric sea surface height is observed independently by the altimeter and *in situ* measurement systems. Assuming that the *in situ* systems are properly calibrated, the difference represents the "altimeter bias."

instrumented experiment site, as shown schematically in Figure 67. Site essentials for this type of "point calibration" are an accurate tide gauge and a means of tying the tide-gauge sea-level readings to the geocenter. In an ideal situation, the experiment site should be located along a repeating ground track, but also where the altimeter and radiometer footprints do not experience significant land intrusion. Small islands (Ménard *et al.*, 1994) or off-shore platforms (Francis, 1992; Christensen *et al.*, 1994a) are excellent candidates. Accurate estimates of the orbit height as the satellite overflies the site are also critical. Traditionally, this has necessitated the presence of a nearby satellite laser-ranging station (Section 4.2.2).

The Bermuda experiments for GEOS-3 (1976) and Seasat (1978) provided the earliest examples of the overhead altimeter calibration concept (Martin and Kolenkiewicz, 1981; Kolenkiewicz and Martin, 1982). Sea-surface-height data from the *in situ* and satellite measurement systems were compared at overflight times to infer the overall bias in the GEOS-3 and Seasat measurement systems. Laser data from the NASA station at Bermuda were the most important source of tracking data available for determining the orbital height of the satellite as it passed overhead.

Dedicated calibration sites continue to serve a vital function in modern altimeter missions. For T/P, calibration sites have been maintained at the Harvest oil platform off the coast of central California (Christensen *et al.*, 1994a) and Lampedusa Island in the Mediterranean Sea (Ménard *et al.*,

1994). Although satellite laser ranging systems remain valuable for determining accurate estimates of the orbit heights over experiment sites (Bonnefond *et al.*, 1995; Murphy *et al.*, 1996), the DORIS (Section 4.2.3) and GPS (Section 4.2.6) systems have emerged as important tools for both satellite- and tide-gauge positioning in calibration experiments. For example, Christensen *et al.* (1994a) used GPS-based estimates for determining the heights of both the Harvest tide-gauge system and the satellite in their validation of the T/P data.

Owing to the collocation of many *in situ* sensors, dedicated calibration sites are useful for measuring environmental conditions that underlie the formation of the fully-corrected sea-surface height. In addition to sea level, a well-instrumented site provides a means of measuring wind, wave, tropospheric and ionospheric conditions, thus permitting a useful segregation of the various potential error sources. The most prominent goal of point calibration, however, has been the determination of the absolute bias in the altimeter measurement system. While it is common to refer to this as simply the "altimeter bias," it actually embraces mean errors from all of the sensors and accompanying corrections that contribute to the measurement of sea-surface-height. Such a bias could have any number of causes, ranging from unmodeled hardware delays onboard the spacecraft to errors in the models used in the ground-processing software. Magnitudes of the biases for recent missions range from about 40 cm for ERS-1 (Francis, 1992; Murphy *et al.*, 1996) to near zero for both the dual-frequency NASA altimeter and the single-frequency CNES altimeter onboard the T/P satellite (Haines *et al.*, 1998). (As discussed in Section 2.4.6, the dual-frequency altimeter originally had a 13-cm bias, but this was later found to be an artifact of a software error in the ground-processing system.) The effect of the bias is to introduce a fictitious elevation or depression of the globally averaged observations of sea-surface height, which impacts specialized studies such as the measurement of the earth's mean radius. More importantly, the biases can negatively impact altimetric studies that combine data from different missions, or even from different altimeter systems flying on the same platform, e.g., the dual-frequency and single frequency altimeters on T/P or the primary and backup dual-frequency altimeter systems on T/P, referred to as Side A and Side B.

The unprecedented accuracy of the T/P data has prompted several studies of the change in the global mean sea level (Chapter 8). Consequently, efforts to calibrate potentially subtle drifts in the measurement system have received significant attention. This problem is quite challenging from the perspective of point calibration, as the eustatic sea-level rise over the last century is estimated to be only 1.8 mm year^{-1} (Douglas, 1995). State-of-the-art point calibration for T/P yields instantaneous bias estimates every 10 days with a repeatability of about 3 cm (Christensen *et al.*, 1994a; Ménard *et al.*, 1994). To estimate measurement-system drift at the 1 mm year^{-1} level requires a minimum of 3–5 years of data, as well as near-perfect control over potential systematic errors such as unmodeled subsidence or uplift of the tide gauges. Harvest is one of the longest operating calibration sites with the potential to support this level of accuracy. One of the most significant challenges at Harvest, however, is measuring the platform subsidence, estimated at close to 1 cm year^{-1}, with sufficient accuracy to support detection of potentially subtle drifts (1 mm year^{-1}) in the T/P measurement system (Haines *et al.*, 1998). Additional long calibration time series have emerged from data collected in Bass Strait (White *et al.*, 1994) and the English Channel (Murphy *et al.*, 1996). CNES is developing a permanent calibration occupation on Corsica Island off the south coast of France in the Mediterranean Sea. These sites should continue to return important measurements for validating data from T/P and future altimeter missions.

An exciting and relatively new technique of point calibration is based on GPS buoys. Data from a precise GPS receiver on drifting or moored buoys can be used to accurately determine local sea surface height (Hein *et al.*, 1990; Rocken *et al.*, 1990; Kelecy *et al.*, 1994). By deploying a moored buoy along a ground track, the altimeter measurement system bias can be estimated. Using data from a GPS-equipped spar buoy near Harvest, Born *et al.* (1994) computed a single-pass bias estimate for the T/P dual-frequency altimeter that agreed with the corresponding figure from the platform tide gauge at the 1-cm level. The advantage of a buoy deployment over a fixed calibration site is that the experiment can be carried out nearly anywhere on the globe. The best determinations of geocentric sea surface height from a GPS buoy are achieved when a terrestrial (fiducial) GPS site is located nearby. However, recent advances in GPS technology enable accurate positioning even for isolated, roving GPS receivers (Zumberge *et al.*, 1998).

8.2.2. Distributed Calibration

While information from dedicated point calibration sites has proven invaluable for detecting biases in the T/P measurement systems, the most reliable *in situ* information on system drift has been obtained by the global tide-gauge network. It has long been common practice to use tide-gauge data, primarily from island stations in the tropics, to validate regional variations in the sea-surface height. In contrast to the overhead calibration approach, the tide-gauge data are processed to expose long-period (e.g., seasonal) sea-level variations that can be compared with smoothed altimeter sea-surface height changes (e.g., Cheney *et al.*, 1994; Mitchum, 1994). The problem of determining a reliable measure of the global stability of the altimeter system is considerably more challenging. Cooperating tide gauges that comprise the global network are rarely found in fortuitous locations along the satellite ground track. Though the numbers are increasing, relatively few of the gauges are presently equipped with GPS or DORIS systems to provide information on vertical land motion. Finally, prior to T/P, it was necessary to reduce the effects of residual orbit errors using various empirical techniques, exacerbating the interpretation of secular changes in the tide-gauge/altimeter comparisons. With the advent of T/P and the availability of precise orbits of nearly uniform global quality, many tide gauges in the global network can serve as ad hoc calibration sites.

Capitalizing on the high quality of the T/P data, Mitchum (1998) devised an innovative approach to monitoring global altimeter stability by comparisons with sea-level time series from selected gauges near the satellite ground track. The drift estimate from the global tide-gauge analysis provides information that is complementary to the calibration estimates from the dedicated point calibration sites. The significance of this complementary information was amply demonstrated with the discovery in 1996 of an algorithm error for the T/P dual-frequency altimeter that introduced both a global bias (13 cm) and a slow drift ($\sim$8 mm year^{-1}) in the global mean sea level (see Section 2.4.6). While the effects of the mean component of the error were readily observed by dedicated calibration sites and a GPS buoy soon after launch (Born et al., 1994; Christensen et al., 1994a; Ménard et al., 1994; White et al., 1994), a multi-year calibration time series from the global tide-gauge network was needed for convincing detection of the slow drift (Mitchum, 1998). In retrospect, the combined results provided a remarkable portrait of the total effect of the algorithm error on the sea surface height. This experience helped spur efforts to further enhance the global network by identifying 30 selected tide gauges where vertical land motion measurements are needed to support improved altimeter stability estimates (e.g., Mitchum, 1997). Some of the stations in the enhanced network are already instrumented with GPS or DORIS (e.g., Cazenave et al., 1999). Additional upgrades are underway and should be in place for the launches of future altimeters.

Other important calibration exercises using distributed sites emerged from the T/P experience. Morris and Gill (1994), for example, examined the performance of the altimeter systems using lake-level stations on the shores of the U.S. Great Lakes. The technique of distributed calibration has also been applied to monitor the measurements from the T/P Microwave Radiometer (TMR) from globally distributed radiosonde observations (Keihm et al., 2000) and from coastal and island GPS stations (Haines and Bar-Sever, 1998). These calibrations of opportunity have detected a slight drift (1–1.5 mm year^{-1}) in the estimated wet-zenith path delays from the TMR (see Section 3.1.2). This spurious drift provides a partial explanation of the residual trend observed in the Mitchum (1998) global tide-gauge calibration time series.

8.3. Measurement Performance

To illustrate how the cal/val results are synthesized, the T/P dual-frequency altimeter can be considered as an example. The estimated accuracies for this measurement system based on information from cal/val studies are summarized in Table 11. The root sum of squares of the environmental errors in the range estimates (i.e., the ionospheric and tropospheric effects and the EM and skewness biases) is 2.7 cm. The single-pass overall accuracy of the sea-surface-height measurements is approximately 4 cm. This exceptional performance is due in large measure to the reduction of the rms orbit error to 2.5 cm although all components of the measurement system meet or exceed expectations. The 4-cm figure is supported by the in situ comparisons discussed previously, and will serve as the benchmark against which the performance of future missions such as Jason-1 (Ménard et al., 1999) will be measured.

By averaging the T/P data over large spatial scales, the performance can be further enhanced. For example, Cheney et al. (1994) showed that altimeter data over the tropical Pacific have an accuracy of approximately 2 cm at wavelengths of a few hundred km. At the global scale, current estimates of the measurement-system bias and drift (Table 11) generally fall in the ranges of ±2 cm and ±2 mm year^{-1}, respectively. At these levels, the estimates are highly sensitive to the exact assumptions used in forming the sea surface heights.

9. OUTLOOK FOR FUTURE ALTIMETER MISSIONS

The unique utility of satellite altimetry for studies of the various branches of earth science as well as practical applications discussed in succeeding chapters underscore the importance of maintaining altimetric observations on a long-term basis. Strategies for such sustained observing systems have been discussed by Nowlin (1999). The highly accurate and long record of sea-surface height variability obtained from the T/P mission has met the requirement for observing the temporal change of the global ocean topography for studying ocean circulation and short-term climate variability. Such capability must be maintained for the understanding and prediction of the oceanic environment and its effects on climate on interannual and longer time scales.

In the near term, T/P will be succeeded by its follow-on mission called Jason-1 (Ménard et al., 1999). Jason-1 is a continuing collaboration between the U.S. and France with an expected launch in early 2001. Jason-1 will observe the sea surface topography in the same orbit as T/P and will carry the same suite of instruments as T/P but with more advanced technology and improved measurement performance. The overall accuracy of Jason-1 measurements of

TABLE 11.   The rms System Measurement Accuracy for Side A of the TOPEX/POSEIDON
Dual-frequency Altimeter, Which was Active from Launch Through February 1999

Single-pass sea surface height accuracy		
NASA radar altimeter noise[a]	1.7 cm	Fu *et al.* (1994)
Ionosphere[b]	0.5 cm	Imel (1994)
EM bias	2.0 cm	Rodriquez and Martin (1994b)
Skewness	1.2 cm	Rodriquez and Martin (1994b)
Dry troposphere	0.7 cm	Fu *et al.* (1994)
Wet troposphere	1.1 cm	Ruf *et al.* (1994)
Orbit	2.5 cm	Ries and Tapley (1999)
Total (RSS) sea surface height	4.1 cm	
Sea-surface height bias and drift errors		
Measurement-system bias[c]	−0.5 cm	Haines *et al.* (1998)
Measurement-system drift[d]	−0.2 cm year^{-1}	Mitchum (1998)
Single-pass wind/wave accuracies		
Wind speed	2 m sec^{-1}	Callahan *et al.* (1994)
Significant wave height[e]	0.2 m	Callahan *et al.* (1994)

[a]Based on 1-sec averages of the range estimates for 2-m significant wave height.

[b]Based on 100-km along-track averages of the dual-frequency altimeter estimates of the ionospheric range correction.

[c]For January 1, 1993, after correcting for the oscillator drift error described in Section 2.4.6. Accounting for systematic *in situ* measurement errors (1–2 cm), the bias is not statistically distinguishable from zero.

[d]Applying a correction for the recently discovered drift (0.1 cm year^{-1}) in the TMR measurements of path delay (Keihm *et al.*, 2000) would decrease the magnitude of this estimate.

[e]Based on data collected prior to degradation of the side A point target response.

sea-surface height is expected to be comparable to and will probably exceed the T/P performance summarized in Section 8.3.

While Jason-1 will demonstrate the operational capability of satellite altimetry, the strategy for making altimetry become part of a long-term operational ocean climate observing system is a pressing issue. A series of Jason-1 follow-on missions has been proposed to maintain continuity of highly accurate, long-term altimetric observations of the global ocean topography. Planning is underway for Jason-2, the first of this proposed series of altimeters for operational climate observations with a proposed launch in late 2004. The performance goal of this operational sequence of altimeter missions is to progress systematically from the high standard set by T/P to an improved overall accuracy of 1 cm through continued dedicated effort to improve the correction algorithms described in this chapter and the introduction of new technologies for improved measurement accuracy. While other systems for observing ocean topography will continue (e.g., the global tide gauge network and other altimeter missions), the T/P-Jason series of altimeters will provide a benchmark for calibrating and integrating other measurements to achieve a globally consistent dataset for climate research.

As discussed in Section 8.1, a key issue in flying operational altimeter missions is orbit selection. Most operational satellite missions are flown in sun-synchronous orbits for a variety of reasons relating to other instruments onboard the satellite. Such orbits are not acceptable for highly accurate altimetry. Tidal signals in sun-synchronous observations of sea surface topography cannot be removed with the accuracy required for monitoring global change of the ocean circulation and mean sea level. Although the accuracies of tide models have greatly improved as a result of extensive data assimilation studies that have utilized the T/P data (see Chapter 6), the unpredictable part of the tides would result in an error of 2–3 cm that is very difficult to remove from altimeter data collected in sun-synchronous orbits. The most notable example is that all of the errors of the solar tidal constituents alias into the zero frequency (see Section 8.1.5) and therefore limit the accuracy of mean sea level estimated from a sun-synchronous altimeter. These residual tidal errors can be reduced only by flying in a carefully selected orbit such as the T/P orbit.

The importance of long-term continuity of altimeter observations along the T/P ground tracks should be emphasized. An 8-year record of sea surface height variability has now been established by T/P. Jason-1 and the proposed Jason-2 will continue the climate record of sea-surface height observations along the same ground tracks. Changing the ground track pattern would disrupt the continuity of the data record, analogous to moving a tide gauge to another location along a coastline. A long data record would have to established along the new ground tracks and cross-calibration against the observational record along the T/P ground tracks would complicate and possibly limit the utility of the merged altimeter datasets for studies of climate variability.

For high-resolution mapping of mesoscale variability of the sea surface height field, the continuity of the altimetric data record along the T/P ground tracks is less of an issue. For such applications, the important issue is the sampling density. A single altimeter is limited by a compromise in sampling the spatial and temporal scales of ocean variability (see Section 8.1.7). Multiple conventional nadir-looking altimeters are needed to map ocean mesoscale eddies (Greenslade *et al.*, 1997, Le Traon and Dibarboure, 1999) and to resolve the full range of scales of variability that are important to large-scale ocean circulation (see Chapter 2). During the Jason-1 time frame, the European ENVISAT will make altimetric measurements in the ERS orbit, continuing the dual-altimeter coverage of the global ocean that is presently being obtained from the simultaneous operation of T/P and the ERS-2 altimeter.

While the cost-effectiveness of traditional satellite altimeter missions is improving, new remote sensing techniques are being developed for mapping ocean topography with high two-dimensional spatial resolution. A radar altimeter in which the full beam-limited footprint (see Section 2.4.1) is utilized for range estimation is described by Raney (1998). Along-track resolution is achieved by utilizing variations of the Doppler shift in the frequencies of the radar returns over the beam-limited footprint area. The number of independent looks for a given target is thus much larger than that of the conventional altimeter, leading to reduced noise level as well as smaller footprint size. Because the same signal-to-noise ratio of conventional altimeters can be achieved with lower power, the weight and cost of this new altimeter design can be significantly reduced. Launching several of these low-cost altimeters into a formation of orbits has been proposed as a method of achieving high cross-track resolution (D. Porter and K. Raney, personal communication, 1999).

A concept for achieving high cross-track resolution from multiple conventional altimeters on a single satellite platform was described by Bush *et al.* (1984). The stringent requirement for the stability of a satellite and its pointing accuracy has been an insurmountable challenge. Recent developments of spacecraft technologies are making multibeam altimetry more feasible. By measuring the interference between signals received by two altimeters separated by a 5-m boom at the T/P altitude of 1336 km, a swath width of about 200 km can be achieved with slight off-nadir pointing angles using an interferometry technique (Rodríguez *et al.*, 1999). Extensive ground-based data processing is required to reduce the measurement errors of such a system to a level of 5 cm.

Another promising technique is the analysis of the reflections of GPS signals off the ocean surface to detect ocean to-pographic features (LaBrecque *et al.*, 1998). A great deal remains to be learned about this technique such as the physics of sea-state bias at large incidence angles as well as the complicated sampling characteristics of the GPS measurements.

Another long-standing issue in satellite altimetry is the lack of high-resolution knowledge of the marine geoid. As summarized in Section 5.2 and Chapter 2, this has long been an impediment to the use of altimetric observations for determining the absolute circulation of the ocean. A substantial improvement in the knowledge of the geoid is expected over the next 5–10 years (see Chapter 10). A joint United States–German satellite mission called the Gravity Recovery and Climate Experiment (GRACE) is under development with a planned launch in 2001 (Davis *et al.*, 1999). GRACE will use a pair of low earth-orbiting satellites whose relative distance is precisely determined (Wahr *et al.*, 1998). The measurement of the distance plus onboard accelerometer measurements will be analyzed to improve the geoid accuracy to better than 1 cm on spatial scales greater than 200 km (half wavelength). In addition, the European Space Agency is developing a high-resolution gravity mission called the Gravity Field and Steady-state Ocean circulation Explorer (GOCE) with a planned launch in mid-2004. GOCE estimates of the gravity field are based on gravity gradiometry and GPS precision tracking in a very low orbit. GOCE observations are expected to extend the 1-cm accuracy of geoid undulations down to spatial scales of 100 km (LeGrand and Minster, 1999).

The accurate high-resolution global geoid that will be constructed from these dedicated gravity missions can be used retroactively to obtain absolute sea-surface topography from past altimeter missions. Aliasing of errors of the solar tidal constituents into the zero frequency will continue limit the accuracy of absolute sea surface topography estimated from sun-synchronous altimeters such as ERS-1, ERS-2 and ENVISAT. However, an accurate global geoid will provide accurate estimates of the absolute sea surface topography and surface geostrophic current velocity from altimeters in non sun-synchronous orbits. The spatial resolution of 100 km or better for the geoid will resolve important details of the mean structures of the boundary currents, the Antarctic Circumpolar Current and mesoscale eddies. Such information will provide powerful constraints for ocean models as well as ocean-atmosphere coupled models for simulation and prediction of the interannual and decadal variability of the global environment.

ACKNOWLEDGMENTS

We thank P. A. M. Abusali, Srinivas Bettadpur, George Born, Frank Lemoine, Xiaoqing Pi, Michael Schlax, Detlef Stammer, and Victor Zlotnicki for their careful reviews and constructive criticisms of various parts of this manuscript. Numerous technical discussions with Michael Schlax,

Michael Freilich, George Hayne, David Hancock, and Ed Walsh were also extremely helpful. Donna Witter provided many helpful comments on an early draft of a portion of this manuscript that was written nearly a decade ago. We are grateful to numerous colleagues for supplying data and figures, as acknowledged in the various figure captions. We also thank Michael Schlax and David Reinert for extensive help with final production of the figures. DBC was supported for this work by contracts 958127 and 1206715 from the Jet Propulsion Laboratory. JCR was supported by contract 956689 from the Jet Propulsion Laboratory. The contributions by LLF, PSC, and BJH were carried out by the Jet Propulsion Laboratory, California Institute of Technology, under contract with the National Aeronautic and Space Administration; support from the TOPEX/POSEIDON Project is gratefully acknowledged.

References

Agnieray, P. (1997). The DORIS system performances and evolutions. Coordination of Space Techniques for Geodesy and Geodynamics Bulletin No. 14, *In* "Advanced Space Technology in Geodesy–Achievements and Outlook," (R. Rummel, C. Reigber, and H. Hornik, Eds.), pp. 73–80, Deutsches Geodätisches Forschungsinstitut, Munich.

Alaka, M. A., and Elvander, R. C. (1972). Optimum interpolation from observations of mixed quality. *Mon. Wea. Rev.*, **100**, 612–624.

Alishouse, J. C. (1983). Total precipitable water and rainfall determinations from the SEASAT scanning multichannel microwave radiometer. *J. Geophys. Res.*, **88**, 1929–1935.

Alishouse, J. C., Snider, J. B., Westwater, E. R., Swift, C. T., Ruf, C. S., Snyder, S. A., Vongsathorn, J., and Ferraro, R. R. (1990a). Determination of cloud liquid water content using the SSM/I. *IEEE Trans. Geosci. Rem. Sens.*, **28**, 817–822.

Alishouse, J. C., Snyder, S., and Ferraro, R. R. (1990b). Determination of oceanic total precipitable water from the SSM/I. *IEEE Trans. Geosci. Rem. Sens.*, **28**, 811–822.

Anderle, R. J. (1986). Doppler satellite measurements and their interpretation, *In* "Space Geodesy and Geodynamics" (A. J. Anderson and A. Cazenave, Eds.), pp. 113–167, Academic Press, London.

Anderson, P. H., Asknes, K., and Skonnord, H. (1998). Precise ERS-2 orbit determination using SLR, PRARE, and RA observations. *J. Geodesy*, **72**, 421–429.

Antreasian, P. G., and Rosborough, G. W. (1992). Prediction of radiant energy forces on the TOPEX/POSEIDON spacecraft. *J. Spacecr. Rockets*, **29**, 81–90.

Apel, J. (1994). An improved model of the ocean surface wave vector spectrum and its effect on radar backscatter. *J. Geophys. Res.*, **99**, 16,269–16,291.

Arnold, D., Kong, J., and Melville, W. (1991). Physical optics prediction of EM bias. *In* "Progress in Electromagnetic Research," IEEE Press, Piscataway, NJ.

Bamber, J. L., Ekholm, S., and Krabil, W. (1998). The accuracy of satellite radar altimeter data over Greenland ice sheet determined from airborne laser data. *Geophys. Res. Lett.*, **25**, 3177–3180.

Barotto, B. (1995). Introduction of stochastic parameters to improve the estimation of the trajectories of a dynamical system by a least-squares method. Application to centimeter-level orbit determination of a satellite. Ph.D. Thesis, L'Universite Paul Sabatier, Toulouse, France, 284 pp.

Barrett, A. H., and Chung, V. K. (1962). A method for the determination of high-altitude water vapor abundance from ground-based microwave observations. *J. Geophys. Res.*, **76**, 4259–4266.

Barrick, D. E. (1968). Rough surface scattering based on the specular point theory. *IEEE Trans. Antennas Propag.*, **13**, 449–454.

Barrick, D. E., and Peake, W. H. (1968). A review of scattering from surfaces with different roughness scales. *Radio Sci.*, **3**, 865–868.

Barrick, D. E., and Lipa, B. J. (1985). Analysis and interpretation of altimeter sea echo, *Adv. Geophys.*, **27**, 60–99.

Bašić, T., and Rapp, R. (1992). "Oceanwide prediction of gravity anomalies and sea surface heights using GEOS-3, Seasat and Geosat altimeter data and ETOPO5U bathymetric data." Rep. 416, Dept. Geod. Sci. and Surv., Ohio State University, Columbus.

Bent, R. B., Llewellyn, S. K., Nesterczuk, G., and Schmid, P. E. (1976). The development of a highly successful worldwide empirical ionospheric model. *In* "Effect of the Ionosphere on Space Systems and Communications" (J. Goodman, Ed.), pp. 13–28, Natl. Tech. Inf. Serv., Springfield, VA.

Berthias, J. P., and Houry, S. (1998). Next day precise orbits for TOPEX/POSEIDON using DORIS. *In* "Advances in the Astronautical Sciences" (T. H. Stengle, Ed.), Vol. 100, pp. 173–184.

Bertiger, W. I., *et al.* (1994). GPS precise tracking of TOPEX/POSEIDON: Results and implications. *J. Geophys. Res.*, **99**, 24,449–24,464.

Bertiger, W. I., *et al.* (1998). A real-time wide area differential GPS system. *Navigation*, **44**, 433–447.

Bettadpur, S., and Eanes, R. J. (1994). Geographical representation of radial orbit perturbations due to ocean tides; Implications for satellites. *J. Geophys. Res.*, **99**, 24,883–24,894.

Beutler, G., Brockman, E., Gurtner, W., Hugentobler, U., Mervart, L., Rothacher, M., and Verdun, A. (1994). Extended orbit modeling techniques at the CODE processing center of the international GPS service for geodynamics (IGS): Theory and initial results. *Man. Geod.*, **19**, 367–386.

Beutler, G., Rothacher, M., Springer, T., Kouba, J., and Neilan, R. E. (1999). The International GPS Service (IGS): An interdisciplinary service in support of Earth sciences. *Adv. Space Res.*, **23**, 631–653.

Biancale, R., Crenne, E., Berger, C., and Barlier, F. (1993). Modeling the upper atmospheric density with the help of satellite data. *In* "Proceedings of the 112th IAG Symposia, Geodesy and Physics of the Earth," (H. Montag and Charles Reigber, Eds.), pp. 124–128, Springer-Verlag.

Bierman, G. J. (1977). *In* "Factorization Methods for Discrete Sequential Estimation," 241 pp. Academic Press, San Diego, CA.

Bilitza, D. (1997). International Reference Ionosphere—Status 1995/96. *Adv. Space Res.*, **20**, 1755–1759.

Bindschadler, R. (1998). Monitoring ice sheet behavior from space. *Rev. Geophys.*, **36**, 79–104.

Bisagni, J. J. (1989). Wet tropospheric range corrections for satellite altimeter-derived dynamic topographies in the western North Atlantic. *J. Geophys. Res.*, **94**, 3247–3254.

Black, H. D., and Eisner, A. (1986). A new technique for monitoring the water vapor in the atmosphere. *J. Geophys. Res.*, **91**, 2331–2337.

Boissier, C., De Mey, P., Dombrowsky, E., Jourdan, D., Ménard, Y., Minster, J. F., Perigaud, C., and Rouquet, M. S. (1989). Progress in mesoscale variability analysis from satellite altimetry data. *Adv. Space Res.*, **9**, 467–472.

Bonnefond, P., Exertier, P., Schaeffer, P., Bruinsma, S., and Barlier, F. (1995). Satellite altimetry from a short-arc technique: Application to the Mediterranean. *J. Geophys. Res.*, **100**, 25,365–25,382.

Bordi, J. J., Ries, J. C., and Tapley, B. D. (2000). Precise orbits for the ERS-2 altimeter satellite. *J. Astron. Sci.*, in press.

Born, G. H., et al. (1994). Calibration of the TOPEX altimeter using a GPS buoy. *J. Geophys. Res.*, **99**, 24,517–24,526.

Born, G. H., Ed. (1995). Special Issue: TOPEX/POSEIDON Calibration/Validation. *Mar. Geod.*, **18**, 1–2.

Born, G. H., Tapley, B. D., and Santee, M. L. (1986). Orbit determination using dual crossing arc altimetry. *Acta Astronautica*, **13**, 157–163.

Bracewell, R. N. (1986). *In* "The Fourier Transform and its Applications," 474 pp. McGraw-Hill, New York.

Brenner, A. C., Koblinsky, C. J., and Beckley, B. D. (1990). A preliminary estimate of geoid-induced variations in repeat orbit satellite altimeter observations. *J. Geophys. Res.*, **95**, 3033–3040.

Bretherton, F. P., Davis, R. E., and Fandry, C. B. (1976). A technique for objective analysis and design of oceanographic experiments applied to MODE-73. *Deep Sea Res.*, **23**, 559-582.

Bretschneider, C. L. (1958). Revisions in wave forecasting: deep and shallow water. *Proc. 6th Conf. Coast. Eng.*, 30–67.

Brown, G. S. (1977). The average impulse response of a rough surface and its applications. *IEEE Trans. Antennas Propag.*, **25**, 67–74.

Brown, G. S. (1990). Quasi-specular scattering from the air-sea interface. *In* "Surface Waves and Fluxes." Vol. 2, pp. 1–39, (G. L. Gernaert and W. J. Plant, Eds.), Kluwer, Norwell, MA.

Brown, G. S., Stanley, H. R., and Roy, N. A. (1981). The wind speed measurement capability of spaceborne radar altimetry. *IEEE J. Oceanic Eng.*, **6**, 59–63.

Brown, O. B., and Cheney, R. E. (1983). Advances in satellite oceanography. *Rev. Geophys. Space Phys.*, **21**, 1216–1230.

Bush, G. B., Dobson, E. B., Matyskiela, R., and Kilgus, C. C. (1984). An analysis of a satellite multibeam altimeter. *Mar. Geod.*, **8**, 345–384.

Cailliau, D., and Zlotnicki, V. (2000). Precipitation detection by the TOPEX/POSEIDON dual-frequency radar altimeter, TOPEX microwave radiometer, special sensor microwave/imager and climatological average shipboard reports. *IEEE Trans. Geosci. Rem. Sens.*, **38**, 205–213.

Callahan, P. S. (1984). Ionospheric variations affecting altimeter measurements: A brief synopsis. *Mar. Geod.*, **8**, 249–263.

Callahan, P. S. (1987). An overview of the ionosphere and combined height for ionospheric corrections algorithm. *Proc. WOCE/NASA Altimeter Algorithm Workshop*, Corvallis, OR.

Callahan, P. S. (1991). "TOPEX gound system science algorithm specification." Jet Propul. Lab. Doc. JPL D-7075, Rev. A, Change 1, Rev. A.

Callahan, P. S. (1993). "TOPEX GDR Users Handbook." Jet Propul. Lab. Doc. JPL D-8944 Rev. A.

Callahan, P. S., and Haub, D. (1995). On-orbit measurement of TOPEX/POSEIDON altimeter antenna pattern. *Mar. Geod.*, **18**, 117–128.

Callahan, P. S., Morris, C. S., and Hsiao, S. V. (1994). Comparison of TOPEX/POSEIDON σ_0 and significant wave height distributions to Geosat. *J. Geophys. Res.*, **99**, 25,015–25,024.

Carter, D. J. T., Challenor, P. G., and Srokosz, M. A. (1992). An assessment of Geosat wave height and wind speed measurements. *J. Geophys. Res.*, **97**, 11,383–11,392.

Cartwright, D. E., and Longuet-Higgins, M. S. (1956). The statistical distribution of the maxima of a random function. *Proc. Roy. Soc.*, **A. 237**, 212–232.

Cazenave, A., Daillet, S., and Lambeck, K. (1977). Tidal studies from perturbations in satellite orbits. *Phil. Trans. R. Soc. Lond.*, **284**, 595–606.

Cazenave, A., Dominh, K., Ponchaut, F., Soudarin, L., Cretaux, J. F., and Le Provost, C. (1999). Sea level changes from TOPEX/POSEIDON altimetry and tide guages, and vertical crustal motions from DORIS. *Geophys. Res. Lett.*, **26**, 2077–2080.

Chelton, D. B. (1994). The sea-state bias in altimeter estimates of sea level from collinear analysis of TOPEX data. *J. Geophys. Res.*, **99**, 24,995–25,008.

Chelton, D. B., and McCabe, P. J. (1985). A review of satellite altimeter measurement of sea surface wind speed: With a proposed new algorithm. *J. Geophys. Res.*, **90**, 4707–4720.

Chelton, D. B., and Wentz, F. J. (1986). Further development of an improved altimeter wind speed algorithm. *J. Geophys. Res.*, **91**, 14,250–14,260.

Chelton, D. B., and Schlax, M. G. (1993a). Spectral characteristics of time-dependent orbit errors in altimeter height measurements. *J. Geophys. Res.*, **98**, 12,579–12,600.

Chelton, D. B., and Schlax, M. G. (1993b). Satellite altimetry: Attempts to progress beyond studies of the statistics of mesoscale variability. *In* "Proceedings of the 'Aha Huliko'a Hawaiian Winter Workshop," pp. 55–101, University of Hawaii, Honolulu.

Chelton, D. B., and Schlax, M. G. (1994). The resolution capability of an irregularly sampled dataset: with application to GEOSAT altimeter data. *J. Atmos. Ocean. Tech.*, **11**, 534–550.

Chelton, D. B., Walsh, E. J., and MacArthur, J. L. (1989). Pulse compression and sea level tracking in satellite altimetry. *J. Atmos. Oceanic Technol.*, **6**, 407–438.

Chelton, D. B., de Szoeke, R. A., Schlax, M. G., El Naggar, K., and Siwertz, N. (1998). Geographical variability of the first baroclinic Rossby radius of deformation. *J. Phys. Oceanogr.*, **28**, 433–460.

Chelton, D. B., Schlax, M. G., Lyman, J. M., and de Szoeke, R. A. (2000). The latitudinal structure of monthly variability of the sea surface height in the tropical Pacific. *J. Phys. Oceanogr.*, submitted.

Chen, G., Chapron, B., Tournadre, J., Katsaros, K., and Vandemark, D. (1997). Global oceanic precipitation: a joint view by TOPEX and the TOPEX microwave radiometer. *J. Geophys. Res.*, **102**, 10,457–10,471.

Cheney, R. E., Douglas, B. C., Sandwell, D. T., Marsh, J. G., and Martin, T. V. (1984). Applications of satellite altimetry to oceanography and geophysics. *Mar. Geophys. Res.*, **7**, 17–32.

Cheney, R. E., Douglas, B. C., McAdoo, D. C., and Sandwell, D. T. (1986). Geodetic and oceanographic applications of satellite altimetry. *In* "Space Geodesy and Geodynamics," pp. 377–405, Academic Press, London.

Cheney, R. E., Douglas, B. C., and Miller, L. (1989). Evaluation of GEOSAT altimeter data with application to tropical Pacific sea level variability. *J. Geophys. Res.*, **94**, 4737–4747.

Cheney, R., Miller, L., Agreen, R., Doyle N., and Lillibridge, J. (1994). TOPEX/POSEIDON: the 2-cm solution. *J. Geophys. Res.*, **99**, 24,555–24,563.

Cheney, R., *et al.* (1997). Operational altimeter data processing and assimilation for El Niño forecasts. *Proc. Monitor. Oceans in the 2000s: An Integrated Approach*, Biarritz, France.

Christensen, E. J., *et al.* (1994a). Calibration of TOPEX/POSEIDON at Platform Harvest. *J. Geophys. Res.*, **99**, 24,465–24,485.

Christensen, E. J., Haines, B. J., McColl, K. C., and Nerem, R. S. (1994b). Observations of geographically correlated orbit errors for TOPEX/POSEIDON using the Global Positioning System. *Geophys. Res. Lett.*, **21**, 2175–2178.

Cleveland, W. S., and Devlin, S. J. (1988). Locally weighted regression: An approach to regression analysis by local fitting. *J. Am. Stat. Assoc.*, **83**, 596–610.

Colombo, O. L. (1984). Altimetry, oceans and tides. NASA Tech. Memo. 86180. 173 pp.

Colombo, O. L. (1989) The dynamics of Global Positioning System orbits and the determination of precise ephemerides. *J. Geophys. Res.*, **94**, 9167–9182.

Coster, A. J., Gaposhkin, E. M., and Thornton, L. E. (1992). Real-time ionospheric monitoring system using the GPS. *Navigation*, **39**, 191.

Cotton, P. D., and Carter, D. J. T. (1994). Cross calibration of TOPEX/ERS-1 and Geosat wave heights. *J. Geophys. Res.*, **99**, 25,025–25,033.

Danby, J. M. A. (1962). *In* "Fundamentals of Celestial Mechanics," 348 pp. Macmillan, New York.

Daniell, R. E., Brown, L. D., Anderson, D. N., Fox, M. W., Doherty, P. H., Decker, D. T., Sojka, J. J., and Schunk, R. W. (1995). Parameterized ionospheric model: A global ionospheric parameterization based on first principles models. *Radio Sci.*, **30**, 1499–1510.

Davies, K., Hartmann, G. K., and Letinger, R. (1977). A comparison of several methods of estimating the columnar electron content of the plasmasphere. *J. Atmos. Terrest. Phys.*, **39**, 571–580.

Davis, E. S., Dunn, C. E., Stanton, R. H., and Thomas, J. B. (1999). The GRACE mission: Meeting the technical challenges, Paper IAF-99-B.2.05 in *Proc. 50th Int. Astronautical Congress*, 4–8 October 1999, Amsterdam, The Netherlands.

Degnan, J. D. (1985). Satellite laser ranging: Current status and future prospects. *IEEE Trans. Geosci. Remote Sens.*, **GE-32**, 398–413.

Degnan, J. D. (1998). SLR2000 project: Engineering overview and status. *Proc. 11th International Workshop on Laser Ranging*, Deggendorf.

Didden, N., and Stammer, D. (1994). Influence of tropospheric water vapor corrections on GEOSAT altimetry in the North Atlantic Ocean. *J. Atmos. Oceanic Technol.*, **11**, 982–993.

Divis, D. A. (1999). Finally! A second signal decision. *GPS World Magazine*, Washington View Column, April.

Dobson, E. B., Monaldo, F., Goldhirsh, J., and Wilkerson, J. (1987). Validation of GEOSAT altimeter derived wind speeds and significant wave heights using buoy data. *J. Geophys. Res.*, **92**, 10,719–10,732.

Dodson, A. H. (1986). Refraction and propagation delays in space geodesy. *Int. J. Rem. Sens.*, **7**, 515–524.

Donelan, M. A., and Pierson, W. J. (1987). Radar scattering and equilibrium ranges in wind-generated waves with application to scatterometry. *J. Geophys. Res.*, **92**, 4971–5029.

Doodson, A. T. (1922). The harmonic development of the tide-generating potential. *Proc. R. Soc. London, Ser. A*, **100**, 305–329.

Dorandeu, J., and Le Traon, P.-Y. (1999). Effects of global mean atmospheric pressure variations on mean sea level changes from TOPEX/Poseidon. *J. Atmos. Oceanic Tech.*, **16**, 1279–1283.

Douglas, B. C. (1995). Global sea level change: Determination and interpretation. *Rev. Geophys.*, Suppl., 1425–1432.

Douglas, B. C., McAdoo, D. C., and Cheney, R. E. (1987). Oceanographic and geophysical applications of satellite altimetry. *Rev. Geophys.*, **25**, 875–880.

Ebuchi, N., and Kawamura, H. (1994). Validation of wind speeds and significant wave heights observed by the TOPEX altimeter around Japan. *J. Oceanogr.*, **50**, 479–487.

Egbert, G. D. (1997). Tidal data inversion: Interpolation and Inference. *Prog. Ocean.*, **40**, 53–80.

Elfouhaily, T., Chapron, B., Katsaros, K., and Vandemark, D. (1999). Weakly nonlinear theory and sea state bias estimations. *J. Geophys. Res.*, **104**, 7641–7647.

Elfouhaily, T., Thompson, D. R., Chapron, B., and Vandemark, D. (2000). Improved electromagnetic bias theory. *J. Geophys. Res.*, **105**, 1299–1310.

Elfouhaily, T., Vandemark, D., Gourrion, J., and Chapron, B. (1998). Estimation of wind stress using dual-frequency TOPEX data. *J. Geophys. Res.*, **103**, 25,101–25,108.

Emery, W. J., Born, G. H., Baldwin, D. G., and Norris, C. L. (1990). Satellite-derived water vapor corrections for Geosat altimetry. *J. Geophys. Res.*, **95**, 2953–2964.

Esbensen, S. K., Chelton, D. B., Vickers, D., and Sun, J. (1993). An analysis of errors in SSMI evaporation estimates over the global oceans. *J. Geophys. Res.*, **98**, 7081–7101.

Eymard, L., Tabary, L., Gérard, E., Boukabara, S.-A., and Le Cornec, A. (1996). The microwave radiometer aboard ERS-1: Part II–Validation of the geophysical products. *IEEE Trans. Geosci. Rem. Sens.*, **34**, 291–303.

Fedor, L. S., and Brown, G. S. (1982). Waveheight and wind speed measurements from the Seasat radar altimeter. *J. Geophys. Res.*, **87**, 3254–3260.

Fedor, L. S., Godbey, T. W., Gower, J. F. R., Guptill, R., Hayne, G. S., Rufenach, C. L., and Walsh, E. J. (1979). Satellite altimeter measurements of sea state — An algorithm comparison. *J. Geophys. Res.*, **84**, 3991–4001.

Fleury, R., Foucher, F., and Lassudrie-Duchesne, P. (1991). Global TEC measurement capabilities of the DORIS system. *Adv. Space Res.*, **11**, 51–54.

Fliegel, H. F., Gallini, T. E., and Swift, E. R. (1992). Global Positioning System radiation pressure force model for geodetic applications. *J. Geophys. Res.*, **97**, 559–568.

Francis, E. (1987). Calibration of the Nimbus-7 Scanning Multichannel Microwave Radiometer (SMMR), 1979–1984. M.S. Thesis, College of Oceanography, Oregon State University, 248 pp.

Francis, C. R. (Ed.) (1992). The calibration of the ERS-1 radar altimeter. ESA Report ER-RP-ESA-RA-0257, ESA/ESTEC Noordwijk, The Netherlands.

Francis, O., and Bergé, M. (1993). Estimate of the radial orbit error by complex demodulation. *J. Geophys. Res.*, **98**, 16,083–16,094.

Freilich, M. H. (1997). Validation of vector magnitude data sets: Effects of random component errors. *J. Atmos. Ocean. Tech.*, **14**, 695–703.

Freilich, M. H., and Dunbar, R. S. (1993). Derivation of satellite wind model functions using operational surface wind analyses: An altimeter example. *J. Geophys. Res.*, **98**, 14,633–14,649.

Freilich, M. H., and Challenor, P. G. (1994). A new approach for determining fully empirical altimeter wind speed model functions. *J. Geophys. Res.*, **99**, 25,051–25,062.

Freilich, M. H., and Vanhoff, B. A. (2000). The accuracy of remotely sensed surface wind speed measurements. *J. Atmos. Oceanic Tech.*, in press.

Fu, L.-L. (1983). Recent progress in application of satellite altimetry to observing the mesoscale variability and general circulation of the oceans. *Rev. Geophys. Space Phys.*, **21**, 1657–1666.

Fu, L.-L., and Glazman, R. (1991). The effect of the degree of wave development on the sea-state bias in radar altimetry measurement. *J. Geophys. Res.*, **96**, 829–834.

Fu, L.-L., and Pihos, G. (1994). Determining the response of sea level to atmospheric pressure forcing using TOPEX/POSEIDON data. *J. Geophys. Res.*, **99**, 24,633–24,642.

Fu, L.-L., and Cheney, R. E. (1995). Application of satellite altimetry to ocean circulation studies: 1987–1994. *Rev. Geophys.*, Suppl., 231–223.

Fu, L.-L., Liu, W. T., and Abbott, M. R. (1990). Satellite remote sensing of the ocean. *In* "The Sea," Vol. 9, pp. 1193–1236, John Wiley & Sons, New York.

Fu, L., Christensen, E., Yamarone, C., Lefebvre, M., Ménard, Y., Dorrer, M., and Escudier, P. (1994). TOPEX/POSEIDON mission overview. *J. Geophys. Res.*, **99**, 24,369–24,381.

Fukumori, I., Raghunath, R., and Fu, L.-L. (1998). Nature of global large-scale sea level variability in relation to atmospheric forcing: A modeling study. *J. Geophys. Res.*, **103**, 5493–5512.

Gandin, L. S. (1965). *In* "Objective Analysis of Meteorological Fields," 242 pp. Israel Program for Scientific Translations, Jerusalem.

Garner, C. S., Tsai, B.-M., and Abshire, J. B. (1983). Remote sensing of atmospheric pressure and sea state from satellites using short-pulse multicolor laser altimeters. *AGARD Conf. on Propagation Factors Affecting Remote Sensing by Radio Waves*, AGARD-CP-345, Paper 46.

Gaspar, P., and Florens, J.-P. (1998). Estimation of the sea state bias in radar altimeter measurements of sea level: Results from a new nonparametric method. *J. Geophys. Res.*, **103**, 15,803–15,814.

Gaspar, P., Ogor, F., Le Traon, P.-Y., and Zanifé, O.-Z. (1994). Estimating the sea state bias of the TOPEX and Poseidon altimeters from crossover differences. *J. Geophys. Res.*, **99**, 24,981–24,994.

Gaspar, P., and Ponte, R. (1997). Relation between sea level and barometric pressure determined from altimeter data and model simulations. *J. Geophys. Res.*, **102**, 961–971.

Gaspar, P., and Ponte, R. (1998). Correction to "Relation between sea level and barometric pressure determined from altimeter data and model simulations." *J. Geophys. Res.*, **103**, 18,809.

Geyling, F. T., and Westerman, H. R. (1971). *In* "Introduction to Orbital Mechanics," pp. 349, Addison-Wesley, Reading, MA.

Gill, A. E. (1982). *In* "Atmosphere-Ocean Dynamics," 662 pp. Academic Press, San Diego, CA.

Ginzburg, V. L. (1964). *In* "The Propagation of Electromagnetic Waves in Plasmas." Addison-Wesley, Reading, MA.

Glazman, R. E., and Pilorz, S. H. (1990). Effects of sea maturity on satellite altimeter measurements. *J. Geophys. Res.*, **95**, 2857–2870.

Glazman, R., and Srokosz, M. (1991). Equilibrium wave spectrum and sea state bias in altimetry. *J. Phys. Oceanogr.*, **21**, 1609–1621.

Glazman, R. E., and Greysukh, A. (1993). Satellite altimeter measurements of surface wind. *J. Geophys. Res.*, **98**, 2475–2483.

Glazman, R. E., Pihos, G. G., and Ip, J. (1988). Scatterometer wind speed bias induced by the large-scale component of the wave field. *J. Geophys. Res.*, **93**, 1317–1328.

Glazman, R., Fabrikant, A., and Srokosz, M. (1996). Numerical analysis of the sea state bias for satellite altimetry. *J. Geophys. Res.*, **101**, 3789–3799.

Goldfinger, A. D. (1980). Refraction of microwave signals by water vapor. *J. Geophys. Res.*, **85**, 4904–4912.

Goldhirsh, J. (1988). Analysis of algorithms for the retrieval of rain-rate profiles from a spaceborne dual-wavelength radar. *IEEE Trans. Geosci. Rem. Sens.*, **26**, 98–114.

Goldhirsh, J., and Rowland, J. R. (1982). A tutorial assessment of atmospheric height uncertainties for high-precision satellite altimeter missions to monitor ocean currents. *IEEE Trans. Geosci. Rem. Sens.*, **20**, 418–434.

Goldhirsh, J., and Dobson, E. B. (1985). A recommended algorithm for the determination of ocean surface wind speed using a satellite-borne radar altimeter. Rep. JHU/APL SIR-85-U005, Johns Hopkins Univ., Appl. Phys. Lab., Laurel, MD.

Goldstein, H. (1951). Attenuation by condensed water. *In* "Propagation of Short Radio Waves," Vol. 13, pp. 671–692, MIT Radar Laboratory Series, McGraw-Hill, New York.

Gorney, D. J. (1990). Solar cycle effects on the near-earth space environment. *Rev. Geophys.*, **28**, 315–336.

Gower, J. F. R. (1996). Intercalibration of wave and winds data from TOPEX/POSEIDON and moored buoys off the west coast of Canada. *J. Geohys. Res.*, **101**, 3817–3829.

Greenslade, D. J. M., Chelton, D. B., and Schlax, M. G. (1997). The mid-latitude resolution capability of sea level fields constructed from single and multiple satellite altimeter datasets. *J. Atmos. Oceanic Technol.*, **14**, 849–870.

Grody, N. C. (1976). Remote sensing of atmospheric water content from satellites using microwave radiometry. *IEEE Trans. Antennas Propag.*, **AP-24**, 155–162.

Grody, N. C., Gruber, A., and Shen, W. C. (1980). Atmospheric water content over the tropical Pacific derived from the Nimbus-6 scanning microwave spectrometer. *J. Appl. Meteor.*, **19**, 986–996.

Hahn, C. J., Warren, S. G., and London, J. (1995). The effect of moonlight on observation of cloud cover at night, and application to cloud climatology. *J. Climate*, **8**, 1429–1446.

Haines, B., and Bar-Sever, Y. (1998). Monitoring the TOPEX microwave radiometer with GPS: Stability of wet tropospheric path delay measurements. *Geophys. Res. Lett.*, **25**, 3563–3566.

Haines, B. J., Born, G. H., Christensen, E. J., Gill, S., and Kubitschek, D. (1998). The Harvest Experiment: TOPEX/POSEIDON absolute calibration results from five years of continuous data. *AVISO Altimetry Newsletter 6:* TOPEX/POSEIDON: 5 Years of Progress, 53–54.

Haines, B. J., *et al.* (1996) Observations of TOPEX/POSEIDON orbit errors due to gravitational and tidal modeling errors using the global positioning system. *In* "GPS Trends in Precise Terrestrial, Airborne and Spaceborne Applications," Vol. 115, IAG Symposia, Springer.

Haines, B. J., Lichten, S. M., Lough, M. F., Muellerschoen, R. J., and Vigue-Rodi, Y. (1999). Determining precise orbits for TOPEX/POSEIDON within one day of real time: results and implications. *Adv. Astronautical Sci.*, (R. Bishop, D. Mackison, R. Culp, and M. Evans, Eds.), **102**, 623–634.

Harrison, I., and Swinerd, G. (1994). Determining a good aerodynamic model for satellite orbit analysis of ERS-1, using laser ranging data. *Proc. 19th Int. Symp. Space Technol. Sci.*, Yokohama, Japan.

Hancock, D. W., Brooks, R. L., and Lockwood, D. W. (1990). Effects of height acceleration on Geosat heights. *J. Geophys. Res.*, **95**, 2843–2848.

Hayne, G. S. (1980). Radar altimeter mean return waveforms from near-normal incidence ocean surface scattering. *IEEE Trans. Antennas Propag.*, **28**, 687–692.

Hayne, G. S., and Hancock, D. W. (1982). Sea-state related altitude errors in the SEASAT radar altimeter. *J. Geophys. Res.*, **87**, 3227–3231.

Hayne, G. S., and Hancock, D. W. (1990). Corrections for the effects of significant wave height and attitude on GEOSAT radar altimeter measurements. *J. Geophys. Res.*, **95**, 2837–2842.

Hayne, G. S., Hancock, D. W., Purdy, C. L., and Callahan, P. S. (1994). The corrections for significant wave height and attitude effects in the TOPEX radar altimeter. *J. Geophys. Res.*, **99**, 24,941–24,955.

Hedin, A. E. (1991). Extension of the MSIS thermospheric model into the middle and lower atmosphere. *J. Geophys. Res.*, **96**, 1159–1172.

Hein, G., Landau, H., and Blomenhofer, H. (1990). Determination of instantaneous sea surface, wave heights and ocean currents using satellite observations of the Global Positioning System. *Mar. Geod.*, **14**, 217–224.

Herrero, F. A. (1988). Satellite drag coefficients and upper atmosphere densities; present status and future directions. Astrodynamics 1987, *Adv. Astronautical Sci.*, (J. K. Solder, A. K. Misra, R. E. Lindberg, and W. Williamson, Eds.), **65**, 1607–1623.

Hevizi, L., Walsh, E., MacIntosh, R., Vandemark, D., Hines, D., Swift, R., and Scott, J. G. (1993). Electromagnetic bias in sea surface range measurements at frequencies of the TOPEX/Poseidon satellite. *IEEE Trans. Geosci. Remote Sens.*, **31**, 367–388.

Ho, C. M., Wilson, B. D., Mannucci, A. J., Lindqwister, U. J., and Yuan, D. N. (1997). A comparative study of ionospheric total electron content measurements using global ionospheric maps of GPS. TOPEX radar and the Bent model. *Radio Sci.*, **32**, 1499–1512.

Hoffman-Wellenhof, B., Lichtenegger, H., and Collins, J. (1993). *In* "GPS: Theory and Practice," 326 pp. Springer-Verlag, New York.

Hopfield, H. S. (1971). Tropospheric effect on electromagnetically measured range: Prediction from surface weather data. *Radio Sci.*, **6**, 357–367.

Houghton, J. T., Taylor, F. W., and Rodgers, C. D. (1984). *In* "Remote Sounding of Atmospheres," Cambridge University Press, Cambridge.

Imel, D. (1994). Evaluation of the TOPEX/POSEIDON dual-frequency ionosphere correction. *J. Geophys. Res.*, **99**, 24,895–24,906.

Jackson, F. C. (1979). The reflection of impulses from a nonlinear random sea. *J. Geophys. Res.*, **84**, 4934–4939.

Jacobs, G. A., Born, G. H., Parke, M. E., and Allen, P. C. (1992). The global structure of the annual and semiannual sea surface height variability from Geosat altimeter data. *J. Geophys. Res.*, **97**, 17,813–17,828.

Ji, M., Reynolds, R. W., and Beringer, D. W. (2000). Use of TOPEX/POSEIDON sea-level data for ocean analyses and ENSO prediction: some preliminary results. *J. Climate*, **13**, 216–231.

Joselyn, J. A., Anderson, J. B., Coffey, H., Harvey, K., Hathaway, D., Heckman, G., Hildner, E., Mende, W., Schatten, K., Thompson, R., Thomson, A. W. P., and White, O. R. (1997). Panel achieves consensus prediction of solar cycle 23. *Eos Trans. Am. Geophys. Union*, **78**, 211–212.

Jourdon, D., Boissier, C., Braun, A., and Minster, J. F. (1990). Influence of wet tropospheric correction on mesoscale dynamic topography as derived from satellite altimetry. *J. Geophys. Res.*, **95**, 17,993–18,004.

Kaula, W. M. (1966). *Theory of Satellite Geodesy*, Blaisdell Press, Waltham, MA, 124 pp.

Keihm, S. J., and Ruf, C. S. (1995). The role of water vapor radiometers in the TOPEX microwave radiometer in-flight calibration. *J. Mar. Geod.*, **18**, 139–156.

Keihm, S. J., Janssen, M. A., and Ruf, C. S. (1995). TOPEX/POSEIDON Microwave Radiometer (TMR): III. Wet tropospheric range correction algorithm and pre-launch error budget. *IEEE Trans. Geosci. Rem. Sens.*, **33**, 147–161.

Keihm, S. J., Zlotnicki, V. L., and Ruf, C. S. (2000). TOPEX microwave radiometer performance evaluation, 1992–1998. *IEEE Trans, Geosci. Rem. Sens.*, **38**, 1379–1386.

Kelecy, T., Parke, M. E., Born, G. H., and Rocken, C. (1994). Precise sea-level measurements using the Global Positioning System. *J. Geophys. Res.*, **99**, 7951–7959.

Killeen, T. L., Burns, A. G., Johnson, R. M., and Marcos, F. A. (1993). Modeling and prediction of density changes and winds affecting spacecraft trajectories. Environmental Effects on Spacecraft Positioning and Trajectories. *Geophys. Monogr. Ser.* 73, (A. Vallence-Jones, Ed.), pp. 83–109, American Geophysical Union, Washington, D.C.

Kim, M. C. (1995). Determination of high resolution mean sea surface and marine gravity field using satellite altimetry. Report CSR-95-02, Center for Space Research, The University of Texas at Austin, 187 pp.

King-Hele, D. G. (1992). *In* "A Tapestry of Orbits," 244 pp. Cambridge University Press.

Klinkrad, H., Koeck, Ch., and Renard, P. (1991). Key features of a satellite skin force modelling technique by means of Monte-Carlo ray tracing. *Adv. Space Res.*, **11**, (6) 147–(6) 150.

Klobuchar, J. A. (1987). Ionoshperic time-delay algorithm for single-frequency GPS users. *IEEE Trans. Aerosp. Electron. Syst.*, **23**, 325–331.

Klokocnik, J., Wagner, C. A., and Kostelecky, J. (1999). Spectral accuracy of JGM-3 from satellite crossover altimetry. *J. Geodesy*, **73**, 138–146.

Knocke, P. C., Ries, J. C., and Tapley, B. D. (1988). Earth Radiation Pressure Effects on Satellites, paper 88–4292. *Proc. AIAA/AAS Astrodynamics Conf.*, 577–587.

Kolenkiewicz, R., and Martin, C. (1982). Seasat altimeter height calibration. *J. Geophys. Res.*, **87**, 3189–3197.

Kuijper, D. C., Ambrosius B. A. C., and Wakker, K. F. (1995). SPOT-2 and TOPEX/POSEIDON precise orbit determination from DORIS Doppler tracking. *Adv. Space Res.*, **6**, 45–50.

LaBrecque, J. L., Lowe, S. T., Young, L. E., Caro, E., Romans, L. J., and Wu, S. C. (1998). The first Spaceborne Observations of GPS Signals Reflected from the Ocean Surface. *Eos Trans. Am. Geophys. Union*, **79**.

Lagerloef, G. S. E. (1987). Comment on "On the joint distribution of surface elevations and slopes for a nonlinear random sea, with an application for radar altimetry" by M. A. Srokosz. *J. Geophys. Res.*, **92**, 2985–2987.

Lambeck, K., Cazenave, A., and Balmino, G. (1974). Solid earth and ocean tides estimated from satellite orbit analyses. *Rev. Geophys. Space Phys.*, **12**, 421–434.

Lanyi, G. E., and Roth, T. (1988). A comparison of mapped and measured total electron content using global positioning system and beacon satellite observations. *Radio Sci.*, **23**, 486–492.

Lawrence, R. S., Little, C. G., and Chivers, H. J. A. (1964). A survey of ionospheric effects upon earth-space radio propagation. *Proc. IEEE*, **52**, 4–27.

Lefebvre, J. M., Barckicke, J., and Ménard, Y. (1994). A significant wave height dependent function for TOPEX/POSEIDON wind speed retrieval. *J. Geophys. Res.*, **99**, 25,035–25,049.

Le Traon, P.-Y., and Hernandez, F. (1992). Mapping the oceanic mesoscale circulation: Validation of satellite altimetry using surface drifters. *J. Atmos. Oceanic Technol.*, **9**, 687–698.

Le Traon, P.-Y., and Dibarboure, G. (1999). Mesoscale mapping capabilities of multiple-satellite altimeter missions. *J. Atmos. Oceanic Technol.*, **16**, 1208–1223.

Le Traon, P.-Y., Rouquet, M. C., and Boissier, C. (1990). Spatial scales of mesoscale variability in the North Atlantic as deduced from Geosat data. *J. Geophys. Res.*, **95**, 20,267–20,285.

Le Traon, P.-Y., Bossier, C., and Gaspar, P. (1991). Analysis of errors due to polynomial adjustment of altimeter profiles. *J. Atmos. Oceanic Technol.*, **8**, 385–396.

Le Traon, P.-Y., Nadal, F., and Ducet, N. (1998). An improved mapping method of multisatellite data. *J. Atmos. Oceanic Technol.*, **15**, 522–534.

LeGrand, P., and Minster, J. (1999). Impact of the GOCE gravity mission on ocean circulation estimates. *Geophys. Res. Lett.*, **26**, 1881–1884.

Leitinger, R., Hartmann, G. K., Lohmar, F.-J., and Putz, E. (1984). Electron content measurements with geodetic Doppler receivers. *Radio Sci.*, **19**, 789–797.

Leitinger, R., Schmidt, G., and Tauriainen, A. (1975). An evaluation method combining the differential Doppler measurements from two stations that enables the calculation of the electron content of the ionosphere. *J. Geophys.*, **41**, 201–213.

Lemoine, F., *et al.* (1998). The development of the joint NASA GSFC and the National Imagery and Mapping Agency (NIMA) geopotential model EGM96. NASA/TP-1998-206861, July, 575 pp.

Lerch, F. J., Wagner, C. A., Klosko, S. M., and Putney, B. H. (1981). Goddard Earth models for oceanographic applications (GEM 10B and 10C). *Mar. Geod.*, **5**, 2–43.

Lerch, F. J., Marsh, J. G., Klosko, S. M., and Williamson, R. G. (1982). Gravity model improvement for SEASAT. *J. Geophys. Res.*, **87**, 3281–3296.

Lerch, F. J., Marsh, J. G., Klosko, S. M., Patel, G. B., Chinn, D. S., Pavlis, E. C., and Wagner, C. A. (1991). An improved error assessment for the GEM-T1 gravitational model. *J. Geophys. Res.*, **96**, 20,023–20,040.

Lerch, F. J., *et al.* (1993). A geopotential model from satellite tracking, altimeter, and surface gravity data: GEM-T3. *J. Geophys. Res.*, **99**, 2815–2839.

Levitus, S., and Boyer, T. P. (1994). *In* "World Ocean Atlas 1994," Volume 4: Temperature. National Oceanic and Atmospheric Administration Atlas NESDIS 4, U.S. Department of Commerce, 117 pp.

Levitus, S., Burgett, R., and Boyer, T. P. (1994). *In* "World Ocean Atlas 1994," Volume 3: Salinity. National Oceanic and Atmospheric Administration Atlas NESDIS 3, U.S. Department of Commerce, 99 pp.

Lichten, S. M. (1990). Estimation and filtering for high-precision GPS positioning applications. *Manuscripta Geodaetica*, **15**, 159–176.

Liebe, H. J. (1985). An updated model for millimeter wave propagation in moist air. *Radio Sci.*, **20**, 1069–1089.

Lipa, B. J., and Barrick, D. E. (1981). Ocean surface height-slope probability density function from SEASAT altimeter echo. *J. Geophys. Res.*, **86**, 10,921–10,930.

Liu, W. T., and Mock, D. (1990). The variability of atmospheric equivalent temperature for radar altimeter range correction. *J. Geophys. Res.*, **95**, 2933–2938.

Liu, W. T., Tang, W., and Niiler, P. P. (1991). Humidity profiles over the ocean. *J. Clim.*, **4**, 1023–1034.

Longuet-Higgins, M. S. (1952). On the statistical distribution of the heights of sea waves. *J. Mar. Res.*, **11**, 245–266.

Longuet-Higgins, M. (1963). The effects of nonlinearities on statistical distribution in the theory of sea waves. *J. Fluid Mech.*, **17**, 459–480.

Lorell, J., Colquitt, E., and Anderle, R. J. (1982). Ionospheric correction for SEASAT altimeter height measurements. *J. Geophys. Res.*, **87**, 3207–3212.

Lough, M. F., *et al.* (1998). Precise orbit determination for low-Earth orbiting satellites using GPS Data: Recent advances. *Proc. 54th Ann. Mtg. of the Inst. of Nav.*, 123–131, Denver.

Lozier, M. S., McCartney, M. S., and Owens, W. B. (1994). Anomalous anomalies in averaged hydrographic data. *J. Phys. Oceanogr.*, **24**, 2624–2638.

Lozier, M. S., Owens, W. B., and Curry, R. G. (1995). The climatology of the North Atlantic. *Prog. Oceanogr.*, **36**, 1–44.

MacArthur, J. L. (1978). Seasat-A radar altimeter design description. Doc. SDO–5232, Appl. Phys. Lab., November, 163 pp.

Mannucci, A. J., Wilson, B. D., Yuan, D. N., Ho, D. H., Lindqwister, U. J., and Runge, T. F. (1998). A global mapping technique for GPS-derived ionospheric total electron content measurements. *Radio Sci.*, **33**, 565–582.

Mannucci, A. J., Iijima, B. A., Linqwister, U. J., Pi, X., Sparks, L., and Wilson, B. D. (1999). GPS and Ionosphere. *In* "Review of Radio Science" (W. Ross Stone, Ed.), pp. 622–655, Oxford University Press.

Marcos, F. A., Baker, C. R., Bass, J. N., Killeen, T. L., and Roble, R. G. (1993). Satellite drag models: Current status and prospects. *Adv. Astronautical Sci.*, **85**, 1253–1274.

Marsh, J. G., *et al.* (1988). A new gravitational model for the earth from satellite tracking data: GEM-T1. *J. Geophys. Res.*, **93**, 6169–6215.

Marsh, J. G., *et al.* (1990). The GEM-T2 Gravitational Model. *J. Geophys. Res.*, **95**, 22,043–22,071.

Marshall, J. A., and Luthcke, S. B. (1994). Modeling radiation forces acting on TOPEX/POSEIDON for precision orbit determination. *J. Spacecr. Rockets*, **31**, 89–105.

Marshall, J. A., Klosko, S. M., and Ries, J. C. (1995b). Dynamics of SLR Tracked Satellites. U.S. National Report to International Union of Geodesy and Geophysics 1991–1994, *Rev. Geophys.*, Suppl., 353–360.

Marshall, J. A., Zelensky, N. P., Klosko, S. M., Chinn, D. S., Luthcke, S. B., Rachlin, K. E., and Williamson, R. G. (1995a). The temporal and spatial characteristics of the TOPEX/POSEIDON radial orbit error. *J. Geophys. Res.*, **100**, 25,331–25,352.

Martin, C., and Kolenkiewicz, R. (1981). Calibration validation of the GEOS-3 altimeter. *J. Geophys. Res.*, **86**, 6369–6381.

Martin, T. V., Zwally, H. J., Brenner, A. C., and Bindschadler, R. A. (1983). Analysis and retracking of continental ice sheet radar altimeter waveforms. *J. Geophys. Res.*, **88**, 1608–1616.

Massmann, F.-H., Neumayer, K. H., Raimando, J. C., Enninghorst, K., and Li, H. (1997). Quality of the D-PAF ERS orbits before and after inclusion of PRARE data. Proc. 3rd ERS Scientific Symposium, ESA SP-414, Florence, Italy.

Maul, G. A. (1985). *In* "Introduction to Satellite Oceanography," 606 pp. Martinus Nijhoff, Dordrecht, The Netherlands.

McCarthy, D. D. (Ed.) (1996). IERS Standards 1996. IERS Tech. Note 21, Observatoire de Paris, France, 150 pp.

Melbourne, W. G., Davis, E. S., Yunck, T. P., and Tapley, B. D. (1994). The GPS flight experiment on TOPEX/POSEIDON. *Geophys. Res. Lett.*, **21**, 2171–2174.

Melville, W. K., Arnold, D. V., Stewart, R. H., Keller, W. C., Kong, J. A., Jessup, A. T., and Lamarre, E. (1990). Measurement of EM bias at K_u and C bands. Proc. Oceans-90 Conf., Washington, D.C.

Melville, W. K., Stewart, R. H., Keller, W. C., Kong, J. A., Arnold, D. V., Jessup, A. T., Loewen, M. R., and Slinn, A. M. (1991). Measurements of electromagnetic bias in radar altimetry. *J. Geophys. Res.*, **96**, 4915–4924.

Ménard, Y., Fu, L.-L., Escudier, P., and Kunstmann, G. (1999). Jason-1, on the tracks of TOPEX/POSEIDON. *Eos Trans. Am. Geophys. Union*, submitted.

Ménard, Y., Jeansou, E., and Vincent, P. (1994). Calibration of the TOPEX/POSEIDON altimeters at Lampedusa: Additional results at Harvest. *J. Geophys. Res.*, **99**, 24,487–24,504.

Milani, A., Nobili, A. M., and Farinella, P. (1987). *In* "Non-gravitational Perturbations and Satellite Geodesy." Adam Hilger Ltd., Bristol, 125 pp.

Minster, J.-F., Jourdan, D., Boissier, C., and Midol-Monnet, P. (1992). Estimation of the sea-state bias in radar altimeter Geosat data from examination of frontal systems. *J. Atmos. Oceanic Technol.*, **9**, 174–187.

Minster, J.-F., Rémy, F., and Normant, E. (1993). Constraints on the repetitivity of the orbit of an altimetric satellite: Estimation of the cross-track slope. *J. Geophys. Res.*, **10**, 410–419.

Mitchum, G. (1994). Comparison of TOPEX sea-surface heights and tide gauge sea levels. *J. Geophys. Res.*, **99**, 24,541–24,553.

Mitchum, G. T. (1995). The source of 90-day oscillations at Wake Island. *J. Geophys. Res.*, **100**, 2459–2475.

Mitchum, G. (1997). A tide-gauge network for altimeter calibration, Methods for Monitoring Sea Level: GPS and tide gauge collocation and GPS altimeter calibration. Proc. IGS/PMSL Workshop, Jet Propulsion Laboratory Publ. 97–17, 45–55.

Mitchum, G. T. (1998). Monitoring the stability of satellite altimeters with tide gauges. *J. Atmos. Oceanic Technol.*, **15**, 721–730.

Moe, M. M., Wallace, S. D., and Moe, K. (1993). Refinements in determining satellite drag coefficients. *J. Guidance Control Dynamics*, **16**, 441–445.

Monaldo, F. (1988). Expected differences between buoy and radar altimeter estimates of wind speed and significant wave height and their implications on buoy-altimeter comparisons. *J. Geophys. Res.*, **93**, 2285–2302.

Monaldo, F. (1990). Path length variations caused by atmospheric water vapor and their effects on the measurement of mesoscale ocean circulation features by a radar altimeter. *J. Geophys. Res.*, **95**, 2923–2932.

Monaldo, F. (1991). Ionospheric variability and the measurement of ocean mesoscale circulation with a spaceborne radar altimeter. *J. Geophys. Res.*, **96**, 4925–4937.

Monaldo, F. (1993). TOPEX ionospheric height correction precision estimated from prelaunch test results. *IEEE Trans. Geosci. Rem. Sens.*, **31**, 371–375.

Monaldo, F. M., Goldhirsh, J., and Walsh, E. J. (1986). Altimeter height measurement error introduced by the presence of variable cloud and rain attenuation. *J. Geophys. Res.*, **91**, 2345–2350.

Montenbruck, O. (1992). Numerical integration methods for orbital motion. *Celest. Mech.*, **53**, 59–69.

Moore, P., and Ehlers, S. (1993). Orbital refinement of ERS-1 using dual crossover arc techniques with TOPEX/POSEIDON. *Manuscripta Geodaetica*, **18**, 249–262.

Moore, P., Ehlers S., and Carnochan, S. (1998). Accuracy assessment and refinement of the JGM-2 and JGM-3 gravity fields for radial positioning of ERS-1. *J. Geodesy*, **72**, 373–384.

Moore, R. K., and Williams, C. S. (1957). Radar terrain return at near-vertical incidence. *Proc. IRE*, **45**, 228–238.

Moran, J. M., and Rosen, B. R. (1981). Estimation of the propagation delay through the troposphere from microwave radiometer data. *Radio Sci.*, **16**, 235–244.

Morris, C. S., and Gill, S. K. (1994). Evaluation of the TOPEX/POSEIDON altimeter system over the Great Lakes. *J. Geophys. Res.*, **99**, 24,527–24,539.

Morris, C., and Gill, S. (1994). Evaluation of the TOPEX/POSEIDON altimeter system over the Great Lakes. *J. Geophys. Res.*, **99**, 24,527–24,539.

Morrow, R., Coleman, R., Church, J., and Chelton, D. (1994). Surface eddy momentum flux and velocity variances in the Southern Ocean from Geosat altimetry. *J. Phys. Oceanogr.*, **24**, 2050–2071.

Muellerschoen, R. J., Bertiger, W. I., Wu, S. C., and Munson, T. (1994). Accuracy of GPS determined TOPEX/POSEIDON orbits during Anti-Spoof periods. Proc. 1994 Natl. Tech. Mtg. of the Inst. of Nav., pp. 607–614, San Diego.

Muellerschoen, R. J., Lichten, S. M., Lindqwister, U., and Bertiger, W. I. (1995). Results of an automated GPS tracking system in support of TOPEX/POSEIDON and GPSMet. Proc. 1995 Intl. Tech. Mtg. of the Inst. of Nav., pp. 183–193, Palm Springs.

Murphy, C. M., Moore P., and Woodworth, P. (1996). Short-arc calibration of the TOPEX/POSEIDON and ERS-1 altimeters utilizing *in situ* data. *J. Geophys. Res.*, **101**, 14,191–14,200.

Naderi, F. M., Freilich, M. H., and Long, D. G. (1991). Spaceborne radar measurement of wind velocity over the ocean — An overview of the NSCAT scatterometer system. *Proc. IEEE*, **97**, 850–866.

Nerem, R. S., *et al.* (1994). Gravity model development for TOPEX/POSEIDON: Joint Gravity Model-1 and 2. *J. Geophys. Res.*, **99**, 24,421–24,447.

Nerem, R. S., *et al.* (1997). Improved determination of global mean sea level variations using TOPEX/POSEIDON altimeter data. *Geophys. Res. Lett.*, **24**, 1331–1334.

Nerem, R. S., Eanes, R. J., Ries, J. C., and Mitchum, G. T. (1998). The use of a precise reference frame in sea level change studies. Proc. International Association of Geodesy Conference, Integrated Global Geodetic Observing System, Munich.

Neumann, G., and Pierson, W. J. (1966). *In* "Principles of Physical Oceanography," 545 pp. Prentice Hall, Englewood Cliffs, NJ.

Nöuel, F., Bardina, J., Jayles, C., Labrune, Y., and Troung, B. (1988). DORIS: A precise satellite positioning doppler system. *Adv. Astron. Sci.*, **65**, 311–320.

Nöuel, F., Berthias, J. P., Deleuze, M., Guitart, A., Laudet, P., Piuzzi, A., Pradines, D., Valorge, C., Dejoie, C., Susini, M. F., and

Taburiau, D. (1994). Precise Centre National d'Etudes Spatiale orbits for TOPEX/POSEIDON: Is reaching 2 cm still a challenge? *J. Geophys. Res.*, **99**, 24,405–24,419.

Nowlin, W. D. (1999). A strategy for long-term ocean observations. *Bull. Am. Meteor. Soc.*, **80**, 621–627.

Okkonen, S. R., and Jacobs, G. A. (1996). Aliased propagating mesoscale features in altimeter data. *J. Atmos. Oceanic Technol.*, **13**, 1311–1316.

Ondrasik, V. J., and Wu, S. C. (1982). A simple and economical tracking system with sub-decimeter Earth satellite and ground receiver position determination capabilities, *In* "Proceedings of the Third Intl. Symposium on the Use of Artificial Satellites for Geodesy and Geodynamics," Ermioni, Greece.

Parke, M. E., and Walsh, E. J. (1995). Altimeter footprint dimensions. *Mar. Geod.*, **18**, 129–137.

Parke, M. E., Stewart, R. H., Farless, D. L., and Cartwright, D. E. (1987). On the choice of orbits for an altimetric satellite to study ocean circulation and tides. *J. Geophys. Res.*, **92**, 11,693–11,707.

Parke, M. E., Born, G., Leben, R., McLaughlin, C., and Tierney, C. (1998). Altimeter sampling characteristics using a single satellite. *J. Geophys. Res.*, **103**, 10,513–10,526.

Parkinson, B. W. (1996). Introduction and heritage of NAVSTAR, the Global Positioning System. *In* "Global Positioning System Theory and Applications," Vol. 1, pp. 3–28, Progress in Aero. and Astro., 164, American Institute of Aerospace and Astro., Washington, D.C.

Partington, K. C., Ridley, J. K., Raply, C. G., and Zwally, H. J. (1989). Observation of the surface properties of ice sheets by satellite radar altimetry. *J. Glaciol.*, **35**, 267–275.

Peckham, G. E., Gatley, C., and Flower, D. A. (1983). Optimizing a remote sensing instrument to measure atmospheric surface pressure. *Int. J. Rem. Sens.*, **4**, 465–478.

Petty, G. W. (1997). An intercomparison of oceanic precipitation frequencies from 10 special sensor microwave/imager rain rate algorithms and shipboard present weather reports. *J. Geophys. Res.*, **102**, 1757–1777.

Phillips, O. M. (1977). *In* "The Dynamics of the Upper Ocean," 2nd ed., 327 pp. Cambridge University Press.

Phillips, O. M. (1980). *In* "The Dynamics of the Upper Ocean," 336 pp. 1st paperback ed., Cambridge University Press.

Phoebus, P. A., and Hawkins, J. D. (1990). The impact of the wet tropospheric correction on the interpretation of altimeter-derived ocean topography in the northeast Pacific. *J. Geophys. Res.*, **95**, 2939–2952.

Pisacane, V. L. (1986). Satellite techniques for determining the geopotential of sea surface elevations. *J. Geophys. Res.*, **91**, 2365–2371.

Plant, W. J. (1986). A two-scale model of short wind-generated waves and scatterometry. *J. Geophys. Res.*, **91**, 10,735–10,749.

Ponte, R. M. (1992). The sea level response of a stratified ocean to barometric pressure forcing. *J. Phys. Oceanogr.*, **22**, 109–113.

Ponte, R. M. (1993). Variability in a homogeneous global ocean forced by barometric pressure. *Dyn. Atmos. Oceans*, **18**, 209–234.

Ponte, R. M., and Gaspar, P. (1999). Regional analysis of the inverted barometer effect over the global ocean using TOPEX/POSEIDON data and model results. *J. Geophys. Res.*, **104**, 15,587–15,602.

Ponte, R. M., Salstein, D. A., and Rosen, R. D. (1991). Sea level response to pressure forcing in a barotropic numerical model. *J. Phys. Oceanogr.*, **21**, 1043–1057.

Press, W. H., Teukolsky, S. A., Vetterling, W. T., and Flannery, B. P. (1992). *In* "Numerical Recipes," 963 pp. Cambridge University Press.

Quartly, G. D. (1998). Determination of oceanic rain rate and rain cell structure from altimeter waveform data. *J. Atmos. Oceanic Technol.*, **15**, 1361–1378.

Quartly, G. D., Guymer, T. H., and Srokosz, M. A. (1996). The effects of rain on TOPEX radar altimeter data. *J. Atmos. Oceanic Technol.*, **13**, 1209–1229.

Quartly, G. D., Srokosz, M. A., and Guymer, T. H. (1999). Global precipitation statistics from dual-frequency TOPEX altimetry. *J. Geophys Res.*, **104**, 31,489–31,516.

Qiu, B. (1999). Seasonal eddy field modulation of the North Pacific Subtropical Countercurrent: TOPEX/POSEIDON observations and theory. *J. Phys. Oceangr.*, **29**, 2471–2486.

Raney, R. K. (1998). The delay/Doppler radar altimeter. *IEEE Trans. Geosci. Rem. Sens.*, **36**, 1578–1588.

Rapp, R. H., and Pavlis, N. K. (1990). The development and analysis of geopotential coefficient models to spherical harmonic degree 360. *J. Geophys. Res.*, **95**, 21,885–21,911.

Rapp, R. H., Yi, Y., and Wang, Y. M. (1994). Mean sea surface and geoid gradient comparisons with TOPEX altimeter data. *J. Geophys. Res.*, **99**, 24,657–24,667.

Ray, J. R., Ma, C., Ryan, J. W., Clark, T. A., Eanes, R. J., Watkins, M. M., Schutz, B. E., and Tapley, B. D. (1991). Comparison of VLBI and SLR geocentric site coordinates. *Geophys. Res. Lett.*, **18**, 231–234.

Ray, R. D., and Koblinsky, C. J. (1991). On the sea-state bias of the GEOSAT altimeter. *J. Atmos. Oceanic Technol.*, **8**, 397–408.

Ray, R. D., and Mitchum, G. T. (1996). Surface manifestation of internal tides generated near Hawaii. *Geophys. Res. Lett.*, **23**, 2101–2104.

Ray, R. D., and Mitchum, G. T. (1997). Surface manifestation of internal tides in the deep ocean: observations from altimetry and tide guages. *Progr. Oceanogr.*, **40**, 135–162.

Reigber, Ch., Luehr, H., and Schwintzer, P. (2000). The CHAMP geopotential mission; Bolletino di Geofisica Teorica ed Applicata, Special section Joint IGC/IGeC meeting, in press.

Reigber, Ch., Massmann, F.-H., and Flechtner, F. (1997). The PRARE system and the data analysis procedure. Coordination of Space Techniques for Geodesy and Geodynamics Bulletin No. 14, Advanced Space Technology in Geodesy — Achievements and Outlook, (R. Rummel, C. Reigber, and H. Hornik, Eds.), 73–80, Deutsches Geodätisches Forschungsinstitut, Munich.

Resch, G. M. (1984). Water vapor radiometers in geodetic applications. *In* "Geodetic Refraction," (F. K. Brunner, Ed.), Springer-Verlag, New York.

Ries, J. C. (1997). Non-gravitational force modelling effects on satellite orbits. Coordination of Space Techniques for Geodesy and Geodynamics Bulletin No. 14, "Advanced Space Technology in Geodesy — Achievements and Outlook," (R. Rummel, C. Reigber, and H. Hornik, Eds.), pp. 73–80, Deutsches Geodätisches Forschungsinstitut, Munich.

Ries, J. C., and Tapley, B. D. (1999). Centimeter level orbit determination for the TOPEX/POSEIDON altimeter satellite. *Adv. Astronautical Sci.*, **102**, 583–598.

Ries, J. C., Shum, C. K., and Tapley, B. D. (1993). Surface force modeling for precision orbit determination, Environmental Effects on Spacecraft Positioning and Trajectories. Geophysical Monograph 73, IUGG Vol. 13, (A. Vallence-Jones, Ed.), pp. 111–124, American Geophysical Union, Washington, D.C.

Rocken, C., Kelecy, T., Born, G., Young, L., Purcell, G., and Wolf, S. K. (1990). Measuring precise sea level from a buoy using the global positioning system. *Geophys. Res. Lett.*, **17**, 2145–2148.

Rodríguez, E. (1988). Altimetry for non-Gaussian oceans: height biases and estimation of parameters. *J. Geophys. Res.*, **93**, 14,107–14,120.

Rodríguez, E., and Chapman, B. (1989). Extracting ocean surface information from altimeter returns: the deconvolution method. *J. Geophys. Res.*, **94**, 9761–9778.

Rodríguez, E., and Martin, D. W. (1994a). Assessment of the TOPEX altimeter performance using waveform retracking. *J. Geophys. Res.*, **99**, 24,957–24,969.

Rodríguez, E., and Martin, J. M. (1994b). Estimation of the electromagnetic bias from retracked TOPEX data. *J. Geophys. Res.*, **99**, 24,971–24,979.

Rodríguez, E., Kim, Y., and Martin, J. M. (1992). The effect of small-wave modulation on the electromagnetic bias. *J. Geophys. Res.*, **97**, 2379–2389.

Rodríguez, E, Pollard, B., and Martin, J. (1999). Wide-Swath Ocean Interferometric Altimetry. Unpublished manuscript, Jet Propulsion Laboratory, Pasadena, Calif.

Rosborough, G. W., and Tapley, B. D. (1987). Radial, transverse and normal satellite position perturbations due to the geopotential. *Celest. Mech.*, **40**, 409–421.

Rosborough, G. W., and Marshall, J. A. (1990). Effect of orbit error on determining sea surface variability using satellite altimetry. *J. Geophys. Res.*, **95**, 5273–5277.

Rossow, W. B., and Schiffer, R. A. (1991). ISCCP cloud data products. *Bull. Am. Meteor. Soc.*, **72**, 2–20.

Rowlands, D. D., Luthcke, S. B., Marshall, J. A., Cox, C. M., Williamson, R. G., and Rowton, S. C. (1997). Space shuttle precision orbit determination in support of the SLA-1 using TDRSS and GPS tracking data. *J. Astron. Sci.*, **45**, 113–129.

Ruf, C. S., Keihm, S. J., and Janssen, M. A. (1995). TOPEX/POSEIDON Microwave Radiometer (TMR): I. Instrument description and antenna temperature calibration. *IEEE Trans. Geosci. Rem. Sens.*, **33**, 125–137.

Ruf, C., Keihm, S., Subramanya, B., and Janssen, M. (1994). TOPEX/POSEIDON microwave radiometer performance and in-flight calibration. *J. Geophys. Res.*, **99**, 24,915–24,926.

Rush, C. M. (1986). Ionospheric radio propagation models and predictions- a mini review. *IEEE Trans. Anten. Propag.*, **34**, 1163–1170.

Saastamoinen, J. (1972). Atmospheric correction for troposphere and stratosphere in radio ranging of satellites. *In* "The Use of Artificial Satellites for Geodesy" (S. Henriksen, A. Mancini and B. Chovitz, Eds.), Vol. 15, pp. 247–251, Geophysics Monograph Series, American Geophysics Union, Washington D.C.

Scharroo, R., and Visser, P.N.A.M. (1998). Precise orbit determination and gravity field improvement for the ERS satellites. *J. Geophys. Res.*, **103**, 8113–8127.

Schlax, M. G., and Chelton, D. B. (1992). Frequency domain diagnostics for linear smoothers. *J. Am. Stat. Assoc.*, **87**, 1070–1081.

Schlax, M. G., and Chelton, D. B. (1994a). Detecting aliased tidal errors in altimeter height measurements. *J. Geophys. Res.*, **99**, 12,603–12,612.

Schlax, M. G., and Chelton, D. B. (1994b). Aliased tidal errors in TOPEX/POSEIDON sea surface height data. *J. Geophys. Res.*, **99**, 24,761–24,775.

Schlax, M. G., and Chelton, D. B. (1996). Correction to "Aliased tidal errors in TOPEX/POSEIDON sea surface height data." *J. Geophys. Res.*, **101**, 18,451.

Schrama, E. J. O. (1992). Some remarks on several definitions of geographically correlated orbit errors: Consequences for satellite altimetry. *Manuscripta Geodaetica*, **17**, 282–294.

Schreiner, W. S., Markin, R. E., and Born, G. H. (1997). Correction of single frequency altimeter measurements for ionosphere delay. *IEEE Trans. Geosci. Rem. Sens.*, **35**, 271–277.

Schutz, B. E., Tapley, B. D., and Shum, C. K. (1985). Precise SEASAT ephemeris from laser and altimeter data. *Adv. Space Res.*, **5**, 155–168.

Schutz, B. E., Tapley, B. D., Abusali, P. A. M., and Rim, H. J. (1994). Dynamic orbit determination using GPS measurements from TOPEX/POSEIDON. *Geophys. Res. Lett.*, **21**, 2179–2182.

Schwintzer, P., *et al.* (1997). Long-wavelength global gravity field models: GRIM4-S4, GRIM4-C4. *J. Geodesy*, **71**, 189–208.

Schwintzer, P., *et al.* (1998). A new global Earth gravity field model from satellite orbit perturbations for support of geodetic/geophysical and oceanographic satellite missions, reference frame and low degree gravity field solution (GRIM5-Step1). Scientific Tech. Rep. STR98/18 Geo-Forschungs Zentrum, Potsdam, Germany.

Seeber, G. (1993). *In* "Satellite Geodesy: Foundations, Methods and Applications," 531 pp. Walter de Gruyter & Co., Berlin.

Shampine, L., and Gordon, M. (1975). *In* "Computer Solution of Ordinary Differential Equations, The Initial Value Problem," W. H. Freeman and Company, San Francisco.

Shapiro, B. E., Bhat, R. S., and Frauenholz, R. B. (1994). Using anomalous along-track forces to control the T/P ground track. *Adv. Astronaut. Sci.*, **87**, 799–812.

Shen, C. Y., McWilliams, J. C., Taft, B. A., Ebbesmeyer, C. C., and Lindstrom, E. J. (1986). The mesoscale spatial structure and evolution of dynamical and scalar properties observed in the northwestern Atlantic Ocean during the POLYMODE Local Dynamics Experiment. *J. Phys. Oceanogr.*, **16**, 454–482.

Shuhy, J. L., Grunes, M. R., Uliana, E. A., and Choy, L. W. (1987). Comparison of GEOSAT and ground-truth wind and wave observations: preliminary results. *Johns Hopkins APL Tech. Digest*, **8**, 219–221.

Shum, C. K., Tapley, B. D., Yuan, D. N., Ries, J. C., and Schutz, B. E. (1990a). An improved model for the Earth's gravity field. *Proc. Int. Assn. of Geod. Symp. 103*, pp. 97–108, Springer-Verlag.

Shum, C. K., Yuan, D. N., Ries, J. C., Smith, J. C., Schutz, B. E., and Tapley, B. D. (1990b). Precise orbit determination for the Geosat exact repeat mission. *J. Geophys. Res.*, **95**, 2887–2898.

Shum, C. K., Zhang, B. H., Schutz, B. E., and Tapley, B. D. (1990c). Altimeter crossover methods for precision orbit determination and the mapping of geophysical parameters. *J. Astron. Sci.*, **38**, 355–368.

Singer, S. F. (1968). Measurement of atmospheric surface pressure with a satellite-borne laser. *Appl. Opt.*, **7**, 1125–1127.

Smith, A. J. E., and Visser, P. N. A. M. (1995). Dynamic and non-dynamic ERS-1 radial orbit improvement from ERS-1/TOPEX dual-satellite altimetry. *Adv. Space. Res.*, **16**, 12,123–12,130.

Smith, A. J. E., Hesper, E. T., Kuijper, D. C., Mets, G. J., Visser, P. N. A. M., Ambrosius, B. A. C., and Wakker, K. F. (1996). TOPEX/POSEIDON orbit error assessment. *J. Geodesy*, **70**, 546–553.

Smith, D. E., Kolenkiewicz, R., Dunn, P. J., Robbins, J. W., Torrence, M. H., Klosko, S. M., Williamson, R. G., Pavlis, E. C., Douglas, N. B., and Fricke, S. K. (1991). Tectonic Motion and Deformation from Satellite Laser Ranging to LAGEOS. *J. Geophys. Res.*, **95**, 22,013–22,041.

Smith, E. A., *et al.* (1998). Results of WetNet PIP-2 Project. *J. Atmos. Sci.*, **55**, 1483–1536.

Smith, E. K., and Weintraub, S. (1953). The constants in the equation for atmospheric refractive index at radio frequencies. *Inst. Radio Eng.*, **41**, 1035–1037.

Smith, S. D., Colles, M. J., and Peckham, G. E. (1972). The measurement of surface pressure from a satellite. *Quart. J. R. Met. Soc.*, **98**, 431–433.

Sofia, S., Fox, P., and Schatten, K. (1998). Forecast update for activity cycle 23 from a dynamo-based method. *Geophys. Res. Lett.*, **25**, 4149–4152.

Soicher, H. (1986). Variability of transionospheric signal time delay at subauroral latitudes. *IEEE Trans. Anten. Propag.*, **34**, 1313–1319.

Sojka, J. J. (1989). Global scale, physical models of the *F*-Region ionosphere. *Rev. Geophys.*, **27**, 371–403.

Spilker, J. J., and Parkinson, B. W. (1996). Overview of GPS operation and design. *In* "Global Positioning System Theory and Applications," Vol. 1, pp. 29–54, Progress in Aero. and Astro., 164, American Institute of Aero. and Astro., Washington, D.C.

Srokosz, M. A. (1986). On the joint distribution of surface elevations and slopes for a nonlinear random sea, with an application for radar altimetry. *J. Geophys. Res.*, **91**, 995–1006.

Srokosz, M. A. (1987). Reply. *J. Geophys. Res.*, **92**, 2989–2990.

Staelin, D. H., Kunzi, K. F., Pettyjohn, R. L., Poon, R. K. L., Wilcox, R. W., and Waters, J. W. (1976). Remote sensing of atmospheric water vapor and liquid water with the Nimbus-5 microwave spectrometer. *J. Appl. Meteor.*, **15**, 1204–1214.

Stammer, D. (1997). Global characteristics of ocean variability estimated from regional TOPEX/POSEIDON altimeter measurements. *J. Phys. Oceanogr.*, **27**, 1743–1769.

Stammer, D., Wunsch, C., and Ponte, R. M. (2000). De-aliasing of global high-frequency barotropic motions in altimeter observations. *Geophys. Res. Lett.*, **27**, 1175–1178.

Stanley, H. R. (1979). The GEOS-3 Project. *J. Geophys. Res.*, **84**, 3779–3783.

Stewart, R. H. (1985). *In* "Methods of Satellite Oceanography," 360 pp. University of California Press, Berkeley, CA.

Stewart, R. H., and Devalla, B. (1994). Differential sea-state bias: A case study using TOPEX/POSEIDON data. *J. Geophys. Res.*, **99**, 25,009–25,013.

Stum, J. (1994). A comparison between TOPEX microwave radiometer, ERS-1 microwave radiometer and European Centre for Medium-Range Weather Forecasting derived wet tropospheric corrections. *J. Geophys. Res.*, **99**, 24,927–24,939.

Sverdrup, H. U., and Munk, W. H. (1947). Wind, sea and swell: theory of relations for forecasting. U.S. Navy Hydrographic Office Pub. No. 601, 44 pp.

Swift, C. T. (1980). Passive microwave remote sensing of the ocean—a review. *Bound. Layer Meteor.*, **18**, 25–54.

Tai, C.-K. (1989). Accuracy assessment of widely used orbit error approximations in satellite altimetry. *J. Atmos. Oceanic Technol.*, **6**, 147–150.

Tai, C.-K. (1991). How to observe the gyre to global-scale variability in satellite altimetry: signal attenuation by orbit error removal. *J. Atmos. Oceanic Technol.*, **8**, 271–288.

Tai, C.-K., and J. Kuhn, J. (1995). Orbit and tide error reduction for the first 2 years of TOPEX/POSEIDON. *J. Geophys. Res.*, **100**, 25,353–25,363.

Tapley, B. D. (1973). Statistical orbit determination theory. *In* "Recent Advances in Dynamical Astronomy," (B. D. Tapley and V. Szebehely, Eds.), D. Reidel, Hingham, MA.

Tapley, B. D. (1989). Fundamentals of orbit determination. *In* "Theory of Satellite Geodesy and Gravity Field Determination," (F. Sansó and R. Rummel, Eds.), Vol. 25, pp. 235–260, Lecture Notes in Earth Sciences, Springer-Verlag, New York.

Tapley, B. D., and Rosborough, G. W. (1985). Geographically correlated orbit error and its effect on satellite altimetry missions. *J. Geophys. Res.*, **90**, 11,817–11,831.

Tapley, B. D., Lundberg, J. B., and Born, G. H. (1982). The SEASAT altimeter wet tropospheric range correction. *J. Geophys. Res.*, **87**, 3213–3220.

Tapley, B. D., Ries, J. C., Davis, G. W., Eanes, R. J., Schutz, B. E., Shum, C. K., Watkins, M. M., Marshall, J. A., Nerem, R. S., Putney, B. H., Klosko, S. M., Luthcke, S. B., Pavlis, D., Williamson, R. G., and Zelensky, N. P. (1994). Precision orbit determination for TOPEX/POSEIDON. *J. Geophys. Res.*, **99**, 24,383–24,404.

Tapley, B. D., Schutz, B. E., and Born, G. H. (2000). *In* "Statistical Orbit Determination." Academic Press, San Diego, in press.

Tapley, B. D., Schutz, B. E., Eanes, R. J., Ries, J. C., and Watkins, M. M. (1993). Lageos Laser Ranging Contributions to Geodynamics, Geodesy, and Orbital Dynamics. *In* "Contributions of Space Geodesy to Geodynamics: Earth Dynamics, Geodynamics Series" (D. E. Smith and D. L. Turcotte, Eds.), Vol. 24, pp. 147–173, American Geophysical Union, Washington, D.C.

Tapley, B. D., Watkins, M. M., Ries, J. C., Davis, G. W., Eanes, R. J., Poole, S. R., Rim, H. J., Schutz, B. E., Shum, C. K., Nerem, R. S., Lerch, F. J., Marshall, J. A., Klosko, S. M., Pavlis, N. K., and Williamson, R. G. (1996). The JGM-3 geopotential model. *J. Geophys. Res.*, **101**, 28,029–28,049.

Tapley, B. D., Shum, C. K., Ries, J. C., Poole, S. R., Abusali, P. A. M., Bettadpur, S. V., Eanes, R. J., Kim, M. C., Rim, H. J., and Schutz, B. E. (1997). The TEG-3 Geopotential Model. Inter. Assoc. Geodesy International Symp. No. 117, Gravity, Geoid and Marine Geodesy, pp. 453–460, (J. Seagwa, H. Fujimoto, and S. Okubo, Eds.), Springer-Verlag.

Teles, J., Pheung, P. B., and Guedeney, V. S. (1980). Tracking and Data Relay Satellite System range and doppler tracking system observation measurements and modeling. NASA Technical Memorandum X-572-80-26.

Thayer, G. D. (1974). An improved equation for the radio refractive index of air. *Radio Sci.*, **9**, 803–807.

Tierney, C., Wahr, J., Bryan, F., and Zlotnicki, V. (2000). Short-period oceanic circulation: Implications for satellite altimetry. *Geophys. Res. Lett.*, **27**, 1255–1258.

Tournadre, J., and Morland, J. C. (1997). The effects of rain on TOPEX/POSEIDON altimeter data. *IEEE Trans. Geosci. Rem. Sens.*, **35**, 1117–1135.

Townsend, W. F. (1980). An initial assessment of the performance achieved by the SEASAT altimeter. *IEEE J. Oceanic Eng.*, **5**, 80–92.

Townsend, W. F., McGoogan, J. T., and Walsh, E. J. (1981). Satellite radar altimeters: present and future oceanographic capabilities. *In* "Oceanography from Space," (J. F. R. Gower, Ed.), pp. 625–636, Plenum Press, New York.

Trenberth, K. E., and Olson, J. G. (1988a). Intercomparison of NMC and ECMWF global analyses: 1980-1986. *NCAR Tech. Note*, NCAR/TN-301+STR, 81 pp.

Trenberth, K. E., and Olson, J. G. (1988b). An evaluation and intercomparison of global analyses from the National Meteorological Center and the European Centre for Medium-Range Weather Forecasts. *Bull. Am. Meteor. Soc.*, **69**, 1047–1057.

Ulaby, F. T., Moore, R. K., and Fung, A. K. (1981). *In* "Microwave Remote Sensing, Active and Passive," Vol. I, Microvawe Remote Sensing Fundamentals and Radiometry, 456 pp. Addison-Wesley, Reading, MA.

Ulaby, F. T., Moore, R. K., and Fung, A. K. (1986a). *In* "Microwave Remote Sensing, Active and Passive," Vol. II, Radar Remote Sensing and Surface Scattering and Emission Theory, 608 pp. Addison-Wesley, Reading, MA.

Ulaby, F. T., Moore, R. K., and Fung, A. K. (1986b). *In* "Microwave Remote Sensing, Active and Passive," Vol. III, From Theory to Applications, 1705 pp. Addison-Wesley, Reading, MA.

Van Leeuwen, A., Rosen, E., and Carrier, L. (1979). The global positioning system and its applications in spacecraft navigation. *Navigation*, **26**, 204–221.

Visser, P. N. A. M., and Ambrosius, B. A. C. (1997). Orbit determination of TOPEX/POSEIDON and TDRSS satellites using TDRSS and BRTS tracking. *Adv. Space Res.*, **19**, 1641–1644.

Wagner, C. A. (1979). The geoid spectrum from altimetry. *J. Geophys. Res.*, **84**, 3861–3870.

Wagner, C. A. (1990). The M_2 tide from GEOSAT altimetry. *Manuscr. Geodet.*, **15**, 283–290.

Wagner, C. A., and Tai, C. K. (1994). Degradation of ocean signals in satellite altimetry due to orbit error removal processes. *J. Geophys. Res.*, **99**, 16,255–16,267.

Wahr, J. M. (1981). Body tides on an elliptical, rotating, elastic and oceanless earth. *Geophys. J. R. Astronom. Soc.*, **64**, 677–703.

Wahr, J., Molenaar, M., and Bryan, F. (1998). Time variability of the Earth's gravity field: hydrological and oceanic effects and their possible detection using GRACE. *J. Geophys. Res.*, **103**, 30,205–30,229.

Wakker, K. F., Zandbergen, R. C. A., Ambrosius, B. A. C. (1991). Seasat precise orbit computation and altimeter data processing. *Int. J. Remote Sensing*, **12**, 1649–1669.

Wallace, J. M., and Hobbs, P. V. (1977). *In* "Atmospheric Science: An Introductory Survey," 457 pp. Academic Press, San Diego.

Walsh, E. J. (1982). Pulse to pulse correlation in satellite radar altimeters. *Radio Sci.*, **17**, 786–800.

Walsh, E. J., Monaldo, F. M., and Goldhirsh, J. (1984). Rain and cloud effects on a satellite dual-frequency radar altimeter system operating at 13.5 and 35 GHz. *IEEE Trans. Geosci. Remote Sens.*, **22**, 615–622.

Walsh, E. J., Jackson, F. C., Uliana, E. A., and Swift, R. N. (1989). Observations on electromagnetic bias in radar altimeter sea surface measurements. *J. Geophys. Res.*, **94**, 14,575–14,584.

Walsh, E. J., Jackson, F. C., Hines, D. E., Piazza, C., Hevizi, L. G., McLaughlin, D. J., Mcintosh, R. E., Swift, R. N., Scott, J. F., Yungel, J. K., and Frederick, E. B. (1991). Frequency dependence of electromagnetic bias in radar altimeter sea surface range measurements. *J. Geophys. Res.*, **96**, 20,571–20,583.

Watkins, M. M., Eanes, R. J., and Ma, C. (1994). Comparison of terrestrial reference frame velocities determined from SLR and VLBI. *Geophys. Res. Lett.*, **21**, 169–172.

Webb, D. J. (1981). A comparison of Seasat-1 altimeter measurements of wave height with measurements made by a pitch-roll buoy. *J. Geophys. Res.*, **86**, 6394–6398.

Webster, W. J., Wilheit, T. T., Ross, D. B., and Gloersen, P. (1976). Spectral characteristics of the microwave emission from a wind-driven foam-covered sea. *J. Geophys. Res.*, **81**, 3095–3099.

Weidman, P. D., Mickler, D. L., Dayyani, B., and Born, G. H. (1999). Analysis of Legeckis eddies in the near-equatorial Pacific, *J. Geophys. Res.*, **104**, 7865–7887.

Wentz, F. J. (1997). A well-calibrated ocean algorithm for special sensor microwave/imager. *J. Geophys. Res.*, **102**, 8703–8718.

Wentz, F. J. , Mattox, L. A., and Peteherych, S. (1986). New algorithms for microwave measurements of ocean winds: Applications to SEASAT and the Special Sensor Microwave Imager. *J. Geophys. Res.*, **91**, 2289–2307.

Wentz, F. J., and Spencer, R. W. (1998). SSM/I rain retrievals within a unified all-weather ocean algorithm. *J. Atmos. Sci.*, **55**, 1613–1627.

White, N. J., Coleman, R., Church, J. A., Morgan, P. J., and Walker, S. J. (1994). A southern hemisphere verification for the TOPEX/POSEIDON altimeter mission. *J. Geophys. Res.*, **99**, 24,505–24,516.

Wilheit, T. T. (1978). A review of applications of microwave radiometry to oceanography. *Bound.-Layer Meteor.*, **13**, 277–293.

Wilheit, T. T., and Chang, A. T. C. (1980). An algorithm for retrieval of ocean surface and atmospheric parameters from the observations of the scanning multichannel microwave radiometer (SMMR). *Radio Sci.*, **15**, 525–544.

Wilheit, T. T., Chang, A. T. C., and Milman, A. S. (1980). Atmospheric corrections to passive microwave observations of the ocean. *Bound. Layer Meteor.*, **18**, 65–77.

Wilheit, T. T., *et al.* (1994). Algorithms for the retrieval of rainfall from passive microwave measurements. *Rem. Sens. Rev.*, **11**, 163–194.

Wilmes, H., Reigber, C., Schafer, W., and Hartl, P. (1987). Precise Range and Range-rate Equipment, PRARE, on-board ERS-1. *In* "Proceedings of the XIX IUGG General Assembly." Vol. II, pp. 586–596, Vancouver.

Witter, D. L., and Chelton, D. B. (1991a). A GEOSAT wind speed algorithm and a method for altimeter wind speed algorithm development. *J. Geophys. Res.*, **96**, 8853–8860.

Witter, D. L., and Chelton, D. B. (1991b). An apparent wave height dependence in the sea-state bias in GEOSAT altimeter range measurements. *J. Geophys. Res.*, **96**, 8861–8867.

Wu, S.-C. (1979). Optimum frequencies of a passive microwave radiometer for tropospheric path-length correction. *IEEE Trans. Antennas Propag.*, **27**, 233–239.

Wu, S.-C., Yunck, T. P., and Thornton, C. L. (1991). Reduced-dynamic technique for precise orbit determination of low Earth satellites. *J. Guid. Control Dynam.*, **14**, 24–30.

Wunsch, C. (1989). Sampling characteristics of satellite orbits, *J. Atmos. Oceanic Technol.*, **6**, 891–907.

Wunsch, C., and Stammer, D. (1997). Atmospheric loading and the oceanic "inverted barometer" effect. *Rev. Geophys.*, **35**, 79–107.

Wunsch, C., and Stammer, D. (1998). Satellite altimetry, the marine geoid, and the oceanic general circulation. *Ann. Rev. Earth Planet. Sci.*, **26**, 219–253.

Yaplee, B. S., Shapiro, A., Hammond, D. L., Au, B. D., and Uliana, E. A. (1971). Nanosecond radar observations of the ocean surface from a stable platform. *IEEE Trans. Geosci. Electron.*, **GE-9**, 171–174.

Yuan, D. N. (1991). *In* "The Determination and Error Assessment of the Earth's Gravity Field Model," Report CSR-91-01, Center for Space Research, The University of Texas at Austin.

Yunck, T. P. (1996). Orbit determination. *In* "Orbit Determination, Global Positioning System Theory and Applications," Vol. II. Progress in Aeronautics and Astronautics, Vol. 164, (B. Parkinson and J. Spilker, Jr., Eds.), pp. 559–592, American Institute of Aeronautics and Astronautics, Washington, D.C.

Yunck, T. P., Wu, S. C., and Lichten, S. M. (1985). A GPS measurement system for precise satellite tracking and geodesy. *J. Astronaut. Soc.*, **33**, 367–380.

Yunck, T. P., *et al.* (1994). First assessment of GPS-based reduced dynamic orbit determination on TOPEX/POSEIDON. *Geophys. Res. Lett.*, **21**, 541–544.

Yunck, T. P., Wu, S. C., Wu, J. T., and Thornton, C. L. (1990). Precise tracking of remote sensing satellites with the Global Positioning System. *IEEE Trans. Geosci. Remote Sens.*, **28**, 108–116.

Yunck, T. P., Bar-Sever, Y. E., Bertigei, W. E., Lichten, S. M., Lindquister, U., Mannucci, A., Muellerschoen, R., Manson, T., Romans, L., and Wu, S. (1996). A prototype WADGPS system for real-time sub-meter positioning worldwide. *Proc. Intl. Tech. Mtg. Inst. Nav.*, 1819–1826.

Zieger, A. R., Hancock, D. W., Hayne, G. S., and Purdy, C. L. (1991). NASA radar altimeter for the TOPEX/Poseidon Project. *Proc. IEEE*, **79**, 810–826.

Zimbelman, D. F., and Busalacchi, A. J. (1990). The wet tropospheric range correction: product intercomparisons and the simulated effect for tropical Pacific altimeter retrievals. *J. Geophys. Res.*, **95**, 2899–2922.

Zlotnicki, V. (1994). Correlated environmental corrections in TOPEX/POSEIDON, with a note on ionospheric accuracy. *J. Geophys. Res.*, **99**, 24,907–24,914.

Zlotnicki, V., Fu, L.-L., and Patzert, W. (1989). Seasonal variability in global sea level observed with GEOSAT altimetry. *J. Geophys. Res.*, **94**, 17,959–17,969.

Zumberge, J., Watkins, M., and Webb, F. (1998). Characteristics and applications of precise GPS clock solutions every 30 s. *Navigation*, **44**, 449–456.

2

Large-Scale
Ocean Circulation

LEE-LUENG FU* and DUDLEY B. CHELTON[†]

*Jet Propulsion Laboratory, California Institute of Technology, Pasadena, California
[†]College of Oceanic and Atmospheric Sciences, Oregon State University, Corvallis, Oregon

1. INTRODUCTION

Understanding the circulation of the ocean has been a slow process because of the ocean's vastness and inaccessibility. Before the time of modern technology, measurement of the ocean on a global scale was an effort that took years to accomplish. The data collected were interpreted in a climatological fashion with the assumption that the ocean did not change much with space and time. The Challenger Expedition conducted during 1872–1876, the first scientific exploration of the global oceans, is such an example. Not until the International Geophysical Year expeditions in the late 1950s did it become apparent that the ocean had variability over relatively short spatial scales (a few hundred kilometers) in a wide range of locations (Fuglister, 1960). With the advent of modern maritime instrumentation in the late 1960s and early 1970s, oceanographers began collecting time series of measurements in the ocean and discovered the richness of ocean variability in both space and time (e.g., the MODE Group, 1978). It was then recognized that it was not feasible to deploy enough instruments in the ocean to produce synoptic maps of the ocean variability.

Space-age technologies have made satellite remote sensing a powerful new tool to study the earth on a global scale. However, the opacity of the ocean to electromagnetic sensing has limited spaceborne measurements to the properties of the surface layer of the ocean (such as sea surface temperature and color). The promise of measuring the height of sea surface using radar altimetry provides a tantalizing opportunity to oceanographers because the sea surface height relative to the geoid, called the ocean topography, is a dy-

namic variable of the ocean and reflects oceanic processes not only at the surface but at depths as well. Using altimetric measurements, oceanographers are able to make inference about the state of the ocean beneath the surface. The challenge is the required measurement accuracy. A simple analysis (e.g., Wunsch and Stammer, 1998) shows that, for an ocean of 4000 m depth at 24° latitude, a 1-cm tilt in the ocean topography is associated with a mass transport of 7 Sv (1 Sv = 1 million tons per second, roughly the transport of all rivers combined), if the entire water column moves at the same velocity. The actual transport varies with latitude and the vertical distribution of current velocity, but this value provides a rough estimate. Such a magnitude is an appreciable fraction of the transport of the Florida Current ($\sim$ 30 Sv), for example. Measuring sea surface height from space with an accuracy of 1 cm is a tall order. As discussed in Chapter 1, it took 2 decades to reach the current level of performance in satellite altimetry with measurement accuracy approaching 4 cm (rms value at 1/sec data rate). After spatial and temporal smoothing, the accuracy is close to 2 cm on monthly time scales (Cheney et al., 1994).

The accuracy of the present geoid models, however, has not yet matched the accuracy of satellite altimetry (see Chapter 10). The derived ocean topography is therefore not sufficiently accurate for determining the details of the absolute ocean circulation. This deficiency has long been recognized by the geodyanmics community (National Research Council, 1997). New missions such as GRACE (Gravity Recovery and Climate Experiment) (Wahr et al., 1998; Davis et al., 1999) and GOCE (Gravity Field and Steady-State Ocean Circulation Explorer) (Legrand and Minster, 1999) have been planned to obtain more accurate measurements of the earth's gravity and hence the knowledge of the geoid. At wavelengths longer than 3000 km, the errors of presently available geoid models are smaller than the magnitude of oceanographic signals. At these large scales, satellite altimetry has provided the first direct measurement of the global ocean topography. The utility of such measurement for the determination of the oceanic general circulation is discussed in Section 2 of this chapter.

Because of the lack of detailed knowledge of the geoid, satellite altimeter data have primarily been used to study the temporal variability of the ocean. The differences in sea-surface height measured along precisely repeating ground tracks or at ground-track crossovers reflect primarily the temporal change plus measurement errors, while the time-invariant geoid is canceled in the sea-surface height differences (see Chapters 3 and 10 for detailed discussions of these methods). Studies based on early altimetry missions had to deal with a host of errors from orbit uncertainties, poor corrections for the various atmospheric effects, and inaccurate knowledge of the tides (e.g., Wunsch and Gaposchkin, 1980; Fu, 1983; Fu and Cheney, 1995). These errors often limited the studies to the energetic mesoscale eddies and boundary currents (Chapter 3). TOPEX/POSEIDON (denoted by T/P hereafter) is the first altimetry mission that produces data sufficiently accurate for studying variabilities on scales larger than the mesoscale without the need for correcting large-scale errors at the expense of distorting oceanographic signals of interests. A recent review of the results from the mission is provided by Wunsch and Stammer (1998). This data set has led to many new discoveries at the large scales where conventional shipboard measurements suffer from inadequate sampling. Results from studies focusing on the large scales (larger than the mesoscale whose upper bound is loosely defined as 500 km) at mid and high latitudes are reviewed in Section 3 of this chapter. Conclusions are given in Section 4.

2. THE OCEAN GENERAL CIRCULATION

The large variability of the ocean in space and time makes direct measurement of the oceanic general circulation, the long-term average of the flow field of the world's oceans, extremely difficult. For example, most current meter records have duration of only a few years and the data often show that the time mean is overwhelmed by the variability and hence is not statistically significant (Müller and Siedler, 1992). The frequency spectra computed from these records reveal that the oceanic spectrum is "red;" namely, the power density increases with decreasing frequency over the recorded period. On the other hand, the large-scale patterns of the ocean's temperature, salinity, and other chemical properties seem to be stable over long periods of time. For instance, the qualitative features from the cross-Atlantic (at 24°N) hydrographic surveys conducted in the 1990s are remarkably similar to those made in the1950s (Parrilla et al., 1994), with temperature changes of only a small fraction of a degree Celsius below the seasonal thermocline. This apparent stable distribution of water properties could result from the existence of a stable general circulation, which is probably best determined from the ocean's density and chemical properties. This is indeed how the description of the oceanic general circulation has been derived in the past. However, the turbulent nature of oceanic flows could also lead to stable distribution of water properties through eddy diffusion (see Section 4.6.2 of Chapter 3) without the need of a mean flow at all. The effect of eddies must be taken into account in the study of the general circulation.

The basic principle allowing the determination of the circulation from the ocean's density field is based on the fact that, in the open ocean at latitudes more than a few degrees from the equator, the large-scale oceanic flows are nearly in geostrophic and hydrostatic balance, expressed by the following equations:

$$fv = \frac{1}{\rho}\frac{\partial p}{\partial x}, \tag{1}$$

$$fu = -\frac{1}{\rho}\frac{\partial p}{\partial y}, \tag{2}$$

$$0 = -\frac{\partial p}{\partial z} - g\rho, \tag{3}$$

where p is the pressure, ρ is the density of sea water, g is the earth's gravity acceleration, u and v are the zonal and meridional velocity components, respectively, f is the Coriolis parameter defined as $f = 2\Omega \sin\varphi$, where Ω is the earth's rotation rate (7.292×10^{-5} rad/sec), and φ is the latitude. If we differentiate Eq. (1) with respect to z and Eq. (3) with respect to x, then Eq. (1) and Eq. (3) can be combined to obtain the following equation:

$$\rho v = -\frac{g}{f}\int_{z_0}^{z}\frac{\partial\rho}{\partial x}dz + v_0, \tag{4}$$

where z_0 is a reference level for the integration and v_0 is the meridional velocity at the reference level. A similar equation can be obtained for u. Historically, the reference level was chosen to be at great depths where the velocity was assumed to be zero, or at least close to zero. Therefore the velocity of the ocean could be determined from the knowledge of the ocean's density field. However, the choice of a "level of no motion" for the reference level has been controversial and problematic (Wunsch, 1996). Determination of the absolute velocity at a given level over the global ocean is one of the most challenging tasks facing physical oceanographers.

Determining the ocean topography from satellite altimetry in principle is a straightforward approach to the prob-

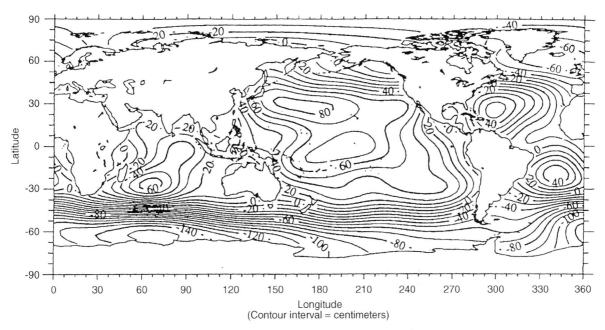

FIGURE 1 Spherical harmonic representation of the ocean topography up to degree and order 10 based on Geosat data. It was obtained as part of a solution to an inverse problem involving simultaneous fitting of ocean topography and a gravity model to altimeter and satellite tracking data. Contour interval is 10 cm. (From Nerem, R. S., Schrama, E. I., Koblmsky, C. J., and Beckley, B. D., 1990. With permission.)

lem of the reference level velocity. When multiplied by g, the ocean topography represents the dynamic pressure at the ocean surface. The surface geostrophic velocity can thus be obtained directly from the gradients of the ocean topography, denoted by η:

$$v = \frac{g}{f}\frac{\partial \eta}{\partial x}, \quad u = -\frac{g}{f}\frac{\partial \eta}{\partial y}. \tag{5}$$

The ocean topography derived from early altimetry missions was of little value for detecting absolute circulation because its uncertainty was larger than the oceanic signal at all scales. The uncertainty is dominated by satellite orbit errors at the scales of the ocean basin and by geoid errors at smaller scales. For instance, the first map of the absolute circulation of the Pacific Ocean derived from the Seasat altimeter showed only a marginal resemblance to the real basin-scale (>6000 km) circulation of the ocean (Tai and Wunsch, 1983). During the preparation for T/P, a systematic gravity model improvement project was conducted in the 1980s, leading to progressively improved gravity and geoid models. These models were first applied to the Geosat altimeter data and the reanalysis of the Seasat data. In addition to straightforward differencing between sea-surface height and a geoid model (e.g., Tai, 1988), ocean topography was often solved as a solution to an inverse problem involving simultaneous adjustment of orbit, geoid, and ocean topography. The resulting ocean topography solutions were extended to spherical harmonics of degree and order 6–10, with wavelengths of 4000–6000 km (Tapley *et al.*, 1988; Marsh *et al.*, 1990;

Denker and Rapp, 1990; Nerem *et al.*, 1990; Visser *et al.*, 1993). The large-scale features of the global ocean surface circulation began to emerge from such efforts (Figure 1), which were quite remarkable given the fact that the orbit errors were of the same order of magnitude as the oceanic signals. The formal estimate of the error in the ocean topography solution was 13 cm for wavelengths longer than 4000 km (Nerem *et al.*, 1990).

Despite the progress in deriving realistic ocean topography from altimetry, the result is not useful scientifically unless it provides new insight into the circulation and improves the estimate of important quantities such as the oceanic heat transport. Wunsch (1981) suggested that the ocean topography based on historical hydrographic data and a conventional mid-depth level of no motion had an accuracy of 10–25 cm. The altimetric ocean topography must therefore be more accurate than 10–25 cm to be useful. In fact, the ocean topography resulting from inverse models based on extensive *in situ* data is more accurate than 10–25 cm and thus places a more stringent test on the utility of altimetry (e.g., Martel and Wunsch, 1993a). Trying to incorporate the Geosat result of Nerem *et al.* (1990) into a North Atlantic model for an improved estimation of the circulation, Martel and Wunsch (1993b) found that the Geosat estimate of the surface velocity was not compatible with the conventional data. The error in the altimetry-derived ocean topography was too large, by at least a factor of two, to allow any new information to be gained about the circulation.

The gravity model improvement activities supported by the T/P Project culminated in a series of models called the Joint Gravity Models (JGM) (see Chapter 10). The geoid derived from the most recent model, JGM-3 (Tapley *et al.*, 1996), has an estimated accuracy of about a factor of two better than the one obtained by Nerem *et al.* (1990). The estimated formal error in the JGM-3 geoid model in terms of a set of functions that are orthonormal over the oceans is shown in Figure 2 (from Rapp *et al.*, 1996), among other information to be discussed later. Displayed in Figure 3a is the ocean topography obtained as the difference between a 2-year averaged T/P mean sea surface (1993–1994) and the JGM-3 geoid, up to degree 14 of the orthonormal expansion (adapted from Rapp *et al.*, 1996). The error in JGM-3 becomes larger than oceanographic signals beyond degree 14 as discussed below. This ocean topography is compared with the simulation of an ocean general circulation model projected onto the same orthonormal expansion (Figure 3b). The model has a spatial resolution of $1/4°$ and 20 vertical levels. The basic formulation of the model was described in Semtner and Chervin (1992). The model simulation was performed for 1987–1994, driven by daily winds and climatological monthly heat flux. The simulation of the global ocean circulation produced by the model has been used by other studies as a benchmark for comparison with observations (e.g., Stammer *et al.*, 1996).

The visual resemblance between the T/P-derived topography and the model is quite encouraging up to this degree and order in the orthonormal expansion. The improvement of the T/P result relative to the Geosat result (Figure 1) is visible, especially in the Atlantic Ocean. Rapp *et al.* (1996) performed a quantitative evaluation of the comparison. The root-mean-square (rms) difference between the T/P topography and the model up to degree 14 is 12.4 cm. The corresponding rms difference in geostrophic velocity speed is 2.5 cm/sec, about 50% of the signal. If the correlation between the model error and the observation error is small, which is not an unreasonable assumption, then the difference can be interpreted in terms of the magnitudes of the observation error and the model error. Since the systematic orbit error (geographically correlated) based on the JGM-3 gravity model is less than 1 cm in the T/P data (Tapley *et al.*, 1996) and other altimetry systematic errors are also at 1 cm level (Chapter 1, Section 8.3), the observation error at the large scales is dominated by the geoid error. Up to degree 14 the JGM-3 geoid model has an estimated rms error of 6.8 cm (Table 1, Rapp *et al.*, 1996). The inferred rms error for the ocean model is then about 10 cm. Fu and Smith (1996) compared the T/P ocean topography with the model of Smith *et al.* (1992), which is similar to the model of Semtner and Chervin (1992) but with a $1/6°$ spatial resolution. This model was developed by the Parallel Ocean Program (POP) of the Los Alamos National Laboratory, sometimes referred to as the POP model. Fu and Smith (1996) found

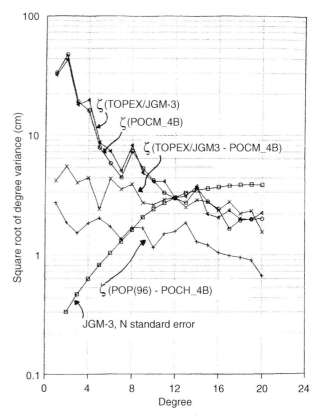

FIGURE 2   The square root of variance in cm as a function of degrees of the orthonormal ocean functions for the following: the JGM-3 geoid undulation standard error, the ocean topography determined from T/P (based on the JGM-3 geoid) and from the ocean model of Semtner and Chervin (1992) (denoted by POCM_4B), the difference between the T/P ocean topography and POCM_4B, as well as the difference between POCM_4B and the POP model (denoted by POP96). The degrees in the abscissa refer to those of a set of expansion functions similar to the spherical harmonics but they are orthonormal over the global oceans. (From Rapp, R. H., Zang, C., and Yi, Y., 1996. With permission.)

that the large-scale error of the POP model was about 10 cm as well. For the first time the errors of global ocean models have been assessed against a global data set with a specified error estimate. One should note, however, that these error estimates are approximate and heavily dependent on the geoid errors, which are often not well determined.

The map of the difference between the T/P ocean topography and the model (T/P minus model) (Figure 3c) reveals a few large discrepancies: −62 cm in the Banda Sea (between Australia and Indonesia and to the east of the Timor Sea) and from 36 cm to −37 cm in the South Pacific sector of the Antarctic Circumpolar Current. The large discrepancy in the Banda Sea is puzzling, because the estimated geoid error in the area is only about 8 cm to degree 14. However, there are large high-wave-number geoid features in the region that are not well modeled and might have aliasing effects on the low-degree representations of the ocean topography (Rapp *et al.*, 1996).

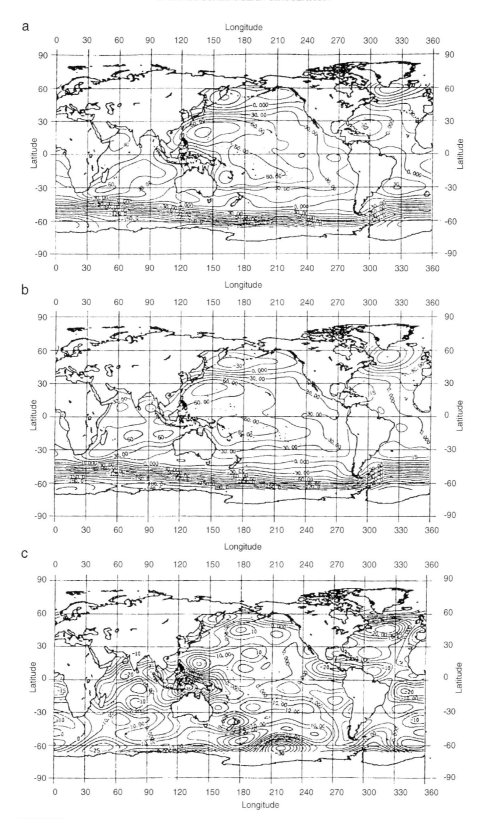

FIGURE 3 (a) Ocean topography estimated from T/P data from 1993–1994 and the JGM-3 geoid. (b) Ocean topography simulated by an ocean general circulation model (POCM_4B) for the same time period. (c) The difference between the top and middle panels (T/P–POCM_4B). Contour interval is 10 cm. (Adapted from Rapp, R. H., Zang, C., and Yi, Y., 1996. With permission.)

There are indications suggesting that the discrepancies in the Southern Ocean might be caused by problems in the ocean model. The difference between the Semtner and Chervin model and the POP model shows large values (20–30 cm) in the same region of the Southern Ocean (Rapp et al., 1996). The rms difference by degree between the two models is also shown in Figure 2. Although the two models are not totally independent, their difference perhaps reflects a lower bound for the errors in both models. The reader is also referred to Park and Gambéroni (1995) for a regional study of the Indian Ocean sector of the Southern Ocean. They reported agreement between the T/P ocean topography and the simulations by the Fine Resolution Antarctic Model (FRAM Group, 1991) in revealing several gyre-scale circulation patterns that were absent in historical hydrographic data.

The error in the JGM-3 geoid becomes larger than the signal in the ocean topography (Figure 2) at degrees higher than 14, where the rms difference between the T/P topography and the model becomes comparable to the signal magnitude. The correlation between the T/P topography and the model drops from the high values of 0.8–0.9 at low degrees to below 0.5 at degrees higher than 14. The utility of the JGM-3 model is thus limited to degrees lower than 14 (corresponding to a wavelengths of about 3000 km). A more recent geoid model, called EGM96 (see Chapter 10; Lemoine et al., 1998), was constructed using a more extensive data base than was used in JGM-3. Preliminary assessment suggests that EGM 96 has about a factor of two improvement over JGM-3 up to degree 20. With this increased accuracy, the utility of the altimetry-derived ocean topography may be extended to degree 18 (Lemoine et al., 1998).

With much improved geoid models, ocean topography derived from altimetry has become quantitatively consistent with what has been obtained from historical data (Williams and Pennington, 1999). Ganachaud et al. (1997) conducted a linear inverse calculation to evaluate whether the T/P-derived ocean topography is useful for improving the estimate of the oceanic general circulation relative to a previous estimate based on historical hydrographic data (Macdonald and Wunsch, 1996). After spatial smoothing to exclude wavelengths shorter than 1600 km, they found that the altimetrically determined velocities were consistent with hydrographic estimates within the error bars of each. After this consistency check, they further combined the two estimates using a recursive inverse procedure to obtain a new solution that was fully consistent with both altimetry and hydrography. However, the resulting solution did not reduce the errors in the circulation and its associated heat transport in any significant way, because the geoid (JGM-3) was still worse than what can be obtained equivalently from the hydrographic data. They further concluded that, in order to obtain geoid models sufficiently accurate to allow improvements upon the hydrographic estimate of the mean ocean

circulation, specifically designed gravity missions meeting very demanding requirements were required (also see National Research Council, 1997 and Chapter 10). The planned missions such as GRACE and GOCE mentioned in the Introduction are designed to fulfill these roles.

The study by Gananchaud et al. (1997) has provided a particular view of the utility of altimetry through the perspective of a linear inverse calculation based on hydrographic data and geostrophic dynamics (also see LeGrand et al., 1998). One could argue that the prior error estimates assigned to the hydrographic solution might be overly optimistic. Given the difference in the sampling (in both space and time) of the circulation between hydrography and altimetry, the combination of the two data sets using a linear inverse model for estimating the ocean general circulation can be quite tricky. When the global altimetry data are continuously assimilated into a state-of-the-art ocean general circulation model (see Chapter 5), one should then be able to obtain a better estimate of the time-evolving global circulation, from which a well-sampled, time-averaged general circulation should thus emerge. As discussed earlier, the errors in the present ocean models are generally comparable to those in the altimetric ocean topography at large scales. The approach of data assimilation should produce an estimate that is better than what can be obtained from either the data or the model by itself. Such an approach was taken by Stammer et al. (1997). Some preliminary results from the study can be found in Wunsch and Stammer (1998). The effects of altimetry on the model simulation are clearly evident. However, it is not yet clear whether such estimates are consistent with the hydrographic estimates as well as other types of observations. The difficulty lies in the determination of the error covariance of the estimates resulting from very complex calculations involving highly non-linear general circulation models.

3. THE TEMPORAL VARIABILITY

The time variation of sea level measured by tide gauges has been a valuable source of information for studying the large-scale dynamics and thermodynamics of the ocean. Long records of tide gauge data are among the few time series available to oceanographers to study oceanic processes over a wide range of time scales: tides, internal waves, fluctuations of current, response to atmospheric thermal and mechanical forcing, planetary waves, global warming, etc (e.g., Pattullo et al., 1955; Wunsch, 1972; Wyrtki, 1974a; Enfield and Allen, 1980; Chelton and Davis, 1982; Douglas, 1991). However, the tide gauges are mostly located along the coasts and their distribution in the open ocean is limited by the available islands. Therefore, the spatial sampling of the tide gauges is extremely poor except perhaps in the western and central tropical Pacific where there are more island gauges than elsewhere. Flying a satellite altimeter is analogous to providing a huge number (on the order of a half million) of tide gauges densely distributed over the global oceans, offering a global coherent view of the various large-scale processes that have been poorly resolved by the sparsely distributed tide gauges in the past.

The largest variability of sea level is due to the ocean tides with a global rms amplitude of about 32 cm (see Chapter 6 as well as Figure 54 of Chapter 1). Because the main rationale for the development of satellite altimetry is to study the ocean circulation, the measurement system must be designed such that the tidal variations can be sampled well to allow them to be separated from the effects of ocean circulation on sea level measurement (see Chapters 1 and 6). The best ocean tide models derived from the T/P data are accurate with an rms error of 2–3 cm. The residual tidal signals appear in the corresponding aliased periods such as 62 days for the M_2 component, and 173 days for the K_1 component in the T/P data. Because the aliased tidal periods are precisely known (Chapter 6, Table 1), these residual errors will not affect the study of ocean circulation in a significant way if the aliased periods do not coincide with those of strong ocean variability. However, the aliased period of K_1, 173 days, is very close to the semi-annual period and could be problematic in regions where the semi-annual variability is strong such as the South Atlantic and the central tropical Pacific (Jacobs et al., 1992). In the tropical regions especially in the Indian Ocean where the ocean has substantial variability at periods near 60 days (Luyten and Roemmich, 1982; Kindle and Thompson, 1989), the M_2 tidal aliasing also presents a problem. Because the phase of the tides are sampled differently by adjacent satellite tracks, aliased tidal signals may appear as traveling waves and create confusions for interpretation (see Section 8.1.5 of Chapter 1; Schlax and Chelton, 1994b, 1996).

After the tidal signals are removed (including the effects of the solid earth tides), the remaining sea-level variability (global rms $\sim$ 12 cm, see Wunsch and Stammer, 1997) can be divided into two categories according to spatial scales. At scales larger than 500 km, the variability is to a large extent caused by the ocean's response to the forcing by the atmosphere through wind stress, pressure, and air-sea exchange of heat and freshwater. This component of sea level variability, which is the subject of this section, has a global rms amplitude of about 9 cm (including the inverted-barometer response, see Section 3.2). This value is derived from the estimates of Wunsch and Stammer (1998) for the variability of large-scale circulation plus the inverted-barometer response. At mesoscales (defined here to be wavelengths shorter than 500 km), sea-level variability has a global rms amplitude of about 8 cm and is mainly caused by ocean currents and eddies, which are the subject of Chapter 3.

Kuragano and Kamachi (2000) investigated the characteristics of spatial and temporal scales of sea-level variability using T/P data. They fitted a three-dimensional Gaussian model to the observed covariance as a joint function (as opposed to separate functions) in latitudinal, longitudinal, and temporal lags. They found that in areas where the mesoscale energy is low, the dominant scales are generally larger than 500 km. At mid and low latitudes, the large-scale variability is generally anisotropic with larger zonal scales and westward phase propagation, exhibiting the presence of Rossby waves (Section 3.4). At high latitudes, the large-scale variability becomes more isotropic, perhaps reflecting locally forced response. In areas where the mesoscale energy is high, the large-scale variability becomes obscured without some space-time filtering.

Changes of sea level at a given location can be caused by two factors: changes in the total mass of the water column, or changes in the density of the water column. Both factors can operate at the same time. The interpretation of sea level change is thus not straightforward. Changes in both mass and density of a water column can be caused by direct atmospheric forcing as well as the internal processes of the ocean such as advection and diffusion. Because advection and diffusion occur primarily at mesoscales, sea level change at scales larger than 500 km is mostly dominated by atmospheric forcing.

Changes in the large-scale distribution of the vertically integrated mass in the ocean are mostly caused by the mechanical forcing of the atmosphere through wind and pressure. The ocean's response involves the movement of the entire water column in unison, called the barotropic mode. Changes in the density distribution of a water column can be caused by two mechanisms. First, the buoyancy forcing by the exchange of heat and freshwater with the atmosphere can change of the density of the water. Second, the constant-density surfaces in the ocean can move vertically, leading to a change of the vertical distribution of density without changing the total columnar mass, called the baroclinic mode. The large-scale baroclinic mode can be excited by both buoyancy and mechanical forcing. The barotropic response occurs on relatively short time scales, generally less than the seasonal scales. The response to buoyancy forcing occurs predominantly on seasonal time scales, namely the annual cycle. The baroclinic response occurs on seasonal and longer time scales. Therefore, the time scales of sea level change offer some clues regarding the underlying physical processes, which in turn determine the extent of the relation between the sea level variability and the variability of the density and velocity at depths.

The organization of the rest of the section is as follows. The buoyancy forcing is discussed in Section 3.1, followed by discussions of the mechanical forcing by atmospheric pressure (Section 3.2) and wind (Section 3.3). Studies of baroclinic Rossby waves, a key dynamical mechanism governing the large-scale low-frequency variability of the ocean, are discussed in Section 3.4. The relation between sea level change and subsurface fields is discussed in Section 3.5. An overview of the phenomena at annual and interannual time scales is given in Section 3.6.

3.1. Buoyancy-Forcing and the Heat Budget of the Ocean

The effects of the ocean's exchange of heat and freshwater (evaporation/precipitation) with the atmosphere change the density of the ocean and hence the sea level (the "steric" effect). The steric variability of sea level is to first order caused by the change of the heat content in the ocean on seasonal time scales, with the effects of freshwater exchange (via salinity) playing a secondary role (Gill and Niiler, 1973). Steric sea-level variability can therefore be used to infer the heat storage in the upper ocean and the rate of heat exchange between the ocean and the atmosphere. However, salinity effects on sea level are important in certain regions (Maes, 1998). Corrections for them generally require simultaneous measurements of salinity because the climatological database is not very useful for the purpose (Sato *et al.*, 2000). On global scales, the effects of freshwater exchange between the ocean and the atmosphere are important on seasonal-to-interannual time scales. At these scales, the variation of the global mean sea level can be used to infer the water balance of the earth (Minster *et al.*, 1999; see Chapter 8).

Steric sea level variability generally occurs over the scales of air-sea heat and freshwater exchange, typically on the order of 1000 km and larger. To study the mechanisms of sea level variability at these scales, it is essential to be able to separate the steric effects from other factors that affect sea level. The steric sea-level anomaly can be computed from the change of the heat and salt content of the water column (Gill and Niiler, 1973) as follows:

$$\eta_s = -\frac{1}{\rho_0}\left(\int_{-h}^{0}\frac{\partial \rho}{\partial T}T'dz + \int_{-h}^{0}\frac{\partial \rho}{\partial S}S'dz\right), \qquad (6)$$

where T' and S' are the temperature and salinity deviations from their mean values and ρ and ρ_0 are the density and its depth average, respectively. As noted above, the steric sea-level variability is dominated by the effects of temperature. The effects of salinity are generally only about 10% of those of temperature. However, salinity effects can be more significant in certain regions such as the western tropical Pacific, where salinity generally accounts for more than 20% of the sea-level anomalies and is sometimes comparable to the thermal effect (Maes, 1998). In any event, because information about salinity is much less available than temperature, steric sea level is normally estimated from temperature information only. With this approximation, steric sea level is

simply related to thermal expansion. To explore the global patterns of thermal effects on sea-level variability, Leuliette and Wahr (1999) conducted a coupled-pattern analysis on simultaneous altimeter data and sea-surface–temperature data. They reported large-scale, coupled patterns of sea level and sea surface temperature on annual, interannual, and secular time scales.

Changes in steric sea level can be computed from air-sea heat flux as follows (Stammer, 1997; Chambers *et al.*, 1997; Wang and Koblinsky, 1997):

$$\frac{\partial \eta_s}{\partial t} = \frac{\alpha Q}{\rho_0 c_p}, \qquad (7)$$

where α is the coefficient of thermal expansion of sea water, c_p is the specific heat at constant pressure and Q is the net air-sea heat flux anomaly. Direct observations of the various components of Q are sparsely distributed in space and time. Time-varying Q is therefore available only from simulations made by models run by meteorological centers. Using Q provided by the ECMWF (European Center for Medium-Range Weather Forecast), Stammer (1997) made estimates of the steric component of sea level and compared them with T/P observations (Figure 4). At mid-latitudes, steric sea level accounts for a major portion of the observed sea level variations (also see Vivier *et al.*, 1999; Ferry *et al.*, 2000). Significant discrepancies are primarily confined to the tropics and subtropics, where the temporal variability of heat flux diminishes and the wind-induced sea level variability (in the form

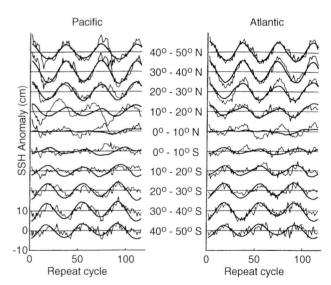

FIGURE 4 Sea surface height anomalies (thin line) observed by T/P. The data were averaged in zonal bands extending 10° meridionally across the width of the Pacific (left panel) and the Atlantic (right panel). Also shown are the steric height anomalies for the same areas (bold line) computed from Eq. (7) based on the ECMWF heat flux. The latitude ranges are marked on the figure. The horizontal axis is T/P repeat cycle. Each cycle has a duration of 9.9 days with Cycle 1 starting on October 2, 1992. (From Stammer, D., 1997. With permission.)

of Rossby and Kelvin waves) dominates (see Section 3.4 and Chapter 4). Note that the steric variability at mid-latitudes is larger in the northern hemisphere than the southern hemisphere by a factor of two, reflecting the contrast between the two hemispheres in the distribution of land mass. The cold air blown from the continents to the warm oceans during winter time is a major contributor to the air-sea heat flux. The lack of land mass in the southern hemisphere leads to less steric variability.

Errors in the model simulation of Q are difficult to quantify. The study by Siefridt (1994) suggests that the errors in the ECMWF heat flux are roughly 20–40 W/m^2 (also see Ferry et al., 2000), larger than the signals of heat flux variability in many regions of the world's oceans. The errors in the southern hemisphere are probably larger because of the lack of data to constrain the model. In some studies, altimetric observations of sea level have actually been used to estimate the heat flux. For example, White and Tai (1995) computed the correlation between sea level anomalies (relative to an annual cycle) from T/P and the upper ocean heat storage anomalies estimated from XBT (expendable bathythermograph) profiles. The correlation obtained over the globe was 0.5–0.8. Based on regression analysis, they were able to estimate interannual changes in heat storage over the global upper ocean (above 400 m) using the T/P data. Although the inferred estimates from altimetry are not as accurate as direct estimates from XBT's, the uniform and frequent global coverage of the altimeter data leads to globally gridded estimates of heat storage anomalies with sampling errors smaller than those obtained from XBT data by a factor of two. When the estimated heat storage anomalies were used to compute the rate of heat storage change integrated over ocean-basin scales, the errors in the estimated basin-wide air-sea heat fluxes are only about 2 W/m^2, about half those obtained from the XBT analyses. Such errors are comparable to the signals of the seasonal-to-interannual variability of basin-wide air-sea heat flux (e.g., integrated over the entire Pacific Ocean), a quantity that is extremely difficult to measure.

Chambers et al. (1997) also made estimates of heat storage and its rate of change directly from Eq. (7) using T/P altimeter data. On interannual time scales, the inferred rates of heat storage change have an error of about 5–10 W/m^2 when compared with the results from the TOGA (Tropical Ocean and Global Atmosphere Program) moorings in the tropical Pacific Ocean. When integrated over an ocean basin, the error in the estimate of interannual heat flux change is essentially dictated by the error in estimating the mean sea-level trends. Based on an uncertainty of 2–3 mm/year for the mean sea-level trends (see Chapter 8), their estimates for the uncertainty in the determination of the basin-wide heat flux variability on interannual time scales are 1–1.5 W/m^2, comparable to the estimate of White and Tai (1995). Given this uncertainty, Chambers et al. (1997) reported that the North

Atlantic and the oceans in the Southern Hemisphere gained heat from the atmosphere at an average rate 0–3 W/m^2 from 1993–1995. However, it is difficult to verify this conclusion from independent sources because of the sparse distribution of in situ observations.

When the altimetric estimates of heat storage change are averaged over ocean-basin scales, the errors introduced by nonthermally forced sea-level signals tend to average out, leading to a fairly accurate estimate of the basin-wide heat flux change. However, such error reduction does not apply to regional analyses. For instance, the effects of heat advection by ocean currents become important in the regions of the Kuroshio Extension and the Gulf Stream. In an effort to investigate the utility of altimetry data for diagnosing the heat budget of the ocean, Qiu and Kelly (1993) used a numerical model of the ocean's mixed-layer driven by wind and heat flux with the geostrophic current velocity estimated from Geosat observations for estimating the mixed-layer depth and temperature of the Kuroshio Extension region. They found that the advection of heat by the surface flows made a substantial contribution to the local heat balance. The advection warms the upstream region of the Kuroshio Extension while cools the downstream region due to the presence of the recirculation gyre. Using a similar approach, Kelly and Qiu (1995a, b) ran the mixed-layer model with assimilation of altimeter data and sea-surface temperature data for studying the heat balance of the Gulf Stream gyre. Rather than forcing the model with heat flux produced by weather centers, heat flux was estimated as the residual of the heat budget that involved horizontal advection, vertical entrainment, and eddy diffusion. The error in the heat flux estimate ranges from 20–100 W/m^2. Based on the heat budget analysis, their conclusions indicate that the seasonal variability of the heat content of the gyre to the south of the Gulf Stream is primarily forced by the air-sea heat flux (also see Wang and Koblinsky, 1996a). Within the Gulf Stream and to the north of it, the cooling by the southward Ekman advection of cold water has a tendency to be balanced by the warming of the Gulf Stream water. A very complex pattern of air-sea interaction is present in the region, showing signatures of the coupling between the air-sea heat flux and the wind-driven circulation (also see Kelly et al., 1999).

3.2. Atmospheric Pressure Forcing

The response of the ocean to atmospheric pressure forcing has received less attention than the response to wind and buoyancy forcing. This is because pressure forcing is a much less effective mechanism for generating oceanic motions. The conventional wisdom is that, to first order, the ocean responds to pressure forcing in a static manner. An increase (decrease) in atmospheric pressure by 1 mbar depresses (raises) sea level by about 1 cm (Chapter 1, Section 5.4). The ocean thus acts like an "inverted barometer."

The horizontal gradient of the atmospheric pressure is completely compensated by the adjustment of sea level, leaving no pressure gradient just below the sea surface. From Eqs. (1) and (2), there is thus no geostrophic movement of water associated with the inverted-barometer response. If not corrected for this static response, sea-level data would be very difficult to analyze for studying the variability caused by ocean circulation. The variability of atmospheric pressure has a global rms magnitude of about 7 mb with a high degree of spatial variability (see Figure 24 of Chapter 1). The resulting inverted barometer (IB) sea-level variations can be as large as 15 cm (rms) in the Southern Ocean where the atmospheric pressure variability is the largest. The inverted barometer correction is thus a very important issue in analyzing altimeter data.

The extent to which the ocean's response can be represented by the IB approximation has recently been reviewed by Wunsch and Stammer (1997). Their conclusion is that an IB response is expected at all frequencies and wavenumbers except those falling on the dispersion curves for the dynamic response in the form of gravity and Rossby waves. Resonance is possible depending on the reflecting properties of the ocean bottom and specific geometry of the ocean basin. Analyses of the T/P altimeter data have indeed shown a near IB response in most of the non-tropical regions. Fu and Pihos (1994) performed a regression analysis between T/P sea-level anomalies (computed as deviations from the time mean) and pressure fluctuations. They found that the sea level response over regions poleward of 30° latitude was weaker than IB with a regression coefficient of -0.84 ± 0.29 (1 standard deviation) cm/mbar. (The minus sign indicates the expected inverse relationship.) They also performed a multivariate regression analysis to remove the effects of wind forcing, which is correlated with the pressure. The result indicated a response much closer to IB at -0.96 ± 0.32 cm/mbar. In the tropics where the pressure forcing is particularly weak, their results showed a complete breakdown of the IB response. However, after the removal of the wind effects using the simulation of a tropical ocean model, the sea-level response became IB-like even in the tropics.

Gaspar and Ponte (1997) and Ponte and Gaspar (1999) used the T/P crossover differences within 10-day repeat cycles to investigate the IB effects at high frequencies (primarily at periods of 3–4 days). Outside the tropics, their results were similar to Fu and Pihos (1994), whose study, based on repeat-track analysis, did not filter out any particular frequencies. The high-frequency sea-level response is generally somewhat weaker than IB with a regression coefficient of about -0.9 cm/mbar. The coefficient drops to -0.7 cm/mbar in a number of regions in the Southern Ocean. They were able to simulate the observed relation between sea level and pressure using a numerical model of a homogeneous ocean driven by both wind and pressure. They

found that the extratropical sea-level response became much closer to IB if the model was driven only by pressure. They concluded that most of the apparent non-IB response in the extratropics was due to the wind effects, consistent with the multivariate regression analysis of Fu and Pihos (1994) and the modeling study of Ponte (1994). However, even the pressure-driven simulations revealed significant non-IB response (~ -0.9 cm/mbar) in the Southern Ocean, where the bottom topography is favorable for resonant response. Another extensive region of pressure-driven non-IB response (from -0.7 to -0.8 cm/mbar) is in the tropics, where the frequency band of free waves is wider than that in the extratropics and is thus more prone to resonant response. However, the atmospheric pressure variability is very small (rms value less than 2 mb) in the tropics and hence the magnitude of the ocean's non-IB response is also extremely small there.

Bryan et al. (2000) also conducted a study of the ocean's response to atmospheric pressure forcing using an ocean general circulation model. They found that the dynamic response of the ocean occurs primarily at periods shorter than 10 days with signatures of strong topographic control. The maximum variance of sea level occurs over closed f/H contours (H is the depth of the ocean bottom), especially in the Southern Ocean where the rms dynamic response reaches 8 cm.

3.3. Wind Forcing

Wind stress exerts forcing on the ocean over a wide range of spatial and temporal scales. The ocean's response is complicated, involving both local adjustment and propagating waves from remote forcing. The vertical structure of the ocean's response is a function of the spatial and temporal scales of the forcing. The relation between wind-forced sea-level variability and the internal structure of the oceanic variability is thus scale-dependent.

Outside of the tropics, the ocean's response to time-varying wind forcing at large scales is primarily through the vertical motion caused by the convergence/divergence of the wind-driven flow in the surface layer (the Ekman flow). This vertical motion is often called Ekman pumping (e.g., Pedlosky, 1987), which is simply proportional to the curl of wind stress with a scaling factor of $1/f$ (f is the Coriolis parameter). Depending on the spatial and temporal scales of the forcing, the ocean's response has different vertical structures. At spatial scales larger than 1000 km and time scales shorter than 300 days, the ocean's response in the extra-tropics is primarily barotropic, or depth independent (Willebrand et al., 1980; Price and Rossby, 1982; Koblinsky et al., 1989; Fukumori et al., 1998). Sea-level variability at these scales should therefore be a good indicator of the motion of the entire water column and can be described to a large extent by the linear barotropic vorticity equation:

$$\frac{\partial}{\partial t}\nabla^2\eta + \beta\frac{\partial\eta}{\partial x} - \frac{f}{H}\left(\frac{\partial\eta}{\partial x}\frac{\partial H}{\partial y} - \frac{\partial\eta}{\partial y}\frac{\partial H}{\partial x}\right)$$
$$= \frac{f}{\rho g}\left[\nabla\times\left(\frac{\tau}{H}\right)\right]_z, \qquad (8)$$

where η is the sea level, β is the meridional derivative of f ($\beta = 2\Omega\cos\varphi/R$, where R is the earth's radius), H is the depth of the ocean bottom, and τ is the wind stress. Fu and Davidson (1995) investigated this vorticity balance from analyzing the T/P sea-level observations at periods shorter than 1 year with the wind stress obtained from NOAA's National Center for Environmental Prediction (NCEP) operational analysis product. They divided the ocean into $10°\times 10°$ boxes and computed the box averages of each term of Eq. (8). Because small-scale variabilities were averaged out in these box averages, the result contained only the large scales for which Eq. (8) should be valid. Over most of the oceans, they found that such computation was too noisy. There were only a few regions where the signal-to-noise ratio was sufficiently high to reveal the dynamic balance described by Eq. (8). These regions are primarily in the central and northeast Pacific and the southeast Pacific Ocean.

Freely propagating barotropic Rossby waves are homogeneous solutions to Eq. (8). According to their dispersion relation, the periods of these waves for wavelengths larger than 1000 km are generally less than 30 days at mid-latitudes. These waves are generally not well-resolved by the data and can therefore be a source of the noise in the calculation of Fu and Davidson (1995). Detection of such high-frequency waves that are marginally resolved by T/P has been reported by Fu et al. (2000). From analyzing the T/P data in the region of the Argentine Basin in the South Atlantic Ocean where high-frequency energy is particularly high [see discussion below and color section (Figure 7)], they reported the finding of large-scale oscillations at a period of 25 days in the region. These oscillations exhibit a dipole pattern of counter-clockwise rotational propagation centered over a seamount in the center of the basin. The scale of the dipole is about 1000 km. The peak-to-trough amplitude is on the order of 10 cm. The amplitude of these oscillations has large seasonal-to-interannual variations. These oscillations are shown to be a homogeneous solution to Eq. (8). Closed f/H contours in the region have apparently provided a mechanism for the confinement of the waves to the seamount. Results from a numerical model simulation have reproduced the observed pattern of the oscillations. The resultant mass transport variability is on the order of 50 Sv. Fu et al. (2000) also reported evidence from deep current meter records in the Argentine Basin revealing signals that were consistent with the altimetry observations.

For time scales longer than 30 days, a time-dependent Sverdrup relation is expected (Willebrand et al., 1980;

Koblinsky et al., 1989). This relation is represented by Eq. (8) without the first term on the right-hand side. In the absence of topographic variability, it can be written

$$Hv(t) = \frac{\nabla\times\tau(t)}{\beta}, \qquad (9)$$

where $v(t)$ is the time-dependent, barotropic meridional velocity. According to Eq. (9), the fluctuations of the barotropic meridional velocity should be coherent and in phase with the fluctuations of the wind stress curl. Stammer (1997) evaluated Eq. (9) for the Pacific Ocean using the ECMWF wind stress and compared the results with the T/P data and an ocean general circulation model. He integrated Eq. (9) zonally across the ocean basin to obtain the total meridional Sverdrup transport and compared it with the transports calculated from both the T/P data and the model based on the zonal sea-surface height differences across the ocean basin, assuming that the transports were barotropic. The steric component of sea-level variability in both the data and model was removed using the calculation from Eq. (7). In the North Pacific north of 40°N, the zonally integrated meridional transports estimated from both the T/P data and the model agree with the wind-induced changes estimated from Eq. (9) (Figure 5). A marginal agreement also occurs in the region between 10°N and 25°N, whereas the comparison shows poor agreement elsewhere. After examination of the model results in detail, Stammer (1997) discovered that the cross-basin sea-level difference has a substantial baroclinic (depth-dependent) contribution, especially in regions where the agreement with Eq. (9) is poor. These "contami-

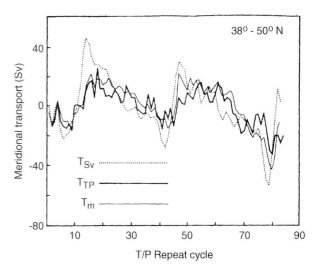

FIGURE 5 Various estimates of the zonally integrated meridional mass transport averaged over 38° to 50°N in the North Pacific. The dotted line (T_{Sv}) represents the calculation based on Eq. (9); the bold solid line (T_{TP}) represents the estimate from the T/P data; the thin solid line (T_m) represents the estimate from a numerical model. The horizontal axis is T/P repeat cycle. Each cycle has a duration of 9.9 days with Cycle 1 starting on October 2, 1992. (Adapted from Stammer, D., 1997.)

nations" by the baroclinic contribution to sea level changes are evidently responsible, at least in part, for the apparent breakdown of Eq. (9) in some regions. The baroclinic contribution was primarily due to the low-frequency (periods much longer than 30 days) part of the signal, which was not filtered in time in the study of Stammer (1997). By filtering out all propagating waves (mostly baroclinic Rossby waves) from the T/P data, Vivier *et al.* (1999) showed that the time-dependent Sverdrup relation, including topographic effects, was valid in most of the subpolar and subtropical regions of the Pacific Ocean.

The validity of Sverdrup dynamics in describing the large-scale, barotropic variability of the North Pacific was also investigated by Chelton and Mestas-Nuñez (1996) using 3 years of T/P data and an ocean general circulation model. They reported that both the T/P-derived and model simulated barotropic transports of the subtropical and subpolar gyres of the North Pacific fluctuated seasonally in phase with the wind-driven Sverdrup transport according to Eq. (9). Based on the assumption of mass conservation for the seasonal barotropic variability, they assert that there is no net mass transport across a zonal section spanning the basin from coast to coast. The transport of a boundary current should thus be equal to (with an opposite sign) the transport of the interior ocean. Because the latter can be estimated from Eq. (9), Chelton and Mestas-Nuñez (1996) estimated the wind-driven, time-dependent transports of the western boundary currents of the North Pacific Ocean, the Kuroshio (a northward warm current from the subtropics), and the Oyashio (a southward cold current from the subpolar region). They also estimated the transport of the Kuroshio Extension (the eastward downstream extension of the Kuroshio) by calculating the difference in the streamfunctions averaged to the north and south of the current. They found that the variability of the transport of the Kuroshio Extension estimated this way was in agreement with the variability of the sum of the transports of the Kuroshio and the Oyashio (Figure 6). On seasonal time scales, the barotropic transport variability of the Kuroshio Extension is thus coherent and in phase with the sum of the variability of the two boundary currents upstream. This study also suggests that the seasonal variability of the barotropic circulation of the North Pacific is largely wind-driven according to Sverdrup dynamics.

Large-scale variability in the T/P data at periods shorter than 100 days was investigated by Fu and Smith (1996) and Chao and Fu (1995) using ocean general circulation models. Geographic distribution of sea-level variance in the period band of 20–100 days is in excellent agreement between the model and the data [see color insert (Figure 7)]. Most of the variance is in the high-latitude regions and the tropics. While the variance in the tropics is mostly caused by baroclinic motion (Fukumori *et al.*, 1998), the variance at the high latitudes is caused by barotropic motion, as suggested by the high degree of coherence between the model simu-

lated sea-level and barotropic stream function in the period band at the high latitudes (Chao and Fu, 1995). Comparisons of the T/P observations with model simulations in these regions show that the models are able to simulate the sea-level variations with respect to the individual high-frequency events, especially in the North Pacific (Figure 8, from Fu and Smith, 1996). This indicates that the wind is the primary forcing for sea level at these scales, because only wind forcing, as opposed to thermal forcing which is also applied to the model simulation, can create such variabilities and an accurate wind field is required to produce faithful simulations of the real events. The relatively poor correlation in the Southern Ocean may be due to the poor accuracy in the wind forcing caused by the scarcity of direct meteorological observations necessary to constrain the ECMWF analysis. Although the analysis of the observations using the simplified barotropic vorticity equation (Eq. 8) leads to mixed results, the general circulation model simulations provide convincing evidence that the large-scale, intraseasonal sea-level variability is caused by wind-forced barotropic motion of the ocean.

Sea-level variability has significant energy at periods shorter than the repeat period of satellite altimeters (e.g., 10 days for T/P, 17 days for Geosat, and 35 days for ERS). Fukumori *et al.* (1998) presented evidence from model simulations driven by wind for the existence of these high-frequency barotropic variabilities in the T/P data. They showed that 12-hr sampled model simulations were able to explain more variance in the T/P data than 3-day sampled simulations, indicating the existence of high-frequency variability and the model's ability to simulate it. These high-frequency variabilities can also be forced by atmospheric pressure as discussed in Section 3.2. If not removed, they would create aliasing problems for studying low-frequency variability using altimetry data. Recent studies of Stammer *et al.* (2000) and Tierney *et al.* (2000) have suggested that dealiasing of this kind may be performed using ocean general circulation models forced by accurate wind and atmospheric pressure (also see Section 5.5 of Chapter 1).

As time scale increases, the ocean's response to wind becomes increasingly baroclinic (depth-dependent). This is because the wavelength of barotropic Rossby waves decreases with increasing period. For periods longer than 300 days, the wavelengths are shorter than 100 km at 45° latitude. The wind forcing has relatively little energy in this frequency/wavenumber band to generate barotropic waves (although a barotropic Sverdrup response is still possible as discussed earlier). The minimum period of baroclinic Rossby waves is a function of latitude (Gill, 1982), approaching 500 days and longer poleward of the 50° latitude. Since there is very little wind energy at such long periods (Fukumori *et al.*, 1998, their Figure 13), there is little wind-forced baroclinic energy at high latitudes. However, due to the limited duration of available observations, we are not able to address

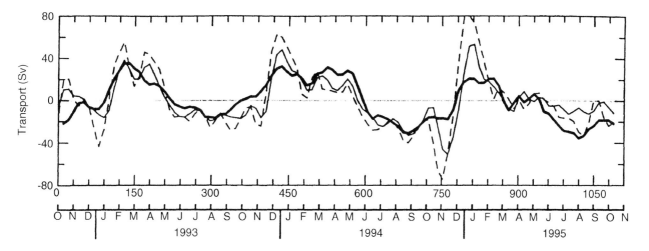

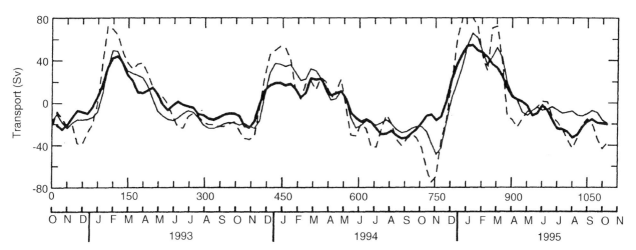

FIGURE 6   Upper panel: The sum of the Oyashio and Kuroshio transports. Dashed lines are from the Sverdrup model represented by Eq. (9); thick solid lines are from the T/P sea surface height; thin solid lines are from the simulation by the Parallel Ocean Program model of the Los Alamos National Laboratory. Lower panel: same as the upper panel except for the Kuroshio Extension transport. (Adapted from Chelton, D. B., and Mestas-Nuñez, A. M., 1996.)

the baroclinic energy at decadal scales when the basin-wide baroclinic adjustment processes become important. On the other hand, the minimum wave period decreases to about 100 days equatorward of 20°, where the wind has substantial energy available in the period band to force the baroclinic waves. Therefore the ocean's baroclinic response to wind forcing is primarily limited to the low and mid latitudes with time scales increasing with latitude. However, waves originating at low latitudes can eventually propagate to high latitudes as coastal-trapped waves along the eastern boundaries of the ocean and then spreading westward as Rossby waves. Such waves may play important roles in ocean-atmosphere interaction on basin-wide scales (Jacobs *et al.*, 1994). A general description of baroclinic Rossby waves is given in the next section.

3.4. Baroclinic Rossby Waves

The relatively slow time scales of baroclinic processes in the ocean make them more amenable to *in situ* observations than faster barotropic variabilities. For example, baroclinic Rossby waves have been studied extensively from temperature profiles of the upper ocean, especially from the XBT program called TRANSPAC (White and Bernstein, 1979). Convincing evidence for Rossby waves has been presented in numerous studies of the North Pacific where upper-ocean thermal data are most abundant. Satellite observations provide a new tool for expanding the studies to the world ocean. Because of the extent of existing literature and the evidence from T/P data that the standard theory is deficient, it is important to review the new findings within the context of pre-

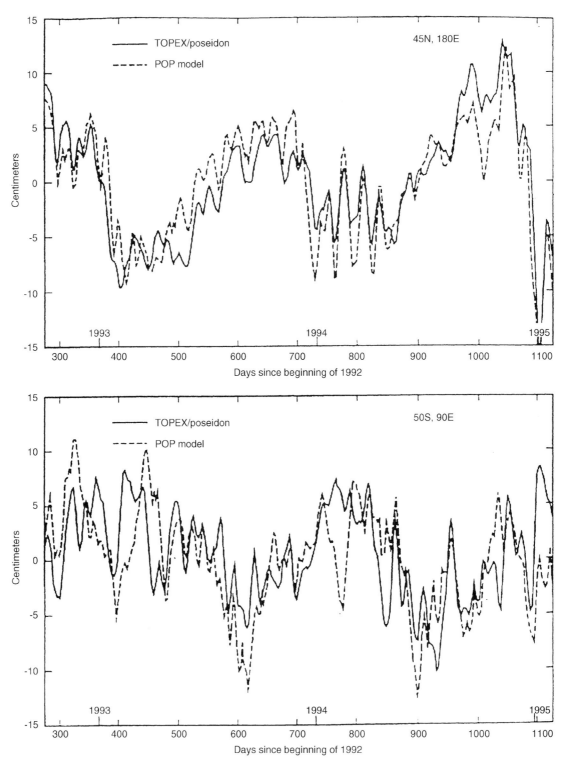

FIGURE 8 Large-scale (>1000 km) sea-level variations in the central North Pacific (45°N, 180°E; top panel) and the South Indian Ocean (50°S, 90°E; bottom panel). T/P observations are shown as solid lines and the model simulations [from the same model as in Figure 7 (see color section)] as dashed lines. (From Fu, L.-L., and Smith, R. D., 1996. With permission.)

vious results from both theory and observations. We therefore start with a comprehensive review of the background of the subject.

3.4.1. The Classical Theory

For small-amplitude motions in a continuously stratified fluid, the standard approach to obtaining low-frequency wave solutions of the quasi-geostrophic equations is to linearize about a state of zero mean background flow. For the case a flat-bottom ocean of depth H, the vertical dependence can be separated from the horizontal and temporal dependence by writing the quasi-geostrophic streamfunction as $\psi = \psi_n(x, y, t) Z_n(z)$. (The significance of the subscript n becomes apparent below.) Outside of the tropical band between about 5°S and 5°N, the linearized equation for the streamfunction can then be separated into the coupled pair of equations,

$$\frac{\partial}{\partial t}\left(\nabla^2 \psi_n - \frac{1}{\lambda_n^2}\psi_n\right) + \beta\frac{\partial\psi_n}{\partial x} = 0, \qquad (10)$$

$$\frac{d}{dz}\left(\frac{f_0^2}{N^2}\frac{dZ_n}{dz}\right) + \frac{1}{\lambda_n^2}Z_n = 0 \qquad (11)$$

(see, for example, Sections 15 and 18 of LeBlond and Mysak, 1978), where λ_n is the separation constant and $N^2(z) = -g\rho^{-1}\partial\rho/\partial z$ is the square of the buoyancy frequency associated with the density stratification $\rho(z)$. For unforced motion with a rigid-lid approximation, the boundary conditions for the ordinary differential Eq. (11) are $dZ_n/dz = 0$ at $z = 0, -H$. Then Eq. (11) is a Sturm-Liouville eigenvalue problem for which there is a countable number of increasing nonnegative eigenvalues λ_n and corresponding eigen-functions Z_n. The subscript n thus labels the different eigen-solutions for the normal modes of the coupled Eqs. (10) and (11). The eigen-function for the vertical structure of mode n has n intersections with the zero axis.

Equation (11) can be satisfied with an eigenvalue of $\lambda_0^{-2} = 0$ and a corresponding eigen-function $Z_0 = 1$ (i.e., uniform vertical structure). In this case, Eq. (10) is identical to Eq. (8) for the case of a flat bottom with no wind forcing. The $n = 0$ mode thus corresponds to unforced barotropic quasi-geostrophic motion. The parameter λ_n is referred to as the Rossby radius of deformation for mode n. The barotropic and first-baroclinic modes are usually the most energetic. Atlases of λ_0 and λ_1 have been published based on the eigenvalues computed from $N^2(z)$ estimated from historical hydrographic data (e.g., Chelton et al., 1998).

For propagating wave solutions of the form

$$\psi \sim \exp i \times (kx + ly - \omega_n t),$$

substitution into Eq. (10) yields a dispersion relation for mode n that can be written in the form

$$c_{nx} = \frac{-\beta}{k^2 + l^2 + \lambda_n^{-2}}, \qquad (12)$$

where $c_{nx} = \omega_n/k$ is the zonal phase speed of the wave. These wave solutions are referred to as Rossby waves or planetary waves. It is evident from Eq. (10) that the restoring mechanism for Rossby waves involves the latitudinal variation of the Coriolis parameter, β. Columns of water deflected from their rest latitude experience a perturbation of vorticity caused by the earth's curvature which a return to the rest latitude tends to cancel. An overshoot of the rest latitude results in wave motion. The negative sign on the right side of Eq. (12) indicates that all Rossby wave solutions have a westward phase speed. In the long-wave limit of zero zonal and meridional wavenumbers k and l, Rossby waves are nondispersive with exactly westward phase and group velocity.

The term "Rossby wave" is correctly applied to westward-propagating disturbances for which the effect of the earth's curvature is the primary restoring mechanism and whose phase speed should be given by Eq. (12). Evidence is presented in later sections for westward phase speeds that differ from Eq. (12), suggesting the likelihood of complications to the simple β-effect restoring mechanism. We will persist in using the term "Rossby wave" to describe such manifestations, despite their evidently more complicated nature.

Baroclinic Rossby waves play a crucial role in large-scale ocean circulation theory. They represent the transient ocean response to low-frequency, large-scale forcing at the sea surface or at an eastern boundary. The westward group velocity of long Rossby waves plays an important role in maintaining the western boundary currents (Anderson et al., 1979; see also Section 12.4 of Gill, 1982). Wind generation of Rossby waves is most efficient at eastern boundaries (e.g., Krauss and Wübber, 1982; Cummins et al., 1986; Reason et al., 1987; Barnier, 1988; Herrmann and Krauss, 1989; Gerdes and Wübber, 1991). Rossby wave generation by buoyancy forcing can also be significant (Magaard, 1977) and Rossby waves can represent the free-wave response to eastern boundary current variations associated with wind-driven upwelling and poleward propagating coastal-trapped waves originating at lower latitudes (e.g., Grimshaw and Allen, 1988; Clarke and Shi, 1991; Mysak, 1983; Shriver et al., 1991; McCalpin, 1995). Modeling studies have found that major topographic features such as mid-ocean ridges act as a barrier to westward propagating baroclinic waves emanating from the eastern boundary (Barnier, 1988; Gerdes and Wübber, 1991). However, the coupling of the barotropic and baroclinic modes over topography generates new baroclinic Rossby waves that propagate west from the topography with larger amplitude than the incident baroclinic Rossby waves from the eastern basin.

3.4.2. Upper-Ocean Thermal Observations

Although the existence of Rossby waves had been accepted since the seminal studies by Rossby et al. (1939) and Rossby (1940), observational verification of oceanic Rossby waves remained elusive until the 1970s. At mid latitudes, the zonal propagation speeds of first-baroclinic Rossby waves are on the order of a few cm/sec (see Figure 14 below). The zonal wavelengths of Rossby waves with annual period, for example, are on the order of 1000 km. Long data records spanning a wide range of longitude are thus essential for detection of Rossby waves. Direct observations of Rossby waves became possible with the accumulation of a sufficiently long and spatially dense collection of profiles of the upper-ocean thermal structure in the North Pacific. Vertical displacements of the thermocline associated with first-baroclinic Rossby waves are about three orders of magnitude greater than sea surface height displacements (e.g., LeBlond and Mysak, 1978; Gill, 1982) and can thus be easily detected from repeated observations of the upper-ocean thermal structure.

Early evidence of first-baroclinic Rossby waves, appearing as variations of tens of meters in the depths of subsurface isotherms, was reported by Emery and Magaard (1976) (see also Magaard and Price, 1977), Roden (1977), White (1977) and Meyers (1979). Over the succeeding decade as the space-time distribution of upper-ocean temperature profiles expanded, a large body of evidence accumulated for the existence of Rossby waves in the North Pacific. These analyses followed two different lines of approach. L. Magaard and co-workers (Price and Magaard, 1980; Kang and Magaard, 1980; Magaard, 1983; Price and Magaard, 1983) modeled the waves as a stochastic process with random phases. Frequencies and wavenumbers were computed based on least-squares fits of model cross spectra to observed cross spectra. In the other approach, W. White and co-workers (Bernstein and White, 1981; White and Saur, 1981, 1983; White, 1982, 1983, 1985; White et al., 1985; White and Tabata, 1987) considered the waves to be deterministic and inferred the frequencies and wavenumbers from the space-time correlation structure of the thermal field. Both approaches found that the variability associated with Rossby waves was about an order of magnitude more energetic west of the Hawaiian Ridge and the Emperor Seamount chain than in the eastern basin (Roden, 1977; Bernstein and White, 1977; White, 1982; Price and Magaard, 1980; Magaard, 1983).

Although less apparent than in the North Pacific, evidence for the existence of low-frequency Rossby waves has been presented from observed depth variations of the subsurface isotherms in the North Atlantic as well (Price and Magaard, 1986). The weaker Rossby wave signals may, at least in part, be due to an inadequate observational dataset because of the lack of a coordinated long-term effort to map the spatial and temporal variability of the upper-ocean thermal structure in the North Atlantic. Alternatively, Rossby wave variability may be obscured in the North Atlantic by energetic variability from other sources such as the mesoscale variability associated with the instability of the Gulf Stream or by the interaction of the flow with the steep topography of

the Mid-Atlantic Ridge. The much wider span of the North Pacific allows a clearer distinction between mesoscale variability that is restricted to the far western basin and westward propagating signals across the rest of the basin.

From analyses of the kinematic features of the westward propagating waves, White (1977), Meyers (1979), and Bernstein and White (1981) concluded that the observed zonal phase speeds across the entire Pacific between 10°N and 20°N and in the western and central Pacific between about 30°N and 40°N were systematically higher by about a factor of two than predicted by Eq. (12) based on the standard theory for free waves with zero background mean flow and a flat bottom. However, most subsequent studies concluded that the observed phase speeds were consistent with the predicted phase speeds (e.g., White, 1982, 1983, 1985; White et al., 1985; Price and Magaard, 1986; White and Tabata, 1987; Shriver et al., 1991). Close inspection of the figures in some of these studies reveals that the observed phase speeds were actually somewhat higher than the predicted phase speeds in some cases, despite the authors' conclusions that the observations were consistent with standard Rossby wave theory.

The significance of the varying conclusions about the relationship between the observed and predicted phase speeds of Rossby waves inferred from upper-ocean thermal data was difficult to assess. Although the historical record of upper-ocean thermal structure is better in the North Pacific than in any other region of the world ocean, Magaard (1983) cautioned that the dataset is quite gappy and only marginally suitable for Rossby wave analyses. Kessler (1990) noted that elaborate space-time interpolation was necessary to construct complete fields from the subsurface thermal data archived at the National Oceanographic Data Center (NODC) that were analyzed in the earlier published investigations of Rossby waves (see, for example, the distributions of observations in Figure 4 of White et al., 1985). Kessler (1990) acquired a large number of temperature profiles from the Japanese Far Seas Fisheries Research Laboratory that expanded the NODC dataset by about 50%. The merged dataset allowed significant improvement in the analysis of Rossby wave variability with minimal need for interpolation over the latitude range from 5°N to 25°N in the North Pacific. From a careful analysis of this dataset, Kessler (1990) presented clear evidence that the observed westward phase speeds were systematically higher than the predicted phase speeds at latitudes poleward of about 7°N. At lower latitudes, the observed phase speeds were slower than predicted by the standard theory for extratropical Rossby waves.

An important point to be noted is that the theoretical phase speeds in most of the historical analyses of upper-ocean thermal data were based on two-layer approximations of the ocean density structure. Notable exceptions are the analyses by White et al. (1985) and Kessler (1990) that compared the observed phase speeds with the theoretical phase speeds for nondispersive first-baroclinic Rossby waves obtained from the eigenvalue solutions of Eqs. (11) and (12) computed by Emery *et al.* (1984) based on buoyancy frequency profiles derived from climatological hydrographic data. The necessity for a rather ad hoc selection of the free parameters of the two-layer approximation (the upper-layer thickness and the density difference between the two layers) allows a considerable range of possible theoretical phase speeds. Tuning these parameters to match the observed phase speeds may account for some of the conclusions that the phase speeds deduced from upper-ocean thermal data are consistent with theoretical phase speeds.

To place the early observational evidence for Rossby waves in a more quantitative context, the observed phase speeds reported from historical analyses of upper-ocean thermal data are compared in Figure 9 with the theoretical phase speeds for nondispersive first-baroclinic Rossby waves determined from Eqs. (2) and (3) based on buoyancy frequency profiles computed from climatological hydrographic data by Chelton *et al.* (1998). Although the range of phase speed estimates at any particular latitude is considerable, it is apparent that the vast majority of the observed phase speeds are faster than predicted from the standard theory. Indeed, the only northern-hemisphere extratropical phase speed estimates that are significantly slower than the theoretical phase speeds are those reported by White (1983) (see the open squares in Figure 9). The sole attempt to estimate Rossby wave phase speeds from the sparsely distributed upper-ocean thermal data in the South Pacific (White *et al.*, 1985; see the open circles in Figure 9) yielded phase speeds that are almost identical to the theoretical phase speeds.

3.4.3. Geosat Observations

The strong tendency in Figure 9 for observed phase speeds to be faster than the predicted westward phase speeds of Rossby waves in the North Pacific cannot be validated globally because of the paucity of suitable upper-ocean thermal data elsewhere in the world ocean. Observations of sea surface height (SSH) variations from a satellite altimeter are ideally suited to such analyses. The measurements must be accurate to better than 5 cm, which is the approximate SSH variation corresponding to a nominal thermocline depth variation of 50 m. The GEOS-3 altimeter observations were too noisy and the sampling was too irregular for investigation of Rossby wave propagation. The measurement accuracy of the Seasat altimeter would have been adequate but the 3-month data record was too short to show the slow westward propagation of baroclinic Rossby waves. The 2.5-year data record from the Geosat altimeter provided the first opportunity for altimetric detection of baroclinic Rossby waves.

Numerous studies reported evidence in the Geosat data of westward propagating SSH signals that were consistent with first-mode baroclinic Rossby waves in a variety of regions

of the world ocean (e.g., White *et al.*, 1990; Jacobs *et al.*, 1992, 1993; Tokmakian and Challenor, 1993; Forbes *et al.*, 1993; Le Traon and Minster, 1993; Van Woert and Price, 1993; Pares-Sierra *et al.*, 1993; Mitchum, 1995; Aoki *et al.*, 1995). These analyses were compromised to some degree by the need to apply elaborate empirical corrections for the large Geosat orbit errors that increased from an rms of about 25 cm at the beginning of the data record to more than 70 cm by the end of the data record (Chelton and Schlax, 1993). In addition, an unfortunate characteristic of the Geosat orbit configuration (see Section 8.1.5 in Chapter 1) is that errors in the lunar semidiurnal M_2 tidal constituent alias into westward propagation with a period of 317 days and a zonal wavelength of 8° of longitude (Jacobs *et al.*, 1992; Schlax and Chelton, 1994a, 1994b, 1996). At latitude θ, large-scale errors in the model for this tidal constituent propagate westward with a phase speed of $3.2 \cos\theta$ cm/sec. Parke *et al.* (1998) have shown the latitudes for which this coincides with the dispersion relation for first-baroclinic Rossby waves in the North Pacific and North Atlantic. Owing to the limitations of the 2.5-year record length, elementary sampling considerations conclude that it is not possible to distinguish aliased M_2 tide errors in Geosat data from baroclinic Rossby waves over a range of latitudes between about 20° and 40° (see Figures 4 and 5 of Parke *et al.*, 1998).

Despite the limitations of the Geosat data for investigation of baroclinic Rossby waves, some of the features previously inferred from upper-ocean thermal data were nonetheless evident in the analyses of Geosat data. In particular, Jacobs *et al.* (1993), Van Woert and Price (1993), and Mitchum (1995) found that the amplitudes of westward propagating SSH signals were much larger to the west of the Hawaiian Ridge and the Emperor Seamount chain than to the east. Tokmakian and Challenor (1993) found a similar increase of energy west of the Mid-Atlantic Ridge between about 30°N and 40°N. In addition, several studies reported westward phase speeds faster than predicted by the standard theory. A synthesis of published estimates of phase speeds deduced from Geosat data is shown in Figure 10. Although the discrepancies between observed and theoretical phase speeds tend to be somewhat smaller than in the case of the phase speed estimates obtained from upper-ocean thermal data in Figure 9, it is apparent that most of the Geosat-based phase speed estimates are faster than predicted from the standard theory for first-baroclinic Rossby waves.

3.4.4. TOPEX/POSEIDON Observations

The T/P orbit configuration was carefully chosen to avoid aliasing of any of the dominant tidal constituents into periods of energetic variability in the ocean (Parke *et al.*, 1987; Schlax and Chelton, 1994b, 1996; see also Section 8.1.5 in Chapter 1). T/P data are therefore much less compromised than Geosat data for studies of baroclinic Rossby waves. At latitudes near 6°, the solar semidiurnal S_2 tide aliases

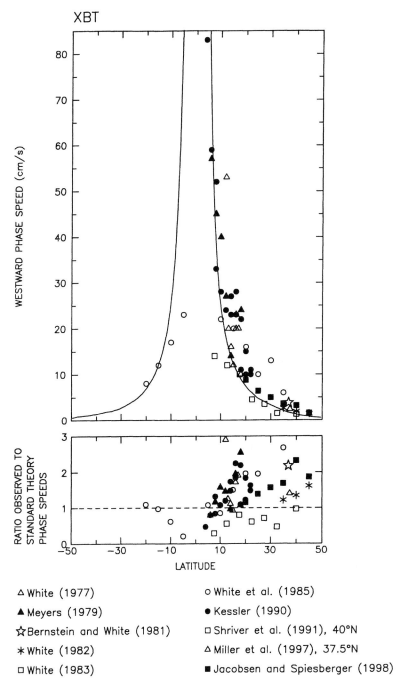

FIGURE 9 The westward phase speed estimates reported in previously published analyses of upper-ocean thermal data. The source of each phase speed estimate is indicated by the various symbols. Note that open squares and open triangles are used to represent two sources each. There should be no confusion, however, because the Shriver *et al.* (1991) and Miller *et al.* (1997) estimates correspond to just the single latitude indicated in the legend. The solid lines in the upper panel indicate the global zonally averaged latitudinal variation of the phase speeds predicted by the standard theory for extratropical freely propagating, nondispersive Rossby waves based on the eigenvalue solutions obtained from climatological hydrographic data by Chelton *et al.* (1998). The bottom panel shows the ratio of each individual observed phase speed estimate to the zonal average of the theoretical phase speeds at 1° intervals across the longitudinal span over which the phase speed was estimated from the upper-ocean thermal data.

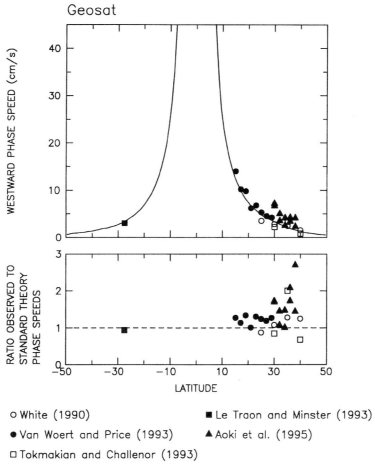

FIGURE 10 The westward phase speed estimates reported in previously published analyses of Geosat altimeter data. The source of each phase speed estimate is indicated by the various symbols. Note that the Van Woert and Price (1993) and Aoki *et al.* (1995) phase speed estimates at 0.5° latitude intervals have been decimated to 2° latitude intervals for display purposes here. The solid lines in the upper panel and the ratios in the bottom panel were obtained as described in Figure 9.

into a period of 59 days with a wavelength that falls on the Rossby wave dispersion curve (see Parke *et al.*, 1998). None of the alias periods and wavelengths of the other tidal constituents are near the Rossby wave dispersion curve at any latitudes. The SSH signatures of extratropical baroclinic Rossby waves can therefore be studied globally from T/P data with immunity from aliasing of tide errors. Moreover, the high accuracy of the T/P data (in particular, the reduction of orbit errors by more than an order of magnitude compared with previous altimeter missions) allows detection of much smaller amplitude SSH signatures of baroclinic Rossby waves.

Because of the large steric effects of annual heating and cooling of the upper ocean (see Section 3.1), westward propagation can be difficult to detect in raw SSH observations at middle and high latitudes. This is illustrated in the left panel of Figure 11 (see color insert) from a time-longitude plot of SSH along 32°N in the North Pacific. Large-amplitude vari-

ability and westward propagation are apparent to the west of the Northwest Hawaiian Ridge at about 175°W. There is no clear evidence of westward propagation to the east of this topographic feature because of the large annual steric sea-level variations. The right panel in Figure 11 shows the same time-longitude plot after zonally high-pass filtering the data to remove the large-scale steric signal. Westward propagation then becomes apparent across the entire basin. Note, however, that the westward propagating signals in the eastern basin are dominated by variability at lower frequencies than to the west of the topographic features.

The typical characteristics of westward propagating SSH signals throughout the world oceans are illustrated in Figure 12 (see color insert) from time-longitude sections of zonally high-pass filtered SSH along 24° latitude in all five oceans. A robust feature of the observed westward propagation is a monotonic increase of the propagation speed from east to west across the ocean basin (Chelton and Schlax,

1996). This westward increase of phase speed is evident in Figure 12 from the decreases of the negative slopes of the trajectories of SSH anomalies from east to west in the time-longitude domain. This is most noticeable in the North and South Pacific where the phase speeds in the far western basins are about double the phase speeds in the eastern basins. Chelton *et al.* (1998) have shown that these geographical variations of the phase speeds are consistent with the increased stratification in the permanent pycnocline in the western basins associated with the deepening of the pycnocline from westward intensification of the subtropical gyres.

Effects apparently associated with topography, and similar to those noted in Figure 11, are also evident in Figure 12. In the North Pacific, the variability along 24°N is much larger to the west of the Hawaiian Ridge at about 160°W. In the South Pacific, SSH variability is similarly larger to the west of the East Pacific Rise at about 120°W. Chelton and Schlax (1996) noted that larger amplitude variability to the west of major topographic features is a common feature in time-longitude sections of T/P data throughout the world ocean. As discussed above, the larger amplitude variability to the west of the Hawaiian Ridge and Emperor Seamount chain was also apparent from in situ observations of upper-ocean thermal structure and from Geosat altimeter data. The larger amplitude variability in the western South Pacific evident in Figure 12 was also found in an analysis of T/P data by Wang *et al.* (1998). Similar increases of SSH variability were observed between latitudes of about 30°N and 40°N to the west of the Mid-Atlantic Ridge from analyses of T/P data by Schlax and Chelton (1994b) and Polito and Cornillon (1997). These T/P results for the North Atlantic are consistent with the earlier conclusions of Tokmakian and Chellenor (1993) deduced from Geosat data. The larger amplitudes of westward propagating signals to the west of major topography are consistent with the previously noted modeling results obtained by Barnier (1988) and Gerdes and Wübber (1991) who found that topography-induced coupling between the barotropic and baroclinic Rossby wave modes incident on meridional topography can generate larger amplitude baroclinic Rossby waves to the west of the topography.

Larger amplitudes of Rossby waves to the west of topography are by no means universal, however. For example, the very large-amplitude signals in the South Indian Ocean and the smaller-amplitude signals in the South Atlantic are seen from Figure 12 to propagate along 24°S across the entire basins oblivious to the major topography. While larger amplitude variability is evident at 24°N in the far western North Atlantic, this does not appear to be associated with topography since the Mid-Atlantic Ridge is located at about 45°W at this latitude.

Another noteworthy feature in Figures 11 and 12 is the energetic variability with a periodicity near the annual cycle at the eastern boundaries of the North and South Pacific. Chelton and Schlax (1996) observed that this variability pen-

etrates only a short distance westward into the ocean interior, often showing no evidence of westward propagation. This rapid decrease in the amplitude of SSH variability near the eastern boundary was also evident in the Geosat data (e.g., White *et al.*, 1990; Pares-Sierra *et al.*, 1993). Qiu *et al.* (1997) suggest that this occurs because of frictional damping of Rossby waves generated at the eastern boundary. Note especially the large positive boundary-trapped SSH anomaly during the second half of 1997 that was associated with the 1997–1998 El Niño (Strub and James, 2000). In the South Pacific, large-amplitude SSH signals associated with this El Niño event propagated westward from the eastern boundary until at least early 1999, albeit with rapidly decreasing amplitude toward the west. Westward propagation of this El Niño signal is also apparent at 32°N in Figure 11, but with even more rapidly decreasing amplitude. There is no evidence in Figure 12 of westward propagation of the boundary manifestation of the El Niño signal at 24°N.

A global synthesis of the phase speeds of westward propagating SSH signals observed during the first three years of T/P data was presented by Chelton and Schlax (1996). An update of that analysis is presented here based on 7 years of T/P data. After zonally high-pass filtering as described above, westward propagation like that shown in Figures 11 and 12 is a very common feature of the T/P data. In some regions, however, the westward propagation appears to be intermittent spatially or temporally, most likely because of the co-existence of processes unrelated to Rossby waves. The locations at which westward propagation is clearly present over longitudinal spans of 30° or more are shown in the upper panel of Figure 13. The phase speeds estimated for these sections by the Radon transform (Jain, 1989; Deans, 1983) are shown in the upper left panel of Figure 14. It can be noted that there are no locations of clear westward propagation north of 40°N or south of 48°S. The critical latitude for annual Rossby waves is about 40° (e.g., Grimshaw and Allen, 1988; Clarke and Shi, 1991). Only interannual Rossby waves can propagate at higher latitudes. The amplitudes of interannual variability are generally small (see the eastern basins in the right panel of Figure 11 and in the North and South Pacific panels of Figure 12) and are therefore easily obscured by more energetic mesoscale variability. The lack of clear evidence of westward propagation at the higher latitudes is thus consistent with the concept of the critical latitude.

The solid line in the upper-left panel of Figure 14 corresponds to the global zonally averaged phase speed of nondispersive first-baroclinic Rossby waves determined from Eqs. (11) and (12) based on $N(z)$ profiles computed from climatological hydrographic data by Chelton *et al.* (1998). It is apparent that nearly all of the phase speed estimates poleward of about 10° of latitude are faster than predicted by the standard theory. The discrepancy between observations and theory is shown point-wise in the middle left

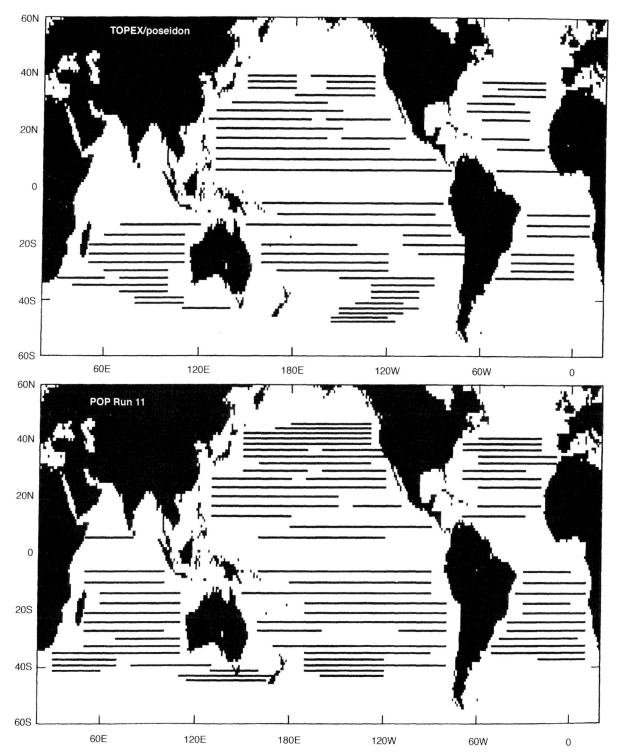

FIGURE 13 The locations of clear westward propagation in the 7-year T/P data record (upper) and the 3-year model simulation from Run 11 of the POP model. SSH fields from the POP simulation were provided courtesy of R. Smith of the Los Alamos National Laboratory.

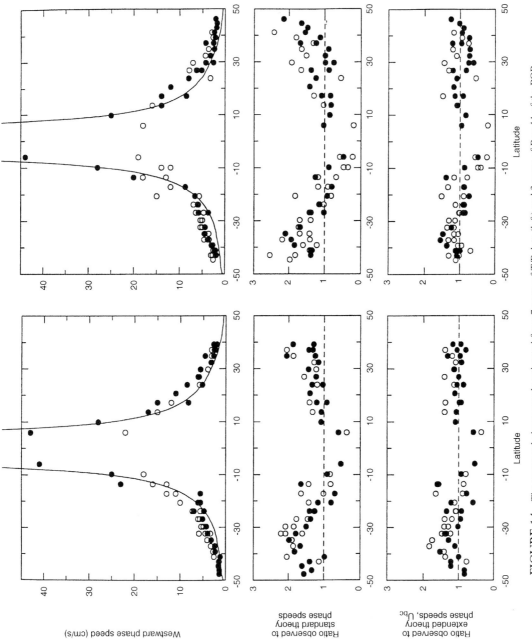

FIGURE 14 The westward phase speeds estimated from 7 years of T/P data (left) and 3 years of Run 11 of the POP model (right) along the respective sections shown in Figure 13. The solid circles correspond to estimates from the Pacific Ocean and the open circles correspond to estimates from the Atlantic and Indian oceans. The solid lines in the upper panels indicate the global zonally averaged latitudinal variation of the phase speeds predicted by the standard theory for extratropical freely propagating, nondispersive Rossby waves in the absence of any background mean flow based on the eigenvalue solutions obtained from climatological hydrographic data by Chelton *et al.* (1998) (left panel) and from the hydrography of the POP model (right panel). The pointwise ratios of each observed phase speed to the phase speeds predicted at the same locations as the observations based on the standard theory and the extended theory that includes the baroclinic background mean flow are shown in the middle and bottom panels, respectively.

panel of Figure 14 from the ratio of each observed phase speed to the average of the theoretical phase speeds over the same longitudinal span as that from which the phase speed estimate was obtained from the T/P data. The ratios smaller than 1 equatorward of 10° of latitude are consistent with the results obtained by Kessler (1990) from upper-ocean thermal data. These slower phase speeds in the equatorial band can apparently be explained by the effects of the vertically sheared mean zonal equatorial currents on the phase speeds of equatorially trapped Rossby waves (Philander, 1979; Chang and Philander, 1989; Zheng et al., 1994). The interest here is in the extratropical region where the observed phase speeds are systematically higher than predicted by the standard theory.

The fast westward phase speeds poleward of 10° latitude are very similar to the previously summarized phase speeds inferred from upper-ocean thermal data in the North Pacific (Figure 9) and from Geosat data at several locations in the northern hemisphere (Figure 10). The fast phase speeds found globally by Chelton and Schlax (1996) from the first three years of T/P data have been verified at selected locations from numerous other analyses of T/P data (e.g., Nerem et al., 1994, see their Figure 2; Wang and Koblinsky, 1995, 1996b; Polito and Cornillon, 1997; Cipollini et al., 1997, 1999; Wang et al., 1998; White et al., 1998; Witter and Gordon, 1999). The fast phase speeds have been further validated from recent analyses of upper-ocean thermal data in the North Pacific between 20°N and 45°N (Miller et al., 1997; Jacobson and Spiesberger, 1998). Phase speeds faster than predicted by the standard theory are evidently ubiquitous in the extratropical world ocean. The discrepancy between the observations and the theory generally increases with increasing latitude, reaching values of more than a factor of 2 difference at 40° latitude (see middle panel of Figure 14).

The fast phase speeds of SSH anomalies inferred from the global T/P data are also evident in the POP model referred to earlier in Section 2. The model SSH fields were interpolated to the locations of the T/P observations and were then smoothed and zonally high-pass filtered in the same manner as the T/P data. The locations of clear westward phase propagation are shown in the lower panel of Figure 13. While these locations are not always precisely the same in the POP model as in the T/P observations, the westward phase speed characteristics are very similar. In particular, the model phase speeds are systematically higher than the phase speeds predicted from the standard theory based on $N(z)$ profiles computed from the model hydrography (see top and middle right panels of Figure 14). Similarly high-phase speeds have been reported by White et al. (1998) from the ocean general circulation model described by Chao and Fu (1995). This may not be true of all ocean general circulation models, however. Miller et al. (1997) report that the westward phase speeds in an eight-layer model

developed by Oberhuber (1993) were significantly slower than the phase speeds observed in upper-ocean thermal data. Jacobson and Spiesberger (1998) similarly report that the phase speeds in a six-layer model developed by Hurlburt et al. (1996) are consistent with the phase speeds predicted from the standard theory. It is noteworthy, however, that the Miller et al. (1997) and Jacobson and Spiesberger (1998) studies both compared the model phase speeds with the theoretical phase speeds computed from the climatological hydrographic dataset rather than from the hydrography of the models.

3.4.5. Recent Theoretical Studies

The T/P observations of fast westward propagation over most of the extratropical world ocean have stimulated a great deal of theoretical interest in Rossby wave propagation. The standard theory given by Eqs. (10) and (11) is evidently incomplete. Extensions of the classical theory for Rossby waves are summarized here.

Following White (1977), Qiu et al. (1997) suggested that the superposition of a monochromatic freely propagating annual Rossby wave and a standing Rossby wave forced by zonally coherent annual variations of the wind stress curl would lead to a factor of two speedup of Rossby wave phase speeds. Qiu et al. (1997) clarified the character of this apparent speedup. Without frictional damping, the Radon transform technique used to estimate the phase speeds in Figure 14 would yield a phase speed equal to the standard theory if applied to the superposed free and forced wave. In this case, the synchronous forced and free wave model clearly cannot account for the observed fast phase speeds in Figure 14. Moreover, the zonal high-pass filtering applied here to remove the steric signal (see Figure 11) eliminates the standing wave component from wind stress curl forcing of large zonal extent, thus isolating the free-wave component that has a phase speed equal to that given by the standard theory. Superposition of a forced and free wave thus fails to account for the observed fast-phase speeds for this reason as well.

When sufficiently strong frictional damping is included in the synchronous forced and free-wave model of Qiu et al. (1997) and when the wind-stress curl forcing has limited zonal extent so that the standing wave signal is not eliminated by the zonal high-pass filtering applied to investigate westward propagation from T/P data, westward propagation at a phase speed double that of the free wave can be deduced over a restricted longitudinal range of about one wavelength of the free Rossby wave. From Figures 11 and 12, however, it is apparent that the observed fast westward propagation persists over longitudinal spans of many wavelengths. The theory of Qiu et al. (1997) therefore cannot account for the observed fast propagation even when frictional damping is included in the model. Furthermore, the spectral content of SSH variability spans a wide range of frequencies [see color insert (Figure 15)]. The mechanism for apparent fast westward propagation from monochromatic resonant forcing suggested by Qiu *et al.* (1997) would therefore have to pertain at each frequency. This mechanism apparently cannot explain the observed fast-phase speeds.

White *et al.* (1998) proposed another mechanism by which coupling between oceanic Rossby waves and the wind-stress curl may increase the westward phase speed of SSH anomalies. Temporal variations of SSH, sea-surface temperature (SST), meridional surface wind and wind-stress curl in the eastern North and South Pacific over the period 1993–1994 were interpreted as one cycle of biennial variability. The dominant complex empirical orthogonal function (CEOF) was remarkably similar for each of the four variables, bearing strong resemblance to the β-refraction patterns of oceanic Rossby waves. Based on the spatial phase relationships in the CEOFs of the various fields, White *et al.* (1998) hypothesize a scenario in which Rossby wave perturbations of the SST field result in air-sea heat flux that modifies the temperature and winds in the lower troposphere in a manner that creates a positive feedback of the wind stress curl on the free Rossby waves. From the linkages of these hypothesized dynamical relationships, they develop an analytical model for coupled ocean-atmosphere Rossby waves that can increase the westward phase speeds. The statistical significance of this scenario for biennial waves deduced from only 2 years of heavily filtered data is marginal at best. Moreover, it remains to be determined whether the same coupling exists over the broad range of frequencies of westward propagating SSH signals exhibited in the T/P data [see color insert (Figure 15)].

Another possible shortcoming of the standard theory for Rossby wave-phase speeds is the assumption of a flat bottom. Killworth and Blundell (1999) investigated the possibility that slowly varying bottom topography could increase the phase speeds of Rossby waves. While they found that topographic effects can alter the phase speed significantly at some locations, the effects are small over most of the world ocean. Hence, topography does not seem a likely explanation for the anomalously fast phase speeds that are observed ubiquitously over the world oceans.

Killworth *et al.* (1997) proposed an explanation for the fast observed phase speeds that is the most promising to date. They suggested that the speedup occurs because of the modification of free Rossby wave modes by vertically sheared mean currents. For travelling wave solutions of the form $\psi = Z_n(z) \exp i(kx + ly - \omega_n t)$, the quasi-geostrophic potential vorticity equation linearized about a mean zonal velocity profile $\bar{u}(z)$ is

$$\left[c_{nx} - \bar{u}(z) \right] \left[\frac{d}{dz} \left(\frac{f_0^2}{N^2} \frac{dZ_n}{dz} \right) - (k^2 + l^2) Z_n \right]$$
$$- \left[\beta - \frac{\partial}{\partial z} \left(\frac{f_0^2}{N^2} \frac{\partial \bar{u}}{\partial z} \right) \right] Z_n = 0. \quad (13)$$

The rigid-lid and flat-bottom boundary conditions of zero vertical velocity at $z = 0, -H$ are

$$\left[c_{nx} - \bar{u}(z) \right] \frac{dZ_n}{dz} + \frac{\partial \bar{u}}{\partial z} Z_n = 0, \quad \text{at } z = 0, -H. \quad (14)$$

The eigenvalues c_{nx} of the Sturm-Liouville eigenvalue problem defined by Eqs. (13) and (14) are the zonal phase speeds of the wave modes. When $\bar{u}(z) = 0$, Eqs. (13) and (14) reduce to the equations for the standard theory that neglects the mean flow. These equations also reduce to the standard theory when $\bar{u}(z)$ is vertically uniform (i.e., barotropic), except that the phase speed is Doppler shifted by $\bar{u}$. When $\bar{u}(z)$ is vertically sheared, Eqs. (13) and (14) are identical to the equations used to investigate baroclinic instability (e.g., Gill *et al.*, 1974; Pedlosky, 1987). The unstable solutions usually have wavelengths near the baroclinic Rossby radius of deformation, which is much shorter than the wavelengths of westward propagating features that are evident in 6° by 6° by 60-day smoothed SSH fields constructed from T/P data (see, for example, Figures 11 and 12). Killworth *et al.* (1997) investigated the stable wave solutions of these equations. Stable solutions of Eqs. (13) and (14) were previously investigated by Kang and Magaard (1979; 1980).

The presence of a vertically sheared mean current can strongly modify the characteristics of the stable baroclinic modes. The number of discrete eigen-solutions is severely truncated (Kang and Magaard, 1979; Killworth *et al.*, 1997). There is always a barotropic mode, and it is only weakly affected by the mean flow. There is usually also a first baroclinic mode, but there is seldom more than one higher-order baroclinic mode. As first noted by Kang and Magaard (1979; 1980), the mean shear can significantly alter the dispersion relation of the baroclinic stable wave solutions. In particular, Kang and Magaard (1979) documented an example in the North Pacific for which the mean shear doubled the zonal phase-speed relative to the standard theory. In another example, the mean shear reduced the phase speed relative to the standard theory.

For comparison with the phase speeds estimated from T/P data, Killworth *et al.* (1997) computed the first baroclinic phase speed eigenvalues c_{nx} from Eqs. (13) and (14) for the long-wave limit $(k, l) \ll \lambda_1$. The calculation was done at each point on a 1° global grid for which estimates of buoyancy frequency $N(z)$ and the baroclinic component of $\bar{u}(z)$ could be computed from climatological hydrographic data. The viewpoint that Eqs. (13) and (14) can be solved locally at each grid point as if the computed profiles of $N(z)$ and $\bar{u}(z)$ typified the entire β-plane ocean has some validity if the mean hydrography varies slowly in space. The phase speeds from the extended theory were found to be higher than the phase speeds from the standard theory over most of the mid-latitude and high-latitude ocean. It should be noted that this speedup is not the Doppler shift since the barotropic component of $\bar{u}(z)$ was not included in this calculation. The mechanism for the speedup is the augmentation of β by the baroclinic vortex stretching term in Eq. (13) (i.e., the last term in the equation) over most of the extratropical ocean, thus increasing the meridional potential vorticity gradient that is the restoring mechanism for the shear-modified Rossby waves.

Dewar (1998) and de Szoeke and Chelton (1999) investigated the properties of the shear-modified Rossby waves from layered models. They showed that a minimum of three layers is required to reproduce the speedup mechanism of the continuously stratified model considered by Killworth *et al.* (1997). For a three-layer model, de Szoeke and Chelton (1999) showed that layer velocities for which the meridional gradient of potential vorticity vanishes in the middle layer yield Rossby wave speedup ratios relative to the standard theory that are very similar to those observed in the T/P data. They suggested that the observed fast phase speeds in the real ocean were at least partially a consequence of the well-known homogenization of potential vorticity in intermediate density layers of the world ocean (Keffer, 1985; Talley, 1988).

The effects of shear modification of the phase speeds of first-baroclinic Rossby waves in the continuously stratified ocean are shown in the bottom panels of Figure 14. The phase speeds predicted from the extended theory for non-dispersive waves were averaged over the same longitudinal spans as those from which each of the T/P phase speed estimates was obtained. The ratios of the T/P estimates to the average of the phase speeds from the extended theory are shown in the lower left panel of Figure 14. The extended theory was similarly used to compute the phase speeds of non-dispersive waves based on $N(z)$ and the baroclinic component of $\bar{u}(z)$ from the POP model. The pointwise ratios of the "observed" POP phase speeds to the predicted phase speeds are shown in the lower right panel of Figure 14. Outside of the tropical band from about 10°S to 10°N, the agreement between the predicted non-dispersive phase speeds and the phase speeds deduced from the T/P data and the POP model is much improved over the standard model. The agreement is especially good in the northern hemisphere T/P observations and in both hemispheres of the POP model simulation. The poorer agreement in the southern hemisphere T/P observations may be an indication of inadequacies of the climatological hydrographic data from which the shear-modified phase speeds were computed.

The shear-modified phase speeds in Figure 14 were computed from only the baroclinic component of $\bar{u}(z)$. Some of the remaining discrepancies between the observed and predicted phase speeds in the bottom panels of Figure 14 may therefore be attributable to the barotropic component of the mean zonal flow. As noted above, the barotropic velocity acts to Doppler shift the zonal phase speed c_{nx}. Over most of the world ocean, the barotropic velocity is probably small enough to justify its neglect in the shear-modified eigen-solutions. In any case, the barotropic velocity is difficult to determine in the real ocean. Using the barotropic velocity obtained from a high-resolution global ocean circulation model as a proxy for the barotropic velocity in the real ocean, Killworth *et al.* (1997) found that the agreement between the theoretical phase speeds and the T/P observed phase speeds was only marginally improved when the shear-modified phase speeds obtained from climatological hydrographic data were adjusted for the Doppler shift.

The conclusion that the effects of the barotropic velocity are negligible is not likely to be valid at high southern latitudes where the barotropic component of the velocity is known to be strong in the core of the Antarctic Circumpolar Current (ACC). Indeed, Hughes (1995; 1996) analyzed the output of the Fine Resolution Antarctic Model (FRAM Group, 1991) and found clear evidence of Doppler shifting of baroclinic Rossby waves from eastward advection by the ACC. The approximate 300-km wavelengths of these features are very difficult to resolve in altimeter data. The zonal spacing of neighboring ground tracks in the 10-day repeat T/P dataset is about the same as the wavelengths of the Doppler-shifted Rossby waves in the FRAM model. Hughes (1995) was nonetheless able to identify features in the T/P data that resembled the expected "chevron-shaped" patterns of Rossby waves refracted by the horizontally sheared ACC in the southeast Pacific. The ground-track spacing for the ERS satellites is adequate to resolve these features but the long 35-day repeat period introduces errors from unresolved temporal variability of the waves. Although they were not able to map the Rossby waves, Hughes *et al.* (1998) presented compelling evidence for eastward propagation in the core of the ACC and westward propagation to the north and south of the ACC from an analysis of a short 4.6-month record of ERS-1 altimeter data. The eastward propagation is consistent with the expected Doppler shifting of westward propagating Rossby waves by a supercritical eastward barotropic flow in the core of the ACC.

3.4.6. Frequency-Wavenumber Characteristics

The long T/P data record allows an investigation of the frequency-wavenumber characteristics of Rossby waves that cannot be addressed from any other observational dataset. Zang and Wunsch (1999) analyzed 4.75 years of T/P data along seven zonal sections in the North Pacific to determine frequency-wavenumber spectra by an array-processing technique that estimates the dominant zonal wavenumber at specified frequencies. These most energetic spectral peaks were compared with the dispersion relation obtained from the standard theory for first-baroclinic Rossby waves neglecting the effects of the background mean flow. They found energetic variability at frequencies higher than can exist according to the standard theory and speculated that this may be an indication of mode coupling between the barotropic and baroclinic modes. At the lower frequencies, many of their spectral estimates were centered at smaller wavenumbers than predicted by the dispersion relation from the standard theory. In nearly all cases, however, the dispersion curve falls within the confidence intervals of the spectral estimates. They concluded that the observed spectral characteristics of the westward propagating signals are indistinguishable at low frequencies and wavenumbers from the standard theory for Rossby waves in the absence of a mean background flow. The validity of this conclusion is clearly sensitive to the widths of the confidence intervals, which are difficult to estimate quantitatively.

As the data record continues to accumulate, the frequency resolution of spectral estimates obtained from T/P data continues to improve. The frequency-wavenumber spectra of SSH obtained by the traditional Fourier transform technique are shown in Figure 15 for subdomains of the five zonally high-pass filtered time-longitude plots along 24° latitude shown in Figure 12. A rich distribution of time scales in the observed westward propagation is apparent in the spectra for these sections; there is no evidence of a dominance of variability at the annual period, for example. The much less energetic variability in the North and South Atlantic is also apparent. The dispersion relations for the standard theory and the extended theory modified by the mean vertical shear are shown by the open and solid circles, respectively. In the North Atlantic, there is no significant difference between the standard and extended theory. In the other basins, however, the two theories diverge significantly with increasing frequency and wavenumber. Consistent with the results obtained by Zang and Wunsch (1999), the T/P data exhibit energetic variability at frequencies higher than are allowed in the standard theory. In all of the cases shown here, however, the dispersion relation from the extended theory fits the observed spectral characteristics at these high frequencies and wavenumbers very well. A noteworthy feature of the spectra and the dispersion relation for the extended theory is that the westward propagating signals are weakly dispersive over a much broader range of frequencies and zonal wavenumbers than is the dispersion relation for the standard theory.

3.4.7. Future Work

The preceding overview summarizes the significant contributions of altimetry toward understanding oceanic baroclinic Rossby waves. There are still many fundamental characteristics of Rossby waves that are not yet understood. Perhaps most notable is that the generation mechanisms for the ubiquitous westward propagation observed in the T/P data are not yet known. It is also not known why the amplitudes of westward propagating signals are often, but not always, larger to the west of major topographic features. Efforts are underway to investigate the meridional propagation of Rossby waves. This is made difficult by the coarse meridional resolution of the SSH fields that can be constructed from T/P data (see Section 8.1.7 in Chapter 1). Meridional propagation is not readily apparent in animations of global

SSH fields smoothed as described in Figures 11 and 12, suggesting that meridional propagation may not be a persistent feature of Rossby waves. A theoretical explanation for the lack of well-defined meridional propagation has not yet been proposed. Efforts are underway, however, to investigate other mechanisms for the observed fast westward phase speeds. It has been suggested, for example, that nonlinearity may be important but it has not yet been shown that this could account for the pervasive fast phase speeds over most of the extratropical world ocean.

The upcoming launches of the Jason-1 altimeter in early 2001 and the Jason-2 altimeter three years later offer exciting prospects for quantitative studies of ocean climate variability from altimeter data. Both of these satellites will sample SSH along the T/P ground tracks, thus extending the T/P data record. The merged T/P and Jason datasets promise to yield new insight into the basin-scale baroclinic adjustment of the world ocean to interannual and decadal variability. A climate signal of particular interest is the North and South Pacific response to the El Niño Southern Oscillation phenomenon. From an analysis of Geosat and ERS-1 data, Jacobs et al. (1994) proposed that poleward propagating Kelvin waves generated along the eastern boundary of the North Pacific by the strong 1982–1983 El Niño radiated offshore as westward propagating Rossby waves that perturbed the Kuroshio Extension a decade later (see Section 3.7 for more discussion). The imprints of the very strong 1997–1998 El Niño event and the succeeding strong La Niña conditions offer an excellent opportunity for quantitative investigation of the scenario hypothesized by Jacobs et al. (1994) by tracking the evolution of the SSH field over the North and South Pacific in the merged T/P, Jason-1, and Jason-2 datasets.

3.5. The Relation Between Sea Level and Subsurface Variability

As noted earlier, a major advantage of altimetry over other ocean remote sensing is the relation between observed sea level and subsurface fields. From the discussions in the preceding sections on the various mechanisms of sea level variations, it is clear that sea-level variations reflect changes in subsurface fields caused by thermal and mechanical processes. Except for steric change of sea level caused by buoyancy forcing, large-scale sea-level variability is predominantly caused by adiabatic, mechanical forcing and thus has a tight relation with subsurface variability as part of an organized motion field of the water column. For the barotropic mode, the surface geostrophic velocity determined from sea level variability through Eq. (5) represents the uniform horizontal velocity of the entire water column. For the baroclinic mode, sea level change is related to the change in the height of the subsurface constant-density surfaces, the isopycnals. Therefore, sea level has been used to estimate the subsurface density field, from which horizontal geostrophic velocity at depths has also been estimated.

In the tropics where the steric sea-level variability is the smallest (Figure 4), there is a tight relationship between sea level and the depth of the thermocline (Wyrtki, 1985). A 1-cm rise (fall) in sea level corresponds roughly to a 2-m fall (rise) of the thermocline. (The ratio of sea level variations to thermocline variations is dependent on geographic location.) Using data from a network of tide gauges in the tropical Pacific, Wyrtki (1985) developed a method for estimating the variation of the volume of warm water in the upper ocean. He was thus able to quantify the heat content change during El Niño by estimating the increase of warm water volume from sea-level changes. Miller and Cheney (1990) applied the method of Wyrtki (1985) to Geosat data for estimating basin-wide meridional transport of the upper-ocean warm water in the tropical Pacific during the 1986–1987 El Niño.

Carnes *et al.* (1990) used Geosat altimeter data to estimate the subsurface temperature field in the Gulf Stream area. They applied statistical regression analysis to a set of AXBT (air-dropped expendable bathythermograph) data deployed along the Geosat ground tracks and derived an empirical relation between subsurface temperature and surface dynamic height. They then applied the empirical relation to the Geosat altimeter data for estimating the subsurface temperature field. They used the ocean topography derived form the altimeter data relative to a geoid model as a surrogate for the dynamic height. The rms difference between the Geosat-derived ocean topography and the dynamic height is 15–19 cm, mostly reflecting the geoid errors and the Geosat altimeter measurement errors. The resulting temperature estimates have an rms error of about 1°C below 200 m. The error increases to 2°C near the surface, where there is significant change in temperature caused by the annual cycle of heat exchange with the atmosphere. The thermally-driven variability of temperature as a function of depth does not have a tight relation with the sea-level variability.

Using T/P data with a series of repeating transects of XBT and XCTD (expendable conductivity, and temperature depth profiler) across the North Pacific Ocean (from Taiwan through Guam to San Francisco), Gilson *et al.* (1998) studied the relationship between altimetric sea-level measurements and subsurface temperature and currents. A total of 5 years of simultaneous *in situ* and satellite data were analyzed, allowing the relation between the two datasets to be examined over a wide range of spatial and temporal scales. The anomalies of the altimetric sea level relative to a 5-year mean were compared with the anomalies of the dynamic height computed from the *in situ* data relative to the same 5-year mean. The altimeter data were interpolated to the times and locations of the in-situ observations using an objective analysis scheme. The overall rms difference between the two anomaly fields is 5.2 cm. At wavelengths longer than 500 km

where lies 65% of the variance of the dynamic height, the two are highly coherent (0.89) with an rms difference of 3.5 cm. This difference is consistent with the measurement errors of T/P (Section 8.3 in Chapter 1) and the errors in the dynamic height estimates, plus the residual dynamic height variability below 800 m, which is the deepest level of the XBT observations. The major source of the dynamic height errors was the variability of salinity, which was sparsely sampled by the XCTD's along the transects (12–25 salinity profiles compared to 300 temperature profiles for each transect from the most recent 14 transects). The salinity-caused rms error in the dynamic height is 1–2 cm. At wavelengths shorter than 500 km, the coarse spacing of the T/P ground tracks led to the underestimation of the mesoscale variability, which, however, was well-sampled by the *in situ* data. The correlation between the two was reduced to 0.56 at these scales. Gilson *et al.* (1998) also compared the basin-wide spatial averages of the two anomaly fields and found an rms difference of 2.4 cm. This difference is puzzling because it cannot be explained by the estimated measurement errors. It is probably due to the barotropic variability present in the altimeter data but not sampled by the *in situ* observations.

From analyzing the XBT/XCTD data, Gilson *et al.* (1998) found that, except for the annual variations, the dynamic height anomalies were largely caused by the vertical motion of the thermocline. The entire thermocline moved coherently in the vertical, causing the temperature at the thermocline depths highly correlated with the surface dynamic height but with an opposite sign. The high correlation between the altimetric sea level anomaly and the surface dynamic height anomaly discussed above thus allows the use of the former to make estimates of the subsurface temperature anomaly. The correlation between the dynamic height and the subsurface temperature was derived after the annual cycle was removed from both fields. This is because the temperature change at the annual period is primarily caused by the heat exchange with the atmosphere rather than the vertical motion of the isopycnals. The vertical distribution of the temperature change at the annual period is thus not highly correlated with the surface dynamic height. Therefore altimeter data were used only to estimate the subsurface temperature anomaly at nonannual time scales. The temperature anomalies derived from the altimeter data were added to the mean and the annual cycle determined from the in-situ data to obtain the absolute temperature field. Such temperature estimates were able to account for 53% of the total variance, with a maximum rms error of 0.7°C at depths of 80–160 m in the central basin. The error increased to 1°C in the thermocline at both the eastern and western ends of the basin. Improvement of such estimates is expected from a longer data set that will allow increased signal-to-noise ratio and more accurate regression analysis.

As reported in Gilson *et al.* (1998), the specific volume anomaly at depths can also be estimated from the surface

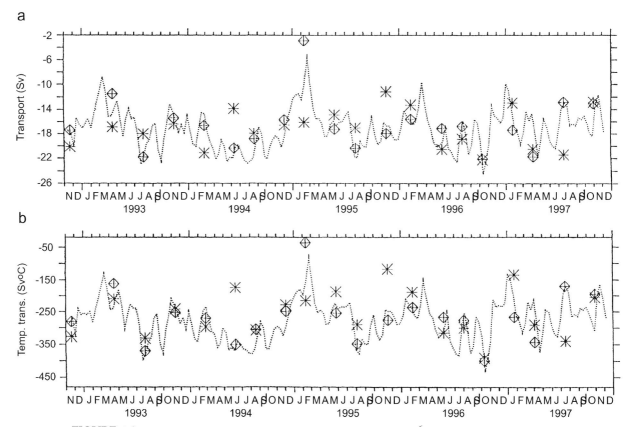

FIGURE 16 Upper panel: Time series of geostrophic transport (in $Sv = 10^6$ tons/sec), integrated from ocean boundary-to-boundary and from 0–800 m. The asterisks represent calculations directly from XBT/XTCD data and are relative to 800 m. The dotted line is the estimate from 10-day T/P data. Diamonds also represent T/P data but interpolated to the positions and times of XBT data. Differences from the 10-day data indicate variability over the 17-day duration of the XBT cruises. Lower panel: same as the upper panel except for temperature transport, which is the integral of velocity times temperature. (From Gilson, J., Roemmich, D., Cornuelle, B., and Fu, L.-L., 1998. With permission.)

dynamic height anomaly and hence from the altimetric sea level anomaly. With the vertical profiles of specific volume anomaly determined, one can then calculate the dynamic height relative to 800 m (the deepest level of the XBT observations) at a given shallower depth, denoted by ΔD. The basin-wide meridional volume transport of the upper 800 m can thus be estimated by

$$\int_0^L \int_{-800}^0 v \, dz dx = \int_0^L \int_{-800}^0 \frac{1}{f} \frac{\partial(\Delta D)}{\partial x} \, dz dx, \quad (15)$$

where x is the along-transect coordinate, z is the vertical coordinate, v is the cross-transect velocity, and L is the length of the transect. The basin-wide meridional temperature transport can also be estimated by

$$\int_0^L \int_{-800}^0 v \, T dz dx = \int_0^L \int_{-800}^0 \frac{T}{f} \frac{\partial(\Delta D)}{\partial x} \, dz dx, \quad (16)$$

where T is the temperature. Displayed in Figure 16 are the volume transport and temperature transport calculated di-

rectly from the XBT/XCTD transects and from T/P altimeter data using Eqs. (15) and (16). The agreement between the two estimates is encouraging. The outstanding disagreements in June 1994 and February 1995 are attributed to the anomalously strong Kuroshio near Taiwan during those periods. Because of its narrow width and a primarily north-south orientation, the Kuroshio near Taiwan is not well sampled by the T/P data due to the coarse ground-track spacing.

The study of Gilson *et al.* (1998) has demonstrated the utility of combined use of altimeter data with *in situ* data for making estimates of subsurface fields and transports that are of importance to the study of climate. The potential of this combination has motivated plans for deploying a network of floats in the ocean to complement future altimetry missions. (The program is called Argo, see Wilson, 2000.) The correlation between sea-level variability and subsurface variability discussed in this section underlies the importance of satellite altimetry as part of a global observing system. In addition to using straightforward statistical techniques discussed above, a more effective approach to the estimation of subsurface variability from altimeter data is provided by the use of an ocean general circulation model through the techniques of data assimilation, which is the subject of Chapter 5.

3.6. The Annual Cycle

The annual cycle is a major component of the large-scale variability over most of the ocean. It is caused by a combination of many different processes discussed in the preceding sections and hence exhibits a very complicated geographic pattern in its amplitude and phase. Owing to its dense spatial coverage, satellite altimetry provides the first detailed description of this complicated pattern. Jacobs et al. (1992) attempted to estimate the annual cycle over the global oceans from Geosat data. The analysis was complicated by large tidal and orbital errors in the data, especially the M_2 tidal variations at an aliased period of 317 days, which is very difficult to distinguish from the annual period in a 3-year record. Despite great efforts to mitigate the effects of these errors, the results inevitably suffered from residual errors because both signals and errors have similar spatial and temporal scales.

Using more accurate data from T/P, Stammer (1997) computed the amplitude and phase of the annual harmonic fit to 3 years (1993–1995) of data (Figure 17 [see color insert]). Also shown are the steric component of the annual cycle computed from Eq. (7) and the residuals (presumably wind-forced) after the steric component has been removed. As discussed in Section 3.1, a major portion of the annual cycle at mid latitudes is due to steric effects with opposite phase in the two hemispheres. The larger amplitude in the Northern Hemisphere caused by steric effects as discussed in Section 3.1 is clearly revealed. This hemispheric asymmetry was not apparent in the Geosat result, probably caused by the effects of the orbit-error removal procedure applied to the Geosat data. In fact, the annual cycle at mid-latitudes away from the western boundary currents was mostly absent in Jacobs et al. (1992).

At high latitudes, where air-sea heat flux has the largest annual variability, the sea-level annual cycle is actually less intense than that at mid latitudes. This is because the coefficient of thermal expansion becomes smaller as the water gets colder. The coefficient at high latitudes is only about 1/3 of its value at low latitudes. In the high-latitude Southern Ocean, there is a sharp change in the phase of the annual cycle (Figure 17b), roughly between 50°S and 60°S where the Antarctic Circumpolar Current (ACC) flows. To the north of the ACC, the annual maximum in sea level occurs in March-April; to the south of the ACC, it occurs in August-September. This front of phase change is consistent with the Geosat study of Chelton et al. (1990) in which the annual cycle accounts for the first empirical orthogonal mode. After the steric component of the annual cycle is removed, the phase of the wind-forced residual variability in the Southern Ocean (Figure 17f) has a maximum in August-October

when the wind is the strongest. It is therefore apparent that the annual cycle to the north of the ACC is dominated by the buoyancy forcing (with a maximum in March/April, see Figure 17d), whereas to the south of the ACC, it is dominated by the wind forcing.

After the removal of the steric component, the residual annual cycle is prominent in the tropics, where the annual cycle is primarily driven by wind, consistent with previous analyses of in situ data (Wyrtki, 1974b; Taft and Kessler, 1991; Reverdin et al., 1994). The annual cycle in the tropical Pacific is characterized by bands of high amplitude: one along 5° to 7°N and the other along 12° to 15°N. The former is associated with the annual variation of the North Equatorial Counter Current, which reaches its maximum around October-November. A westward phase propagation is observed across the entire extent of the varying current and is consistent with the interpretation of Meyers (1979) in terms of wind-driven Rossby waves. The variability along 12° to 15°N is indicative of the annual variations of the North Equatorial Current. Westward phase propagation is observed only in the eastern end of the band. There is little phase propagation in the western part of the band, where Meyers's study suggested a near in-phase relationship between sea level and the wind stress curl (also see Vivier et al., 1999). The maximum in the Atlantic (along 5° to 7°N) is associated with the annual variations of the Atlantic North Equatorial Counter Current (Richardson and Reverdin, 1987). The zonal phase variation in the Indian Oceans shows westward propagation. Previous studies (Woodbury et al., 1989; Perigaud and Delecluse, 1992) have suggested that the annual cycle in the southern tropical Indian Ocean (10° to 20°S) is associated with Rossby waves driven by the annual cycle of the trade winds (also see Masumoto and Meyers, 1998). It is also interesting to note the 180° phase change across the Arabian Sea, representing the ocean's response to the annual monsoon wind cycle during its two opposite phases (Bruce et al., 1994). Significant semiannual signals are found in the Indian Ocean (Basu et al., 2000; Jacobs et al., 1992), as well as the central equatorial Pacific and the southwest Atlantic.

The ability of numerical models to simulate the oceanic annual cycle is an important test of the usefulness of the models for climate studies, since the annual cycle is the result of the complicated interaction between the ocean and the atmosphere. Comparisons of ocean general circulation models with the T/P altimeter data have revealed the strengths and weaknesses of the models (Stammer et al., 1996; Fu and Smith, 1996; Jacobs et al., 1996). In general, the models have good skills in reproducing the wind-driven component of the variability, but not the steric component. The model-simulated amplitude at mid latitudes where the steric component dominates the annual cycle is generally too weak (by as much as 3 cm), reflecting possible problems in the model's mixing mechanism as well as the poor quality of the heat flux fields used to force the models. Although the error in the ECMWF heat flux could cause an error of 2 cm in the annual cycle (Stammer *et al.*, 1996), there is no evidence for a systematic low bias in the ECMWF heat flux. The major culprit should be the model's lack of a good mixed layer. Li *et al.* (2000) demonstrated that the use of a state-of-the-art mixing scheme significantly improved the simulation of the annual cycle by an ocean model.

3.7. Interannual Variability

Although the annual variation of the sun angle is the dominant external source for the variability in the circulation of the ocean and the atmosphere, the inherently nonlinear response of the coupled ocean-atmosphere system is not purely periodic at the annual period, but has a rich spectrum with power density generally increasing with period (a red spectrum as noted earlier). There are large sea-level variations at periods longer than annual. As the record length of altimetry data is increasing, the large-scale patterns of long-period variability becomes increasingly observable. The current 8-year data records from T/P and ERS-1 and -2 provide an unprecedented opportunity to study the interannual variability of the ocean.

The most prominent sea-level variations on interannual scales are associated with ENSO (El Niño Southern Oscillation) events (Hendricks *et al.*, 1996; Stammer *et al.*, 1996; Chambers *et al.*, 1999; Potemra and Lukas, 1999). Because these events are strongly tied to the variability of the tropical oceans, they are discussed mainly in Chapter 4. On a global scale, the variation of the mean sea level is also significantly affected by ENSO. This subject is discussed in Chapter 8. In this section we focus on basin-wide variations. We begin with a description of the evolution of large-scale sea level features over a 4-year span to provide a qualitative view of the extent of the interannual variability of the ocean, followed by brief discussions of basin-wide effects of ENSO and low-frequency interactions between gyres and eddies.

Shown in Figure 18 (see color insert) are yearly mean sea-level anomalies obtained from the T/P data during 1996–1999. The anomalies were computed by first subtracting from the data a 4-year mean averaged over the period of 1993–1996 and then averaging the residuals over yearly intervals. The 4-year period for estimating the mean includes a weak warm event (1994–1995) and a weak cold event (1996) in the tropical Pacific Ocean and is thus considered a reasonable period for computing a mean somewhat close to the norm. The yearly averaging filters out the annual cycle and other high-frequency signals. The resulting yearly anomalies thus reveal primarily the variability on time scales longer than a year. However, the mesoscale variability is still visible in many places after the averaging, especially along the Indian Ocean sector of the ACC as well as in the Brazil/Malvinas Confluence region of the South Atlantic.

The large-scale features in Figure 18 clearly illustrate the effects of the dramatic 1997–1998 El Niño and its transition into La Niña in 1998–1999. The high sea levels in the western Pacific and the eastern Indian Ocean in 1996 set the buildup stage for El Niño. During El Niño, the tropical Pacific Ocean and the Indian Ocean exhibit large-scale changes, with elevated sea levels in the eastern Pacific Ocean (the maximum reached over 30 cm in December 1997) and the western Indian Ocean. Chambers *et al.* (1999) computed the empirical orthogonal functions for the T/P data during 1992–1998 after removing an annual and a semiannual sinusoid as well as a linear trend (also see Potemra and Lukas, 1999). They found that the ENSO-related variability (appearing as the leading mode) in the Indian and Pacific Oceans are highly correlated with no significant phase lags. They also demonstrated that the wind anomalies (from the analysis of the Florida State University) in the western tropical Pacific and the eastern Indian Ocean were significantly correlated with each other and opposite in direction. As the wind anomalies (westerly anomaly in the western Pacific and easterly anomaly in the eastern Indian Ocean) became intensified during the initial phase of the 1997–1998 El Niño, downwelling (thermocline depressed/sea level elevated) Kelvin waves started propagating eastward in the Pacific, whereas downwelling Rossby waves started propagating westward in the Indian Ocean. Such processes led to the buildup of the high sea levels in the western Indian Ocean and the eastern Pacific Ocean during the peak of El Niño. More discussions of the roles of the Kelvin and Rossby waves in the development of El Niño are given in Chapter 4.

After El Niño evolved into La Niña in mid-1998, the entire North Pacific Ocean became progressively abnormal. In 1999, the western part of the basin was characterized by a large pool of high sea level north of New Guinea and west of the Philippines, as well as by a series of bands of high sea level extending from the western boundary of the basin to the central basin from subtropical to high latitudes. The two bands of high sea level emanating from east of Japan are most notable. The eastern part of the basin exhibits low sea levels from the Gulf of Alaska to the southern tip of Baja California. This strip of low sea-level connects to the huge pool of low sea level associated with La Niña in the central equatorial Pacific. This pattern of sea level change has drawn significant attention from the climate research community, because it bears a strong resemblance to a phase of ENSO-like long-term variability of the Pacific Ocean (Zhang *et al.*, 1997). The record is obviously too short to lead to any conclusions, but something apparently very interesting has been occurring in the Pacific Ocean.

The 1997–1999 ENSO event is probably the most intensively observed to date in comparison with previous events. Satellite altimetry has played a significant role in providing a unique global perspective to complement *in situ* observations, which are mostly concentrated in the equatorial regions. In fact, T/P data have been incorporated into the data stream used by NOAA's National Center for Environment Prediction for short-term climate forecast. The merits of altimeter data in improving the skills of prediction model have been demonstrated in hindcast experiments and reported in Ji *et al.* (2000).

At interannual and longer time scales, the interactions between the tropical oceans and the mid- and high-latitude oceans are important in determining the large-scale ocean–atmosphere coupled climate system (e.g., Gu and Philander, 1997). By patching the Geosat and ERS-1 altimeter data together, Jacobs *et al.* (1994) conducted an interesting study in which suggestive evidence was shown for the long-term and basin-wide influence of El Niño. They performed a simulation of the Pacific Ocean circulation (from 1981–1993) using a numerical model and demonstrated that the sea level rise in the eastern tropical Pacific Ocean during the 1982–1983 El Niño had triggered northward-propagating coastal-trapped waves along the coast of North America. The sea-level anomalies caused by these relatively fast coastal waves (50 km/day) eventually transformed into slow westward-propagating (5 km/day) Rossby waves (see Section 3.4), which reached the region of the Kuroshio Extension in 1992–1993. These Rossby waves in the model simulation influenced the path of the Kuroshio Extension and hence the sea surface temperature 10 years after the onset of El Niño. They suggested that the resulting sea surface temperature change may be an important factor in determining the weather and climate of North America. The basin-wide temporal variations of the sea surface height in the North Pacific Ocean simulated by the model were compared with the differences between the Geosat (1986–1989) data and the ERS-1 data (1992–1993) with favorable results. Observations of sea surface temperature were also in agreement with the altimeter data and the model simulation. Earlier theoretical and numerical studies of the remote influence of El Niño in terms of coastal Kelvin waves along the eastern boundary of the Pacific Ocean can be found in McCreary (1976), Hurlburt *et al.* (1976), and Johnson and O'Brien (1990). Observational evidence for such coastal-trapped waves has been obtained from tide gauge data (Enfield and Allen, 1980; Chelton and Davis, 1982).

There is evidence that ENSO may also affect the remote Southern Ocean (Peterson and White, 1998) via the so-called Antarctic Circumpolar Wave, which is an ocean-atmosphere-coupled wave that evidently propagates eastward around the Antarctica. This wave originates in the western subtropical South Pacific Ocean in association with ENSO activities, creating a slow oceanic teleconnection that spreads the influence of ENSO to other ocean basins 6–8 years later. Jacobs and Mitchell (1996) showed earlier evidence suggestive of such waves in the relatively short Geosat altimeter data.

At interannual and longer time scales, there is evidence for the interaction of oceanic variabilities at different spatial and temporal scales, in particular the interactions between gyre and eddy scales (see Chapter 3 for more discussions). Qiu (1995) found that the eddy kinetic energy in the Kuroshio Extension and its southern recirculation gyre had undergone significant variations. The eddy energy in the recirculation region steadily increased in 1993–1994, while the flow strengths of the Kuroshio Extension and the recirculation steadily decreased in 1993–1994. Analysis of the energetics suggested that there was a transfer of energy from the mean flow to the eddy field of the recirculation region via barotropic instability

Witter and Gordon (1999) analyzed 4 years of T/P data and discovered an interannual basin-scale mode in the South Atlantic. This mode indicates that the state of the eastern South Atlantic underwent a transition from high sea level and enhanced gyre-scale circulation in 1993 and 1994, to a state of lower sea level and weaker circulation in 1996. They also noted that the dominant mode of basin-scale zonal wind had a similar temporal signature. In the meantime, the migration path of the Agulhas eddies from the Indian Ocean to the Atlantic became more constricted and narrower in 1996 than in 1993–1994. These observations suggest that the salt and vorticity inputs from the Indian Ocean to the Atlantic are probably affected by interannual variations of wind-forced large-scale circulation.

4. CONCLUSIONS

Satellite altimetry has revolutionized the study of large-scale ocean circulation by providing a capability of observing the sea surface height with global coverage and frequent sampling. As noted earlier, sea surface height is a dynamic variable of the physical state of the ocean. It contains information about the state of the entire water column. Although an accurate geoid model is required to obtain the ocean topography from which absolute surface geostrophic current velocity can be derived, a great deal has been learned about the temporal variability of ocean circulation in spite of the lack of a desirable knowledge of the geoid.

Errors in the geoid models presently available are smaller than the magnitude of oceanographic signals at wavelengths longer than 3000 km. Atimetrically-determined ocean topography has provided the first test bed for examining the performance of global ocean general circulation models at large scales. The rms large-scale (wavelengths longer than 3000 km) errors in model-simulated ocean topography have been estimated to be at the level of 10 cm. At these scales, the ocean topography derived from altimetry is generally consistent with what is required by *in situ* data in an inverse calculation. However, to obtain significantly new information from altimetry on absolute ocean topography and hence the oceanic general circulation, more accurate geoid knowledge over a wide range of scales is required.

A substantial improvement of the knowledge of the geoid is expected in the next 5–10 years. GRACE, a United States/Germany joint mission, will utilize a pair of low-earth-orbiting satellites whose relative distance is precisely determined (Wahr *et al.*, 1998; Davis *et al.*, 1999). The measurement of the distance plus onboard accelerometer measurements will be analyzed to improve the geoid accuracy to better than 1 cm at spatial scales greater than 200 km (half wavelength). GRACE is planned for launch in 2001. The GOCE mission, currently planned by the European Space Agency, will fly a system based on gradiometry and GPS precision tracking in a low orbit. GOCE is expected to extend the 1-cm geoid accuracy to a scale of 100 km. With such accurate knowledge of the geoid, absolute surface geostrophic current velocity will be determined from altimeter data with details including the swift boundary currents and mesoscale eddies (LeGrand and Minster, 1999). Such new knowledge of the geoid can be used retrospectively with existing and future altimeter data to improve the knowledge of the oceanic general circulation. With the use of ocean models that are constrained by these new observations, a significant improvement down to 100 km scale is anticipated in the next 5–10 years.

Based on the combined 8-year record of altimetry from T/P and ERS-1 and -2, significant progress has been made in the knowledge of temporal variations of large-scale ocean variability. Many of the limitations in early data from Seasat and Geosat have been mitigated in the new data due to improved measurement systems. New insights of ocean dynamics have been obtained over a wide range of scales. Large-scale high-frequency barotropic variability, which has not been sampled adequately by past *in situ* data, has emerged in the T/P data. The distribution of such variability has been mapped globally and its dynamics are understood through model simulations. Without the new observations, we would not be able to test ocean models at these scales and appreciate their surprisingly good performance.

At seasonal-to-interannual scales, a new look into the dynamics of baroclinic Rossby waves has been made possible by the high quality and the 7-year T/P data record. The zonal phase speeds of the waves have been a focus of investigation. The estimated westward phase speeds are generally higher than those of nondispersive, long waves predicted by the standard linear wave theory. Inspired by these findings, a number of theories have been proposed for the anomalous phase speeds, including the effects of the vertical shear of mean currents, wind-forced waves, and ocean-atmosphere-coupled waves. Among these, the enhancement of the wave restoring mechanism (the beta effect) by the vertical mean shear seems to be the most promising explanation for the fast phase speeds. Spectral analysis reveals that, especially

at high frequencies and wavenumbers, the most energetic variability deviates significantly from the dispersion relation deduced from the standard theory that neglects the effects of background mean flow. The dispersion relation derived from the new theory that includes the effects of a vertically sheared mean flow yields a much improved fit to the observations.

The distribution of the amplitude and phase of the global annual cycle of sea-level variations has been derived from T/P data. However, the published results are based on only 3 years of data and are anticipated to be updated with more recent data. Nevertheless, improvements over previous estimates from Geosat data are notable at large-scales, especially on the hemispheric asymmetry of the annual amplitude. The dominance of steric effects at mid and high latitudes and the transition to wind forcing at low latitudes has been clearly shown. A barotropic wind-driven Sverdrup balance at seasonal scales across ocean basins has been demonstrated in the subtropical and subpolar regions of the Pacific Ocean.

The exceptionally strong ENSO event from 1997–1999 provided a central stage for demonstrating the utility of satellite altimetry in providing a global perspective of large-scale year-to-year changes in the ocean. The data have clearly showed that ENSO is not limited to the tropical Pacific where most of the *in situ* observations are concentrated. Significant changes in the Indian Ocean also took place during the same period, seemingly driven by the same system of anomalies in the atmospheric circulation over the western Pacific and eastern Indian Ocean. The large expanse of high sea-level anomalies in the western mid-latitude North Pacific after El Niño evolved into La Niña is intriguing. Understanding of the evolution of these anomalies and their impact on climate requires consistent long-term observations on global scales. Satellite altimetry is particularly amenable to observing such large-scale, low-frequency phenomena and useful for improving the capability of climate prediction. For example, the utility of T/P data in enhancing the capability of predicting ENSO has been demonstrated.

Satellite altimetry is a truly global observing system that measures a dynamical variable of the ocean of importance to climate change. The workings of such a system and its utility have been demonstrated for the past 8 years, long enough to show its impacts on climate studies but too short to address the "red" spectrum of the climate system. It is clear that the dedicated, precision system of T/P ought to be continued. Determination of the ocean circulation to an extent that is useful for climate studies requires the level of performance achieved by T/P. The joint France/U.S. follow-on mission to T/P, called Jason-1, will be launched in early 2001 to continue T/P measurement along the same ground tracks with measurement performance potentially even better than T/P. A series of such missions, coupled with complementary *in situ* observing systems such as the Argo floats (Wilson,

2000), is deemed an important element of a long-term global ocean observing system for climate and other maritime applications. Incorporation of such an element into national and international operational systems is a pressing issue (Nowlin, 1999). Also of importance is to improve the spatial resolution of ocean topography measurement for monitoring the energetic mesoscale eddies and currents, a task that cannot be adequately carried out by a single, nadir-looking altimeter. New technologies and measurement techniques amenable to addressing this issue are emerging. Potential scenarios include launching low-cost multiple altimeters, new radar instruments with wide swaths, and special GPS receivers for collecting ocean reflections of GPS signals. Such new developments should be an integral part of a long-term strategy for the measurement of ocean topography.

ACKNOWLEDGMENTS

We thank Carl Wunsch, Roland de Szoeke, and Michael Schlax for helpful comments on various portions of this manuscript. We also thank Michael Schlax for extensive help with the figures in Section 3.4 and Rick Smith for providing the model simulations of sea level from the Parallel Ocean Program (POP) at the Los Alamos National Laboratory.

The research described in the paper was carried out in part by the Jet Propulsion Laboratory, California Institute of Technology, under contract with the National Aeronautics and Space Administration. Support from the TOPEX/POSEIDON Project is acknowledged. DBC was supported for this work by contracts 958127 and 1206715 from the Jet Propulsion Laboratory.

References

Anderson, D. L. T., Bryan, K., Gill, A. E., and Pacanowski, R. C. (1979). The transient response of the North Atlantic: Some model studies. *J. Geophys. Res.*, **84**, 4795–4815.

Aoki, S., Imawaki, S., and Ichikawa, K. (1995). Baroclinic disturbances propagating westward in the Kuroshio Extension region as seen by a satellite altimeter and radiometers. *J. Geophys. Res.*, **100**, 839–855.

Barnier, B. (1988). A numerical study on the influence of the mid-Atlantic ridge on nonlinear first-mode baroclinic Rossby waves generated by seasonal winds. *J. Phys. Oceanogr.*, **18**, 417–433.

Basu, S., Meyers, S. D., and O'Brien, J. J. (2000). Annual and interannual sea-level variations in the Indian Ocean from TOPEX/Poseidon observations and ocean model simulations, *J. Geophys. Res.*, **105**, 975–994.

Bernstein, R. L., and White, W. B. (1977). Zonal variability in the distribution of eddy energy in the mid-latitude North Pacific Ocean. *J. Phys. Oceanogr.*, **7**, 123–126.

Bernstein, R. L., and White, W. B. (1981). Stationary and traveling mesoscale perturbations in the Kuroshio Extension Current. *J. Phys. Oceanogr.*, **11**, 692–704.

Bruce, J. G., Johnson, D. R., and Kindle, J. C. (1994). Evidence for eddy formation in the eastern Arabian Sea during the northeast monsoon. *J. Geophys. Res.*, **99**, 7651–7664.

Bryan, F. O., Wang, D., and Holland, W. R. (2000). Response of a global ocean general circulation model to atmospheric pressure loading. *J. Geophys. Res.*, in press.

Carnes, M. R., Mitchell, J. L., and de Witt P. W. (1990). Synthetic Temperature Profiles Derived from Geosat Altimetry: Comparison with Airdropped Expendable Bathythermograph Profiles. *J. Geophys. Res.*, **95**, 17979–17992.

Chambers, D. P., Tapley, B. D., and Stewart, R. H. (1997). Long-period ocean heat storage rates and basin-scale heat fluxes from TOPEX. *J. Geophys. Res.*, **102**, 10525–10533.

Chambers, D. P., Tapley, B. D., and Stewart, R. H. (1999). Anomalous warming in the Indian Ocean coincident with El Niño. *J. Geophys. Res.*, **104**, 3035–3047.

Chang, P., and Philander, S. G. H. (1989). Rossby wave packets in baroclinic mean currents, *Deep-Sea Res.*, **36**, 17–37.

Chao, Y., and Fu, L.-L. (1995). A comparison between the TOPEX/POSEIDON data and a global ocean general circulation model during 1992–1993. *J. Geophys. Res.*, **100**, 24965–24976.

Chelton, D. B., and Davis, R. E. (1982). Monthly mean sea-level variability along the west coast of North America. *J. Phys. Oceanogr.*, **12**, 757–784.

Chelton, D. B., and Mestas-Nuñez, A. M. (1996). The large-scale, wind-driven response of the North Pacific. *Int. WOCE Newsletter*, **25**, 3–6.

Chelton, D. B., and Schlax, M. G. (1993). Spectral characteristics of time-dependent orbit errors in altimeter height measurements. *J. Geophys. Res.*, **98**, 12,579–12,600.

Chelton, D. B., and Schlax, M. G. (1996). Global observations of oceanic Rossby waves. *Science*, **272**, 234–238.

Chelton, D. B., Schlax, M. G., Witter, D. L., and Richman, J. G. (1990). Geosat altimeter observations of the surface circulation of the Southern Ocean. *J. Geophys. Res.*, **95**, 17877–17903.

Chelton, D. B., de Szoeke, R. A., Schlax, M. G., El Naggar, K., and Siwertz, N. (1998). Geographical variability of the first-baroclinic Rossby radius of deformation. *J. Phys. Oceanogr.*, **28**, 433–460.

Cheney, R. E., Miller, L., Agreen, R., Doyle, N., and Lillibridge, J. (1994). TOPEX/POSEIDON: The 2-cm solution. *J. Geophys. Res.*, **99**, 24555–24564.

Cipollini, P., Cromwell, D., Jones, M. S., Quartly, G. D., and Challenor, P. G. (1997). Concurrent altimeter and infrared observations of Rossby wave propagation near 34°N in the northeast Atlantic. *Geophys. Res. Lett.*, **24**, 889–892.

Cipollini, P., Cromwell, D., and Quartly, G. D. (1999). Observations of Rossby-wave propagation in the northeast Atlantic with TOPEX/POSEIDON altimetry. *Adv. Space Res.*, **22**, 1553–1556.

Clarke, A. J., and Shi, C. (1991). Critical frequencies at ocean boundaries. *J. Geophys. Res.*, **96**, 10731–10738.

Cummins, P. F., Mysak, L. A., and Hamilton, K. (1986). Generation of annual Rossby waves in the North Pacific by the wind stress curl. *J. Phys. Oceanogr.*, **16**, 1179–1189.

Davis, E. S., Dunn, C. E., Stanton, R. H., and Thomas, J. B. (1999). The GRACE mission: Meeting the technical challenges, Paper IAF-99-B.2.05 in Proc. 50th Int. Astronautical Congress, 4–8 October 1999, Amsterdam, The Netherlands.

Deans, S. R. (1983). *In* "The Radon Transform and Some of its Applications," pp. 289. John Wiley & Son, New York.

Denker, H., and Rapp, R. H. (1990). Geodetic and Oceanographic Results from the Analysis of One Year of Geosat Data. *J. Geophys. Res.*, **95**, 13151–13168.

de Szoeke, R. A., and Chelton, D. B. (1999). The modification of long planetary waves by homogeneous potential vorticity layers. *J. Phys. Oceanogr.*, **29**, 500–511.

Dewar, W. K. (1998). On "too fast" baroclinic planetary waves in the general circulation. *J. Phys. Oceanogr.*, **28**, 1739–1758.

Douglas, B. (1991). Global sea-level rise. *J. Geophys. Res.*, **96**, 6981–6992.

Emery, W. J., and Magaard, L. (1976). Baroclinic Rossby waves as inferred from temperature fluctuations in the eastern Pacific. *J. Mar. Res.*, **34**, 365–385.

Enfield, D. B., and Allen, J. S. (1980). On the structure and dynamics of monthly mean sea-level anomalies along the Pacific coast of North and South America. *J. Phys. Oceanogr.*, **10**, 557–578.

Ferry, N., Reverdin, G., and Oschlies, A. (2000). Seasonal sea surface height variability in the North Atlantic Ocean, *J. Geophys. Res.*, **105**, 6307–6326.

Forbes, C., Leaman, K., Olson, D., and Brown, O. (1993). Eddy and wave dynamics in the South Atlantic as diagnosed from Geosat altimeter data. *J. Geophys. Res.*, **98**, 12297–12314.

FRAM Group (1991). An eddy resolving model of the Southern Ocean. *Eos. Trans. AGU*, **72**, 169, 174–175.

Fu, L.-L. (1983). Recent progress in the application of satellite altimetry to observing the mesoscale variability and general circulation of the oceans. *Rev. Geophys. Space Phys.*, **21**, 1657–1666.

Fu, L.-L., and Cheney, R. E. (1995). Applications of satellite altimetry to ocean circulation studies: 1987-1994. *Rev. Geophys.*, **32 (Supplement)**, 213–223.

Fu, L.-L., and Davidson, R. A. (1995). A note on the barotropic response of sea level to time-dependent wind forcing. *J. Geophys. Res.*, **100**, 24955–24963.

Fu, L.-L., and Pihos, G. (1994). Determining the response of sea level to atmospheric pressure forcing using TOPEX/POSEIDON data. *J. Geophys. Res.*, **99**, 24633–24642.

Fu, L.-L., and Smith, R. D. (1996). Global ocean circulation from satellite altimetry and high-resolution computer simulation. *Bull. Am. Meteorolog. Soc.*, **77**, 2625–2636.

Fu, L.-L., Cheng, B., and Qiu, B. (2000). 25-Day Period Large-Scale Oscillations in the Argentine Basin Revealed by the TOPEX/POSEIDON Altimeter, *J. Phys. Oceanogr.*, in press.

Fu, L.-L., Christensen, E. J., Yamarone, C. A., Lefebvre, M., Menard, Y., Dorrer, M., and Escudier, P. (1994). TOPEX/POSEIDON Mission Overview. *J. Geophys. Res.*, **99**, 24369–24381.

Fuglister, F. C. (1960). *In* "Atlantic Ocean Atlas of Temperature and Salinity Profiles and Data from the International Geophysical Year of 1957–1958." Woods Hole Oceanographic Institution, Woods Hole, Mass.

Fukumori, I., Raghunath, R., and Fu, L.-L. (1998). The nature of global large-scale sea-level variability in relation to atmospheric forcing: A modeling study. *J. Geophys. Res.*, **103**, 5493–5512.

Ganachaud, A., Wunsch, C., Kim, M.-C., and Tapley, B. (1997). Combination of TOPEX/POSEIDON data with a hydrographic inversion for determination of the oceanic general circulation and its relation to geoid accuracy. *Geophys. J. Int.*, **128**, 708–722.

Gaspar, P., and Ponte, R. M. (1997). Relation between sea level and barometric pressure determined from altimeter data and model simulations. *J. Geophys. Res.*, **102**, 961–971.

Gerdes, R., and Wübber, C. (1991). Seasonal variability of the North Atlantic Ocean—a model intercomparison. *J. Phys. Oceanogr.*, **21**, 1300–1322.

Gill, A. E. (1982). *In* "Atmosphere-Ocean Dynamics." Academic Press, New York.

Gill, A. E., and Niiler, P. P. (1973). The theory of the seasonal variability in the ocean. *Deep-Sea Res.*, **20**, 141–177.

Gill, A. E., Green, J. S. A., and Simmons, A. J. (1974). Energy partition in the large-scale ocean circulation and the production of mid-ocean eddies. *Deep-Sea Res.*, **21**, 499–528.

Gilson, J., Roemmich, D., Cornuelle, B., and Fu, L.-L. (1998). Relationship of TOPEX/POSEIDON altimetric height to steric height and circulation in the North Pacific. *J. Geophys. Res.*, **103**, 27947–27965.

Grimshaw, R., and Allen, J. S. (1988). Low-frequency baroclinic waves off coastal boundaries. *J. Phys. Oceanogr.*, **18**, 1124–1143.

Gu, D. F., and Philander, S. G. H. (1997). Interdecadal climate fluctuations that depend on exchanges between the tropics and extratropics. *Science*, **275**, 805–807.

Hendricks, J. R., Leben, R. R., Born, G. H., and Koblinsky, C. J. (1996). Empirical orthogonal function analysis of global TOPEX/POSEIDON altimeter data and implications for detection of global sea-level rise. *J. Geophys. Res.*, **101**, 14131–14145.

Herrmann, P., and Krauss, W. (1989). Generation and propagation of annual Rossby waves in the North Atlantic. *J. Phys. Oceanogr.*, **19**, 727–744.

Hughes, C. W. (1995). Rossby waves in the Southern Ocean: A comparison of TOPEX/POSEIDON altimetry with model predictions. *J. Geophys. Res.*, **100**, 15933–15950.

Hughes, C. W. (1996). The Antarctic Circumpolar Current as a waveguide for Rossby waves. *J. Phys. Oceanogr.*, **26**, 1375–1387.

Hughes, C. W., Jones, M. S., and Carnochan, S. (1998). Use of transient features to identify eastward currents in the Southern Ocean. *J. Geophys. Res.*, **103**, 2929–2943.

Hurlburt, H. E., Kindle, J. C., and O'Brien, J. J. (1976). A numerical study of the onset of El Niño. *J. Phys. Oceanogr.*, **6**, 621–631.

Hurlburt, H. E., Wallcraft, A. J., Schmitz, W. H., Hogan, P. J., and Metzger, E. J. (1996). Dynamics of the Kuroshio/Oyashio current system using eddy-resolving models of the North Pacific Ocean. *J. Geophys. Res.*, **101**, 941–976.

Jacobs, G. A., and Mitchell, J. L. (1996). Ocean circulation variations associated with the Antarctic Circumpolar Wave. *Geophys. Res. Lett.*, **23**, 2947–2950.

Jacobs, G. A., Born, G. H., Parke, M. E., and Allen, P. C. (1992). The global structure of the annual and semiannual sea surface height variability from Geosat altimeter data. *J. Geophys. Res.*, **97**, 17813–17828.

Jacobs, G. A., Emery, W. J., and Born, G. H. (1993). Rossby waves in the Pacific Ocean extracted from Geosat altimeter data. *J. Phys. Oceanogr.*, **23**, 1155–1175.

Jacobs, G. A., Hurlburt, H. E., Kindle, J. C., Metzger, E. J., Mitchell, J. L., Teague, W. J., and Wallcraft, A. J. (1994). Decade-scale trans-Pacific propagation and warming effects of an El Niño anomaly. *Nature*, **370**, 360–363.

Jacobs, G. A., Teague, W. J., Mitchell, J. L., and Hurlburt, H. E. (1996). An examination of the North Pacific Ocean in the spectral domain using Geosat altimeter data and a numerical ocean model. *J. Geophys. Res.*, **101**, 1025–1044.

Jacobson, A. R., and Spiesberger, J. L. (1998). Observations of El Niño-Southern Oscillation induced Rossby waves in the northeast Pacific using in situ data. *J. Geophys. Res.*, **103**, 24585–24596.

Jain, A. K. (1989). *In* "Fundamentals of Digital Image Processing," 569 pp. Prentice-Hall, Englewood Cliffs, NJ.

Ji, M., Reynolds, R. W., and Beringer, D. W. (2000). Use of TOPEX/Poseidon sea-level data for ocean analyses and ENSO prediction: some preliminary results. *J. Climate*, **13**, 216–231.

Johnson, M. A., and O'Brien, J. J. (1990). The northeast Pacific Ocean responses to the 1982-1983 El Niño. *J. Geophys. Res.*, **95**, 7155–7166.

Kang, Y. Q., and Magaard, L. (1979). Stable and unstable Rossby waves in the North Pacific Current as inferred from the mean stratification. *Dyn. Atmos. Oceans*, **3**, 1–14.

Kang, Y. Q., and Magaard, L. (1980). Annual baroclinic Rossby waves in the central North Pacific. *J. Phys. Oceanogr.*, **10**, 1159–1167.

Keffer, T. (1985). The ventilation of the world's oceans: Maps of the potential vorticity field. *J. Phys. Oceanogr.*, **15**, 509–523.

Kelly, K. A., and Qiu, B. (1995a). Heat flux estimates for the western North Atlantic. Part I: assimilation of satellite data into a mixed layer model. *J. Phys. Oceanogr.*, **25**, 2344–2360.

Kelly, K. A., and Qiu, B. (1995b). Heat flux estimates for the western North Atlantic. Part II: the upper ocean heat balance. *J. Phys. Oceanogr.*, **25**, 2361–2373.

Kelly, K. A., Singh, S., and Huang, R. X. (1999). Seasonal variations of sea surface height in the Gulf Stream region, *J. Phys. Oceanogr.*, **29**, 313–327.

Kessler, W. S. (1990). Observations of long Rossby waves in the northern tropical Pacific. *J. Geophys. Res.*, **95**, 5183–5217.

Killworth, P. D., and Blundell, J. R. (1999). The effect of bottom topography on the speed of long extratropical planetary waves. *J. Phys. Oceanogr.*, **29**, 2689–2710.

Killworth, P. D., Chelton. D. B., and de Szoeke, R. A. (1997). The speed of observed and theoretical long extra-tropical planetary waves. *J. Phys. Oceanogr.*, **27**, 1946–1966.

Kindle, J. C., and Thompson, J. D. (1989). The 26- and 50-day oscillations in the western Indian Ocean: model results. *J. Geophys. Res.*, **94**, 4721–4736.

Koblinsky, C. J., Niller, P. P., and Schmitz, W. J. (1989). Observations of wind-forced deep ocean currents in the North Pacific. *J. Geophs. Res.*, **94**, 10773–10790.

Krauss, W., and Wübber, C. (1982). Response of the North Atlantic to annual wind variations along the eastern coast. *Deep-Sea Res.*, **29**, 851–864.

Kuragano, T., and Kamachi, M. (2000). Global statistical space-time scales of oceanic variability estimated from the TOPEX/POSEIDON altimeter data. *J. Geophys. Res.*, **105**, 955–974.

LeBlond, P. H., and Mysak, L. A. (1978). *In* "Waves in the Ocean," 602 pp. Elsevier Scientific Publishing Company, Amsterdam.

LeGrand, P., and Minster, J. (1999). Impact of the GOCE gravity mission on ocean circulation estimates. *Geophys. Res. Lett.*, **26**, 1881–1884.

LeGrand, P., Mercier, H., and Reynaud, T. (1998). Combining T/P altimetric data with hydrographic data to estimate the mean dynamic topography of the North Atlantic and improve the geoid. *Ann. Geophys.*, **16**, 638–650.

LeTraon, P.-Y., and Minster, J.-F. (1993). Sea level variability and semiannual Rossby waves in the South Atlantic subtropical gyre. *J. Geophys. Res.*, **98**, 12315–12326.

Lemoine, F., *et al.* (1998). *In* "The Development of the Joint NASA GSFC and the National Imagery and Mapping Agency (NIMA) Geopotential Model EGM96," NASA/TP-1998-206861, NASA Goddard Space Flight Center, Greenbelt, MD.

Leuliette, E. W., and Wahr, J. M. (1999). Coupled pattern analysis of sea surface temperature and TOPEX/Poseidon sea surface height, *J. Phys. Oceanogr.*, **29**, 599–611.

Li, X., Chao, Y., Mcwilliams, J. C., and Fu, L.-L. (2000). A comparison of two vertical mixing schemes in a Pacific Ocean general circulation model. *J. Climate*, in press.

Luyten, J. R., and Roemmich, D. H. (1982). Equatorial currents at semiannual period in the Indian Ocean. *J. Phys. Oceanogr.*, **12**, 406–413.

Macdonald, A., and Wunsch, C. (1996). The global ocean circulation and heat flux. *Nature*, **382**, 436–439.

Maes, C. (1998). Estimating the influence of salinity on sea-level anomaly in the ocean. *Geophys. Res. Lett.*, **25**, 3551–3554.

Magaard, L. (1977). On the generation of baroclinic Rossby waves in the ocean by meteorological forces. *J. Phys. Oceanogr.*, **7**, 359–364.

Magaard, L. (1983). On the potential energy of baroclinic Rossby waves in the North Pacific. *J. Phys. Oceanogr.*, **13**, 38–42.

Magaard, L., and Price, J. M. (1977). Note on the significance of a previous Rossby wave fit to internal temperature fluctuations in the eastern Pacific. *J. Mar. Res.*, **35**, 649–651.

Marsh, J. G., *et al.* (1990). Dynamic sea surface topography, gravity, and improved orbit accuracies from the direct evaluation of seasat altimeter data. *J. Geophys. Res.*, **95**, 13129–13150.

Martel, F., and Wunsch, C. (1993a). The North Atlantic circulation in the early 1980's—an estimate from inversion of a finite difference model. J. Phys. *Oceanogr.*, **23**, 898–924.

Martel, F., and Wunsch, C. (1993b). Combined inversion of hydrography, current meter data and altimetric elevations for the North Atlantic circulation. *Manuscripta Geodaetica*, **18**, 219–226.

Masumoto, Y., and Meyers, G. (1998). Forced Rossby waves in the southern tropical Indian Ocean. *J. Geophys. Res.*, **103**, 27589–27602.

McCalpin, J. D. (1995). Rossby wave generation by poleward propagating Kelvin waves: the midlatitude quasigeostrophic approximation. *J. Phys. Oceanogr.*, **25**, 1415–1425.

McCreary, J. (1976). Eastern tropical ocean response to changing wind systems: with application to El Niño. *J. Phys. Oceanogr.*, **6**, 632–645.

Meyers, G. (1979). On the annual Rossby wave in the tropical North Pacific Ocean. *J. Phys. Oceanogr.*, **9**, 663–674.

Miller, A. J., White, W. B., and Cayan, D. R. (1997). North Pacific thermocline variations on ENSO timescales. *J. Phys. Oceanogr.*, **27**, 2023–2039.

Miller, L., and Cheney, R. (1990). Large-scale meridional transport in the tropical Pacific Ocean during the 1986–1987 El Niño from Geosat. *J. Geophys. Res.*, **95**, 17905–17919.

Minster, J.-F., Cazenave, A., Serafini, Y. V., Mercier, F., Gennero, M. C., and Rogel, P. (1999). Annual cycle in mean sea level from TOPEX/Poseidon and ERS-1: Inference on the global hydrological cycle. *Glob. Planet. Change*, **20**, 57–66.

Mitchum, G. T. (1995). The source of 90-day oscillations at Wake Island. *J. Geophys. Res.*, **100**, 2459–2475.

Müller, T. J., and Siedler, G. (1992). Multi-year current time series in the eastern North Atlantic Ocean. *J. Mar. Res.*, **50**, 63–98.

Mysak, L. A. (1983). Generation of annual Rossby waves in the North Pacific. *J. Phys. Oceanogr.*, **13**, 1908–1923.

National Research Council (1997). *In* "Satellite Gravity and the Geosphere," National Academy Press, Washington, D.C.

Nerem, R. S., Tapley, B. D., and Shum, C.-K. (1990). Determination of the Ocean Circulation using Geosat Altimetry. *J. Geophys. Res.*, **95**, 3163–3180.

Nerem, R. S., Schrama, E. J., Koblinsky, C. J., and Beckley, B. D. (1994). A preliminary evaluation of ocean topography from the TOPEX/POSEIDON mission. *J. Geophys. Res.*, **99**, 24565–24583.

Nowlin, W. D. (1999). A strategy for long-term ocean observations. *Bull. Am. Meteorolog. Soc.*, **80**, 621–627.

Oberhuber, J. M. (1993). Simulation of the Atlantic circulation with a coupled sea ice-mixed layer-isopycnal general circulation model. Part I: Model description. *J. Phys. Oceanogr.*, **23**, 808–829.

Pares-Sierra, A., White, W. B., and Tai, C.-K. (1993). Wind-driven coastal generation of annual mesoscale eddy activity in the California current. *J. Phys. Oceanogr.*, **23**, 1110–1121.

Parke, M. E., Born, G., Leben, R., McLaughlin, C., and Tierney, C. (1998). Altimeter sampling characteristics using a single satellite. *J. Geophys. Res.*, **103**, 10513–10526.

Park, Y.-H., and Gambéroni, L. (1995). Large-scale circulation and its variability ion the south Indian Ocean from TOPEX/POSEIDPON altimetry. *J. Geophys. Res.*, **100**, 24911–24929.

Parke, M. E., Stewart, R. H., Farless, D. L., and Cartwright, D. E. (1987). On the choice of orbits for an altimetric satellite to study ocean circulation and tides. *J. Geophys. Res.*, **92**, 11693–11707.

Parrilla, G., Lavin, A., Bryden, H., Garcia, M., and Millard, R. (1994). Rising temperatures in the subtropical North Atlantic Ocean over the past 35 years. *Nature*, **369**, 48–51.

Pattullo, J., Munk, W., Revelle, R., and Strong, E. (1995). The seasonal oscillation in sea-level. *J. Mar. Res.*, **14**, 88–155.

Pedlosky, J. (1987). *In* "Geophysical Fluid Dynamics," 2nd ed., Springer-Verlag, New York.

Perigaud, C., and Delecluse, P. (1992). Annual sea level variations in the southern tropical Indian Ocean from Geosat and shallow water simulations. *J. Geophys. Res.*, **97**, 20169–20178.

Peterson, R. G., and White, W. B. (1998). Slow oceanic teleconnections linking the Antarctic circumpolar wave with the tropical El Niño southern oscillation. *J. Geophys. Res.*, **103**, 24573–24583.

Philander, S. G. H. (1979). Equatorial waves in the presence of the equatorial undercurrent. *J. Phys. Oceanogr.*, **9**, 254–262.

Polito, P. S., and Cornillon, P. (1997). Long baroclinic Rossby waves detected by TOPEX/POSEIDON. *J. Geophys. Res.*, **102**, 3215–3235.

Ponte, R. M. (1994). Understanding the relation between wind- and pressure-driven sea-level variability. *J. Geophys. Res.*, **99**, 8033–8039.

Ponte, R. M., and Gaspar, P. (1999). Regional analysis of the inverted barometer effect over the global ocean using TOPEX/POSEIDON data and model results. *J. Geophys. Res.*, **104**, 15587–15602.

Potemra, J. T., and Lukas, R. (1999). Seasoanl to interannual modes of sea-level variability in the western Pacific and eastern Indain Oceans, *Geophys. Res. Lett.*, **26**, 365–368.

Price, J. F., and Rossby, H. T. (1982). Observations of a barotropic planetary waves in the western North Atlantic. *J. Mar. Res.*, **40 (Suppl)**, 543–558.

Price, J. M., and Magaard, L. (1980). Rossby wave analysis of the baroclinic potential energy in the upper 500 meters of the North Pacific. *J. Mar. Res.*, **38**, 249–264.

Price, J. M., and Magaard, L. (1983). Rossby wave analysis of subsurface temperature fluctuations along the Honolulu-San Francisco great circle. *J. Phys. Oceanogr.*, **13**, 258–268.

Price, J. M., and Magaard, L. (1986). Interannual baroclinic Rossby waves in the midlatitude North Atlantic. *J. Phys. Oceanogr.*, **16**, 2061–2070.

Qiu, B. (1995). Variability and energetics of the Kuroshio extension and its recirculation gyre from the first two-year TOPEX data. *J. Phys. Oceanogr.*, **25**, 1827–1842.

Qiu, B., and Kelly, K.A. (1993). Upper-ocean heat balance in the Kuroshio Extension region. *J. Phys. Oceanogr.*, **23**, 2027–2041.

Qiu, B., Miao, W., and Müller, P. (1997). Propagation and decay of forced and free baroclinic Rossby waves in off-equatorial oceans. *J. Phys. Oceanogr.*, **27**, 2405–2417.

Rapp, R. H., Zhang, C., and Yi, Y. (1996). Analysis of dynamic ocean topography using TOPEX data and orthonormal functions. *J. Geophys. Res.*, **101**, 22583–22598.

Reason, C. J. C., Mysak, L. A., and Cummins, P. F. (1987). Generation of annual-period Rossby waves in the South Atlantic Ocean by the wind stress curl. *J. Phys. Oceanogr.*, **17**, 2030–2042.

Reverdin, G., Frankingnoul, C., Kestenare, E., and McPhanden, M. J. (1994). Seasonal variability in the surface currents of the equatorial Pacific. *J. Geophys. Res.*, **99**, 20323–20344.

Richardson, P. L., and Reverdin, G. (1987). Seasonal cycle of velocity in the Atlantic North Equatorial Countercurrent as measured by surface drifters, current meters and ship data. *J. Geophys. Res.*, **92**, 3691–3708.

Roden, G. I. (1977). On long-wave disturbances of dynamic height in the North Pacific. *J. Phys. Oceanogr.*, **7**, 41–49.

Rossby, C. G. (1940). Planetary flow patterns in the atmosphere. *Q. J. Roy. Meteor. Soc.*, **66**, 68–87.

Rossby, C. G., and Collaborators (1939). Relations between variations in the intensity of the zonal circulation of the atmosphere and the displacements of the semi-permanent centers of action. *J. Mar. Res.*, **2**, 38–55.

Sato, O. T., Polito, P. S., and Liu, W. T. (2000). Importance of salinity measurements in the heat storage estimation from TOPEX/POSEIDON. *Geophys. Res. Lett.*, **27**, 549–551.

Schlax, M. G., and Chelton, D. B. (1994a). Detecting aliased tidal errors in altimeter height measurements. *J. Geophys. Res.*, **99**, 12603–12612.

Schlax, M. G., and Chelton, D. B. (1994b). Aliased tidal errors in TOPEX/POSEIDON sea surface height data. *J. Geophys. Res.*, **99**, 24761–24775.

Schlax, M. G., and Chelton, D. B. (1996). Correction to "Aliased tidal errors in TOPEX/POSEIDON sea surface height data." *J. Geophys. Res.*, **101**, 18451.

Shriver, J. F., Johnson, M. A., and O'Brien, J. J. (1991). Analysis of remotely forced oceanic Rossby waves off California. *J. Geophys. Res.*, **96**, 749–757.

Semtner, A. J., Jr., and Chervin, R. M. (1992). Ocean general circulation from a global eddy resolving model. *J. Geophys. Res.*, **97**, 5493–5550.

Siefridt, L. (1994). Validation des donne'es de vent ERS-1 et des flux de surface du CEPMMT dans le context de la mede'lisation des circulations oce'aniques 'a l'e'chelle d'un bassin, Ph.D. thesis, Univ. Joseph Fourier, Grenoble, France.

Smith, R. D., Dukowicz, J. K., and Malone, R. C. (1992). Parallel ocean general circulation modeling. *Physica D*, **60**, 38–61.

Stammer, D. (1997). Steric and wind-induced changes in TOPEX/POSEIDON large-scale sea surface topography observations. *J. Geophys. Res.*, **102**, 20987–21009.

Stammer, D., Tokmakian, R., Semtner, A., and Wunsch, C. (1996). How well does a 1/4 degree global circulation model simulate large-scale oceanic observations? *J. Geophys. Res.*, **101**, 25779–25811.

Stammer, D., Wunsch, C., Giering, R., Zhang, Q., Marotzke, J., Marshall, J., and Hill, C. (1997). The global ocean circulation estimated from TOPEX/POSEIDON altimetry and the MIT general circulation model, Rep. N. 49, pp. 40. MIT, Cambridge, MA.

Stammer, D., Wunsch, C., and Ponte, R. M. (2000). De-aliasing of global high frequency barotropic motions in altimeter observations. *Geophys. Res. Lett.*, **27**, 1175–1178.

Strub, P. T., and James, C. (2000). Altimeter-derived surface circulation in the large-scale Northeast Pacific gyres: Part 2. 1997–1998 El Niño anomalies. *Prog. Oceanogr.*, submitted.

Taft, B. A., and Kessler, W. S. (1991). Variations of zonal currents in the central tropical Pacific during 1970 to 1987. *J. Geophys. Res.*, **96**, 12599–12618.

Tai, C.-K. (1988). Estimating the basin-scale circulation from satellite altimetry. Part1: straightforward spherical harmonic expansion. *J. Phys. Oceanogr.*, **18**, 1398–1413.

Tai, C.-K., and Wunsch, C. (1983). Absolute measurement by satellite altimetry of the dynamic topography of the Pacific Ocean. *Nature*, **301**, (5899), 408–410.

Talley, L. D. (1988). Potential vorticity distribution in the North Pacific. *J. Phys. Oceanogr.*, **18**, 89–106.

Tapley, B. D., Nerem, R. S., Shum, C. K., Ries, J. C., and Yuan, D. N. (1988). Determination of the general ocean circulation from a joint gravity field solution. *Geophys. Res. Lett.*, **15**, 1109–1112.

Tapley, B. D., Watkins, M. M., Ries, J. C., Davis, G. W., Eanes, R. J., Poole, S. R., Rim, H. J., Schutz, B. E., Shum, C. K., Nerem, R. S., Lerch, F. J., Marshall, J. A., Klosko, S. M., Pavlis, N. K., and Williamson, R. G. (1996). The joint gravity model 3. *J. Geophys. Res.*, **101**, 28029–28049.

The MODE Group (1978). The mid-ocean dynamics experiment. *Deep-Sea Res.*, **25**, 859–910.

Tierney, C., Wahr, J., Bryan, F., and Zlotnicki, V. (2000). Short-period oceanic circulation: implications for satellite altimetry. *Geophys. Res. Lett.*, **27**, 1255–1258.

Tokmakian, R. T., and Challenor, P. G. (1993). Observations in the Canary Basin and the Azores frontal region using Geosat data. *J. Geophys. Res.*, **98**, 4761–4773.

Van Woert, M. L., and Price, J. M. (1993). Geosat and Advanced Very High Resolution Radiometer observations of oceanic Rossby waves adjacent to the Hawaiian Islands. *J. Geophys. Res.*, **98**, 14619–14631.

Visser, P. N. A. M., Wakker, K. F., and Ambrosius, B. A. C. (1993). Dynamic sea surface topography from GEOSAT altimetry. *Mar. Geod.*, **16**, 215–239.

Vivier, F., Kelly, K. A., and Thompson, L. (1999). Contributions of wind forcing, waves, and surface heating to sea surface height observations in the Pacific Ocean. *J. Geophys. Res.*, **104**, 20767–20788.

Wahr, J., Molenaar, M., and Bryan, F. (1998). Time variability of the Earth's gravity field: hydrological and oceanic effects and their possible detection using GRACE. *J. Geophys. Res.*, **103**, 30205–30229.

Wang, L., and Koblinsky, C. J. (1995). Low-frequency variability in regions of the Kuroshio Extension and the Gulf Stream. *J. Geophys. Res.*, **100**, 18313–18331.

Wang, L., and Koblinsky, C. J. (1996a). Annual variability of the subtropical recirculations in the North Atlantic and North Pacific: a TOPEX/Poseidon study. *J. Phys. Oceanogr.*, **26**, 2462–2479.

Wang, L., and Koblinsky, C. J. (1996b). Low-frequency variability in the region of the Agulhas Retroflection. *J. Geophys. Res.*, **101**, 3597–3614.

Wang, L., and Koblinsky, C. (1997). Can the Topex/Poseidon altimetry data be used to estimate air-sea heat flux in the North atlantic? *Geophys. Res. Lett.*, **24**, 139–142.

Wang, L., Koblinsky, C., Howden, S., and Beckley, B. (1998). Large-scale Rossby wave in the mid-latitude South Pacific from altimetry data. *Geophys. Res. Lett.*, **25**, 179–182.

White, W. B. (1977). Annual forcing of baroclinic long waves in the tropical North Pacific Ocean. *J. Phys. Oceanogr.*, **7**, 50–61.

White, W. B. (1982). Traveling wave-like mesoscale perturbations in the North Pacific Current. *J. Phys. Oceanogr.*, **12**, 231–243.

White, W. B. (1983). Westward propagation of short-term climatic anomalies in the western North Pacific Ocean from 1964–1974. *J. Mar. Res.*, **41**, 113–125.

White, W. B. (1985). The resonant response of interannual baroclinic Rossby waves to wind forcing in the eastern midlatitude North Pacific. *J. Phys. Oceanogr.*, **15**, 403–415.

White, W. B., and Bernstein, R. L. (1979). Design of an oceanographic network in the mid-latitude North Pacific. *J. Phys. Oceangr.*, **9**, 592–606.

White, W. B., and Saur, J. F. T. (1981). A source of annual baroclinic waves in the eastern subtropical North Pacific. *J. Phys. Oceanogr.*, **11**, 1452–1462.

White, W. B., and Saur, J. F. T. (1983). Sources of interannual baroclinic waves in the eastern subtropical North Pacific. *J. Phys. Oceanogr.*, **13**, 531–543.

White, W. B., and Tabata, S. (1987). Interannual westward-propagating baroclinic long-wave activity on Line P in the eastern midlatitude North Pacific. *J. Phys. Oceanogr.*, **17**, 385–396.

White, W. B., and Tai, C.-K. (1995). Inferring interannual changes in global upper ocean heat storage from TOPEX altimetry. *J. Geophys. Res.*, **100**, 24943–24954.

White, W. B., Meyers, G. A., Donguy, J. R., and Pazan, S. E. (1985). Short-term climatic variability in the thermal structure of the Pacific Ocean during 1979-82. *J. Phys. Oceanogr.*, **15**, 917–935.

White, W. B., Tai, C.-T., and DiMento, J. (1990). Annual Rossby wave characteristics in the California current region from the Geosat exact repeat mission. *J. Phys. Oceanogr.*, **20**, 1297–1311.

White, W. B., Chao, Y., and Tai, C.-K. (1998). Coupling of biennial oceanic Rossby waves with the overlying atmosphere in the Pacific basin. *J. Phys. Oceanogr.*, **28**, 1236–1251.

Williams, R. G., and Pennington, M. (1999). Combining altimetry with a thermocline model to examine the transport of the North Atlantic, *J. Geophys. Res.*, **104**, 18269–18280.

Willebrand, J., Philander, S. G. H., and Pacanowski, R. C. (1980). The oceanic response to large-scale atmospheric disturbances, *J. Phys. Oceanogr.*, **10**, 411–429.

Wilson, S. (2000). Launching the Argo armada: taking the ocean's pulse with 3000 free-ranging floats, *Oceanus*, **42**, 17–19.

Witter, D. L., and Gordon, A. L. (1999). Interannual variability of South Atlantic circulation from four years of TOPEX/POSEIDON satellite altimeter observations. *J. Geophys. Res.*, **104**, 20927–20948.

Woodbury, K., Luther, M., and O'Brien, J. J. (1989). The wind-driven seasonal circulation in the southern tropical Indian Ocean. *J. Geophys. Res.*, **94**, 17985–18002.

Wunsch, C. (1972). Bermuda sea level in relation to tides, weather, and baroclinic fluctuations. *Rev. Geophys. Space Phys.*, **10**, 1–49.

Wunsch, C. (1981). An interim relative sea surface for the North Atlantic Ocean. *Mar. Geod.*, **5**, 103–119.

Wunsch, C. (1996). *In* "The Ocean Circulation Inverse Problem," Cambridge University Press, Cambridge.

Wunsch, C., and Gaposchkin, E. M. (1980). On using satellite altimetry to determine the general circulation of the oceans with application to geoid improvement. *Rev. Geophys. Space Phys.*, **18**, 725–745.

Wunsch, C., and Stammer, D. (1997). Atmospheric loading and the oceanic "inverted barometer" effect. *Rev. Geophys.*, **35**, 79–107.

Wunsch, C., and Stammer, D. (1998). Satellite altimetry, the marine geoid, and the oceanic general circulation. *Annu. Rev. Earth Planet. Sci.*, **26**, 219–253.

Wyrtki, K. (1974a). Sea level and the seasonal fluctuations of the equatorial currents in the western Pacific Ocean. *J. Phys. Oceanogr.*, **4**, 91–103.

Wyrtki, K. (1974b). Equatorial currents in the Pacific 1950 to 1970 and their relations to the trade winds. *J. Phys. Oceanogr.*, **4**, 372–380.

Wyrtki, K. (1985). Water displacements in the Pacific and the genesis of El Niño. *J. Geophys. Res.*, **90**, 7129–7132.

Zang, X., and Wunsch, C. (1999). The observed dispersion relationship for North Pacific Rossby wave motions. *J. Phys. Oceanogr.*, **29**, 2183–2190.

Zhang, Y., Wallace, J. M., and Battisti, D. S. (1997). ENSO-like interdecadal variability: 1900–1993. *J. Climate*, **10**, 1004–1020.

Zheng, Q., Yan, X.-H., Ho, C.-R., and Tai, C.-K. (1994). The effects of shear flow on propagation of Rossby waves in the equatorial oceans. *J. Phys. Oceanogr.*, **24**, 1680–1686.

3

Ocean Currents and Eddies

P. Y. LE TRAON[*] and R. MORROW[†]

[*]*CLS Space Oceanography Division*
8–10 rue Hermes
Parc Technologique du Canal
31526 Ramonville St Agne, France
[†]*LEGOS, 14 avenue Edouard Belin*
31055 Toulouse Cedex, France

1. INTRODUCTION

The large-scale ocean circulation can be described by large-scale gyres, with slow, diffuse equatorward currents at the eastern boundary and fast, intense, poleward flows at the western boundary—the so-called western boundary currents. In the zonally unbounded Southern Ocean, the Antarctic Circumpolar Current (ACC) flows continuously around the globe and provides a major link for water-property exchanges between the Atlantic, Indian, and Pacific oceans. Time scales of this large-scale horizontal circulation are typically of a few years. Superimposed on (but also interacting with) this mainly wind-driven circulation is the thermohaline vertical circulation, which allows the exchange of heat between the equatorial and high-latitude regions. Time scales of this vertical circulation (the famous, but oversimplified, conveyor belt representation) are typically of one to several hundred years. Characterizing the ocean circulation through theories of large-scale wind-driven and thermohaline ocean circulation (see Pedlosky, 1996, for a review), while often illuminating, is, however, only a crude and first-order approximation of the real ocean. The ocean is indeed a turbulent system and the ocean circulation is not a large-scale and temporally stable phenomenon; it varies over an almost continuous frequency/wavenumber spectrum, with space and time scales ranging from tens to thousands of kilometers and from days to years (e.g., Wunsch, 1981). An instantaneous view of the ocean circulation would thus reveal areas of in-

tense and small-scale ocean currents almost everywhere and would be dominated by mesoscale variability. In particular, western boundary currents and the ACC are areas of intense mesoscale variability. Open ocean currents, which are part of the large scale gyre circulation, are also often intense and narrow currents embedded with mesoscale eddies. At the eastern boundaries, superimposed on the broad equatorward flow are energetic currents and coastal upwelling currents, which can be highly variable in space and time.

The mesoscale variability is the dominant signal in the ocean circulation. There is not a precise definition of mesoscale variability, but it usually refers to a subclass of energetic motions with typical space and time scales of 50–500 km and 10–100 days. Phenomenological representations of this variability include eddies, vortices, fronts, narrow jets, meanders, rings, filaments, and waves.

Over the past 30 years, a major effort has been made to observe and model eddy variability. Our understanding has progressed greatly, and the eddy field has been visualized qualitatively (e.g., Holland and Lin, 1975; Robinson, 1982). Ocean eddies can be observed almost everywhere at mid and high latitudes, their energy generally exceeding the energy of the mean flow by an order of magnitude or more (e.g., Wyrtki *et al.*, 1976; Richardson, 1983; Schmitz and Luyten, 1991). As in the atmosphere, baroclinic instability processes are thought to be the principal source of eddy energy in the global oceans. There is a distinct concentration of eddy kinetic energy (EKE) along the mean frontal zones in the global oceans, which suggests that variability is primarily generated by instabilities of the mean flow. The baroclinic instability process depends on the presence of vertical shear in the mean flow, implying sloping density gradients, and an available store of potential energy. Under certain conditions, instabilities form, releasing this potential energy and converting it to eddy potential and kinetic energies. If the mean flow has significant horizontal shear as in strong narrow jets, barotropic instabilities can also occur. In this case, there is a release of mean kinetic energy to EKE. In the real ocean, the mean flow has both vertical and horizontal shear, so potentially both baroclinic and barotropic (mixed) instabilities can coexist. Through these instability-forcing mechanisms, eddy activity is thus maximum in regions of major oceanic currents. This relationship between ocean currents and mesoscale variability explains, in particular, why the chapter includes a discussion of both ocean currents and eddies.

Mesoscale variability occurs everywhere in the ocean, not just in regions of strong currents. In regions of weak mesoscale activity, mesoscale variability may also be directly forced by fluctuating wind (Frankignoul and Müller, 1979; Müller and Frankignoul, 1981). There are also several examples of eddies that are directly forced by strong small-scale wind stress curl (e.g., caused by orographic effects). The interaction of currents with bottom topography is also an important (local) mechanism for eddy energy generation. Eddies downstream of small islands are thus often observed (e.g., Aristegui *et al.*, 1994; Heywood *et al.*, 1996). The highest eddy energy is actually also often found adjacent to major topographic obstacles, which suggests the importance of topography in the stability of the mean flow. From these source regions, eddy energy can be redistributed throughout the oceanic gyre by radiation of Rossby waves (e.g., Pedlosky, 1977; Talley, 1983).

Mesoscale eddies, or more exactly vortices (i.e., characterized by a persistent closed circulation), have their own dynamics which are dominated by nonlinear effects (Nof, 1981; McWilliams and Flierl, 1979; McWilliams, 1985; Cushman-Roisin *et al.*, 1990). In contrast to waves, they may transport mass (closed streamlines in the moving fluid) and properties such as heat, salt, and chemical tracers including nutrients over long distances. They also have different propagation velocities. McWilliams and Flierl (1979), using a numerical quasigesotrophic model, showed, for example, that a vortex moves westward at the greatest linear Rossby wave speed. Nof (1981) and Cushman-Roisin *et al.* (1990) have derived more general formulas for westward eddy motion. Anticyclones also have a tendency for southward propagation and cyclones for northward propagation (e.g., Cushman-Roisin, 1994). The relationship between vortices and waves is intricate, and the distinction between eddy and wave signals may be difficult. Whether the mid-latitude westward propagation observed in altimeter data (see Chapter 2) is due

to Rossby waves or to eddies is thus still an open question. Eddies lose energy to smaller scales by dissipation, but they also interact, merge, and grow in the framework of quasi-geostrophic turbulence (e.g., McWilliams, 1989). They can also feed energy back to the mean flow and drive deep circulation (Holland *et al.*, 1982; Lozier, 1997). Their climate role in terms of heat and salt transport is, however, not yet well established. Eddies and, more generally, mesoscale variability thus form a crucial component of the ocean circulation dynamics, including the large-scale and time-mean ocean circulation.

Despite all this progress, our understanding of eddy and ocean current dynamics remains incomplete. We still need to quantify the energy sources and sinks of eddies and the eddy energy radiation mechanisms and to better rationalize their horizontal and vertical structures. Their contribution to the total heat transport and their interaction with the general circulation should also be better established: What part of the ocean circulation is eddy-driven? What is the role of eddies in the ventilation of the thermocline? Can we parametrize their effect on the general circulation? Similarly, we need to better understand the ocean current dynamics, their seasonal/interannual variations, and their forcing and instability mechanisms. A better understanding of these mesoscale phenomena requires an extensive, global observing system. In this respect, satellite altimetry has unique capabilities for producing a global and synoptic view of the ocean surface, even in remote areas. Spatial sampling along the satellite ground track is well suited to mesoscale studies, while a temporal sampling period of 10–20 days should avoid aliasing of most of the mesoscale signal. This description is, of course, limited to the surface oceanic circulation. However, given the high vertical coherence of mesoscale variability, this information reflects more than just surface conditions. For a stratified rotating fluid, simple scaling arguments show that the ratio of horizontal scale/vertical scale should be on the order of N/f, where N is the Brunt Väisälä frequency and f the Coriolis parameter (e.g., Wunsch, 1982). The influence of a mid-latitude eddy of 100-km radius should be seen to a depth of several hundred meters. Altimeter data, when assimilated in ocean models, can thus be a strong constraint for inferring the three-dimensional ocean mesoscale circulation (see Chapter 5). Complementary information on the vertical structure of the mesoscale variability (e.g., barotropic and baroclinic components) needs to be obtained, however, from *in situ* data.

The mesoscale variability in high-variability regions was one of the first ocean signals to be observed with Skylab, GEOS-3, and Seasat altimetric missions (Fu, 1983a). Seasat, in particular, despite its short duration, provided the first global description of mesoscale variability (Fu, 1983a). The U.S. Navy's Geosat altimeter opened new horizons for the quantitative use of satellite altimetry. Geosat was particularly suitable for mesoscale observations because of its long duration—the satellite operated on a near-repeat orbit (17.05-day cycle) for almost 3 years (November 1986 through June 1989)—and its lower noise level (Douglas and Cheney, 1990; Le Traon, 1992). TOPEX/POSEIDON (hereafter T/P) and ERS-1/2 now provide an improved description of ocean currents and mesoscale variability with better accuracy and an improved space/time sampling (e.g., Fu and Cheney, 1995). Up to now, however, this enhanced capability has not been fully exploited mainly because most T/P data analysis has focused on the large-scale signals (see Chapters 2 and 4). We now have more than 10 years of good mesoscale variability measurements from Geosat (1986–1989), ERS-1 and ERS-2 (1991–2000+), and T/P (1992–2000+), so we can learn more about the seasonal/interannual variations in mesoscale variability intensity.

Most of the discussion in this chapter centers on results obtained from Geosat, T/P, and ERS-1/2 data and shows how satellite altimetry contributes to the description of ocean currents and mesoscale variability. The chapter is organized as follows. Section 2 discusses the specific issues for altimeter data processing for mesoscale studies. Section 3 contains the description of ocean currents. Section 4 is focused on a global description of mesoscale eddies and of their dynamics. Section 5 provides the main conclusions and prospects for the future.

2. ALTIMETER DATA PROCESSING FOR MESOSCALE STUDIES

We proceed here with a brief description of altimeter data processing issues, focusing on the aspects that are relevant to mesoscale studies. These include signal extraction and measurement errors, mapping and merging, velocity estimation, and sampling issues.

2.1. Ocean Signal Extraction

The altimetric observation of the sea surface topography S can be described by

$$S = N + \eta + \varepsilon, \tag{1}$$

where N is the geoid, η the dynamic topography, and ε the measurement errors. Present geoids are not generally accurate enough to estimate globally the absolute dynamic topography η except at very long wavelengths. The variable part of the dynamic topography η' ($\eta' = \eta - \langle \eta \rangle$) or sea-level anomaly (hereafter SLA) is, however, easily extracted since the geoid is stationary on the time scale of an altimetric mission. The most commonly used method is the repeat-track method (collinear analysis). This method is suitable for satellites whose orbits repeat their ground tracks (to within ±1 km) at regular intervals. For a given track, the variable part of the signal is obtained by removing a mean profile (e.g., over the mission duration), which contains the geoid and the quasi-permanent dynamic topography from each profile.

Most mesoscale studies have used the repeat-track method to extract the SLA. The error variance induced by removing the mean is typically $\langle \eta'^2 \rangle T_I / T$, where $\langle \eta'^2 \rangle$ is SLA variance, T is duration of observation (e.g., 1 year), and T_I is decorrelation time of SLA. Assuming a decorrelation time T_I of 20 days, this amounts to a relative quadratic error between 5% (T/P) and 10% (ERS-1/2) for a 1-year mean. Note also that the mean profile calculation procedure is complicated by the fact that altimeter data are generally gappy. This means that only partial profiles are used to estimate the mean profile to avoid a contamination in the mean sea level from the orbit errors. However, there are several ways to minimize this problem (e.g., Chelton et al., 1990). The sampling rate also induces aliasing of frequencies higher than the Nyquist frequency [$(20 \text{ days})^{-1}$ for T/P, $(70 \text{ days})^{-1}$ for ERS-1/2, and $(34 \text{ days})^{-1}$ for Geosat].

Measurement errors ε can significantly affect the estimation of altimeter SLA (see Chapter 1). Most of these errors are of large-scale although there are probably mesoscale components which are not well quantified. Spectral analysis of known error sources thus shows that for wavelength shorter than 1000 km (mesoscale range), the oceanic signal dominates the errors (e.g., Fu, 1983b; Le Traon et al., 1990). The only exception concerns the short wavelengths (below 100 km) where altimeter instrumental noise and small-scale ageostrophic variability (e.g., internal waves) start to dominate the altimeter SLA spectrum. Cross-track geoid errors may also affect the small wavelengths, but the effect is small and limited to very specific areas (e.g., fracture zones) (e.g., Minster et al., 1993).

The altimeter data processing for mesoscale studies should thus include a correction for long-wavelength errors and a reduction of instrumental noise. An approach commonly used approach to correct the long-wavelength errors is to approximate them by a first- or second-degree polynomial over a given arc length. Although the polynomial adjustment method induces nonnegligible errors in the mesoscale signal (Le Traon et al., 1991), it is still necessary to use it for satellites or regions where the large-scale errors are large (e.g., orbit error) and/or are not well-known. Alternatively, to minimize the removal of the oceanic signal, one can use more complex methods such as global crossover minimization or inverse techniques. The instrumental and small-scale noise can be reduced using appropriate low-pass filtering of along-track data (see also discussion in Section 2.3).

2.2. Mapping and Merging of Multiple Altimeter Missions

For mesoscale studies, there are two predominant approaches. The first approach uses along-track SLA data, which provide a very good one-dimensional spatial sampling. Along-track altimeter data have been used extensively for computing mesoscale statistics. The second approach uses maps derived from one or preferably several altimeters. Maps can provide a synoptic view of mesoscale eddies allowing a much better visualization and understanding of eddy dynamics. However, particular processing and sampling issues must be considered when dealing with maps. Sampling issues will be detailed in Section 2.4, and we will focus here on the processing issues.

The mapping can be performed with optimal interpolation methods that use a priori knowledge of the space and time scales of the ocean signal. When data are mapped onto a regular space/time grid using such interpolation methods, the along-track long-wavelength errors (or high frequency ocean signals) can induce artificial cross-track gradient at smaller scales and thus spurious eddy signals (Le Traon et al., 1998). The effect is particularly important in low-eddy-energy regions and when several altimeter data sets are merged. Along-track long-wavelength errors can also lead to other serious problems. A well-known example comes from Geosat for which errors in the tide correction were aliased to produce spurious Rossby wave-like signals in the mapped data (Chelton and Schlax, 1994). To minimize these problems, either the mapping method should take into account an along-track long-wavelength error (i.e., a correlated noise caused by orbit, tidal, or inverse barometer residual errors) or the long-wavelength errors should be removed before the mapping (Le Traon et al., 1998).

The merging of multisatellite altimeter data sets is generally necessary to map the mesoscale variability. This is not an easy task. To merge multisatellite altimetric missions, it is first necessary to have homogeneous and intercalibrated data sets. Homogeneous means that the same geopotential model and reference systems for the orbit and the same (as far as possible) instrumental and geophysical corrections should be used (e.g., same tidal models, same meteorological models). Intercalibrated means that relative biases and drifts must be corrected and also that the orbit error must be reduced. An effective methodology is to use the most precise mission (T/P, Jason-1) as a reference for the other satellites (Le Traon and Ogor, 1998). When altimetric data have been homogenized and intercalibrated, the next step is to extract the SLA for the different missions. It is preferable that the SLAs from different missions be calculated relative to the same ocean mean using a common reference surface (e.g., either a very precise mean sea surface or mean profiles consistent between the different missions). The final step is to merge the SLAs via a mapping or assimilation technique.

Ducet et al. (2000) provide a detailed analysis of the merging of T/P and ERS-1/2 over a 5-year period, in particular to quantify the contribution of merging for the description of the ocean mesoscale circulation.

2.3. Surface Geostrophic Velocity Calculations

Away from the equator, sea-surface velocities can be calculated from the along-track sea-surface slope using the geostrophic relation, i.e.,

$$fu = -g\partial\eta/\partial y, \qquad (2)$$

where u is the surface velocity normal to the track (positive eastward), g is gravitational acceleration, f is the Coriolis parameter, y is the along-track direction, and η is the SLA. That is, the ocean geostrophic currents are associated with a sea-level slope across the current. The sea-surface slope can be 1 m across the Gulf Stream, for example, and as such it is easily detected by altimetry.

Note that this calculation gives only the cross-track component of the residual flow. In many altimetric studies, we assume that the flow is isotropic so that the variance of the cross-track component is taken as representative of the variance of the total surface velocity. This isotropic approximation is valid for a turbulent field, but fluctuations become anisotropic in the presence of mean shear, close to bathymetry, or at larger scales and in the equatorial regions (β effect) where the zonal fluctuations tend to dominate meridional fluctuations. Another point is that the direction of the cross-track component changes with the curvature of the ground track: the component measures more zonal variability near the equator while the part caused by meridional flow becomes equal or larger at high latitudes.

The geostrophic relation is a good first-order approximation for estimating the flow field, and it is used extensively with hydrographic data, as well as altimetry. However, approximately 10% of the flow field is not in geostrophic balance, including the frictional surface wind-driven Ekman transports which flow across isobars. Ageostrophic flow also occurs in frontal zones and in regions of meander formation and is associated with cyclostrophic effects in small eddies.

As the calculation of the derivative acts as a high-pass filter, cross-track geostrophic velocities are much less sensitive to large-scale signals and are thus more representative of the mesoscale signals. The computation of geostrophic velocity is sensitive to measurement noise and small-scale ageostrophic ocean signals and needs a careful filtering of altimeter data (e.g., Zlotnicki et al., 1993; Morrow et al., 1994; Strub et al., 1997). Previous estimates obtained with Geosat data used filtering over at least 100 km to significantly reduce the contamination of noise. Even with the more accurate T/P data, the measurement noise influences the smaller spatial scales out to 50–70 km. As an illustration, Figure 1

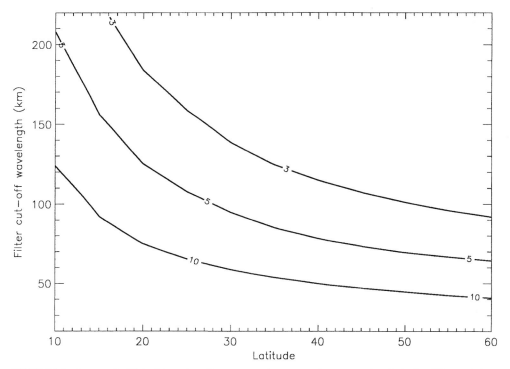

FIGURE 1 Impact of a SLA white noise of 3-cm rms on the error on the cross-track velocity for different choices of low-pass filtering cutoff wavelength and for different latitude bands. Velocities are calculated from slopes of filtered SLA over 14 km. Units are cm sec^{-1}.

gives the estimation of the impact of a white noise of 3-cm rms on the error on the cross-track velocity for different choices of Lanczos low-pass filtering cutoff wavelengths of SLA and for different latitude bands. This shows, for example, that in order to have velocity errors below 5 cm/s rms, it is necessary to filter the SLA wavelengths shorter than 200 km at 10°N and 80 km at 40°N. In altimetric studies, a satisfactory trade-off between noise reduction and ocean signal removal needs to be found; otherwise, the filtered SLA data may miss a significant fraction of the eddy energy. This holds, in particular, at high latitudes, where eddies can have spatial correlation scales below 50 km. An analysis of the effect of different filter cutoff wavelengths on the estimation of cross-track velocity variance can be found in Stammer (1997).

2.3.1. Estimating Orthogonal Velocity Components at Crossover Points

The magnitude and direction of the residual velocity can be resolved at crossover points by a simple geometric transformation. There, we have available two (generally nonorthogonal) components of geostrophic velocity from each of the ascending and descending passes $(V_{a'}, V_{d'})$ and a known angle (ϕ) between the ground-track and the north meridian, which varies as a function of latitude only. The residual velocity can be calculated in any orthogonal projection, e.g., in east/north components (u', v') following Parke

et al. (1987):

$$u' = \frac{V_{a'} + V_{d'}}{2\cos\phi}, \qquad (3)$$

$$v' = \frac{V_{a'} - V_{d'}}{2\sin\phi}. \qquad (4)$$

The errors associated with the geometric transformation in Eqs. (3) and (4) can also be determined. If the residual sea-surface slope measurements from the ascending and descending passes have a variance error of $\sigma_{s'}^2$, then using the law of propagation of variances and assuming $V_{a'}$ and $V_{d'}$ are independent, the variance of the errors in the u' and v' components will be

$$\sigma_{u'}^2 = \frac{1}{2}\left(\frac{g\sigma_{s'}}{f\cos\phi}\right)^2, \qquad (5)$$

$$\sigma_{v'}^2 = \frac{1}{2}\left(\frac{g\sigma_{s'}}{f\sin\phi}\right)^2. \qquad (6)$$

If there were no measurement errors in the corrected altimeter data and no time lag between the ascending and descending passes, the geometrical transformation of Eqs. (3) and (4) would resolve the orthogonal velocities perfectly. However, Eqs. (5) and (6) show that the error in each velocity component depends on the estimated measurement error, $\sigma_{s'}$, i.e., how well the geostrophic sea-surface slopes can be estimated from the filtered along-track residuals. The error in u' and v' will also vary with latitude, depending on three factors: the Coriolis parameter f, the crossover angle ϕ, and the time difference between the ascending and the descending passes (Morrow *et al.*, 1994). For example, at low latitudes as the crossover angle, ϕ, decreases, the geometrical transformation tends to map more of the error into the northward velocity component. Particular care needs to be taken in applying this technique in low-eddy-energy regions, where measurement errors can dominate the ocean signal.

2.3.2. Estimating Orthogonal Velocity Components from Mapped Data

The velocity anomaly field can also be derived from gridded fields of SLA or can be directly mapped from along-track SLA data (e.g., Le Traon and Dibarboure, 1999; Ducet *et al.*, 2000). Mapping has the advantage of combining the information from ascending, descending, and neighboring tracks. Even at crossover points, they can thus provide a slightly better estimation than the previous method because they take into account the sea-level gradients between the tracks. They also allow a more rigorous treatment of measurement errors and are actually the only means of extracting velocity from a combination of several altimeters. The drawback is that mapping errors are not homogeneous (see Section 2.4) and this should be taken into account in the interpretation. The combination of T/P and ERS-1/2 yields, however, rather homogeneous U and V mapping errors, generally below 25% of the signal variance (Le Traon and Dibarboure, 1999). An analysis of velocity statistics derived from T/P and ERS-1/2 combined maps can be found in Ducet *et al.* (2000).

2.3.3. Reynolds Stresses and Velocity Variance Ellipses

Velocity variance statistics can be calculated from the time series of orthogonal velocity components, at either each crossover point or each grid point of the mapped data. From these time series we can compute the Reynolds stress terms, which include the north/east velocity variance terms $(\overline{u'^2}, \overline{v'^2})$ and the covariance term $\overline{u'v'}$. The magnitude and the direction of the eddy variability can be represented using variance ellipses or axes (see Morrow *et al.*, 1994; after Preisendorfer, 1988). The direction, θ, of the axis of principal variability, measured anticlockwise from east, is

$$\tan\theta = \frac{\sigma_{11} - \overline{u'^2}}{\overline{u'v'}}, \qquad (7)$$

where the magnitude of the variance along the major axis is given by

$$\sigma_{11} = \frac{1}{2}\left(\overline{u'^2} + \overline{v'^2} + \sqrt{\left(\overline{u'^2} - \overline{v'^2}\right)^2 + 4\left(\overline{u'v'}\right)^2}\right) \quad (8)$$

and along the minor axis by

$$\sigma_{22} = \left(\overline{u'^2} + \overline{v'^2}\right) - \sigma_{11}. \qquad (9)$$

Anisotropic flow is represented by an elongated ellipse, with the principal direction of the velocity variance aligned with the direction of the major axis. The orientation of the ellipse depends on the covariance term: the major and minor ellipse axes define the coordinate system in which u' and v' are uncorrelated. Ellipses with a major axis oriented in the northeast quadrant have a positive $\overline{u'v'}$; ellipses oriented toward the southeast quadrant have a negative $\overline{u'v'}$. As such, the direction of horizontal eddy momentum flux can be inferred from the ellipse orientation. Isotropic flow is represented by circular ellipses and zero covariance.

2.4. Sampling Issues

The along-track sampling of altimeters is quite well-suited to resolve mesoscale eddies. A combination of several altimeter missions is required, however, for a 3D (x, y, t) mesoscale variability mapping. Le Traon and Dibarboure (1999) have quantified the mesoscale mapping capability when combining various existing or future altimeter missions in terms of SLA (Figure 2) and zonal (U) and meridional (V) velocities. Their main results, which only take into account the different sampling characteristics for each satellite, are as follows:

1. The Geosat (or Geosat Follow On) 17-day orbit provides the best sea-level and velocity mapping for the single-satellite case. The Jason-1 + T/P (interleaved T/P–Jason-1 tandem orbit scenario) provides the best mapping for the two-satellite case. There is only minor improvement, however, with respect to the T/P + ERS (or Jason-1 + ENVISAT) scenario.

2. There is a large improvement in sea-level mapping when two satellites are included. For example, compared to T/P alone, the combination of T/P and ERS has a mean mapping error reduced by a factor of 4 and a standard deviation reduced by a factor of 5. Compared to ERS, the reduction is smaller but still a factor of more than 2. The improvement in sea-level mapping is not as large when going from two to three or from three to four satellites.

3. The velocity field mapping is more demanding in terms of sampling. The U and V mean mapping errors are two to four times larger than the SLA mapping error. The contribution from a third satellite is also more significant than for SLA. Only a combination of three satellites can actually provide a velocity field mapping error below 10% of the signal variance. Mapping of the meridional velocity is less accurate but by only 10–20% even at low latitudes. This suggests that the criterion of a satellite orbit with a rather low inclination for a better estimation of the velocity

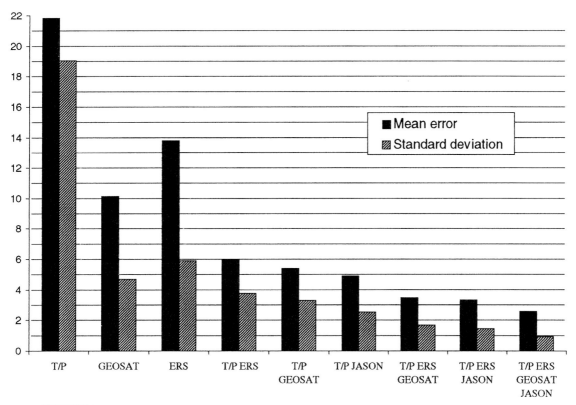

FIGURE 2   Mean and standard deviation of sea-level anomaly (SLA) mapping error for single and multiple altimeter missions. The calculation assumes a space scale of 150 km (zero crossing of the correlation function) and an e-folding time scale of 15 days. Units are percentages of signal variance. (From Le Traon, P. Y., and Dibarboure, G., 1999. With permission.)

field is not really relevant, in particular for a multisatellite configuration.

The main conclusions of that study are that at least two (and preferably three) missions are needed to map the mesoscale ocean circulation and that existing and future two-satellite configurations (T/P and ERS and later Jason-1 and ENVISAT) will provide a rather good mapping of mesoscale variability. Note that these conclusions differ from those of Greenslade et al. (1997), who analyzed the resolution capability of multiple altimeter missions and concluded that the mesoscale signal cannot be mapped with an acceptable accuracy. This can be explained by the definition of resolution chosen by Greenslade et al. (1997). They required a very homogeneous mapping error which cannot be achieved with two, three, or even four satellites (in particular, the mesoscale signal will be always better estimated along the tracks). A resolution definition is, however, a fundamentally subjective choice and, although the mesocale mapping errors are not homogeneous, Le Traon and Dibarboure (1999) argue that they remain sufficiently small relative to the signal (see Figure 2). All three satellite configurations they analyzed always have, for example, a mapping error below 10% of the signal variance.

3. OCEAN CURRENTS

In Chapter 2, the global, large-scale ocean circulation was described in detail. In this chapter, we will take another approach and look at each individual current system (Figure 3) and show how altimetry has improved our understanding of the dynamics of these ocean currents. Although all open ocean currents are essentially driven by external influences (principally wind stress and heating/cooling), their responses are quite different. Not only are there regional differences in the atmospheric forcing, but the circulation in each ocean basin can be strongly controlled by local bathymetry and interocean exchanges. Altimetry provides a regular sampling of all of these ocean currents, from the fast, energetic currents at the western boundaries and in the Antarctic Circumpolar region to the slow, more quiescent flow in the center and east of most ocean basins.

Ocean currents are also highly variable. Shallow surface currents can be associated with seasonal changes in the mixed layer. The sea-surface slope associated with these currents can also increase or decrease as the volume transport changes seasonally. The current's mean position also changes, influenced by meanders, by the formation and sep-

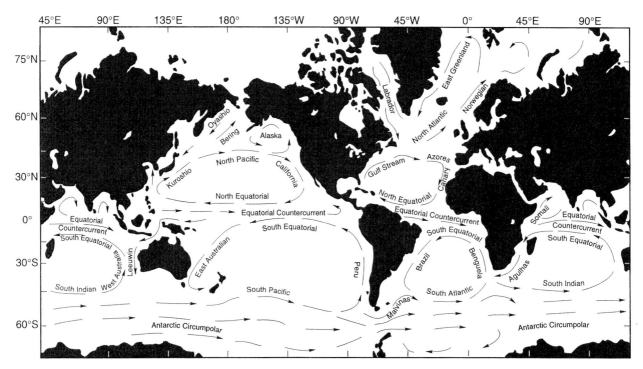

FIGURE 3   Schematic showing the major current systems described in this section.

aration of rings and eddies, and by the interaction with westward propagating waves and eddies. Altimetry allows us to monitor this variability, a task that is nearly impossible with *in situ* hydrographic and moored current meter data alone. However, to estimate absolute velocities requires an accurate estimate of the marine geoid, or an independent estimate of the mean ocean dynamic topography, and thus the mean current. Transport estimates also require a knowledge of the vertical structure or subsurface measurements.

In the following section, we review some of the main techniques used to estimate the absolute velocities and transports of the major current systems from altimeter data. We then look at each major current system separately and review the considerable progress made by altimetric studies in the main western boundary currents, eastern boundary currents, open ocean currents (such as the Circumpolar and Agulhas Currents), and the semi-enclosed seas. Studies on near-equatorial currents and eddies will be addressed separately in Chapter 4, Tropical Circulation.

3.1. Estimating the Absolute Velocities and Transports from Altimetry

Accessing the time-mean component of the flow is necessary for transport calculations, but a difficult task with altimetry given the large geoid errors at small spatial scales of $O(100–200$ km). One technique that has been developed to estimate the surface transport is to combine altimeter residual heights with a simple analytical model of the mean sur-

face velocity profile (Kelly and Gille, 1990; Kelly, 1991, Tai, 1990). The analytical surface velocity model they use has a Gaussian shape that describes an isolated jet; the shape is based on previous studies of the Gulf Stream with subsurface moorings and ADCP data. The amplitude, width, and position of the jet are allowed to vary as shown in Figure 4, and they are adjusted to fit the altimetric residual heights data, using a least-squares technique. Kelly and Gille (1990) analyzed a single pass and found maximum surface velocities between 1.2 and 2.0 m/sec. Surface transports were also estimated by integrating the velocities along each track from 34°N to 40°N, across the Gulf Stream axis. Comparisons with simultaneous ADCP velocity measurements were favorable (Joyce *et al.*, 1990), suggesting that the technique could provide a means for monitoring the surface Gulf Stream transport from altimetry alone.

Qiu *et al.* (1991) have refined the Kelly and Gille (1990) method. They estimate the two-dimensional mean surface height by combining synthetic profiles along ascending and descending tracks through an inverse method. Their method also accommodates possible modifications of the Gaussian shape such as recirculation gyres. This method was applied to 2.5 years of Geosat data in the Kuroshio Extension. The mean sea-surface height agrees well with the climatological mean obtained from hydrographic data. This type of technique worked well in the Gulf Stream and the Kuroshio Extension not only because the horizontal mean velocity profile is well known from *in situ* data and the mean velocity is strong so the signal-to-noise ratio is high, but also be-

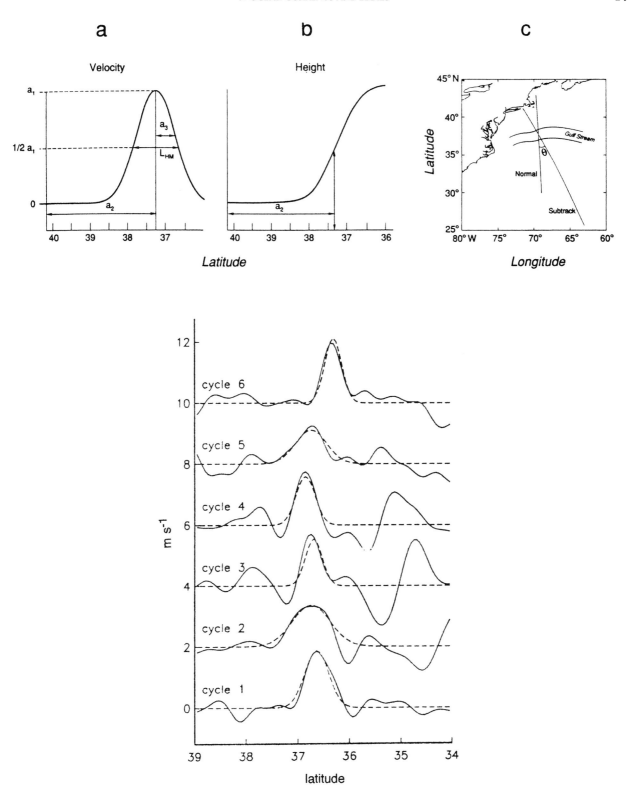

FIGURE 4 Sketch of the Gaussian jet model. (a) The Gaussian velocity profile is characterized by the maximum velocity a_1, the central position, a_2, and the jet width, a_3. These parameters are allowed to vary, and are adjusted to fit the altimetric residual heights data, using a least-squares technique. (b) Sea-surface height is obtained by integrating the velocity profile. (c) The angle between the groundtrack and a perpendicular to the Gulf Stream axis. The bottom panels show geostrophic velocity profiles from altimetry residuals and synthetic mean (solid) and the Gaussian model (dashed) in the Gulf Stream region for the first six Geosat cycles. (From Kelly, K. A., and Gille, S. T., 1990. With permission.)

cause the current existed as a single intense jet throughout the 2-year observation period. However, this method is difficult to apply in other regions where the mean jet is weak or too stable in time, bifurcates, or has seasonal reversals.

An alternative is to estimate the mean height profile using climatological hydrographic data. The advantage is that the mean profile is available anywhere in the ocean (with varying degrees of accuracy). A disadvantage is that the mean is often weak and the oceanic fronts quite diffuse, especially in sparsely sampled regions where a large degree of spatial smoothing is necessary. Another problem is the uncertainty in specifying the reference level. For the ACC and most western boundary currents, this problem can be serious because these currents have a significant barotropic component. Despite these problems, composite maps of altimetric anomalies with a hydrographic mean have been used to derive surface geostrophic velocities in the Gulf Stream Extension (Willebrand *et al.*, 1990) and Kuroshio (Ichikawa *et al.*, 1995). These mapped velocities were significantly correlated with available drifter velocities and independent hydrographic sections, although the mapped velocities were generally weaker than the drifter velocities. This is mainly because of the objective analysis mapping, which removed energy on scales less than 100 km. Another similar technique is to use a model mean dynamic topography.

Some investigators have also attempted to estimate the marine geoid itself as accurately as possible. In the Gulf Stream, Rapp and Wang (1994) and Rapp and Smith (1994) applied a gravimetric geoid model combined with simple models of the meandering jet to estimate absolute Gulf Stream parameters (locations, width, velocity, and height jump across the current). The gravimetric geoid undulations were calculated using a 360 potential coefficient model; land, ship, and altimeter-derived gravity anomalies; and bathymetric data. The spatial resolution of the potential model is of order 1° (100 km), which can be improved locally by the available *in situ* data. Hwang (1996) has also estimated the mean baroclinic transports from a combination of a marine geoid model, climatology, and altimetry in the upstream region of the Kuroshio. This application may underestimate the transport estimates since the geoid and hydrographic mean do not include the smaller spatial scales, and the calculation does not include the barotropic component of the flow. In both these examples, removing the geoid estimate does not really improve the transport calculations, since we are in regions where there are sufficient hydrographic data to estimate the mean flow, and the simple model of a meandering jet already works well. However, the geoid approach may work better in regions where there is little current meandering and less dense hydrographic data. We note though, that in sparse gravity data regions, the errors in the geoid at the mesoscale still tend to obscure the mean oceanic component. In the future, gravimetric missions such as CHAMP, and especially GRACE and GOCE, will provide a geoid improvement by one or two orders of magnitude. This should allow an estimation of a mean dynamic topography with accuracy of about 1–2 cm rms for scales larger than 100–200 km.

A final technique for estimating absolute velocities and transports is to combine the altimeter SLAs with simultaneous *in situ* data to estimate a "synthetic geoid" or more exactly a mean dynamic topography. The technique proceeds as follows. *In situ* data can provide estimates of the absolute dynamic topography η (although the barotropic part maybe more difficult to estimate) and satellite altimetry gives η'; the combination of the two estimates can thus yield the mean dynamic topography $\langle \eta \rangle$. This may be a very powerful methodology but it requires a large number of simultaneous data. This technique has been applied in the Gulf Stream (Mitchell *et al.*, 1990; Glenn *et al.*, 1991; Porter *et al.*, 1992; Howden *et al.*, 1999), the Kuroshio (Imawaki and Uchida, 1997), and the Azores Current (Hernandez, 1998).

The next logical step for monitoring the transport of ocean currents is to assimilate all available altimetric and *in situ* data into high-resolution ocean models. This will be discussed more completely in Chapter 5.

3.2. Western Boundary Currents

The western boundary currents are a crucial element in the ocean circulation system. They play an important role in the poleward transport of heat, and they provide a boundary between the less dense surface water of the subtropical gyres and the denser surface water of the subpolar gyres. The western boundary currents are also regions of large vorticity dissipation, which act to balance the total vorticity input into the subtropical gyres by the wind field.

The western boundary current in the North Atlantic is the Gulf Stream; its partner in the North Pacific is the Kuroshio (Figure 3). Both currents are typically 100-km wide and have surface velocities of 1–2 m/sec. They are associated with a large signal in mean transport and in variability. In the Southern Hemisphere, the western boundary currents are less intense but more complicated. They are strongly influenced by interactions with the Antarctic Circumpolar Current in the Atlantic (for the Brazil-Malvinas Confluence zone) and the Indian oceans (Agulhas/Mozambique Currents). In the South Pacific, the lack of intensity in the East Australian Current is perhaps caused by the existence of the Indonesian Throughflow at the western boundary at 10°S, and complicated by the presence of New Zealand further east.

Historically, the western boundary currents have attracted the largest number of hydrographic measurements in each ocean basin, but even so, their variability and true mean structure were not well-defined. Now, with nearly a decade of altimetric measurements available, we have made considerable progress in understanding the dynamics of western boundary currents. The wealth of altimetric studies in the western boundary current regions exists for a number of reasons. Not only are the dynamics interesting, but the signal-to-noise ratio of variability is high so that even the early generation altimeters with larger errors (GEOS3, Seasat, and Geosat) can detect coherent signals. The western boundary currents also have the largest density of *in situ* measurements to validate the altimetric signals, so new techniques are often validated in the Gulf Stream and Kuroshio regions before they are applied in more remote ocean areas.

The variability in these intense western boundary currents tends to be dominated by meanders of the current, by the separation and reabsorption of eddies, by the interaction of the mean jet with bathymetry, and by the interaction of the eastward current with westward propagating Rossby waves. These combined mechanisms generate a very complicated surface-height signal. Early descriptive studies concentrated on the statistics of the variability and on validating the observed variations with concurrent *in situ* data. More recent studies have analyzed a combination of altimetry, *in situ* data, and other satellite data and models and as such have made much progress in understanding the physical processes governing the observed variability.

3.2.1. The Gulf Stream

The Gulf Stream, the intense western boundary current of the North Atlantic Ocean, is perhaps the best-known current of the world. The Gulf Stream is fed to the south by the Florida Current, flows north along the North American continental slope with an increasing transport reaching 70–100 Sv, and then separates around 35°N and continues eastward as an open-ocean inertial jet. The Gulf Stream transport increases downstream because of the continual inflow of Sargasso Sea recirculation water, reaching a maximum transport of 150 Sv near 65°W. The Gulf Stream Extension is defined east of the Newfoundland Rise at 50°W, where the flow separates into three parts: the northern branch eventually becomes the North Atlantic Current, the eastward branch feeds into the Azores Current, and a southward branch forms part of the Sargasso Sea recirculation. Current speeds reach a maximum of 1–2 m sec^{-1} at the surface and decrease rapidly with depth. Instabilities of the flow form meanders, which can eventually separate into eddies or rings. Approximately 8–11 warm-core rings pinch off poleward of the main flow, and 5–8 cold-core rings form equatorward of the current per year (Hummon and Rossby, 1998). These rings generally drift westward with speeds of a few kilometers per day. The Gulf Stream is very energetic, and at any time up to 15% of the Sargasso Sea is covered by cold-core rings originating from Gulf Stream instabilities.

The seasonal variations of the Gulf Stream have been the subject of numerous altimetric studies. An analysis of climatological data showed maximum dynamic height differences across the current occurs in November (Zlotnicki, 1991), in contrast to the maximum volume transport which

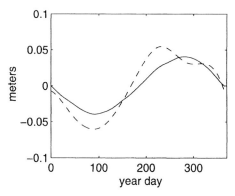

FIGURE 5 Seasonal variations of sea-level difference across the Gulf Stream from TOPEX/POSEIDON data (dashed line) and estimated from ECMWF net surface heat fluxes (solid line). The component representing Gulf Stream seasonal position variations has been removed. (From Kelly, K. A. *et al.*, 1999. With permission.)

occurs in spring (Sato and Rossby, 1995). Various studies based on Geosat data during the period November 1986 to December 1988 suggested that maximum sea-level slopes (Zlotnicki, 1991) and surface transports (Kelly and Gille, 1990) occurred in late autumn (September/October), somewhat earlier than in the climatological data. The annual cycle of sea-level differences across the Gulf Stream during this period explained 60% of the observed variance, with an amplitude of 9 cm. Kelly and Gille (1990) noted that the maximum surface transports occurred in late autumn when the Gulf Stream was north of its mean position. Minimum surface transports occurred in the late spring, with the Gulf Stream south of its mean position. T/P data during the period 1992–1997 confirm that the maximum surface transport and most northerly position occur in autumn, with a minimum in meandering in summer (Wang and Koblinsky, 1995; Kelly *et al.*, 1999).

The cause for these annual variations is most likely the upper-ocean seasonal heating cycle. Vazquez *et al.* (1990) found the annual signal to be dominated by meanders. However, if the SSH fluctuations caused by meandering are removed, most of the seasonal transport variations can be explained by seasonal heating differences across the Gulf Stream (Figure 5). Hydrographic data confirm that the seasonal variations in sea level come from the upper 250 m (Kelly *et al.*, 1999). This surface seasonal signal, with a fall maximum, obscures the seasonal variation in total volume transport, which has a spring maximum and is dominated by the larger volume transport below 250 m. Although there are some phase differences between the steric effect and the observed SSH, Kelly *et al.* (1999) suggest that these minor differences may be caused by seasonal variations in advection.

The idea of a seasonal modulation in the Gulf Stream meanders has been refuted by Lee and Cornillon (1995), who examined the Gulf Stream path from a much longer series of AVHRR data and found a dominant 9-month cycle instead. T/P data also reveal a dominant 9-month signal propagating southwestward from the Gulf Stream jet (Rogel, 1995) which appears to be related to incoming westward-propagating Rossby waves of that frequency interacting with the jet.

Kelly (1991) described in detail the Gulf Stream meandering and structure, using the technique of a simple analytical model to estimate the mean surface velocity profile. Two different flow regimes, separated by a transition region coinciding with the New England Seamount Chain, are found west of 64°W and east of 58°W. East of 58°W, increased meandering makes the mean current twice as wide as west of 64°W. Eulerian time scales also drop by a factor of 3 and peak velocities and surface transport drop by 25%. Tai (1990) has used a similar method on an ensemble of ascending tracks in the Gulf Stream and Kuroshio extensions. Surface transport is found to increase from 90×10^3 m^2 sec^{-1} (after leaving the coast) to reach a maximum of 130×10^3 m^2 sec^{-1} at 63°W for the Gulf Stream and at 150°E for the Kuroshio. These results compare well with surface transport deduced from drifter data (Richardson, 1985). The variability maximum coincides with the mean position of the jet.

Time series of the Gulf Stream transport have also been calculated near 71°W from T/P altimetry, a high-resolution geoid, and an empirical model for the Gulf Stream vertical structure by Howden et al. (1998). The vertical structure is a function of main thermocline depth, which is then calibrated to the sea-surface topography using historical AXBT data and reprocessed Geosat data. The transport is corrected for seasonal steric height changes. Altimetric sea-surface changes across the Gulf Stream of about 1.2 m are comparable to in situ measurements from a weekly repeated XBT and ADCP line between Newark, New Jersey, and Bermuda. These results suggest that altimetry, combined with a tight relation for the vertical density structure, can be used to monitor changes in the Gulf Stream position and transport.

A decade of altimetric studies have provided a better statistical description of the Gulf Stream transport, its seasonal and mesoscale variability, and the importance of surface heating in generating its annual cycle. With longer time series, we will also be able to investigate interannual changes, its interaction with the subpolar and subtropical gyres and eventually learn more of its role in interdecadal variations such as those for the North Atlantic Oscillation.

3.2.2. Kuroshio

The Kuroshio Current system is the strong western boundary current of the North Pacific subtropical gyre. The characteristic feature of the Kuroshio is that it has several quasi-stationary paths, controlled partly by topographic features. Upstream, the Kuroshio passes east of Taiwan and then flows along the eastern boundary of the East China Sea,

where its path undergoes strong seasonal meanders. Around 30°N, the Tsushima Current separates from the Kuroshio to the west of Japan. The main branch passes eastward through the Tokara Strait, south of Kyushu, and undergoes large meanders before reaching the Izu Ridge south of Honshu. The Kuroshio separates from the Japan coast near 35°N and becomes the Kuroshio Extension, where it again undergoes large meanders before passing over the Shatsky Rise at 157°E and the Emperor Seamounts at 170°E. The transport of the Kuroshio also increases along its path from about 60 Sv near 135°E (Imawaki and Uchida, 1997) to up to 130 Sv near 145°W (Wijffels et al., 1998). The Kuroshio Extension is also very energetic; in periods where the current follows a stable path, around 5 rings form each year, increasing to 10 rings per year during transition periods.

The seasonal cycle in the Kuroshio and its Extension is more variable than that in the Gulf Stream, with a large interannual modulation. Zlotnicki (1991) analyzed sea-surface differences across the Kuroshio Extension for the first 2 years of Geosat (1986–1988) and found a maximum peak in September/October, similar to the Gulf Stream, although he noted that the annual cycle in climatological dynamic height differences was negligible. Qiu et al. (1991) found annual variations of the Kuroshio Extension in agreement with Zlotnicki's results. They also observed that the maximum surface-height difference in the Kuroshio Extension lagged by 2 months the maximum sea-level difference across Tokara Strait. The Kuroshio Extension annual cycle calculated by Wang and Koblinsky (1995) from T/P data in 1993 also shows a maximum in October. As in the Gulf Stream, the maximum surface transport occurs when the current axis is north of its mean position. Annual variations are found to be significant in the upstream region (141°E to 153°E). On average, they explain only 15% of the total variance whereas interannual variations explain 23%. Qiu (1992) found that the seasonal transport variations were closely related to the seasonal change in intensity of the southern recirculation gyre. However, other authors suggest that the principal cause of this seasonal cycle is differential heating over the Kuroshio Extension (Zlotnicki, 1991; Wang and Koblinsky, 1995).

Wang et al. (1998) have separated the annual and intra-annual sea-level variability into separate spatial scales. They find the large-scale annual variability (>2000 km) is essentially a standing oscillation, related to seasonal heating. The shorter-scale annual and intra-annual variability (800–1500 km) is found to have strong, westward phase propagation and is associated with large bathymetric features. In particular, maximum low-frequency wave activity occurs west of the Shatsky Rise (~160°E).

Pronounced interannual fluctuations are evident in the Kuroshio Extension region. Qiu (1992) suggests that interannual variations during the Geosat period were possibly influenced by the intensification of the subtropical wind gyre

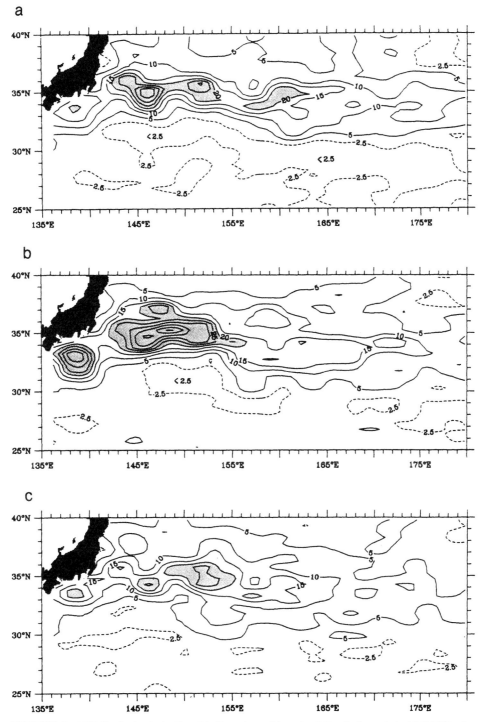

FIGURE 6 EKE distribution in the Kuroshio Extension and its recirculation for the periods: (a) Oct 92 to Apr 93, (b) Apr 93 to Oct 93, (c) Oct 93 to Jun 94. The contour unit is $0.01 \ m^2 \ sec^{-2}$, areas with energy levels greater than $0.2 \ m^2 \ sec^{-2}$ are shaded. (From Qiu, B., 1995. With permission.)

with the 1986 to 1987 ENSO event. During the first 2 years of T/P, Qiu (1995) also noted significant variations in EKE levels in the Kuroshio Extension and its recirculation gyre (Figure 6). Significant interannual changes in EKE were also observed by Adamec (1998). Qiu (1995) and Adamec (1998) analyzed those changes in terms of eddy–mean flow interaction (see discussion in Section 4.6).

Mean velocities and transports have been estimated in the Kuroshio region using a number of different methods. The technique of combining a simple Gaussian model of the mean jet with the variable flow characteristics derived from altimetry has been used to map the mean sea surface and variations in the Kuroshio Extension from 2.5 years of Geosat data (Qiu et al., 1991), and including the associated recirculation gyres (Qiu, 1992, 1995). Using the derived absolute surface velocities, the authors investigated quasi-stationary meanders and the relative eddy and mean kinetic energies. The ratio of EKE to mean kinetic energy is high, with a nearly constant value of 1.5–2.0 along the Kuroshio Extension path. Propagation of mesoscale fluctuations is generally westward except for the upstream region of the Kuroshio Extension.

Recently, the ASUKA Group[1] undertook an ambitious program to monitor the absolute volume and heat transports of the Kuroshio and its recirculation south of Shikoku, Japan. ASUKA carried out repeated hydrographic surveys under a T/P satellite groundtrack near 135°E during 1993–1995 (Imawaki and Uchida, 1997); the intensive survey also included 9 current meter moorings with 33 current meters, 2 upward-looking ADCPs, and 10 inverted echo sounders (Figure 7a). The volume transport in the top 1000 m showed a very tight relation with the surface dynamic topography (Figure 7c). This means that variations in the sea-surface slopes across the Kuroshio measured by altimetry can be used to monitor the variations in total transport. For absolute transport estimates, a mean sea-surface profile was derived from the along-track hydrographic data and combined with the 10-day altimetric anomalies. The average volume transport was estimated to be 63 ± 13 Sv, with no apparent seasonal cycle although interannual and intraseasonal cycles were very strong (Figure 7b). The latter were probably influenced by fluctuations in local stationary eddies. With such an accurate along-track estimate of the mean profile, the surface geostrophic velocities compared extremely well with surface drifters (Uchida et al., 1998) and highlighted the drifter sampling bias as they tend to converge in the high-velocity core of the Kuroshio.

These altimeter studies of the Kuroshio Current System have improved our understanding of seasonal heating variations on the seasonal transport cycle and the important role of topographic forcing in generating meanders, Rossby

waves, and other low-frequency variations. With longer time series, altimetry has also revealed a strong interannual signal, which may be directly influenced by tropical Pacific variations but may also have a direct effect on the dynamics of the subpolar gyre. The Kuroshio appears an important conduit between the subtropical and subpolar ocean regions, and future altimetric studies will certainly examine its role in modulating the circulation and climate of the North Pacific.

3.2.3. Brazil–Malvinas Confluence Region

The two western boundary currents of the South Atlantic are the Brazil Current and the Malvinas Current. In this region, the Brazil Current flows poleward along the continental slope of South America. The Malvinas Current originates as a branch of the Antarctic Circumpolar Current. Between 35°S and 39°S, the two current systems, which have strongly contrasting water types, converge in the Brazil–Malvinas Confluence region and form a strong frontal structure. The Confluence region is associated with an intense and complex mesoscale variability field characterized by rings, eddies, and filaments. Several studies have used Geosat data to analyze the mesoscale variability of the Brazil–Malvinas Confluence area (Provost and Le Traon, 1993; Forbes et al., 1993; Goni et al., 1996). The mesoscale field is highly inhomogeneous, with low sea-level variability and EKE in the Malvinas Current (<8 cm and 150 $cm^2 sec^{-2}$, respectively), intermediate values in the Brazil Current (16 cm and 800 $cm^2 sec^{-2}$), and high values in the Confluence region (30 cm and 1700 $cm^2 sec^{-2}$). Provost and Le Traon (1993) showed that the mesoscale variability has a marked anisotropy with larger meridional velocity variance and is dominated by relatively large spatial scales and low-frequency fluctuations. Contrary to what is found in most western boundary and open ocean currents, they found little energy at the annual frequency, but a significant semi-annual signal (Figure 8), probably indirectly related to atmospheric forcing. Note that semi-annual Rossby waves were also observed in the interior of the South Atlantic subtropical gyre (Le Traon and Minster, 1993). More recently, Vivier and Provost (1999) showed that T/P data can be used to monitor the volume transport of the Malvinas Current. The transport is estimated from T/P data using a priori statistical information on the vertical structure of the flow. T/P transport compares favorably (correlation of 0.8) with estimates from current meter data. Transport time series derived from 3 years of T/P data show salient periodicities between 50 and 80 days and at the semi-annual period. Little energy is again found at the annual period. They suggest that the semi-annual variability may be remotely forced by winds at Drake Passage. The semi-annual periodicity of the Malvinas Current may explain the importance of the semi-annual signal in the Brazil–Malvinas Confluence region.

---

[1] Affiliated Surveys of the Kuroshio off Cape Ashizuri.

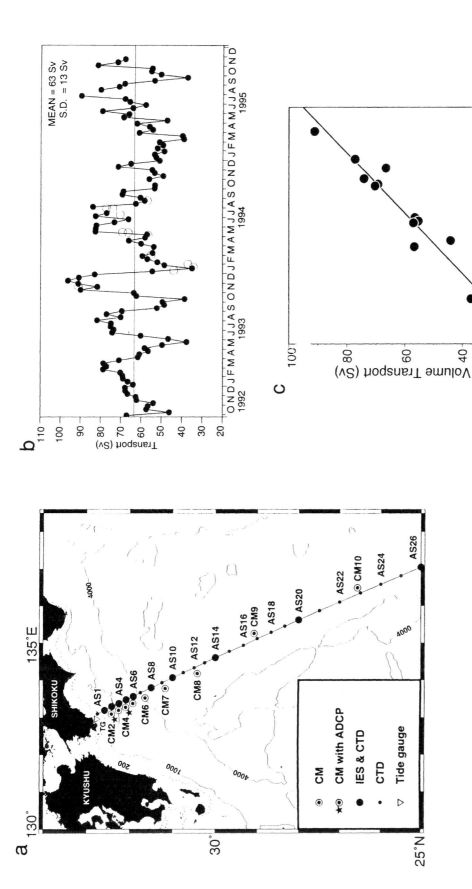

FIGURE 7   (a) The ASUKA Group's *in situ* observations along a TOPEX/POSEIDON satellite groundtrack in the Kuroshio, deployed from October 1993 to November 1995. (b) Time series of the absolute volume transport estimated from TOPEX/POSEIDON data, using (c) the regression relation between transport variations and SLA derived from the *in situ* data. (From Imawaki, S., and Uchida, H., 1997. With permission.)

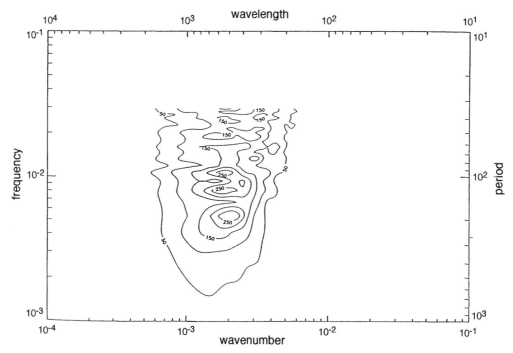

FIGURE 8   Mean frequency/wavenumber spectrum of SLA in the Brazil–Malvinas Confluence area as derived from Geosat data. Units are in cm^2. There is a peak at the semi-annual frequency while little energy is found at the annual frequency. (From Provost, C., and Le Traon, P. Y., 1993. With permission.)

3.2.4. East Australian Current and East Auckland Current

The western boundary current of the South Pacific, the East Australian Current (EAC), flows south along the Australian coast, then separates near 32°S to form the Tasman Front, reattaches to the shelf break near North Cape, New Zealand, and continues southeast alongshore as the East Auckland Current. Although the mean flow associated with the EAC is weaker than other western boundary currents, the eddy field is quite energetic. Using a combination of T/P altimeter data and AVHRR sea-surface temperature data, Ridgway *et al.* (1998) observed that the eddies, both cyclonic and anticyclonic, are advected southward in the EAC about three times per year. These eddies appear phase-locked with westward-propagating Rossby waves generated in the eastern Pacific. They suggest that the eddies develop as instabilities when the Rossby waves interact with the eastward mean flow in a region of sharply deepening bathymetry. Wilkin and Morrow (1994) calculated eddy kinetic energy and Reynolds stresses at Geosat crossover points in the Tasman Sea region. They found that high EKE associated with instabilities of the EAC extended from the coast to the Lord How Rise at 160°E. Velocity variance ellipses from the altimeter and drifter data showed distinct anisotropic structure (Figure 9, see color insert), with a strong alongshore component at the coast, with offshore Reynolds stresses converging around the first semi-permanent meander of the Tasman

Front. Farther offshore, the ellipses are orientated north–south, consistent with meridional velocities associated with the incoming Rossby waves.

Roemmich and Sutton (1998) have investigated the East Auckland Current with a combination of T/P altimetric data, broad-scale XBT data, two repeating high-resolution XBT transects, and neutrally buoyant floats. The mean transport of the East Auckland Current is about 9 Sv, with an additional 10 Sv in the recirculation of three permanent warm-core eddies. Their study further compares the extensive time-series data with hydrographic climatology, and it is one of the first studies to quantify the error in the estimated climatological mean field, in particular from the significant mesoscale contribution.

3.3. Eastern Boundary Currents

The eastern boundary currents are generally characterized as diffuse, equatorward flows, that extend from the eastern boundary into the ocean interior, as part of the return flow of the wind-forced gyres. Superimposed on this broad equatorward flow are narrow coastal currents and often upwelling currents, driven either by the wind-stress curl or by the equatorward wind stress that exists along the eastern boundaries. These coastal currents are noted for their equatorward surface currents and poleward undercurrents, but they can be highly variable, even reversing under certain conditions. Altimetric studies of the broad eastern boundary currents are rather limited, essentially for the same reasons that western boundary current studies are so numerous. From an altimetric perspective, they have lower signal-to-noise ratios, especially in the earlier missions; they have fewer *in situ* measurements for validation; and at first order, their variability has less influence on the overall dynamics than on western boundary currents. The narrow coastal boundary currents have also been difficult to measure from altimetry, since they are sampled by few data points, they are subject to various coastal errors (tidal errors, nonisostatic pressure responses, altimeter or radiometer data dropouts close to land), and the wind-forced currents can have a strong ageostrophic component that is not seen by altimetry.

Despite these early measurement difficulties, the new generation of more precise altimeters such as T/P have allowed us to monitor the variability associated with these eastern boundary currents, and in some cases measure their velocities and transports. This is important not only for understanding the dynamical processes at the eastern boundaries, but also because perturbations and instabilities at the eastern boundary can then propagate westward as Rossby waves, and so influence the ocean interior at later periods. In the following section, we consider each eastern boundary region separately and look at how altimetry has aided our understanding of eastern boundary current variability.

3.3.1. California Current

The California Current System on the eastern boundary of the North Pacific is one of the few eastern boundary regions to be extensively studied with altimetry. This region is one of strong seasonal currents forced by the large wind-stress curl variability. The mainly equatorward current system also reverses seasonally, with periods of poleward flow over the shelf from October to March (also known as the Davidson Current). A combination of Geosat altimeter data and satellite AVHRR sea-surface temperature data has been used to describe the large-scale summer circulation (Strub and James, 1995). The circulation is characterized by a large-scale equatorial jet and temperature front that lies close to the Oregon coast (20–50 km) and extends farther offshore from the Californian coast in a convoluted, meandering jet. Eddies associated with this jet may persist for 3–6 months and play an active role in offshore transport. A comparison between T/P altimetry and subsurface current meter data shows that the altimeter resolves horizontal scales of 50–80 km in the along-track direction (Strub *et al.*, 1997). The rms difference between the altimeter and current meters was around 7–8 cm sec^{-1}, and was mostly caused by strong and persistent small-scale variability in the currents. Velocity variances, eddy kinetic energy levels, and the major axis of variance ellipses were in good agreement, and demonstrated the anisotropic nature of the variability. Most of the high eddy energy within 500 km of the coast is associated with the the seasonal jet which develops each spring and

moves offshore, rather than with deep-ocean eddies moving onshore.

A comparison of EKE measured by different instruments has also been made in the Californian Current region (Kelly *et al.*, 1998). Surface drifters, T/P altimetry data, and moored current data all show similar seasonal cycles in EKE, with maximum values in late summer/autumn, a month or so earlier than for the western boundary currents. The velocity fluctuations were typically 240–370 km in wavelength and lasted several months. In the northern section (36° to 41°N), the maximum EKE migrates zonally on a seasonal time scale, with maximum EKE associated with increased equatorward flow. A simple model was used to show that the seasonal variations in EKE amplitude were related to wind-stress curl forcing, but this did not explain the offshore movement of the front.

The zonal migration in EKE is probably related to annual Rossby waves. White *et al.* (1990) first studied these waves using Geosat altimetry. They found that altimetric sea-level residuals showed local maxima at three key locations off the Californian Coast: at Point Eugenia at 27°N SW of Point Conception at 32°N, and between Monterey and Cape Mendicino (37° to 40°N). These local maxima were mainly associated with annual Rossby waves (53% variance), which originated near the coast, with wavelength scales of 400–800 km, periods of 6–12 months, and with westward phase speeds at 2–5 cm sec^{-1}, faster at lower latitudes which were consistent with Rossby wave theory. In a later study, Pares-Sierra *et al.* (1993) compared two numerical models to test whether these waves were generated by a coastal response to wind forcing or baroclinic instabilities of the California Current. Only the wind-forced model simulated correctly the distribution of eddy variance, suggesting that the dominant source of this mesoscale eddy activity is the wind forcing adjacent to the coast, modified by both Rossby and Kelvin wave dynamics.

These studies have concentrated on the offshore dynamics. The narrow coastal currents are more difficult to access with altimetry, and statistical gridding techniques often give unrealistic results when extrapolated towards the data-sparse coastal region. An alternate method is to combine offshore altimeric sea-level measurements with coastal-tide gauge data. This has been done for the California Current region using T/P and ERS-1 altimeter data (Strub and James, 1997) and a "successive correction" method, which results in a smoother field in data sparse regions and a more detailed field around the data. The tide gauge data are filtered to remove both tides and coastal trapped waves, using a 20-day filter. Including the tide gauge data lowers the sea level along the coast and reveals an alongshore jet close to the coast which is not sampled by altimetry. Satellite SST data confirm that the jet flows along the outer edges of colder upwelled water, which is an expected pattern for this region of coastal upwelling. However, several cold filaments with horizontal scales of 50 km are apparent in the SST data and are not resolved by the gridded altimetric/tide gauge data. This reaffirms that features with scales less than 100 km are not adequately resolved by altimetry, even when T/P and ERS-1 data are combined.

3.3.2. Alaska Current and Gulf of Alaska

North of the California Current system is the Alaska Current, which forms the eastern component of the Pacific subpolar gyre. The current is constrained to the shelf region by a strong cross-shelf pressure gradient that is maintained by the freshwater input from Alaska's rivers, which reduces the coastal density. Most energetic eddy activity is confined to the shelf/slope and boundary current region, with little eddy energy in the gyre interior (Lagerloef *et al.*, 1994). The coastal mesoscale eddy activity is around 100 km in wavelength (Matthews *et al.*, 1992), the most notable being the Sitka eddy around 57°N, which has an approximate annual cycle interior (Lagerloef *et al.*, 1994). Two-dimensional spectral analysis across the Gulf of Alaska reveals both westward- and eastward-propagating features at this latitude, with mainly annual period and longer wavelength of 1000 km. Using a longer time series of nearly 4 years of Geosat data, Bhaskaran *et al.* (1993) also identified significant interannual variability, which was closely linked to baroclinic variations detected by hydrographic data. Meyers and Basu (1995) have more recently analyzed 6 years of T/P data. They show that eddy activity in the Gulf of Alaska fluctuates with the ENSO cycle.

3.3.3. The Peru/Chile Current

The Peru/Chile current is the eastern boundary current of the South Pacific subtropical gyre. The equatorward current transports enough cooler water from higher latitudes to lower the SST along South America by several degrees from the zonal average. Superimposed is a vigorous upwelling circulation, which lowers the coastal SST by another 2 to 4°C. Strub and James (1997) used maps of T/P and ERS-1 altimetry to study the nearshore circulation off Chile. They described the nearshore equatorward jet that was aligned along the SST front between February and March 1993. The jet appeared to break up into a series of eddies in April, which widened the region of cooler coastal water by 100 km. Longer-term monitoring of this coastal jet was difficult because of the poor spatial resolution (100 km) of the gridded altimeter data. Strub *et al.* (1995) have also monitored the Peru/Chile Countercurrent, which flows from 10°S to 35–40°S and lies approximately 100–300 km offshore. Although the temporal mean current was not available in the altimetric observations, they assumed the mean countercurrent was poleward, based on historical observations. The altimeter variability allowed them to deduce maximum velocities in spring and minimum in autumn.

3.3.4. Leeuwin Current

The dynamics of the eastern boundary current in the southern Indian Ocean are quite unusual, since it flows poleward against the prevailing equatorward winds. The current is forced by the large alongshore pressure gradient along the West Australian coast, which is five times greater than that for other eastern boundary regions. This large pressure gradient is maintained by the lower-density water which enters to the north from the Indonesian Throughflow. The alongshore pressure gradient drives an eastward geostrophic flow toward the coast; when geostrophy breaks down at the coast, the flow accelerates poleward down the pressure gradient and overcomes the equatorward wind-driven flow. Morrow and Birol (1999) have used T/P altimetry to monitor the variability of this alongshore pressure gradient; the mean pressure gradient is derived from climatological data. The altimeter data confirm previous studies that the alongshore pressure gradient is maximum in May, when the coastal boundary current, the Leeuwin Current, is strongest (Figure 10). However, the 3-year time series also reveals a semi-annual signal, with a secondary peak in November which is not evident in the climatology, although it is present for this period in XBT data. There is also significant interannual variability in the pressure gradient, and also in the structure of instabilities associated with the Leeuwin Current. Away from the coast, T/P data reveal a band of energetic Rossby waves between 20 and 35°S which propagate westward across the basin, originating near the eastern boundary, with near semi-annual periods of 100–200 days. Birol and Morrow (2000) used a simple vorticity model forced by ERS-1 scatterometer winds and an eastern boundary condition derived from XBT data to determine that these semi-annual waves are not locally wind-forced, but appear remotely forced from the north.

3.3.5. Benguela Current

The Benguela Current is the broad equatorward flow in the southeast Atlantic Ocean. As in other eastern boundary currents, the flow is dominated by geostrophic eddies and is associated with a strong upwelling system. This current system is also instrumental in the Indian/Atlantic exchange of mass, heat, and salt, since eddies from the Agulhas region (see below) are often entrained by the Benguela Current and transported north. Altimeter data have been combined with *in situ* data to monitor the transport in the Benguela Current system by Garzoli *et al.* (1997). From the *in situ* data, they derive a correlation function between sea-surface elevation and thermocline depth, indicating a dominant baroclinic flow. With a combination of these correlations and a linear stratification geostrophic relation, they can calculate baroclinic transports. During the 3 years from 1993–1995, the mean northward baroclinic transport was between 12 and 15 Sv, with fluctuations up to 25 Sv because of high

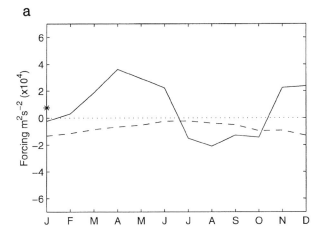

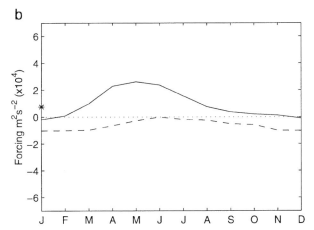

FIGURE 10 (a) Forcing terms for the poleward Leeuwin Current for the 3-year period 1993–1995: The alongshore pressure gradient from TOPEX/POSEIDON data (solid line) drives onshore geostrophic flow, which is balanced by the offshore wind forced flow from the alongshore wind stress from ERS-1 (dashed line). Positive forcing indicates conditions for a poleward coastal current period. (b) Similar fields from climatology, with no semi-annual signal. (From Morrow, R. A., and Birol, F., 1999. With permission.)

velocities associated with ring shear. The interannual variability in transport was small, although the source regions changed from year to year. In 1993/1994, the main contribution came from the South Atlantic, with equal partitions of Indian Ocean and tropical Atlantic contributions. In 1995, the main contribution was from the Indian Ocean, for reasons yet unknown.

3.3.6. Canary Current

The easternmost branch of the Azores Current (see below) feeds the Canary Current, an essential part of the North Atlantic eastern boundary current system. Because of the seasonal shift of the trade winds, coastal upwelling north of 25°N is strongest in summer and fall. Filaments of cold water from the upwelling regime often extend several hundred kilometers offshore. Most of them develop near the capes

at the African coast and can be traced in satellite images of sea-surface temperature or pigments (Hernandez-Guerra and Nykjaer, 1997). An analysis of more than 5 years of T/P and ERS-1/2 data in the Canary basin (Hernandez and Le Traon, 1999) clearly shows the seasonal variation of the Canary Current and the associated upwelling. There is a remarkably strong variability south of the Canary Islands (7–9 cm rms, 150–200 $cm^2 sec^{-2}$), which is related to the eddy field generated by instabilities of the Canary Current as it crosses the archipelago and by the wind-stress curl forcing downstream from the Islands (Arístegui *et al.*, 1994). The signal has a very clear seasonal modulation with a maximum EKE observed in summer and fall.

3.4. Open Ocean Currents

3.4.1. Antarctic Circumpolar Current

The Antarctic Circumpolar Current is unique in many respects. It is the only ocean current to flow continuously around the globe in the zonally unbounded Southern Ocean, and it provides a major link for water-property exchanges between the Atlantic, Indian, and Pacific oceans. It is a region of large heat and momentum exchanges between the ocean and the atmosphere. It is also a region of high eddy variability, and ocean eddies may play a role in the southward oceanic heat transport and in transporting momentum in the circumpolar region. Before the advent of satellite altimetry, there had been few long-term observations of the ACC and its variability because of the difficulty and expense in making hydrographic or current-meter measurements in this remote and inhospitable environment. Altimetry allows us to monitor the surface signature of these remote currents and their variability.

One of the early findings from altimetry was that a vigorous eddy field existed along the path of the ACC, with large geographical variations (Zlotnicki *et al.*, 1989). This had never been fully quantified from sporadic hydrographic sections or the limited current meter moorings, and eddy-resolving models were rather limited at this time. The spatial sampling of Geosat was particularly useful for Southern Ocean mesoscale studies. Regions of high mesoscale variability were significantly correlated with the mean circulation, as determined from historical hydrographic data (Chelton *et al.*, 1990). This indicated the importance of hydrodynamic instabilities in the dynamics of the ACC. The geographical distribution of both the mean currents and their variability was also shown to be strongly controlled by bathymetry (Sandwell and Zhang, 1989; Chelton *et al.*, 1990). Gille and Kelly (1996) analyzed the spatial and temporal characteristics of the mesoscale variability and also concluded that local instability mechanisms were more important for the ACC dynamics than basin-scale processes.

The dynamical significance of this mesoscale variability was quantified by Morrow *et al.* (1994), who calculated

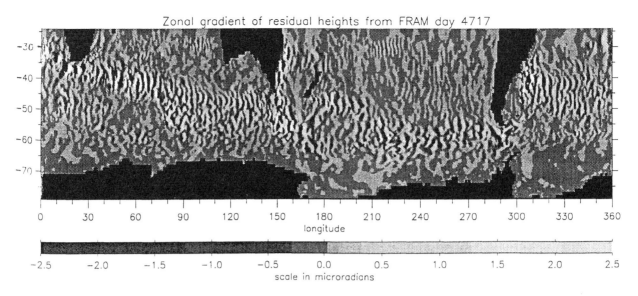

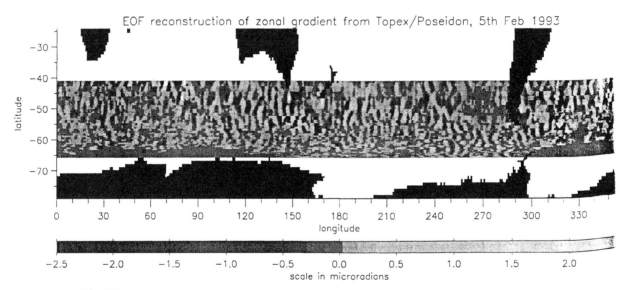

FIGURE 11   (a) Zonal gradients of residual sea-surface heights from the fine-resolution Antarctic model (FRAM) compared with (b) an EOF reconstruction of similar fields from TOPEX/POSEIDON, for Feb 5, 1993. The pattern highlights Rossby waves, with wavelengths of 200–400 km. (From Hughes, C. W., 1995. With permission.)

velocity variance ellipses and Reynolds stresses at Geosat crossover points in the Southern Ocean. They demonstrated that the eddy variability was often anisotropic, in particular close to bathymetric features and in the vicinity of the mean current. This anisotropy is associated with horizontal eddy momentum fluxes. Calculations made along the path of the ACC showed that eddies tend to converge momentum into the mean current and, on average, tend to accelerate the frontal jets. The Geosat data reveal a surprisingly complex geographical distribution of this eddy momentum flux convergence (and in some regions divergence), which was con-

firmed by comparisons with numerical models (Wilkin and Morrow, 1994).

Part of this mesoscale eddy variability is associated with propagating Rossby waves, with wavelength of about 300 km and periods of 4 to 12 months (Hughes, 1995). These waves are advected eastward in the mean axis of the ACC, and westward elsewhere, and are evident in both T/P data and the fine-resolution Antarctic model (FRAM), as shown in Figure 11. Morrow et al. (1994) estimated the mean axis of the ACC to coincide with the meridional maximum of $\overline{v'v'}$, which they interpreted as the mean position of the time-varying meanders. The finer-resolution FRAM model supports this, but it also shows that the meanders are often wave-like, with a steady eastward propagation. The characteristic patterns of eddy momentum flux convergence form an arrowhead pointing eastward (Morrow *et al.*, 1994); this is also the hallmark of Rossby waves traveling along a waveguide formed by the eastward jet (Hughes, 1995).

The dynamics of the eddy-mean interactions near steep topographic slopes in the Southern Ocean was evaluated by Witter and Chelton (1998) using Geosat altimetry and a wind-forced quasigeostrophic channel model. They found that largest eddy energies occur downstream of zonal modulations of bottom topography, in particular when topographic steering forces the flow into regions of reduced ambient potential vorticity. The types of instability processes are also modified. With zonally uniform topography, baroclinic instability and recycling of eddy energy occurs, whereas in regions with strong zonal modulations in topography, mixed baroclinic-barotropic instabilities occur with an associated strong downward transfer of eddy energy. This is an important mechanism in the transfer of wind-input momentum from the surface to deeper levels, where it can be balanced by form drag over topographic ridges. Thus the Geosat observations showed enhanced variability near steep topographic slopes in the ACC; the model results detail the vertical structure and suggest that these high-eddy-energy regions make a large contribution to the dynamics of the ACC.

The large-scale sea-level variability was examined by Chelton *et al.* (1990) in the Southern Ocean using 2 years of Geosat data, smoothed to a resolution of 12° longitude by 6° latitude by 9 days. The variability was dominated by the seasonal cycle, with a zonally coherent annual component and a semi-annual component that showed large amplitude and phase changes over the three ocean bases. Interannual variability of the seasonal cycle was also strongly regional. Park and Gambéroni (1995) considered the large-scale circulation in the southern Indian Ocean sector of the ACC, using 18 months of T/P data. During this period and in this location, they find no significant interannual variations and find that the variability is strongly constrained by the current-topography interactions.

Monitoring the transport of the ACC with altimetry is a tricky task. The frontal regions carry up to 75% of the transport because of their enhanced horizontal density gradients and associated geostrophic currents. However, because these fronts are so narrow and variable, they are also difficult to sample with altimetry, and mapped altimetry data tend to smooth the fronts and reduce their gradients and associated currents. The ACC fronts also have a large barotropic component which is measured by altimetry, but they are difficult to estimate from hydrographic data. Despite these difficulties, a number of altimetric studies have estimated the mean circulation and transport of the ACC. Gille (1994) applied a model of a meandering Gaussian jet to estimate the

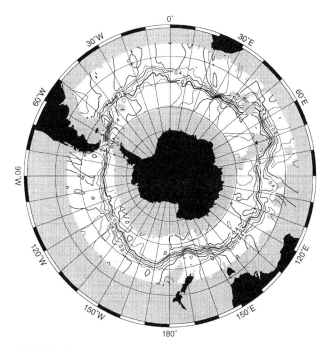

FIGURE 12 Mean sea-surface height across the Antarctic Circumpolar Current reconstructed from Geosat height variability using a meandering Gaussian jet, after Gille (1994). Contour interval is 0.2 m.

mean sea surface across the Subantarctic Front and the Polar Front (Figure 12). The meandering jet model explains between 40 and 70% of the height variance along the jet axes; the remaining variance represents rings and eddies that separate from the jets. The fronts are substantially steered by topography, with the mean jet width of 44 km, and meanders around 75 km either side of their mean position. The latitude of the ACC axis varies from 40°S to 60°S, and although the flow retains its sharp frontal structure, the frontal width varies with latitude because of to the changing Rossby radius. The average height difference across the Subantarctic Front was 0.7 m, with a 0.6-m drop across the Polar Front. In the southern Indian Ocean sector of the ACC, Park and Gaméroni (1995) calculated the mean circulation using a combination of altimetry and a geoid model. Their mean altimetric circulation indicated the presence of an anticyclonic subtropical gyre north of the ACC and two cyclonic subpolar gyres south of the current, either side of the Kerguelen Plateau. These gyre structures were in good agreement with a fine-resolution numerical model, but they were not evident in the climatological data, probably because of the strong barotropic component of these currents.

As a choke point, Drake Passage has been the site of numerous hydrographic campaigns, including extensive current meter measurements, so the ACC transport and its variability are reasonable well known here. Challenor *et al.* (1996) have estimated the surface geostrophic currents in Drake Passage from a combination of altimetry, hydrography, and current data. Their method depends on having *in situ* density and current data along an altimeter ground-track to specify the absolute surface current for one altimeter pass. This *in situ* surface current is then used as a reference for the altimetric data, providing a time series of absolute surface current from altimetry. Their technique finds two main jets at 56.6°S and 58.5°S, which remain consistently in position. Woodworth *el al.* (1996) compared altimetry with bottom-pressure measurements (for the barotropic component) and FRAM model sections. Although sea-level and bottom-pressure measurements were in good agreement in the northern part of Drake Passage, the FRAM results showed that they were virtually uncorrelated with transport fluctuations. In the southern section, which is more important for transport studies, sea level and bottom pressure show significant rms differences, explained by a larger baroclinic variability. Their conclusion is that altimetry will need to be combined with extensive *in situ* measurements to properly understand the dynamics of Drake Passage.

These Southern Ocean altimetric studies have highlighted the large regional differences in eddy energy and fluxes, the existence of seasonal and interannual variations, and the very strong dependence on bottom topography. However, dynamical studies have also shown the difficulty in establishing a strong relationship between the altimetric surface heights and the subsurface velocities in this region of weak vertical density gradients and a strong barotropic component. Future altimetric studies will almost certainly rely on a combination of satellite surface measurements with *in situ* data and numerical models.

3.4.2. Agulhas Current and Retroflection

The Agulhas Current system south of Africa is actually the strongest western boundary current in the Southern Hemisphere, fed to the north by the Mozambique Current and the East Madagascar Current and from recirculation in the southwest corner of the Indian Ocean. As for most western boundary currents, the Agulhas increases its transport along its path from around 70 Sv near 31°S to 100–135 Sv near 35°S (Tomczak and Godfrey, 1994). The reason it is included in this section on open ocean currents is that as it moves southwest into the Atlantic, the current retroflects and turns eastward in the direction of the prevailing winds and eastward flowing Antarctic Circumpolar Current. So the Agulhas Current System, including the Retroflection, is closer to an open-ocean current than a western boundary current.

The Agulhas Retroflection is one of the more highly variable oceanic regions, with mean variance levels measured by Geosat altimetry of around 550 cm^2, compared with levels around 350 cm^2 for the Gulf Stream (Zlotnicki *et al.*, 1989). To the west, the high variability is mainly caused by eddy shedding; the Retroflection moves westward until its western part pinches off and forms an eddy and the loop retreats to its most eastern position. Further east, the high eddy variability is caused by instability processes as the current inter-

acts with topography and the eastward flowing Circumpolar Current.

Seasonal variations in the Agulhas current and mesoscale activity have been reported by a number of studies, which used up to 2 years of altimeter data (Zlotnicki *et al.*, 1989; Quartly and Srokosz, 1993). There is no clear seasonal signal either in climatological data or in the FRAM model (Quartly and Srokosz, 1993). Given the short duration of these time series, and the fact that intermittent ring shedding and Retroflection meanders may bias the seasonal results, it is difficult to quantify any true seasonal cycle—a much longer time-series analysis may be necessary.

The separation of Agulhas rings from the Retroflection region and their westward propagation into the South Atlantic have been extensively studied using altimetry. These rings may be one of the important mechanisms for the inter-ocean transport of mass, heat, and salt between the Indian and Atlantic oceans. However, given their limited spatial extent and sporadic detachment, they have been difficult to measure using hydrographic sections, and so their effects on the total mass, heat, and salt budgets were largely uncertain. Altimetry has allowed us to monitor the separation and propagation of these eddies away from the Retroflection region and to build a catalogue of their surface characteristics. Gordon and Haxby (1990) describe one such eddy using a combination of Geosat altimetry and *in situ* data. Dynamic height calculated from both data sets was equivalent; furthermore, this eddy had distinct signs of Indian Ocean stratification. Most studies find that between five and six rings are shed from the Retroflection region each year, whether during the Geosat period (Gordon and Haxby, 1990; Feron *et al.*, 1992; Byrne *et al.*, 1995) or the T/P period (Grundlingh, 1995; Goni *et al.*, 1997). Eddy shedding can be detected from altimetry by sharp changes in the stability of the sea-level patterns (Feron *et al.*, 1992). An analysis of 3 years of T/P by Goni *et al.* (1997) found that ring shedding was neither continuous nor periodic, and for long periods there were no ring formations. The eddies detected were mostly anticyclonic, and they drifted west-northwest across the South Atlantic with speeds of 3–7 cm sec^{-1} (Grundlingh, 1995), with a slowed translation across strong topographic slopes. The residence time of the eddies in the South Atlantic was around 3–4 years, and their amplitude decayed slowly with an e-folding distance of O (1700–3000 km) (Byrne *et al.*, 1995).

A combination of altimetry and concurrent hydrographic sections has been used to estimate the annual mean property fluxes from the Indian to the Atlantic Ocean that are associated with these Agulhas eddies. In terms of water mass transfer, estimates range from a minimum of 5 Sv (Byrne *et al.*, 1995) to a maximum of 15 Sv (Gordon and Haxby, 1990); van Ballegooyen *et al.* (1994) estimate a transport of 6.3 Sv of water warmer than 10°C, and 7.3 Sv warmer than 8°C. The total upper layer transport across 32°S in the south Atlantic has been estimated at 10 Sv northward (Schmitz, 1995), so the contribution from Agulhas eddies is a substantial component of the total interocean exchange. The energy flux associated with these eddies is on the order of 10^{17} J (Byrne *et al.*, 1995: Geosat; Goni *et al.*, 1997: T/P). Annual heat fluxes estimates vary between 3.8×10^{13} W (Grundlingh, 1995: T/P) and 4.5×10^{13} W (van Ballegooyen *et al.*, 1994: Geosat). Annual salt fluxes are estimated at 78×10^{12} kg per year (van Ballegooyen *et al.*, 1994).

Interannual changes in the behavior of these Agulhas eddies have also been investigated. Witter and Gordon (1999) used an empirical orthogonal function analysis of T/P data and found a transition in basin-scale sea level in the eastern South Atlantic, from higher sea level and enhanced gyre-scale circulation in 1993/1994 to lower sea level and sluggish circulation in 1996. The Agulhas eddy trajectory also changed; eddies dispersed over a broad region in 1993/1994 and remained in a narrow corridor in 1996 (Figure 13). This suggests that in different years, the injection of mass, heat, and salt from Agulhas eddies may occur in different regions of the subtropical gyre, and they may be partially controlled by interannual variations in the large-scale circulation.

The mechanisms that cause ring-shedding in the Retroflection region have also been investigated using altimetry. One of these mechanisms is the Natal Pulse, an occasional large meander that forms in the Agulhas Current off the African coast. An analysis of Geosat, ERS-1, and T/P data has been used to monitor the Natal Pulses from Durban to the Agulhas Bank (de Ruijter *et al.*, 1999), and after a time lag of around 100 days an Agulhas ring is shed. Rings also appear to be shed by a combination of Natal Pulses and the arrival of Rossby waves in the Agulhas Return Current. Possible triggering mechanisms for these Natal Pulses have been investigated from altimetry and cross-shelf velocity measurements (de Ruijter *et al.*, 1999). The intermittent formation of these pulses is argued to be related to barotropic instabilities of the Agulhas Current. Occasionally, the correct conditions for these instability processes occur in the Natal Bight, particularly when the intensity of the incoming jet exceeds the average. The combined velocity structure and altimetry observations show that the increased sharpness is related to offshore anomalies, upstream and eastward of the Natal Bight.

3.4.3. Azores Current

The Azores Current is part of the North Atlantic subtropical gyre recirculation. West of the Mid-Atlantic Ridge, it is the extension of the southeast branch of the Gulf Stream. It crosses the ridge near the Oceanographer and Hayes fracture zones between 32°N and 37°N and then flows across the Canary Basin at about 34°N (Gould, 1985). In the eastern basin, it splits into three main southward branches. These branches, which vary seasonally and interannually, are found just east of the Mid-Atlantic Ridge, in the central basin near

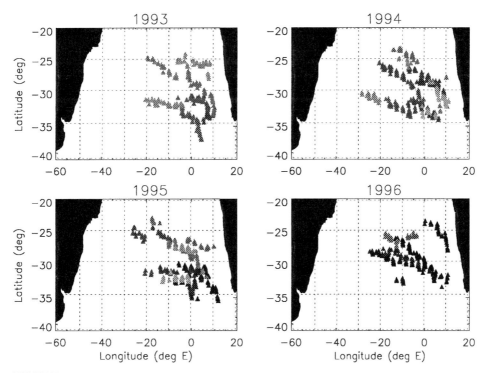

FIGURE 13 Agulhas eddy trajectories computed from the T/P data for 1993 to 1996. Note the changes from a broad Agulhas eddy corridor in 1993/1994 to a narrower corridor in 1996. Different graytones represent individual eddy tracks. (From Witter, D. C., and Gordon, A. L., 1999. With permission.)

23°W and near the coast of West Africa feeding the Canary Current, respectively (Stramma and Siedler, 1988; Klein and Siedler, 1989; Siedler and Onken, 1996). The mean transport of the Azores current is about 10 Sv in the eastern basin while it is about 30 Sv west of the ridge (Gould, 1985; Klein and Siedler, 1989). As observed by Pingree and Sinha (1997), however, the instantaneous transport in the Azores front in the eastern basin (near 32°W) can reach up to 30 Sv, but the net transport is much smaller because of north and south recirculations. The Azores current is related to a thermohaline front with strong meanders, which are embedded in a field of mesoscale eddies. It is poorly reproduced in most model simulations, and its associated mesoscale variability is largely underestimated. This makes the analysis of altimeter data particularly useful.

The Azores Current/Front mesoscale variability was first analyzed with Geosat data by Tokmakian and Challenor (1993) and Le Traon and De Mey (1994). They showed a tongue of high variability along 34°N associated with the Azores Front with rms SLA larger than 8 cm and EKE between 100 and 200 $cm^2 sec^{-2}$. Dominant space and time scales are greater than 300 km and 100 days (Le Traon and De Mey, 1994). Propagation velocities were westward, and they also noted the wavelike structure of the mesoscale field, which they attributed to Rossby waves generated near the eastern coast. Cippolini et al. (1997) also found westward propogation in both T/P and ATSR (SST) data, with Rossby wave characteristics consistent with the theory of Killworth et al. (1997). An analysis of Reynolds stresses and eddy–mean flow interaction (see Section 4.6.1) from Geosat data by Le Traon and De Mey (1994) showed that the eddy field tends to accelerate the mean flow and its recirculations.

Hernandez et al. (1995) have analyzed T/P and ERS-1/2 data in the Azores front region east of the Mid-Atlantic ridge, together with Semaphore in situ measurements (Eymard et al., 1996; Tychensky et al., 1998). There is very good agreement between hydrographic and drifter measurements and altimetry (Figure 14). Both altimetry and in situ data reveal the development of large Azores front meanders from July 1993 to November 1993. In situ data show that this development is caused by the interaction of the Azores Front with Mediterranean lenses (Meddies) coming from the east (Tychensky et al., 1998). Meddies are often observed in the Canary basin, and an early study with Geosat data (Stammer et al., 1991) showed that their surface signature could be detected by altimetry. In the Azores Frontal region just west of the Mid-Atlantic Ridge, a large cyclonic eddy associated with a sea-level depression of 40 cm (a "Storm") was observed with both ERS-1/2 and hydrographic data (Pingree and Sinha, 1998). The in situ data showed that the eddy signature penetrated down to 4000 m. The eddy was likely associated with a cyclonic meander of the Azores Front. Pingree and Sinha (1997) conclude that the observed westward-propagating features in the Canary basin are cyclonic and

anticyclonic eddies rather than Rossby waves. From their field data, the eddy separation was ~500 km and the eddies move at a mean westward speed of 3 km per day (in the eastern basin). Cromwell et al. (1996) applied a synthetic geoid technique using ERS-1 and hydrographic data on the northern side of the Azores Front near 28°W to analyze absolute velocities for the ERS-1 3-day repeat periods in spring 1992 and spring 1994. They found a persistent westward flow at 35°N, with velocities of typically 25 cm sec^{-1}, which is probably related to recirculation north of the Azores Front. The Azores Front recirculation has also been observed from hydrographic data and is probably eddy-driven through rectification mechanisms (Alves and Colin de Verdière, 1998).

Most of these results are confirmed by a recent analysis of more than 5 years of T/P, ERS-1, and ERS-2 data in the Canary basin (Hernandez and Le Traon, 1999). Marked seasonal to interannual variations exist, both in the relative position of the current and in the sea-level variability. However, the seasonal behavior of the Azores Current and recirculation is less clear because the signal is dominated by meoscale variability and Rossby wave signals. In agreement with Geosat results (Le Traon and De Mey, 1994), altimeter data show the existence of Rossby wave signals with wavelengths larger than 500 km. They correspond to Azores Front meanders which propagate upstream (westward) (see Cushman-Roisin, 1994). While altimetry has provided a much better visualization of the Azores front, the mechanisms responsible for its mesoscale variability are not yet completely understood. Baroclinic instability is probably one important mechanism, but Rossby waves, meddies, Mediterranenan outflow, wind forcing, and bathymetry also have to be taken into account to explain the dynamics of the region. In that respect, recent numerical and theoretical evidences of the role of Mediterranean outflow as a possible source of forcing of the Azores Current (Jia, 2000) are partculary interesting.

3.4.4. Iceland-Faeroe Front and Norwegian Atlantic Current

The Iceland-Faeroe Front region, lying between Iceland and the Faraoe Islands, is a permanent but highly variable frontal region, separating the warm, salty North Atlantic water from the colder, fresher Arctic waters coming from the Greenland and Norwegian seas. Although this is a dynamically important region, there are limited in situ observations which do not resolve the large spatial and temporal variability of the front, and the frequent cloud cover limits satellite SST measurements. Altimetry provides a means for monitoring the position and strength of the front, which has cross-frontal slopes of 10–35 cm, current speeds of 25–50 cm sec^{-1}, and meanders with radii of curvature of around 25 km (Robinson et al., 1989, Pistek and Johnson, 1992a). The 3-day repeat ERS-1 data show that the variability in the frontal zone is on time scales of days (Tokmakian, 1994),

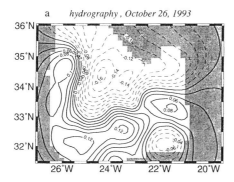

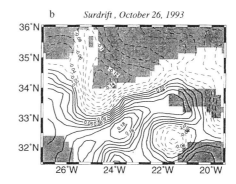

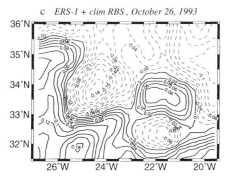

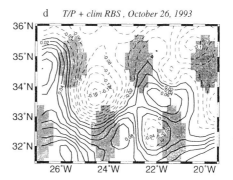

FIGURE 14 The meandering of the Azores front east of the mid-Atlantic ridge as observed from (a) hydrographic and (b) drifter data from the Semaphore experiment and (c) ERS-1 and (d) TOPEX/POSEIDON data. (From Hernandez, F., *et al.*, 1995. With permission.)

although the time scales are longer away from the main jet. This suggests that significant frontal variability may be missed by longer repeat missions, and it also indicates that combinations of satellite missions (ERS-1 and T/P, for example) are necessary for better monitoring of surface frontal characteristics. Altimetric studies have also been used to monitor the occurrence of cold cross-frontal jets, related to the formation of cold eddies south of the frontal boundary (Scott and McDowall, 1990).

The transport of the Norwegian Atlantic Current has also been determined from a combination of altimetry and along-track and climatological CTD data by Pistek and Johnson (1992b). Mean transports of volume, heat, and salt were calculated as 2.9 Sv, 8×10^{11} Kcal sec^{-1}, and 1×10^8 kg sec^{-1}. Samuel *et al.* (1994) used a different technique of combining Geosat residuals with a mean sea surface determined from an isopycnal ocean model, but found similar mean transports with a significant seasonal variation of 1.8 Sv, with maximum transports in winter (February to March) and minimum in summer (July to August).

3.5. Semi-Enclosed Seas

A major problem in studies of semi-enclosed seas is that the standard altimetric corrections for the tides and atmo-

spheric pressure response have larger errors in the coastal or semienclosed seas than in the open ocean. These local tidal and inverse-barometer correction errors are often reduced by applying specific regional corrections.

3.5.1. Mediterranean Sea

The Mediterranean Sea circulation is characterized by a complex combination of mesoscale, subbasin, and basin scale signals (Millot, 1991; Robinson *et al.*, 1991) (Figure 15). In the western basin, the Atlantic Water first flows eastward along the Spanish coast, veers south across the Alboran sea, and forms the unstable Algerian current whose instabilities generate intense mesoscale activity in the form of anticyclonic eddies and meanders (e.g., Millot, 1991). Other components of the western circulation are the cyclonic gyres in the Tyrrhenian Sea and in the Balearic basin and the Liguro–Provençal current. The Atlantic Water then enters the eastern basin, via the Sicily Straits, meanders in the Ionian basin, and forms the Mid-Mediterranean jet. This current is described as being persistent and weak. On both sides of the jet, subbasin gyres can be identified, such as the rather stable cyclonic Rhodes and Ionian gyres to the north and the seemingly transient anticyclonic Mersa-Matruh and Shikmona gyres to the south (e.g., Robinson *et al.*, 1991). The seasonal and interannual variability of the circulation is thought to be mainly caused by wind-stress forcing, although heat-flux variations and changes in inflow/outflow at Gibraltar are also important mechanisms.

Satellite altimetry has provided an important contribution to analyzing the Mediterranean Sea circulation variability. Use of satellite altimetry in the Mediterranean Sea is particularly difficult, however, because the signal is small and the geometry (narrowness, straits, islands) of the basin is complex. The latter makes the orbit error reduction more difficult and means that the specific problems in coastal areas (radiometer correction, tides) need to be addressed. Larnicol et al. (1995) analyzed T/P data in the entire Mediterranean Sea and showed that the surface circulation was more complex in the eastern basin than in the western basin. The eastern basin is composed of subbasin-scale gyres, such as the so-called Mersa-Matruh and Shikmona gyres, which do not have an obvious recurrence period. They observed a winter intensification of the large-scale cyclonic circulation in the western and in the Ionian basins. Several mesoscale structures, such as the Alboran gyres (see also Vazquez et al., 1996) and the Ierepetra gyre, also showed a clear seasonal cycle with a maximum in summer. The good qualitative and quantitative agreement of the results with in situ data from the Mediterranean illustrated the improved accuracy of T/P over its predecessors (see, for example, Fuda et al., 1999). Iudicone et al. (1998) also analyzed 2 years of T/P data and AVHRR data and confirmed the conclusions of Larnicol et al. (1995). They also suggested that the variations in mesoscale variability intensity between 1993 and 1994 were related to wind-forced interannual variations.

Algerian eddies were analyzed in several studies (Ayoub, 1997; Vignudelli, 1997; Ayoub et al., 1998; Iudicone et al., 1998; Bouzinac et al., 1998). These studies showed the development of Algerian eddies, their eastward propagation, and their detachment from the coast and propagation to the northwest. Bouzinac et al. (1998) deduced from a complex EOF analysis of combined T/P and ERS-1 maps (from Ayoub et al., 1998) that Algerian eddies may detach from the coast at two locations (4°E and 8°E). Nardelli et al. (1999) performed a detailed analysis of the Sicily channel using T/P and ERS-2 along-track data, AVHRR data, and in situ measurements. They found a good match between altimeter and hydrographic data.

Ayoub (1997) and Ayoub et al. (1998) extended the Larnicol et al. (1995) work using ERS-1 and ERS-2 data and T/P over 5 years. SLA maps were systematically calculated from T/P and ERS-1/2 separately and T/P and ERS-1/2 combined. They showed that the improved sampling given by the combination of T/P and ERS-1/2 is vital for monitoring mesoscale signals in the Mediterranean. Over 5 years, the 3-month variations of combined SLA maps show the main characteristics of the Mediterranean circulation (Alboran gyres; Algerian eddies; Ionian, Ierepetra, Mersa-Matruh, and Shikmona gyres, etc.) and the seasonal variations in these features (strengthening of cyclonic circulation in winter, strengthening of anticyclonic Alboran and Ierepetra gyres in summer and fall) (Figure 16). T/P and ERS-1/2 reveal some of the particularly strong signals very well: Alboran gyres east of Gibraltar and Ierepetra gyre southeast of Crete. The strong seasonal signal from the Ierepetra gyre (its

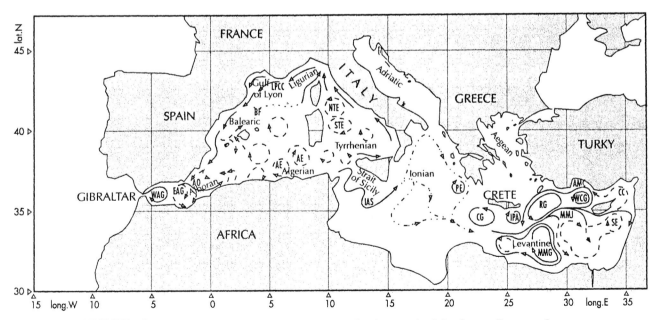

FIGURE 15 General circulation in the Mediterranean Sea and main ocean circulation features. Permanent features are solid lines and recurrent features are dashed lines. (From Iudicone, D., et al., 1998. With permission.)

diameter is about 100–200 km, which is much larger than the local internal Rossby radius, hence the gyre appellation) is probably linked to direct forcing by strong Etesian winds, which interact with the Cretan topography (orographic effect). There is also a large interannual variability, in particular in the Levantine basin. The interannual variability in the Levantine might be related to a change in the Etesian winds and the switching from a state with a well-developed Ierepetra gyre to a state with a large anticyclonic system in the central Levantine basin (with the development of the Mersa-Matruh and Shikmona anticyclonic gyres) (Ayoub, 1997).

These results provide, for the first time, a global view of the intraseasonal, seasonal, and interannual variations of

the circulation in the Mediterranean Sea. This is crucial for model validation and for a better understanding of the Mediterranean Sea circulation (e.g., Pinardi and Navarra, 1993).

3.5.2. Yellow and East China Seas

The Yellow and East China seas form a vast expanse of continental shelf bounded by the Chinese mainland and limited offshore by the Kuroshio, with Kyushu to the north and Taiwan to the south. The circulation in this region is strongly determined by the proximity to the Kuroshio and by the monsoon winds, which are northerly in winter and southeasterly in summer. Altimetry has been used to describe the seasonal variations in surface circulation by Yanagi *et*

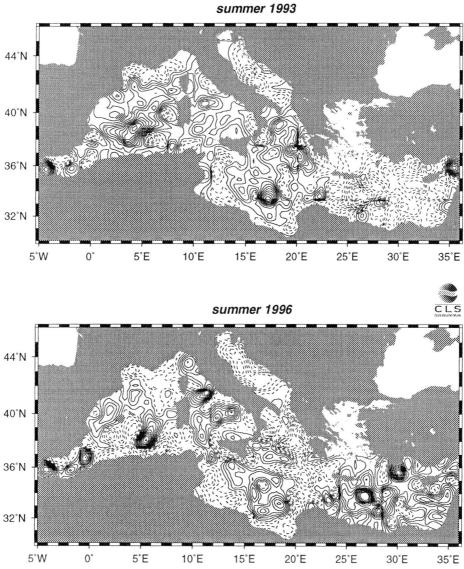

FIGURE 16 Seasonal/interannual variations of the Mediterranean circulation as derived from the combination of T/P and ERS-1/2. The maps are 3-month averages and correspond to the the summer 1993 and the summer 1996. Units are in cm. (Courtesy of G. Larnicol.)

al. (1997), who apply a regional tidal model to correct the T/P residuals. For the Yellow Sea, they find that a cyclonic circulation develops in summer, with an anticyclonic downwelling circulation in winter. In the East China Sea, the wintertime circulation is cyclonic. The sea-surface response to regional wind forcing in the Yellow and East China seas has also been studied at intraseasonal time scales by Jacobs (1999). The residual tidal variations are removed by a least-squares fit of the eight main tidal frequencies at each point along the ground-track. Although the instantaneous wind stress sets up local ageostrophic Ekman transports, this explains only a small fraction of the sea-level response. Most of the sea-level variability is remotely forced, and an extended EOF analysis of the wind field reveals the origins. The principal wind mode is significantly related to variations and intrusions of the Kuroshio along the shelf break. Another mode is caused by northerly wind bursts in the winter, which produces large drops in sea-surface height in the Bohai Bay and northern Yellow Sea regions. The third and fourth modes are associated with typhoon passages in the autumn, which are related to positive SSH anomalies moving from the northern Yellow Sea to the Chinese coast.

3.5.3. Gulf of Mexico

In the Gulf of Mexico, Geosat altimetry has been used to investigate the shedding of eddies from the Loop Current. Johnson *et al.* (1992) tracked two major rings shed from the Loop Current, which drifted southwestward across the Gulf; they also monitored the build-up of the Loop Current during 1985/1986. Jacobs and Leben (1990) found eddies shed from the Loop Current with period of approximately 10.5 months, consistent with results from numerical models and satellite infrared data. More recently, various U.S. institutions have used a combination of T/P and ERS-1/ERS-2 data in near-real time, to monitor Loop Current eddies and other circulation variability in the Gulf of Mexico. As an example, Biggs *et al.* (1996) describe the cleavage of an anticyclonic Loop Current eddy, Eddy Triton, by a deepwater cyclonic circulation. The altimeter data allowed them to track both pieces: the major part drifted northwest to the "eddy graveyard" in the northwest corner of the Gulf; the minor part drifted southwest to the continental margin, turned north, later coalesced with the major fragment, and was ultimately entrained into another eddy. This level of detailed surveillance is important in this region for oil exploration, for fisheries, and for tracking marine mammals.

3.5.4. Arabian Sea and Bay of Bengal

The ocean variability in the semi-enclosed seas of the northern Indian Ocean is strongly controlled by the monsoons. The western part of each basin is highly energetic, with root mean square values of 15–17 cm in the Somali Current and 13–15 cm in the western part of the Bay of Bengal (Jensen *et al.*, 1997). Altimetry has been used to monitor the three dominant gyres off the coast of Somalia, i.e., the Great Whirl, the Southern Gyre, and the Socotra Eddy (Subrahmanyam *et al.*, 1997). The strength of these eddies varies significantly with the strength of the monsoon. Propagating signals with annual and semi-annual periods have also been examined from altimetry using complex principal component analyses. In the eastern Arabian Sea, Rossby waves that radiate from the west coast of India are associated with the passage of Kelvin waves along the coastline (Subrahmanyam *et al.*, 1997). Altimetry has also been used to study the Arabian Sea Laccadive High—an anticyclonic circulating feature that forms off the southwest coast of India during the northeast monsoon. Bruce *et al.* (1998) find that the Laccadive High consists of multiple eddies, and it is forced by local and remote seasonal monsoon forcing, as well as being influenced by an intraseasonal signal that originates in the Bay of Bengal. Finally, Subrahmanyam *et al.* (1996) noted the existence of the poleward East Indian Coastal Current (EICC) along the east coast of India, using T/P data. Although the current is not detected in climatological hydrographic data, the T/P analysis suggests the current persists for the entire year, though it changes direction from north (January to August) to south (September to December) and is dominated by eddies in January to March. When fully developed in April, the current extends some 800 km, and the speed of the residual current is around 20–30 cm sec^{-1}.

4. MESOSCALE EDDIES

The first part of the chapter focused on the description of the main ocean currents and on their associated mesoscale variability (meandering, eddy shedding) as observed by satellite altimetry. The second part deals with a global statistical description of the mesoscale variability. The global space-time sampling of satellite altimetry is very well-suited to statistically describing mesoscale phenomena. Such a description can reveal geographical and temporal variations in mesoscale eddy statistics. This information can then be used to interpret the eddy structures, identify sources of eddy energy, and analyze energy transfer from these sources. This can help us understand eddy dynamics, particularly the mechanisms that generate and dissipate eddies. Global statistical descriptions are also a means of testing and validating eddy-resolving models. They are also necessary for inverse modeling and altimeter data assimilation studies.

In the first section, we start with a description of the rms sea-level variability, cross-track geostrophic velocity variance, and EKE (Section 4.1). We then review in Section 4.2 the analysis of seasonal to interannual variations in eddy intensity. Section 4.3 summarizes the findings on space and time scales of mesoscale variability and their relationship with the internal Rossby radius while Section 4.4 deals with SLA frequency and wavenumber spectra and their relationship with quasi geostrophic turbulence theory. Section 4.5 focusses on the comparison of these eddy statistics with eddy-resolving model simulations. Eddy dynamics (eddy-mean flow interaction and eddy transport) are finally discussed in Section 4.6.

4.1. Global Statistical Description

The simplest description of mesoscale variability is that obtained by the global mapping of the rms of SLA. While the first estimates derived from Seasat revealed only one-tenth of the total mesoscale energy (because of the short—3-month—duration of the repeat mission), estimates obtained from Geosat were quite accurate (e.g., Koblinsky, 1988; Sandwell and Zhang, 1989; Shum *et al.*, 1990). T/P and ERS-1/2 values are comparable to those from Geosat with some important differences. For example, T/P results contain more of the large-scale signal (which was generally removed from Geosat by the orbit-error adjustment) and also the longer time-scale signal (depending on the duration of the data set). T/P also has a reduced noise level and improved altimetric corrections. The satellites also have different repeat period and do not alias or observe the same frequencies (see also Section 2.1). Using the 3-day repetitive ERS-1 data over a 3-month period, Minster and Gennero (1995) have estimated that between 5 and 10% of the energy in the Gulf Stream and Kuroshio regions was at periods shorter than 20 and 34 days, respectively, and are thus aliased in T/P and Geosat sampling. Wunsch and Stammer (1995) have estimated the contribution of mesoscale variability to the total energy. In low-eddy-energy regions, the contribution of the large-scale and long-time-scale signals (e.g., steric signals) can be as high as half of the total energy.

Global statistics from T/P can be found in Wunsch and Stammer (1995) while ERS-1/2 results can be found in Le Traon and Ogor (1998). Figure 17 (see color insert) shows the rms SLA derived from almost 5 years of T/P and ERS-1/2 combined maps (Ducet *et al.*, 2000). The rms sea-level variability in western boundary currents (Gulf Stream, Kuroshio, Brazil–Malvinas Confluence, Agulhas) and the Antarctic Circumpolar Current is higher than 30 cm rms and can reach up to 50 cm rms. The maximum eddy variability is actually observed in the Agulhas region, followed by the Gulf Stream, Kuroshio, and Confluence regions and finally by the ACC. In very low-eddy-energy regions, i.e., in regions with no or weak mean currents, the mesoscale signal after instrumental noise filtering is typically 3 cm rms and may be influenced by other small-scale signals (e.g., internal waves). ERS-1/2 results are in excellent agreement with T/P results when orbit error has been corrected (Le Traon and Ogor, 1998). In high-latitude areas (which are not sampled by T/P), the signal is generally weak (5–8 cm rms).

Because the calculation of derivative acts as a high-pass filter, cross-track geostrophic velocities are much less sensitive to large-scale signals and are thus more representative of the mesoscale signals. The computation of velocity is more sensitive, however, to measurement noise and needs a careful filtering of sea-level data (see Section 2.3). The map of cross-track geostrophic variance is equivalent to EKE [$EKE = 1/2\,(\langle u'^2 \rangle + \langle v'^2 \rangle)$] if the field is isotropic. An analysis of global SLA maps obtained from the combination of T/P and ERS-1/2 shows that, to a first order, this approximation is valid outside the tropics (Ducet *et al.*, 2000). Shum *et al.* (1990) produced a global map of cross-track geostrophic velocity variance using Geosat data. More than 65% of the ocean is shown to have EKE values less than $300\ cm^2\ sec^{-2}$. The maximum EKE exceeds $2000\ cm^2\ sec^{-2}$ for most of the western boundary currents but reaches only $500\ cm^2\ sec^{-2}$ in the ACC. Sandwell and Zhang (1989) have produced a global map of the variance of dynamic topography slope, which translates directly into variance of geostrophic velocity outside equatorial areas. They established a correlation between the intensity of the variability and the ocean depth: areas of highest variability are in deep basins (>4 km). In the ACC, there is a close relationship between the geographical distribution of mesoscale variability and the strength of the mean circulation. This is not surprising since these narrow currents are likely to be baroclinically and barotropically unstable. There is also strong topographic control of the eddy field (and the mean field) consistent with numerical simulations (e.g., Treguier and Mc Williams, 1990). Most of these areas are also characterized by intense mean currents (western boundary currents and ACC), which are the main source of eddy energy through instability.

Ducet *et al.* (2000) provide a more recent estimation of the EKE based on the combination of 5 years of T/P and ERS-1/2. They show that the maximum levels of EKE can reach values of up to $4500\ cm^2\ sec^{-2}$ in the western boundary currents. Because of the higher resolution provided by the combination of T/P and ERS-1/2 and a low background noise variance (about $15\ cm^2\ sec^{-2}$), their estimation reveals many details that cannot be accessed with along-track data only. Stammer (1997) analyzed the correlation between the T/P-derived EKE (assuming isotropy) and the mean kinetic energy (MKE) (0/1000 dbar geostrophic current) as derived from Levitus historical data. As expected, there is good correspondence between T/P EKE and MKE maxima, as the currents are the main sources of eddies. There are a few noteworthy exceptions in the Agulhas retroflection, the East and West Australian currents, and the Brazil–Malvinas Confluence regions. In the ACC, rather high mesoscale variability areas are also found in regions of abrupt changes of bottom topography and do not appear to be associated with strong mean currents. This may imply that in these areas, baroclinic instability (caused by vertical shear) is not likely to occur and that other mechanisms are to be sought (e.g., barotropic instability, mean current/bottom topography interaction). Finally, it should be noted that altimeter estimates generally agree well with *in situ* measurements (in particular, drifting buoys). Differences can almost always be explained by differences in sampling and/or differences in measurement content (ageostrophic signals, differences in mean signal removal–time mean or space/time mean for drifting buoy) (Le Traon *et al.*, 1990; Hernandez *et al.*, 1995; Stammer, 1997).

4.2. Seasonal Variations of Mesoscale Variability Intensity

Several mechanisms could explain seasonal or interannual variations in mesoscale variability intensity. In regions of intense currents (e.g., western boundary currents), where baroclinic/barotropic instabilities of the mean currents are the main forcing mechanisms, a seasonal/interannual intensification of the current may induce a growth of the instability rates. A change in the current position could also imply a similar shift of the eddy field. These changes could be indirectly related to wind forcing, which is the main forcing of the ocean circulation. Strass *et al.* (1992) also suggest that the development of the seasonal pycnocline during the heating season can generate strong potential vorticity gradients favoring baroclinic instability. In such a case, heat flux forcing would be indirectly responsible for an enhanced mesoscale variability during the summer season. In regions of low eddy energy, fluctuating wind forcing could be a possible direct source of mesoscale variability (e.g., Frankignoul and Müller, 1979; Willebrand *et al.*, 1980; Müller and Frankignoul, 1981). At high and intraseasonal frequencies, the wind forcing generates predominantly barotropic fluctuations in the form of forced waves and Rossby waves and has been invoked to explain the coherence between wind-stress curl fluctuations and deep-ocean mooring sites in the North Atlantic and North Pacific (Willebrand *et al.*, 1980; Samelson, 1990; Luther, 1990) and the increase in deep eddy energy in winter (Dickson *et al.*, 1992; Koblinsky *et al.*, 1989). As emphasized by Lippert and Müller (1995) and Müller (1997) and in qualitative agreement with observations, local and nonlocal coherence between ocean response and wind forcing is expected because the ocean response integrates forcing from different places. Model simulations show also high-frequency large-scale barotropic signals in high-latitude regions consistent with T/P observations (Fu and Davidson, 1995). Except in a few localized regions, the barotropic signals are small, however, and will not induce an important change in the surface mesoscale signals as seen by an altimeter, which are dominated by baroclinic signals (Wunsch, 1997). Wind fluctuations also generate baroclinic signals (e.g., Frankignoul and Müller, 1979; Müller and Frankignoul, 1981). Wind is thus one of the forcing

mechanisms of the low-frequency (mainly seasonal to interannual) baroclinic Rossby waves that are ubiquitous in altimetric data (see Chapter 2). In the mesoscale band, the ocean response to stochastic wind fluctuations can also be baroclinic when nonlinearities and bottom topography are taken into account (Treguier and Hua, 1987, 1988). Wind forcing is thus a possible candidate for direct surface mesoscale signal generation.

Several attempts have been made to analyze seasonal variations of mesoscale intensity with Geosat data (Fu *et al.*, 1988; Zlotnicki *et al.*, 1989). Zlotnicki *et al.* (1989) showed that some regions (North-East Atlantic and North-East Pacific) had higher mesoscale energy during winter, when wind intensity is stronger. This suggests that wind forcing may be a source of eddy energy there. Small seasonal variations were also found in the western boundary currents where a maximum of eddy energy was generally found during summer and fall. Stammer and Böning (1996) reexamined results from Geosat and concluded that the maximum of mesoscale energy was in fall rather than in winter, thus before the maximum of wind-stress curl energy supporting the thermodynamic mechanism proposed by Strass *et al.* (1992). Most of these results were based on only 2 years of Geosat data and should be interpreted with caution. In addition, to analyze the mesoscale variability variations, it would be preferable to analyze the EKE rather than the rms sea-level variability, as the latter includes more large-scale signals.

Garnier and Schopp (1999) have analyzed the effect of wind on the mesoscale activity in the North Atlantic using 2 years of T/P data and ERS-1 scatterometer winds. Their study suggests that the wind plays an important role in the time evolution of the mesoscale variability. The mesoscale variability associated with the Gulf Stream and the North Atlantic Drift increases when the wind induces an intensified eastward Sverdrupian velocity. In this case, the baroclinic instability mechanism could be responsible for the mesoscale variability intensification. Seasonal and interannual variations in eddy energy were analyzed by Stammer and Wunsch (1999) using 4 years of T/P data (see also Ducet *et al.*, 2000). Over most of the subtropical region and along major mean fronts, the variations are weak and often negligible. Western boundary current extensions show an annual cycle, with a maximum occurring in late summer/early fall. A pronounced annual cycle in eddy energy was also apparent in the eastern North Pacific and the northern and eastern North Atlantic. In these regions, Stammer and Wunsch (1999) found a significant correlation between EKE and wind-stress forcing at seasonal and interannual scales (Figure 18) EKE seasonal variations related to wind forcing are also observed in a North Atlantic high resolution model simulation (Stammer *et al.*, 2000). Note also that seasonal modulations of the eddy energy are clearly observed in other regions (e.g., California Current, Canary Current, Mediterranean Sea) (see previous

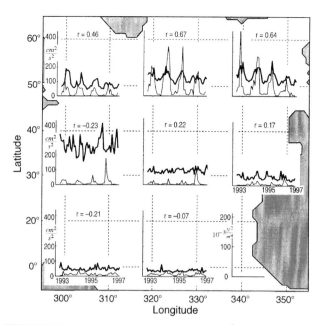

FIGURE 18 Time series of monthly KS ($K_S = \sin^2(\phi)\, K_E$, where K_E is the T/P cross-track geostrophic velocity variance and ϕ is the latitude) and τ^2 (where τ is the wind stress) in the North Atlantic (from Stammer and Wunsch, 1999).

section). In the California Current, Kelly *et al.* (1998) found, in particular, maximum EKE values along fronts which coincided with regions of maximum Ekman pumping (caused by wind-stress curl) suggesting an indirect relationship between the eddies, presumably formed by baroclinic instability, and winds.

Qiu (1999) has made a detailed analysis of the North Pacific Subtropical Countercurrent (STCC) using more than 5 years of T/P data. The mean EKE of the STCC is about 330 cm^2 sec^{-2} and reaches half the energy of the Kuroshio Extension. The unique characteristics of this current system is that it has a very clear EKE seasonal cycle with a maximum of EKE in April/May and a minimum in December/January (Figure 19). Using a 2 1/2 layer reduced-gravity model, representing the vertically sheared STCC/North Equatorial Current (NEC) system, Qiu (1999) convincingly shows that the seasonal modulation of the STCC eddy field is related to seasonal variations in the intensity of baroclinic instability. The seasonal cooling/heating of the upper thermocline modifies the vertical velocity shear of the STCC/NEC and the density difference between the STCC and NEC layers. As a result, the spring time condition is considerably more favorable for baroclinic instability than the fall-time condition. The theoretically predicted e-folding time scale of the instability is 60 days and matches the time lag between the EKE maximum and the maximum shear of the STCC/NEC (Figure 20).

4.3. Space and Time Scales of Mesoscale Variability

Space and time scales can be derived from altimeter covariance functions, which are mathematically equivalent to frequency and wavenumber spectra (through a Fourier transform). We will focus here on the main space and time scales as defined by the zero crossing of the correlation function while the next section on frequency/wavenumber spectra allows a more detailed discussion of the frequency and wavenumber content of altimeter signals.

Results obtained with Geosat in the Atlantic (Le Traon *et al.*, 1990; Le Traon, 1991; Stammer and Böning, 1992) showed a clear latitudinal variation of the space scales, decreasing toward the poles. Although a relationship with internal Rossby radius (IRR) was observed, the spatial scales seen by altimetry varied by a factor of 2 from the equator to the pole, while the Rossby radius varied by more than a factor 4 (Le Traon, 1993). Time scales were shorter in high-variability areas while longer time scales were observed above the Mid-Atlantic Ridge and in low-eddy-energy regions. They were generally not proportional to space scales (Le Traon, 1991). Bottom topography appeared to play an important role in the temporal coherence of mesoscale structures. Stammer (1997) recently repeated the calculation using T/P data. Since T/P SLA data also contain the large-scale and low-frequency signals (which were in large part removed during the Geosat orbit error removal), T/P space and time scales are much larger than those for Geosat. If the large-scale signal is removed, however, or if the analysis is performed on geostrophic velocities, the scales are much more representative of the mesoscale variability. Time and space scales are then consistent with Geosat values. Globally, they vary between 5 and 20 days and between 60 and 200 km, respectively. A relationship with the IRR similar to Geosat results is noted by Stammer (1997) (Figures 21 and 22). Spatial scales in the high latitudes seen by T/P (or Geosat) are too large, however, compared to the IRR. Because of low stratification and larger Coriolis parameter, the ocean is expected to be more barotropic and baroclinic instability may not be the main eddy-generation mechanism. A relationship with IRR would be valid only if one could separate barotropic and baroclinic signals in altimeter SLA measurements. While baroclinic signals dominate SLA signals (Wunsch, 1997), they may not be negligible at high latitudes. This was confirmed by an analysis of CME model simulations (Stammer, 1997).

The relationship of spatial scales with IRR certainly points to the baroclinic instability as the main generation mechanism for mesoscale variability. Indeed, in baroclinic instability, the most unstable perturbation has the IRR scale (e.g., Gill, 1982). However, the dependence is not expected to follow a simple law. It depends, in particular, on the further evolution of ocean eddies through turbulent cascade,

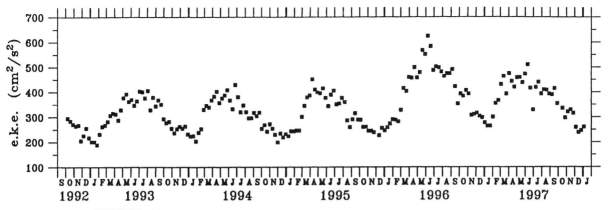

FIGURE 19   Time series of the EKE in the STCC region. (From Qiu, B., 1999. With permission).

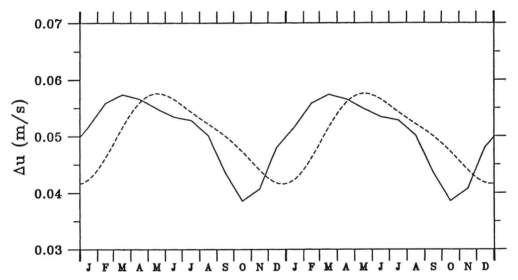

FIGURE 20   Seasonal change in the vertical shear between the STCC and its underlying NEC (solid line) versus the seasonal change in the EKE (dashed line). (From Qiu, B., 1999. With permission).

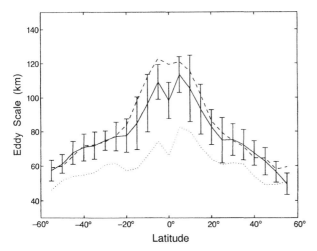

FIGURE 21 Eddy scales estimated from TOPEX data as the integral scale of the sea-level anomaly autocorrelation function (up to the first zero crossing) and averaged zonally between 0° and 360° in longitude. (From Stammer, D., 1997. With permission.)

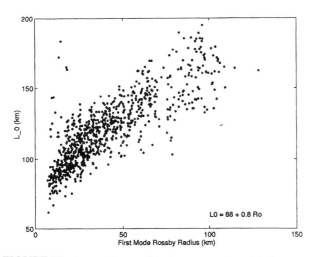

FIGURE 22 Scatter diagram of the first zero crossing of the SLA autocorrelation function against the corresponding first internal Rossby radius. (From Stammer, D., 1997. With permission.)

which may not exist everywhere (see discussion below). Different relationships with IRR can also be observed in the case of other forcing mechanisms. For wind forcing, the energy input is on larger wavelengths than the IRR, and it is the ratio of the largest forced wavelength to the IRR that determines the space scales of the ocean response (Treguier and Hua, 1987).

The relationship with IRR is more difficult to estimate from *in situ* data because of the lack of adequate space/time sampling. Paillet (1999), in an analysis of scales of vortices from XBT data, reports no significant relationship in the eastern North Atlantic. He argues that in the baroclinic instability mechanism the most unstable perturbation is not necessarily the one that is excited and that grows. Vortices are also long-lived structures and may be observed far

away from their initial formation region. Finally, vortices are known to coalesce with neighbors (quasigeostrophic turbulence—see below) but the turbulence may not be well developed in quiet regions because it requires a large number of eddies. There, eddies would keep their initial size, while in regions where turbulence is active they would grow. Paillet (1999) points to this as a possible explanation for rather large eddies in the North Atlantic Current compared to regions more to the South. On the contrary, Krauss *et al.* (1990) report a good relationship between eddy spatial scales and the IRR from drifting buoy data.

4.4. Frequency/Wavenumber Spectral Analysis

4.4.1. Frequency Spectra

SLA frequency spectra have been computed by Stammer (1997) in different areas of the world ocean with T/P data. All spectra show a peak at the annual frequency (mainly related to the dilatation/contraction of surface waters in response to heat flux forcing and to a less extent to wind forcing). Slopes of the spectra generally range between -1 and -2, the steeper slopes being found in high-mesoscale-variability areas. Slopes are slightly lower than slopes derived from subsurface currentmeter data. Stammer (1997) suggests that this may be because of near-surface signals not seen in subsurface currentmeter data.

4.4.2. Wavenumber Spectra

4.4.2.1. Observations
Whereas frequency spectra were already well known from *in situ* (currentmeter) measurements (although not with the same global coverage but with a much better time sampling), the wavenumber spectra are a unique contribution of altimetry. Following the pioneering work of Fu (1983) with Seasat, a detailed analysis was conducted by Le Traon *et al.* (1990) in the North Atlantic with Geosat data (Figure 23). They found spectral slopes in the mesoscale band of -4 in the western part of the basin and between -2 and -3 in the other areas between 50–200 km and 200–600 km. After the break in the slope, which showed a decrease according to latitude, the spectra remain red in the eastern part of the basin. An increased energy at smaller scales (below 100 km) is also observed there. Forbes *et al.* (1993) systematically studied of the wavenumber spectra from Geosat data in the South Atlantic and found similar shapes of the spectra. A different view of the same spectra was, however, proposed by Stammer and Böning (1992). They suggested that Geosat spectra were not significantly different from a k^{-5} law and that noise was probably responsible for the weaker slopes in the eastern basin. This led to some controversy (Le Traon, 1993; Stammer and Böning, 1993). Results obtained with the more precise T/P data (Stammer, 1997) are actually in excellent agreement with Geosat results and thus confirmed that noise

was not responsible for the observed shape of the Geosat spectra. Surprisingly, slopes as weak as k^{-1} can also be observed in very high latitudes from T/P and/or ERS-1/2 data. More recently, Paillet (1999) used an ensemble of 102 high-resolution XBT tracks in the North-East Atlantic and showed that the dynamic height wavenumber spectrum was following a k^{-3} law consistent with Geosat and T/P results.

4.4.2.2. Interpretation of Wavenumber Spectra The interpretation of wavenumber spectra is closely related to theories of geostrophic turbulence. Geostrophic turbulence is a complex and evolving topic, and the reader is referred to Kraichnan (1967), Charney (1971), Rhines (1979), Colin de Verdière (1982), or McWilliams (1989) for a much more detailed review. Geostrophic turbulence is close to 2D turbulence and is characterized by a red cascade of energy toward larger scales while there is an inverse cascade (i.e., toward smaller scales) for the enstrophy. In pure 2D homogeneous turbulence, the cascade would continue up to the scale of the basin, which is obviously not occurring. By taking into account the presence of a background planetary vorticity gradient (β-plane turbulence), however, a competition between Rossby waves (dominant at large-scales) and turbulence (dominant at small-scales) occurs. The cascade of energy is severely reduced at large-scales by the dispersion of Rossby waves which prevents the merging of eddies. Rhines defines the β-arrest scale as $L_\beta = (\beta/2U)^{-1/2}$, which is the scale for which the β effect becomes more important than nonlinearities. At scales larger than L_β, the turbulence is strongly reduced and a linear regime consisting of Rossby waves is observed. While the turbulent regime is characterized by an isotropy, the more linear regime becomes anisotropic (owing to the dispersion relation of Rossby waves). At the end state of the cascade, flow is mainly zonal ($u > v$) (because of the form of the Rossby wave dispersion relation at long wavelengths). Stratified quasigeostrophic turbulence theories yield a similar cascade in the horizontal (Hua and Haidvogel, 1986; McWilliams, 1989). In addition, there is also a red cascade in the vertical scale. In a two-layer system, eddies in the two layers start to interact when the horizontal scale is close to the internal Rossby radius and to lock in the vertical leading to a barotropic signal. Then the energy cascade proceeds as in the 2D case. More generally, the turbulent cascade in a rotating stratified fluid leads to a 3D isotropy once the vertical coordinate has been rescaled by N/f and for wavenumbers higher than the wavenumber at which the energy is input (McWilliams, 1989). Those are highly idealized views of the interaction of eddies in the ocean. There are many obstacles to the development of the energy cascade in either the horizontal or vertical scales (coastal boundaries, emergence of localized vortices, bottom topography) (McWilliams, 1989).

Through dimensional arguments, Kraichnan (1967) predicted for 2D turbulence a $k^{-5/3}$ energy spectrum at scales larger than the energy input scale, while the enstrophy cascade to smaller scales leads to a k^{-3} energy spectrum. In extending these ideas to 3D quasigeostrophic turbulence, Charney (1971) also obtained a k^{-3} energy spectrum which has been confirmed in numerical simulations (e.g., Herring, 1980; Hua and Haidvogel, 1986). A k^{-3} energy spectrum would yield a k^{-5} one-dimensional (along-track) sea-surface height spectrum assuming isotropy (Fu, 1983b; Le Traon *et al.*, 1990). As noted by Le Traon *et al.* (1990) and Stammer (1997), altimeter wavenumber spectra in high-eddy-energy regions compare qualitatively with quasigeostrophic turbulence models although the observed slopes are only in k^{-4} rather than the expected k^{-5} law. In these regions, eddies are mainly generated through baroclinic instability, and the dynamics are highly nonlinear. In low-eddy-energy regions, where the altimeter slopes are significantly lower (k^{-3}, k^{-2}, or even k^{-1}), the turbulence may not be as active and the ocean may be in a more linear regime. The ubiquitous observation of Rossby waves in the subtropical oceans (see Chapter 2) favor this interpretation of a more linear regime outside major ocean currents.

The shape of the spectra can also be different if there is significant wind forcing. Treguier and Hua (1987), in simulations of quasigeostrophic turbulence forced by wind, found spectra more in agreement with altimeter observations in low-eddy-energy regions. In particular, they found that the spectra remain red for wavelengths larger than those at the break of the slope. Altimetric spectral slopes may also be weaker than expected because of nongeostrophic effects related to the mesoscale variability of the mixed layer (Klein and Hua, 1988; Le Traon *et al.*, 1990). Glazman *et al.* (1996) and Glazman and Cheng (1999) thus suggest that internal waves may affect the high-wavenumber part of the altimeter spectra. Note that internal tide signals were clearly detected in altimeter data in the tropics (e.g., Ray and Mitchum, 1996) and can also be clearly seen as peaks in the T/P wavenumber and frequency/wavenumber spectra. Filtering of those signals may increase the spectral slopes in low-eddy-energy regions for intermediate wavelengths.

4.4.3. Frequency/Wavenumber Spectra

Frequency/wavenumber spectra allow us to associate space and time scales of mesoscale variability as well as the propagation characteristics. Le Traon (1991) calculated frequency/wavenumber spectra in the Atlantic from Geosat. The dominant wavelengths of around 200–600 km (depending on latitude) are associated with long periods (>150 days) in the eastern part of the basin, while near the Gulf Stream significant energy is also found at shorter periods. In the Gulf Stream area, propagation velocities can be either westward or eastward for short periods (<80 days). At longer periods and in the eastern North Atlantic, they are mainly westward. Seasonal signals associated with westward propagation are also observed in these frequency/wavenumber spec-

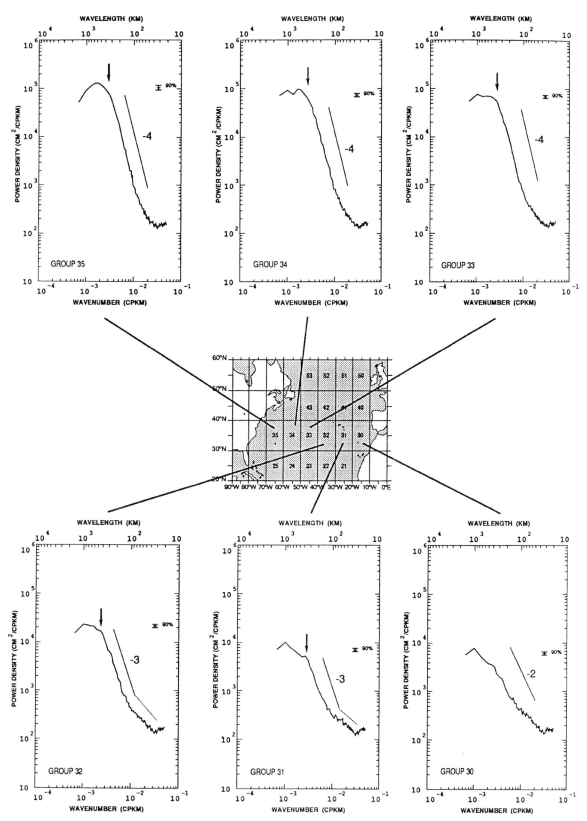

FIGURE 23 Along-track sea-level anomaly wavenumber spectra in the North Atlantic as derived from 2 years of Geosat data. (From Le Traon, P. Y., *et al.*, 1990. With permission.)

tra (Le Traon, 1991). These results confirm the interpretation of wavenumber spectra by Le Traon *et al.* (1990), who suspected a change in dynamic regime after the break in the wavenumber spectrum slope. Indeed, pseudo-dispersion relations deduced from Geosat data point to two distinct dynamic regimes, as in numerical models: a turbulent regime for smaller scales (<300–500 km), where there is proportionality between space and time scales, and an apparently more linear regime after the spectral peak in wavenumber where an inverse dispersion relation is found in the eastern part of the basin. This latter feature is in agreement with quasigeostrophic models forced by fluctuating winds (Treguier and Hua, 1987). Several investigators have used these analyses to characterize the mesoscale variability in regional areas (e.g., Le Traon and Minster, 1993; Provost and Le Traon, 1993; Morrow and Birol, 1999).

4.5. Comparison with Eddy-Resolving Models

The comparison of statistical descriptions such as spectra and EKE with results from eddy-resolving models is useful for model validation and for a better understanding of eddy dynamics. It should also help us to better understand the observed frequency/wavenumber of mesoscale variability and its relation with the forcing. Altimeter data have been compared with model data in numerous studies (e.g., Stammer and Böning, 1992; Treguier, 1992; Hughes, 1995; Fu and Smith, 1996; Stammer *et al.*, 1996, the DYNAMO group, 1997; McClean *et al.*, 1997). Comparisons have been made between the model and altimetry rms sea-level variability and EKE, but a few studies include higher-order statistics (space and time scales, wavenumber spectra, Reynolds stresses, and eddy–mean flow interaction). These studies show that eddy-resolving models still largely underestimate the mesoscale variability. Recent numerical simulations in the North Atlantic with a $0.1°$ resolution by Smith *et al.* (1999) show remarkable improvements over previous lower-resolution experiments for both the time–mean circulation and mesoscale variability. With a $0.1°$ resolution, the rms sea-level variability compares favorably with T/P observations over most of the domain, including the Gulf Stream region, the North Atlantic Current to the east of the Grand Banks, and the Azores Current (Smith *et al.*, 1999; Ducet, 2000). Such a good degree of qualitative agreement should allow a much more detailed comparison with altimetry, including higher-order statistics.

4.6. Eddy Dynamics

4.6.1. Eddy–Mean Flow Interactions

Eddy–mean flow interactions include the generation of eddies through barotropic and baroclinic instabilities of the mean flow, as well as the convergence of eddy momentum

fluxes (Reynolds stresses) to accelerate the mean current. Before the advent of altimetry, these eddy–mean flow interactions had mostly been investigated using analytical or numerical models. Some *in situ* studies based on current meter moorings at particular locations had revealed the complexities of the local eddy–mean field. However, there are very few locations with *in situ* records longer than 2 years, necessary to derive stable eddy statistics (see Wunsch, 1997, for a summary). Even though we cannot derive an accurate mean current from altimetry, the eddy–mean interactions can still be estimated using a mean derived from hydrography or from numerical simulations.

In terms of an energy budget, the eddy–mean flow interactions can be separated into four components:

1. The transfer from the mean kinetic energy to mean potential energy
2. The transfer of mean potential energy to eddy potential energy
3. The transfer of eddy potential energy to EKE
4. The transfer of EKE to mean kinetic energy (MKE)

The first three components require details of the vertical density structure to define the mean and eddy potential energy fields, which are clearly not accessible from altimetry. Only the final component includes the purely kinematic terms. This barotropic energy conversion of EKE to MKE can be calculated from the horizontal momentum equations (see Wilkin and Morrow, 1994, for details),

$$\overline{u'u'}\frac{\partial \overline{u}}{\partial x} + \overline{u'v'}\frac{\partial \overline{v}}{\partial x} + \overline{u'v'}\frac{\partial \overline{u}}{\partial y} + \overline{v'v'}\frac{\partial \overline{v}}{\partial y}, \qquad (10)$$

where the overbar denotes a time mean, and the prime denotes the departure from this mean. When this term is positive, EKE is converted to MKE at this location via the Reynolds stresses doing work on the mean shear so as to accelerate the mean flow. There is also a contribution from the vertical Reynolds stresses ($\overline{u'w'}$, $\overline{v'w'}$), which are neglected since the vertical velocities are assumed weak near the sea surface.

A similar calculation can be made to derive the eddy momentum flux divergence from the horizontal momentum equations. Again, neglecting the contribution from vertical velocity perturbations, the eddy momentum flux divergence is given by

$$-\frac{\overline{u}}{|\overline{u}|}\left(\frac{\partial}{\partial x}\overline{u'u'} + \frac{\partial}{\partial y}\overline{u'v'}\right) - \frac{\overline{v}}{|\overline{u}|}\left(\frac{\partial}{\partial x}\overline{u'v'} + \frac{\partial}{\partial y}\overline{v'v'}\right).$$
$$(11)$$

Here $(\overline{u}, \overline{v})/|\overline{u}|$ denotes the unit vector parallel to the coordinates of the mean flow$(\overline{u}, \overline{v})$, so the first term represents a contribution to the zonal momentum balance, and the second contributes to the meridional momentum balance.

Altimetry has provided the first global measurement of the surface eddy Reynolds stress terms, ($\overline{u'u'}$, $\overline{v'v'}$, $\overline{u'v'}$), which has been useful for improving our understanding of ocean dynamics, as well as validating estimates from numerical models. Note that energy and momentum budgets also rely on vertically averaged quantities over the water column, so estimates of the full barotropic conversion terms require a relationship for the vertical velocity structure, as well as an estimate of the mean flow.

One of the first studies to investigate the spatial distribution of eddy momentum fluxes from altimetry was made by Tai and White (1990) in the Kuroshio region. They created 0.5° latitude/longitude maps of Geosat ascending track sea-level anomalies for every 17-day repeat cycle, using a spatial decorrelation radius of 200 km. Velocity anomalies are calculated from the mapped SLA assuming geostrophy, and the eddy flux of zonal momentum is estimated from the $\overline{u'v'}$ component. Their maps show positive eddy momentum flux values south of the jet and negative values to the north, implying a convergence of eddy momentum flux which tends to accelerate the eastward mean jet. The north-south slope in eddy momentum flux $\partial/\partial y(\overline{u'v'})$ shows convergence and eastward acceleration between 35 and 37°N (the jet axis), and divergence or westward acceleration farther north from 37.5 to 38.5°N and south from 30 to 33°S. This pattern is consistent with theories of nonlinear baroclinic instabilities occurring near the jet axis, which favor meander growth, and westward Rossby wave propagation dominating to the north and south of the mean jet. Geosat measurements also confirm the convergence of eddy momentum flux at zonal frontal zones in the eastern North Atlantic (Beckman et al., 1994), at both the subpolar front and the Azores Front (see also Le Traon and De Mey, 1994).

The interannual variations in eddy–mean flow interactions have been investigated in the Kuroshio Extension region by Qiu (1995) and Adamec (1998). During the first 2 years of T/P, Qiu (1995) noted significant variations in EKE levels in the Kuroshio Extension and its recirculation gyre (see Figure 6). The EKE in the southern recirculation gyre increased during this period, and energetic analyses showed an energy transfer from the mean field to the eddy field, because of barotropic instabilities. This energy transfer also led to a reduction in the intensity of both the eastward-flowing Kuroshio Extension and the westward recirculation gyre. The relation between the barotropic energy conversion and eddy heat fluxes was also examined by Adamec (1998) using a combination of T/P and AVHRR data. During summer 1994, the convergence of surface Reynolds stresses in the Kuroshio Extension was much weaker than normal, leading to low values of EKE 3 months later. The time difference is consistent with the characteristic time scales of the eddy energy (about 90–100 days) in this region. The upper-ocean heat content was then calculated using a combination of AVHRR sea-surface temperature data and XBT data. The lower EKE in summer 1994 was accompanied by a cooler southern recirculation and warming north of the Kuroshio Extension. This decreased the large-scale baroclinicity across the jet and therefore reduced the near surface transport and EKE. Calculations of the baroclinic zonally symmetric circulation indicated that the fronts were also substantially weakened during summer 1994, consistent with the reduced EKE. The primary reason for this change in baroclinicity was the changes in the convergence of eddy heat fluxes (Figure 24; see color insert). Adamec (1998) suggested that interannual variations in eddy activity, including eddy heat fluxes, may play a crucial role in modulating the seasonal signal in the Kuroshio Extension.

Another technique that involves less spatial smoothing of the eddy signals is calculating the surface geostrophic velocity vectors at altimeter crossover points (Morrow et al., 1992, Johnson et al., 1992), as described in Section 2.3.1. Morrow et al., (1992, 1994) applied this technique using 2 years of Geosat data in the Southern Ocean, and they found anisotropic geostrophic velocity variability in the vicinity of the major currents and strong bathymetric features. The variance ellipse orientation also has important implications for the eddy flux of horizontal momentum: where the ellipse axes are aligned parallel or perpendicular to the mean flow, there is no cross-stream transfer of momentum. Eddy momentum fluxes in the Southern Ocean were calculated by Morrow et al. (1992, 1994) using this crossover technique and streamwise coordinates. They found a net convergence of along-stream momentum along the mean axis of the ACC, suggesting that eddies tend to accelerate the mean jet. The Geosat data revealed a very complex geographical distribution of Reynolds stress convergence and divergence, suggesting that other instability mechanisms may be important locally. The altimeter data also provided the first quantitative proof that the Southern Ocean eddy momentum flux divergence was too small, and in the wrong direction, to balance the momentum input by the wind.

4.6.2. Eddy Transports, Eddy Viscosity, and Eddy Diffusivity

Eddies are an important mechanism for the transport of heat, salt, and momentum in the ocean. However, the eddy transport of different properties has remained a large uncertainty for estimating the global property budgets, because of the difficulty and expense in measuring eddies *in situ*. In addition, eddy effects must be parameterized in coarse-resolution ocean models, and the parameterizations are tested on a very limited eddy data base. Altimetry allows us to monitor the global surface eddy characteristics of the flow and has the potential to improve our understanding of the geographical distribution of eddy transports of different properties.

In turbulence theory, the turbulent Reynolds stresses can be taken as proportional to the local gradient of the mean flow, via a proportionality term given as the eddy viscosity, ν. For example, for the zonal component of the flow, u, the horizontal and vertical eddy viscosity terms are defined from

$$-\overline{u'u'} = v_x \frac{\partial \overline{u}}{\partial x}; \quad -\overline{u'v'} = v_y \frac{\partial \overline{u}}{\partial y}; \quad -\overline{u'w'} = v_z \frac{\partial \overline{u}}{\partial z}. \quad (12)$$

In coarse-resolution ocean models, the eddy viscosity is often assumed constant. Altimetry allows us to investigate the global distribution of the surface eddy viscosity. The formulation given above has been used by Johnson *et al.* (1992) to estimate eddy viscosity locally in the Pacific sector of the ACC. Using a combination of Geosat altimetry for the horizontal Reynolds stresses and climatological data for the zonal mean flow, they estimated an average eddy viscosity of $v_y = 8 \times 10^3$ m^2 sec^{-1}, consistent with other estimates for oceanic flows. They note, however, that point measurements of eddy viscosity undergo much larger fluctuations. Note that this parametrization is obviously oversimplified, as eddies or waves can drive mean motions yielding negative viscosity.

The eddy transport associated with any passive tracer can be represented in the form of a Fickian diffusion,

$$v'\lambda' = -\kappa \nabla_h \overline{\lambda}, \quad (13)$$

where $\overline{\lambda}$ is the time-mean passive tracer field (such as salt or temperature) and κ is an isotropic diffusivity tensor. Again, in coarse-resolution ocean models, κ is often taken as spatially uniform in all three spatial dimensions, so estimating its geographical distribution is useful for improving model parameterizations.

A number of different techniques have been used to estimate the eddy diffusivity, k, for passive tracers directly from altimetry. Assuming conditions of statistically homogeneous, barotropic, β-plane turbulence, Holloway (1986) and Keffer and Holloway (1988) estimated $\kappa \cong C\tau\psi$, where C is a proportionality constant ($\sim$0.4) and τ is an O(1) anisotropy tensor induced by Rossby wave propagation. ψ is the rms value of the streamfunction derived from altimetric sea-level anomalies h', where $\psi = gh'/f$, g is the acceleration of gravity, and f is the Coriolis parameter. The calculations by Holloway (1986) and Keffer and Holloway (1988) are based on 3 months of Seasat data and include various ad hoc scaling factors to take account of the short time series and the fact that the surface variability overestimates the depth-averaged eddy fluxes. To derive meridional eddy heat and salt fluxes in the Southern Ocean, they apply Eq. (13) with the meridional gradient in mean temperature and salinity derived from Levitus climatological data. Given the data limitations, they derive surprisingly good estimates of poleward heat flux, sufficient to supply the heat lost to the atmosphere south of the Antarctic Polar Front. The salinity fluxes were plausible at most latitudes, but poleward at high latitudes and thus in the wrong direction to compensate for the observed excess of P-E south of the Polar Front. (The errors in the salinity calculations are not surprising, given that the

Seasat data spanned the austral winter from July to October, when *in situ* salinity observations are almost nonexistent.)

An alternative technique developed by Stammer (1998) is to calculate the eddy diffusivity, κ, as proportional to the typical horizontal turbulent velocity scale, u', and length scale, l', i.e., $\kappa \propto u'l'$, both derived from altimeter data. This is rewritten in terms of altimetric eddy kinetic energy, K_E, and an averaged eddy integral time scale derived from altimetry on a 5° geographical grid, T_{alt}:

$$\kappa = 2\alpha K_E T_{alt}. \quad (14)$$

Here, α is a scaling factor, $\alpha = 0.7 \times (L_{R_0}/L)^2$, where L_{R_0} and L are the first internal Rossby radius and an eddy mixing scale; α is further scaled by 0.1 to convert altimetric surface EKE values to an averaged EKE over the top 1000 m. The factor of 2 is included to convert the integral time scale, T_{alt}, to the equivalent of the first zero crossing of the autocorrelation function, which is closer to the decorrelation scales observed from current meter data.

Stammer (1998) calculates the eddy diffusivity κ globally from T/P altimetry using equation (14) and compares this with the method of Holloway (1986). Both techniques give similar geographical distributions, but the Holloway calculation is twice as large as Stammer's. This may be because the streamfunction calculation includes more large-scale variability from seasonal steric changes and planetary wave motion, which are unrelated to eddy transfer processes. Eddy diffusivity values reach 2500 m^2 sec^{-1} in the energetic western boundary currents, decreasing to 250 m^2 sec^{-1} in the interior and eastern parts of the basins, and the geographical distribution is quite inhomogeneous. Stammer (1998) also calculates global eddy heat and salt transports using mean meridional gradients of temperature and salinity from Levitus mean climatology. Strong poleward eddy heat and salt transports occur in the energetic western boundary currents: the Gulf Stream, the Kuroshio, and the Agulhas Current (Figure 25). In the equatorial band 5°N to 5°S, eddy heat and salt fluxes are northward in the Pacific and southward in the Indian Ocean, with equatorward eddy transports occurring between 5 and 20° latitude.

These altimeter-based eddy transports are consistent with local *in situ* estimates [see Wunsch (1999) for a review] but demonstrate the strong spatial inhomogeneity of the eddy diffisivity field. One should note, however, that numerous *ad hoc* scaling factors have been included, in particular for the vertical distribution of properties, and as such they provide a first-order and qualitative approximation of eddy transports. In addition, these transport estimates are also based on climatological mean meridional gradients, and so should be considered a lower bound on instantaneous eddy-transports in the ocean. In the future, better estimates should be possible using eddy-resolving models with assimilation of remote sensing and *in situ* data.

5. CONCLUSIONS

Satellite altimetry has made a unique contribution to observing and understanding ocean currents and eddies, which stems from its excellent space-time coverage, providing both a quasi-synoptic description and a statistical description of the ocean surface circulation. We now have a much better description of the upper-ocean current systems and their associated mesoscale variability. Altimeter data analyses have produced global, quantitative estimates of eddy energy with high spatial resolution, revealing details such as the correlation of EKE with the mean currents and the role of the bathymetry. They have provided, for the first time, a global description of the seasonal/interannual variations in eddy energy. In most regions, significant mesoscale variations appear to be related to changes in the intensity of the mean current instabilities. An additional feature revealed by studying these variations has been the possible role of forcing by fluctuating winds in a few regions of the ocean. The frequency/wavenumber mesoscale circulation spectrum has been characterized, as have been the corresponding time and space scales. The eddy–mean flow interactions have also

been mapped for the first time and provide an important ingredient for understanding the western boundary current and ACC dynamics. Our examples also show the potential of synoptic mapping of mesoscale variability, for example, for monitoring Agulhas eddies. Such studies are useful for explaining the structure of eddies and for better understanding eddy dynamics. They provide a good means of testing and validating models and theories.

Despite all this progress, there is still much to learn from altimeter data for mesoscale variability studies. Some suggestions for future work are given below. Recent improvements of eddy-resolving model simulations are impressive and models have now a high degree of realism. More detailed comparisons of altimetry (including comparison of higher-order statistics such as frequency/wavenumber spectra and Reynolds stresses) with eddy-resolving models should now be very instructive. The global frequency/wavenumber characteristics of altimeter sea-level variability could also be explained and possibly related to different kinds of forcing. It may also be complemented by a 3D (x, y, t) spectral analysis of combined T/P + ERS-1/2 maps, which would allow us in addition to better characterize the anisotropy of the eddy field. A better understanding of altimeter

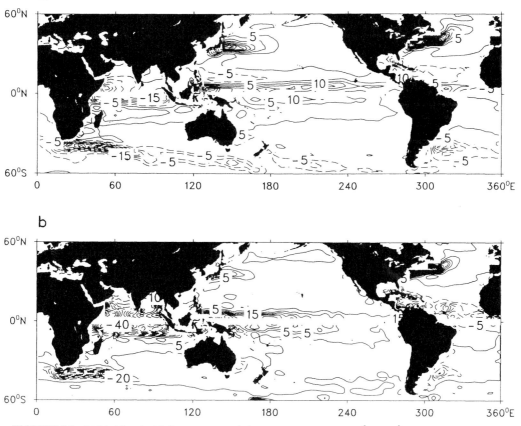

FIGURE 25   (a) Meridional eddy heat transport, $v'T'$; contour interval is 5×10^6 W m^{-1}. (b) Meridional eddy salt transport, $v'S'$; contour interval is 5 kg m^{-1}. Negative values are dashed. (From Stammer, D., 1998. With permission.)

wavenumber spectra will also require more detailed comparisons with models and quasigeostrophic turbulence simulations and theories. This is important for interpretation of altimeter data in regions of weak wavenumber spectral slopes. Systematic studies of individual eddies (such as studies of Agulhas eddies) should also be performed and the relation between eddies and Rossby waves should be better analyzed. Finally, the seasonal/interannual variations of the major current systems and the seasonal/interannual variations in eddy energy can be analyzed further since we now have almost 10 years of good altimeter data sets (at least for mesoscale studies). All these studies should take advantage of the sampling provided by the combination of several altimeter missions (two and possibly three missions).

The combined use of altimeter data with other data sets such as *in situ* hydrographic data (*T* and *S* profiles), current meters and drifters, SST from infrared imagery, and forcing fields should also be developed further. The comparison/combination of altimeter data with TOGA and WOCE surface drifter data should provide, for example, a means of estimating the ageostrophic component of the surface circulation. The vertical structure of mesoscale variability should also be obtained (at least in a statistical sense—see Wunsch, 1997) from an analysis of *in situ* data such as the ARGO profiling float array (ARGO Science Team, 1998). More generally, the integration of altimeter data with *in situ* data (*T* and *S* profiles, drifters, floats) and forcing data is crucial for a better understanding of ocean currents and mesoscale variability dynamics. Assimilation into models is a powerful means for performing such an integration. This aspect is developed in the last chapter of this book and is obviously highly relevant to the above discussion. It is related to the development of operational oceanography which will benefit scientific studies on ocean currents and eddies by providing (as it is today for the atmosphere) a regular description of the three-dimensional ocean dynamically consistent with altimeter data, *in situ* data, and forcing data. This is one of the challenges of the future Global Ocean Data Assimilation Experiment (GODAE) (Smith and Lefebvre, 1997).

ACKNOWLEDGMENTS

We thank the two anonymous reviewers for their careful review of the chapter; the detailed comments were very useful in improving the quality of the chapter. We also thank D. Adamec, N. Ducet, F. Hernandez, B. Qiu, and G. Reverdin for their comments/suggestions on the chapter.

References

Adamec, D. (1998). Modulation of the seasonal signal of the Kuroshio Extension during 1994 from satellite data. *J. Geophys. Res.* **103**, 10,209–10,222.

Alves, M., and de Verdière, A. C. (1999). Instability dynamics of a subtropical jet and applications to the Azores front-current system – eddy driven mean flow. *J. Phys. Oceanogr.* **29**, 837–864.

Argo Science Team (1998). On the design and implementation of ARGO – an initial plan for a global array of profiling floats. ICPO Report N°21. GODAE Report N°5. Published by the GODAE International Project Office, Melbourne, 32 pp.

Arístegui, J. P., Sangrá, S., Hernández-León, Cantón, M. A., Hernández-Guerra, and Kerling, J. L. (1994). Island eddies in the Canary Islands. *Deep-Sea Res.* **41**, 1509–1525.

Ayoub, N. (1997). Variabilité du niveau de la mer et de la circulation en Méditerranée à partir des données altimétriques et de champs de vent. Comparaison avec des simulations numériques. Phd Thesis, Université Paul Sabatier.

Ayoub, N., Le Traon, P. Y., and De Mey, P. (1998). Combining ERS-1 and TOPEX/POSEIDON data to observe the variable oceanic circulation in the Mediterranean sea. *J. Mar. Sys.* **18**, 3–40.

Beckman, A., Böning, C. W., Brügge, B., and Stammer, D. (1994). On the generation and role of eddy variability in the central North Atlantic Ocean. *J. Geophys. Res.* **99**, 20,381–20,391.

Bhaskaran, S., Lagerloef, G. S. E., Born, G. H., Emery, W. J., and Leben, R. R. (1993). Variability in the Gulf of Alaska from Geosat altimetry data. *J. Geophys. Res.* **98**, 16,311–16,330.

Biggs, D. C., Fargion, G. S., Hamilton, P., and Leben. R. R. (1996). Cleavage of a Gulf of Mexico Loop Current eddy by a deep water cyclone. *J. Geophys. Res.* **101**, 20,629–20,641.

Birol, F., and Morrow, R. A. (2000). Source of the baroclinic waves in the Southeast Indian Ocean, *J. Geophys. Res.* (in press).

Bouzinac, C., Vasquez, J., and Font, J. (1998). Complex empirical orthogonal function analysis of ERS-1 and TOPEX/POSEIDON combined altimetric data in the region of the Algerian current. *J. Geophys. Res.* **103**, 8059–8072.

Bruce, J. G., Kindle, J. C., Kantha, L. H., Kerling, J. L., and Bailey, J. F. (1998). Recent observations and modeling in the Arabian Sea Laccadive High region. *J. Geophys. Res.* **103**, 7593–7600.

Byrne, D. A., Gordon, A. L., and Haxby. W. F. (1995). Agulhas eddies: A synoptic view using Geosat ERM data. *J. Phys. Oceanogr.* **25**, 902–917.

Challenor, P. G., Read, J. F., Pollard, P. T., and Tokmakian, R. T. (1996). Measuring surface currents in Drake Passage from altimetry and hydrography. *J. Phys. Oceanogr.* **26**, 2748–2759.

Charney. (1971). Geostrophic turbulence. *J. Atmos. Sci.* **28**, 1087–1095.

Chelton, D. B., Schlax, M. G., Witter, D. L., Richman, J. G. (1990). Geosat altimeter observations of the surface circulation of the southern ocean. *J. Geophys. Res.* **95**, 17,877–17,903.

Chelton, D. B., and Schlax, M. (1994). The resolution capability of an irregularly sampled data set: With application to Geosat altimeter data. *J. Atm. Ocean. Tech.* **95**, 17,877–17,903.

Cippolini, P., Cromwell, D., Jones, M. S., Quartly, G. D., and Challenor, P. (1997). Concurrent altimeter and infrared observations of Rossby wave propagation near 34°N in the Northeast Atlantic. *Geophys. Res. Lett.* **24**, 889–892.

Colin de Verdière, A. (1982). Dynamics of low frequency motions. In "Space Oceanology," pp.179–247. Cepadues-Editions, Toulouse.

Cromwell, D., Challenor, P. G., New, A. L., and Pingree, R. D. (1996). Persistent Westward Flow in the Azores Current as seen from altimetry and hydrography. *J. Geophys. Res.* **101**, 423–933.

Cushman-Roisin, B. C. (1994). *Introduction to Geophysical Fluid Dynamics*, p. 320. Prentice Hall, Englewood Cliffs, NJ.

Cushman-Roisin, B. C., Chassignet, E. P., and Tang, B. (1990). Westward motion of mesoscale eddies. *J. Phys. Oceanogr.* **20**, 758–768.

De Ruijter, W. P. M., Van Leeuwen, P. J., and Lutjeharms, J. R. E. (1999). Generation and evolution of notal pulses: solitary meanders in the Algulhas current. *J. Geophys. Res.* **29**, 3043–3055.

Dickson, R. R., Gould, W. J., Gurbutt, P. A., and Killworth, P. D. (1982). A seasonal signal in ocean currents to abyssal depths. *Nature* **295**, 193–198.

Douglas, B. C., and Cheney, R. E. (1990). Geosat: Beginning a new era in satellite oceanography. *J. Geophys. Res.* **95**, 2833–2836.

Ducet, N. (2000). Combinaison des données altimétriques TOPEX/POSEIDON et ERS-1/2 pour l'étude de la variabilité mésoéchelle. Comparaison avec un modèl à haute résolution de l'Atlantique Nord, Ph.D. Thesis. Paul Sabatier University, Toulouse.

Ducet N., Le Traon, P.Y., and Reverding, G. (2000). Global high resolution mapping of ocean circulation from TOPEX/POSEIDON and ERS-1/2. *J. Geophys. Res.* (in press).

DYNAMO group, (1997). DYNAMO Dynamics of North Atlantic Models: Simulation and assimilation with high resolution models, Berichte IfM Kiel Nr. 294.

Eymard *et al.* (1996). The Semaphore-93 mesoscale air/sea experiment. *Ann. Geophys.* **14**, 986–1015.

Feron, R. C. V., De Ruijter, W. P. M., and Oskam, D. (1992). Ring shedding in the Agulhas Current system. *J. Geophys. Res.* **97**, 9467–9477.

Forbes C., Leaman, K., Olson, D., and Brown, O. (1993). Eddy and wave dynamics in the South Atlantic as diagnosed from Geosat altimeter data. *J. Geophys. Res.* **98**, 12297–12314.

Frankignoul, C., and Müller, P. (1979). Quasi-geostrophic response of an infinite β-plane ocean to stochastic forcing by the atmosphere. *J. Phys. Oceanogr.* **9**, 104–127.

Fu, L.-L. (1983a). Recent progress in the application of satellite altimetry to observing the mesoscale variability and general circulation of the oceans. *Rev. Geophys. Space Phys.* **21**, 1657–1666.

Fu, L.-L. (1983b). On the wave number spectrum of oceanic mesoscale variability observed by the Seasat altimeter. *J. Geophys. Res.* **88**, 4331–4341.

Fu, L.-L., Zlotnicki, V., and Chelton D. B. (1988). Satellite altimetry—observing ocean variability from space. *Oceanography* **1**, 4–11.

Fu, L.-L., and Davidson R. A. (1995). A note on the barotropic response of sea level to time-dependent wind forcing. *J. Geophys. Res.* **100**, 24,955–24,963.

Fu, L.-L., and Cheney R.E. (1995). Applications of satellite altimetry to ocean circulation studies: 1987–1994. *Rev. Geophys.* **32**, Supplement, 213–223, 1995.

Fu, L.-L., and Smith R. D. (1996). Global ocean circulation from satellite altimetry and high-resolution computer simulations. *Bull. Amer. Meteor. Soc.* **77**, 2625–2636.

Fuda, J. L., Millot, C., Taupier-Letage, I., Send, U., and Bocognano, J. M. (1999). XBT monitoring of a meridian section across the Western Mediterranea Sea. *Deep-Sea Res.* (submitted).

Garnier, V., and Schopp R. (1999). Wind forcing and mesoscale activity along the Gulf Stream and the North Atlantic current. *J. Geophys. Res.* (in press).

Garzoli, S. L., Goni, G. J., Mariano, A. J., and Olson, D. B. (1997). Monitoring the upper southeastern Atlantic transports using altimeter data. *J. Mar. Res.* **55**, 453–481.

Gill, A. E. (1982). *In* "Atmosphere-Ocean Dynamics." Academic Press, London.

Gille, S. T. (1994). Mean sea surface height of the Antarctic Circumpolar Current from Geosat data: Method and application. *J. Geophys. Res.* **99**, 18,255–18273.

Gille, S. T., and Kelly, K. A. (1996). Scales of spatial and temporal variability in the Southern Ocean. *J. Geophys. Res.* **101**, 8759–8773.

Glazman, R. E., and Cheng, B. (1998). Altimeter observations of baroclinic oceanic inertia-gravity wave turbulence. *Proc. Roy. Soc. London* (in press).

Glazman, R. E., Fabrikant, A., and Greysukh, A. (1996). Statistics of spatial-temporal variations of sea surface height based on Topex altimeter measurements. *Int. J. Remote Sensing* **17**, 2647–2666.

Glenn, S. M., Porter, D. L., and Robinson, A. R. (1991). A synthetic geoid validation of Geosat mesoscale dynamic topography in the Gulf Stream region. *J. Geophys. Res.* **96**, 7145–7166.

Goni, G., Kamholz, S., Garzoli, S., and Olson, D. (1996). Dynamics of the Brazil-Malvinas Confluence based on inverted echo sounders and altimetry. *J. Geophys. Res.* **101**, 16,273–16,289.

Goni, G. J., Garzoli, S. L., Olson, D., and Brown, O. (1997). Agulhas ring dynamics from TOPEX/POSEIDON satellite altimeter data. *J. Mar. Res.* **55**, 861–883.

Gordon, A. L., and Haxby, W. F. (1990). Agulhas eddies invade the south Atlantic: evidence from Geosat altimeter and shipboard conductivity temperature-depth survey. *J. Geophys. Res.* **95**, 3117–3125.

Gould, W. J. (1985). Physical oceanography of the Azores front. *Progress in Oceanography* **14**, 167–190.

Greenslade, D. J. M., Chelton, D. B., and Schlax, M. G. (1997). The Midlatitude resolution capability of sea level fields constructed from single and multiple satellite altimeter datasets, *J. Atm. Ocean. Tech.* **14**, 849–870.

Gründlingh, M. L. (1995). Tracking eddies in the southeast Atlantic and southwest Indian oceans with TOPEX/POSEIDON. *J. Geophys. Res.* **100**, 24,977–24,986.

Hernandez, F., Le Traon, P. Y., and Morrow, R. (1995). Mapping mesoscale variability of the Azores current using TOPEX/POSEIDON and ERS-1 altimetry, together with hydrographic and Lagrangian measurements. *J. Geophys. Res.* **100**, 24,995–25,006.

Hernandez, F., and Le Traon, P. Y. (1999). Variability of the Canary basin circulation from satellite altimetry. In CANIGO (MAS3-CT96-0060). Final scientific report, pp. 45–55.

Hernandez, F. (1998). Estimation d'un géoïde synthétique sur la zone Semaphore. Rapport final. Contrat SHOM/CLS N° 97.87.003.00.470.29.45.

Hernandez-Guerra, A., and Nykjaer, L. (1997). Sea surface temperature variability off north-west Africa: 1981–1989. *Int. J. Remote Sensing* **18**, 2539–2558.

Herring. (1980). Statistical theory of quasigeostrophic turbulence. *J. Atmos. Sci.* **37**, 969–977.

Heywood, K. J., Stevens, D. P., and Bigg, G. R. (1996). Eddy formation behind the tropical island of Aldabra. *Deep-Sea Res.* **43**, 555–578.

Holland, W. R., Harrison, P. E., and Semtner, A. J. (1982). Eddy resolving numerical models of large scale ocean circulation, *In* "Eddies in Marine Science," 609 pp., (A. R. Robinson ed.), Springer-Verlag, Berlin.

Holland, W. R., and Lin, L. B. (1975). On the generation of mesoscale eddies and their contribution to the oceanic general circulation. I. A preliminary numerical experiment. *J. Phys. Oceanogr.* **5**, 642–657.

Holloway, G. (1986). Estimation of oceanic eddy transports from satellite altimetry. *Nature* **323**, 243–244.

Howden, S. D., Gottlieb, E., Koblinsky, C., Rossby, T., and Wang, L. (1998). Gulf Stream observations from TOPEX/POSEIDON and in comparison with in-situ measurements near 71°W. *J. Geophys. Res.* (submitted).

Hua, B. L., and Haidvogel, D. B. (1986). Numerical simulations of the vertical structure of quasi-geostrophic turbulence. *J. Atmos. Sci.* **43**, 2923–2936.

Hughes, C. W. (1995). Rossby waves in the Southern Ocean: A comparison of TOPEX/POSEIDON altimetry and model predictions. *J. Geophys. Res.* **100**, 15,933–15,950.

Hummon, J. M., and Rossby T. (1998). Spatial and temporal evolution of a Gulf Stream crest-warm core ring interaction. *J. Geophys. Res.* **103**, 2795–2810.

Hwang, C. (1996). A study of the Kuroshio's seasonal variabilities using an altimetric-gravimetric geoid and TOPEX/POSEIDON altimeter data. *J. Geophys. Res.* **101**, 6313–6335.

Ichikawa, K., Imawaki, S., and Ishii, H. (1995). Comparison of surface velocities determined from altimeter and drifting buoy data. *J. Oceanogr.* **51**, 729–740.

Imawaki, S., and Uchida, H. (1997). Time series of the Kuroshio transport derived from field observations and altimetry data. *Intl. WOCE Newsletter* **25**, 15–18.

Iudicone, D., Santolieri, R., Marullo, S., and Gerosa, P. (1998). Sea level variability and surface eddy statistics in the Mediterranean sea from TOPEX/POSEIDON data. *J. Geophys. Res.* **103**, 2995–3012.

Jacobs, G. A. (1999). Sea surface height response to regional wind forcing in the Yellow and East China Seas. 1. Linear response to local wind stress. *J. Geophys. Res.* **103**, 18459–18478.

Jacobs, G. A., and Leben, R. R. (1990). Loop Current eddy shedding estimated using Geosat altimeter data. *Geophys. Res. Lett.* **17**, 2385–2388.

Jia, Y. (2000). On the formation of an Azore current due to Mediterranean outflow in a modelling study of the North Atlantic. *J. Phys. Oceanogr.* (in press).

Jensen, V. E., Samuel, P., and Johannessen, O. M. (1997). Studying the monsoon circulation in the Indian Ocean using altimeter data, TOPEX/POSEIDON SWT Abstracts, Biarritz, France.

Johnson, D. R., Thompson, J. D., and Hawkins, J. D. (1992). Circulation in the Gulf of Mexico from Geosat altimetry during 1985–1986. *J. Geophys. Res.* **97**, 2201–2214.

Johnson, T. J., Stewart, R. H., Shum, C. K., and Tapley, B. D. (1992). Distribution of Reynolds stress carried by mesoscale variability in the Antarctic Circumpolar Current. *Geophys. Res. Lett.* **19**, 1201–1204.

Joyce, T. M., Kelly, K. A., Schubert, D. M., and Caruso, M. J. (1990). Shipboard and altimetric studies of rapid Gulf Stream variability between Cape Cod and Bermuda. *Deep Sea Res.* **37**, 897–910.

Keffer, T., and Holloway, G. (1988). Estimating Southern Ocean eddy flux of heat and salt from satellite altimetry. *Nature* **332**, 624–626.

Kelly, K. A., and Gille, S. T. (1990). Gulf stream surface transport and statistics at 69°W from the Geosat altimeter. *J. Geophys. Res.* **95**, 3149–3161.

Kelly, K. A. (1991). The meandering Gulf Stream as seen by the Geosat Altimeter: surface transport, position, and velocity variance from 73° to 46°W. *J. Geophys. Res.* **96**, 16,721–16,738.

Kelly, K. A., Beardsley, R. C., Limeburner, R., Brink, K. H., Paduan, J. D., and Cherekin, T. K. (1998). Variability of the near-surface eddy kinetic energy in the California Current based on altimetric, drifter, and moored current data. *J. Geophys. Res.* **103**, 13,067–13,084.

Kelly, K. A., Singh, S., and Huang R. X. (1999). Seasonal variations of sea surface height in the Gulf Stream region, *J. Phys. Oceanogr.*, **29**, 313–327.

Killworth, P. D., Chelton, D. B., and De Szoeke, R. A. (1997). The speed of observed and theoretical long extra-tropical planetary waves. *J. Phys. Oceanogr.* **27**, 1,946–1,966.

Klein, P., and Hua, B. L. (1988). Mesoscale heterogeneity of the wind driven mixed layer: influence of a quasi-geostrophic flow. *J. Mar. Res.* **46**, 495–525.

Klein, B., and Siedler, G. (1989). On the origin of the Azores Current. *J. Geophys. Res.* **94**: 6,159–6,168.

Koblinsky, C. (1988). Geosat versus Seasat. *EOS, Transactions of the American Geophysical Union* **69**, 1026.

Koblinsky, C. J., Niiler, P. P., and Schmitz, W. J., Jr. (1989). Observations of wind-forced deep ocean currents in the North Pacific. *J. Geophys. Res.* **94**, 10,773–10,790.

Kraichnan. (1967). Inertial ranges in two-dimensional turbulence. *Phys. Fluids* **10**, 1417–1423.

Krauss W., Döscher, R., Lehmann, A., and Viehoff, T. (1990). On eddy scales in the eastern and northern North Atlantic ocean as a function of latitude. *J. Geophys Res.* **95**, 18,049–18,056.

Lagerloef, G. S. E., Born, G. H., Leben, R. R., Royer, T. C., and Musgrave, D. L. (1994). Mesoscale variations in the Alaska Gyre from ERS-1 altimeter data, Proceedings Second ERS-1 Symposium, Hamburg, Germany. *ESA SP* **361**, 1217–1219.

Larnicol, G., Le Traon, P. Y., Ayoub, N., and De Mey, P. (1995). Sea Level Variability in the Mediterranean Sea from two years of TOPEX/POSEIDON data. *J. Geophys. Res.* **100**, 25,163–25,177.

Lee, T., and Cornillon, P. (1995). Temporal variation of the meandering intensity and domain-wide lateral oscillations of the Gulf Stream. *J. Geophys. Res.* **100**, 13,603–13,613.

Le Traon, P. Y. (1991). Time scales of mesoscale variability and their relationship with spatial scales in the North Atlantic. *J. Mar. Res.* **49**, 467–492.

Le Traon, P. Y. (1992). Contribution of satellite altimetry to the observation of oceanic mesoscale variability. *Oceanologica Acta* **15**, 441–457.

Le Traon, P. Y. (1993). Comment on "Mesoscale variability in the Atlantic ocean from Geosat altimetry and Woce high resolution numerical modeling by D. Stammer and C.W. Böning." *J. Phys. Oceanogr.* **23**, 2729–2732.

Le Traon, P. Y., and Minster, J. F. (1993). Sea level variability in the South-Atlantic Subtropical gyre: semi-annual Rossby waves and large scale signal. *J. Geophys. Res.* **98**, 12,315–12,326.

Le Traon, P. Y., and De Mey, P. (1994). The eddy field associated with the Azores Front east of the Mid-Atlantic Ridge as observed by the Geosat altimeter. *J. Geophys. Res.* **99**, 9,907–9,923.

Le Traon, P. Y., and Ogor, F. (1998). ERS-1/2 orbit improvement using TOPEX/POSEIDON: the 2 cm challenge. *J. Geophys. Res.* **95**, 8045–8057.

Le Traon, P. Y. and G. Dibarboure (1999). Mesoscale mapping capabilities from multiple altimeter missions. *J. Atm. Ocean. Tech.* **16**, 1208–1223.

Le Traon, P. Y., Rouquet, M. C., and Boissier, C. (1990). Spatial scales of mesoscale variability in the North Atlantic as deduced from Geosat data. *J. Geophys. Res.* **95**, 20,267–20,285.

Le Traon, P. Y., Boissier, C., and Gaspar, P. (1991). Analysis of errors due to polynomial adjustments of altimeter profiles. *J. Atm. Ocean. Tech.* **8**, 385–396.

Le Traon, P. Y., Nadal, F., and Ducet, N. (1998). An improved mapping method of multisatellite altimeter data. *J. Atm. Ocean. Tech.* **15**, 522–533.

Lippert, A., and Müller, P. (1995). Direct atmospheric forcing of geostrophic eddies. Part II: coherence maps. *J. Phys. Oceanogr.* **25**, 106–121.

Lozier, M. S. (1997). Evidence for large-scale eddy-driven gyres in the North Atlantic. *Science* **277**, 361–364.

Luther, D. S., Chave, A. D., Filloux, J. H., and Spain, P. F. (1990). Evidence for local and non-local barotropic responses to atmospheric forcing during BEMPEX. *Geophys. Res. Lett.* **17**, 949–952.

Matthews, P. E., Johnson, M. A., and O'Brien, J. J. (1992). Observation of mesoscale ocean features in the northeast Pacific using Geosat radar altimetry data. *J. Geophys. Res.* **97**, 17,829–17,840.

McClean, J. L., Semtner, A. J., and Zlotnicki, V. (1997). Comparisons of mesoscale variability in the Semtner-Chervin 1/4° model, the Los Alamos Parallel Ocean Program 1/6° model, and TOPEX/POSEIDON data. *J. Geophys. Res.* **11**, 25,203–25,226.

McWilliams, J. C., and Flierl, G. R. (1979). On the evolution of isolated, nonlinear vortices. *J. Phys. Oceanogr.* **9**, 1155–1182.

McWilliams, J. C. (1985). Submesoscale, coherent vortices in the ocean. *Rev. Geophys.* **23**, 165–182.

McWilliams, J. C. (1989). Statistical properties of decaying gesotrophic turbulence. *J. Fluid Mech.* **108**, 199–230.

Meyers, S. D., and Basu, S. (1999). Eddies in the eastern Gulf of Alaska from TOPEX/POSEIDON altimetry. *J. Geophys. Res.* **104**, 13333–13343.

Millot, C. (1991). Mesoscale and seasonal variabilities of the circulation in the western Mediterranean. *Dyn. Atmos. Oceans.* **15**, 179–214, 1991.

Mitchell, J. L., Dastugue, J. M., Teague, W. J., and Hallock, Z. R. (1990). The estimation of geoid profiles in the northwest Atlantic from simultaneous satellite altimetry and airborne expendable bathythermograph sections. *J. Geophys. Res.* **95**, 17,965–17,977.

Minster, J. F., Remy, F., and Normant, E. (1993). Constraints on the repetitivity of the oribt of an altimetric satellite: estimation of the cross-track slope. *J. Atm. Ocean. Tech.* **10**, 410–419.

Minster, J. F., and Gennero, M. C. (1995). High-frequency variability of western boundary currents using ERS-1 three-day repeat altimeter data. *J. Geophys. Res.* **100**, 22,603–22,612.

Morrow, R. A., and Birol, F. (1999). Variability in the south-east Indian Ocean from altimetry: Forcing mechanisms for the Leeuwin Current. *J. Geophys. Res.* **103**, 18289–18544.

Morrow, R. A., Church, J. A., Coleman, R., Chelton, D. B., and White, N. (1992). Eddy momentum flux and its contribution to the Southern Ocean momentum balance. *Nature* **357**, 482–484.

Morrow, R. A., Coleman, R., Church, J. A., and Chelton, D. B. (1994). Surface eddy momentum flux and velocity variance in the Southern Ocean from Geosat altimetry. *J. Phys. Oceanogr.* **24**, 2050–2071.

Müller, P. (1997). Coherence maps for wind-forced quasigeostrophic flows. *J. Phys. Oceanogr.* **27**, 1927–1936.

Müller, P., and Frankignoul, C. (1981). Direct atmospheric forcing of geostrophic eddies. *J. Phys. Oceanogr.* **11**, 287–308.

Nardelli, B. B., Santolieri, R., Marullo D., Iudicone D., and Zoffoli S. (1999). Altimetric signal and three dimensional structure of the sea in the channel of Sicily. *J. Geophys. Res.* **104**, 20585–20603.

Nof, D. (1981). On the β-induced movement of isolated baroclinic eddies. *J. Phys. Oceanogr.* **11**, 1662–1672.

Paillet, J., (1999). Central water vortices of the eastern North Atlantic, *J. Phys. Oceanogr.* **29**, 2487–2503.

Pares-Sierra, A., White, W. B., and Tai, C. K. (1993). Wind-driven coastal generation of annual mesoscale eddy activity in the California Current. *J. Phys. Oceanogr.* **23**, 1,110–1,121.

Park, Y. H., and Gambéroni, L. (1995). Large-scale circulation and its variability in the south Indian Ocean from TOPEX/POSEIDON altimetry. *J. Geophys. Res.* **100**, 24,911–24,929.

Parke, M. E., Stewart, R. L., Farless, D. L., and Cartwright, D. E. (1987). On the choice of orbits for an altimetric satellite to study ocean circulation and tides. *J. Geophys. Res.* **92**, 11,693–11,707.

Pedlosky, J. (1977). On the radiation of mesoscale energy in the mid-ocean. *Deep Sea Res.* **24**, 591–600.

Pedlosky, J. (1996). Ocean Circulation Theory. Springer-Verlag Berlin, 453 pp.

Pinardi, N., and Navarra, A. (1993). Baroclinic wind adjustment processes in the Mediterranean Sea. *Deep Sea Res.* **40**, 1299–1326.

Pingree, R. D., and Sinha B. (1998). Dynamic topography (ERS-1/2 and seatruth) of subtropical ring (STORM 0) in the STORM corridor (32°N-34°N, Eastern basin, North Atlantic Ocean). *J. Mar. Biol. Ass. of the UK* **78**, 351–376.

Pistek, P., and Johnson, D. R. (1992a). A study of the Iceland-Faeroe Front using Geosat altimetry and current-following drifters. *Deep-Sea Res.*, Part A, **39**, 2029–2051.

Pistek, P., and Johnson, D. R. (1992b). Transport of the Norwegian Atlantic Current as determined from satellite altimetry. *Geophys. Res. Lett.* **19**, 1379–1382.

Porter, D. L., Dobson, E. B., and Glenn, S. M. (1992). Measurements of dynamic topography during SYNOP utilizing a Geosat synthetic geoid. *Geophys. Res. Lett.* **19**, 1847–1850.

Preisendorfer, R. W. (1988). "Principal component analysis in meteorology and oceanography." Elsevier, New York.

Provost, C., and Le Traon, P. Y. (1993). Spatial and temporal scales in altimetric variability in the Brazil-Malvinas Current Confluence region: dominance of the semiannual period and large spatial scales. *J. Geophys. Res.* **98**, 18,037–18,051.

Qiu, B. (1992). Recirculation and seasonal change of the Kuroshio from altimetry observations. *J. Geophys. Res.* **97**, 17,801–17,813.

Qiu, B. (1995). Variability and energetics of the Kuroshio Extension and its recirculation gyre from the first two-year TOPEX data. *J. Phys. Oceanogr.* **25**, 1827–1842.

Qiu, B. (1999). Seasonal eddy field modulation of the North Pacific subtropical Countercurrent: TOPEX/POSEIDON observations and theory. *J. Phys. Oceanogr.* **29**, 2471–2486.

Qiu., B., Kelly, K. A., and Joyce, T. M. (1991). Mean flow and variability in the Kuroshio Extension from Geosat altimetry data. *J. Geophys. Res.* **96**, 18,491–18,507.

Quartly, G. D., and Srokosz, M. A. (1993). Seasonal variations in the region of the Agulhas Retroflection: Studies with Geosat and FRAM. *J. Phys. Oceanogr.* **23**, 2107–2122.

Rapp, R. H., and Smith, D. A. (1994). Preliminary estimates of Gulf Stream characteristics from TOPEX data and a precise gravimetric geoid. *J. Geophys. Res.* **99**, 24,707–24,723.

Rapp, R. H., and Wang, Y. M. (1994). Dynamic topography estimates using Geosat data and a gravimetric geoid in the Gulf Stream region. *Geophys. J. Int.* **117**, 511–528.

Ray, R. D., and Mitchum, G. T. (1996). Surface manifestation of internal tides generated near Hawai. *Geophys. Res. Lett.* **23**, 2101–2014.

Rhines, P. B. (1979). Geostrophic turbulence. *Ann. Rev. Fluid Mech.* **69**, 417–443.

Richardson, P. L. (1983). Eddy kinetic energy in the North Atlantic from surface drifters. *J. Geophys. Res.* **88**, 4355–4367.

Richardson, P. L. (1985). Average velocity and transport of the Gulf Stream near 55°W. *J. Mar. Res.* **43**, 83–111.

Ridgway, K. R., Dunn, J. R., Wilkin, J. L., and Walker, A. E. (1998). A satellite based ocean analysis system for Australian waters, Proceedings of the TOPEX/POSEIDON SWT Meeting, Keystone, Co, U.S.

Robinson, A. R. ed. (1982). "*Eddies in Marine Science.*" 609 pp., Springer-Verlag, Berlin.

Robinson, A. R., Golnaraghi, M., Leslie, W. G., Artegiani, A., Hecht, A., Lazzoni, E., Michelato, A., Sansone, E., Theocharis, A., and Ünlüata, Ü. (1991). The eastern Mediterranean general circulation: features, structure and variability. *Dyn. Atmos. Oceans* **15**, 215–240.

Robinson, A. R., Walstad, L. J., Calman, J., Dobson, E. B., Dembo, D. W., Glenn, S. M., Porter, D. L., and Goldhirsh, J. (1989). Frontal signals east of Iceland from the Geosat altimeter. *Geophys. Res. Lett.* **16**, 77–80.

Roemmich, D., and Sutton, P. (1998). The mean and variability of ocean circulation past northern New Zealand: determining the representativeness of hydrographic climatologies. *J. Geophys. Res.* **103**, 13,041–13,054.

Rogel, P. (1995). Variabilite mésochelle et saisonnière du gyre Nord-Atlantique, PhD Thesis, Université Paul Sabatier, Toulouse, 210 pp.

Samelson, R. M. (1990). Evidence for wind-driven current fluctuations in the eastern North Atlantic, *J. Geophys. Res.* **95**, 11,359–11,368.

Samuel, P., Johannessen, J. A., and Johannessen, O. M. (1994). A study of Atlantic water in the GIN Sea using Geosat altimeter data. *In* "The Polar Oceans and Their Role in Shaping the Global Environment." *Geophysical Monograph 85, AGU* **95**, 108.

Sandwell, D. T., and Zhang, B. (1989). Global mesoscale variability from the Geosat Exact Repeat Mission—Correlation with ocean depth. *J. Geophys Res.* **94**, 17,971–17,984.

Sato, O. T., and Rossby T. (1995). Seasonal and low frequency variations in dynamic height anomaly and transport of the Gulf Stream, *Deep-Sea Res.*, **42**, 149–162.

Schmitz, W. J., Jr. (1995). On the interbasin-scale thermohaline circulation. *Reviews of Geophys.* **33**, 151–173.

Schmitz, W. J., Jr., and Luyten, J. R. (1991). Spectral time scales for mid-latitude eddies. *J. Mar. Res.* **49**, 75–105.

Scott, J. C., and McDowall, A. L. (1990). Cross-frontal cold jets near Iceland: In-water, satellite infrared and Geosat altimeter data. *J. Geophys. Res.* **95**, 18,005–18,014.

Shum, C. K., Werner, R. A., Sandwell, D. T., Zhang, B. H., Nerem, R. S., and Tapley, B. D. (1990). Variations of global mesoscale eddy energy observed from Geosat. *J. Geophys. Res.* **95**, 17,865–17,876.

Siedler, G., and Onken, R. (1996). Eastern Recirculation. *In* "The Warmwatersphere of the North Atlantic Ocean." (W. Krauss, Ed.), pp. 339–364. Gebrueder Borntraeger, Berlin.

Smith, N., and Lefebvre M. (1997). The Global Ocean Data Assimilation Experiment (GODAE). Monitoring the oceans in the 2000s: an integrated approach, International Symposium, Biarritz, October 15–17 1997.

Smith, R. D., Maltrud, M. E., Bryan, F. O., and Hecht, M. W. (1999). Numerical simulation of the North Atlantic ocean at 1/10°. *J. Phys. Oceanogr.* (submittted).

Stammer, D. (1997). Global characteristics of ocean variability estimated from regional TOPEX/POSEIDON altimeter measurements. *J. Phys. Oceanogr.* 27, 1743–1769.

Stammer, D. (1998). On eddy characteristics, eddy transports, and mean flow properties. *J. Phys. Oceanogr.* 28, 727–739.

Stammer, D., and Böning, C. W. (1992). Mesoscale variability in the Atlantic ocean from Geosat altimetry and WOCE high resolution numerical modelling. *J. Phys. Oceanogr.* 22, 732–752.

Stammer, D., and Böning, C. W. (1993). Reply. *J. Phys. Oceanogr.* 23, 2733–2735.

Stammer, D., and Böning, C. W. (1996). Generation and distribution of Mesoscale eddies in the North Atlantic Ocean. In "The warmwatersphere of the North Altantic Ocean," (W. Krauss, Ed.), pp. 159–193, Gebruder Borntraeger, Berlin.

Stammer, D., and Wunsch, C. (1999). Temporal changes in eddy energy of the oceans. *Deep Sea Res.* 46, 77–108.

Stammer, D., Hinrichsen, H. H., and Käse, R. H. (1991). Can meddies be detected by satellite altimetry? *J. Geophys. Res.* 96, 7005–7014.

Stammer D., Tokmakian R., Semtner A., and Wunsch C. (1996). How well does a 1/4° global ocean circulation model simulate large scale oceanic observations. *J. Geophys. Res.* 101, 25,779–25,811.

Stammer, D., Böning, C., and Dieterich, C. (2000). The role of variable wind forcing in generating eddy energy in the North Atlantic. *Progr. Oceanogr.* (in press).

Stramma, L., and Siedler, G. (1988). Seasonal changes in the North Atlantic subtropical gyre. *J. Geophys. Res.* 93, 8111–8118.

Strass, V. H., Leach, H., and Woods, J. D. (1992). On the seasonal development of mesoscale variability: the influence of the seasonal pycnocline formation. *Deep-Sea Res.* 29, 1627–1639.

Strub, P. T., and James, C. (1995). The large-scale summer circulation of the California Current. *Geophys. Res. Lett.* 22, 207–210.

Strub, P. T., and James, C. (1997). Satellite comparisons of eastern boundary currents: Resolution of circulation features in "coastal" oceans, Proceedings of the TOPEX/POSEIDON SWT Meeting, Biarritz, France.

Strub, P. T., Mesias, J. M., and James, C. (1995). Altimeter observations of the Peru-Chile countercurrent. *Geophys. Res. Lett.* 22, 211–214.

Strub, P. T., Chereskin, T. K., Niiler, P. P., James, C., and Levine, M. D. (1997). Altimeter-derived variability of surface velocities in the California Current system; 1. Evaluation of Topex altimeter velocity resolution. *J. Geophys. Res.* 102, 12,727–12,748.

Subrahmanyam, B., Snaith, H. M., Jones, M. S., Challenor, P. G., and Robinson, I. S. (1996). Identification of the East India Coastal Current during 1993 using Topex/Poseidon altimeter data. *TOPEX/POSEIDON SWT Abstracts.* (Southampton) UK.

Subrahmanyam, B., Robinson, I. S., and Challenor, P. G. (1997). Seasonal variability of the Arabian Sea circulation from TOPEX/POSEIDON altimeter data. *TOPEX/POSEIDON SWT Abstracts.* (Biarritz) France.

Tai, C. K. (1990). Estimating the surface transport of meandering oceanic jet streams from satellite altimetry: surface transport estimates for the Gulf Stream and Kuroshio Extension. *J. Phys. Oceanogr.* 20, 860–879.

Tai, C. K., and White, W. B. (1990). Eddy variability in the Kuroshio Extension as revealed by Geosat altimetry: Energy propagation away from the jet, Reynolds stress and seasonal cycle. *J. Phys. Oceanogr.* 20, 1761–1777.

Talley, L. D. (1983). Radiating instabilities in thin baroclinic jets. *J. Phys. Oceanogr.* 13, 2161–2181.

Tokmakian, R. (1994). The Iceland-Faeroe Front: A synergistic study of hydrography and altimetry. *J. Phys. Oceanogr.* 24, 2245–2262.

Tokmakian, R. T., and Challenor, P. G. (1993). Observations in the Canary Basin and the Azores frontal region using Geosat data. *J. Geophys. Res.* 98, 4761–4773.

Tomczak, M., and Godfrey, J. S. (1994). In "Regional Oceanography: An Introduction," 422 pp. Pergamon Press.

Treguier, A. M. (1992). Kinetic energy analysis of an eddy resolving, primitive equation model of the North Atlantic. *J. Geophys. Res.* 97, 687–701.

Treguier, A. M., and Hua, B. L. (1987). Oceanic quasi-geostrophic turbulence forced by stochastic wind fluctuations. *J. Phys. Oceanogr.* 17, 397–411.

Treguier, A. M., and Hua, B. L. (1988). Influence of bottom topography on stratified quasi-geostrophic turbulence in the ocean. *Geophys. Astrophys. Fluid Dynamics* 43, 265–305.

Treguier, A. M., and McWilliams, J. C. (1990). Topographic influences on wind-driven, stratified flow in a b-plane channel: an idealized model for the Antarctic circumpolar current. *J. Phys. Oceanogr.* 20, 321–343.

Tychensky, A., Le Traon, P. Y., Hernandez, F., and Jourdan, D. (1998). Structure and temporal change in the Azores Front during the SEMAPHORE experiment. *J. Geophys. Res.* 103, 25,009–25,027.

Uchida, H. S., Imawaki, S., and Hu, J. H. (1998). Comparison of Kuroshio surface velocities derived from satellite altimeter and drifting buoy data. *J. Oceanogr.* 54, 115–122.

Van Ballegooyen, R. C., Grundlingh, M. L., and Lutjeharms, J. R. E. (1994). Eddy fluxes of heat and salt from the southwest Indian Ocean into the southeast Atlantic Ocean: A case study. *J. Geophys. Res.* 99, 14,053–14,070.

Vazquez, J., Zlotnicki, V., and Fu, L. L. (1990). Sea level variabilities in the Gulf Stream between Cape Hatteras and 50 W: A Geosat study. *J. Geophys. Res.* 95, 17,957–17,964.

Vasquez, J., Font, J., and Martinez-Benjamin, J. J. (1996). Observations on the circulation in the Alboran sea using ERS-1 altimetry and sea surface temperature data. *J. Phys. Oceanogr.* 26, 1426–1439.

Vivier, F., and Provost, C. (1999). Volume transport of the Malvinas current. Can the flow be monitored by TOPEX/POSEIDON. *J. Geophys. Res.* 104, 21105–21122.

Wang, L., and Koblinsky, C. J. (1995). Low-frequency variability in regions of the Kuroshio Extension and the Gulf Stream. *J. Geophys. Res.* 100, 18,313–18,331.

Wang, L., Koblinsky, C. J., and Howden, S. (1998). Annual and intra-annual sea level variability in the region of the Kuroshio Extension from TOPEX/POSEIDON and Geosat altimetry. *J. Phys. Oceanogr.* 28, 692–711.

White, W. B., Tai, C. K., and DiMento, J. (1990). Annual Rossby wave characteristics in the California Current region from the Geosat Exact Repeat Mission. *J. Phys. Oceanogr.* 20, 1298–1311.

Wijffels, S. E., Hall, M. M., Joyce, T., Torres, D. J., Hacker, P., and Firing, E. (1998). Multiple deep gyres of the western North Pacific: A WOCE section along 149°E. *J. Gephys. Res.* 103, 12,985–13,009.

Wilkin, J., and Morrow, R. A. (1994). Eddy kinetic energy and momentum flux in the Southern Ocean: Comparison of a global eddy-resolving model with altimeter, drifter and current-meter data. *J. Geophys. Res.* 99, 7903–7916.

Willebrand, J., Philander, S. G. H., and Pacanowski, R. C. (1980). The oceanic response to large-scale disturbances. *J. Phys. Oceanogr.* 10, 411-429.

Willebrand, J., Käse, R. H., Stammer, D., Hinrichsen, H. H., and Krauss, W. (1990). Verification of Geosat sea surface topography in the Gulf Stream Extension with surface drifting buoys and hydrographic measurements. *J. Geophys. Res.* 95, 3007–3014.

Witter, D. L., and Gordon, A. L. (1999). Interannual variability of the South Atlantic Circulation from four years of TOPEX/POSEIDON altimeter observations. *J. Geophys. Res.* 104, 20885–20910.

Witter, D. L., and Chelton, D. B. (1998). Eddy-mean flow interaction in zonal oceanic jet flow along zonal ridge topography. *J. Phys. Oceanogr.* 28, 2019–2039.

Woodworth, P. L., Vassie, J. M., Hughes, C. W., and Meredith, M. P. (1996). A test of the ability of TOPEX/POSEIDON to monitor flows through the Drake Passage. *J. Geophys. Res.* **101** 11,935–11,947.

Wunsch, C. (1981). Low frequency variability in the sea. *In "Evolution of physical oceanography,"* (B. Warren and C. Wunsch, Eds.). MIT Press, Cambridge, MA, 342–374.

Wunsch, C. (1982). The ocean circulation and its measurement from space. *In "Space Oceanology"*, pp.13–67. Cepadues-Editions, Toulouse.

Wunsch, C. (1997). The vertical partition of oceanic horizontal kinetic energy. *J. Phys. Oceanogr.* **27**, 1770–1794.

Wunsch, C. (1999). Where do ocean eddy heat fluxes matter? *J. Geophys. Res.* **104**, 13235–13250.

Wunsch, C., and Stammer, D. (1995). The global frequency-wavenumber spectrum of oceanic variability estimated from TOPEX/POSEIDON altimetric measurements. *J. Geophys. Res.* **100**, 24,895–24,910.

Wyrtki, K., Magaard, L., and Hager, J. (1976). Eddy energy in the oceans. *J. Geophys. Res.* **15**, 2641–2646.

Yanagi, T., Morimoto, A., and Ichikawa, K. (1997). Seasonal variation in surface circulation of the East China Sea and the Yellow Sea derived from satellite altimetric data. *Cont. Shelf Res.* **17**, 655–664.

Zlotnicki, V. (1991). Sea level differences across the Gulf Stream and Kuroshio Extension. *J. Phys. Oceanogr.* **21**, 599–609.

Zlotnicki, V., Fu, L. L., and Patzert, W. (1989). Seasonal variability in a global sea level observed with Geosat altimetry. *J. Geophys. Res.* **94**, 17,959–17,969.

Zlotnicki, V., Siedler, G., and Klein, B. (1993). Can the weak surface currents of the Cape Verde frontal zone be measured with altimetry. *J. Geophys. Res.* **98**, 2485–2493.

4

Tropical Ocean Variability

JOEL PICAUT* and ANTONIO J. BUSALACCHI†

*Institut de Recherche pour le Développement (ORSTOM)
LEGOS/GRGS
Toulouse, France
†Laboratory for Hydrospheric Processes
NASA/Goddard Space Flight Center
Greenbelt, Maryland

1. INTRODUCTION

The tropical oceans, with a surface area more than one-third that of the whole earth, are a major source of heat, with roughly equal contributions to the oceanic and atmospheric circulations. Air-sea interactions, which originate in the tropics, affect the climate of the earth on monthly to decadal time scales. The tropical Pacific Ocean, as the largest of the three tropical oceans, is at the origin of the El Niño-Southern Oscillation (ENSO) phenomenon (Philander, 1990). ENSO impacts across the world were particularly notable during the warm phases (El Niño) of the 1982–1983 and 1997–1998 events. The tropical Indian Ocean is an important part of the monsoon system, which affects nearly the entire Asian continent (Webster et al., 1998). The tropical Atlantic Ocean is subject to the so-called Atlantic sea surface temperature (SST) dipole that influences the rainfall of the African and South American continents on interannual to decadal time scales (Moura and Shukla, 1981). Moreover, there exists an El Niño-like phenomenon in the Atlantic (Hisard, 1979; Merle et al., 1980), but it is less important than its Pacific counterpart. While it is the SST variability that forces the atmosphere, the enormous heat content of the upper layer of the tropical oceans provides the thermal inertia for these various coupled phenomena and offers prospects for prediction. ENSO, as the most important coupled mode, was proved to be predictable at the be-

ginning of the TOGA (Tropical Ocean and Global Atmosphere) program (Cane et al., 1986). In essence, the memory of the short-term climate of the earth resides in the tropical ocean.

Most of the heat content in the tropical ocean is situated in the upper layer, and as a result, low-latitude oceans are often approximated as a two-layer system. This attribute implies that variations in the depth of the thermocline, and hence of the heat content, are directly reflected in sea-level height (Rébert et al., 1985). For example, an inverse ratio of 200 between the thermocline depth and sea-level height was found as being typical of the equatorial band. The two-layer approximation implies that the first baroclinic mode is dominant. This fact, together with the vanishing Coriolis force toward the equator, renders the tropical oceans to a very specific dynamic. As shown by Matsuno (1966) and Moore (1968), the latitude band of the tropical oceans is the site of a singular set of equatorial waves that propagate eastward (equatorial Kelvin waves) or westward (equatorial Rossby waves) at a much faster phase speed than the midlatitude Rossby waves. A key result is that the influence of wind forcing can extend thousands of kilometers away from the forcing region and be felt a few months or a year later at the other side of a tropical ocean basin. These equatorial waves can be detected by their notable sea-level signature (order 10 cm). However, with their fast-phase speed (order 1 m sec^{-1}) and very specific meridional signature, they can be hard to detect by conventional *in situ* measurements. Satellite altimeters are now capable of measuring sea level to a few centimeters. Hence, these satellite platforms appear to be the best means for studying these fundamental large-scale signals of the tropical oceans.

Simple ocean models were a significant aid to the simulation and understanding of the sea level variations associated with ENSO events because of the applicability of linear equatorial dynamics (e.g., McCreary, 1976; Busalacchi and O'Brien, 1981). Interestingly, some of these linear models were used to validate the first altimeters such as Geosat, in regions where there were not many *in situ* observations (e.g., Arnault *et al.*, 1992a). The use of oceanic general circulation models (OGCMs) is still plagued by the inaccuracy of the heat and wind forcing, and biases in the subsurface thermal structure. Data assimilation into these models is a promising way to correct such deficiencies. Because of the two-layer approximation of the tropical oceans and the remotely forced effect induced by equatorial waves, sea-level information can be easily projected into the thermocline and influence other parts of the ocean. The assimilation of altimeter data into ocean models is treated in another chapter of this book, and only very specific results will be used in the present chapter. For additional information on tropical assimilation, the reader can refer to Busalacchi (1996). Since the tropical oceans are the source of short-term climate signals that are potentially predictable, assimilation of altimeter data may prove to be an important means to improve the initial conditions of coupled forecast models, and ultimately their prediction skill.

Satellite altimetry has been supported by several *in situ* measurements. During the 1985–1994 TOGA program, various means of *in situ* measurement were developed, with many of them transmitting their data in near-real time (McPhaden *et al.*, 1998). Tide gauge sea level has been a useful way to validate altimetric measurements in the tropics, especially in the western half of the tropical Pacific where there are a number of island stations. Fortuitously, the Pacific tide gauge network was operational for the launch of Geosat. Moreover, the XBT and TAO (Tropical Atmosphere Ocean) networks of expendable bathythermograph and instrumented moorings were able to provide reliable sea surface dynamic height measurements. Hence, these TOGA networks were very useful for validating the Geosat, ERS-1, ERS-2, and TOPEX/POSEIDON altimeters. The tropical oceans and, in particular, the tropical Pacific where the TOGA measurements were concentrated, were an ideal location for testing the usefulness of altimetric measurements. This was particularly true considering the insignificance of the barotropic signal and the usually weak variation of sea-level pressure over the tropical oceans.

A series of altimetry missions captured several ENSO events of particular interest. Geosat observed the successive 1986–1989 El Niño-La Niña, while ERS-1, ERS-2,

and TOPEX/POSEIDON observed the series of weak 1992–1995 El Niño followed by the weak 1996 La Niña. This class of altimeters performed well during the powerful 1997–1998 El Niño, which abruptly shifted into the 1998–1999 La Niña (Figure 1; see color insert). TOPEX/POSEIDON was dedicated to measuring the circulation of the global ocean, and within the tropics it benefited from the complete installation of the TOGA observing system. The vanishing of the Coriolis force at the equator, which makes the tropical oceans so interesting, seemed at first to be a serious limitation for such circulation monitoring in the equatorial oceans. We will see in a subsequent section that this problem was overcome.

This chapter on altimetry and the tropical oceans is organized as follows. The tropical Pacific will be presented first since it is so important for short-term climate and thus the most studied. With some discussion on the Indo-Pacific throughflow, the section on Indian Ocean is next, followed by the tropical Atlantic and the conclusion.

2. TROPICAL PACIFIC

With a width nearly half the circumference of the earth, the tropical Pacific is by far the largest of the three tropical oceans. It is subject to a self-sustained air-sea coupled oscillation on interannual time-scales, which is the basis for the El Niño-Southern Oscillation (ENSO). From an oceanic point of view, the equatorial Pacific is separated into a cold tongue in the east and a warm pool in the west. The cold tongue is the signature of the wind-driven equatorial and coastal upwelling. The warm pool exists because of atmospheric heat fluxes at low latitudes and the lack of cold water entrainment from below. It is maintained in the western Pacific as a result of the trade winds and the associated westward surface currents. ENSO is manifested by a notable change in the intensity of the equatorial upwelling, and therefore of the SST in the cold tongue, and also by the change in the zonal extension of the cold tongue and warm pool. The warm pool is characterized by SSTs among the warmest in the world's oceans, and because of its size (larger than the continental United States), it is the world's most important source of heat into the atmosphere and to higher latitudes in the Pacific Ocean.

For these reasons the tropical Pacific has been the most studied ocean during the 1985–1994 TOGA program. In particular, various ocean-observing systems were developed or installed during this program (McPhaden *et al.*, 1998), and they were of great use for the validation of the altimetry satellites launched during the corresponding decade.

2.1. Sea-level Validation

Wyrtki (1975) was the first to realize the importance of island tide gauge measurements to study and monitor the equatorial surface currents and upper tropical ocean heat content. With the presence of a significant number of islands, especially in the western tropical Pacific, he developed a network of sea-level measurements with real-time data transmission, which was fully operational at the beginning of TOGA and therefore at the launch of the Geosat satellite.

In preparation for the Geosat mission, Miller *et al.* (1986) built up 3-month time series of Seasat sea-level anomalies from crossover differences and compared them with tide gauge records from two islands of the central Pacific. This technique was applied to the first 18 months of the primary Geosat Geodetic Mission, and the resulting sea-level anomalies were compared to an island tide gauge in the central Pacific with an rms difference of 4 cm (Miller *et al.*, 1988). A more complete validation study of the non-repeat Geosat geodetic sea-level data was done by the same group of authors (Cheney *et al.*, 1989) by comparing the complete 18 months of Geosat data with monthly sea-level measurements from 14 island tide gauges and surface dynamic heights from two equatorial moorings. The average correlation of 0.68 and rms difference of 3.7 cm were surprisingly good, especially knowing the deficiency in the various corrections (e.g., orbit, electromagnetic bias and water vapor), the crossover technique used, and the inherent imprecision of *in situ* sea-level measurements. A comparison by Tai *et al.* (1989) of a similar 18-month Geosat product with 23 island sea-level records resulted in comparable correlation. In the same study, a lower correlation between maps of surface dynamic height derived from XBT and maps of Geosat sea level confirmed the superiority of island sea level for altimeter validation purposes. In any case, the favorable evaluation from these various studies was very encouraging for the scientific use of the data of the primary Geosat Geodetic Mission in the tropical Pacific Ocean.

The Geosat Exact Repeat Mission (ERM) began in November 1986 and enabled sea-level time series to be computed using the preferred method of collinear differences. Miller and Cheney (1990), combining the preceding 18-month Geosat product with 2.5 years of Geosat ERM, obtained a similar comparison with the same set of 14 tide gauges. The correlation coefficient was higher (0.86), but the rms difference went up to 5.3 cm, probably because of the greater sea-level changes during the ERM, with the presence of the 1986–1989 El Niño-La Niña, as compared to the Geodetic Mission. The presence of the Australian coasts in the western Pacific caused Cheney and Miller (1990) to improve the orbit corrections over short arcs, trying to keep as much as possible of the oceanographic signal. Despite such improvements, uncertainty in the wet tropospheric corrections remained a problem for Geosat, especially during ENSO events (Zimbelman and Busalacchi, 1990). Spatially and temporally coherent shifts between the Geosat sea-level product and tide-gauge measurements on time scales longer than a year limited the use of this altimetry product

for monitoring large-scale low-frequency variability (Wyrtki and Mitchum, 1990). This need of datum for altimetry measurements confirmed the necessity to maintain the TOGA tide-gauge network in parallel with altimetry satellites.

In the first years of TOGA, an experimental array of TAO moorings equipped with thermistor chains was installed progressively in the equatorial Pacific (Hayes *et al.*, 1991). This permitted time series of surface dynamic height relative to 500 dbar to be derived from the temperature measurements and a T-S (Temperature-Salinity) relationship. Delcroix *et al.* (1991) were the first to successfully use such validation techniques over the first year of Geosat ERM data. By doing additional comparisons with XBT and CTD (Conductivity-Temperature-Depth) measurements, these authors pointed out the deficiency in the use of a climatological T-S relationship, at least in the warm pool where the salinity structure can change drastically. More comprehensive comparisons between the complete set of Geosat ERM, a dozen TAO moorings, most of the island tide gauges and OGCM sea level, confirmed the quality of the Geosat data with an average correlation of 0.7 and a rms difference of 5 cm (e.g., Chao *et al.*, 1993; Delcroix *et al.*, 1994). In addition, such comparison studies proved very useful for understanding the dynamics of the 1986–1989 El Niño-La Niña.

By the time of the launch of the TOPEX/POSEIDON satellite, 56 TAO moorings were installed, and at the end of TOGA in December 1994, the full array of 70 moorings extending from 137°E to 95°W and 8°N to 8°S was completed (McPhaden, 1995). At the end of the verification phase of the mission, the first 500 days of TOPEX/POSEIDON altimetry data were compared to TAO-derived surface dynamic heights (Busalacchi *et al.*, 1994). Given the large 15° longitude × 3° latitude decorrelation scale intrinsic to the tropics (White *et al.*, 1985), this intercomparison was facilitated through the optimal interpolation of both surface height fields onto a common, 10-day, 1° × 1° grid. Cross-correlation was generally higher than 0.7 and the rms difference smaller than 4 cm, except off the equator in the eastern Pacific. The observed sea-level signal in both TAO and TOPEX/POSEIDON fields appeared generally larger than the expected measurement errors. Furthermore, the unresolved systematic underestimation of altimeter signals in the equatorial Pacific, as observed by Geosat (Cheney *et al.*, 1989; Delcroix *et al.*, 1994), did not appear to be an issue with TOPEX/POSEIDON. This initial intercomparison between TOPEX/POSEIDON and TAO derived sea level was extended by Menkes *et al.* (1995). It was found that for periods greater than 35 days, the two fields compared quite well over the entire equatorial basin, with a mean correlation of 0.79 and mean rms difference of 2.6 cm. In particular, the low correlation found off the equator in the east by Busalacchi *et al.* (1994) appeared to be caused by the presence of instability waves, truncated TAO time series, and low oceanic signals which render the correlation inadequate.

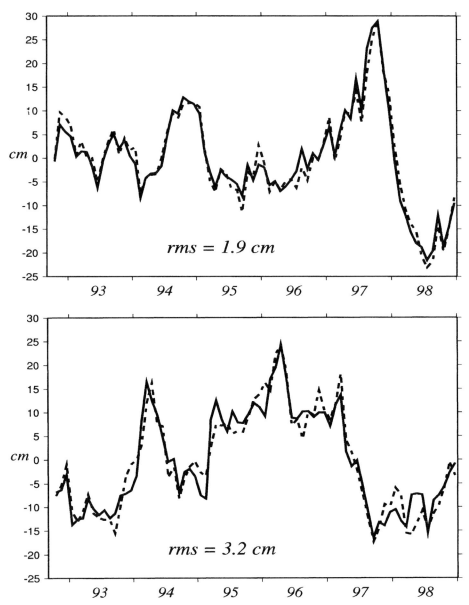

FIGURE 2 Comparison between monthly sea-level anomalies from tide gauge (dashed line) and TOPEX/POSEIDON (solid line). Upper panel: Christmas Island, 0°, 158°W; lower panel: Honiara, 10°S, 158°E. (Courtesy of Bob Cheney.)

Mitchum (1994) carefully compared the first 300 days of the TOPEX/POSEIDON data with 71 tide gauge measurements over the three oceans. Of the total number of tide gauges used, three-fourth were situated in the Pacific, with 31 on tropical Pacific islands. A comparison between the island tide gauges and the four nearest smoothed alongtrack altimeter data resulted in a 0.58 correlation coefficient and a 5.8 cm rms difference. With an additional 10-day filter, the rms difference went down to 4 cm. Using monthly means, Cheney *et al.* (1994) found that the comparison between 17 island sea levels within 10°N to 10°S and the first 51 10-day

cycles of TOPEX/POSEIDON data resulted in a correlation of 0.85 and rms difference of 2.3 cm (Figure 2).

All such previous intercomparisons in the tropical Pacific were limited, since individual *in situ* sea level and surface dynamic height measurements have environmental and instrumental errors of order 3–7 cm. A dedicated open-ocean validation experiment was conducted in the western equatorial Pacific during the first 6 months of the verification phase of the TOPEX/POSEIDON mission, in order to determine the effective accuracy of the two altimeters over the tropics. Two TOGA-TAO moorings at 2°S, 156°E and 2°S,

164°E were outfitted with additional temperature, salinity, and pressure sensors to measure, within 1 cm, the dynamic height from the surface to the bottom at 5-min intervals directly beneath two TOPEX/POSEIDON crossovers; bottom pressure sensors and inverted echo sounders were deployed as well (Picaut, *et al.*, 1995; Katz *et al.*, 1995). Instantaneous comparisons of the 1-sec TOPEX and Poseidon altimeter retrievals with the 5-min surface dynamic height resulted in a rms difference as low as 3.7 cm at 2°S, 156°E and 3.3 cm at 2°S, 164°E (upper part of Figure 3). After the use of a 3-D interpolation technique and a 30-day low-pass filter, *in situ* and satellite data were found to be highly correlated. The correlation coefficient and rms difference were 0.96 and 1.8 cm at 2°S, 156°E and 0.94 and 1.9 cm at 2°S, 164°E, respectively (lower part of Figure 3).

The stability of the TOPEX/POSEIDON altimeters was checked by Mitchum (1998) using tide-gauge measurements mostly from the tropical Pacific. The method used was able to detect a small drift caused by an algorithm error. It confirmed an overall stability of the satellite heights of less than 10 mm over the first 4 years of TOPEX/POSEIDON, which is more than satisfactory for most of the altimetry applications.

2.2. Altimetry-derived Surface Current

Many maps of currents derived from satellite altimetry avoid the equatorial band, since the geostrophic approximation is very sensitive to small sea-level variations near the equator. For example, a 1-cm difference within the equator and 1°N or 1°S, caused by high-frequency variations or instrumental errors, would lead to an erroneous estimate for zonal geostrophic current of 0.7 m sec^{-1}. However, it was shown from TOGA-TAO dynamic heights and current meters that geostrophy can be extended to the equator for periods equal to or greater than a month from the use of the second derivative of the meridional pressure field (Picaut *et al.*, 1989). This was successfully applied to the Geosat sea-level data, with an additional linear and nonlinear filter along the tracks in order to remove small-scale sea-level variability (Picaut *et al.*, 1990). The along-track filtering is apparently not necessary for deriving surface currents at the equator from TOPEX/POSEIDON, given the quality of the altimetry data (Menkes *et al.*, 1995). Away from the equator, the use of the classical geostrophic relationship (i.e., the first derivative of the meridional pressure field) was successfully tested with TOPEX/POSEIDON data in the western tropical Pacific through a comparison with near-surface drifters (Yu *et al.*, 1995). However, very close to the equator the zonal currents deduced from the first derivative rarely match the second derivative, and a correction factor within 5°N to 5°S was proposed by Picaut and Tournier (1991) to ensure the continuity of the zonal currents in the equatorial band. This

technique was applied to the Geosat data in order to calculate geostrophic surface currents over the whole tropical Pacific (Delcroix *et al.*, 1991). A similar technique was developed with TOPEX/POSEIDON data, with the addition of Ekman flow in order to estimate the total near-surface currents from altimetry and surface wind measurements (Lagerloef *et al.*, 1999). Geostrophic currents appear important over most of the tropics, but Ekman current divergence is the dominant feature around the cold tongue (Figure 4). Knowing the sensitivity of the geostrophic approximation to small sea-level variations near the equator, these successful studies on altimetry-derived surface currents represent the most stringent test of using altimetry observations to estimate sea level and surface currents anywhere in the world ocean.

2.3. Evidence of Equatorial Waves

With the seminal theoretical work of Matsuno (1966) and Moore (1968), it was revealed that the tropical oceans are subject to a particular dynamic, in which equatorial waves and their reflection on ocean boundaries are very important. Wyrtki (1975) was the first to suggest that equatorial waves can be an important part of the ENSO mechanism, through remote wind forcing from the central-western part of the equatorial basin into the eastern part and, in particular, through the surge of warm water from the warm pool into the east. The successful simulations of El Niño with linear equatorial ocean models (e.g., McCreary, 1976; Busalacchi and O'Brien, 1981) led observationalists to search for the existence of these waves. Knox and Halpern (1982) were able to infer the propagation of a first vertical mode equatorial Kelvin wave, through the variations of transport as observed from several current meters on two equatorial moorings at 110 and 152°W. Sea-level observations were tentatively used to detect equatorial waves, since the predominant effect on thermocline depth is directly reflected in sea-level variations. Hence, Ripa and Hayes (1981) were able to infer equatorial Kelvin waves from their meridional structure, and Lukas *et al.* (1984) identified, from propagating sea-level signals, equatorial Kelvin waves and possible Rossby waves issued from Kelvin wave reflection on the eastern ocean boundary. Arguments about the existence of these waves and their importance in the physics of ENSO were perpetuated by the difficulties in observing them from sparse *in situ* measurements. Satellite altimetry appeared to be an excellent system to observe equatorial waves, as they have a broad sea-level signal, with a specific meridional signature in sea level. Furthermore, the sea-level signal propagates zonally at a speed easily detectable with repeat orbit satellites.

Cheney *et al.* (1987) and Miller *et al.* (1988) were the first to clearly document equatorial Kelvin waves during the onset of the 1986–1987 El Niño from the Geosat data. As seen from Figure 5, westerly wind bursts generate downwelling Kelvin waves which propagate eastward within the range

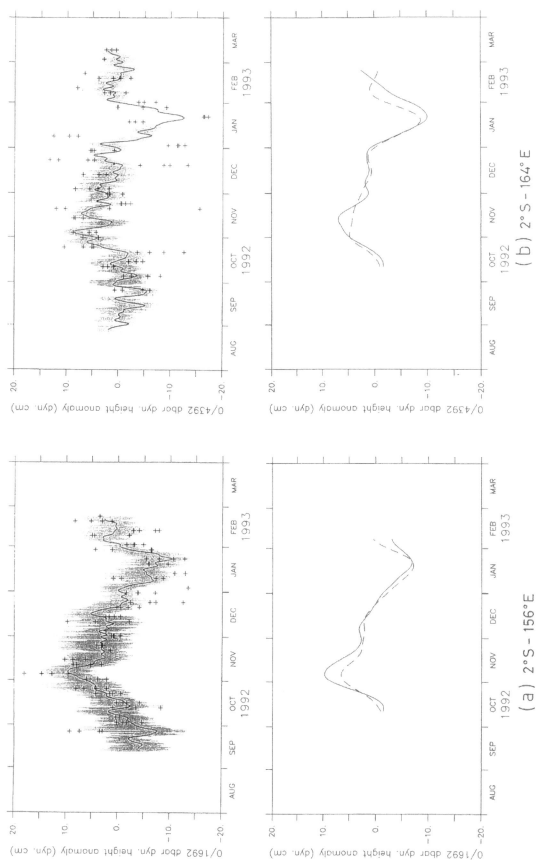

FIGURE 3 Open-ocean validation of the TOPEX/POSEIDON altimeters in the western equatorial Pacific. *Upper panel*: The surface to bottom dynamic height, calculated from a suite of instruments along the two mooring lines, are represented in two ways. The thin line (seen as gray shading) corresponds to the original 5-min time series, and the thick line is a low-pass filtered (5-day Hanning filter) version. Superimposed are the 1-sec sea-level data of the eight points of TOPEX/POSEIDON measurements (pulse) nearest to the two mooring sites. All time series are anomalies relative to the period of measurements. The larger high-frequency variability of the thin line at the shallowest of the two sites (1739 m depth versus 4400 m) reflects the existence of strong internal tides. *Lower panel*: Low-frequency (30-day Hanning filter) comparison between the sea surface dynamic height (dashed line) relative to the bottom and the TOPEX/POSEIDON sea level (solid line) obtained from a three-dimensional interpolation. (From Picaut, J., *et al.*, 1995. With permission.)

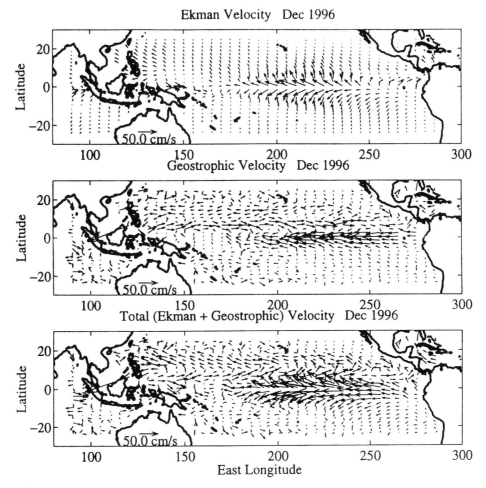

FIGURE 4 The Ekman term, geostrophic term and their sum (top, middle, and bottom panels respectively) for the monthly near-surface velocity composite during December 1996. NSCAT satellite scatterometer data were used to compute the Ekman term, and the geostrophic surface currents were derived from TOPEX/POSEIDON sea level. (Courtesy of G. Lagerloef.)

of the theoretical phase speed (2.4–2.8 m sec^{-1}). In particular, the beginning of the 1986–1987 El Niño is marked by a well-defined downwelling Kelvin wave. This wave appeared as a 10–15 cm rise in sea level trapped to the equator with an equatorial radius of deformation corresponding to the first vertical mode (Delcroix *et al.*, 1991). As seen from Geosat and TOPEX/POSEIDON measurements (e.g., Delcroix *et al.*, 1994; Boulanger and Menkes, 1995; Zheng *et al.*, 1995; Chelton and Schlax, 1996) and also recently by the fully operational TAO array (e.g., Kessler *et al.*, 1995), eastward propagating Kelvin waves are present most of the time along the equator. They appear to be associated frequently with variations of zonal winds in the western part of the basin, and most commonly of downwelling (upwelling) type during El Niño (La Niña). Once equatorial Kelvin waves reach the coast of South America, they reflect into coastal Kelvin waves. These coastal waves propagate poleward and excite a full set of equatorial Rossby

waves into the ocean interior with distinct meridional structures (Moore, 1968). The coastal Kelvin wave propagation was evidenced from coastal tide-gauge measurements all the way to Alaska and Chile (Enfield and Allen, 1980), and with the improvements of altimetry it can now be detected from satellites (Born *et al.*, 1998).

Equatorial Rossby waves emanating from the reflection of equatorial Kelvin waves on the eastern boundary can be tracked toward the central part of the basin from TOPEX/POSEIDON data (Boulanger and Fu, 1996; Susanto *et al.*, 1998). Equatorial Rossby waves can also be generated in the ocean interior by wind forcing. Such forcing happens to be situated predominantly in the western part of the basin on interannual time scales (Mantua and Battisti, 1994) and in the central-eastern part of the basin on annual time scales (du Penhoat *et al.*, 1992). As seen from Figure 6 and Table 1, first meridional Rossby waves of the first vertical mode can be identified with Geosat sea-level data from their

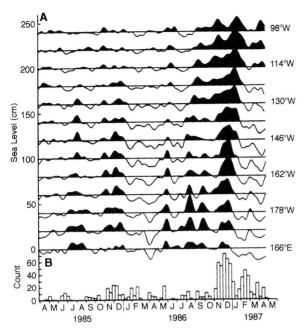

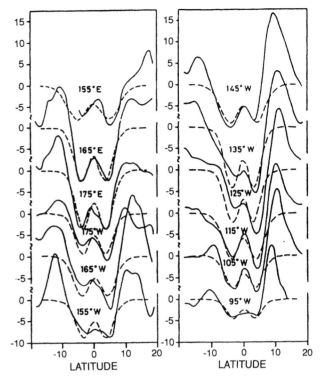

FIGURE 5   (A) Geosat-derived sea level time series at 13 locations (8° interval between 166°E and 98°W) along the equator, April 1985 through April 1987. Horizontal lines indicate zero mean values for the first 12 months (April 1985 through April 1986) with 20-cm offset between pairs of series. (B) Histogram along the x-axis indicating the time intervals during which westerly wind bursts were persistently observed in the far western equatorial Pacific. The histogram was constructed by summing the number of days for which analyzed westerly winds greater than 5 knots were found in each 5° by 5° box in the region 0° to 5°S, 130°E to 170°E. (From Miller, L., et al., 1988. With permission.)

FIGURE 6   Least-square fits of the Geosat sea-level anomalies (in cm, solid curves) to the theoretical meridional structures (dashed curved) of the first horizontal equatorial upwelling Rossby wave, at different longitudes (155°E to 95°W), and at the time reported in Table 1. (From Delcroix, T., et al., 1991. With permission.)

TABLE 1.   Mean Date of the Passage of the Maximum of Negative Geosat Sea Level Anomalies at 4°N and 4°S at Various Longitudes Associated with the Upwelling Equatorial Rossby Wave as Shown in Figure 6. The Third Column Gives the Phase Speed Estimated from the Least Square Fits of the Geosat Sea Level Meridional Structures to the Theoretical First Meridional Upwelling Equatorial Rossby Wave Shape, as Plotted on Figure 6. The Mean and Standard Deviation 0.86 ± 0.22 m sec^{-1} of These Phase Speeds Are Comparable to the 1.02 ± 0.32 m sec^{-1} Propagation Estimated From Time-lag Correlation Analysis (Adapted from Delcroix et al., 1991)

Longitude	Date (1987)	Phase speed (m sec^{-1})
155°E	Sept. 3	1.44
165°E	Aug. 29	0.81
175°E	Aug. 24	0.77
175°W	July 20	0.77
165°W	July 10	0.95
155°W	June 10	0.78
145°W	May 21	1.08
135°W	May 16	0.75
125°W	May 11	0.67
115°W	May 6	0.91
105°W	Apr. 25	0.76
95°W	March 31	0.67

specific meridional structure and their zonal phase propagation (one-third of the phase speed of Kelvin wave). However, because of the complicated wind forcing in the equatorial basin, reflected and wind-forced Rossby waves can be mixed together or can be hidden from one another. This has been studied through a combined analysis of Geosat data and linear ocean-model output (du Penhoat et al., 1992). Furthermore, the downward leakage of energy from equatorial Rossby waves (Kessler and McCreary, 1993; Dewitte et al., 1999) complicates their evidence from altimetry far into the ocean interior, as suggested by Boulanger and Fu (1996) from TOPEX/POSEIDON data analysis. The westward extent of equatorial Rossby waves, issued from reflection of equatorial Kelvin waves on the eastern boundary, is of interest not only in the tropics. For example, Rossby waves issued from the reflection of strong Kelvin waves during the 1982–1983 El Niño seemed to propagate westward well into the mid-latitude Pacific. Ten years later, they may have reached and affected the Kuroshio, as indicated by Geosat and TOPEX/POSEIDON data and model analyses (Jacobs et al., 1994).

Once the equatorial Rossby waves reach the western ocean boundary, they are transformed into short Rossby waves that bring mass into the equator and result in equatorial Kelvin waves. This reflection of equatorial Rossby waves into equatorial Kelvin waves was suggested by Geosat analysis on annual and interannual time scale (White *et al.*, 1990; White and Tai, 1992) and carefully analyzed from TOPEX/POSEIDON by Boulanger and Menkes (1995, 1999). In their latter paper these authors indicated a nearly perfect reflection of the first meridional Rossby waves on the western boundary into equatorial Kelvin waves, and a rather good reflection of these equatorial waves on the eastern boundary. They also identified from altimetry measurements equatorial Rossby waves of the second- and third-meridional modes. These waves are, respectively, antisymmetric and symmetric and propagate at one-fifth and one-seventh of the speed of the Kelvin waves.

2.4. Testing Theories of ENSO, Improving Its Prediction

From an oceanographic point of view, theoreticians and modelers like McCreary (1976) were among the first to imagine that equatorial waves are a distinctive part of ENSO mechanisms, through remote forcing. During El Niño, equatorial downwelling Kelvin waves seemed to be the mechanism by which the decreasing trades or westerly winds in the central-western part of the basin could induce SST warming further east. El Niño remote forcing was gradually accepted by the scientific community through numerous model simulations and limited basin-scale observations. Ultimately, satellite altimetry confirmed this basin-wide aspect of ENSO.

Based on Pacific island sea-level analysis, Wyrtki (1985) suggested that a build-up might be necessary in order to initiate an El Niño. Prior to such an event, warm waters accumulate into the 15°N to 15°S band, especially in the warm pool, and are released poleward during El Niño. As a consequence, the time between two successive ENSO events may correspond to the time it takes for the equatorial region to be refilled with warm waters. This concept was tested by Miller *et al.* (1990) using Geosat data. They suggested that the equatorial band over which the accumulation of warm water occurred was much narrower than Wyrtki assumed, a result suggested previously in a model sampling study by Springer *et al.* (1990). The build-up was not obvious from Geosat data prior to the 1986–1987 El Niño (Miller *et al.*, 1990), but it seemed to occur from TOPEX/POSEIDON data prior to the strong 1997–1998 El Niño. More needs to be done with longer altimetry data from several satellites to fully test Wyrtki's hypothesis on the onset of El Niño.

Bjerknes (1966) was the first to recognize the coupling between the El Niño oceanic signal and the Southern Oscillation atmospheric signal. Although this ocean-atmosphere coupling contributed a large piece to the ENSO puzzle, if left alone this positive feedback would lead to a never ending El Niño. For the shift from El Niño into La Niña (or

vice versa), Schopf and Suarez (1988) and Battisti (1988) proposed a theory, known as the delayed-action oscillator. Assuming an initial warming in the central-eastern equatorial Pacific, the feedback mechanism envisioned by Bjerknes (1966) results in westerly winds. These winds generate equatorial downwelling Kelvin waves that enhance the warming further east. West of the westerly winds, equatorial upwelling Rossby waves are generated which propagate westward and hit the western ocean boundary after 6–8 months. These waves are reflected into equatorial upwelling Kelvin waves that propagate eastward and act to damp the growing SST-wind-coupled system in the east. It is the successful arrival of these delayed equatorial upwelling Kelvin waves that erodes this growth, stops it, and eventually shifts El Niño into La Niña. This fascinating theory was built from theoretical considerations and model simulations, but it needed to be validated from basin-wide data analysis.

From extended EOF analysis of Geosat data during 1986–1989, White and Tai (1992) suggested that an equatorial Rossby wave reflected into an equatorial Kelvin wave at the western boundary. However, detailed projections of Geosat and TOPEX/POSEIDON data on individual equatorial wave modes indicated little evidence of first meridional Rossby wave reflection into Kelvin waves during either the 1986–1989 El Niño-La Niña or during the 1992–1993 El Niño (Delcroix *et al.*, 1994; Boulanger and Menkes, 1995). These authors suggested instead that wind forcing rather than reflected Rossby waves was the main trigger of Kelvin waves during these events. However, Boulanger and Fu (1996) suggested that first-mode downwelling Rossby waves were reflected into downwelling Kelvin waves at the western boundary at the onset of the 1994–1995 warm event. With the improved and longer duration TOPEX/POSEIDON data, Boulanger and Menkes (1999) succeeded in showing a nearly perfect reflection of first meridional Rossby waves into equatorial Kelvin waves over the 1992–1998 TOPEX/POSEIDON period. These recent results substantiate the delayed-action oscillator theory, but more needs to be done to see how much of this theory of ENSO is at work in the real ocean.

An additional perspective on the ENSO oscillation was derived from the recent evidence of an oceanic zone of convergence at the eastern edge of the western Pacific warm pool, which moves in phase with the Southern Oscillation Index over thousands of kilometers, eastward during El Niño and westward during La Niña (Figure 7; see color insert). Using hypothetical drifters displaced by zonal current fields derived from Geosat sea level, Picaut and Delcroix (1995) demonstrated that zonal advection is the dominant mechanism for the zonal displacement of the eastern edge of the warm pool during the 1986–1989 El Niño-La Niña. This analysis of altimeter data was extended to TOPEX/POSEIDON derived currents, two *in situ* near-current fields, and three classes of models over the 1982–

1994 period (Picaut *et al.*, 1996). In addition, it was found from Geosat, rainfall, and sea-surface salinity data, that the eastern edge of the warm pool is characterized by a well-defined salinity front, through the encounter of fresh water from the warm pool into saltier water from the upwelling region further east (Delcroix and Picaut, 1998). Zonal displacement of this oceanic zone of convergence at the eastern edge of the warm pool is the reason why SST in the central equatorial Pacific varies within 26–29°C on ENSO time scales. Given that an SST of 28°C is the threshold for organized atmospheric convection, the resulting variations of the atmospheric heating is associated with local wind-SST coupling and leads to global ENSO teleconnection (Graham and Barnett, 1996). From altimetry data analyses, Picaut and Delcroix (1995) noted that equatorial wave reflection on the eastern ocean boundary contributes to the shift in the zonal advection of the oceanic zone of convergence at the eastern edge of the warm pool. Hence, a revised delayed-action oscillator theory of ENSO was proposed in which zonal advection and equatorial wave reflection on the eastern boundary are at least as important as vertical advection and equatorial wave reflection on the western boundary (Picaut *et al.*, 1997). It must be pointed out that this revised theory of ENSO, where the air-sea interaction occurs in the central-western equatorial Pacific, as observed, could not have been imagined without the use of altimetry data.

Altimetry data also have the potential for improving the skill of ENSO prediction. Experiments with the NCEP (National Centers for Environmental Prediction) coupled model for ENSO prediction suggest that the assimilation of TOPEX/POSEIDON sea level in addition to subsurface temperature observations could have positive impact on the skill of ENSO prediction (Ji *et al.*, 2000). Indeed the assimilation of altimetric sea level may improve the initialization of large scale sea level variabilities and therefore of upper surface temperature variabilities. As discussed in the next section, sea level also reflects the variations of the salinity structure, and presently, the assimilation of TOPEX/POSEIDON sea level in the NCEP coupled operational model is done by correcting the temperature field alone. More studies are needed to clearly demonstrate the impact that altimetry may have on short-term climate prediction.

2.5. Changes of Mass, Heat and Salt of the Upper Ocean

The comparison of Rébert *et al.* (1985) between temperature profiles and island sea level in the tropical Pacific relates the changes in the upper heat content from sea-level variations. White and Tai (1995) extended this relationship to the global ocean from 30°S to 60°N through the use of XBT data and the first 2 years of TOPEX/POSEIDON data. They found that the relationship is more solid in the tropics than in mid latitude because of the two-layer charac-

ter of the tropical ocean. This implies that altimetry is able to detect the significant interannual changes in upper heat content in the tropical ocean. Using more than 4 years of TOPEX/POSEIDON data and most of the TAO moorings, Chambers *et al.* (1998) refined this relationship in the equatorial Pacific and estimated the error in heat storage anomaly to be about 50×10^7 J m^{-2}. However, in the western equatorial Pacific these authors found a difference as much as 30% between the heat storage anomalies inferred from the TOPEX/POSEIDON and those calculated from TAO. This appeared to be because of the specifics of the salinity structure of the warm pool. This finding was also noted from the differences between two sets of sea level (up to 8 cm over the warm pool) simulated by the NCEP oceanic model, one with assimilation of TOPEX/POSEIDON data only and the other with assimilation of subsurface temperature only. The importance of salinity variations in sea-surface dynamic heights, in particular in regions like the warm pool, was confirmed by further analyses of TOPEX/POSEIDON data and CTDs over the tropical Pacific (Vossepoel *et al.*, 1999; Maes *et al.*, 2000). An important aspect of these studies is that it may be possible to infer the salinity structure of the upper layers from altimetry measurements and temperature profile. Such a method may be useful for estimating the transport of salt in the tropics, especially in view of the difficulty in acquiring *in situ* salinity measurements.

Altimetry measurements are nevertheless an adequate means to study the important redistribution of heat and mass during ENSO, especially those associated with the generation and propagation of equatorial waves. One of the most notable effects of these waves in the equatorial band is the zonal seesaw-like pattern in thermocline depth during the warm and cold phases of ENSO. Wyrtki (1984) tentatively described such changes in the zonal slope of sea level from tide-gauge measurements in the equatorial band during the 1982–1983 El Niño. Altimetry measurements described such basin-scale changes in thermocline depth in a more detailed way (e.g., Figure 7a). In particular, it was confirmed, in agreement with equatorial wave theory, that the warm phase of ENSO is always marked by a 10- to 30-cm rise in sea level (i.e., 20–60 m deepening of the thermocline) in the narrow equatorial band of the eastern Pacific, followed slightly in time by a drop in sea level of comparable amount (i.e., rising of the thermocline) over the broader band of the warm pool. The seesaw pattern is inverted during the cold phase of ENSO, but usually with smaller amplitude. As illustrated in Figure 1, the changes from El Niño to La Niña during 1997–1998 were over 40 cm on either side of the basin.

These changes in thermocline depth and sea-level height in the equatorial band are associated with a huge transfer of water mass and heat. As seen from Figure 7b, the surface zonal current averaged in the equatorial band can vary from 40 cm sec^{-1} westward during the 1996 La Niña to

50 cm sec^{-1} eastward during the 1997 El Niño. Similar surface current variations were found from Geosat during the 1986–1989 El Niño-La Niña (Picaut and Delcroix, 1995). Such current variations are not restricted to the near-surface. Direct current measurements along 165°E (McPhaden and Picaut, 1990) indicate the current transports over the first 100 m varied from 47 Sverdrup (Sv) to the east during 1986 to −7 Sv to the west during 1988, a change of transport nearly equivalent to the total transport of the Kuroshio. These changes along the equatorial band are associated with an important redistribution of mass and heat to higher latitudes. Miller and Cheney (1990) have looked at the meridional redistribution of the upper-layer volume above the 20°C isotherm from Geosat sea-level data within 20°N to 20°S. They found that the notable seasonal exchange of warm water across the 8°N nodal line was roughly doubled during the 1986–1989 ENSO events. This meridional seesaw, with its pivot line roughly under the Intertropical Convergence Zone (ITCZ), was confirmed on ENSO time scales with temperature profiles and surface dynamic heights by Delcroix (1998) over the 1979–1995 period. Since the relation between the upper heat content and altimetry sea level is imperfect, a careful analysis of an OGCM that assimilates altimetric sea level is needed to clarify these meridional and zonal exchanges of mass and heat and define their role in the mechanisms of ENSO.

2.6. High-frequency Oscillations

From Seasat sea level data, Musman (1986) and Malardé et al. (1987) were able to detect an important source of high-frequency variability in the eastern tropical Pacific, known as the Legeckis waves or tropical instability waves. Theses waves, which propagate westward, are coincident with the shear zone between the South Equatorial Current (SEC) and the North Equatorial Counter Current (NECC) and between the NECC and the North Equatorial Current (NEC). Using 26 months of Geosat ERM data, Perigaud (1990) evidenced two distinct zones of energetic variability in the northeastern tropical Pacific. The first zone, between 4 and 7°N along the southern shear front of the NECC, showed up as an 8-cm maximum sea-level signal of 1000–2200 km wavelength, and 28–40 days period. The second zone, along 12°N and the northern shear front of the NECC, corresponded to a fluctuation of 6 cm amplitude, 50–90 days period, and 630–950 km wavelength. Despite the difficulty in resolving these high-frequency signals from a 17-day repeat orbit satellite, this study was also able to show that these signals do not extend west of the dateline and they are seasonally modulated.

In their intercomparison between TOPEX/POSEIDON sea level and TAO surface dynamic heights, Busalacchi et al. (1994) found that the satellite sensor is better than the array of moorings for detecting such instability waves. Giese

et al. (1994), using both means of measurements, found two regions of high-frequency variability similar to those found by Perigaud (1990) with Geosat data. The source between 5 and 7°N is clearly identified as an instability wave. It appears modulated by the strength of the equatorial current system and extends between 110 and 165°W. The variability at 11°N has the form of anticyclonic eddies, but its source appears difficult to establish. Hansen and Maul (1991) have observed an anticyclonic current ring propagating westward in the same location from Geosat and in situ data, and they suggest a retroflection mechanism in the NECC. On the other hand, Giese et al. (1994) thought that these eddies are generated by strong wind-stress curl variations off the coast of Central America during northern fall and winter, and are therefore distinct from the instability waves.

During their open-ocean validation experiment of TOPEX/POSEIDON altimeters, Picaut et al. (1995) found quasi-permanent semi-diurnal internal tides in the western equatorial Pacific. These noticeable features (rms amplitude around 2 cm) showed up in the intercomparison between the 5 min surface to bottom dynamic height measured on two TAO moorings and the 1-sec TOPEX and Poseidon altimetric sea level (upper part of Figure 3). At the shallowest of the two mooring sites (2°S, 156°E on the Ontong Java Plateau), abrupt changes of the entire vertical structure (up to 100 m) and of sea level (up to 30 cm) associated with the strongest internal tides, occurred in less than 1 hr. These manifestations of solitary waves, generated in tandem with internal tides during spring tides, were studied by other means of measurements around 2°S, 156°E (Pinkel et al., 1997), during the simultaneous TOGA-Coupled Ocean-Atmosphere Response Experiment (COARE). From the sea-level traces of the internal tides along the descending and ascending TOPEX/POSEIDON tracks, Gourdeau (1998) was able to determine the main direction of propagation of these internal tides and their source. TOPEX/POSEIDON sea level combined with island tide gauges proved very useful for detecting internal tides around the Hawaiian Islands (Ray and Mitchum, 1997). These waves appeared to propagate over more than 1000 km before decaying significantly. However, these authors warned that it is premature to expect to study internal tides and their dissipation rate on the global ocean from altimetry measurements alone.

3. INDIAN OCEAN

In contrast to the Pacific Ocean, the large-scale changes in the sea-surface topography in the tropical Indian Ocean were relatively unknown on seasonal to interannual time scales prior to the advent of satellite altimetry. Whereas the network of tide gauges and TAO moorings in the Pacific Ocean helped define, in near real time, the variability of sea level and dynamic height on a day-to-day basis, the small number of instrumented islands in the Indian Ocean and limited access to coastal tide gauge data meant that Geosat and TOPEX/POSEIDON observations would rapidly advance our understanding of the Indian Ocean variability. Beginning in the mid-1980s, radar altimeter observations began to be used to define the basin-scale circulation in the tropical Indian Ocean, as well as regional scale circulation features such as the Great Whirl and the Indonesian Throughflow.

The first clues that altimeter data could be used to study the Indian Ocean circulation were provided by Perigaud *et al.* (1986) using Seasat altimeter observations for September to October 1978. Equatorial variations in the zonal slope of sea level were examined using crossover differences and careful consideration of signal degradation due to geoid error, tidal error, sea state bias, and orbit error. As a result of this processing, a signal of ±10 cm was extracted in the western equatorial Indian Ocean and found to control the sea-level slope along the equator during the months analyzed.

With the launch of Geosat, Perigaud and Zlotnicki (1992) focused on the orbit and tidal errors that needed to be reduced in the Geosat ERM data for basin-scale analyses of the tropical Indian Ocean. They were able to identify large annual variations in sea level in the southern tropical Indian Ocean along the path of the SEC. The source of this variability was attributed to westward propagating Rossby waves generated by wind forcing to the east. Subsequently, in a series of papers by Perigaud and Delecluse (1992a, b; 1993), the seasonal and interannual variability for various time segments between 1985–1989 were analyzed in more detail together with XBT observations and wind-forced reduced-gravity ocean model simulations. The greatest sea-level variability was observed to be between 8°S to 20°S. On seasonal time scales the maximum amplitude was of order 12 cm. Root mean square errors between the altimeter sea level data and dynamic heights along three XBT sections were between 3–5 cm. Complex Empirical Orthogonal Functions were used to highlight the westward propagating Rossby wave signals whose speed was found to be in general agreement with theoretical estimates. Comparison with numerical model solutions suggested that nonlinear interaction with the mean flow in the vicinity of the SEC may increase the propagation speed of the Rossby waves. On interannual time scales, a significant increase in basin-wide average sea level of 0.9 cm was found to be coincident with the 1986–1987 El Niño.

On regional scales, Subrahmanyam *et al.* (1996) used Geosat altimeter observations to describe the variability of the surface circulation in the western Indian Ocean. Hydrographic data in the western Indian Ocean and off the coast of Somali, taken during cruises in August to September 1988, formed the basis for comparisons of the mesoscale variabil-

ity in the region. The altimeter and *in situ* observations were in good agreement with regards to their mutual description of a two-gyre system for the Somali Current consisting of the Great Whirl near 9°N and a secondary gyre to the south. Geosat data were also used by Ali (1993) and Sharma and Ali (1996) to infer the mixed-layer depth variability along the equator. Climatological data were used to construct a relation between sea level and mixed-layer depth. In turn, this relation was applied to the Geosat observations. Equatorial mixed layer depths were estimated to be between 30 m in the west, down to 115 m in the east. Comparisons with *in situ* temperature profiles suggested a rms error of 20 m.

The improved accuracy of TOPEX/POSEIDON data increased the emphasis on regional scale variability and the gradients in sea level that were difficult to estimate with Geosat data because of contamination problems with orbit error. For example, Kumar *et al.* (1998) used TOPEX/POSEIDON observations to describe the sea-level variability in the northern Indian Ocean in response to the winter and summer monsoons. Approximately 3 years of TOPEX/POSEIDON data were used to characterize the order 40 cm sea-level excursions in the Arabian Sea and Bay of Bengal in response to the northeast and southwest monsoon wind forcing. The space-based perspective of the northern Indian Ocean allowed the effects of local wind forcing to be contrasted with those attributed to remote forcing of sea-level variability into the interior of the Indian Ocean.

The sea-level variability described by TOPEX/POSEIDON in the southeast Indian Ocean was considered by Morrow and Birol (1998), with emphasis on the pressure gradient variability associated with the Leeuwin Current. Their analyses, similar to the previous work of Perigaud and Delecluse with Geosat, identified a region of energetic variability between 10°S to 15°S that had a maximum amplitude of 15 cm at 90°E. Consistent with previous findings, this was attributed to the annual generation of Rossby waves. A second band of lesser variability was noted at 25°S to 30°S with propagation characteristics suggestive of semi-annual Rossby waves. However the phase speed of this propagation, in a region of mean westward flow, was nearly two times the theoretical estimate. A third region of high variability was found along the west coast of Australia. In this region, the Leeuwin Current, a poleward boundary current, is known to flow against the prevailing equatorward wind. It was determined that TOPEX/POSEIDON data could provide a good description of the alongshore pressure gradient. In this analysis, the pressure gradient term was found to be larger than, and in opposition to, the direct wind forcing estimated from ERS-1 scatterometer data, suggesting that TOPEX/POSEIDON data were a valuable means for monitoring the variability of the Leeuwin Current.

Most of the early model-altimeter intercomparison studies involved limited-duration simulations with the one and a half layer-reduced gravity class of models. In Murtugudde and Busalacchi (1999), a reduced gravity, primitive equation, sigma coordinate OGCM was coupled to an advective atmospheric mixed-layer model to study the seasonal and interannual variability of the dynamics and thermodynamics of the tropical Indian Ocean for 1980–1995. The sea-level variability and Rossby wave propagation along 12°S were shown to be in good agreement with Geosat and TOPEX/POSEIDON observations upstream of the mid-ocean ridge. However, the agreement broke down downstream of 70°E, suggesting the influence of bottom topography as pointed out by Morrow and Birol (1998). In a companion study by Murtugudde *et al.* (1998), the interannual transport of the Indonesian Throughflow was shown to be capable of being described in terms of the sea-level difference between the western Pacific and eastern Indian Oceans by a judicious choice of the end points guided by both model simulations and altimeter data. This followed the work of Potemra *et al.* (1997) who were the first to show that the interbasin gradient in sea level, from the first 3 years of TOPEX/POSEIDON data, was highly correlated ($r = 0.78$) with the throughflow transport anomalies with the output from an OGCM. The main effect of the throughflow in the study of Murtugudde *et al.* (1998) was to warm the Indian Ocean and cool the Pacific. Analysis of the model and observed sea level indicated a decrease in the throughflow during El Niño years that was governed by Pacific Ocean winds. However, a significant fraction of the interannual throughflow variability was not associated with El Niño, and was, in fact, governed by the sea-level variability induced by wind changes in the Indian Ocean.

It has already been stated how fortuitous it was that TOPEX/POSEIDON was flying during the extreme El Niño of 1997–1998. What is less well known, however, is how anomalous the Indian Ocean was during this same time period. Figure 8 (see color insert) shows the sea-level and SST anomalies for February 1998. During this extreme Indian Ocean event, SST was more than 2°C warmer than normal. Sea level was more than 20 cm higher than normal in the west and more than 15 cm lower than normal in the east. Heavy rainfall and extreme flooding occurred over Somalia and the neighboring region. Although the Indian Ocean is normally subject to the seasonally reversing monsoon circulation, during this event the Indian Ocean was subject to easterly wind forcing that was more reminiscent of the trade-wind circulation and oceanic response in the tropical Pacific Ocean. Limited SST and tide-gauge observations indicate that the last time the Indian Ocean experienced an interannual event of this nature was in 1961. The fact that TOPEX/POSEIDON has provided a tandem depiction of the evolution of the El Niño event in the Pacific and a coincident extreme event in the Indian Ocean will undoubtedly lead to the study of anomalous events in the Indian Ocean and the bearing these may have on the coupled climate system (R. Murtugudde, personal communication).

4. TROPICAL ATLANTIC

The first analysis of sea-level measurements from space dedicated to the tropical Atlantic was performed by Ménard (1988) through the use of GEOS 3 and Seasat data. A reasonable agreement was found between the seasonal variability of the sea surface topography deduced from these altimetric measurements and with historical hydrographic data analysis (Merle and Arnault, 1985). The tropical Atlantic Ocean was somewhat forgotten by the 1985–1994 TOGA program. However, just before this decade, an important French–U.S. experiment (FOCAL/SEQUAL for Français Océan et Climat en Atlantique Tropical/Seasonal Equatorial Atlantic) focused on the seasonal cycle of the upper layer, which was considered as the dominant mode of variability of this tropical ocean. As a consequence, the analyses of the subsequent Geosat altimetric data focused at first on topics of interest for this experiment, e.g., the seasonal upwelling in the Gulf of Guinea, the variation of the slope along the equator, the importance of remote forcing for these phenomena, and the variation of the NECC. The improvement from Geosat to TOPEX/POSEIDON satellites was most notable in the tropical Atlantic. Truncated passes, with the presence of the northern coast in the Gulf of Guinea, resulted in serious orbit errors and a limited use of Geosat in this gulf (Arnault *et al.*, 1992a). Further west, the unprecedented duration of the TOPEX/POSEIDON mission gave the possibility of studying various temporal scales of the western boundary current system. It confirmed the improvement brought by this dedicated mission for the understanding of an important feature of the global circulation of the Atlantic and thus of the World Ocean Circulation Experiment (WOCE). Because of the limited number of islands and the absence, until very recently, of moored thermistor chain measurements in the tropical Atlantic, various authors have validated altimetric sea level with historical surface dynamic height field. For the same reasons, comparison with modeled sea-surface topography is more common in this ocean for validation purpose than in the tropical Pacific.

Using the anomalous along-track data of the first year of the Geosat ERM (November 1986 to November 1987), a gridded altimetric sea-level product over the tropical Atlantic was compared to a historical seasonal dynamic height field (Arnault *et al.*, 1990). The discrepancies between these two surface fields and sea level obtained from a linear model, stressed the importance of orbit and wet tropospheric corrections for altimetry and of interannual variations in this basin. Simultaneous comparisons with scattered *in situ* data were therefore done with improved altimetric measurements taken over several years.

A comparison was made in 1993 between two tide-gauge measurements at Principe Island and Pointe Noire (respectively off and on the south-east coast of the Gulf of Guinea) and an old and a new version of the Geosat data, over the April 1985 to September 1989 period (Arnault and Cheney, 1994). Such analyses, complemented with a comparison with sea level simulated with an OGCM, confirmed the difficulties in using coastal sea-level measurements for the validation of such satellite data. Overall, these analyses corroborated the improvement of the new Geosat product, especially in the Gulf of Guinea. Tide gauge measurements from some of the few islands in the tropical Atlantic were successfully used to validate the TOPEX/POSEIDON altimeters (Verstraete and Park, 1995; Arnault and Le Provost, 1997). Another source of validation was provided by inverted echo sounders. Six of them were deployed in the western equatorial Atlantic from February 1987 to June 1988. Although some IES time series compared well with the nearby Geosat alongtrack data, the discrepancies pointed out the necessity to improve the water-vapor correction in a region of important moisture (Carton and Katz, 1990; Arnault et al., 1990).

A dedicated validation experiment was completed at the end of September-early October 1988 from a merchant ship. About 200 XBTs were launched within 40°N to 20°S along a route following a cross-equatorial Geosat track in the tropical Atlantic (Arnault et al., 1992b). Surface-dynamic height and Geosat sea-level anomalies agreed quite well, except in the southern part of the track where salinity changes and meso-scale variability may have been important. An extensive comparison with in situ data was possible from numerous hydrographic data collected as part of the Soviet Union SECTIONS (acronym for Energetically Active Zones of Ocean and Climate Variability) program in the tropical Atlantic in the 1980s. During the lifetime of the Geosat ERM, eight cruises were conducted and regional surface-dynamic height maps were compared with their counterpart from altimetric measurements (Carton et al., 1993). However, the absence of a mean Geosat sea-level estimate for each individual cruise rendered the validation task difficult.

The availability of the precise TOPEX/POSEIDON data during two WOCE cruises in the western equatorial Atlantic, respectively in September to October 1985 and April to May 1996, enabled Arnault et al. (1999) to pursue this type of comparison with sea-surface dynamic heights. Encouraging results led the authors to claim that appropriate objective analysis of TOPEX/POSEIDON could be used to detect the meso-scale variability of the surface layer of the western equatorial Atlantic, despite the large scale of the satellite measurements. Of most interest is the good comparison between the surface current anomalies derived from altimetry and the near-surface measurements of the ship-mounted Acoustic Doppler Current Profiler during these two cruises. The example provided in Figure 9 shows that the distinctive meandering and eddy-like structures off Brazil and French Guyana can be described with TOPEX/POSEIDON data.

In contrast with the Pacific Ocean, a moored array of thermistor chains was not available in the tropical Atlantic for sea-level validation. However, TOPEX/POSEIDON data were utilized to optimize the deployment of a Pilot Research Moored Array in the Tropical Atlantic (PIRATA) (Servain et al., 1998). Fields of TOPEX/POSEIDON sea-level anomalies were subsampled with the goal being to reconstruct the original fields through the use of reduced-space Kalman filter data assimilation at a limited number of potential moorings sites (Hackert et al., 1998). This study was limited to the use of sea level as a proxy for upper heat content and subsurface thermal structure. Nevertheless, this was the first time that an array of moorings benefited from a series of observing system simulation experiments incorporating altimetry observations before being installed.

Basin-scale comparisons between the data of the first year of the Geosat ERM and historical hydrographic data proved that altimetric data can depict the large-scale low-frequency variability of surface dynamic heights (Carton, 1989; Arnault et al., 1990). In particular, it was suggested that this first year corresponds to a period when the equatorial upwelling was weak and occurred earlier than the regular cycle. A subsequent analysis of 2 years of Geosat ERM data indicated that boreal Spring 1988 was warmer in the Gulf of Guinea than its counterpart in 1987 (Arnault et al., 1992a). The difficulty in fully detecting sea-level signals in the Gulf of Guinea by the first Geosat product was obviously a limit to the detection of interannual variations. However a linear model that assimilates Geosat data confirmed that the seasonal upwelling arrived in the Gulf of Guinea 1 month earlier in 1987 than in 1988 (Gourdeau et al., 1992). The use of better GDRs, with more accurate orbit and water vapor corrections over the full Geosat period (April 1985 to September 1989) appeared to improve the weak signal problem in the Gulf of Guinea (Arnault and Cheney, 1994). Basin scale analysis of this new product confirmed the 1987–1988 warming in the Gulf of Guinea. This warming was a consequence of the slackened trades in the western equatorial Atlantic, following the remote forcing mechanism proposed by Moore et al. (1978) and confirmed by Servain et al. (1982). Interestingly this basin-scale analysis of the Geosat product evidenced a depletion of water mass in the equatorial band associated with the warming in the Gulf of Guinea similar to Wyrtki's study of the Pacific El Niño sea level (Wyrtki, 1985). These features observed by sea-level altimetry confirmed the existence of El Niño-like interannual variations in the equatorial Atlantic (Hisard, 1979; Merle et al., 1980) and their possible connection with the Pacific El Niño. The basin-wide Geosat sea-level analysis of Arnault and Cheney (1994) also confirmed the dominance of the seasonal cycle in tropical Atlantic sea level and the presence of two seesaws, zonally along the equator (Merle, 1980) and meridionally under the mean position of the ITCZ. The presence of

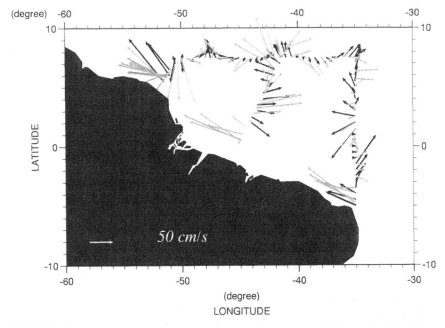

FIGURE 9 Comparison between the geostrophic surface currents (dark arrows) derived from TOPEX/POSEIDON sea level and the near surface ship-mounted Acoustic Doppler Current Profiler measurements (gray arrows) taken during the ETAMBOT 1 cruise in September to October 1995. (From Arnault, S., *et al.*, 1999. With permission.)

a semi-annual cycle in the Gulf of Guinea and in the northwestern part of the basin was also corroborated.

All these basin-wide sea-level features and associated readjustment of mass emphasized the need to monitor, through altimetry, the response to winds of the upper ocean, especially along the equator, and the variability of the major surface equatorial currents. Zonal changes in the pressure gradient along the equator were already approached by some of the previous authors from Geosat data. Using the more precise TOPEX/POSEIDON data, Katz *et al.* (1995) found that this zonal slope along the equator is, on a mean, in equilibrium with the zonal wind stress in the western half of the basin. Further east, such a mean relationship cannot hold given the remote forcing from the winds in the west, and thus the presence of Kelvin waves propagating into the Gulf of Guinea, as suggested from altimetric data (Arnault and Cheney, 1994; Gourdeau *et al.*, 1997; Katz, 1997).

Concerning zonal surface circulation, a qualitative approach was done with the first years of Geosat data by looking at the change of the meridional gradient of sea level (Carton, 1989; Carton and Katz, 1990; Arnault *et al.*, 1990, 1992a). Special attention was directed by these authors to the NECC variations, given the good comparison with *in situ* data around this current. In particular, it was found that the NECC was subject to important changes during the 1986–1988 period, especially in the west. During its fall decrease the longitudinal variations evidenced by altimetry established that the NECC can be separated in two parts, east

and west of about 32°W. The variations of the major equatorial currents in the tropical Atlantic can be estimated with reasonable confidence from Geosat altimetry (Carton, 1989; Arnault *et al.*, 1990). Maps of total surface geostrophic currents, over the 3°N to 15°N and 30°W to 60°W region and from November 1986 to November 1988, were derived from the superposition of Geosat sea-level anomalies and an estimate of mean sea level (Carton and Katz, 1990). The mean sea level came from an objective correction of the 0/1000 dbar Levitus climatology with about 120 contemporaneous hydrographic casts. These first-ever maps of surface currents derived from altimetry in the tropics provided useful information on the major equatorial currents in the region. In particular, the seasonal increase of the NECC in fall is notable with geostrophic currents of up to 130 cm sec^{-1} in November 1987. Other maps of altimetry-derived currents were determined by Didden and Schott (1992) in the northwestern tropical Atlantic in a meso-scale analysis of geostrophic current anomaly from the first 2 years of Geosat ERM. Without any estimate of the mean sea surface, such maps were limited to the annual harmonic of geostrophic velocity. Surprisingly good comparisons with similar calculations from a high-resolution OGCM indicate that such a technique can be extrapolated to the whole tropical basin and usefully utilized for the detection of the variations of equatorial current. Another effort to derive maps of surface currents from altimetry was done by Arnault *et al.* (1999) in the same northwestern region using TOPEX/POSEIDON sea-level anomalies, the surface dynamic height climatology of Levitus and hydrographic data from two WOCE cruises. As noted earlier, the favorable comparison between *in situ* and altimetry-derived currents (Figure 9) is very encouraging for studying the upper circulation in the tropical Atlantic from TOPEX/POSEIDON. Without current meter measurements to validate the second derivative of the meridional pressure field, the previous geostrophic studies were not extended to the equatorial band, contrary to the Pacific, therefore limiting the use of such maps of surface currents and associated transports.

The altimetric studies in the northwestern tropical Atlantic also proved useful for the detection and analysis of a very important feature in the exchange of heat between the northern and southern Atlantic hemispheres, i.e., the low-latitude western boundary surface current, its retroflection into the NECC, and the associated propagating eddies which apparently feed the Florida Current (Nystuen and Andrade, 1993). As shown by Arnault *et al.* (1999) the western boundary current system can be better resolved by TOPEX/POSEIDON than Geosat. In spite of the geostrophic limitations, such as the nonlinear radial acceleration, altimetric data analyses showed the formation, evolution, and northwestern propagation of these eddies, especially during the breakdown of the retroflection process (Didden and Schott, 1993). Similar analyses of Geosat and TOPEX/POSEIDON data have also detected a systematic, but not yet fully understood, 40–50 day oscillation in the same region. The understanding of these various features of the western tropical Atlantic by means of altimetry data analyses are an important goal of the WOCE and CLIVAR (Climate Variability and predictability) DecCen (Decadal-Centennial) programs.

To close this chapter on the tropical Atlantic, it is worth noting the detection of tropical instability waves from altimetry. Using six Geosat tracks within 8°N to 8°S in the center of the basin and a composite mean-wave pattern, Musman (1992) was able to detect instability waves between May and November 1987. These waves appear as two sea-level maxima of 3 cm at 4°N and 4°S, and propagate westward at about 45 cm sec^{-1}, with a period around 25 days. A different approach using 3 years of TOPEX/POSEIDON by Katz (1997) resulted in a lower 23 cm sec^{-1} westward propagation and a similar 28-day period. The discrepancy between the two studies reflects the difficulties in resolving the short time-space structure of tropical instability waves from altimetry data.

5. CONCLUSION

Satellite radar altimetry has proven to be a significant enhancement to our ability to describe and understand the tropical ocean circulation and significant short-term climatic

events such as El Niño. Space-based altimeter platforms have provided an unprecedented perspective of the global tropics. An excellent example of this has been the extreme changes that have been observed in the tropical Pacific and Indian Oceans during the evolution of the 1997–1998 El Niño and the La Niña that followed. Geosat, ERS1, ERS-2, and TOPEX/POSEIDON altimeter data constitute a valuable complement to observations in the tropical Pacific where there is a rather extensive *in situ* network. In other low-latitude regions where there is not a comprehensive *in situ* observing system, and in particular the Indian Ocean, altimeter data are one of the most, if not the most, important data streams for describing the large-scale ocean circulation. Meridionally, altimeter data in the Pacific Ocean have permitted studies of the exchanges of mass and heat between the tropics and higher latitudes. Zonally, altimeter data have been used to monitor the Indonesian Throughflow between the Pacific and Indian Oceans. In the Indian and Atlantic Oceans, altimeter data have been used to characterize the basin-scale variability on seasonal to interannual time scales. It is worth noting that essentially all the previous applications of satellite altimetry to studies of the tropical ocean have relied on Geosat and TOPEX/POSEIDON, in contrast to ERS-1 and ERS-2. This has occurred for a variety of reasons, including access to the data, quality of data, and the formulation of science teams devoted to analyzing the altimeter data.

In view of the two-layer approximation for tropical oceans, altimeter data were relatively easy to apply to studies of tropical ocean circulation. A majority of such applications have been in the tropical Pacific Ocean because of the importance of the ENSO phenomenon, the broad expanse of the Pacific basin, and the presence of the TOGA observing system. The island sea-level and dynamic-height measurements from the tide-gauge network and TAO array of 70 surface moorings, respectively, allowed Geosat and TOPEX/POSEIDON data to be rigorously validated in the tropical Pacific. The moored current meters of the TOGA observing system enabled the calculation and validation of altimetry derived estimates of the zonal geostrophic flow field right down to the equator using the second derivative of the sea-level fields. The lack of similar *in situ* current measurements in the Atlantic and Indian Oceans has hindered the estimation of the equatorial currents in these oceans.

During the late 1970s and early 1980s there was considerable debate within the equatorial theoretical community regarding the existence and importance of equatorial wave modes and the role they play in remote forcing and El Niño. The space-time coverage of the altimeter data unequivocally identified the existence of equatorial waves via their propagation speed and horizontal structure. Going back to the mid-1980s, satellite altimeters have now observed several ENSO events in succession. Geosat data during the 1986–1989 El Niño and La Niña clearly demonstrated the presence of equatorially trapped Kelvin waves. Subsequently, Geosat and later TOPEX/POSEIDON data were used to analyze the existence of westward propagating Rossby waves and their role in mechanisms of ENSO such as the delayed action oscillator. Access to altimeter data has also shown that the reflection of equatorial Kelvin waves into Rossby waves at the eastern boundary and associated perturbations to zonal advection may be more important in the evolution of the ENSO cycle and the related displacements of the warm pool than previously thought.

In the Indian Ocean, Geosat and TOPEX/POSEIDON altimeter data highlighted an energetic band of westward propagating variability between 8°S to 20°S in response to the wind forcing of annual Rossby waves in the southeastern portion of the basin. TOPEX/POSEIDON data have been used to describe the response of the Bay of Bengal and Arabian Sea to monsoon forcing and the degree to which the interior circulation is a combination of local and remote forcing. The greater precision of the TOPEX/POSEIDON data also permitted the description and monitoring of important circulation features such as the Leeuwin Current and the Indonesian Throughflow. The exceptional deviations in sea level in the Indian Ocean captured by TOPEX/POSEIDON in 1998 were some of the largest anomalies believed to have occurred in the Indian Ocean since 1961.

The application of Geosat data within the Atlantic Ocean was hindered by orbit error problems and altimeter retrievals related to the configuration of the Gulf of Guinea coastline. As in the Indian Ocean, the lack of a network of *in situ* observations in the tropical Atlantic limited most of the early validation studies to be comparisons with ocean model solutions. The availability of refined estimates of Geosat altimeter data and the higher quality TOPEX/POSEIDON data has led to a number of studies of the tropical Atlantic Ocean circulation such as the seasonal upwelling in the Gulf of Guinea, its remote forcing, variations in the zonal pressure gradient along the equator, changes in the meridional pressure gradient across the NECC, and the variability of narrow, energetic western boundary currents and associated retroflection.

The perspective of the tropical ocean circulation that altimetry has offered to date is really just the beginning of what can be anticipated. The incorporation of these observations into ocean general circulation models has just started recently. Although the altimeter data have been used to infer the surface geostrophic flow field, ocean data assimilation techniques are required to advance our understanding of the three-dimensional circulation at low latitudes. Altimeter data, as a proxy for the upper ocean heat content, represents a valuable integral constraint on the subsurface thermal structure for such studies. Already, we have seen the importance of subsurface salinity highlighted as a result of confronting ocean general circulation model solutions with altimeter observations. In addition, the subsurface thermal structure provides the thermal inertia, or memory as it were, for the coupled climate system in the tropics. In this regard, satellite altimeter data hold considerable potential for advancing the prediction skill of the coupled climate system on seasonal to interannual time scales. Should this potential be fulfilled, it will generate an operational demand for satellite altimetry observations that will need to be sustained. As was seen during the TOGA program, altimeter data proved to be a very valuable complement to the *in situ* observing system in the equatorial Pacific Ocean. As we expand our focus out and beyond this region, satellite altimeter data will prove to be a rich source of data for the CLIVAR GOALS (Global Ocean-Atmosphere-Land System) and DecCen programs for years to come.

ACKNOWLEDGMENTS

We would like to thank the many scientists who answered our requests and sent reprints or pre-prints of their work on altimetry. Eric Hackert and Guoqing Li contributions for several of the figures are appreciated. Comments on a first version of the manuscript by Sabine Arnault, Bob Cheney, and Claire Perigaud are greatly acknowledged. We thank Sabine Arnault, Bob Cheney, Gary Lagerloef, and Laury Miller for their authorization to reproduce some of the figures. This work was supported by NASA (Grant 621-55-04), CNES, and IRD-ORSTOM.

References

Ali, M. M. (1993). Inference of the reversal of mixed layer zonal slope along the Equatorial Indian Ocean using Geosat altimeter data. *Int. J. Remote Sensing*, **14**, 2043–2049.

Arnault, S., and R. E. Cheney (1994). Tropical Atlantic sea-level variability from Geosat (1985–1989). *J. Geophys. Res.*, **99**, 18,207–18,223.

Arnault, S., and C. Le Provost (1997). Regional identification in the tropical Atlantic Ocean of residual tide errors from an empirical orthogonal function analysis of TOPEX/POSEIDON altimetric data. *J. Geophys. Res.*, **102**, 21,011–21,036.

Arnault, S., Y. Ménard, and J. Merle (1990). Observing The Tropical Atlantic-Ocean in 1986–1987 from altimetry. *J. Geophys. Res.*, **95**, 17,921–17,945.

Arnault, S., A. Morliere, J. Merle J., and Y. Ménard (1992a). Low-frequency variability of the Tropical Atlantic surface-topography, altimetry and model comparison. *J. Geophys. Res.*, **97**, 14,259–14,288.

Arnault, S., L. Gourdeau, and Y. Ménard (1992b). Comparison of the altimetric signal with in situ measurements in the Tropical Atlantic Ocean. *Deep-Sea Res.*, **39**, 481–499.

Arnault, S., B. Bourles, Y. Gouriou, and R. Chuchla (1999). Intercomparison of the upper layer circulation of the western equatorial Atlantic Ocean: in-situ and satellite data. *J. Geophys. Res*, **104**, 21,171–21,194.

Battisti, D. S. (1988). Dynamics and thermodynamics of a warming event in a coupled atmosphere-ocean model. *J. Atmos. Sci.*, **45**, 2889–2919.

Bjerknes, J. (1966). A possible response of the atmospheric Hadley circulation to equatorial anomalies of ocean temperature. *Tellus*, **18**, 820–829.

Born, G., R. Leben, C. Fox, and G. Tierney (1998). Wave monitoring and analysis in the Pacific. *AVISO Altimetry*, **6**, 16–17.

Boulanger, J.-P., and C. Menkes (1995). Propagation and reflection of long equatorial waves in the Pacific ocean during the 1992–1993 El Niño. *J. Geophys. Res.*, **100**, 25,041–24,059.

Boulanger, J.-P., and L.-L. Fu (1996). Evidence of boundary reflection of Kelvin and first-mode Rossby waves from TOPEX/POSEIDON sea level data. *J. Geophys. Res.*, **101**, 16,361–16,371.

Boulanger, J.-P., and C. Menkes (1999). Long equatorial wave reflection in the Pacific Ocean during the 1992–1998 TOPEX/POSEIDON period. *Clim. Dynam.*, **15**, 205–225.

Busalacchi, A. J. (1996). Data assimilation in support of tropical Ocean circulation studies. *In* "Modern Approaches to Data Assimilation in Ocean Modeling," pp. 235–269. P. Malanotte-Rizoli ed., Elesevier Sc.

Busalacchi, A. J., and J. J. O'Brien (1981). Interannual variability of the equatorial Pacific in the 1960s. *J. Geophys. Res.*, **86**, 10,901–10,907.

Busalacchi, A. J., M. J. McPhaden, and J. Picaut (1994). Variability in the equatorial Pacific sea surface topography during the Verification Phase of the TOPEX/POSEIDON mission. *J. Geophys. Res.*, **99**, 24725–24738.

Cane M. A., S. E. Zebiak, and S. C. Dolan (1986). Experimental Forecasts of El-Niño. *Nature*, **321**, 827–832.

Carton, J. A. (1989). Estimates of sea level in the tropical Atlantic Ocean using Geosat altimetry . *J. Geophys. Res.*, **94**, 8029–8039.

Carton J. A., and E. J. Katz (1990). Estimates of the zonal slope and seasonal transport of the Atlantic North Equatorial Countercurrent. *J. Geophys. Res.*, **95**, 3091–3100.

Carton, J., G. A. Chepurin, G. K. Korotaev, and T. Zhu (1993). Comparison of dynamic height variations in the tropical Atlantic during 1987–1989 as viewed in Sections hydrography and Geosat altimetry. *J. Geophys. Res.*, **98**, 14,369–14,377.

Chambers, D. P., B. D. Tapley, and R. H. Steward (1998). Measuring heat storage changes in the equatorial Pacific: a comparison between TOPEX altimetry and Tropical-Atmosphere-Ocean buoys. *J. Geophys. Res.*, **103**, 18,591–18,597.

Chao, Y., D. Halpern, and C. Perigaud (1993). Sea-surface height variability during 1986–1988 in the tropical Pacific-Ocean. *J. Geophys. Res.*, **98**, 6947–6959.

Chelton, D. B., and M. G. Schlax (1996). Global observations of Oceanic Rossby waves. *Science*, **272**, 234–238.

Cheney, R. E., and L. Miller (1990). Recovery of the sea level signal in the western tropical Pacific from Geosat altimetry. *J. Geophys. Res.*, **17**, 2977–2984.

Cheney, R. E., L. Miller, B. C. Douglas, and R. W. Agreen (1987). Monitoring equatorial Pacific sea level with Geosat. *Johns Hopkins APL Tech. Dig.*, **8**, 245–250.

Cheney, R. E., B. C. Douglas, and L. Miller (1989). Evaluation of Geosat data with application to tropical Pacific sea level variability. *J. Geophys. Res.*, **94**, 4737–4747.

Cheney, R., L. Miller, R. Agreen, N. Doyle, and J. Lillibridge (1994). TOPEX/POSEIDON: the 2-cm solution. *J. Geophys. Res.*, **99**, 24,555–24,563.

Delcroix, T. (1998). Observed surface oceanic and atmospheric variability in the tropical Pacific at seasonal and ENSO time scales: A tentative overview. *J. Geophys. Res.*, **103**, 18,611–18,633.

Delcroix, T., and J. Picaut (1998). Displacement of the western Pacific fresh pool. *J. Geophys. Res.*, **103**, 1087–1098.

Delcroix, T., J. Picaut, and G. Eldin (1991). Equatorial Kelvin and Rossby waves evidenced in the Pacific Ocean through Geosat sea level and surface current anomalies. *J. Geophys. Res.*, **96**, 3249–3262.

Delcroix, T., J.-P. Boulanger, F. Masia, and C. Menkes (1994). Geosat-derived sea-level and surface-current anomalies in the equatorial Pacific, during the 1986–89 El Niño and La Niña. *J. Geophys. Res.*, **99**, 25,093–25,107.

Dewitte, B., G. Reverdin, and C. Maes (1999). Vertical structure of an OGCM simulation of the equatorial Pacific I 1985–1994. *J. Phys. Oceanogr.*, **29**, 2363–2388.

Didden N., and F. Schott (1992). Seasonal variations in the western tropical Atlantic surface circulation from Geosat altimetry and WOCE model results. *J. Geophys. Res.*, **97**, 3529–3541.

Didden N., and F. Schott (1993). Eddies in the North Brazil Current retroflection region observed by Geosat altimetry . *J. Geophys. Res.*, **98**, 20121–20131.

du Penhoat, Y., T. Delcroix, and J. Picaut (1992). Interpretation of Kelvin/Rossby waves in the equatorial Pacific from model-Geosat data intercomparison during the 1986–1987 El Niño. *Oceanol. Acta*, **15**, 545–554.

Enfield, D., and J. Allen (1980). On the structure and dynamics of monthly mean sea level anomalies along the Pacific coast of North and South America. *J. Phys. Oceanogr.*, **10**, 557–578.

Giese, B. S., J. A. Carton, and L. J. Holl (1994). Sea level variability in the eastern tropical Pacific as observed by TOPEX and Tropical Ocean-Global Atmosphere Tropical Atmosphere-Ocean Experiment. *J. Geophys. Res.*, **99**, 24,739–24,748.

Gourdeau, L. (1998). Internal tides observed at 2°S–156°E by in situ and TOPEX/POSEIDON data during the coupled Ocean-Atmosphere Response Experiment (COARE). *J. Geophys. Res.*, **103**, 12629–12638.

Gourdeau, L., S. Arnault, Y. Ménard, and J. Merle (1992). Geosat sea-level assimilation in a tropical Atlantic model using Kalman filter. *Oceanol Acta*, **15**, 567–574.

Gourdeau, L., J.-F. Minster, and M. C. Gennero (1997). Sea level anomalies in the tropical Atlantic from Geosat data assimilated in a linear model, 1986–1988. *J. Geophys. Res.*, **102**, 5583–5594.

Graham, N. E., and T. P. Barnett (1996). ENSO and ENSO-related predictability. Part II: Northern-hemisphere 700-mb height predictions based on a hybrid coupled ENSO model. *J. Climate*, **8**, 544–549.

Hackert, E. C., R. N. Miller, and A. J. Busalacchi (1998). An optimized design for a moored instrument array in the tropical Atlantic. *J. Geophys. Res.*, **103**, 7491–7509.

Hansen, D. V., and G. A. Maul (1991). Anticyclonic current rings in the eastern tropical Pacific Ocean. *J. Phys. Oceanogr.*, **10**, 1168–1196.

Hayes, S. P., L. J. Mangum, J. Picaut, A. Sumi, and K. Takeuchi (1991). TOGA-TAO: A moored array for real-time measurements in the tropical Pacific Ocean. *Am. Met. Soc.*, **72**, 339–347.

Hisard, P. (1979). Observations de réponse de type El Niño dans l'Atlantique tropical oriental Golfe de Guinée. *Oceanol. Acta*, **3**, 69–78.

Jacobs, G., H. Hurlburt, J. Kindle, E. Metzger, J. Mitchell, W. Teague, and A. Wallcraft (1994). Decade-scale trans-Pacific propagation and warming effects of an El Niño anomaly. *Nature*, **370**, 360–363.

Ji, M., R. W. Reynolds, and D. Behringer (2000). Use of TOPEX/POSEIDON sea level data for ocean analyses and ENSO prediction: some early results. *J. Climate*, **13**, 216–231.

Katz, E. J. (1997). Waves along the equator in the Atlantic. *J. Phys. Oceanogr.*, **27**, 2536–2544.

Katz, E. J., J. A. Carton, and A. Chakraborty (1995). Dynamics of the equatorial Atlantic from altimetry *J. Geophys. Res.*, **100**, 25061–25067.

Katz, E. J., A. J. Busalacchi, M. Bushnell, F. I. Gonzalez, L. Gourdeau, M. J. McPhaden, and J. Picaut (1995). A comparison of coincidental time series of the ocean sea surface height by satellite altimeter, mooring and inverted echo sounder. *J. Geophys. Res.*, **100**, 25101–25108.

Kessler, W. S., and J. P. McCreary (1993). The annual wind-driven Rossby wave in the subthermocline equatorial Pacific. *J. Phys. Oceanogr.*, **23**, 1192–1207.

Kessler, W. S., M. J. McPhaden, and K. M. Weickmann (1995). Forcing of intraseasonal Kelvin waves in the equatorial Pacific. *J. Geophys. Res.*, **100**, 10,613–10,631.

Knox, R., and D. Halpern (1982). Long range Kelvin wave propagation of transport variations in Pacific Ocean equatorial currents. *J. Mar. Res.*, **40**, 329–339.

Kumar, S.-P., H. Snaith, P. Challenor, and H. T. Guymer (1998). Seasonal and inter-annual sea surface height variations of the northern Indian Ocean from the TOPEX/POSEIDON altimeter. *Ind. J. Marine Sci.*, **27**, 10–16.

Lagerloef, S. E., G. T. Mitchum, R. B. Lukas, and P. P. Niiler (1999). Tropical Pacific near-surface currents estimated from altimetry, wind and drifter data. *J. Geophys. Res.*, **104**, 23,313–23,326.

Lukas, R., S. Hayes, and K. Wyrtki (1984). Equatorial sea level response during 1982–83 El Niño. *J. Geophys. Res.*, **89**, 10,425–10,430.

Malardé J.-P., P. De Mey, C. Perigaud, and J.-F. Minster (1987). Observation of long equatorial waves in the Pacific Ocean by Seasat altimetry. *J. Phys. Oceanogr.*, **17**, 2273–2279.

Maes, C., D. Behringer, R. W. Reynolds, and M. Ji (2000). Retrospective analysis of the salinity variability in the western tropical Pacific Ocean using and indirect minimization approach. *J. Atmos. Ocean. Tech.*, **17**, 512–524.

Mantua, N. J., and D. S. Battisti (1994). Aperiodic variability in the Zebiak-Cane coupled ocean-atmosphere model: Air-sea interactions in the western equatorial Pacific *J. Climate*, **8**, 2897–2827.

Matsuno, T. (1966). Quasi-geostrophic motions in the equatorial area. *J. Meter. Soc. Jpn*, **44**, 25–43.

McCreary, J. P. (1976). Eastern tropical ocean response to changing wind systems: with application to El Niño. *J. Phys. Oceanogr.*, **6**, 632–645.

McPhaden, M. J. (1995). The Tropical Atmosphere-Ocean Array is completed. *Bull. Am. Meteorol. Soc.*, **76**, 739–741.

McPhaden, M. J., and J. Picaut (1990). Zonal displacement of the western Pacific warm pool associated with the 1986–87 ENSO event. *Science*, **250**, 1385–1388.

McPhaden, M. J., A. J. Busalacchi, R. Cheney, J.-R. Donguy, K. S. Gage, D. Halpern, M. Ji, P. Julian, G. Meyers, G. T. Mitchum, P. P. Niiler, J. Picaut, R. W. Reynolds, N. Smith, and K. Takeuchi (1998). The tropical ocean global atmosphere observing system: A decade of progress. *J. Geophys. Res.*, **103**, 14,169–14,240.

Ménard, Y. (1988). Observing the seasonal variability in tropical Atlantic from altimetry. *J. Geophys. Res.*, **93**, 13967–13978.

Menkes, C., J.-P. Boulanger, and A. J. Busalacchi (1995). Evaluation of TOPEX/POSEIDON and TOGA-TAO sea level topographies and their derived currents. *J. Geophys. Res.*, **100**, 25087–25099.

Merle, J. (1980). Seasonal heat budget in the equatorial Atlantic Ocean. *J. Phys. Res.*, **10**, 464–469.

Merle, J., and S. Arnault (1985). Seasonal variability of the surface dynamic topography in the tropical Atlantic Ocean. *J. Mar. Res.*, **43**, 267–288.

Merle, J., P. Hisard, and M. Fieux (1980). Annual signal and interannual anomalies of SST in the eastern tropical Atlantic Ocean. *Deep-Sea Res.*, **26**, GATE Sup., 2–5, 77–102.

Miller, L., and R. E. Cheney (1990). Large-scale meridional transport in the tropical Pacific ocean during the 1986–1987 El Niño. *J. Geophys. Res.*, **95**, 17,905–17,920.

Miller, L., R. E. Cheney, and D. Milbert (1986). Sea level time series in the equatorial Pacific from satellite altimetry. *Geophys. Res. Lett.*, **13**, 475–478.

Miller, L., R. E. Cheney, and B. C. Douglas (1988). Geosat altimeter observations of Kelvin waves and the 1986–87 El Niño. *Science*, **239**, 52–54.

Mitchum, G. T. (1994). Comparison of TOPEX sea surface heights and tide gauge sea level. *J. Geophys. Res.*, **99**, 24,541–24,553.

Mitchum, G. T. (1998). Monitoring the stability of satellite altimeters with tide gauges. *J. Atmos. Ocean Tech.*, **15**, 721–730.

Moore, D. W. (1968). Planetary-gravity waves in an equatorial ocean. Ph. D. thesis, Harvard University, Cambridge, Mass.

Moore, D. W., P. Hisard, J. P. McCreary, J. Merle, J. J. O'Brien, J. Picaut, J.-M. Verstraete, and C. Wunsch (1978). Equatorial adjustment in the eastern Atlantic. *Geophys. Res. Let.*, **5**, 637–640.

Morrow, R., and F. Birol (1998). Variability in the southeast Indian Ocean from altimetry: Forcing mechanisms for the Leeuwin Current. *J. Geophys. Res.*, **103**, 18,529–18,544.

Moura, A., and J. Shukla (1981). On the dynamics of droughts in northeast Brazil: Observations, theory, and numerical experiments with a general circulation model, *J. Atmos. Sci.*, **38**, 2653–2675.

Murtugudde, R., and A. J. Busalacchi (1999). Interannual variability of the dynamics and thermodynamics of the tropical Indian Ocean. *J. Climate*, **12**, 2300–2326.

Murtugudde, R., A. J. Busalacchi, and J. Beauchamp (1998). Seasonal-to-Interannual effects of the Indonesian throughflow on the tropical Indo-Pacific basin. *J. Geophys. Res.*, **103**, 21,425–21,441.

Musman, S. (1986). Sea level changes associated with westward propagating equatorial temperature fluctuations. *J. Geophys. Res.*, **91**, 10,753–10,757.

Musman, S. (1992). Geosat altimeter observations of long waves in the equatorial Atlantic. *J. Geophys. Res.*, **97**, 3573–3579.

Nystuen, J. A., and C. A. Andrade (1993). Tracking mesoscale ocean features in the Caribbean Sea using Geosat altimetry. *J. Geophys. Res.*, **98**, 8389–8394.

Perigaud, C. (1990). Sea level oscillations observed with Geosat along the two shear fronts of the Pacific North Equatorial Countercurrent. *J. Geophys. Res.*, **95**, 7239–7248.

Perigaud, C., and P. Delecluse (1992a). Annual sea level variations in the southern tropical Indian Ocean from Geosat and shallow-water simulations. *J. Geophys. Res.*, **97**, 20,169–20,178.

Perigaud, C., and P. Delecluse (1992b). Low-frequency sea level variations in the Indian Ocean from Geosat altimeter and shallow-water simulations. *Trends Phys. Oceanogr.*, **1**, 85–110.

Perigaud, C., and V. Zlotnicki (1992). Importance of Geosat orbit and tidal errors in the estimation of large-scale Indian Ocean variations. *Oceanogr. Acta*, **15**, 491–505.

Perigaud, C., and P. Delecluse (1993). Interannual sea-level variations in the tropical Indian-Ocean from Geosat and shallow-water simulations. *J. Phys. Oceanogr.*, **12**, 1916–1934.

Perigaud, C., J.-F. Minster, and G. Reverdin (1986). Zonal slope variability of the tropical Indian-Ocean studied from Seasat altimetry. *Mar. Geodesy*, **10**, 53–68.

Philander, S. G. (1990). *In* "El Niño, La Niña, and the Southern Oscillation," 293 pp. Academic Press, San Diego.

Picaut, J., and R. Tournier (1991). Monitoring the 1979–1985 Equatorial Pacific current transports with expendable bathythermography data. *J. Geophys. Res.*, **96**, 3263–3277.

Picaut, J., and T. Delcroix (1995). Equatorial wave sequence associated with warm pool displacements during the 1986–1989 El Niño-La Niña. *J. Geophys. Res.*, **100**, 18393–18408.

Picaut, J., M. J. McPhaden, and S. P. Hayes (1989). On the use of the geostrophic approximation to estimate time-varying zonal currents at the equator. *J. Geophys. Res.*, **94**, 3228–3236.

Picaut, J., A. J. Busalacchi, M. J. McPhaden, and B. Camusat (1990). Validation of the geostrophic method for estimating zonal currents at the equator from Geosat altimeter data. *J. Geophys. Res.*, **95**, 3015–3024.

Picaut, J., A. J. Busalacchi, M. J. McPhaden, L. Gourdeau, F. I. Gonzalez, and E. C. Hackert (1995). Open-ocean validation of TOPEX/POSEIDON sea level in the western equatorial Pacific. *J. Geophys. Res.*, **100**, 25109–25127.

Picaut, J., M. Ioualalen, C. Menkes, T. Delcroix, and M. J. McPhaden (1996). Mechanism of displacements of the Pacific warm pool: implications for ENSO. *Science*, **274**, 1486–1489.

Picaut J., F. Masia, and Y. du Penhoat (1997). An advective-reflective conceptual model for the oscillatory nature of ENSO. *Science*, **277**, 663–666.

Pinkel, R., M. Merrifield, M. McPhaden, J. Picaut, S. Rutledge, D. Siegel, and L. Washburn, 1997: Solitary waves in the western equatorial Pacific Ocean. *Geophys. Res. Lett.*, **24**, 1603–1606.

Potemra, J. T., R. Lukas, and G. Mitchum (1997). Large-scale estimation of transport from the Pacific to the Indian Ocean. *J. Geophys. Res.*, **102**, 27,795–27,812.

Ray, R. D., and G. T. Mitchum (1997). Surface manifestation of internal tides in the deep ocean: observations from altimetry and island gauges. *Prog. Oceanogr.* **40**, 135–162.

Rébert, J., J. Donguy, G. Eldin, and K. Wyrtki, (1985). Relations between sea level, thermocline depth, heat content and dynamic height in the tropical Pacific Ocean, *J. Geophys. Res.*, **90**, 11,719–11,725.

Ripa, P., and S. P. Hayes (1981). Evidence of for equatorial trapped waves at the Galapagos. *J. Geophys. Res.*, **86**, 6509–6516.

Schopf, P. S., and M. J. Suarez (1988). Vacillations in a coupled ocean-atmosphere model, *J. Atmos. Sci.*, **45**, 549–566.

Servain J., J. Picaut, and J. Merle (1982). Evidence of remote forcing in the equatorial Atlantic Ocean. *J. Phys. Oceanogr.*, **12**, 457–463.

Servain J., A. J. Busalacchi, M. J. McPhaden, A. D. Moura, G. Reverdin, M. Vianna, and S. E. Zebiak (1998). A Pilot Research Moored Array in the Tropical Atlantic (PIRATA). *Bull. Am. Meteor. Soc.*, **79**, 2019–2031.

Sharma, R., and M. M. Ali (1996). Variation of mixed layer depth obtained from Geosat altimeter observations in the equatorial Indian Ocean. *Int. J. Remote Sens.*, **17**, 1539–1546.

Springer, S. R., M. J. McPhaden, and A. J. Busalacchi (1990). Oceanic heat content variability in the tropical Pacific during the 1982–1983 El Niño. *J. Geophys. Res.*, **95**, 22,089–22,101.

Subrahmanyam, B., V. R. Babu, V. S. N. Murty, and L. V. G. Rao (1996). Surface circulation off Somalia and western equatorial Indian Ocean during summer monsoon of 1988 from Geosat. *Int. J. Remote Sens.*, **17**, 761–770.

Susanto, R. D., Q. N. Zheng, and X. H. Yan (1998). Complex singular value decomposition analysis of equatorial waves in the Pacific observed by TOPEX/POSEIDON. *J. Atmos. Oceanogr. Tech.*, **15**, 764–774.

Tai, C.-K., W. B. White, and S. E. Pazan (1989). Geosat crossover analysis in the tropical Pacific. Part II: Verification analysis of altimetric sea-level maps with expendable bathythermograph and island sea-level data. *J. Geophys. Res.*, **94**, 897–908.

Verstraete, J. M., and Y. H. Park (1995). Comparison of TOPEX/POSEIDON altimetry and in situ sea level data at Sao Tome Island, Gulf of Guinea. *J. Geophys. Res.*, **100**, 25129–25134.

Vossepoel, F. C., R. W. Reynolds, and L. Miller (1999). Use of sea level observations to estimate salinity variability in the tropical Pacific. *J. Atmosph. Ocean. Technol.*, **16**, 1401–1415.

Webster, P. J., V. O. Magana, T. N. Palmer, J. Shukla, R. A. Tomas, M. Yanai, and T. Yasunari (1998). Monsoons: Processes, predictability, and the prospects for prediction. *J. Geophys. Res.*, **103**, 14451–14510.

White, W. B., and C.-K. Tai (1992). Reflection of interannual Rossby waves at the maritime western boundary of the tropical Pacific. *J. Geophys. Res.*, **97**, 14,305–14,322.

White, W. B., and C.-K. Tai (1995). Inferring interannual changes in global upper ocean heat storage from TOPEX altimetry. *J. Geophys. Res.*, **100**, 24,943–24,954.

White, W. B., G. Meyers, J. R. Donguy, and S. E. Pazan (1985). Short term climate variability in the thermal structure of the Pacific Ocean during 1979–1982. *J. Phys. Oceanogr.*, **15**, 917–935.

White, W. B., N. Graham, and C.-K. Tai (1990). Reflection of annual Rossby waves at the maritime western boundary of the tropical Pacific. *J. Geophys. Res.*, **95**, 3101–31116.

Wyrtki, K. (1975). El Niño The dynamic response of the equatorial Pacific Ocean to atmospheric forcing. *J. Phys. Oceanogr.*, **5**, 572–584.

Wyrtki, K. (1984). The slope of sea level along the equator during the 1982/83 El Niño. *J. Geophys. Res.*, **89**, 10,419–10,424.

Wyrtki, K. (1985). Water displacements in the Pacific and the genesis of El Niño cycles. *J. Geophys. Res.*, **90**, 7129–7132.

Wyrtki, K., and G. Mitchum (1990). Interannual differences of Geosat altimeter heights and sea-level:the importance of a datum. *J. Geophys. Res.*, **95**, 2969–2975.

Yu, Y., W. J. Emery, and R. R. Leben (1995). Satellite altimeter derived geostrophic currents in the western tropical Pacific during the 1992–1993 and their validation with drifting buoy trajectories. *J. Geophys. Res.*, **100**, 25,069–25,085.

Zheng, Q., X.-H. Yan, C.-R. Ho, and C.-K. Tai (1995). Observations of equatorially trapped waves in the Pacific using Geosat data. *Deep Sea Res.*, **42**, 797–817.

Zimbelman, D. F., and A. J. Busalacchi (1990). The wet tropospheric range correction - product intercomparisons and the simulated effect for tropical Pacific altimeter retrievals. *J. Geophys. Res.*, **95**, 2899–2922.

5

Data Assimilation by Models

ICHIRO FUKUMORI

Jet Propulsion Laboratory
California Institute of Technology
Pasadena CA 91109

1. INTRODUCTION

Data assimilation is a procedure that combines observations with models. The combination aims to better estimate and describe the state of a dynamic system, the ocean in the context of this book. The present article provides an overview of data assimilation with an emphasis on applications to analyzing satellite altimeter data. Various issues are discussed and examples are described, but presentation of results from the non-altimetric literature will be limited for reasons of space and scope of this book.

The problem of data assimilation belongs to the wider field of estimation and control theories. Estimates of the dynamic system are improved by correcting model errors with the observations on the one hand and synthesizing observations by the models on the other. Much of the original mathematical theory of data assimilation was developed in the context of ballistics applications. In earth science, data assimilation was first applied in numerical weather forecasting.

Data assimilation is an emerging area in oceanography, stimulated by recent improvements in computational and modeling capabilities and the increase in the amount of available oceanographic observations. The continuing increase in computational capabilities have made numerical ocean modeling a commonplace. A number of new ocean general circulation models have been constructed with different grid structures and numerical algorithms, and incorporating various innovations in modeling ocean physics (e.g., Gent and McWilliams, 1990; Holloway, 1992; Large *et al.*, 1994). The fidelity of ocean modeling has advanced to a stage where models are utilized beyond idealized process studies and are now employed to simulate and study the actual circulation of the ocean. For instance, model results are operationally produced to analyze the state of the ocean (e.g., Leetmaa and Ji, 1989), and modeling the global ocean circulation at eddy resolution is nearing a reality (e.g., Fu and Smith, 1996).

Recent oceanographic experiments, such as the World Ocean Circulation Experiment (WOCE) and the Tropical Ocean and Global Atmosphere Program (TOGA), have generated unprecedented amounts of *in situ* observations. Moreover, satellite observations, in particular satellite altimetry such as TOPEX/POSEIDON, have provided continuous synoptic measurements of the dynamic state of the global ocean. Such extensive observations, for the first time, provide a sufficient basis to describe the coherent state of the ocean and to stringently test and further improve ocean models.

However, although comprehensive, the available *in situ* measurements and those in the foreseeable future are and will remain sparse in space and time compared with the energy-containing scales of ocean circulation. An effective means of synthesizing such observations then becomes essential in utilizing the maximum information content of such observing systems. Although global in coverage, the nature of satellite altimetry also requires innovative approaches to effectively analyze its measurements. For instance, even though sea level is a dynamic variable that reflects circulation at depth, the vertical dependency of the circulation is not immediately obvious from sea-level measurements alone. The nadir-pointing property of altimeters also limits sampling in the direction across satellite ground tracks, making analyses of meso-scale features problematic, especially with a single satellite. Furthermore, the complex space-time sampling pattern of satellites caused by orbital dynamics makes analyses of even large horizontal scales nontrivial, especially for analyzing high-frequency variability such as tides and wind-forced barotropic motions.

Data assimilation provides a systematic means to untangle such degeneracy and complexity, and to compensate for the incompleteness and inaccuracies of individual observing systems in describing the state of the ocean as a whole. The process is effected by the models' theoretical relationship among variables. Data information is interpolated and extrapolated by model equations in space, time, and into other variables including those that are not directly measured. In the process, the information is further combined with other data, which further improves the description of the oceanic state. In essence, assimilation is a dynamic extrapolation as well as a synthesis and averaging process.

In terms of volume, data generated by a satellite altimeter far exceeds any other observing system. Partly for this reason, satellite altimetry is currently the most common data type explored in studies of ocean data assimilation. (Other reasons include, for example, the near real-time data availability and the nontrivial nature of altimetric measurements in relation to ocean circulation described above.) This chapter introduces the subject matter by describing the issues, particularly those that are often overlooked or ignored. By so doing, the discussion aims to provide the reader with a perspective on the present status of altimetric assimilation and on what it promises to accomplish.

An emphasis is placed on describing what exactly data assimilation solves. In particular, assimilation improves the oceanic state *consistent* with both models and observations. This also means, for instance, that data assimilation does not and cannot correct every model error, and the results are not altogether more accurate than what the raw data measure. This is because, from a pragmatic standpoint, models are always incomplete owing to unresolved scales and physics, which in effect are *inconsistent* with models. Overfitting models to data beyond the model's capability can lead to inaccurate estimates. These issues will be clarified in the subsequent discussion.

We begin in Section 2 by reviewing some examples of data assimilation, which illustrate its merits and motivations. Reflecting the infancy of the subject, many published studies are of relatively simple demonstration exercises. However, the examples describe the diversity and potential of data assimilation's applications.

The underlying mathematical problem of assimilation is identified and described in Section 3. Many of the issues, such as how best to perform assimilation, what it achieves, and how it differs from improving numerical models and/or data analyses per se, are best understood by first recognizing the fundamental problem of combining data and models.

Many of the early studies on ocean data assimilation center on methodologies, whose complexities and theoretical nature have often muddied the topic. A series of different assimilation methods are heuristically reviewed in Section 4

with references to specific applications. Mathematical details are minimized for brevity and the emphasis is placed instead on describing the nature of the approaches. In essence, most methods are equivalent to each other so long as the assumptions are the same. A summary and recommendation of methods is also presented at the end of Section 4.

Practical Issues of Assimilation are discussed in Section 5. Identification of what the model-data combination resolves is clarified, in particular, how assimilation differs from model improvement per se. Other topics include prior error specifications, observability, and treatment of the time-mean sea level. We end this chapter in Section 6 with concluding remarks and a discussion on future directions and prospects of altimetric data assimilation.

The present pace of advancement in assimilation is rapid. For other reviews of recent studies in ocean data assimilation, the reader is referred to articles by Ghil and Malanotte-Rizzoli (1991), Anderson *et al.* (1996), and by Robinson *et al.* (1998). The books by Anderson and Willebrand (1989) and Malanotte-Rizzoli (1996) contain a range of articles from theories and applications to reviews of specific problems. A number of assimilation studies have also been collected in special issues of *Dynamics of Atmospheres and Oceans* (1989, vol 13, No 3–4), *Journal of Marine Systems* (1995, vol 6, No 1–2), *Journal of the Meteorological Society of Japan* (1997, vol 75, No 1B), and *Journal of Atmospheric and Oceanic Technology* (1997, vol 14, No 6). Several papers focusing on altimetric assimilation are also collected in a special issue of *Oceanologica Acta* (1992, vol 5).

2. EXAMPLES AND MERITS OF DATA ASSIMILATION

This section reviews some of the applications of data assimilation with an emphasis on analyzing satellite altimetry observations. The examples here are restricted because of limitation of space, but are chosen to illustrate the diversity of applications to date and to point to further possibilities in the future.

One of the central merits of data assimilation is its extraction of oceanographic signals from incomplete and noisy observations. Most oceanographic measurements, including altimetry, are characterized by their sparseness in space and time compared to the inherent scales of ocean variability; this translates into noisy and gappy measurements. Figure 1 (see color insert) illustrates an example of the noise-removal aspect of altimetric assimilation. Sea-level anomalies measured by TOPEX (left) and its model equivalent estimates (center and right) are compared as a function of space and time (Fukumori, 1995). The altimetric measurements (left panel) are characterized by noisy estimates caused by measurement errors and gaps in the sampling, whereas the assimilated estimate (center) is more complete, interpolating

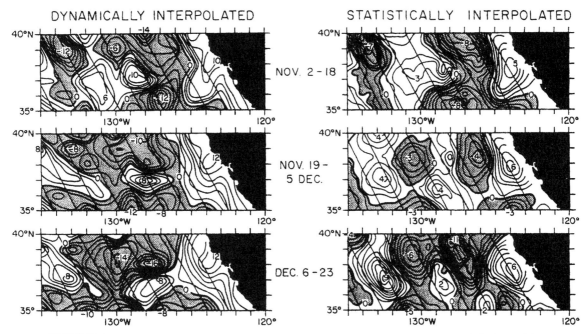

FIGURE 2   A time sequence of sea-level anomaly maps based on Geosat data; (Left) model assimilation, (Right) statistical interpolation of the altimetric data. Contour interval is 2 cm. Shaded (unshaded) regions indicate negative (positive) values. The model is a 7-layer quasi-geostrophic (QG) model of the California Current, into which the altimetric data are assimilated by nudging. (Adapted from White *et al.* (1990a), Fig. 13, p. 3142.)

over the data dropouts and removing the short-scale temporal and spatial variabilities measured by the altimeter. In the process, the assimilation corrects inaccuracies in model simulation (right panel), elucidating the stronger seasonal cycle and westward propagating signals of sea-level variability.

The issue of dynamically interpolating sea level information is particularly critical in studying meso-scale dynamics, as satellites cannot adequately measure eddies because the satellite's ground-track spacing is typically wider than the size of the eddy features. Figure 2 compares a time sequence of dynamically (i.e., assimilation; left column) and statistically (right column) interpolated synoptic maps of sea level by White *et al.* (1990a). The statistical interpolation is based solely on spatial distances between the analysis point and the data point (e.g., Bretherton *et al.*, 1976), whereas the dynamical interpolation is based on assimilation with an ocean model. While the statistically interpolated maps tend to have maxima and minima associated with meso-scale eddies along the satellite ground-tracks, the assimilated estimates do not, allowing the eddies to propagate without significant distortion of amplitude, even between satellite ground tracks. An altimeter's resolving power of meso-scale variability can also significantly improve variabilities simulated by models. For instance, Figure 3 shows distribution of sea-surface height variability by Oschlies and Willebrand (1996), comparing measurements of Geosat (middle) and an eddy-resolving primitive equation model. The bottom and top panels show model results with and without assimilation,

respectively. The altimetric assimilation corrects the spatial distribution of variability, especially north of 30°N, reducing the model's variability in the Irminger Sea but enhancing it in the North Atlantic Current and the Azores Current.

The virtue of data assimilation in dynamically interpolating and extrapolating data information extends beyond the variables that are observed to properties not directly measured. Such an estimate is possible owing to the dynamic relationship among different model properties. For instance, Figure 4 shows estimates of subsurface temperature (left) and velocity (right) anomalies of an altimetric assimilation (gray curve) compared against independent (i.e., non-assimilated) *in situ* measurements (solid curve) (Fukumori *et al.*, 1999). In spite of the assimilated data being limited to sea-level measurements, the assimilated estimate (gray) is found to resolve the amplitude and timing of many of the subsurface temperature and velocity "events" better than the model simulation (dashed curve). The skill of the model results are also consistent with formal uncertainty estimates (dashed and solid gray bars) that reflect inaccuracies in data and model. Such error estimates are by-products of assimilation that, in effect, quantify what has been resolved by the model (see Section 5.3 for further discussion).

Although uncertainties in our present knowledge of the marine geoid (cf., Chapter 10) limit the direct use of altimetric sea-level measurements to mostly that of temporal variabilities, the nonlinear nature of ocean circulation allows estimates of the mean circulation to be made from measure-

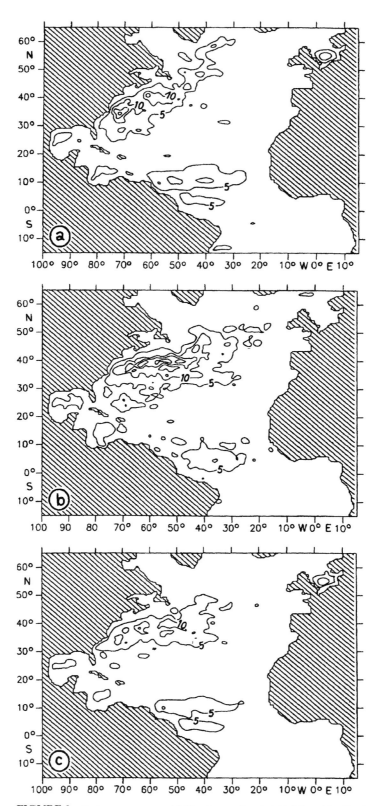

FIGURE 3   Root-mean-square variability of sea surface height; (a) model without assimilation, (b) Geosat data, (c) model with assimilation. Contour interval is 5 cm. The model is based on the Community Modeling Effort (CME; Bryan and Holland, 1989). Assimilation is based on optimal interpolation. (Adapted from Oschlies and Willebrand (1996), Fig. 7, p. 14184.)

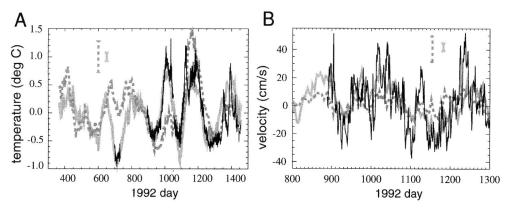

FIGURE 4 Comparison of model estimates and *in situ* data; (A) temperature anomaly at 200 m 8°N 180°E, (B) zonal velocity anomaly at 120 m 0°N 110°W. The different curves are data (black), model simulation (gray dashed), and model estimate by TOPEX/POSEIDON assimilation (gray solid). Bars denote formal uncertainty estimates of the model. The model is based on the GFDL Modular Ocean Model, and the assimilation scheme is an approximate Kalman filter and smoother. This model and assimilation are further discussed in Sections 5.1.2, 5.1.4, and 5.2. (Adapted from Fukumori *et al.* (1999), Plates 4 and 5.)

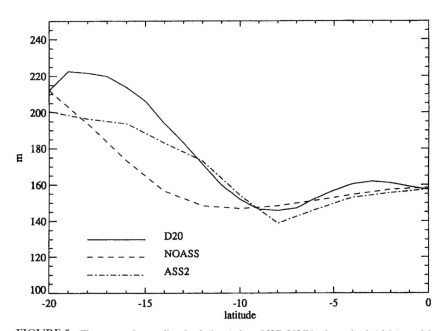

FIGURE 5 Time-mean thermocline depth (in m) along 95°E; 20°C isotherm depth (plain), model simulation (dashed), and model with assimilating Geosat data (chain-dashed). The model is a nonlinear 1.5-layer reduced gravity model of the Indian Ocean. Geosat data are assimilated over 1-year (November 1986 to October 1987) employing the adjoint method. The 20°C isotherm is deduced from an XBT analyses (Smith, 1995). (Adapted from Greiner and Perigaud (1996), Fig. 10, p. 1744.)

ments of variabilities alone. Figure 5 compares such an estimate by Greiner and Perigaud (1996) of the time-mean depth of the thermocline in the Indian Ocean, based solely on assimilation of temporal variabilities of sea level measured by Geosat. The thermocline depth of the altimetric assimilation (chain-dash) is found to be significantly deeper between 10°S and 18°S than without assimilation (dash) and is in closer agreement with *in situ* observations based on XBT measurements (solid).

Data assimilation's ability to estimate unmeasured properties provides a powerful tool and framework to analyze data and to combine information systematically from multiple observing systems simultaneously, making better estimates that are otherwise difficult to obtain from measurements alone. Stammer *et al.* (1997) have begun the process of synthesizing a wide suite of observations with a general circulation model, so as to improve estimates of the complete state of the global ocean. Figure 6 illustrates im-

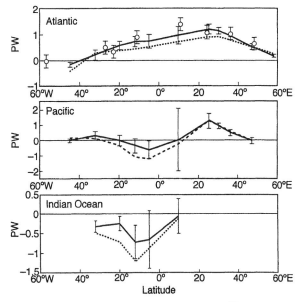

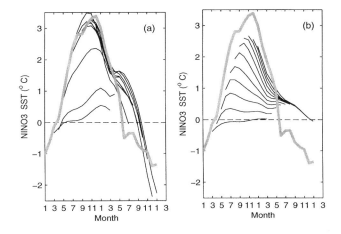

FIGURE 7   Hindcasts of Niño3 index of sea surface temperature (SST) anomaly with (a) and without (b) assimilation. The gray and solid curves are observed and modeled SSTs, respectively. The model is a simple coupled ocean-atmosphere model, and the assimilation is of altimetry, winds, and sea surface temperatures, conducted by the adjoint method. (Adapted from Lee *et al.* (2000), Fig. 10.)

FIGURE 6   Mean meridional heat transport (in 10^{15} W) estimate of a constrained (solid) and unconstrained (dashed lines) model of the Atlantic, the Pacific, and the Indian Oceans, respectively. The model (Marshall *et al.*, 1997) is constrained using the adjoint method by assimilating TOPEX/POSEIDON data in addition to a hydrographic climatology and a geoid model. Bars on the solid lines show root-mean-square variability over individual 10-day periods. Open circles and bars show similar estimates and their uncertainties of Macdonald and Wunsch (1996). (Adapted from Stammer *et al.* (1997), Fig. 13, p. 28.)

provements made in the time-mean meridional heat transport estimate from assimilating altimetric measurements from TOPEX/POSEIDON, along with a geoid estimate and a hydrographic climatology. For instance, in the North Atlantic, the observations require a larger northward heat transport (solid curve) than an unconstrained model (dashed curve) that is in better agreement with independent estimates (circles). Differences in heat flux with and without assimilation are equally significant in other basins.

One of the legacies of TOPEX/POSEIDON is its improvement in our understanding of ocean tides. Refer to Chapter 6 for a comprehensive discussion on tidal research using satellite altimetry. In the context of this chapter, a significant development in the last few years is the emergence of altimetric assimilation as an integral part of developing accurate tidal models. The two models chosen for reprocessing TOPEX/POSEIDON data are both based on combining observations and models (Shum *et al.*, 1997). In particular, Le Provost *et al.* (1998) give an example of the benefit of assimilation, in which the data assimilated tidal solution (FES95.2) is shown to be more accurate than the pure hydrodynamic model (FES94.1) or the empirical tidal estimate (CSR2.0) used in the assimilation. That is, assimilated estimates are more accurate than analyses based either on data or model alone.

Data assimilation also provides a means to improve prediction of a dynamic system's future evolution, by providing optimal initial conditions and other model parameters from which forecasts are issued. In fact, such applications of data assimilation are the central focus in ballistics applications and in numerical weather forecasting. In recent years, forecasting has also become an important application of data assimilation in oceanography. For example, oceanographic forecasts in the tropical Pacific are routinely produced by the National Center for Environmental Prediction (NCEP) (Behringer *et al.*, 1998; Ji *et al.*, 1998), with particular applications to forecasting the El Niño–Southern Oscillation (ENSO). Of late, altimetric observations have also been utilized in the NCEP system (Ji *et al.*, 2000). Lee *et al.* (2000) have explored the impact of assimilating altimetry data into a simple coupled ocean-atmosphere model of the tropical Pacific. For example, Figure 7 shows improvements in their model's skill in predicting the so-called Niño3 sea-surface temperature anomaly as a result of assimilating TOPEX/POSEIDON altimeter data. The model predictions (solid curves) are in better agreement with the observed index (gray curve) in the assimilated estimate (left panel) than without data constraints (right panel).

Apart from sea level, satellite altimetry also measures significant wave height (SWH), which is another oceanographic variable of interest. In particular, the European Centre for Medium-Range Weather Forecasting (ECMWF) has been assimilating altimetric wave height (ERS1) in producing global operational wave forecasts (Janssen *et al.*, 1997). Figure 8 shows an example of the impact of assimilating altimetric SWH in improving predictions made by this wave model up to 5-days into the future (Lionello *et al.*, 1995).

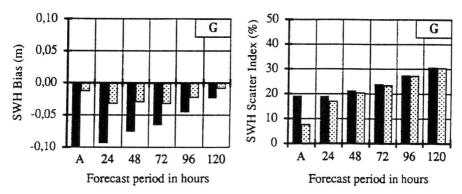

FIGURE 8 Bias and scatter index of significant wave height (SWH) analysis (denoted A on the abscissa) and various forecasts. Comparisons are between model and altimeter. Full (dotted) bars denote the reference experiment without (with) assimilating ERS-1 significant wave height data. The scatter index measures the lack of correlation between model and data. The model is the third generation wave model WAM. Assimilation is performed by optimal interpolation. (Adapted from Lionello *et al.* (1995), Fig. 12, p. 105.)

The figure shows the assimilation (dotted bars) resulting in a smaller bias (left panel) and higher correlation (i.e., smaller scatter) (right panel) with respect to actual wave-height measurements than those without assimilation (full bars). Further discussions on wave forecasting can be found in Chapter 7.

In addition to the state of the ocean, data assimilation also provides a framework to estimate and improve model parameters, external forcing, and open boundary conditions. For instance, Smedstad and O'Brien (1991) estimated the phase speed in a reduced-gravity model of the tropical Pacific Ocean using sea-level measurements from tide gauges. Fu *et al.* (1993) and Stammer *et al.* (1997) estimated uncertainties in winds, in addition to the model state, from assimilating altimetry data. (The latter study also estimated errors in atmospheric heat fluxes.) Lee and Marotzke (1998) estimated open boundary conditions of an Indian Ocean model.

Data assimilation in effect fits models to observations. Then, the extent to which models can or cannot be fit to data gives a quantitative measure of the model's consistency with measurements, thus providing a formal means of hypothesis testing that can also help identify specific deficiencies of models. For example, Bennett *et al.* (1998) identified inconsistencies between moored temperature measurements and a coupled ocean-atmosphere model of the tropical Pacific Ocean, resulting from the model's lack of momentum advection. Marotzke and Wunsch (1993) found inconsistencies between a time-invariant general circulation model and a climatological hydrography, indicating the inherent nonlinearity of ocean circulation. Alternatively, excessive model-data discrepancies found by data assimilation can also point to inaccuracies in observations. Examples of such analysis at present can be best found in meteorological applications (e.g., Hollingsworth, 1989).

Lastly, data assimilation has also been employed in evaluating merits of different observing systems by analyzing model results with and without assimilating particular observations. For instance, Carton *et al.* (1996) found TOPEX/POSEIDON altimeter data having larger impact in resolving intra-seasonal variability of the tropical Pacific Ocean than data from a mooring array or a network of expendable bathythermographs (XBTs). Verron (1990) and Verron *et al.* (1996) conducted a series of numerical experiments (observing system simulation experiments, OSSEs, or twin experiments) to evaluate different scenarios of single- and dual-altimetric satellites. OSSEs and twin experiments are numerical experiments in which a set of pseudo observations are extracted from a particular numerical simulation and are assimilated into another (e.g., with different initial conditions and/or forcing, etc.) to examine the degree to which the former results can be reconstructed. The relative skill of the estimate among different observing scenarios provides a measure of the observation's effectiveness. From such an analysis, Verron *et al.* (1996) conclude that a 10–20 day repeat period is satisfactory for the spatial sampling of mid-latitude meso-scale eddies but that any further gain would come from increased temporal, rather than spatial, sampling provided by a second satellite that is offset in time. Twin experiments are also employed in testing and evaluating different data assimilation methods (Section 4).

3. DATA ASSIMILATION AS AN INVERSE PROBLEM

Recognizing the mathematical problem of data assimilation is essential in understanding what assimilation could achieve, where the difficulties exist, and where the issues arise from. For example, there are theoretical and practical difficulties involved in solving the problem, and various assumptions and approximations are necessarily made, oftentimes implicitly. A clear understanding of the problem is critical in interpreting the results of assimilation as well as in identifying sources of inconsistencies.

Mathematically, as will be shown, data assimilation is simply an inverse problem, such as,

$$\mathcal{A}(\mathbf{x}) \approx \mathbf{y} \tag{1}$$

in which the unknowns, vector $\mathbf{x}$, are estimated by inverting some functional $\mathcal{A}$ relating the unknowns on the left-hand-side to the knowns, $\mathbf{y}$, on the right-hand-side. Equation (1) is understood to hold only approximately (thus $\approx$ instead of $=$), as there are uncertainties on both sides of the equation. Throughout this chapter, bold lowercase letters will denote column vectors.

The unknowns $\mathbf{x}$ in the context of assimilation, are independent variables of the model that may include the state of the model, such as temperature, salinity, and velocity over the entire model domain, and various model parameters as well as unknown external forcing and boundary conditions. The knowns, $\mathbf{y}$, include all observations as well as known elements of the forcing and boundary conditions. The functional $\mathcal{A}$ describes the relationships between the knowns and unknowns, and includes the model equations that dictate the temporal evolution of the model state. All variables and functions will be assumed discretized in space and time as is the case in most practical numerical model implementations.

The data assimilation problem can be identified in the form of Eq. (1) by explicitly noting the available relationships. Observations of the ocean at some particular instant (subscript i), $\mathbf{y}_i$, can be related to the state of the model (including all uncertain model parameters), $\mathbf{x}_i$, by some functional $\mathcal{H}_i$:

$$\mathcal{H}_i(\mathbf{x}_i) \approx \mathbf{y}_i. \tag{2}$$

(The functional $\mathcal{H}_i$ is also dependent on i because the particular set of observations may change with time i.) In case of a direct measurement of one of the model unknowns, $\mathcal{H}_i$ is simply a functional that returns the corresponding element of $\mathbf{x}_i$. For instance, if $\mathbf{y}_i$ were a scalar measurement of the jth element of $\mathbf{x}_i$, $\mathcal{H}_i$ would be a row vector with zeroes except for its jth element being one:

$$\mathcal{H}_i = (0, \quad \ldots, \quad 0, \quad 1, \quad 0, \quad \ldots, \quad 0). \tag{3}$$

Functional $\mathcal{H}$ would be nontrivial for diagnostic quantities of the model state, such as sea level in a primitive equation model with a rigid-lid approximation (e.g., Pinardi *et al.*, 1995). However, even for such situations, a model equivalent of the observation can be expressed by some functional $\mathcal{H}$ as in Eq. (2), be it explicit or implicit.

In addition to the observation equations (Eq. [2]), the model algorithm provides a constraint on the temporal evolution of the model state, that could be brought to bear upon the problem of determining the unknown model states $\mathbf{x}$:

$$\mathbf{x}_{i+1} \approx \mathcal{F}_i(\mathbf{x}_i). \tag{4}$$

Equation (4) includes the initialization constraint,

$$\mathbf{x}_0 = \mathbf{x}_{\text{first guess}}. \tag{5}$$

Function $\mathcal{F}_i$ is, in practice, a discretization of the continuous equations of the ocean physics and embodies the model algorithm of integrating the model state in time from one observed instant i to another $i + 1$. The function generally depends on the state at i as well as any external forcing and/or boundary condition. (For multi-stage algorithms that involve multiple time-steps in the integration, such as the leap-frog or Adams-Bashforth schemes, the state at i could be defined as concatenated states at corresponding multiple time-steps.)

Combining observation Eq. (2) and model evolution Eq. (4), the assimilation problem as a whole can be written as,

$$\begin{pmatrix} \vdots \\ \mathcal{H}_i(\mathbf{x}_i) \\ \vdots \\ \mathbf{x}_{i+1} - \mathcal{F}_i(\mathbf{x}_i) \\ \vdots \end{pmatrix} \approx \begin{pmatrix} \vdots \\ \mathbf{y}_i \\ \vdots \\ 0 \\ \vdots \end{pmatrix}. \tag{6}$$

By solving the data and model equations simultaneously, assimilation seeks a solution (model state) that is consistent with both data and model equation.

Eq. (6) defines the assimilation problem and can be recognized as a problem of the form Eq. (1), where the states in Eq. (6) at different time steps $(\ldots, \mathbf{x}_i^T, \mathbf{x}_{i+1}^T, \ldots)^T$ define the unknown $\mathbf{x}$ on the left-hand side of Eq. (1). Typically, the number of unknowns far exceed the number of independent equations and the problem is ill-posed. Thus, data assimilation is mathematically equivalent to other inverse problems such as the classic box model geostrophic inversion (Wunsch, 1977) and the beta spiral (Stommel and Schott, 1977). However, what distinguishes assimilation problems from other oceanic inverse problems is the temporal evolution and the sophistication of the models involved. Instead of simple constraints such as geostrophy and mass conservation, data assimilation employs more general physical principles applied at much higher resolution and spatial extent. The intervariable relationship provided by the model equations solved together with the observation equations allows data information to affect the model solution in space and time, both with respect to times that formally lie in the future and past of the observed instance, as well as among different properties.

From a practical standpoint, the distinguishing property of data assimilation is its enormous dimensionality. Typical ocean models contain on the order of several million independent variables at any particular instant. For example, a global model with 1° horizontal resolution and 20 vertical levels is a fairly coarse model by present standards, yet it would have 1.3 million grid points ($360 \times 180 \times 20$) over the globe. With four independent variables per grid node (the two components of horizontal velocity, temperature, and salinity), such as in a primitive equation model with the rigid-lid approximation, the number of unknowns would equal 5 million globally or approximately 3 million when counting points only within the ocean.

The amount of data is also large for an altimeter. For TOPEX/POSEIDON, the Geophysical Data Record provides a datum every second, which over its 10-day repeat cycle amount to approximately 500,000 points over the ocean, which is an order of magnitude larger than the number of horizontal grid points of the $1°$ model considered above. In light of the redundancy the data would provide for such a coarse model, the altimeter could be thought of as providing sea level measurements at the rate of one measurement at every grid point per repeat cycle. Then, assuming for simplicity that all observations within a repeat cycle are coincident in time, each observation equation of form Eq. (2) would have approximately 50,000 equations, and there would be 180 such sets (time-levels or different i's) over a course of a 5-year mission amounting to 9 million individual observation equations. The number of time-levels involved in the observation equations would require at least as many for the model equations in Eq. (6), amounting to 540 million (180×3 million) individual model equations.

The size of such a problem precludes any direct approach in solving Eq. (6), such as deriving the inverse of the operator on the left-hand side even if it existed. In practice, there is generally no solution that exactly satisfies Eq. (6), because of inaccuracies of models and uncertainties in observations. Instead, an approximate solution is sought that solves the equations as "close" as possible in some suitably defined manner. Several ingenious inverse methods are known and/or have been developed, and are briefly reviewed in the section below.

4. ASSIMILATION METHODOLOGIES

Because of the problem's large computational task, devising methods of assimilation has been one of the central issues in data assimilation. Many assimilation methods have been put forth and explored, and they are heuristically reviewed in this section. The aim of this discussion is to elucidate the nature of different methods and thereby allow the reader familiarity with how the problems are approached. Rigorous descriptions of the methods are deferred to references herein.

Assimilation problems are in practice ill-posed, in the sense that no unique solution satisfies the problem Eq. (6). Consequently, many assimilation methodologies are based on "classic" inverse methods. Therefore, for reference, we will begin the discussion with a simple review of the nature of inverse methods. Different assimilation methodologies are then individually described, preceded by a brief overview so as to place the approaches into a broad perspective. A Summary and Recommendation is given in Section 4.11.

4.1. Inverse Methods

Comprehensive mathematical expositions of oceanographic inverse problems and inverse methods can be found, for example, in the textbooks of Bennett (1992) and Wunsch (1996). Here we will briefly review their nature for reference.

Inverse methods are mathematical techniques that solve ill-posed problems that do not have solutions in the strict mathematical sense. The methods seek solutions that approximately satisfy constraints, such as Eq. (6), under suitable "optimality" criteria. These criteria include, various least-squares, maximum likelihood, and minimum-error variance (Bayesian estimates). Differences among the criteria lie in what are explicitly assumed.

Least-squares methods seek solutions that minimize the weighted sum of differences between the left- and right-hand sides of an inverse problem (Eq. [1]):

$$\mathcal{J} = (\mathbf{y} - \mathcal{A}(\mathbf{x}))^T \mathbf{W}^{-1} (\mathbf{y} - \mathcal{A}(\mathbf{x})) \tag{7}$$

where $\mathbf{W}$ is a matrix defining weights.

Least-squares methods do not have explicit statistical or probabilistic assumptions. In comparison, the maximum likelihood estimate seeks a solution that maximizes the a posteriori probability of the right-hand side of Eq. (6) by invoking particular probability distribution functions for $\mathbf{y}$. The minimum variance estimate solves for solutions $\mathbf{x}$ with minimum a posteriori error variance by assuming the error covariance of the solution's prior expectation as well as that of the right-hand side.

Although seemingly different, the methods lead to identical results so long as the assumptions are the same (see for example Introduction to Chapter 4 of Gelb [1974] and Section 3.6 of Wunsch [1996]). In particular, a lack of an explicit assumption can be recognized as being equivalent to a particular implicit assumption. For instance, a maximum likelihood estimate with no prior assumptions about the solution is equivalent to assuming an infinite prior error covariance for a minimum variance estimate. For such an estimate, any solution is acceptable as long as it maximizes the a posteriori probability of the right-hand side (Eq. [6]).

Based on the equivalence among "optimal methods," Eq. (7) can be regarded as a practical definition of what various inverse methods solve (and therefore assimilation). Furthermore, the equivalence provides a statistical basis for prescribing weights used in Eq. (7). In particular, $\mathbf{W}$ can be identified as the error covariance among individual equations of the inverse problem Eq. (6).

When the weights of each separate relation are uncorrelated in time, Eq. (7) may be expanded as,

$$\mathcal{J} = \Sigma_{i=0}^{M}(\mathbf{y}_i - \mathcal{H}_i(\mathbf{x}_i))^T \mathbf{R}_i^{-1}(\mathbf{y}_i - \mathcal{H}_i(\mathbf{x}_i))$$
$$+ \Sigma_{i=0}^{M}(\mathbf{x}_{i+1} - \mathcal{F}_i(\mathbf{x}_i))^T \mathbf{Q}_i^{-1}(\mathbf{x}_{i+1} - \mathcal{F}_i(\mathbf{x}_i)) \quad (8)$$

where $\mathbf{R}$ and $\mathbf{Q}$ denote weighting matrices of data and model equations, respectively, and M is the total number of observations of form Eq. (2). Most assimilation problems are formulated as in Eq. (8), i.e., uncertainties are implicitly assumed to be uncorrelated in time.

The statistical basis of optimal inverse methods allows explicit a posteriori uncertainty estimates to be derived. Such estimates quantify what has been resolved and is an integral part of an inverse solution. The errors identify what is accurately determined and what remains indeterminate, and thereby provide a basis for interpreting the solution and a means to ascertain necessary improvements in models and observing systems.

4.2. Overview of Assimilation Methods

Many of the so-called "advanced" assimilation methods originate in estimation and control theories (e.g., Bryson and Ho, 1975; Gelb, 1974), which in turn are based on "classic" inverse methods. These include the adjoint, representer, Kalman filter and related smoothers, and Green's function methods. These techniques are characterized by their explicit assumptions under which the inverse problem of Eq. (6) is consistently solved. The assumptions include, for example, the weights $\mathbf{W}$ used in the problem identification (Eq. [7]) and specific criteria in choosing particular "optimal" solutions, such as least-squares, minimum error variance, and maximum likelihood. As with "classic" inverse methods, these assimilation schemes are equivalent to each other and result in the same solution as long as the assumptions are the same. Using specific weights allows for explicitly accounting for uncertainties in models and data, as well as evaluation of a posteriori errors. However, because of significant algorithmic and computational requirements in implementing these optimal methods, many studies have explored developing and testing alternate, simpler approaches of combining model and data.

The simpler approaches include optimal interpolation, "3D-var," "direct insertion," "feature models," and "nudging." Many of these approaches originate in atmospheric weather forecasting and are largely motivated in making practical forecasts by sequentially modifying model fields with observations. The methods are characterized by various ad hoc assumptions (e.g., vertical extrapolation of altimeter data) to effect the simplification, but the results are at times obscured by the nature of the choices made without a clear understanding of the dynamical and statistical implications. Although the methods aim to adjust model fields towards observations, it is not entirely clear how the solution relates to the problem identified by Eq. (6). Many of the simpler approaches do not account for uncertainties, potentially allowing the models to be forced towards noise, and data that are formally in the future are generally not used in the estimate except locally to yield a temporally smooth result. However, in spite of these shortcomings, these methods are still widely employed because of their simplicity, and, therefore, warrant examination.

4.3. Adjoint Method

Iterative gradient descent methods provide an effective means of solving minimization problems of form Eq. (7), and a particularly powerful method of obtaining such gradients is the so-called adjoint method. The adjoint method transforms the unconstrained minimization problem of Eq. (7) into a constrained one, which allows the gradient of the "cost function" (Eq. [7]), $\partial \mathcal{J}/\partial \mathbf{x}$, to be evaluated by the model's adjoint (i.e., the conjugate transpose [Hermitian] of the model derivative with respect to the model state variables [Jacobian]). Namely, without loss of generality, uncertainties of the model equations (Eq. [4]) are treated as part of the unknowns and moved to the left-hand side of Eq. (6). The resulting model equations are then satisfied identically by the solution that also explicitly includes errors of the model as part of the unknowns. As a standard method for solving constrained optimization problems, Lagrange multipliers are introduced to formally transform the constrained problem back to an unconstrained one. The Lagrange multipliers are solutions to the model adjoint, and in turn give the gradient information of $\mathcal{J}$ with respect to the unknowns. The computational efficiency of solving the adjoint equations is what makes the adjoint method particularly useful. Detailed derivation of the adjoint method can be found, for example, in Thacker and Long (1988).

Methods that directly solve the minimization problem (7) are sometimes called variational methods or 4D-var (four-dimensional variational method). Namely, four-dimensional for minimization over space and time and variational because of the theory based on functional variations. However, strictly speaking, this reference is a misnomer. For example, Kalman filtering/smoothing is also a solution to the four-dimensional optimization problem, and to the extent that assimilation problems are always rendered discrete, the adjoint method is no longer variational but is algebraic.

Many applications of the adjoint are of the so-called "strong constraint" variety (Sasaki, 1970), in which model equations are assumed to hold exactly without errors making initial and boundary conditions the only model unknowns. As a consequence, many such studies are of short duration because of finite errors in $\mathcal{F}$ in Eq. (4) (e.g., Greiner *et al.*, 1998a, b). However, contrary to common misconceptions, the adjoint method is not restricted to solving only "strong constraint" problems. As described above, by explicitly incorporating model errors as part of the unknowns (so-called controls), the adjoint method can be applied to solve Eq. (7) with nonzero model uncertainties **Q**. Examples of such "weak constraint" adjoint may be found in Stammer *et al.* (1997) and Lee and Marotzke (1998). (See also Griffith and Nichols, 1996.)

Adjoint methods have been used to assimilate altimetry data into regional quasi-geostrophic models (Moore, 1991; Schröter *et al.*, 1993; Vogeler and Schröter, 1995; Morrow and De Mey, 1995; Weaver and Anderson, 1997), shallow water models (Greiner and Perigaud, 1994, 1996; Cong *et al.*, 1998), primitive equation models (Stammer *et al.*, 1997; Lee and Marotzke, 1998), and a simple coupled ocean-atmosphere model (Lee *et al.*, 2000), de las Hera *et al.* (1994) explored the method in wave data assimilation.

One of the particular difficulties of employing adjoint methods has been in generating the model's adjoint. Algorithms of typical general circulation models are complex and entail on the order of tens of thousands of lines of code, making the construction of the adjoint technically challenging. Moreover, the adjoint code depends on the particular set of control variables that varies with particular applications. The adjoint compiler of Giering and Kaminski (1998) greatly alleviates the difficulty associated with generating the adjoint code by automatically transforming a forward model into its tangent linear approximation and adjoint. Stammer *et al.* (1997) employed the adjoint of the MITGCM (Marshall *et al.*, 1997) constructed by such a compiler.

The adjoint method achieves its computational efficiency by its efficient evaluation of the gradient of the cost function. Yet, typical application of the adjoint method requires several tens of iterations until the cost function converges, which still requires a significant amount of computations relative to a simulation. Moreover, for nonlinear models, integration of the Lagrange multipliers requires the forward model trajectory which must be stored or recomputed during each iteration. Approximations have been made by saving such trajectories at coarser time levels than actual model time-steps ("checkpointing"), recomputing intermediate time-levels as necessary or simply approximating them with those that are saved (e.g., Lee and Marotzke, 1997). In the "weak constraint" formalism, the unknown model errors are estimated at fixed intervals as opposed to every time-step, so as to limit the size of the control. Although efficient, such computational overhead still makes the adjoint method too costly to apply directly to global models at state-of-the-art resolution (e.g., Fu and Smith, 1996).

To alleviate some of the computational cost associated with convergence, Luong *et al.* (1998) employ an iterative scheme in which the minimization iterations are conducted over time periods of increasing length. This progressive strategy allows the initial decrease in cost function to be achieved with relatively small computational requirements than otherwise. In comparison, D. Stammer (personal communication, 1998) employs an iterative scheme in space. Namely, assimilation is first performed by a coarse resolution model. A finer-resolution model is used in assimilation next, using the previous coarser solution interpolated to the fine grid as the initial estimate of the adjoint iteration. It is anticipated that the resulting distance of the fine-resolution model to the optimal minimum of the cost function $\mathcal{J}$ is closer than otherwise and that the convergence can therefore be achieved faster.

Courtier *et al.* (1994) instead put forth an incremental approach to reducing the computational requirements of the adjoint method. The approach consists of estimating modifications of the model state (increments) based on a simplified model and its adjoint. The simplifications include the tangent linear approximation, reduced resolution, and approximated physics (e.g., adiabatic instead of diabatic). Motivated in part to simplify coding the adjoint model, Schiller and Willebrand (1995) employed an approximate adjoint in which the adjoint of only the heat and salinity equations were used in conjunction with a full primitive equation ocean general circulation model.

The adjoint method is based on accurate evaluations of the local gradient of the cost function (Eq. [7]). The estimation is rigorous and consistent with the model, but could potentially lead to suboptimal results should the minimization converge to a local minimum instead of a global minimum as could occur with strongly nonlinear models and observations (e.g., convection). Such situations are typically assessed by perturbation analyses of the system near the optimized solution.

A posteriori uncertainty estimates are an integral part of the solution of inverse problems. The a posteriori error covariance matrix of the adjoint method is given by the inverse of the Hessian matrix (second derivative of the cost function $\mathcal{J}$ with respect to the control vector) (Thacker, 1989). However, computational requirements associated with evaluating the Hessian render such calculation infeasible for most practical applications. Yet, some aspects of the error and sensitivity may be evaluated by computations of the dominant structures of the Hessian matrix (Anderson *et al.*, 1996). Practical evaluations of such error estimates require further investigation.

4.4. Representer Method

The representer method (Bennett, 1992) solves the optimization problem Eq. (6) by seeking a solution linearly expanded into data influence functions, called representers, that correspond to each separate measurement. The assimilation problem then becomes one of determining the optimal coefficients of the representers. Because typical dimensions of observations are much smaller than elements of the model state (two orders of magnitude in the example above), the resulting optimization problem becomes much smaller in size than the original problem (Eq. [6]) and is therefore easier to solve.

Representers are functionals corresponding to the effects of particular measurements on the estimated solution, viz., Green's functions to the data assimilation problem (Eq. [6]). Egbert *et al.* (1994) and Le Provost *et al.* (1998) employed the representer method in assimilating T/P data into a model of tidal constituents. Although much reduced, representer methods still require a significant amount of computational resources. The largest computational difficulty lies in deriving and storing the representer functions; the computation requires running the model and its adjoint N-times spanning the duration of the observations, where N is the number of individual measurements. Although much smaller than the size of the original inverse problem (Eq. [6]), the number of representer coefficients to be solved, N, is also still fairly large.

Approximations are therefore necessary to reduce the computational requirements for practical applications. Egbert *et al.* (1994) employed a restricted subset of representers noting that representers are similar for nearby measurement functionals. Alternatively, Egbert and Bennett (1996) formulate the representer method without explicitly computing the representers.

Theoretically, the representer expansion is only applicable to linear models and linear measurement functionals, because otherwise a sum of solutions (representers) is not necessarily a solution of the original problem. Bennett and Thorburn (1992) describe how the method can be extended to nonlinear models by iteration, linearizing nonlinear terms about the previous solution.

4.5. Kalman Filter and Optimal Smoother

The Kalman filter, and related smoothers, are minimum variance estimators of Eq. (6). That is, given the right-hand side and the relationship in Eq. (6), the Kalman filter and smoothers provide estimates of the unknowns that are optimal, defined as having the minimum expected error variance,

$$\langle (\mathbf{x} - \bar{\mathbf{x}})^T (\mathbf{x} - \bar{\mathbf{x}}) \rangle. \tag{9}$$

In Eq. (9), $\bar{\mathbf{x}}$ is the true solution and the angle brackets denote statistical expectation. Although not immediately obvious, minimum variance estimates are equivalent to least-squares solutions (e.g., Wunsch, 1996, p. 184). In particular, the two are the same when the weights used in Eq. (7) are prior error covariances of the model and data constraints. That is, the Kalman filter assumes no more (statistics) than what is assumed (i.e., choice of weights) in solving the least-squares problem (e.g., adjoint and representers). When the statistics

are Gaussian, the solution is also the maximum likelihood estimate.

The Kalman filter achieves its computational efficiency by its time recursive algorithm. Specifically, the filter combines data at each instant (when available) and the state predicted by the model from the previous time step. The result is then integrated in time and the procedure is repeated for the next time-step. Operationally, the Kalman filter is in effect a statistical average of model state prior to assimilation and data, weighted according to their respective uncertainties (error covariance). The algorithm guarantees that information of past measurements are all contained within the predicted model state and therefore past data need not be used again. The savings in storage (that past data need not be saved) and computation (that optimal estimates need not be recomputed from the beginning of the measurements) is an important consideration in real-time estimation and prediction.

The filtered state is optimal with respect to measurements of the past. The smoother additionally utilizes data that lie formally in the future; as future observations contain information of the past, the smoothed estimates have smaller expected uncertainties (Eq. [9]) than filtered results. In particular, the smoother literally "smoothes" the filtered results by reducing the temporal discontinuities present in the estimate due to the filter's intermittent data updates. Various forms and algorithms exist for smoothers depending on the time window of observations used relative to the estimate. In general, the smoother is applied to the filtered results (which contains the data information) backwards in time. The occasional references to "Kalman smoothers" or "Kalman smoothing" are misnomers. They are simply smoothers and smoothing.

The computational difficulty of Kalman filtering, and subsequent smoothing, lies in evaluating the error covariances that make up the filter and smoother. The state error evolves in time according to model dynamics and the information gained from the observations. In particular, the error covariances' dynamic evolution, which assures the estimate's optimality, requires integrating the model the equivalent of twice-the-size-of-the-model times more than the state itself, and is the most computationally demanding step of Kalman filtering.

Although the availability of a posteriori error estimates are fundamental in estimation, the large computational requirement associated with the error evaluation makes Kalman filtering impractical for models with order million variables and larger. For this reason, direct applications of Kalman filtering to oceanographic problems have been limited to simple models. For instance, Gaspar and Wunsch (1989) analyzed Geosat altimeter data in the Gulf Stream region using a spectral barotropic free Rossby wave model. Fu *et al.* (1991) detected free equatorial waves in Geosat measurements using a similar model.

More recently, a number of approximations have been put forth aimed directly at reducing the computational requirements of Kalman filtering and smoothing, and thereby making it practical for applications with large general circulation models. For example, errors of the model state often achieve near-steady or cyclic values for time-invariant observing systems or cyclic measurements (exact repeat missions of satellites are such), respectively. Exploiting such a property, Fukumori et al. (1993) explored approximating the model state error covariance by its time-asymptotic limit, thereby eliminating the need for the error's continuous time-integration and storage. Fu et al. (1993), assimilating Geosat data with a wind-driven spectral equatorial wave model, demonstrated that estimates made by such a time-asymptotic filter are indistinguishable from those obtained by the unapproximated Kalman filter. Gourdeau et al. (1997) employed a time-invariant model state error covariance in assimilating Geosat data with a second baroclinic mode model of the equatorial Atlantic.

A number of studies have explored approximating the errors of the model state with fewer degrees of freedom than the model itself, thereby reducing the computational size of Kalman filtering while still retaining the original model for the assimilation. Fukumori and Malanotte-Rizzoli (1995) approximated the model-state error with only its large-scale structure, noting the information content of many observing systems in comparison to the number of degrees of freedom in typical models. Fukumori (1995) and Hirose et al. (1999) used such a reduced state filter and smoother in assimilating TOPEX/POSEIDON data into shallow water models of the tropical Pacific Ocean and the Japan Sea, respectively. Cane et al. (1996) employed a limited set of empirical orthogonal functions (EOFs) arguing that model errors are insufficiently known to warrant estimating the full error covariance matrix. Parish and Cohn (1985) proposed approximating the model-error covariance with only its local structure by imposing a banded approximation of the covariance matrix. Based on a similar notion that model errors are dominantly local, Chin et al. (1999) explored state reductions using wavelet transformation and low-order spatial regression.

In comparison, Menemenlis and Wunsch (1997) approximated the model itself (and consequently its error) by a state reduction method based on large-scale perturbations. Menemenlis et al. (1997) used such a reduced-state filter to assimilate TOPEX/POSEIDON data in conjunction with acoustic tomography measurements in the Mediterranean Sea.

For nonlinear models, the Kalman filter approximates the error evolution by linearizing the model about its present state, i.e., the so-called extended Kalman filter. (Error covariance evolution is otherwise dependent on higher order statistical moments.) For example, Fukumori and Malanotte-Rizzoli (1995) employed an extended Kalman filter with both time-asymptotic and reduced-state approximations. In many situations, such linearization is found to be adequate.

However, in strongly nonlinear systems, inaccuracies of the linearized error estimates can be detrimental to the estimate's optimality (e.g., Miller et al., 1994). Evensen (1994) proposed approximating the error evaluation by integrating an ensemble of model states. The covariance among elements of the ensemble is then used in assimilating observations into each member of the ensemble, thus circumventing the problems associated with explicitly integrating the error covariance. Evensen and van Leeuwen (1996) used such an ensemble Kalman filter in assimilating Geosat altimeter data into a quasi-geostrophic model of the Agulhas current.

Pham et al. (1998) proposed a reduced-state filter based on a time-evolving set of EOFs (Singular Evolutive Extended Kalman Filter, SEEK) with the aim of reducing the dimension of the estimate at the same time as taking into account the time-evolving direction of a model's most unstable mode. Verron et al. (1999) applied the method to analyze TOPEX/POSEIDON data in the tropical Pacific Ocean.

4.6. Model Green's Function

Stammer and Wunsch (1996) utilized model Green's functions to analyze TOPEX/POSEIDON data in the North Pacific. The approach consists of reducing the dimension of the least-squares problem (Eq. [6]) into one that is solvable by expanding the unknowns in terms of a limited set of model Green's functions, corresponding to the model's response to impulse perturbations. The amplitudes of the functions then become the unknowns. Stammer and Wunsch (1996) restricted the Green's functions to those corresponding to large-scale perturbations so as to limit the size of the problem. Bauer et al. (1996) employed a similar technique in assimilating altimetric significant wave height data into a wave model.

The expansion of solutions into a set of limited functions is similar to the approach taken in the representer method, albeit with different basis functions, while the method's identification of the large-scale corrections is closely related to the approach taken in the reduced-state Kalman filters (e.g., Menemenlis and Wunsch [1997]).

4.7. Optimal Interpolation

Optimal interpolation (OI) is a minimum variance sequential estimator that is algorithmically similar to Kalman filtering, except OI employs prescribed weights (error covariances) instead of ones that are theoretically evaluated by the model over the extent of the observations. Sequential methods solve the assimilation problem separately at different instances, i,

$$\begin{pmatrix} \mathcal{H}_i(\mathbf{x}_i) \\ \mathbf{x}_i \end{pmatrix} \approx \begin{pmatrix} \mathbf{y}_i \\ \mathcal{F}_{i-1}(\mathbf{x}_{i-1}) \end{pmatrix} \qquad (10)$$

given the observations $\mathbf{y}_i$ and the estimate at the previous instant, $\mathbf{x}_{i-1}$. The main distinction between Eqs. (10) and (6) is the lack of time dimension in the former. Observed temporal evolution provides an explicit constraint in Eq. (6), whereas it is implicit in Eq. (10), contained supposedly within the past state and its uncertainties (weights). Although optimal interpolation provides "optimal" instantaneous estimates under the particular weights used, the solution is in fact suboptimal over the entire measurement period due to lack of the time dimension from the problem it solves.

OI is presently one of the most widely employed assimilation methods; Marshall (1985) examined the problem of separating ocean circulation and geoid from altimetry using OI with a barotropic quasi-geostrophic (QG) model. Berry and Marshall (1989) and White *et al.* (1990b) explored altimetric assimilation with an OI scheme using a multilevel QG model, but assumed zero vertical correlation in the stream function, modifying sea surface stream function alone. A three-dimensional OI method was explored by Dombrowsky and De Mey (1992) who assimilated Geosat data into an open domain QG model of the Azores region. Ezer and Mellor (1994) assimilated Geosat data into a primitive equation (PE) model of the Gulf Stream using an OI scheme described by Mellor and Ezer (1991), employing vertical correlation as well as horizontal statistical interpolation. Oschlies and Willebrand (1996) specified the vertical correlations so as to maintain deep temperature-salinity relations, and applied the method in assimilating Geosat data into an eddy-resolving PE model of the North Atlantic.

The empirical sequential methods that include OI and others discussed in the following sections are distinctly different from the Kalman filter (Section 4.5), which is also a sequential method. The Kalman filter and smoother algorithm allows for computing the time-evolving weights according to model dynamics and uncertainties of model and data, so that the sequential solution is the same as that of the whole time domain problem, Eq. (6). The weights in the empirical methods are specified rather than computed, often neglecting the potentially complex cross covariance among variables that reflects the information's propagation by the model (see Section 5.1.4). Some applications of OI, however, allow for the error variance of the model state to evolve in time as dictated by the model-data combination and intrinsic growth, but still retain the correlation unchanged (e.g., Ezer and Mellor, 1994). The Physical-Space Statistical Analysis System (PSAS) (Cohn *et al.*, 1998), is a particular implementation of OI that solves Eq. (10) without explicit formulation of the inverse operator.

4.8. Three-Dimensional Variation Method

The so-called three-dimensional variational method (3D-var) solves Eq. (10) as a least-squares problem, mini-

mizing the residuals:

$$\mathcal{J}' = (\mathbf{y}_i - \mathcal{H}_i(\mathbf{x}_i))^T \mathbf{R}_i^{-1}(\mathbf{y}_i - \mathcal{H}_i(\mathbf{x}_i))$$
$$+ (\mathbf{x}_i - \mathcal{F}_{i-1}(\mathbf{x}_{i-1}))^T \mathbf{Q}_{i-1}^{-1}(\mathbf{x}_i - \mathcal{F}_{i-1}(\mathbf{x}_{i-1})). \quad (11)$$

This is similar to the whole domain problem (Eq. 8) except without the time dimension. Thus the name "three-dimensional" as opposed to "four-dimensional" (Section 4.3). However, as with 4D-var, 3D-var is a misnomer, and the method is merely least-squares. Because there is no model integration of the unknowns involved, the gradient of $\mathcal{J}'$ is readily computed, and is used in solving the minimum of $\mathcal{J}'$.

Bourles *et al.* (1992) employed such an approach in assimilating Geosat data in the tropical Atlantic using a linear model with three vertical modes. The approach described by Derber and Rosati (1989) is a similar scheme, except the inversion is performed at each model time-step, reusing observations within a certain time window, which makes the method a hybrid of 3D-var and nudging (Section 4.9).

4.9. Direct Insertion

Direct insertion replaces model variables with observations, or measurements mapped onto model fields, so as to initialize the model for time-integration. Direct insertion can be thought of as a variation of OI in which prior model state uncertainties are assumed to be infinitely larger than errors in observations. Hurlburt (1986), Thompson (1986), and Kindle (1986) explored periodic direct insertions of altimetric sea level using one- and two-layer models of the Gulf of Mexico. Using the same model, Hurlburt *et al.* (1990) extended the studies by statistically initializing deeper pressure fields from sea level measurements. De Mey and Robinson (1987) initialized a QG model by statistically projecting sea surface height into the three-dimensional stream function. Gangopadhyay *et al.* (1997) and Gangopadhyay and Robinson (1997) performed similar initializations by the so-called "feature model." Instead of using correlation in the data-mapping procedure, which tends to smear out short-scale gradients, feature models effect the mapping by assuming analytic horizontal and vertical structures for coherent dynamical features such as the Gulf Stream and its rings. "Rubber sheeting" (Carnes *et al.*, 1996) is another approach aimed at preserving "features" by directly moving model fields towards observations in spatially correlated displacements. Haines (1991) formulated the vertical mapping of sea level based on QG dynamics, keeping the subsurface potential vorticity unchanged while still directly inserting sea level data into the surface stream function. Cooper and Haines (1996) examined a similar vertical extension method preserving subsurface potential vorticity in a primitive equation model.

4.10. Nudging

Nudging blends data with models by adding a Newtonian relaxation term to the model prognostic equations (Eq. [4]) aimed at continuously forcing the model state towards observations (Eq. [2]),

$$\mathbf{x}_{i+1} = \mathcal{F}_i(\mathbf{x}_i) - \gamma(\mathcal{H}_j(\mathbf{x}_j) - \mathbf{y}_j). \qquad (12)$$

The nudging coefficient, γ, is a relaxation coefficient that is typically a function of distance in space and time $(i - j)$ between model variables and observations. Nudging is equivalent to the so-called robust diagnostic modeling introduced by Sarmiento and Bryan (1982) in constraining model hydrographic structures. While other sequential methods intermittently modify model variables at the time of the observations, nudging is distinct in modifying the model field continuously in time, re-using data both formally in the future and past at every model time-step, aimed at gradually modifying the model state, avoiding "undesirable" discontinuities due to the assimilation. The smoothing aspect of nudging is distinct from optimal smoothers of estimation theory (Section 4.5); whereas the optimal smoother propagates data information into the past by the model dynamics (model adjoint), nudging effects a smooth estimate by using data interpolated backwards in time based solely on temporal separation.

Verron and Holland (1989) and Holland and Malanotte-Rizzoli (1989) explored altimetric assimilation by nudging surface vorticity in a multi-layer QG model. Verron (1992) further explored other methods of nudging surface circulation including surface stream function. These studies were followed by several investigations assimilating actual Geosat altimeter data using similar models and approaches in various regions; examples include White *et al.* (1990a) in the California Current, Blayo *et al.* (1994, 1996) in the North Atlantic, Capotondi *et al.* (1995a, b) in the Gulf Stream region, Stammer (1997) in the eastern North Atlantic, and Seiss *et al.* (1997) in the Antarctic Circumpolar Current. In particular, Capotondi *et al.* (1995a) theoretically examined the physical consequences of nudging surface vorticity in terms of potential vorticity conservation. Most recently, Florenchie and Verron (1998) nudged TOPEX/POSEIDON and ERS-1 data into a QG model of the South Atlantic Ocean.

Other studies explored directly nudging subsurface fields in addition to surface circulation by extrapolating sea level data prior to assimilation. For instance, Smedstad and Fox (1994) used the statistical inference technique of Hurlburt *et al.* (1990) to infer subsurface pressure in a two-layer model of the Gulf Stream, adjusting velocities geostrophically. Forbes and Brown (1996) nudged Geosat data into an isopycnal model of the Brazil-Malvinas confluence region by adjusting subsurface layer thicknesses as well as surface geostrophic velocity. The monitoring and forecasting system developed for the Fleet Numerical Meteorology and Oceanography Center (FNMOC) nudges three-dimensional fields generated by "rubber sheeting" and OI (Carnes *et al.*, 1996).

4.11. Summary and Recommendation

Innovations in estimation theory, such as developments of adjoint compilers and various approximate Kalman filters, combined with improvements in computational capabilities, have enabled applications of optimal estimation methods feasible for many ocean data assimilation problems. Such developments were largely regarded as impractical and/or unlikely to succeed even until recently. The virtue of these "advanced" methods, described in Sections 4.3 to 4.6 above, are their clear identification of the underlying "four-dimensional" optimization problem (Eq. [6]) and their objective and quantitative formalism. In comparison, the relation between the "four-dimensional" problem and the approach taken by other ad hoc schemes (Sections 4.7 to 4.10) is not obvious, and the nature and consequence of their particular assumptions are difficult to ascertain. Arbitrary assumptions can lead to physically inconsistent results, and therefore analyses resulting from ad hoc schemes must be interpreted cautiously. For instance, nudging subsurface temperature can amount to assuming heating and/or cooling sources within the water column.

As a result of the advancements, ad hoc schemes used in earlier studies of assimilation are gradually being superseded by methods based on estimation theory. For example, even though operational requirements often necessitate efficient methods to be employed, thus favoring simpler ad hoc schemes, the European Center for Medium-Range Weather Forecasting has recently upgraded their operational meteorological forecasting system from "3D-var" to the adjoint method.

Differences among the "advanced" methods are largely of convenience. As in "classic" inverse methods, solutions by optimal estimation are identical so long as the assumptions, explicit and implicit, are the same. Some approaches may be more effective in solving nonlinear optimization problems than others. Others may be more computationally efficient. However, published studies to date are inconclusive on either issue.

Given the equivalence, accuracy of the assumptions is a more important issue for estimation rather than the choice of assimilation method. In particular, the form and weights (prior covariance) of the least-squares "cost function" (Eq. [8]) require careful selection. Different assimilations often make different assumptions, and the adequacy and implication of their particular suppositions must properly be assessed. These and other practical issues of assimilation are reviewed in the following section.

5. PRACTICAL ISSUES OF ASSIMILATION

As described in the previous section, assimilation techniques are equivalent as long as assumptions are the same, although very often those assumptions are not explicitly recognized. Identifying the assumptions and assessing their appropriateness are important issues in assuring the reliability of assimilated estimates. Several other issues of practical importance exist that warrant careful attention when assimilating data, including some that are particular to altimetric data. These issues are discussed in turn below, and include: the weights used in defining and solving the assimilation problem in Eq. (7), methods of vertical extrapolation, determination of subsurface circulation (observability), prior data treatment such as horizontal mapping and conversion of sea level to geostrophic velocities, and the treatment of the unknown geoid and reference sea level.

5.1. Weights, A Priori Uncertainties, and Extrapolation

The weights **W** in Eq. (7) define the mathematical problem of data assimilation. As such, suitable specification of weights is essential to obtaining sensible solutions, and is the most fundamental issue in data assimilation. While advancements in computational capabilities will directly solve many of the technical issues of assimilation (Section 4), they will not resolve the weight identification. Different weights amount to different problems, thereby leading to different solutions. Misspecification of weights can lead to overfitting or underfitting of data, and/or the failure of the assimilation altogether.

On the one hand least-squares problems are deterministic in the sense that, mathematically, weights could be chosen arbitrarily, such as minimum length solutions and/or solutions with minimum energy (e.g., Weaver and Anderson, 1997). On the other hand, the equivalence of least-square solutions with minimum error variance and maximum likelihood estimates, suggests a particularly suitable choice of weights being a priori uncertainties of the data and model constraints, Eqs. (2) and (4). Specifically, the weights can be identified as the inverse of the respective error covariance matrices.

5.1.1. Nature of Model and Data Errors

Apart from the problem of specifying values of a priori errors (Section 5.1.2), it is important first to clarify what the errors correspond to, as there are subtleties in their identification. In particular, the a priori errors in Eqs. (7) and (8) should be regarded as errors in model and data *constraints* rather than merely model and data errors. A case in point is the so-called representation error (e.g., Lorenc, 1986), that corresponds to real processes that affect measurements but are not represented or resolvable by the models. Representation "errors" concern the *null space* of the model, as opposed to *errors* within the model range space. For instance, inertial oscillations and tides are not included in the physics of quasi-geostrophic models and are therefore within the models' null space. To the extent that representation errors are inconsistent with models but contribute to measurements, errors of representativeness should be considered part of the uncertainties of the data constraint (Eq. [2]) instead of the model constraint (Eq. [4]). Cohn (1997) provides a particularly lucid explanation of this distinction, which is summarized in the discussion below.

Several components of what may be regarded as "model error" exist, and a careful distinction is required to define the optimal solution. In particular, three types of model error can be distinguished; these could be called *model state error*, *model equation error*, and *model representation error*. First, it is essential to recognize the fundamental difference between the ocean and the models. Models have finite dimensions whereas the real ocean has infinite degrees of freedom. The model's true state ($\bar{\mathbf{x}}$) can be mathematically defined by a functional relationship with the real ocean (**w**):

$$\bar{\mathbf{x}} \equiv \mathcal{P}(\mathbf{w}). \tag{13}$$

Functional $\mathcal{P}$ relates the complete and exact state of the ocean to its representation in the finite and approximate space of the model. Such an operator includes both spatial averaging as well as truncation and/or approximation of the physics. For instance, finite dimensional models lack scales smaller than their grid resolution. Quasi-geostrophic models resolve neither inertial waves nor tides as mentioned above and reduced gravity shallow water models (e.g., 1.5-layer models) ignore high-order baroclinic modes. The difference between a given model state and the true state defined by Eq. (13),

$$\mathbf{x} - \bar{\mathbf{x}} = \mathbf{x} - \mathcal{P}(\mathbf{w}) \tag{14}$$

is the *model state error*, and its expected covariance, **P**, forms the basis of Kalman filtering and smoothing (Section 4.5).

The errors of the model constraint (Eq. 4) or *model equation error*, **q**, can be identified as,

$$\mathbf{q}_i = \bar{\mathbf{x}}_{i+1} - \mathcal{F}_i(\bar{\mathbf{x}}_i). \tag{15}$$

The covariance of $\mathbf{q}_i$, $\mathbf{Q}_i$, is the inverse of the weights for the model constraint Eq. (8) in the maximum likelihood estimate. *Model equation error* (Eq. 15) is also often referred to as system error or process noise. Apart from its dependence on errors of the initial condition and assimilated data, the *model state error* **P** is a time-integral by the model equation $\mathcal{F}$ of process noise (*model equation error*) **Q**. Process noise includes inaccuracies in numerical algorithms (e.g., integration errors caused by finite differencing) as well as errors in external forcing and boundary conditions.

The third component of model error is *model representation error* and arises in the context of comparing the model with observations (reality). Observations **y** measure properties of the real ocean and can be described symbolically as:

$$\mathbf{y} = \mathcal{E}(\mathbf{w}) + \varepsilon \tag{16}$$

where $\mathcal{E}$ represents the measurements' sampling operation of the real ocean **w**, and ε denotes the measuring instruments' errors. Functional $\mathcal{E}$ is generally different from the model's equivalent, $\mathcal{H}$ in Eq. (2), owing to differences between **x** and **w** (Eq. [13]). Measuring instrumentation errors are strictly errors of the observing system and represent quantities unrelated to either the model or the ocean. For satellite altimetry, ε includes, for example, errors in the satellite's orbit and ionospheric corrections (cf. Chapter 1).

In terms of quantities in model space, Eq. (16) can be rewritten as:

$$\mathbf{y} = \mathcal{H}(\bar{\mathbf{x}}) + \{\mathcal{E}(\mathbf{w}) - \mathcal{H}(\mathcal{P}(\mathbf{w}))\} + \varepsilon. \tag{17}$$

Assimilation is the inversion of Eq. (2), which can be identified as the first term in Eq. (17) that relates model state to observations rather than a solution of Eq. (16). The second term in { } on the right-hand side of Eq. (17) describes differences between the observing system and the finite dimension of the model, and is the *representation error*.

Representation errors arise from inaccuracies or incompleteness in both model and observations. *Model representation errors* are largely caused by spatial and physical truncation errors caused by its approximation $\mathcal{P}$ (Eq. [13]). For example, coarse-resolution models lack sea level variabilities associated with meso-scale eddies, and reduced gravity shallow water models are incapable of simulating the barotropic mode. Such inaccuracies constitute model representation error when assimilating altimetric data to the extent that an altimeter measures sea level associated with such missing processes of the model.

Data representation error is primarily caused by the observing system not exactly measuring the intended property. For instance, errors in altimetric sea state bias correction may be considered data representation errors. Sea-state bias arises because altimetric measurements do not exactly represent a uniformly averaged mean sea level, but an average depending on wave height (sea state) and the reflecting characteristics of the altimetric radar, a process that is not exactly known. Some island tide gauge stations, because of their geographic location (e.g., inlet), do not represent sea levels of the open ocean and thus can also be considered as contributing to data representation error. (Alternatively, such geographic variations can be ascribed to the model's lack of spatial resolution and thus identified as model representation error, but such distinctions are moot.)

Representation errors are inconsistent with model physics, and therefore are not correctable by assimilation. As far as the model inversion is concerned, representation error, whether of data or model origin, is indistinguishable from instrument error ε. Representation error and instrument noise together constitute uncertainties relating data and the model state, viz., data constraint error, whose covariance is **R** in Eq. (8). Data constraint error is often referred to merely as data error, which can be misleading as there are components in **R** that are unrelated to observations **y**. The data constraint error covariance **R** is identified as the inverse of the weights for the data constraint in the maximum likelihood estimate (Eq. [8]) as well as the data uncertainty used in sequential inversions. In effect, representation errors downweight the data constraint (Eq. [2]) and prevent a model from being forced too close to observations that it cannot represent, thus guarding against model overfitting and/or "indigestion," i.e., a degradation of model estimate by insisting models obey something they are not meant to.

The fact that part of the model's inaccuracies should contribute to downweighting the data constraint is not immediately obvious and even downright upsetting for some (especially for those who are closest to making the observations). However, as it should be clear from discussions above, the error of the data constraint is in the accuracy of the relationship in Eq. (2) and not about deficiencies of the observations **y** per se.

On the one hand, most error sources can readily be identified as one of the three error types of the assimilation problem; measurement instrumentation error, model process noise (or equivalently model equation error), and representation error. Specific examples of instrument and representation errors were given above. Process noise include errors in external forcing and boundary conditions, inaccuracies of numerical algorithms (finite differencing), and errors in model parameterizations. These and other examples are summarized in Table 1.

On the other hand, representation errors are sometimes also sources of process noise. For example, while meso-scale variabilities themselves are representation error for non-eddy resolving models, the *effects* of meso-scale eddies on the large-scale circulation that are not accurately modeled, contribute to process noise (e.g., uncertainties in eddy parameterization such as that of Gent and McWilliams, 1990). Furthermore, some model errors (but not all) can be categorized either as process noise or representation error depending on the definition of the true model state, viz., operator $\mathcal{P}$ in Eq. (13).[1] For instance, $\mathcal{P}$ may be defined alternatively as including or excluding certain forced responses of the

---

[1] What strictly constitutes $\mathcal{P}$ is in fact ambiguous for many models. For instance, variables in finite difference models are loosely understood to represent averages in the vicinity of model grid points. However the exact averaging operator is rarely stated.

TABLE 1. Examples of Error Sources in Altimetric Assimilation

Error type	Source
Instrument error	Orbit determination, ionospheric correction, dry and wet tropospheric corrections, instrument noise (bias and random noise), earth tide
Representation error	Sea-state bias, atmospheric pressure loading[a], ocean tides[a], sub-grid scale variability, diabatic effects for adiabatic models (such as QG models), barotropic modes in 1.5-layer models (or any other model lacking the barotropic mode), baroclinic modes higher than what a particular model can resolve
Model equation error (process noise)	Numerical truncation (inaccuracies in numerical algorithm, e.g., finite differencing), parameterization error including effects of subgridscale processes, errors in external forcing and boundary conditions

[a]Could be regarded as either process noise or representation error, depending on definition of model state. See text for discussion.

ocean. A particular example is tides (and residual tidal errors) in altimetric measurements. While typically treated as representation error, for free-surface models, the lack of tidal forcing (or inaccuracies thereof) could equally be regarded as process noise as well. (External tides are always representation errors for rigid lid models which lack the physics of external gravity waves.) Other examples of similar nature include effects of baroclinic instability in 1.5-layer models (e.g., Hirose *et al.*, 1999), and external variability propagating in through open boundaries (e.g., Lee and Marotzke, 1998). Although either model lacks the physics of the respective "forcing," the resulting variability such as propagating waves within the model domain could be resolved as being a result of process noise.

5.1.2. Prescribing Weights

Instrument errors (ε in Eq. [16]) and data representation errors are relatively well known from comparisons among different observing systems. Discussions of errors in altimetric measurements can be found in Chapter 1. Model errors, including errors of the initial condition, process noise, and model representation error, are far less accurately known. In practice, prior uncertainties of data and model are often simply guesses, whose consistency must be examined based on results of the assimilation (cf. Section 5.2). In particular, error *covariances* (i.e., off-diagonal elements of the covariance matrix including temporal correlations and biases) are often assumed to be nil for simplicity or for lack of sufficient knowledge that suggests alternatives.

One of the largest sources of model error is considered to be forcing error. While some knowledge exists of the accuracy of meteorological forcing fields, estimates are far from complete; geographic variations are not well known and estimates particularly lack measures of error *covariances*. In fact, an accurate assessment of atmospheric forcing errors has been identified as one of the most urgent needs for ocean state estimation (WOCE International Project Office, 1998).

The problem of estimating a priori error covariances is generally known in estimation theory as adaptive filtering. Many of these methods are based on statistics of the so-

called innovation sequence, i.e., the difference between data and model estimates based on past observations. Prior errors are chosen and/or estimated so as to optimize certain properties of the innovation sequence. For instance, Gaspar and Wunsch (1989) adjusted the model process noise so as to minimize the innovation sequence. Blanchet *et al.* (1997) compared several adaptive Kalman filtering methods in a tropical Pacific Ocean model using maximum likelihood estimates for the error. Hoang *et al.* (1998) put forth an alternate adaptive approach, whereby the Kalman gain matrix (the filter) itself is estimated parametrically as opposed to the errors. Such an approach is effective because the filter is in effect only dependent on the ratio of data and model constraint errors and not on the absolute error magnitude, but the resulting state lacks associated error estimates.

Fu *et al.* (1993) introduced an "off-line" approach in which a priori errors are estimated prior to assimilation based on comparing observations with a model *simulation*, i.e., a model run without assimilation. The method is similar to a class of adaptive filtering methods termed "covariance matching" (e.g., Moghaddamjoo and Kirlin, 1993). The particular estimate assumes stationarity and independence among different errors and the signal, and is described below with simplifications suggested by R. Ponte (personal communication, 1997). First we identify data $\mathbf{y}$ and its model equivalent $\mathbf{m} = \mathcal{H}(\mathbf{x})$ (simulation) as being the sum of the true signal $\mathbf{s} = \mathcal{H}(\bar{\mathbf{x}})$ plus their respective errors $\mathbf{r}$ and $\mathbf{p}$:

$$\mathbf{y} = \mathbf{s} + \mathbf{r} \tag{18}$$

$$\mathbf{m} = \mathbf{s} + \mathbf{p}. \tag{19}$$

Then, assuming the true signal and the two errors are mutually uncorrelated with zero means, the covariance among data and its model equivalent can be written as:

$$\langle \mathbf{y}\mathbf{y}^T \rangle = \langle \mathbf{s}\mathbf{s}^T \rangle + \langle \mathbf{r}\mathbf{r}^T \rangle \tag{20}$$

$$\langle \mathbf{m}\mathbf{m}^T \rangle = \langle \mathbf{s}\mathbf{s}^T \rangle + \langle \mathbf{p}\mathbf{p}^T \rangle \tag{21}$$

$$\langle \mathbf{y}\mathbf{m}^T \rangle = \langle \mathbf{s}\mathbf{s}^T \rangle \tag{22}$$

where angle brackets denote statistical expectation. By substituting the brackets with temporal and/or spatial averages (assuming ergodicity), one can estimate the left-hand sides of Eqs. (20) to (22) and solve for the individual terms on the right-hand sides. In particular, the error covariances of the data constraint and the simulated model state can be estimated as,

$$\langle \mathbf{rr}^T \rangle = \langle \mathbf{yy}^T \rangle - \langle \mathbf{ym}^T \rangle \qquad (23)$$

$$\langle \mathbf{pp}^T \rangle = \langle \mathbf{mm}^T \rangle - \langle \mathbf{ym}^T \rangle. \qquad (24)$$

Equation (24) implicitly provides an estimate of model process noise $\mathbf{Q}$ (Eq. [15]) since the model state error of the simulation $\mathbf{p}$ is a function of the former. (The state error can be regarded as independent of initial error for sufficiently long simulations.) Therefore, Eq. (24) can be used to calibrate process noise $\mathbf{Q}$.

An example of error estimates based on Eqs. (23) and (24) is shown in Figure 9 (see color insert) (Fukumori et al., 1999). The data are altimetric sea level from TOPEX/POSEIDON (T/P), and the model is a coarse resolution ($2° \times 1° \times 12$ vertical levels) global general circulation model based on the NOAA Geophysical Fluid Dynamics Laboratory's Modular Ocean Model (Pacanowski et al., 1991), forced by National Center for Environmental Prediction winds and climatological heat fluxes (Comprehensive Ocean-Atmosphere Data Set, COADS).

Errors of the data constraint (Figure 9a, Eq. [23]) and those of the simulated model state (Figure 9b, Eq. [24]) are both spatially varying, reflecting the inhomogeneities in the physics of the ocean. In particular, the data constraint error (Figure 9a) is dominated by meso-scale variability (e.g., western boundary currents) that constitutes representation error for the particular model, and is much larger than the corresponding model state error estimate (Figure 9b) and the instrumental accuracy of T/P ($2 \sim 3$ cm). Process noise was modeled in the form of wind error (Figure 9c) and calibrated such that the resulting simulation error (Figure 9d) (solution of the Lyapunov Equation, which is the time-asymptotic limit of the Riccati Equation with no observations; see for example, Gelb, 1974) is comparable to the estimate based on Eq. (24), i.e., Figure 9b. Similar methods of calibrating errors were employed in assimilating Geosat data by Fu et al. (1993) and TOPEX measurements by Fukumori (1995).

Menemenlis and Chechelnitsky (2000) extended the approach of Fu et al. (1993) by using only model-data differences (residuals),

$$\langle (\mathbf{y} - \mathbf{m})(\mathbf{y} - \mathbf{m})^T \rangle = \langle \mathbf{rr}^T \rangle + \langle \mathbf{pp}^T \rangle, \qquad (25)$$

and not assuming uncorrelated signal and model errors. (The two errors, $\mathbf{r}$ and $\mathbf{p}$, are assumed to be uncorrelated.) To separately estimate $\mathbf{R}$ and $\mathbf{Q}$ (equivalently $\mathbf{P}$) in Eq. (25), the time-lagged covariance of the residuals is further employed,

$$\langle (\mathbf{y}(t) - \mathbf{m}(t))(\mathbf{y}(t + \Delta t) - \mathbf{m}(t + \Delta t))^T \rangle = \langle \mathbf{p}(t)\mathbf{p}(t + \Delta t)^T \rangle \qquad (26)$$

where data constraint error, $\mathbf{r}$, is assumed to be uncorrelated in time. Menemenlis and Chechelnitsky (2000) estimate the a priori errors by matching the empirical estimates of Eqs. (25) and (26) with those based on theoretical estimates using the model and a parametrically defined set of error covariances.

Temporally correlated data errors and/or model process noise require augmenting the problem that is solved. For instance, the expansion in Eq. (8) assumes temporal independence among the constraints in the assimilation problem, Eq. (7). Time correlated errors include biases, caused for example, by uncertainties in model parameters and errors associated with closed passageways in the ocean. The augmentation is typically achieved by including the temporally correlated error as part of the estimated state and by explicitly modeling the temporal dependence of the noise, for instance, by persistence or by a low-order Gauss-Markov process (e.g., Gelb, 1974). The modification amounts to transforming the problem (Eq. [7]) into one with temporally uncorrelated errors at the cost of increasing the size of the estimated state. Dee and da Silva (1998) describe a reformulation allowing estimation of model biases separately from the model state in the context of sequential estimation. Derber (1989) and Griffith and Nichols (1996) examine the problem of model bias and correlated model process noise in the framework of the adjoint method.

Finally, it should be noted that the significance of different weights depend entirely on whether or not those differences are resolvable by models and available observations. To the extent that different error estimates are indistinguishable from each other, further improvement in modeling a priori uncertainties is a moot point. The methods described above provide a simple means of estimating the errors, but their adequacy must be assessed through examination of individual results. Issues of verifying prior errors and the goodness of resulting estimates are discussed in Section 5.2.

5.1.3. Regularization and the Significance of Covariances

The data assimilation problem, being a rank-deficient inverse problem (see, for example, Wunsch, 1996), requires a criterion for choosing a particular solution. To assure the solution's regularity (e.g., spatial smoothness), specific regularization or background constraints are sometimes imposed in addition to the minimization of Eq. (8). For instance, Sheinbaum and Anderson (1990), in investigating assimilation of XBT data, used a smoothness constraint of the form,

$$(\nabla_H \mathbf{x})^2 + \left(\nabla_H^2 \mathbf{x} \right)^2 \qquad (27)$$

where ∇_H is a horizontal gradient operator. The gradient and Laplacian operators are linear operators and can be expressed by some matrix, $\mathbf{G}$ and $\mathbf{L}$, respectively. Then Eq. (27) can be written

$$\mathbf{x}^T \mathbf{G}^T \mathbf{G} \mathbf{x} + \mathbf{x}^T \mathbf{L}^T \mathbf{L} \mathbf{x} = \mathbf{x}^T \left(\mathbf{G}^T \mathbf{G} + \mathbf{L}^T \mathbf{L} \right) \mathbf{x}. \qquad (28)$$

The weighting matrix $\mathbf{G}^T \mathbf{G} + \mathbf{L}^T \mathbf{L}$ is a symmetric nondiagonal matrix, and Eq. (27) can be recognized as a particular weighting of $\mathbf{x}$. Namely, regularization constraints can be specified in the weights already used in Eq. (8) by appropriately prescribing their elements, particularly their off-diagonal values. Alternatively, regularization may be viewed as correcting inadequacies in the explicit weighting factors, i.e., the prior covariance weighting, used in defining the assimilation problem, Eq. (8).

Other physical constraints also render certain diagonal weighting matrices unphysical. For instance, mass conservation in the form of velocity nondivergence requires model velocity errors to be nondivergent as well,

$$\mathbf{D}(\mathbf{x} - \bar{\mathbf{x}}) = 0 \qquad (29)$$

where $\mathbf{D}$ is the divergence operator for the velocity components of model state $\mathbf{x}$. Then the covariance of the initial model state error $\mathbf{P}_0$ as well as the process noise $\mathbf{Q}$ should be in the null space of $\mathbf{D}$, e.g.,

$$\mathbf{D} \mathbf{Q} \mathbf{D}^T = 0 \qquad (30)$$

which a diagonal $\mathbf{Q}$ will not satisfy.

Data constraint *covariances*, in particular the off-diagonal elements of the weighting matrices, are equally as important in determining the optimality of the solution as are the error *variances*, i.e., the diagonal elements. For example, Fu and Fukumori (1996) examined effects of the differences in covariances of orbit and residual tidal errors in altimetry. Orbit error is a slowly decaying function of time following the satellite ground track, and is characterized by a dominating period of once per satellite revolution around the globe. Geographically, errors are positively correlated along satellite ground tracks, and weakly so across-track. While precision orbit determination has dramatically decreased the magnitude of orbit errors, it is still the dominating measurement uncertainty of altimetry (Table 1). Tidal error covariance is characterized by large positive as well as negative values about the altimetric data points, because of the narrow band nature of tides and the sampling pattern of satellites. Consequently, tidal errors have less effect on the accuracy of estimating large-scale circulation than orbit errors of comparable variance, because of the canceling effect of neighboring positive and negative covariances.

5.1.4. Extrapolation and Mapping of Altimeter Data

How best to process or employ altimeter data in data assimilation has been a long-standing issue. The problems include, for example, vertical extrapolation (Hurlburt *et al.*,

1990; Haines, 1991), horizontal mapping (Schröter *et al.*, 1993), and data conversion such as sea level to geostrophic velocity (Oschlies and Willebrand, 1996). (Issues concerning reference sea level are discussed in Section 5.4.) Many of these problems originate in utilizing simple ad hoc assimilation methods and in altimetric measurements not directly being a prognostic variable of the models. For instance, many primitive equation models utilize the rigid lid approximation for computational efficiencies. For such models, sea level is not a prognostic variable but is diagnosed instead from pressure gradients against the sea surface, which is dependent on stratification (dynamic height) and barotropic circulation (e.g., Pinardi *et al.*, 1995). Altimeters also measure significant wave height whereas the prognostic variable in wave models is spectral density of the waves (e.g., Bauer *et al.*, 1992).

From the standpoint of estimation theory, there is no fundamental distinction between assimilating prognostic or diagnostic quantities, as both variables can be defined and utilized through explicit forward relationships of similar form, Eq. (2). That is, no explicit mapping of data to model grid is required, and free surface models provide no more ease in altimetric assimilation than do rigid-lid models. What enables estimation theory to translate observations into unique modifications of model state in effect are the weights in Eq. (8). For instance, specifying data and model uncertainties uniquely defines the Kalman filter which sequentially maps data to the entire model state (Section 4.5). The Kalman filter determines the optimal extrapolation/interpolation by time-integration of the model state error covariance. The covariance defines the statistical relation between uncertainties of an arbitrary model variable and that of another variable, either being prognostic or diagnostic. The covariance computed in Kalman filtering, by virtue of model integration, is dynamically consistent and reflects the propagation of information in space, time, and among different properties. Least-squares methods achieve the equivalent implicitly through direct optimization of Eq. (8). To the extent that model state errors are correlated, as they realistically would be by the continuous dynamics, the optimal weights necessarily extrapolate surface information instantaneously in space (vertically and horizontally) and among different properties.

Figures 10 and 11 show examples of some structures of the Kalman gain corresponding to that based on the model and errors of Figure 9. Reflecting the inhomogeneous nature of wind-driven large-scale sea level changes (Fukumori *et al.*, 1998), Figure 10 shows sea-level differences between model and data largely being mapped to baroclinic changes (model state increments) (black curve) in the tropics and barotropic changes (gray curve) at higher latitudes. Horizontally, the modifications reflect the dynamics of the background state (Figure 11). For instance, the effect of a sea-level difference at the equator (Figure 11B) is similar to the effects of local wind-forcing (the assumed error source), that is a Kelvin wave with temperature and zonal velocity anomalies centered on the equator and an associated Rossby wave of opposite phase to the west of the Kelvin wave with off-equatorial maxima. The Antarctic Circumpolar Current and the presence of the mid-ocean ridge elongates stream function changes in the Southern Ocean in the east-west direction (Figure 11A). The ocean physics render structures of the model error covariance, and thus the optimal filter, spatially inhomogeneous and anisotropic. Such complexity makes it difficult to directly specify an extrapolation scheme for al-

timetry data, as done in ad hoc schemes of data assimilation (Section 4).

Because mapping is merely a combination of data and statistical information (e.g., Bretherton *et al.*, 1976), the information content of a mapped sea level should be no more than what is already available from data along satellite ground tracks and the weights used in mapping the data. However, mapping procedures can potentially filter out or alias oceanographic signals if the assumed statistics are inaccurate. In particular, sea level at high latitudes contain variabilities with periods of a few days, that is shorter than the Nyquist period of most altimetric satellites (Fukumori *et al.*, 1998). Therefore a mapping of altimetric measurements must be carefully performed to avoid possible aliasing of high frequency variability. The simplest and most prudent approach would be to assimilate along-track data directly.

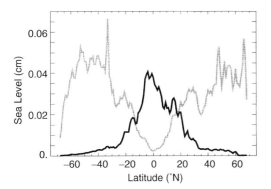

FIGURE 10 Property of a Kalman gain. The figure shows zonally averaged sea level change (cm) as a function of latitude associated with Kalman filter changes in model state (baroclinic displacement [black], barotropic circulation [gray]) corresponding to an instantaneous 1 cm model-data difference. The estimates are strictly local reflecting sea-level differences at each separate grid point. The model is a global model based on the GFDL MOM. The Kalman filter assumes process noise in the form of wind error (Figure 9). (Adapted from Fukumori *et al.* (1999), Plate 2.)

5.2. Verification and the Goodness of Estimates

Improvements achieved by data assimilation not only require accurate solution of the assimilation problem (Section 4), but also depend on the accuracy of the assumptions underlying the definition of the problem itself (Eq. [7]), in particular the a priori errors of the model and data constraints (Section 5.1). The validity of the assumptions must be carefully assessed to assure the quality and integrity of the estimates. At the same time, the nature of the assumptions must be fully appreciated to properly interpret the estimates.

If a priori covariances are correct and the problem is solved consistently, results of the assimilation should necessarily be an improvement over prior estimates. In particular, the minimum variance estimate by definition should become more accurate than prior estimates, including simulations

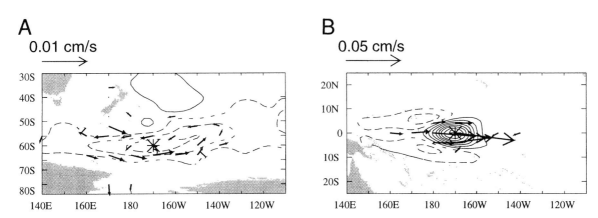

FIGURE 11 Examples of a Kalman gain's horizontal structure. The figures describe changes in a model corresponding to assimilating a 1 cm sea level difference between data and model at the asterisks. The model and errors are those in Figure 9. The figures are, (A) barotropic mass transport stream function (c.i. 2×10^{-10} cm^3/sec) and (B) temperature at 175 m (c.i. 4×10^{-4} °C). Positive (negative) values are shown in solid (dashed) contours. Arrows are barotropic (A) and baroclinic (B) velocities. To reduce clutter, only a subset of vectors are shown where values are relatively large. The assumed data locations are (A) 60°S 170°W and (B) 0°N 170°W. Corresponding effects of the changes on sea level are small due to relatively large magnitudes of data error with respect to model error; changes are 0.02 and 0.03 cm at the respective data locations for (A) and (B). (Adapted from Fukumori *et al.* (1999), Fig. 4.)

without assimilation or the assimilated observations themselves. Mathematically, the improvement is demonstrated, for example, by the minimum variance estimate's accuracy (inverse of error covariance matrix, $\mathbf{P}$) being the sum of the prior model and data accuracies (e.g., Gelb, 1974),

$$\mathbf{P}^{-1} = \mathbf{P}(-)^{-1} + \mathbf{H}^T \mathbf{R}^{-1} \mathbf{H} \qquad (31)$$

where the minus sign in the argument denotes the model state error prior to assimilation. Consequently, the trace of the model state error covariance matrix is a nonincreasing function of the amount of assimilated observations. Matrices with smaller trace define smaller inner products for arbitrary vectors, $\mathbf{h}$; i.e.,

$$\mathbf{h}^T \mathbf{P} \mathbf{h} \leq \mathbf{h}^T \mathbf{P}(-) \mathbf{h}. \qquad (32)$$

Equation (32) implies that not only diagonal elements of $\mathbf{P}$ but errors of any linear function of the minimum variance estimate are smaller than those of non-assimilated estimates. Therefore, for linear models at least, assimilated estimates will not only have smaller errors for the model equivalent of the observations but will also have smaller errors for model state variables not directly measured as well as the model's future evolution. In the case of altimetric assimilation, unless incorrect a priori covariances are used, the model's entire three-dimensional circulation will be improved from, or should be no worse than, prior estimates. (For nonlinear models, such improvement cannot be proven in general, but a linear approximation is a good approximation in many practical circumstances.) Given the equivalence of minimum variance solutions with other assimilation methods (Section 4), these improvements apply equally as well to other estimations, provided the assumptions are the same.

Various measures are used to assess the adequacy of a priori assumptions. For instance, the particular form of Eq. (8), as in most applications, assumes a priori errors being uncorrelated in time. Then, if a priori errors are chosen correctly, the optimal estimate will extract all the information content from the observations except for noise, making the innovation sequence uncorrelated in time. Blanchet et al. (1997) used such measure to assess the adequacy of adaptively estimated uncertainty estimates. However, in practice, representation errors (Section 5.1.1) often dominate model and data differences, such that strict whiteness in residuals cannot always be anticipated. As in the definition of the assimilation problem, the distinction of signal and representation error is once again crucial in assessing the goodness of the solution. The improvement that is expected of the model estimate is that of the signal as defined in Section 5.1.1, and not of the complete state of the ocean.

Another quantitative measure of assessing adequacies of prior assumptions is the relative magnitude of a posteriori model-data differences with respect to their a priori expectations. For instance, the Kalman filter provides formal uncertainty estimates with which to measure magnitudes of actual model-data differences. Figure 12 (see color insert) shows an example comparing residuals (i.e., model-data differences; Figure 12A) and their expectations (Figure 12B) from assimilating TOPEX/POSEIDON data using the Kalman filter described by Figure 9. The comparable spatial structures and magnitudes over most regions demonstrate the consistency of the a priori assumptions with respect to model and data. For least-squares estimates, the equivalent would be for each term in Eq. (8) being of order one (or of comparable magnitude) after assimilation (e.g., Lee and Marotzke, 1998).

The model-data misfit should necessarily become smaller following an assimilation because assimilation forces models towards observations. What is less obvious, however, is what becomes of model properties not directly constrained. If solved correctly, assimilated estimates are necessarily more accurate regardless of property. Then, comparisons of model estimates with independent observations withheld from assimilation provide another, and possibly the strongest, direct measure of the goodness of the particular assimilation and are one of the common means utilized in assessing the quality of the estimates. For instance, Figure 4 in Section 2 compared an altimetric assimilation with *in situ* measurements of subsurface temperature and velocity; it showed not only improvements made by assimilation but also their quantitative consistency with formal error estimates. Others have compared results of an altimetric assimilation with measurements from drifters (e.g., Schröter et al., 1993; Morrow and De Mey, 1995; Blayo et al., 1997), current meters (e.g., Capotondi et al., 1995b; Fukumori, 1995; Stammer, 1997, Blayo et al., 1997), hydrography (e.g., White et al., 1990a; Dombrowsky and De Mey, 1992; Oschlies and Willebrand, 1996; Greiner and Perigaud, 1996; Stammer, 1997), and tomography (Menemenlis et al., 1997). To the extent that future observations contain information independent of past measurements, forecasting skills also provide similar measures of the assimilation's reliability (e.g., Figure 7, see also Lionello et al., 1995; Morrow and De Mey, 1995). The so-called innovation vector in sequential estimation, i.e., the difference of model and data immediately prior to assimilation (Section 5.1.2), provides a similar measure of forecasting skill albeit generally over a short period (e.g., Figure 12; see also Gaspar and Wunsch, 1989; Fu et al., 1993).

The comparative smallness of model-data differences, on the one hand, does not by itself verify or validate the estimation, but it does demonstrate a lack of any outright inadequacies in the calculation. On the other hand, an excessively large difference can indicate an inconsistency in the calculation, but the presence of representation error precludes immediate judgment and requires a careful analysis as to the cause of the discrepancy. For instance, Figure 13 shows an altimetric assimilation (gray curve) failing to resolve subsurface temperature variability (solid curve) at two depths

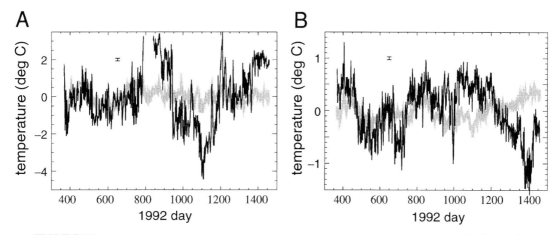

FIGURE 13 An example of model representation error. The example compares temperature anomalies (°C) at 2°S 165°E; (A) 125 m, (B) 500 m. Different curves are *in situ* measurements (black; Tropical Atmosphere and Ocean array) and altimetric assimilation (gray solid). The simulation is hardly different from the assimilation and is not shown to reduce clutter. Bars denote formal error estimates. Model and assimilation are based on those described in Figure 9. (Adapted from Fukumori *et al.* (1999), Plate 5.)

with error estimates being much smaller than actual differences. However, the lack of vertical coherence in the *in situ* measurements suggests the data being dominated by variations with a vertical scale much smaller than the model's resolution (150 m). Namely, the comparison suggests that the model-data discrepancy is caused by model representation error instead of a failure of assimilation. The formal error estimates are much smaller than actual differences as the estimate only pertains to the signal consistent with model and data, and excludes effects of representation error (Section 5.1.1).

Withholding observations is not necessarily required to test consistencies of an assimilation. In fact, the optimal estimate by its very nature requires that all available observations be assimilated simultaneously. Equivalent tests of model-data differences can be performed with respect to properties of a posteriori differences of the estimate. However, from a practical standpoint, when inconsistencies are found it may be easier to identify the source of the inaccuracy by assimilating fewer data and therefore having fewer assumptions at a time.

5.3. Observability

Observability, as defined in estimation theory, is the ability to determine the state of the model from observations in the absence of both model process noise and data constraint errors. Weaver and Anderson (1997) empirically examined the issue of observability from altimetry using twin experiments. Mathematically, the degree of observability is measured by the rank of the inverse problem, Eq. (6). In the absence of errors, the state of the model is uniquely determined by the initial condition, $\mathbf{x}_0$, in terms of which the left-handside of Eq. (6) may be rewritten,

$$\begin{pmatrix} \vdots \\ \mathcal{H}_i(\mathbf{x}_i) \\ \vdots \\ \mathbf{x}_{j+1} - \mathcal{F}_j(\mathbf{x}_j) \\ \vdots \end{pmatrix} = \begin{pmatrix} \vdots \\ \mathcal{H}_i \mathcal{F}_0^i \\ \vdots \\ \mathcal{F}_0^{j+1} - \mathcal{F}_0^{j+1} \\ \vdots \end{pmatrix} \mathbf{x}_0 \quad (33)$$

where the model $\mathcal{F}$ was assumed to be linear, and $\mathcal{F}_i^j$ denotes integration from time i to j. The process noise being zero, the model equations are identically satisfied, and therefore the rank of Eq. (33) is equivalent to that of the equations regarding observations alone; viz.,

$$\begin{pmatrix} \mathcal{H}_M \mathcal{F}_0^M \\ \vdots \\ \mathcal{H}_i \mathcal{F}_0^i \\ \vdots \\ \mathcal{H}_0 \end{pmatrix} \mathbf{x}_0 \quad (34)$$

where M denotes the total incidences of observations. The rank and the range space of the coefficient matrix respectively determine how many and what degrees of freedom are uniquely determined by the observations. In particular, when the rank of the coefficient matrix equals the dimension of $\mathbf{x}$ (i.e., full rank), all components of the model can be uniquely determined and the model state is said to be completely observable.

Hurlburt (1986) and Berry and Marshall (1989), among others, have explored the propagation of surface data into subsurface information. While on one hand, sequential assimilation transfers surface information into the interior of the ocean, on the other hand, future observations also contain information of the past state. That is, the entire temporal evolution of the measured property, viz., indices $i = 0, \ldots, M$ in Eq. (34), provides information in determining the model state and thus the observability of the assimilation problem. Webb and Moore (1986) provide a physical illustration of the significance of the measured temporal evolution in the context of altimetric observability. Namely, as baroclinic waves of different vertical modes propagate at different speeds, the phase among different modes will become distinct over time and thus distinguishable, by measuring the temporal evolution of sea level. Thus dynamics allows different model states that cannot be distinguished from each other by observations alone to be differentiated (Miller, 1989). Mathematically, the "distinguishability" corresponds to the rows of Eq. (34) being independent from each other. In fact, most components of a model are theoretically observable from altimetry, as any perturbation in model state will eventually lead to some numerical difference in sea level, even though perhaps with a significant time-lag and/or with infinitesimal amplitude. Miller (1989) demonstrated observability of model states from measurements of temporal differences, such as those provided by an altimeter (see also Section 5.4). Fukumori et al. (1993) demonstrated the complete observability (i.e., observability of the entire state) of a primitive equation model from altimetric measurements alone.

Observability, as defined in estimation theory, is a deterministic property as opposed to a stochastic property of the assimilation problem. In reality, however, data and model errors cannot be ignored and these errors restrict the degree to which model states can be improved even when they are mathematically observable, and thus limit the usefulness of the strict definition and measure of observability. What is of more practical significance in characterizing the ability to determine the model state is the estimated error of the model state, in particular the difference of the model state error with and without assimilation. For example, Fukumori et al. (1993) show that the relative improvement by altimetric assimilation of the depth-dependent (internal or baroclinic mode) circulation is larger than that of the depth-averaged (external or barotropic mode) component caused by differences in the relative spin-up time-scales. Actual improvements of unmeasured quantities are also often used to measure the fidelity in assimilating real observations (e.g., Figure 4 and the examples in Section 5.2).

5.4. Mean Sea Level

Because of our inadequate knowledge of the marine geoid, altimetric sea level data are often referenced to their time-mean, that is, the sum of the mean dynamic sea surface topography and the geoid. The unknown reference surface makes identifying the model equivalent of such "altimetric

residuals" (Eq. [2]) somewhat awkward, necessitating consideration as to the appropriate use of altimetric measurements. One of several approaches has been taken in practice, including direct assimilation of temporal differences, using mean model sea level in place of the unknown reference, and estimating the mean from separate observations.

The temporal difference of model sea level is a direct equivalent of altimetric variability. Miller (1989), therefore, formulated the altimetric assimilation problem by directly assimilating temporal differences of sea level at successive instances by expanding the definition of the model state vector to include model states at corresponding times. Alternatively, Verron (1992), modeling the effect of assimilation as stretching of the surface layer, reformulated the assimilation problem into assimilating the tendency (i.e., temporal change) of model-data sea level differences, thereby eliminating the unknown time-invariant reference surface from the problem.

The mean sea level of a model simulation is used in many studies to reference altimetric variability (e.g., Oschlies and Willebrand, 1996), which asserts that the model sea level anomaly is equivalent to the altimetric anomaly. Using the model mean to reference model sea level affirms that there is no direct information of the mean in the altimetric residuals. In fact, for linear models, the model mean is unchanged when assimilating altimetric variabilities (Fukumori et al., 1993). Yet for nonlinear physics, the model mean can be changed by such an approach. Using a nonlinear QG model, Blayo et al. (1994) employed the model mean sea level but iterated the assimilation process until the resulting mean converges between different iterations.

Alternatively, a reference sea level can also be obtained from in situ measurements. For instance, Capotondi et al. (1995b) and Stammer (1997) used dynamic height estimates based on climatological hydrography in place of the unknown time-mean altimetric reference surface. Morrow and De Mey (1995) and Ishikawa et al. (1996) utilized drifter trajectories as a means to constrain the absolute state of the ocean.

In spite of their inaccuracies, geoid models have skills, especially at large-spatial scales, which information may be exploited in the estimation. For instance, Marshall (1985) theoretically examined the possibility of determining mean sea level and the geoid simultaneously from assimilating altimetric measurements, taking advantage of differences in spatial scales of the respective uncertainties. Thompson (1986) and Stammer et al. (1997) further combined independent geoid estimates in conjunction with hydrographic observations.

Finally, Greiner and Perigaud (1994, 1996), noting nonlinear dependencies of the oceanic variability and the temporal mean, estimated the time-mean sea level of the Indian Ocean by assimilating sea level variabilities alone measured by Geosat, and verified their results by comparisons with hydrographic observations (Figure 5).

6. SUMMARY AND OUTLOOK

The last decade has witnessed an unprecedented series of altimetric missions that includes Geosat (1985–1989), ERS-1 (1991–1996), TOPEX/POSEIDON (1992–present), ERS-2 (1995–present), and Geosat Follow-On (1998–present), whose legacy is anticipated to continue with Jason-1 (to be launched in 2001) and beyond. At the same time, advances in computational capabilities have prompted increasingly realistic ocean circulation models to be developed and used in studies of ocean general circulation. These developments have led to the recognition of the possibilities of combining observations with models so as to synthesize the diverse measurements into coherent descriptions of the ocean; i.e., data assimilation.

Many advances in data assimilation have been accomplished in recent years. Assimilation techniques first developed in numerical weather forecasting have been explored in the context of oceanography. Other assimilation schemes have been developed or modified, reflecting properties of ocean circulation. Methods based on estimation and control theories have also been advanced, including various approximations that make the techniques amenable to practical applications. Studies in ocean data assimilation are now evolving from demonstrations of methodologies to applications. Examples can be found in practical operations, such as in studies of weather and climate (e.g., Behringer *et al.*, 1998), tidal modeling (e.g., Le Provost *et al.*, 1998), and wave forecasting (e.g., Janssen *et al.*, 1997).

Data assimilation provides an optimal estimate of the ocean consistent with *both* model physics and observations. By doing so, assimilation improves on what either a given model or a set of observations alone can achieve. For instance, although useful for theoretical investigations, modeling alone is inaccurate in quantifying actual ocean circulation, and observations by themselves are incomplete and limited in scope.

Yet, data assimilation is not a panacea for compensating all deficiencies of models and observing systems. A case in point is representation error (Section 5.2). Mathematically, data assimilation is an estimation problem (Eq. [7]) in which the oceanic state is sought that satisfies a set of simultaneous constraints (i.e., model and data). Consequently, the estimate is limited in what it can resolve (or improve) by what observations and models represent *in common*. While errors caused by measuring instruments and numerical schemes can be reduced by data assimilation, model and data representation errors cannot be corrected or compensated by the process. Overfitting models to data beyond what the models represent can have detrimental consequences leading the assimilation to degrade rather than to improve model estimates.

To recognize such limits and to properly account for the different types of errors are imperative for making accurate estimates and for interpreting the results. The a priori errors of model and data in effect define the assimilation problem (Eq. [7]), and a misspecification amounts to solving the wrong problem (Section 5.1.1). However, in spite of adaptive methods (Section 5.1.2), in practice, weights used in assimilation are often chosen more or less subjectively, and a systematic effort is required to better characterize and understand the a priori uncertainties and thereby the weights. In particular, the significance of representation error is often under-appreciated. Quantifying what models and observing systems respectively do and do not represent is arguably the most urgent and important issue in estimation.

In fact, identifying representation error is a fundamental problem in modeling and observing system assessment and is the foundation to improving our understanding of the ocean. Moreover, improving model and data representation can only be achieved by advancing the physics in numerical models and conducting comprehensive observations. Such limitations and requirements of estimation exemplify the relative merits of modeling, observations, and data assimilation. Although assimilation provides a new dimension to ocean state estimation, the results are ultimately limited to what models and observations resolve and our understanding of their nature.

A wide spectrum of assimilation efforts presently exist. For example, on the one hand, there are fine-resolution state-of-the-art models using relatively simple assimilation schemes, and on the other there are near optimal assimilation methods using simpler models. The former places a premium on minimizing representation error while the latter minimizes the error of the resolved state. The differences in part reflect the significant computational requirements of modeling and assimilation and the practical choices that need to be made. Such diversity will likely remain for some time. Yet, differences between these opposite ends of the spectrum are narrowing and should eventually become indistinguishable as we gain further experience in applications.

In spite of formal observability, satellite altimetry, as with other observing systems, cannot by itself accurately determine the complete state of the ocean because of finite model errors, and to a lesser extent data uncertainties. Various other data types must be analyzed and brought together in order to better constrain the estimates. Several efforts have already begun in such an endeavor of simultaneously assimilating *in situ* observations with satellite altimetry. Field experiments such as the World Ocean Circulation Experiment (WOCE) and the Tropical Ocean Global Atmosphere Program (TOGA) have collected an unprecedented suite of *in situ* observations. In particular, the analysis phase of WOCE specifically calls for a comprehensive synthesis of its measurements. The Global Ocean Data Assimilation Experiment (GODAE) plans to demonstrate the utility of global ocean observations through near-real-time analyses by data assimilation. The task of simultaneously assimilating a diverse set of observations is a formidable one, both in terms of computation and analysis and in the assessment of the results. Yet the results of such a synthesis will be far-reaching, leading to exciting new applications and discoveries. Satellite altimetry, being the only presently available means of synoptically measuring the global ocean circulation, will be critical to the success of such effort.

ACKNOWLEDGMENTS

Comments by Jacques Verron and an anonymous reviewer were most helpful in improving this chapter. The author is also grateful to Lee-Lueng Fu, Ralf Giering, Tong Lee, Dimitris Menemenlis, Van Snyder, and Carl Wunsch for their valuable suggestions on an earlier version of the manuscript. This research was carried out in part by the Jet Propulsion Laboratory, California Institute of Technology, under contract with the National Aeronautics and Space Administration.

References

Anderson, D. L. T., and J. Willebrand, (1989). *In* "Oceanic Circulation Models: Combining Data and Dynamics," 605 pp. Proceedings of the NATO Advanced Study Institute on "Modelling the Ocean General Circulation and Geochemical Tracer Transport," Les Houches, France, February 1988, Kluwer.

Anderson, D. L. T., J. Sheinbaum, and K. Haines, (1996). Data assimilation in ocean models, *Rep. Progr. Phys.*, **59**, 1209–1266.

Bauer, E., S. Hasselmann, K. Hasselmann, and H. C. Graber, (1992). Validation and assimilation of Seasat altimeter wave heights using the WAM wave model, *J. Geophys. Res.*, **97**, 12671–12682.

Bauer, E., K. Hasselmann, I. R. Young, and S. Hasselmann, (1996). Assimilation of wave data into the wave model WAM using an impulse response function method, *J. Geophys. Res.*, **101**, 3801–3816.

Behringer, D. W., M. Ji, and A. Leetmaa, (1998). An improved coupled model for ENSO prediction and implications for ocean initialization. Part I: The ocean data assimilation system, *Mon. Weather Rev.*, **126**, 1013–1021.

Bennett, A. F., (1992). *In* "Inverse methods in physical oceanography," 346 pp. Cambridge University Press, Cambridge, U.K.

Bennett, A. F., and M. A. Thorburn, (1992). The generalized inverse of a nonlinear quasi-geostrophic ocean circulation model, *J. Phys. Oceanogr.*, **22**, 213–230.

Bennett, A. F., B. S. Chua, D. E. Harrison, and M. J. McPhaden, (1998). Generalized inversion of tropical atmosphere-ocean data and a coupled model of the tropical Pacific, *J. Climate*, **11**, 1768–1792.

Berry, P., and J. Marshall, (1989). Ocean modelling studies in support of altimetry, *Dyn. Atmos. Oceans*, **13**, 269–300.

Blanchet, I., C. Frankignoul, and M. A. Cane, (1997). A comparison of adaptive Kalman filters for a tropical Pacific Ocean model, *Mon. Weather Rev.*, **125**, 40–58.

Blayo, E., J. Verron, and J. M. Molines, (1994). Assimilation of TOPEX/POSEIDON altimeter data into a circulation model of the North Atlantic, *J. Geophys. Res.*, **99**, 24,691–24,705.

Blayo, E., J. Verron, J. M. Molines, and L. Testard, (1996). Monitoring of the Gulf Stream path using Geosat and TOPEX/POSEIDON altimetric data assimilated into a model of ocean circulation, *J. Marine Syst.*, **8**, 73–89.

Blayo, E., T. Mailly, B. Barnier, P. Brasseur, C. Le Provost, J. M. Molines, and J. Verron, (1997). Complementarity of ERS 1 and TOPEX/POSEIDON altimeter data in estimating the ocean circulation: Assimilation into a model of the North Atlantic, *J. Geophys. Res.*, **102**, 18,573–18,584.

Bourles, B., S. Arnault, and C. Provost, (1992). Toward altimetric data assimilation in a tropical Atlantic model, *J. Geophys. Res.*, **97**, 20,271–20,283.

Bretherton, F. P., R. E. Davis, and C. B. Fandry, (1976). A technique for objective analysis and design of oceanographic experiments applied to MODE–73, *Deep-Sea Res.*, **23**, 559–582.

Bryan, F. O., and W. R. Holland, (1989). A high-resolution simulation of the wind- and thermohaline-driven circulation in the North Atlantic Ocean, *In* "Parameterization of Small–Scale Processes," Proceedings 'Aha Huliko'a, Hawaiian Winter Workshop, University of Hawaii at Manoa, 99–115.

Bryson, A. E., Jr., and Y.-C. Ho, (1975). "Applied Optimal Control," Rev. ed. Hemisphere, New York, 481 pp.

Cane, M. A., A. Kaplan, R. N. Miller, B. Tang, E. C. Hackert, and A. J. Busalacchi, (1996). Mapping tropical Pacific sea level: Data assimilation via a reduced state space Kalman filter, *J. Geophys. Res.*, **101**, 22,599–22,617.

Capotondi, A., P. Malanotte-Rizzoli, and W. R. Holland, (1995a). Assimilation of altimeter data into a quasigeostrophic model of the Gulf Stream system, Part I: Dynamical considerations, *J. Phys. Oceanogr.*, **25**, 1130–1152.

Capotondi, A., P. Malanotte-Rizzoli, and W. R. Holland, (1995b). Assimilation of altimeter data into a quasigeostrophic model of the Gulf Stream system, Part II: Assimilation results, *J. Phys. Oceanogr.*, **25**, 1153–1173.

Carnes, M. R., D. N. Fox, R. C. Rhodes, and O. M. Smedstad, (1996). Data assimilation in a North Pacific Ocean monitoring and prediction system, *In* "Modern Approaches to Data Assimilation in Ocean Modeling," (P. Malanotte-Rizzoli, Ed.) Elsevier, 319–345.

Carton, J. A., B. S. Giese, X. Cao, and L. Miller, (1996). Impact of altimeter, thermistor, and expendable bathythermograph data on retrospective analyses of the tropical Pacific Ocean, *J. Geophys. Res.*, **101**, 14147–14159.

Chin, T. M., A. J. Mariano, and E. P. Chassignet, (1999). Spatial regression and multiscale approximations for sequential data assimilation in ocean models, *J. Geophys. Res.*, **104**, 7991–8014.

Cohn, S. E., (1997). An introduction to estimation theory, *J. Meteorol. Soc. Jpn.*, **75**, 257–288.

Cohn, S. E., A. da Silva, J. Guo, M. Sienkiewicz, and D. Lamich, (1998). Assessing the effects of data selection with the DAO Physical-space Statistical Analysis System, *Mon. Weather Rev.*, **126**, 2913–2926.

Cong, L. Z., M. Ikeda, and R. M. Hendry, (1998). Variational assimilation of Geosat altimeter data into a two-layer quasi-geostrophic model over the Newfoundland ridge and basin, *J. Geophys. Res.*, **103**, 7719–7734.

Cooper, M., and K. Haines, (1996). Altimetric assimilation with water property conservation, *J. Geophys. Res.*, **101**, 1059–1077.

Courtier, P., J.-N. Thépaut, and A. Hollingsworth, (1994). A strategy for operational implementation of 4D-Var, using an incremental approach, *Q. J. Roy. Meteorol. Soc.*, **120**, 1367–1387.

de las Heras, M. M., G. Burgers, and P. A. E. M. Janssen, 1994. Variational wave data assimilation in a third-generation wave model, *J. Atmosph. Oceanic Technol.*, **11**, 1350–1369.

De Mey, P., and A. R. Robinson, (1987). Assimilation of altimeter eddy fields in a limited-area quasi-geostrophic model, *J. Phys. Oceanogr.*, **17**, 2280–2293.

Dee, D. P., and A. M. da Silva, (1998). Data assimilation in the presence of forecast bias, *Q. J. Roy. Meteorol. Soc.*, **124**, 269–295.

Derber, J., and A. Rosati, (1989). A global oceanic data assimilation system, *J. Phys. Oceanogr.*, **19**, 1333–1347.

Derber, J., (1989). A variational continuous assimilation technique, *Mon. Weather Rev.*, **117**, 2437–2446.

Dombrowsky, E., and P. De Mey, (1992). Continuous assimilation in an open domain of the Northeast Atlantic 1. Methodology and application to AthenA-88, *J. Geophys. Res.*, **97**, 9719–9731.

Egbert, G. D., A. F. Bennett, and M. G. G. Foreman, (1994). TOPEX/POSEIDON tides estimated using a global inverse model, *J. Geophys. Res.*, **99**, 24,821–24,852.

Egbert, G. D., and A. F. Bennett, (1996). Data assimilation methods for ocean tides, *In* "Modern Approaches to Data Assimilation in Ocean Modeling," (P. Malanotte-Rizzoli, Ed.), Elsevier, 147–179.

Evensen, G., (1994). Sequential data assimilation with a nonlinear quasi-geostrophic model using Monte Carlo methods to forecast error statistics, *J. Geophys. Res.*, **99**, 10143–10162.

Evensen, G., and P. J. van Leeuwen, (1996). Assimilation of Geosat altimeter data for the Agulhas Current using the ensemble Kalman filter with a quasi-geostrophic model, *Mon. Weather Rev.*, **124**, 85–96.

Ezer, T., and G. L. Mellor, (1994). Continuous assimilation of Geosat altimeter data into a three-dimensional primitive equation Gulf Stream model, *J. Phys. Oceanogr.*, **24**, 832–847.

Florenchie, P., and J. Verron, (1998). South Atlantic Ocean circulation: Simulation experiments with a quasi-geostrophic model and assimilation of TOPEX/POSEIDON and ERS 1 altimeter data, *J. Geophys. Res.*, **103**, 24,737–24,758.

Forbes, C., and O. Brown, (1996). Assimilation of sea-surface height data into an isopycnic ocean model, *J. Phys. Oceanogr.*, **26**, 1189–1213.

Fu, L.-L., J. Vazquez, and C. Perigaud, (1991). Fitting dynamic models to the Geosat sea level observations in the Tropical Pacific Ocean. Part I: A free wave model, *J. Phys. Oceanogr.*, **21**, 798–809.

Fu, L.-L., I. Fukumori, and R. N. Miller, (1993). Fitting dynamic models to the Geosat sea level observations in the Tropical Pacific Ocean. Part II: A linear, wind-driven model, *J. Phys. Oceanogr.*, **23**, 2162–2181.

Fu, L.-L., and I. Fukumori, (1996). A case study of the effects of errors in satellite altimetry on data assimilation, *In* "Modern Approaches to Data Assimilation in Ocean Modeling," (P. Malanotte-Rizzoli, Ed.), Elsevier, 77–96.

Fu, L.-L., and R. D. Smith, (1996). Global ocean circulation from satellite altimetry and high-resolution computer simulation, *Bull. Am. Meteorol. Soc.*, **77**, 2625–2636.

Fukumori, I., J. Benveniste, C. Wunsch, and D. B. Haidvogel, (1993). Assimilation of sea surface topography into an ocean circulation model using a steady-state smoother, *J. Phys. Oceanogr.*, **23**, 1831–1855.

Fukumori, I., and P. Malanotte-Rizzoli, (1995). An approximate Kalman filter for ocean data assimilation; An example with an idealized Gulf Stream model, *J. Geophys. Res.*, **100**, 6777–6793.

Fukumori, I., (1995). Assimilation of TOPEX sea level measurements with a reduced-gravity shallow water model of the tropical Pacific Ocean, *J. Geophys. Res.*, **100**, 25027–25039.

Fukumori, I., R. Raghunath, and L. Fu, (1999). Nature of global large-scale sea level variability in relation to atmospheric forcing: A modeling study, *J. Geophys. Res.*, **103**, 5493–5512.

Fukumori, I., R. Raghunath, L. Fu, and Y. Chao, (1998). Assimilation of TOPEX/POSEIDON altimeter data into a global ocean circulation model: How good are the results?, *J. Geophys. Res.*, **104**, 25647–25655.

Gangopadhyay, A., A. R. Robinson, and H. G. Arango, (1997). Circulation and dynamics of the western North Atlantic. Part I: Multiscale feature models, *J. Atmosph. Oceanic Technol.*, **14**, 1314–1332.

Gangopadhyay, A., and A. R. Robinson, (1997). Circulation and dynamics of the western North Atlantic. Part III: Forecasting the meanders and rings, *J. Atmosph. Oceanic Technol.*, **14**, 1352–1365.

Gaspar, P., and C. Wunsch, (1989). Estimates from altimeter data of barotropic Rossby waves in the northwestern Atlantic ocean, *J. Phys. Oceanogr.*, **19**, 1821–1844.

Gelb, A., (1974). *In* "Applied Optimal Estimation," M.I.T. Press, Cambridge, MA, 374 pp.

Gent, P. R., and J. C. McWilliams, (1990). Isopycnal mixing in ocean circulation models, *J. Phys. Oceanogr.*, **20**, 150–155.

Ghil, M., and P. Malanotte-Rizzoli, (1991). Data assimilation in meteorology and oceanography, *Adv. Geophys.*, **33**, 141–266.

Giering, R., and T. Kaminski, (1998). Recipes for adjoint code construction, *ACM Trans. Mathematical Software*, **4**, 437–474.

Gourdeau, L., J. F. Minster, and M. C. Gennero, (1997). Sea level anomalies in the tropical Atlantic from Geosat data assimilated in a linear model, 1986–1988, *J. Geophys. Res.*, **102**, 5583–5594.

Greiner, E., and C. Perigaud, (1994). Assimilation of Geosat altimetric data in a nonlinear reduced-gravity model of the Indian Ocean Part1: adjoint approach and model-data consistency, *J. Phys. Oceanogr.*, **24**, 1783–1804.

Greiner, E., and C. Perigaud, (1996). Assimilation of Geosat altimetric data in a nonlinear shallow-water model of the Indian Ocean by adjoint approach, Part2: Some validation and interpretation of the assimilated results, *J. Phys. Oceanogr.*, **26**, 1735–1746.

Greiner, E., S. Arnault, and A. Morlière, (1998a). Twelve monthly experiments of 4D-variational assimilation in the tropical Atlantic during 1987: Part 1: Method and statistical results, *Progr. Oceanogr.*, **41**, 141–202.

Greiner, E., S. Arnault, and A. Morlière, (1998b). Twelve monthly experiments of 4D-variational assimilation in the tropical Atlantic during 1987: Part 2: Oceanographic interpretation, *Progr. Oceanogr.*, **41**, 203–247.

Griffith, A. K., and N. K. Nichols, (1996). Accounting for model error in data assimilation using adjoint models, *In* "Computational Differentiation: Techniques, Applications, and Tools," pp. 195–204, Proceedings of the Second International SIAM Workshop on Computational Differentiation, Santa Fe, New Mexico, 1996, Society of Industrial and Applied Mathematics, Philadelphia, PA.

Haines, K., (1991). A direct method for assimilating sea surface height data into ocean models with adjustments to the deep circulation, *J. Phys. Oceanogr.*, **21**, 843–868.

Hirose, N., I. Fukumori, and J.-H. Yoon, (1999). Assimilation of TOPEX/POSEIDON altimeter data with a reduced gravity model of the Japan Sea, *J. Oceanogr.*, **55**, 53–64.

Hoang, S., R. Baraille, O. Talagrand, X. Carton, and P. De Mey, (1998). Adaptive filtering: application to satellite data assimilation in oceanography, *Dyn. Atmos. Oceans*, **27**, 257–281.

Holland, W. R., and P. Malanotte-Rizzoli, (1989). Assimilation of altimeter data into an ocean model: Space verses time resolution studies, *J. Phys. Oceanogr.*, **19**, 1507–1534.

Hollingsworth, A., (1989). The role of real-time four-dimensional data assimilation in the quality control, interpretation, and synthesis of climate data, *In* "Oceanic Circulation Models: Combining Data and Dynamics," (D. L. T. Anderson and J. Willebrand, Eds.), Kluwer, 303–343.

Holloway, G., (1992). Representing topographic stress for large-scale ocean models, *J. Phys. Oceanogr.*, **22**, 1033–1046.

Hurlburt, H. E., (1986). Dynamic transfer of simulated altimeter data into subsurface information by a numerical ocean model, *J. Geophys. Res.*, **91**, 2372–2400.

Hurlburt, H. E., D. N. Fox, and E. J. Metzger, (1990). Statistical inference of weakly correlated subthermocline fields from satellite altimeter data, *J. Geophys. Res.*, **95**, 11,375–11,409.

Ishikawa, Y., T. Awaji, K. Akitomo, and B. Qiu, (1996). Successive correction of the mean sea-surface height by the simultaneous assimilation of drifting buoy and altimetric data, *J. Phys. Oceanogr.*, **26**, 2381–2397.

Janssen, P. A. E. M., B. Hansen, and J.-R. Bidlot, (1997). Verification of the ECMWF wave forecasting system against buoy and altimeter data, *Weather Forecasting*, **12**, 763–784.

Ji, M., D. W. Behringer, and A. Leetmaa, (1998). An improved coupled model for ENSO prediction and implications for ocean initialization. Part II: The coupled model. *Mon. Weather Rev.*, **126**, 1022–1034.

Ji, M., R. W. Reynolds, and D. W. Behringer, (2000). Use of TOPEX/POSEIDON sea level data for ocean analyses and ENSO prediction: Some early results, *J. Climate*, **13**, 216–231.

Kindle, J. C., (1986). Sampling strategies and model assimilation of altimetric data for ocean monitoring and prediction, *J. Geophys. Res.*, **91**, 2418–2432.

Large, W. G., J. C. McWilliams, and S. C. Doney, (1994). Oceanic vertical mixing: A review and a model with a nonlocal boundary-layer parameterization, *Rev. Geophys.*, **32**, 363–403.

Lee, T., and J. Marotzke, (1997). Inferring meridional mass and heat transports of the Indian-Ocean by fitting a general-circulation model to climatological data, *J. Geophys. Res.*, **102**, 10585–10602.

Lee, T., and J. Marotzke, (1998). Seasonal cycles of meridional overturning and heat transport of the Indian Ocean, *J. Phys. Oceanogr.*, **28**, 923–943.

Lee, T., J.-P. Boulanger, L.-L. Fu, A. Foo, and R. Giering, (2000). Data assimilation by a simple coupled ocean-atmosphere model: Application to the 1997–'98 El Niño, *J. Geophys. Res.* (in press).

Leetmaa, A., and M. Ji, (1989). Operational hindcasting of the tropical Pacific, *Dyn. Atmos. Oceans*, **13**, 465–490.

Le Provost, C., F. Lyard, J. M. Molines, M. L. Genco, and F. Rabilloud, (1998). A hydrodynamic ocean tide model improved by assimilating a satellite altimeter-derived data set. *J. Geophys. Res.*, **103**, 5513–5529.

Lionello, P., H. Günther, and B. Hansen, (1995). A sequential assimilation scheme applied to global wave analysis and prediction, *J. Marine Syst.*, **6**, 87–107.

Lorenc, A. C., (1986). Analysis methods for numerical weather prediction, *Q. J. Roy. Meteorol. Soc.*, **112**, 1177–1194.

Luong, B., J. Blum, and J. Verron, (1998). A variational method for the resolution of a data assimilation problem in oceanography, *Inverse Probl.*, **14**, 979–997.

Macdonald, A. M., and C. Wunsch, (1996). An estimate of global ocean circulation and heat fluxes, *Nature*, **382**, 436–439.

Malanotte-Rizzoli, P., (1996). *In* "Modern Approaches to Data Assimilation in Ocean Modeling," Elsevier, Amsterdam, The Netherlands, 455 pp.

Marotzke, J., and C. Wunsch, (1993). Finding the steady state of a general circulation model through data assimilation: Application to the North Atlantic Ocean, *J. Geophys. Res.*, **98**, 20149–20167.

Marshall, J. C., (1985). Determining the ocean circulation and improving the geoid from satellite altimetry, *J. Phys. Oceanogr.*, **15**, 330–349.

Marshall, J. C., C. Hill, L. Perelman, and A. Adcroft, (1997). Hydrostatic, quasi-hydrostatic, and nonhydrostatic ocean modeling, *J. Geophys. Res.*, **102**, 5733–5752.

Mellor, G. L., and T. Ezer, (1991). A Gulf Stream model and an altimetry assimilation scheme, *J. Geophys. Res.*, **96**, 8779–8795.

Menemenlis, D., T. Webb, C. Wunsch, U. Send, and C. Hill, (1997). Basin-scale ocean circulation from combined altimetric, tomographic, and model data, *Nature*, **385**, 618–621.

Menemenlis, D., and C. Wunsch, (1997). Linearization of an oceanic general circulation model for data assimilation and climate studies, *J. Atmos. Oceanic Technol.*, **14**, 1420–1443.

Menemenlis, D., and M. Chechelnitsky, (2000). Error estimates for an ocean general circulation model from altimeter and acoustic tomography data, *Mon. Weather Rev.* **128**, 763–778.

Miller, R. N., (1989). Direct assimilation of altimetric differences using the Kalman filter, *Dyn. Atmos. Oceans*, **13**, 317–333.

Miller, R. N., M. Ghil, and F. Gauthiez, (1994). Advanced data assimilation in strongly nonlinear dynamical models, *J. Atmos. Sci.*, **51**, 1037–1056.

Moghaddamjoo, R. R., and R. L. Kirlin, (1993). Robust adaptive Kalman filtering, *In* "Approximate Kalman Filtering," (G. Chen, Ed.), World Scientific, Singapore, 65–85.

Moore, A. M., (1991). Data assimilation in a quasi-geostrophic open-ocean model of the Gulf Stream region using the adjoint method, *J. Phys. Oceanogr.*, **21**, 398–427.

Morrow, R., and P. De Mey, (1995). Adjoint assimilation of altimetric, surface drifter and hydrographic data in a QG model of the Azores Current, *J. Geophys. Res.*, **100**, 25007–25025.

Oschlies, A., and J. Willebrand, (1996). Assimilation of Geosat altimeter data into an eddy-resolving primitive equation model of the North Atlantic Ocean, *J. Geophys. Res.*, **101**, 14175–14190.

Pacanowski, R., K. Dixon, and A. Rosati, (1991). *In* "Modular Ocean Model Users' Guide," Ocean Group Tech. Rep. 2, Geophys. Fluid Dyn. Lab., Princeton, N.J.

Parish, D. F., and S. E. Cohn, (1985). A Kalman filter for a two-dimensional shallow water model: formulation and preliminary experiments, *Office Note 304*, National Meteorological Center, Washington DC 20233, 64 pp.

Pham, D. T., J. Verron, and M. C. Roubaud, (1998). A singular evolutive extended Kalman filter for data assimilation in oceanography, *J. Marine Syst.*, **16**, 323–340.

Pinardi, N., A. Rosati, and R. C. Pacanowski, (1995). The sea surface pressure formulation of rigid lid models. Implications for altimetric data assimilation studies, *J. Marine Syst.*, **6**, 109–119.

Robinson, A. R., P. F. J. Lermusiaux, and N. Q. Sloan III, (1998). Data Assimilation, *In* "The Sea, Vol 10," (K. H. Brink and A. R. Robinson, Eds.), John Wiley & Sons, New York, NY.

Sarmiento, J. L., and K. Bryan, (1982). An ocean transport model for the North Atlantic, *J. Geophys. Res.*, **87**, 394–408.

Sasaki, Y., (1970). Some basic formalisms in numerical variational analysis, *Mon. Weather Rev.*, **98**, 875–883.

Schiller, A., and J. Willebrand, (1995). A technique for the determination of surface heat and freshwater fluxes from hydrographic observations, using an approximate adjoint ocean circulation model, *J. Marine Res.*, **53**, 453–497.

Schröter, J., U. Seiler, and M. Wenzel, (1993). Variational assimilation of Geosat data into an eddy-resolving model of the Gulf Stream extension area, *J. Phys. Oceanogr.*, **23**, 925–953.

Seiss, G., J. Schröter, and V. Gouretski, (1997). Assimilation of Geosat altimeter data into a quasi-geostrophic model of the antarctic circumpolar current, *Mon. Weather Rev.*, **125**, 1598–1614.

Sheinbaum, J., and D. L. T. Anderson, (1990). Variational Assimilation of XBT data. Part II: Sensitivity studies and use of smoothing constraints, *J. Phys. Oceanogr.*, **20**, 689–704.

Shum, C. K., P. L. Woodworth, O. B. Andersen, G. D. Egbert, O. Francis, C. King, S. M. Klosko, C. Le Provost, X. Li, J. M. Molines, M. E. Parke, R. D. Ray, M. G. Schlax, D. Stammer, C. C. Tierney, P. Vincent, C. I. Wunsch, (1997). Accuracy assessment of recent ocean tide models, *J. Geophys. Res.*, **102**, 25,173–25,194.

Smedstad, O. M., and J. J. O'Brien, (1991), Variational data assimilation and parameter estimation in an equatorial Pacific ocean model, *Progr. Oceanogr.*, **26**, 179–241.

Smedstad, O. M., and D. N. Fox, (1994). Assimilation of altimeter data in a two-layer primitive equation model of the Gulf Stream, *J. Phys. Oceanogr.*, **24**, 305–325.

Smith, N. R., (1995). An improved system for tropical ocean subsurface temperature analyses, *J. Atmos. Oceanic Technol.*, **12**, 850–870.

Stammer, D., and C. Wunsch, (1996). The determination of the large-scale circulation of the Pacific Ocean from satellite altimetry using model Green's functions, *J. Geophys. Res.*, **101**, 18,409–18,432.

Stammer, D., (1997). Geosat data assimilation with application to the eastern North-Atlantic, *J. Phys. Oceanogr.*, **27**, 40–61.

Stammer, D., C. Wunsch, R. Giering, Q. Zhang, J. Marotzke, J. Marshall, and C. Hill, (1997). The global ocean circulation estimated from TOPEX/POSEIDON altimetry and the MIT general circulation model, Report No. 49, Center for Global Change Science, Massachusetts Institute of Technology, Cambridge, MA, 40 pp.

Stommel, H., and F. Schott, (1977). The beta spiral and the determination of the absolute velocity field from hydrographic station data, *Deep-Sea Res.* **24**, 325–329.

Thacker, W. C., and R. B. Long, (1988). Fitting dynamics to data, *J. Geophys. Res.*, **93**, 1227–1240.

Thacker, W. C., (1989). The role of the Hessian matrix in fitting models to measurements, *J. Geophys. Res.*, **94**, 6177–6196.

Thompson, J. D., (1986). Altimeter data and geoid error in mesoscale ocean prediction: Some results from a primitive equation model, *J. Geophys. Res.*, **91**, 2401–2417.

Verron, J., and W. R. Holland, (1989). Impacts de données d'altimétrie satellitaire sur les simulations numériques des circulations générales océaniques aux latitudes moyennes, *Ann. Geophys.*, **7**, 31–46.

Verron, J., (1990). Altimeter data assimilation into an ocean circulation model—sensitivity to orbital parameters, *J. Geophys. Res.*, **95**, 11443–11459.

Verron, J., (1992). Nudging satellite altimeter data into quasi-geostrophic ocean models, *J. Geophys. Res.*, **97**, 7479–7491.

Verron, J., L. Cloutier, and P. Gaspar, (1996). Assessing dual-satellite altimetric missions for observing the midlatitude oceans, *J. Atmos. Oceanic Technol.*, **13**, 1071–1089.

Verron, J., L. Gourdeau, D. T. Pham, R. Murtugudde, and A. J. Busalacchi, (1999), An extended Kalman filter to assimilate satellite altimeter data into a nonlinear numerical model of the tropical Pacific Ocean: Method and validation, *J. Geophys. Res.*, **104**, 5441–5458.

Vogeler, and J. Schröter, (1995). Assimilation of satellite altimeter data into an open-ocean model, *J. Geophys. Res.*, **100**, 15,951–15,963.

Weaver, A. T., and D. L. T. Anderson, (1997). Variational assimilation of altimeter data in a multilayer model of the tropical Pacific Ocean, *J. Phys. Oceanogr.*, **27**, 664–682.

Webb, D. J., and A. Moore, (1986). Assimilation of altimeter data into ocean models, *J. Phys. Oceanogr.*, **16**, 1901–1913.

White, W. B., C.-K. Tai, and W. R. Holland, (1990a). Continuous assimilation of Geosat altimeter sea level observations into a numerical synoptic ocean model of the California Current, *J. Geophys. Res.*, **95**, 3127–3148.

White, W. B., C.-K. Tai, and W. R. Holland, (1990b). Continuous assimilation of simulated Geosat altimetric sea level into an eddy-resolving numerical ocean model, 1, Sea level differences, *J. Geophys. Res.*, **95**, 3219–3234.

WOCE International Project Office, (1998). Report of a GODAE/WOCE meeting on large-scale ocean state estimation, Johns Hopkins University, Baltimore, MD, USA, 9–11 March 1998. WOCE International project Office, WOCE Report No. 161/98, GODAE Report No. 2, 21 pp.

Wunsch, C., (1977). Determining the general circulation of the oceans: A preliminary discussion, *Science*, **196**, 871–875.

Wunsch, C., (1996). *In* "The Ocean Circulation Inverse Problem," Cambridge University Press, New York, NY, 442 pp.

6

Ocean Tides

C. LE PROVOST

LEGOS/GRGS, UMR CNES-CNRS-UPS
14 Avenue Edouard Belin
31401 Toulouse Cedex
France

1. INTRODUCTION

Ocean tides are one of the most fascinating natural events in the world. Each day, the sea rises and falls along the coasts around the world oceans with amplitudes that can reach several meters. Extremes up to 18 m occur in the Bay of Fundy, Canada, and up to 14 m in the Bay of Mt. St. Michel, France. However, it is only since Newton (1687) that ocean tides are explained by the gravitational attraction of the sun and the moon. Since then it has taken nearly one century to move from Newton's equilibrium theory to the dynamic response concept of the ocean tides formulated in Laplace's (1776) Tidal Equations (LTE). The solutions of these LTE strongly depend on the bathymetry and the shape of the ocean's boundaries. Moreover, we know that the oceans have clusters of natural resonance in the same frequency bands as the gravitational forcing function (Platzman, 1981), so that friction, determining the Quality Factor of the resonance, is a critical factor. This explains why all attempts to analytically solve the LTE is hopeless. This is also the reason why their numerical resolution is still not fully satisfactory. One major step of the nineteenth century had been the development by Darwin (1883) of the harmonic techniques for tidal predictions, based on known astronomical frequencies of relative motions of the earth, moon, and sun. This allowed to extract empirical "harmonic constants" from a year's tide-gage record, which in turn could be used to provide reliable predictions for future tides at the same site. The accuracy of these predictions gave too many people the impression that the tides were well understood. Unfortunately the reality is that during the first three-quarters of the last century, our understanding of how the tide behaves in the ocean remained at best conjectural. Knowledge of ocean tides remained confined to the vicinity of the coastlines and of mid-ocean islands where they have been observed. The very irregular variations in amplitude and phase of the tides around the coasts of all oceans let us easily imagine how complex they are in the open seas. The "single point" or "tide gage" measurement approach to map ocean tides at the global scale is therefore doomed to fail because of the complexity of the tides.

In this context, the advent of satellite altimetry has been totally revolutionary: it offers for the first time a means to estimate tides everywhere over the global oceans. The aim of this chapter is to point out the major progress in tidal sciences since the beginning of high-precision satellite altimetry.

Tides are indeed an important mechanism, which have many impacts in geophysics and oceanography. The demands for tidal information have become more exacting in recent years. For earth rotation studies, knowledge of the total dissipation in the tides is needed. In the 1960–1970 decade, this was a totally open question. Since then, this quantity has been derived indirectly from satellite orbit determination and lunar laser ranging. As will be shown in this chapter, these values are now confirmed by direct altimetric measurements of the tidal field. In geodesy, tidal loading of the lithosphere needs to be taken into account thereby requiring a good model of the ocean tides, which was lacking up until recently at the level of precision of modern space techniques. In oceanography, new needs have recently emerged. For example, in ocean acoustic tomography tidal currents can be calculated from the gradients of surface elevation, but this requires still higher precision. Tidal energy dissipation, where and how it takes place, and the evaluation of the horizontal flux of tidal energy are still basically open questions which need tidal currents to be known. But probably the most critical need in these recent years came from the use of satellite altimetry to monitor changes in the slope of the sea surface caused by ocean circulation (Wunsch and Gaposchkin, 1980). The tidal variation of the surface represents more than 80% of the sea-surface variability. Tides must therefore be removed from the altimeter signal for ocean-current monitoring from altimetry, hopefully to a few centimeter precision. It is this requirement of high-quality tidal prediction which has spurred the scientific community to strive for new and better methods for tidal analysis and modeling during the last decade or so, especially for the TOPEX/POSEIDON (T/P) satellite project. We shall review the spectacular improvements of our knowledge of ocean tides, which have resulted from the exploitation of satellite altimetry.

2. MATHEMATICAL REPRESENTATION OF OCEAN TIDES

This section provides a brief review of the usual mathematical representations of ocean tides.

2.1. The Harmonic Expansion

The ocean tide response $\xi_k(\mathbf{x}, t)$, at location $\mathbf{x}$ and time t, of a tidal component k with frequency ω_k and astronomical phase V_k originating from the tide generating potential is generally expressed in terms of an amplitude $A_k(\mathbf{x})$ and Greenwich phase lag $G_k(\mathbf{x})$, so that the sea-surface tidal elevation ξ is expressed as:

$$\xi(\mathbf{x}, t) = \sum_{k=1, \text{Nc}} A_k(\mathbf{x}) \cos[\omega_k t + V_k - G_k(\mathbf{x})]. \quad (1)$$

Argument numbers like $d_1 d_2 d_3 d_4 d_5 d_6$ were introduced by Doodson (1921) and define the frequency and astronomical phase angle of each of the tidal components using the six principal astronomical arguments:

$$\omega_k t + V_k = d_1 \tau + (d_2 - 5)s + (d_3 - 5)h + (d_4 - 5)p \\ + (d_5 - 5)N' + (d_6 - 5)p'. \quad (2)$$

τ, s, h, p, N', p' are the mean lunar time, mean longitude of the moon, sun, lunar perigee, lunar node, and solar perigee, respectively.

2.2. The Response Formalism

The series in Eq. (1) is usually truncated to a limited number of constituents by assuming that the oceanic response to the tide-generating potential varies smoothly with frequency (Munk and Cartwright, 1966).

This truncation is usually done through two steps:

1. The introduction of nodal corrections in amplitude $f_k(t)$ and phase $u_k(t)$ accounts for slow modulations of the tidal forcing over the nodal period of 18.61 years. The nodal modulation factors ensure that the side lines and main lines of the fully explicit development of Doodson (1921) are properly put together in the so-called "constituents." This procedure allows the Doodson series to be reduced from about 400 constituents to only a few tens, say N_s (Schureman, 1958).

$$\xi(\mathbf{x}, t) = \sum_{k=1, \text{Ns}} f_k(t) A_k(\mathbf{x}) \cos[\omega_k t + V_k \\ + u_k(t) - G_k(\mathbf{x})]. \quad (3)$$

2. The further reduction of the number of unknowns from N_s to N, with $N < N_s$ is obtained by relating the complex characteristics of the minor constituents to a limited number of major constituents, through linear or more complex interpolations and extrapolations, called admittance functions (Cartwright and Ray, 1990; Le Provost et al., 1991). They are generally defined as complex functions $Z(\omega_k, \mathbf{x})$ with real and imaginary components, $X(\omega_k, \mathbf{x})$ and $Y(\omega_k, \mathbf{x})$, with $Z = X + iY$. The ocean tide height expressed in terms of admittance function is:

$$\xi_k(\mathbf{x}, t) = H_k \operatorname{Re}\{Z^*(\omega_k, \mathbf{x}) \exp[-i(\omega_k t + V_k)]\}. \quad (4)$$

H_k is the normalized forcing tide potential amplitude at frequency ω_k, $\operatorname{Re}\{f\}$ denotes the real part of f and the asterisk denotes the complex conjugate of a complex number. Linear interpolation can be applied to a minor constituent k located between the major constituents k_1 and k_2:

$$Z(\omega_k, \mathbf{x}) = Z(\omega_{k1}, \mathbf{x}) + [(\omega_k - \omega_{k1})/(\omega_{k2} - \omega_{k1})] \\ \times [Z(\omega_{k2}, \mathbf{x}) - Z(\omega_{k1}, \mathbf{x})]. \quad (5)$$

This enables the problem to be reduced to a determination of the characteristics of a very limited number of major constituents.

This is the way along which some of the models, presented below, limit the direct modeling of the major constituents to five semidiurnal ($M_2, S_2, N_2, K_2, 2N_2$) and three diurnal (K_1, O_1, Q_1), although the associated prediction models include a much larger number of constituents. As an example, in the model of Le Provost et al. (1998) which includes 26 constituents (listed in Table 1), the eight above-mentioned major constituents are computed from the hydrodynamic model, and corrected by assimilation, and the other 18 are deduced by admittance:

- $\mu_2, \nu_2, L_2, \lambda_2,$ and T_2, are estimated from splines based on $M_2, N_2,$ and K_2
- ε_2 and η_2 are extrapolated from a linear admittance based on respectively $2N_2 - N_2$, and $M_2 - K_2$
- P_1 is estimated from a spline based on $Q_1, O_1,$ and K_1

TABLE 1.   Tidal Periods (in hours) of the 26 Ocean Tide Constituents Included in the FES95.1 Prediction Code and their Aliasing Periods (in days) for TOPEX/POSEIDON, ERS1, and Geosat Altimetric Missions

Tides	Tidal period (hours)	Aliased periods (days)		
		Topex/Poseidon 10-day repeat orbit	ERS1 35-day repeat orbit	Geosat 17-day repeat orbit
M_2	12.42	62.11	94.49	317.11
S_2	12.00	58.74	∞	168.82
N_2	12.66	49.53	97.39	52.07
K_2	11.98	86.60	182.62	87.72
K_1	23.93	173.19	365.24	175.45
O_1	25.82	45.71	75.07	112.95
P_1	24.07	88.89	365.24	4466.67
ε_2	13.13	77.31	3166.10	98.75
$2N_2$	12.91	22.54	392.55	58.52
μ_2	12.87	20.32	135.06	81.76
ν_2	12.63	65.22	74.37	41.56
λ_2	12.22	21.04	129.53	35.36
L_2	12.19	20.64	349.25	39.20
T_2	12.02	50.60	365.26	313.89
S_1	24.00	117.48	∞	337.63
Q_1	26.87	69.36	132.81	74.05
OO_1	22.31	29.92	102.31	49.38
J_1	23.10	32.77	95.62	60.03
$2Q_1$	28.01	19.94	5250.89	43.88
σ_1	27.85	21.81	214.30	55.77
ρ_1	26.72	104.61	80.73	54.46
χ_1	24.71	26.88	7974.60	38.86
π_1	24.13	71.49	182.63	397.79
ϕ_1	23.80	3354.43	121.75	89.48
θ_1	23.21	38.97	178.53	46.48
M_1	24.83	23.77	200.71	35.64

[a] The constituents are ordered with increasing periods.

- $2Q_1$, σ_1, and ρ_1 rely on linear admittance estimates based on Q_1 and O_1
- M_1, χ_1, π_1, ϕ_1, θ_1, J_1, and OO_1 come from linear admittance estimates based on O_1 and K_1

2.3. The Orthotide Formalism

Groves and Reynolds (1975) introduced an orthogonalized form of the response formalism by defining a set of functions $\xi_l^{mn}(t)$, called orthotides, which are orthogonal over all time. n and m are the degrees and orders of the development of the tide generating potential. The ocean tide elevation is then expressed as:

$$\xi(\mathbf{x}, t) = \sum_{n=2,3} \sum_{m=0,n} \sum_{l=1,L} \alpha_l^{mn}(\mathbf{x}) \xi_l^{mn}(t) \qquad (6)$$

where the orthoweights, $\alpha_l^{mn}(\mathbf{x})$, are estimated from the observation time series. The orthotide functions are formed as linear combinations of the tidal potential terms, $a_{nm}(t)$ and $b_{mn}(t)$ (Cartwright and Taylor, 1971).

$$\xi_1^{mn} = \sum_{s=-S,+S} [U_{ls}^{mn} a_{nm}(t + s\Delta\tau) - V_{ls}^{mn} b_{nm}(t + s\Delta\tau)] \qquad (7)$$

with $a_{nm}(t) = \sum_j H_{nmj} \cos(\omega_{nmj}t + V_{nmj})$
$b_{nm}(t) = -\sum_j H_{nmj} \sin(\omega_{nmj}t + V_{nmj})$
and where $\Delta\tau = 2$ days (Munk and Cartwright, 1966). U_{ls}^{mn} and V_{ls}^{mn} are the orthoweights. This formalism has been used in several of the models, which will be introduced later. For example, Desai and Wahr (1995) computed their orthoweights using 161 tidal components in the diurnal band and 116 in the semidiurnal band.

3. STATUS BEFORE HIGH-PRECISION SATELLITE ALTIMETRY

3.1. *In Situ* Observations

In situ observations have long been restricted along the coasts, because their motivation was for shipping and the access to harbors. This explains why the geographical distribution of these sites of measurement is mainly concentrated along the coasts of intense commercial activity, cf. Figure 1a. Some 4000 shore-based tide gauges have had their harmonic constants compiled by the International Hydrographic Organization (IHO) for over a century. However, quality is variable. Many harmonic analyses are based on only 1 month or less duration of record; some are from very old and poorly recorded data; others are from estuary sites, where the local tide is not representative of the open sea. If we restrict data to 1-year records less than 50 years old on well-exposed coasts and islands, the total number of well-analyzed stations available worldwide are less than a few hundred.

From about 1965 onwards, deep-pressure recorders were developed which may be left on the ocean floor for several months. They opened up new, long-desired possibilities of obtaining pelagic tidal data from the open ocean (Eyries, 1968; Snodgrass, 1968; Cartwright *et al.*, 1980). Pressure records have a much lower noise level than conventional coastal-surface gauges and their harmonic constants are usually very accurate even if derived from rather short records. Pelagic tidal constants have been compiled by the International Association for the Physical Sciences of the Ocean (IAPSO) (Smithson, 1992), independently of the IHO. At present, about 350 pelagic stations have been operating, but many of them are clustered within a few hundred kilometers of the coasts of Europe and North America, thereby leaving large unrecorded areas in the Indian and South Pacific Oceans. Deployment of the instruments is limited to a few specialized laboratories and to areas frequented by research vessels. Recent 1-year deployments in the Southern Ocean associated with the World Ocean Circulation Experiment (WOCE) have usefully extended the coverage.

3.2. Hydrodynamic Numerical Modeling

As we said above, empirical charting of ocean tides from pelagic data alone is impossible, and because of the complexity of tides in real ocean basins, analytical approaches are hopeless. Hence numerical modeling has long been the most objective way to map the tides. Global ocean tide numerical modeling started in the late 1960s (Bogdanov and Magarik, 1967; Pekeris and Accad, 1969). These models were based on the LTE, but complemented by dissipation, which is indeed critical. It is commonly admitted that bottom friction is very weak in the deep ocean, but is the major contributor to tidal energy budget over the continental shelves and shallow-water seas where tidal currents are amplified. Some models used linear or quadratic parameterization of bottom friction and included the shallow areas in their domain of integration, as far as the spatial resolution of their grid allowed them to do so (Pekeris and Accad, 1969; Zahel, 1977). Others treated the ocean as frictionless, but with energy radiating through boundaries opening on the shallow water areas where energy is dissipated (Accad and Pekeris, 1978; Parke and Hendershott, 1980). A strong improvement of the numerical tidal models resulted from the introduction of earth tides, ocean tide loading, and self-attraction (Hendershott, 1977; Zahel, 1977; Accad and Pekeris, 1978; Parke and Hendershott, 1980).

Although these hydrodynamic numerical models brought very significant contributions to our understanding of the tidal regimes and their dependency on specific parameters like topography, friction, tidal loading, and self-attraction, their solutions only qualitatively agreed with *in situ* observations. Their accuracy was not at the level required for geophysical applications. Hence the need to compensate the deficiencies of these unconstrained models by additional empirical forcing. In this way, solutions fit to observed data at coastal boundaries, on islands, and even in the deep ocean. This was the approach developed by Schwiderski (1980) with his "hydrodynamic interpolation" method. His solutions were much closer to reality, but they depended on the quality of the observations used, some of the data being erroneous and some others representative of local coastal effects not resolved by the model grid. Moreover they suffered from the same weakness as the purely hydrodynamic models over the areas where data were not available (Woodworth, 1985). Nevertheless, these Schwiderski solutions (1980, 1983) have been used as the best available through the last decade. With a resolution of $1° \times 1°$, they cover the world ocean, except for some semi-enclosed basins like the Mediterranean. They include 11 cotidal maps: four semi-diurnal (M_2, S_2, N_2, K_2), four diurnal (K_1, O_1, P_1, Q_1), and three long periods (S_{sa}, M_m, M_f).

One of the difficulties for hydrodynamic models to realistically reproduce the ocean tides at the global scale is their inadequacy to correctly simulate energy dissipation. To improve this shortcoming, it is particularly necessary to reproduce the details of the tidal motions over the shelf areas and the marginal seas, which control the turbulent momentum exchanges. One way to do so is to increase the resolution. Models have been developed with grids of variable size: $4°$ over the deep ocean, $1°$ over some continental shelves, and $0.5°$ in particular shallow seas (Krohn, 1984). Another approach used the finite element (FE) method which improves the modeling of rapid changes in ocean depth, the refinement of the grid in shallow waters, and the description of the irregularities of the coastlines (Le Provost and Vincent, 1986; Kuo, 1991). The FE tidal model of Le Provost *et al.* (1994) used a mesh size of the order of 200 km over the deep

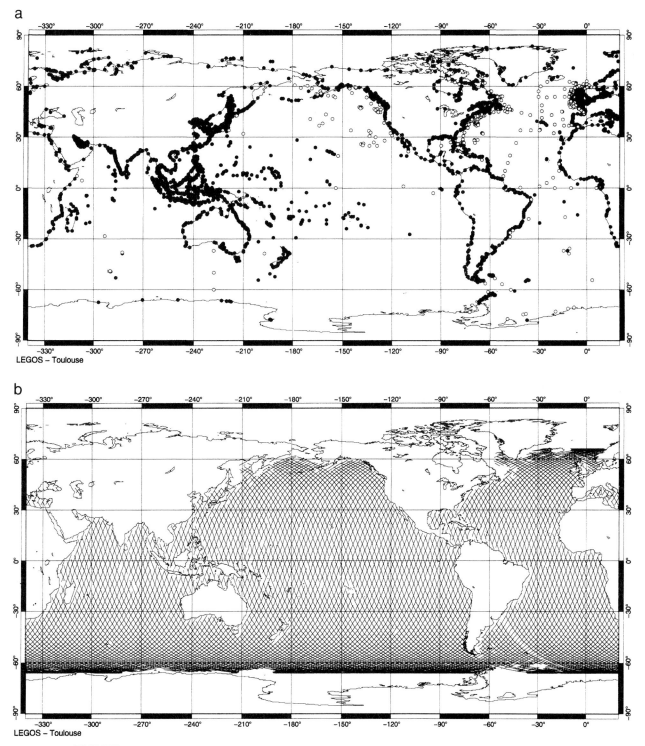

FIGURE 1 (a) Location of the *in situ* observations collected since the end of the last century. Harmonic constants from coastal and island sites have been archived by the International Hydrographic Organisation (IHO) (black dots). Pelagic harmonic constants have been compiled by the International Association for the Physical Sciences of the Oceans (IAPSO)(white dots). (b) Distribution of the ground tracks of the TOPEX/POSEIDON (T/P) Mission (cycle 126) along which sea-level heights are measured every 10 days.

oceans, but reduced to 10 km near the coasts. Qualitatively, their solutions look similar to the one produced by Schwiderski, although they did not force their solutions to agree with data, except for some tuning along the open boundaries of the subdomains of integration (Arctic, North Atlantic, South Atlantic, Indian Ocean, North Pacific, and South Pacific). Compared to the available observed data, these FE solutions were at many places closer to reality than the Schwiderski's solutions. They offered a new set of improved hydrodynamic solutions, which has been considered during these recent years as the best solutions independent of any altimetric data. However, discrepancies remained in these solutions, even with this more sophisticated modeling approach. Comparisons to the first T/P-derived solutions of Schrama and Ray (1994) revealed that they contained large-scale errors of the order of up to 6 cm in amplitude for M_2, in the deep ocean (Le Provost et al., 1995). Major discrepancies are in the South Atlantic Ocean, the mid-Indian Ocean, south west of Australia, east of Asia, and over large areas in the North and Equatorial Pacific Ocean. Referring to observations, it was clear that these large-scale differences were mainly due to inaccuracies in the FE solution. The differences for the diurnal components are lower, of the order of 3 cm for K_1, partly because this wave is globally weaker than M_2. But very high local discrepancies were noted (up to 6 cm) in the south of the Indian Ocean along Antarctica. Uncertainties on the bathymetry appear to be one major limiting factor for hydrodynamic modeling.

3.3. Modeling With Data Assimilation

Given the difficulty in reducing the remaining weakness of the hydrodynamic models, one solution to again improve the precision of the numerical models is to take advantage of the increasing quality of the in situ tidal data set. The idea is the one of Schwiderski (1980), but the methods are based on the assimilation approach considered as an inverse problem (Bennett and McIntosh, 1982; McIntosh and Bennett, 1984; Zahel, 1991; Bennett, 1992). Admitting the uncertainties of the hydrodynamic equations and in the data, the methods seek fields of tidal elevation and currents, which provide the best fit to the dynamic equations and to the data. The fundamental scientific challenge is the choice of weights for the various information: for the dynamic equations, for the boundary conditions, and for the data. Most techniques require explicit inversion of the covariance matrices or operators for the various unknown errors, dynamic and observation, in order to derive the weights. This inversion is a serious computational challenge. The size of the least-squares tidal problem is formidable. With the coarsest acceptable 1° spatial resolution, the number of real unknowns is about 10^6 (even when limiting the problem to the four major constituents: M_2, S_2, K_1, O_1, each involving amplitude and phase of elevation and two velocity components). And there

can be hundreds of pelagic and coastal tide gauge data, and hundreds of thousands altimetric data to assimilate. So the size of the inverse problem precludes direct minimization. To make it feasible, one is forced to adopt oversimplified forms for the covariance. Some less direct methods, such as representer expansions, only require the error covariance themselves even though the same weighted penalty function is minimized. The representer method finds the unique solution of the Euler-Lagrange equations, which are obeyed by minima of the penalty function. The solution of the Euler-Lagrange equations consists of a prior solution, which is an exact solution of the LTE, plus a finite linear combination of the representers (one per data site). These functions are obtained by solving the adjoint LTE and the LTE, once per data site. Many integrations are required, but these are stable inversions of the LTE (Egbert et al., 1994).

Several tentative solutions with data assimilation have been reported in the literature over the recent years. A global tidal inverse at 1° degree resolution using the LTE plus 55 gauge data (pelagic, island, and coastal) and 15 loading gravity data has been successfully constructed by Zahel, 1991. In this application, the size of the problem was reduced by imposing exact conservation of mass, elevation acted as a dependant variable. Separate inversions were reported for M_2 and O_1 constituents: although quantitative comparisons with other available solutions were not presented, it was confirmed that the inversions do lead to much better agreement with data. After Zahel, Grawunder (unpublished results) constructed a global tidal inverse at 0.5° resolution for the semidiurnal S_2 constituent. His LTE model included full loading and self-attraction Green's functions as well as atmospheric tides. Inversion was realized by the representer method. Assimilating only 41 pelagic constants reduced the rms error by more than 50% when compared to other available pelagic data. More recently, Egbert et al. presented in their 1994 paper (leading to their altimeter assimilated solution TPXO.1) global tidal inversions at resolution of $0.7° \times 0.7°$ on the basis of the representer method. Careful analysis of the representer matrix allowed the authors to substantially reduce the number of independent variables. An inverse for the four main constituents (M_2, S_2, K_1, and O_1), using 80 pelagic and island gauge data, gave a solution similar to Schwiderski, but much smoother. All these applications have demonstrated at least qualitatively the feasibility of the assimilation approach for tidal modeling.

4. METHODOLOGIES FOR EXTRACTING OCEAN TIDES FROM ALTIMETRY

As said in the introduction, the advent of satellite altimetry has brought the way to observe tides at the world ocean scale. Suddenly, since the beginning of the era of high precision satellite altimetry, we have moved from the situation visualized in Figure 1a to the one in Figure 1b. In Figure 1a, after more than a century of *in situ* tide gauges measurements, the distribution of available observations was still very spotty over the deep ocean and essentially concentrated along the coasts and in ocean islands. In Figure 1b, after a few years of satellite altimetry, tidal measurements are available along the many tracks of the altimeter satellites. However, this revolutionary technique does not provide the exact equivalent of thousands of tide gauges for a number of reasons. First, one major difficulty is the unusual time sampling of the signal (in terms of tidal analysis), as the repeat cycle of the satellites are ranging from a few days to tens of days. As will be shown later, the consequence is that the semidiurnal and diurnal tides are aliased into periods of several months to years. As the background spectrum of the ocean increases sharply at longer periods (Wunsch and Stammer, 1995), this aliasing results in a considerable increase of the noise-to-signal ratio in terms of tidal signal extraction. Second, altimeter instrument errors and other associated errors are numerous and complex. In the earlier missions, orbit errors were particularly problematic: even after specific orbit error corrections, the presence of systematic residual inaccuracies in the tidal solutions extracted from Geosat by Cartwright and Ray (1990) were observed (Molines *et al.*, 1994). With the advent of T/P, the improvements of the budget error of this mission have greatly facilitated the exploitation of the data, including for tidal studies (Fu *et al.*, 1994). This is particularly true for the precision of the orbit determination, which allowed to directly use the T/P data without orbit correction, without any noticeable impact, at least to a first order (Ma *et al.*, 1994). A third limiting factor of satellite altimetry, as an observing technique for mapping ocean tides, is its spatial sampling, which varies inversely with the length of the repeat period. We will later see the impact of this limitation in purely altimetric solutions and how it has been overcome by combining a priori hydrodynamic solutions with altimetric data, through empirical approaches or more sophisticated assimilation methods. We will also point out the synergy of T/P and ERS altimeter data for improving tidal solutions over continental shelves.

4.1. Tidal Aliasing in Altimeter Data

A discussion of tidal aliasing for satellite altimeters in repeat orbit has been given by Parke *et al.* (1987). Tidal aliasing is dependent on orbital characteristics and tidal frequencies. All the tidal components with periods less than twice the satellite repeat period, $2\,\Delta T$, are aliased into a period longer than $2\,\Delta T$. The alias period T_a of a tidal constituent of frequency f_T is given by the relation:

$$T_a = (2\pi\,\Delta T / \Delta\Phi)$$

where

$$\Delta\Phi = 2\Pi(f_T\,\Delta T - [f_T\,\Delta T + 0.5])$$

$[x]$ is integer part of x, and $\Delta\Phi$ is the phase difference between two consecutive samples by the satellite of the tidal component with frequency f_T. In Table 1, the aliased periods for the 26 most important tidal constituents are given for the Geosat, ERS, and T/P missions, of respectively 17.05-, 35-, and 9.9-day repeat periods. And Figure 2 shows the original and aliased spectra with T/P sampling, for the major tidal components.

For T/P, unlike Geosat and ERS, there are no frozen tides or aliased tidal periods larger than half a year, except for the very small Φ_1 and Ψ_1 components. T/P has been effectively designed to allow the best possible observation of ocean tides: the three major semidiurnal constituents are aliased at nearly 2 months, only K_1 is aliased at 173 days. Note however that several main constituents are aliased close together. If we follow the Rayleigh's criterion, it implies that at least 3 years of T/P observations are necessary to separate M_2 and S_2, 1.5 years to separate N_2 from O_1, and 9 years to separate K_2 from P_1, and K_1 from the semi-annual S_{sa}. Note also that, unfortunately, it is K_1, Φ_1, and Ψ_1 that are of geophysical interest for free core nutation resonance studies.

ERS has more problematic alias periods for tidal mapping. Because of its sun-synchronous orbit, S_2 is always observed with the same phase, so that it is removed as part of the stationary sea surface topography. Also, K_1 and P_1 have aliased periods at exactly 1 year, so that they are not separable from the annual oceanic signal.

For Geosat, apart from P_1, which has an 11-year alias period, all constituents alias to periods smaller than a year. However several main constituents alias to about half a year (K_1 and S_2) or a year (M_2). Noteworthy is that K_1 aliases to 175 days. If referring to the Rayleigh criterion, more than 12 years are needed for a full separation of K_1 from the semi-annual cycle, which is worse than for T/P.

Ascending and descending ground tracks provide additional phase information, which may aid in the estimation of ocean tides. Schrama and Ray (1994) have carefully studied this question. Although the time intervals between intersecting tracks are a complicated function of latitude, they have tabulated for T/P the tidal phase advance between ascending and descending tracks at crossover points for the eight major tidal constituents (see their Table 2). They showed that all tides have one or more latitude bands where the use of intersecting tracks add little information. For the solar P_1 and K_1, this band is in the highest latitudes (at 66°N and S), and for S_2 and K_2, both in the maximum and minimum latitudes. But elsewhere, throughout most of the globe, these intersecting tracks help to solve the aliasing problems.

Also, with some sacrifice in the spatial resolution, advantage can be taken of the fact that the tidal phase can change significantly at the neighboring track: T/P passes over a point

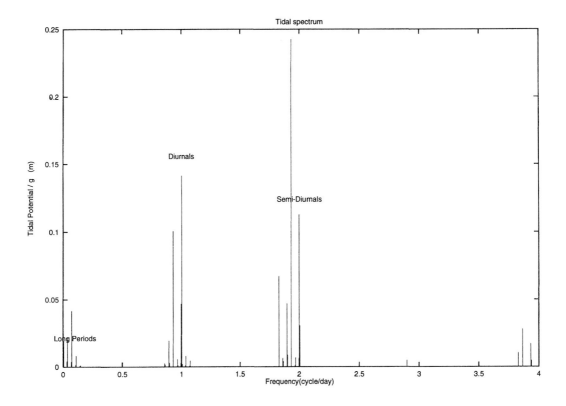

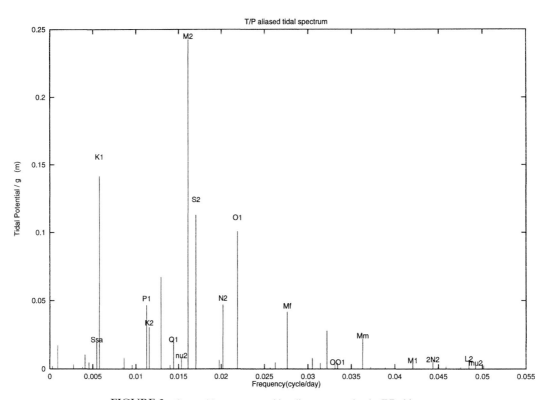

FIGURE 2   Ocean tide spectrum and its alias spectrum by the T/P altimeter.

TABLE 2. List of Models Considered by the T/P SWT Tide Subgroup

Model	Origin (year)	Altimetry data included	Number of constituents	Resolution
SCH80	Schwiderski (1980)	No	11	$1° \times 1°$
CR91	Cartwright-Ray (1991)	Geosat	60^1	$1° \times 1.5°$
FES94.1	Le Provost et al. (1994)	No	$8 + 5^2$	$0.5° \times 0.5°{}^3$
RSC94	Ray-Sanchez-Cartwright (1994)	T/P	60^1	$1° \times 1°$
GSFC94A	Pavlis-Sanchez (1994)	T/P	8	$2° \times 2°$
AG95.1	Andersen (1995)	T/P	$2 + 6^4 + 5^2$	$0.5° \times 0.5°$
CSR3.0	Eanes (1995)	T/P	60^1	$0.5° \times 0.5°$
DW95.0/.1	Desai-Wahr (1995)	T/P	364	$1° \times 1°$
FES95.1/.2	Le Provost et al. (1995)	T/P	$8 + 18^2$	$0.5° \times 0.5°{}^3$
Kantha.1/.2	Kantha (1995)	T/P	60^1	$0.2° \times 0.2°$
ORI	Matsumoto et al. (1995)	T/P	$8 + 8^2$	$1° \times 1°$
SR95.0/.1	Schrama-Ray (1994/1995)	T/P	$4/5 + 2^4 + 16^2$	$1° \times 1°$
TPX0.2	Egbert et al. (1994)	T/P	$8 + 9^3$	$0.58° \times 0.7°$

[1] Number of constituents included in the tide generating potential for the orthotide formulation.
[2] Additional constituents induced by admittance.
[3] The full resolution of this solution is the one of the finite Element grid of Le Provost et al., 1994.
[4] Constituents are adopted from Le Provost et al., 1994.
From Shum, C. K. et al., (1997). With permission.

at $360°/127 = 2.835°$ (to the east of a given track) 38 orbital revolutions later. By choosing to estimate tides in bins of a given size, typically around $3°$, a little larger than the T/P longitudinal sampling, it allows to combine information at several crossover points, and thus the problems of aliasing and closeness of some of the aliased frequencies can be considerably reduced.

4.2. Methods for Estimating Ocean Tides from Satellite Altimetry

The history of satellite altimetry started in 1973 when the Skylab platform flew the first altimeter. Then came the three missions Geos3 (1975–1978), Seasat (1978), and Geosat (1985–1989). They brought significant improvements in the precision of the altimetric measurements (from 1 m for the Skylab altimeter to 4 cm for Geosat), and in the orbit determination (from 5 m in 1973, to 0.5 m at the beginning of the 1990s). The evidence of a tidal signal in altimeter data was first showed from Seasat (Le Provost, 1983; Cartwright and Alcock, 1983). Mazzega (1985) demonstrated that it was indeed feasible to extract tides at the global scale from the Seasat data set, even though this mission ended prematurely. His approach was through a spherical harmonic representation and was limited to the M_2 tide. Geosat provided the first altimetric data set for extended global tide studies, enabling the derivation of models of practical utility for oceanography and tidal science. The Exact Repeat Mission (ERM) of Geosat from 1986–1989, with its 17-day repeat cycle, pro-

vided 2.5 years of altimetry data, which were extensively analyzed by Cartwright and Ray (1990). Their approach was based on binning the data into grid boxes of $1°$ by $1.4754°$ (the Geosat ground track spacing at the equator) and on the analysis of this data set through a response method based on the orthotide representation [see Eq. (6) with $L = 6$ for the semi-diurnal and diurnal admittance functions]. They produced a new set of solutions for the eight major constituents (M_2, S_2, N_2, K_2, K_1, O_1, P_1, Q_1). Molines et al. (1994) provided an analysis indicating that this model was more accurate than the Schwiderski model, considered as the best one available at the time of the launch of T/P.

But it is since the launch of ERS1 in 1991, and most of all T/P in 1992, that an impressive effort started to develop new ocean tides models. By mid-1995, after a little more than 2 years of T/P data, 10 new global ocean tide models were made available to the international scientific community (see Table 2). The methods developed to produce these models can be classified in four groups:

1. Direct analysis of the altimeter data
2. Direct analysis of the altimeter residuals after a preliminary first-guess correction of the data relying on an a priori tidal model
3. Analysis of the altimeter data or their residuals (as in 2-) but after an expansion of the tidal solutions in term of physical modes (typically Proudman functions)
4. The use of inverse methods which incorporate hydrodynamic equation resolution constrained by altimetric data assimilation

Some of these methods are directly based on the harmonic representation of the tides, and others primarily rely on the orthotide formulation. Each method has its advantages and drawbacks. In the following, we will classify the 10 models listed in Table 2 into these four categories and give the major characteristics of the versions available by 1995, when they have been compared by Shum *et al.* (1997).

4.2.1. Direct Analysis of the Altimeter Data

The Desai and Wahr, version 95.0 (DW95.0), is an empirical solution. Their analysis relies on the orthotide formulation for the semidiurnal and diurnal tidal bands. They used Eq. (6) with $L = 6$, like Cartwright and Ray (1990), with 161 and 116 tidal components in each respective band, and additional constant admittance functions in the monthly (M_m), fortnightly (M_f), and ter-mensual (M_t) bands including, respectively 22, 25, and 40 tidal components. This model can be considered as the most empirical one: no reference to any a priori tidal model and no direct or indirect information from the dynamics of tides. It is also among the ones which have the finer initial resolution. It is based on T/P data binned in boxes of 2.8347° in longitude by 1° in latitude, the minimum to ensure that observations from at least one ascending and one descending ground track are included in the tide estimates.

4.2.2. Direct Analysis of the Altimeter Residuals after Correction from an A Priori Tidal Model

Four of the ten models have used an a priori tidal solution coming from previous studies: these are the ones of Andersen (AG95.1), of Schrama and Ray (SR95.0/.1), of Eanes and Bettadpur (CSR3.0), and of Sanchez and Pavlis (GSFC94A). One major interest of this approach is that it takes benefit in the final solutions from the short wavelength structures of the models used as a priori. The first three studies are based on direct harmonic or response methods; only the fourth one used, in addition, an expansion of the orthoweights in term of Proudman functions.

The Andersen-Grenoble, version 95.1 model (AG95.1) is a long-wavelength adjustment to the FES94.1 hydrodynamic solution (Le Provost *et al.*, 1994), for the M_2 and S_2 constituents, using the first 2 years of T/P crossover data (70 cycles). These corrections are estimated using an orthotide approach and interpolated adjustments onto regular grids using collocation with a half width of 3500 km. The final solutions for M_2 and S_2 are given on a 0.5° × 0.5° grid within the latitude range 65°N to 65°S. Outside of these limits, the solutions are the same as other major constituents of FES94.1.

The details of the data processing for the Schrama-Ray solution version 95.0/.1 (SR95.0/.1 model) have been described by Schrama and Ray (1994). In this preliminary paper, they developed solutions as corrections to the Schwiderski and the Cartwright and Ray models. They followed a simple harmonic analysis on altimetric residuals binned in clusters of 3° radius assuring at least two ascending and two descending repeating tracks for each point of analysis (repeated on a grid of 1° × 1°). The final version was computed as a correction to the Finite Element purely hydrodynamic model FES94.1 of Le Provost *et al.* (1994), with T/P altimetric data from cycle 9–71. Only five constituents were solved: M_2, S_2, N_2, K_1, and O_1. The Q_1 and K_2 constituents were adopted directly from FES94.1 and 16 minor constituents were added in the prediction code, by linear response inference.

The Centre for Space Research, version 3.0 model (CSR3.0) of Eanes and Bettadpur (1996) is also a long-wavelength adjustment to the a priori solution of FES94.1. First, diurnal orthoweights were fitted to the Q_1, O_1, P_1, and K_1 constituents of FES94.1 and semidiurnal orthoweights to the N_2, M_2, S_2, and K_2 constituents of AG95.1. Then 89 cycles of T/P altimetry were used to solve for corrections to these orthoweights in 3° × 3° bins. These corrections were then smoothed by convolution with a two-dimensional gaussian for which the full-width-half-maximum was 7°. The smoothed orthoweight corrections were finally output on the standard 0.5° × 0.5° grid of the FES94.1 gridded solution, and combined with them to obtain the new model over the global world ocean domain.

4.2.3. Analysis of the Altimeter Data or Residuals through an Expansion in Terms of Physical Modes

We have already noticed that the first global ocean tide model has been produced by Mazzega (1985) who succeeded in extracting an M_2 solution from the short Seasat data set by using spherical harmonics developments. Two recent studies have produced valuable complete tidal solutions with T/P data, following a more physical approach based on Proudman functions, which form a natural orthogonal basis for the dynamic LTE equations.

The **RSC94** model has been produced by Ray, Sanchez, and Cartwright: the method has been presented in only an abstract (RSC 1994). It is derived from the response approach, with the response weights expressed by expansions in Proudman functions (up to a maximum of 700). These Proudman functions were computed on a 1° grid, but over an ocean limited to 68°N and S, excluding several marginal seas such as the Mediterranean and the Hudson Bay. The method was applied directly on the altimeter data, from cycles 1 to 64. Like for the Desai and Wahr model, this one is totally independent of any previous model. Additionally, the harmonic constants of 20 tide gauges were used in the inversion, mainly located around the Labrador Sea (to supply the lack of altimeter data due to ice cover) and the North Sea.

The Goddard Space Flight Center (GSFC94A) model of Sanchez and Pavlis (1995) is based on corrections to the Schwiderski model for eight major constituents (M_2, S_2, N_2, K_2, K_1, O_1, P_1, Q_1). The residuals of the first 40 cycles of TOPEX data were analysed in terms of Proudman functions.

The resulting solutions were defined on a $2° \times 2°$ grid, from 76,75°S to 69,25°N.

4.2.4. The Use of Inverse Methods Combining Hydrodynamic Equations and Altimetric Data Assimilation

The four last models listed in Table 2 are based on hydrodynamic modeling and data assimilation. Two of them are similar in their approach to the "hydrodynamic interpolation" method of Schwiderski. The two others are based on the more sophisticated "representer" method initially formulated by Bennett and McIntosh (1982) (see also Egbert and Bennett, 1996).

Kantha.1/2: These models are based on the fully nonlinear barotropic shallow water equations (Kantha, 1995). The computational grid is $0,2° \times 0,2°$, expected to provide more accurate tides in coastal oceans and marginal seas. For the assimilation step, Kantha used the two versions of Desai and Wahr (95.0/.1), plus coastal tide gauges. These models are extended to 80°S to cover the Antarctic Ocean.

ORI: This model was developed by the Ocean Research Institute at the University of Tokyo by Matsumoto et al. (1995). The hydrodynamic model has a resolution of $1° \times 1°$. The data assimilated are obtained from a harmonic analysis of the first 77 TOPEX cycles data at the crossover points. The eight major constituents are directly computed, while eight additional terms are deduced by admittance in the semi-diurnal and diurnal bands.

TPX0.2: This model corresponds to a best fit of the dynamics of the shallow water equations and of the crossover data of the first 38 cycles of T/P (Egbert et al., 1994). The method is based on the representer approach, which requires solving the hydrodynamic equations and their adjoint system once for each representer, i.e., each point where one datum is assimilated. This is a tremendous computation, which leads Egbert et al. to linearize their hydrodynamic equations and consider the altimetric data to be assimilated only at the crossover points. The resolution used was a $0,7° \times 0,7°$ grid. The eight major constituents were computed, with nine minor constituents included through admittance interpolation.

FES95.1/.2: The Finite Element Solutions versions 95 are issued from the hydrodynamic solution FES94.1, improved through an assimilation of the a priori altimeter derived solution CSR2.0 of Eanes and Bettadpur. The computational grid of the hydrodynamic model and its adjoint is the Finite Element grid developed by Le Provost et al. (1994), with, as said above, grid sizes of the order of 200 km in the deep ocean, but refined down to 10 km along the coasts. The assimilation method is the one presented by Egbert et al. (1994) and Lyard (1998). The fact that this model is formulated in the time-frequency domain considerably reduces the computational constraints. The data subset assimilated is also limited to a $5° \times 5°$ gridding of the altimeter database. In FES95.1, only the main two constituents M_2 and S_2 are

adjusted by assimilation, and the 11 other constituents are kept from the hydrodynamic solution FES94.1. The version FES95.2 includes on one side three other components adjusted by assimilation (N_2, K_1, and O_1), and the spectrum is extended to 26 constituents by admittance (Le Provost et al., 1998).

5. THE SEMI-DIURNAL AND DIURNAL TIDES OVER THE DEEP OCEAN

The different models made available by the mid-1990s, after the extensive exploitation of the high-precision altimetric data issued from ERS-1 and T/P, have reached a high level of accuracy. To characterize their respective qualities and defaults, a very careful analysis is necessary. In this chapter, we will first briefly describe the characteristics of the ocean tides, as they are seen now, restricting this description to the major components in each species: the semidiurnal components M_2 and S_2, and the diurnal components K_1 and O_1. We will then review the accuracy assessment provided by Shum et al. (1997) for the 10 models that we have briefly introduced in the preceding section. This accuracy assessment is based on a comparison exercise conducted by the Tide Subgroup of the T/P Science Working Team in 1995. We will briefly review the conclusions of the different evaluation tests applied: (1) comparison to pelagic and island gauge in situ data, (2) crossover residual analysis, (3) T/P and tide gauge time-series analysis, and (4) comparison to gravity-loading measurements.

5.1. Characteristics of the Ocean Tides at the Global Scale

5.1.1. The Semidiurnal M_2 and S_2 Tides

Figure 3a shows the global cotidal map for the M_2 constituent. As said before, the different models introduced above propose very similar solutions. In this section, we have chosen to display the FES95.2 solutions. Qualitatively, the patterns look even, very like the solution of Schwiderski (1980), the one of Cartwright and Ray (1990), and the one of Le Provost et al. (1994), hereafter denoted SCH80, CR91, and FES94.1. All the well-known amphidromic systems are distributed over the World Ocean in almost the same positions. It is really difficult for a non-expert to note the differences. The major differences are noticeable in the southern part of the Pacific Ocean, where the amphidromic system north of the Ross Sea is centered at 210°E in FES95.2 whereas that of SCH80 is located at 165°E, on the western side of the Ross Sea. This position is in better agreement with the ones of CR91 and FES94.1.

Figure 3b shows the global cotidal map for the S_2 constituent. As expected from the credo of smoothness in the

a

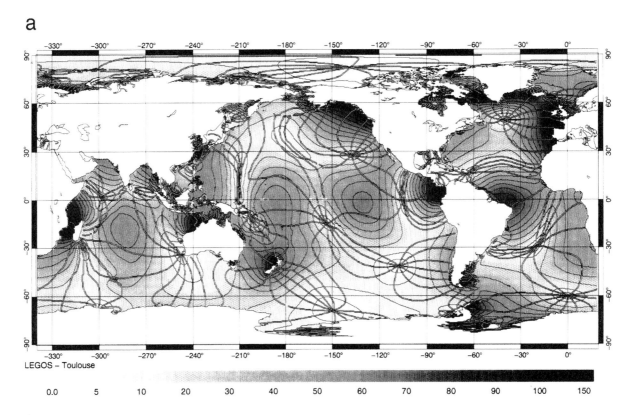

LEGOS – Toulouse

| 0.0 | 5 | 10 | 20 | 30 | 40 | 50 | 60 | 70 | 80 | 90 | 100 | 150 |

b

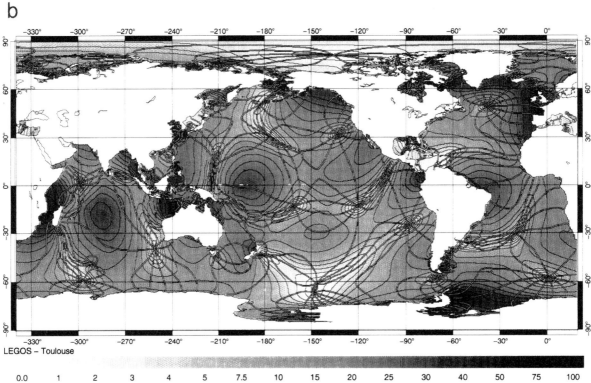

LEGOS – Toulouse

| 0.0 | 1 | 2 | 3 | 4 | 5 | 7.5 | 10 | 15 | 20 | 25 | 30 | 40 | 50 | 75 | 100 |

FIGURE 3   (a) Cotidal maps of the lunar semidiurnal component M_2. Co-amplitudes are in centimeters, co-phases are drawn with an interval of $30°$. (b) As (a) but for the solar semidiurnal component S_2.

response of the ocean to the astronomical forcing, and although S_2 is subject to a radiational contribution, the global pattern of the main solar semidiurnal component looks similar to the lunar component M_2. However, a significant number of shifts in the position of the amphidromic points are observed, because of the shorter wavelength of S_2. The most remarkable changes occur in the South Pacific Ocean and the Southern Ocean. In the South Pacific Ocean, the number and distribution of the amphidromic points is significantly different, with one new amphidromic point in the Drake Passage, and the appearance of a doublet north of the Ross Sea. Also, systematically, the amphidromic points around Antarctica are shifted northward.

5.1.2. The Diurnal K_1 and O_1 Tides

Figure 4a shows the global cotidal map for the main diurnal K_1 constituent. This component is the third one in terms of intensity at the scale of the World Ocean, although there are regions where it is the first or second, mainly at high latitudes. The solution represented here looks very similar to the solutions of SCH80, CR91, and FES94.1. The prominent characteristics of this wave, and more generally of the diurnal constituents, are the existence of a Kelvin wave propagating westward around Antarctica, coupled with a resonance of the Pacific Ocean (Platzman, 1984; Cartwright and Ray, 1991). Another major characteristic of the diurnal components is the existence of energy-trapping processes along the continental shelves and over the main bathymetric features leading to topographic Rossby waves at latitude larger than the critical latitudes (30° for K_1 and 28° for O_1). Although quasi-nondivergent and thus of only a few centimeters amplitude, these waves induce signatures with typical wavelengths of a few hundred kilometers. These waves are not resolved by coarse-grid hydrodynamic models, like SCH80, TPXO.2, and ORI; they are in the FES models, thanks to the high resolution used in the FE model over the continental shelves. Such features are observable in the North Atlantic, over Rockall Plateau, west of the United Kingdom, and along the continental shelf of Greenland and Canada. Similar features are present in all the altimetric models that have adopted FES94.1 as a priori solution (SR95.1, AG95.1, CSR3.0, FES95.2), but they are there as heritage of the hydrodynamic model. Validation of these small-scale structures is not easy, although some have been observed and intensively studied, like over the Rockall Plateau (Pingree and Griffiths, 1984; Kowalik, 1994). Figure 4b shows the cotidal map of the second most important diurnal O_1 constituent. This component is of particular interest, because of the resonant behavior of the diurnal waves around Antarctica. Cartwright and Ray (1991, Figure 6) noted that this resonance peaks at frequencies lower than O_1 and K_1. The solution here presented confirms effectively the bigger amplification of the O_1 constituent south of 60°S.

5.2. Coherency and Accuracy Assessment

5.2.1. Coherency and Differences

An analysis was conducted by Shum *et al.* (1997) on eight of the models listed in Table 3 (GSFC94A and Kantha.1 were not available at the time of the study). The standard deviation between the different models for the semi-diurnal M_2 and the diurnal K_1 constituents were computed. The conclusion is that these solutions are within less than 1 cm difference for the deep ocean (>1000 m) for both components. By contrast, the disagreement is quite large in shallow waters. The standard deviation is 9.8 cm for M_2. Amplitudes of the tidal waves are indeed much bigger there, which partly explains these larger differences. But also tidal scales are shorter than in the deep ocean: it is then more difficult there to extract accurate information from altimetric data, especially because of the binning procedures. Note that in this analysis, the standard deviation for M_2 reduced to 7.9 cm if differences larger than 50 cm were ignored. This is explained by the fact that the FES95.1 solution included in this comparison contains spurious and totally abnormal high values over some coastal areas (the Bass Straits in Australia for example). While it is expected that the Finite Element solutions are better over shallow water areas, as indeed it is the case for the hydrodynamic solution FES94.1, it is not true for the assimilated solution FES95.1. This is because of uncontrolled resonance in the computation of some of the representers involved in the assimilation procedure.

The standard deviation is 0.7 cm for K_1, although this component is much smaller than M_2, by a factor of three in amplitude. This is mostly because the preliminary version FES95.1 was used for these evaluations. This version, by contrast to the FES95.2 version, does not include adjustment (assimilation) of the diurnal constituents, but instead retains the purely hydrodynamic solution which has large scale discrepancies in the southern part of the Indian Ocean, as noted above (Le Provost *et al.*, 1995).

5.2.2. Comparison to In Situ Data

One major test allowing to check the accuracy of the tidal models is to refer to the set (ST102) of *in situ* data carefully assembled from pelagic and island data by Cartwright and Ray (1991) and Le Provost (1994). The location of these stations is shown on Figure 5. It must be reminded that among these 102 sites, 53 are from bottom pressure recorders, but atmospheric tides have been removed from the data by using an analytical model from Harwitz and Cooley (1973) following Cartwright and Ray (1994). Differences in rms between tidal constants from *in situ* data and altimeter-derived solutions and global RSS of these differences were computed by the T/P SWT tidal group. These numbers are given in Table 3. The best score is obtained for the Schrama and Ray SR95.1 solution (RSS of 2.53 cm), closely followed by CSR3.0 (2.61 cm), FES95.2 (2.65 cm),

a

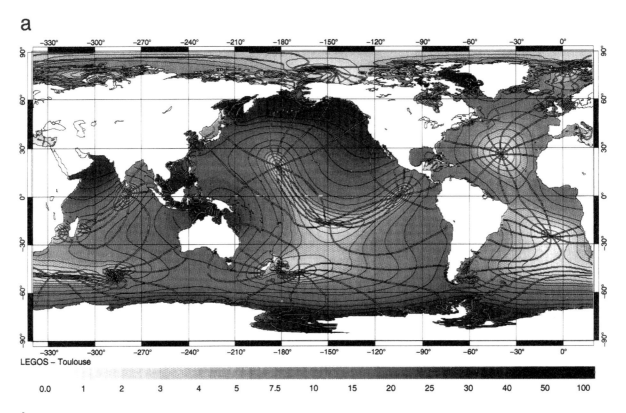

LEGOS – Toulouse

0.0   1   2   3   4   5   7.5   10   15   20   25   30   40   50   100

b

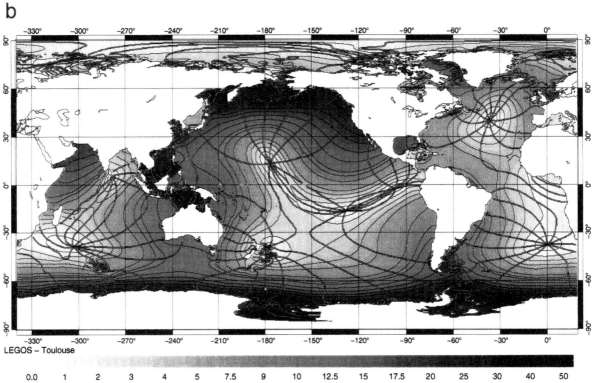

LEGOS – Toulouse

0.0   1   2   3   4   5   7.5   9   10   12.5   15   17.5   20   25   30   40   50

FIGURE 4   (a) Cotidal maps of the luni-solar diurnal component K_1. Co-amplitudes are in centimeters, co-phases are drawn with an interval of $30°$. (b) As (a) but for the lunar diurnal component O_1.

TABLE 3. rms Difference Between Several Global Model Results and the 102 Tide Gauge Data Set (cm)

Model	M_2	S_2	N_2	K_2	K_1	O_1	Q_1	RSS
SCH80	3.86	1.66	1.22	0.59	1.43	1.22	0.35	4.84
CR91	3.20	2.20	0.98	0.67	1.89	1.23	0.46	4.59
FES94.1	2.80	1.59	0.83	0.48	1.22	1.04	0.29	2.73
RSC94	1.89	1.18	0.78	0.49	1.26	0.99	0.37	2.94
GSFC94A	2.18	1.21	0.87	0.63	1.41	1.06	0.35	3.29
AG95.1	1.64	1.05	0.83	0.48	1.22	1.04	0.29	2.75
CSR3.0	1.64	1.01	0.67	0.52	1.12	0.95	0.30	2.61
DW95.1	1.85	1.07	0.70	0.54	1.23	0.96	0.33	2.77
FES95.2	1.65	0.98	0.74	0.48	1.15	1.00	0.29	2.65
Kantha.2	2.08	1.13	0.64	0.56	1.40	1.04	0.57	3.16
ORI	1.93	1.38	0.87	0.90	1.34	1.02	0.42	3.25
SR95.1	1.55	0.99	0.70	0.48	1.04	0.96	0.29	2.53
TPX0.2	2.16	1.19	0.75	0.55	1.31	0.98	0.29	3.14

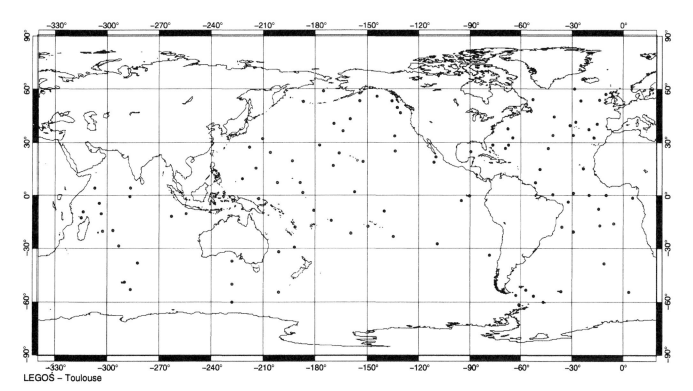

LEGOS – Toulouse

FIGURE 5   Location of the 103 tide gauges (ST103) from the standard sea truth data set agreed by the T/P Scientific Working Team for comparison of the tidal solutions to observations. This data set include island and pelagic sites.

and AG95.1 (2.75 cm). Compared to the RSS of Schwiderski (4.84 cm) and Cartwright and Ray (4.59 cm), the gain is of more than 2 cm in RSS, which is very significant. Compared to FES94.1 (3.80 cm), the gain is a little less, but still of the order of 1.2 cm. A detailed analysis of the improvements on the different tidal constituents shows that the main gain is on the major component M_2 (more than 2 cm rms between the altimetric solutions and SCH80). Also it must be noted

that the score of SCH80 is probably not representative of the level of accuracy of the solution at the global scale, because several of the data included in the ST102 set were used by Schwiderski in his "hydrodynamic interpolation."

5.2.3. Crossover Residual Analysis

Another major test considered by the T/P SWT tidal group has been to examine the variance reduction of the

T/P altimeter signal at crossover points resulting from the tidal corrections issued from each of the models. This reduction of the variance after tidal correction is a measure of the ability of the model used to remove the tidal signal. The crossover measurements were edited for values larger than 60 cm and crossover time difference exceeding 3.5 days to minimize aliasing of ocean signal. The analysis has been carried for the three last altimetric missions: over 4 cycles of Geosat (68 days), 4 cycles of ERS1 (140 days) and 13 cycles of T/P (130 days). The crossover residual statistics have been computed for the global ocean, but also separately in the deep ocean and over the shallow areas (respectively, deeper or shallower than 800 m). These statistics are given in Table 4, adapted from Li et al. (1996). Models are ordered following the lowest crossover residuals obtained on the T/P data set. Although the cycles considered in this study are not the ones used by the different groups to build their solutions, the test can be somewhat biased for T/P, because these models are in one way or another best fitted along the T/P tracks. Nevertheless, the order of efficiency of the tidal models, in terms of residual variance reduction, is almost the same for the three missions. These residuals decrease in amplitude in agreement with the increased accuracy of the missions: 9.7 cm rms crossover residuals at best for Geosat, 8.6 cm for ERS1, 5.8 cm for T/P. The differences in rms residuals between the models are very small: at the millimeter level for T/P. Molines et al. (1994) suggested using the statistical test of Fischer-Snedecor to estimate whether or not the differences are significant. Following this test, it appears that Geosat does not discriminate between four models: CSR3.0, RSC94, SR95.1, and FES95.2. For ERS, it is even less clear; TPX0.2 is not significantly different from the four cited above. By contrast, T/P allows some selection, with two models above the others, with regard to this test: CSR3.0 and SR95.1.

Another indication given by this study is the much higher residual variance in the shallow waters: these altimetric models have effectively been developed primarily for open ocean studies. We will present later some on-going studies seeking new improvement of these models over shelves and coastal seas.

5.2.4. Altimeter Signals and Tide Gauge Time-Series Analysis

Another study which contributed to assess the accuracy of these models analyzed the consistency of altimetric sea surface height data after removal of ocean tides from the measured heights (King et al., 1995). The idea was that a perfect tidal model will allow removal of the whole tidal energy and lead to smooth spectra power distribution where energy on tidal energy aliasing frequencies falls into the overall continuum background. Once again, two models emerge from this test: CSR30 and SR95.1. The second model leads to the smoothest results at the M_2 and S_2 aliased frequencies. By contrast, this approach clearly illustrated the presence of residual energy in tidal bands like M_2 and K_1 for the models not well-ranked in the preceding tests. It also convincingly demonstrated the interest in including in the prediction software very minor constituents generally ignored in the usual models. This led Schrama and Ray (1994) to add 16 minor tides in their model, and Le Provost et al. (1998) to include 13 minor constituents to their usual 13 major ones they considered in FES94.1. As noted by Shum et al. (1997) "that T/P can so clearly detect these often neglected tides is a testament to its unprecedented accuracies."

Comparisons to tide gauge time series allows characterization of how close these altimetric prediction models are to the usual predictions based on in situ analysis. The test was based on 69 tide gauge time series at islands and pelagic locations distributed around the world and provided by the World Ocean Circulation Experiment and Tropical Atlantic

TABLE 4.   Crossover Residuals Statistics with Geosat, ERS1, and T/P Data, for Different Tidal Models

Tide models	Geosat			ERS1			T/P		
	Global	Deep	Shallow	Global	Deep	Shallow	Global	Deep	Shallow
CSR3.0	9.74	9.73	15.17	8.58	8.52	11.55	5.82	5.74	10.05
SR95.1	9.77	9.75	15.14	8.67	8.63	11.35	5.91	5.83	10.16
DW95.1	9.88	9.89	14.87	8.65	8.64	11.54	5.98	5.87	11.78
RSC94	9.75	9.75	13.49	8.65	8.64	10.84	6.06	5.98	10.78
FES95.2	9.80	9.82	14.69	8.69	8.72	11.10	6.12	6.02	11.51
TPX0.2	9.94	9.92	13.94	8.69	8.68	11.35	6.17	6.07	10.72
Kantha.2	9.86	9.87	14.00	8.73	8.69	11.72	6.28	6.21	11.30
AG95.1	10.04	10.03	15.87	8.81	8.77	10.59	6.58	6.49	11.80
ORI	10.20	10.18	16.31	8.94	8.91	14.70	6.65	6.52	13.05
GSFC94	10.25	10.22	14.96	9.28	9.21	13.31	7.07	6.95	13.11

Global Atmosphere programs. Predictions of the tides at these 69 sites with the traditional harmonic method (using harmonic constants obtained from an harmonic analysis of the same data set) leads to an rms residual variance of 2.6 cm in the high frequency. The use, instead, of the different altimetric models gave at best rms residuals of 3.5 cm for CSR3.0, 3.7 cm for SR95.1, and 3.9 cm for FES95.2, i.e., typically 1 cm more than the best possible predictions.

5.2.5. Comparison to Gravity Loading Measurements

Five models have been studied in detail from the point of view of their implied loading tide effects by Melchior and Francis (1996). These are SCH80, AG95.1, CSR3.0, FES95.1, and ORI. SR95.0 was also considered, but its lack of global coverage precluded any conclusion. A set of 286 gravimetric stations were used for this test, taken from the International Centre for Earth Tides (ICET) data bank (Melchior, 1994). At each station, earth-loading tide parameters were determined for the major tidal gravity constituents by subtracting the body tide from the station observations, and comparing them to predictions of loading from each ocean tide model. This study demonstrates a significant improvement in the quality of the altimetric models, compared to SCH80, with an overall gain in the standard deviation of 0.05 mGal (0.65 mGal for the altimetric models against 0.7 mGal for SCH80). And the conclusion was that AG95.1 allows the best agreement, with very good results for FES95.1 and CSR3.0.

A similar comparison presented by Llubes and Mazzega (1997) has extended this test to all the recent tidal models (up to fifteen). Some models have been completed in the high latitudes, in some marginal seas, and near the coasts by using the FES94.1 model. Stations closer than 10 km from the coasts, and those with residual vectors larger than three times the standard deviation were also eliminated from the study. The conclusion of this study is that the best score for the M_2 constituent is obtained with CSR3.0, with a standard deviation reduced to 0.425 mGal by reference to a subset of 220 stations.

5.2.6. Conclusions on the Accuracy Tests and Possible Improvements

The synthesis of these accuracy assessment investigations leads us to conclude that, although all these models are now very similar, some of them seem to better fit reality: SR95.1, CSR3.0, and FES95.2 according to test 1, CSR3.0, SR95.1, and DW95.1 for test 2, CSR3.0, SR95.1, and FES95.2 for test 3, and CSR3.0 for test 4. The overall level of accuracy of the tidal predictions obtained from these models must be of the order of 3.5 cm rms (estimated over periods of typically a year's duration).

Differences remain at the centimeter level. Vector differences (i.e., maximum possible differences) between some of these solutions are presented for M_2 and K_1 on Figures 6

and 7. The misfits between the two "best solutions" CSR3.0 and SR95.1, illustrated in Figure 6a, let one think at first that, except over the shallow water areas, these solutions give the truth, as they differ by less than 1 cm. In fact, they are both issued from similar analysis (method 2 commented in Section 4.2), i.e., direct analysis of binned residuals after removing a first guess of the ocean tide contribution predicted on the basis of the hydrodynamic model FES94.1. And they contain similar systematic errors, which are revealed in Figure 6b showing the vector differences between SR95 and DW95, a purely altimetric solution. Some of the areas of noticeable misfits are certainly caused by DW95. But some features are clearly identified as errors in both SR95 and CSR3.0. A band of difference ranging up to 2 cm is visible SE of Hawaii Ridge: this signal is a "bad souvenir" of the FES94.1 solution related to inter-basin boundary condition problems (FES94.1 is continuous along these boundaries, but not its normal derivative). Other similar signatures are visible between Brazil and Africa, and south of South Africa. These bands are also visible in Figure 6c presenting the differences between SR95.1 and FES95.2: for this new version of the FES model, the boundary problem has been resolved, so that the signatures observed in Figure 6c probably result from inaccuracies in SR95.1 there.

Similar pictures are given on Figure 7 for K_1. The major differences are located at the high latitudes, ranging up to 3 cm around Antarctica. This is not surprising when we remember that this component is very difficult to separate from the semi-annual signal at high latitudes (Schrama and Ray, 1994).

Although these differences are very small, within the accuracy limits of both the altimetric measurements, and the range of variability of harmonic tidal constituents (caused by meteorological effects, internal tides, . . .), one might ask if further improvements could be still expected. Two related questions can be addressed: (1) what can be the gain of the increasing length of altimetric time series? and (2) what can be obtained from analysis of combined ERS1, T/P and Geosat data sets? Desai et al. (1997) have investigated the impact of a continuous update of their empirical ocean tide models along with the increase of the T/P data set. They used a subset of 86 stations (selected from the ST102 standard tide gauge data set) to evaluate the effect of including additional T/P data into their computations, i.e., on seven estimates obtained with data from cycles 10–50, 10–60, up to 10–110. They have concluded that their semidiurnal and diurnal solutions have converged, except for the smaller components such as N2, where small residual variance reduction are more easily distinguishable, and for the long period tides which will be introduced later.

In the same study, they proposed an error budget for the T/P empirical ocean tide models presented above. They have evaluated the individual contribution of the errors in the altimeter range corrections, orbit errors caused by the back-

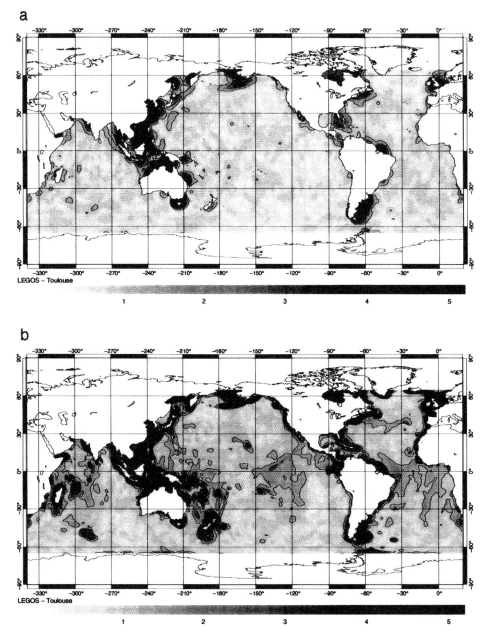

FIGURE 6 Vector amplitude difference between (a) SR95.1 and CSR3.0, (b) SR95.1 and DW95.0, (c) SR95.1 and FES95.2, and (d) DW95.0 and FES95.2 for the M_2 solutions. Units are in centimeters.

ground ocean tide potential, and errors caused by the ocean general circulation, to the errors in the eight principal diurnal and semidiurnal components and the two principal long-period components. Their conclusion is that the dominant source of errors appears to be caused by the general ocean circulation variability. To examine how the general ocean circulation of the ocean may be aliasing into the ocean tides observed by T/P, they used a simulation provided by the Los Alamos National Laboratory (Smith *et al.*, 1992). Such simulations have been shown by Fu and Smith (1996) to produce a fairly good geographical distribution of eddy energy,

although weaker than observed by T/P by a factor of 2. The major errors are in K_1 and S_2, with rms amplitudes of 4 and 2.4 mm (possibly underestimated by a factor of 2). They provided maps of these error distributions for M_2, S_2, K_1, M_f, M_m (see their plate 3). It is indeed interesting to compare these maps with Figures 6b and 7b of this section, related to the misfits between DW95 and SR95. The disagreements between the two models are very coherent with these error maps induced by the ocean circulation variability, in particular for the major differences in the South Indian Ocean and the South Pacific Ocean. Ocean circulation variability might

c

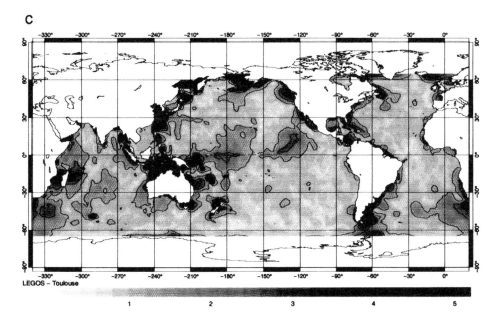

LEGOS – Toulouse

d

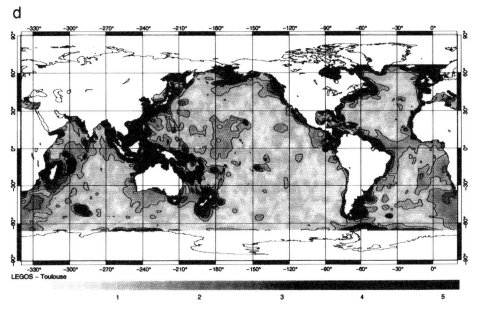

LEGOS – Toulouse

FIGURE 6 (Continued).

be partly responsible of the differences observed between the different new models. Consequently, Desai *et al.* conclude that a gain will result from additional data by reducing the impact of the ocean variability on the aliased tidal signals.

Another possibility to further improve tide solutions at the global scale over the deep ocean areas could be to combine observations from several altimeter satellites. Andersen and Knudsen (1997) have addressed the question. They have focused their study on the diurnal K_1, which is among the less well-determined components up to now from T/P data, because of the proximity of its alias period to the semi-annual oceanic signal. They have also focused their investigations on the high latitudes, where we have noted in Section 4.1 that the inclusion from crossing T/P tracks does not help separate the semi-annual cycle and K_1. They have shown that, by entering ERS1 and Geosat into the computation, the correlation between K_1 and the semi-annual cycle is lowered significantly, however at the price of a slight increase of the interconstituent correlations compared with T/P alone, because these correlations are larger for both ERS1 and Geosat. They have also investigated the impact of combining these different data sets on the quality of the three main constituent solutions M_2, S_2, and K_2, computed fol-

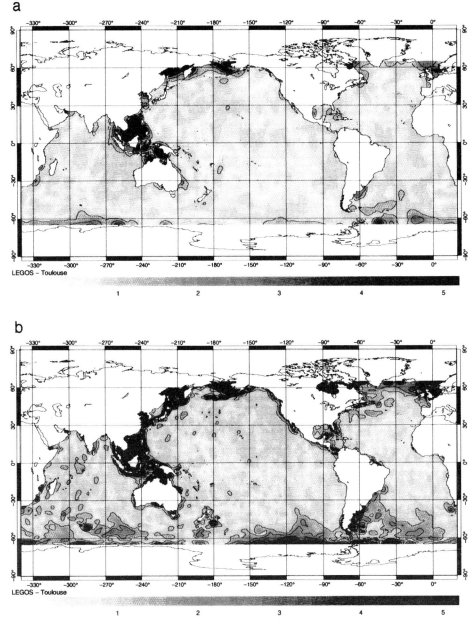

FIGURE 7 Vector amplitude difference between (a) SR95.1 and CSR3.0, (b) SR95.1 and DW95.0, (c) SR95.1 and FES95.2, and (d) DW95.0 and FES95.2 for the K_1 solutions. Units are in centimeters.

lowing either the binning method or the spherical harmonic expansion method, with FES95.1 as a priori (as presented in Section 4.2). They have computed solutions based on T/P only, T/P+ERS1, T/P+Geosat and T/P+ERS1+Geosat. Comparisons of rms to the standard ST102 tide gauge data set used in Shum *et al.* (1997) lead to the conclusion that the inclusion of ERS1 and Geosat data degrades the solutions. This might be caused by the signal-to-noise level of ERS1 and Geosat, and the alias periods (especially S_2 for ERS1). This signal-to-noise level is probably too high

to correct an already very accurate T/P model. However, from their correlation study, they were expecting possible improvement for at least K_1 at high latitudes when including ERS1 and Geosat. They have thus suspected the distribution of the tide gauges in ST102, denser at medium and low latitudes (see Figure 5). Effectively, another comparison with a different sea truth data set of 29 tide gauges, adding new stations (mostly coastal stations) to the ones in ST102 south of 50°S, demonstrate that the combined solutions improve the T/P solutions. Part of these improvement

c

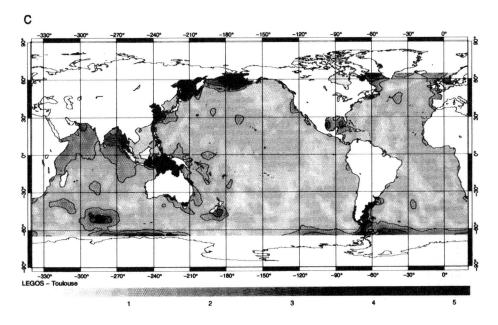

LEGOS - Toulouse

d

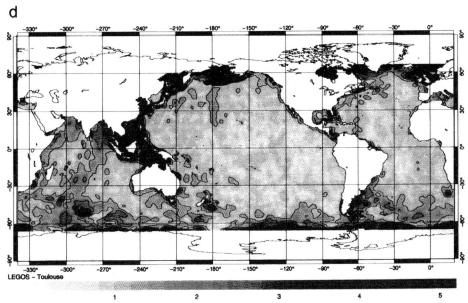

LEGOS - Toulouse

FIGURE 7 (Continued).

certainly comes from data south of 66°S, the limit of the T/P coverage.

6. THE LONG PERIOD OCEAN TIDES

The oceanic response to long-period tidal forcing has be a subject of controversy since more than two centuries after Laplace (1976). A reviews can be found in Wunsch (1967) of the analytical works developed at the end of the nineteenth and the beginning of the twentieth centuries by Darwin (1886), Hough (1897), Poincaré (1910), Proudman (1913) and others. The works of the last 30 years have been reviewed by Miller, Luther, and Hendershott (1993), including Wunsch (1967), Kagan, Rivkind, and Chernayev (1976), Agnew and Farell (1978), Luther (1980), Schwiderski (1982), and Carton (1983). The problem has been recently revisited by Wunsch *et al.* (1997) from the theoretical and modeling point of view, through altimeter data analysis by Ray and Cartwright (1994), Desai and Wahr (1995), and through modeling and data assimilation by Kantha *et al.* (1998) and Lyard *et al.* (1999).

In term of tidal representation, the question of whether long-period tides can be approximated by their equilibrium

a

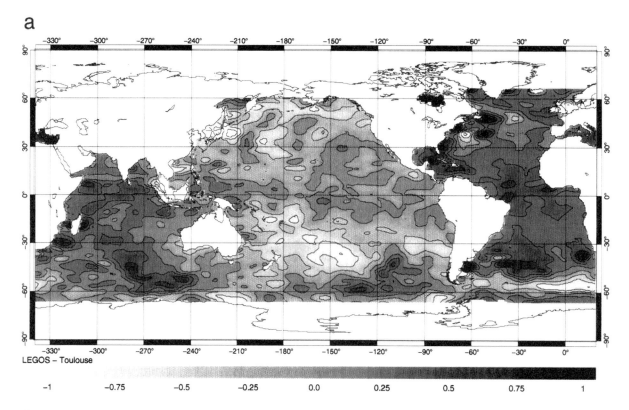

LEGOS – Toulouse

b

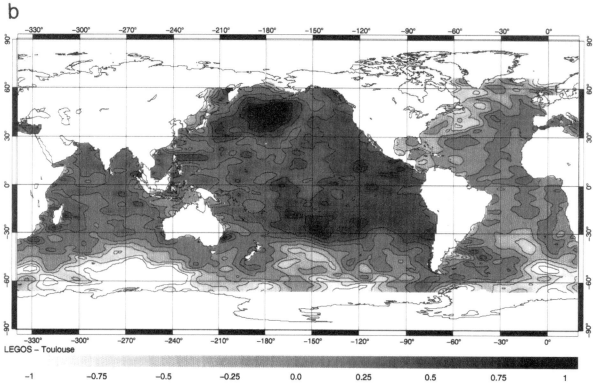

LEGOS – Toulouse

FIGURE 8   Fortnightly M_f ocean tide from T/P model DW95.0 differences with the equilibrium ocean tide. (a) In-phase component, (b) quadrature component.

a

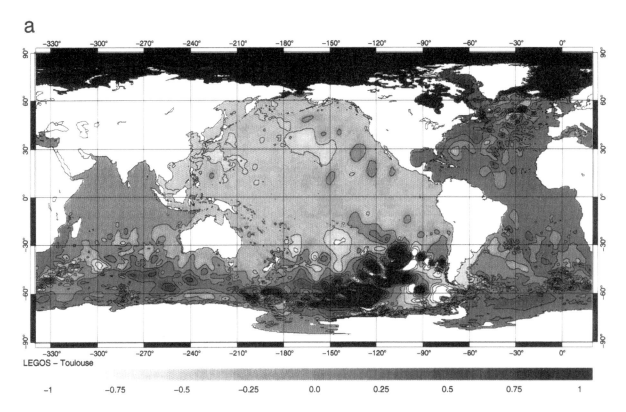

LEGOS – Toulouse

b

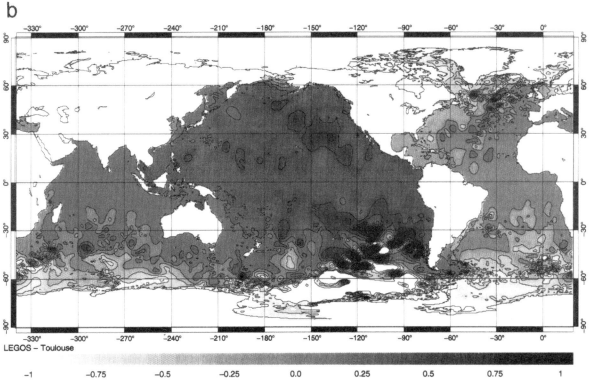

LEGOS – Toulouse

FIGURE 9 Fortnightly M_f ocean tide from FE hydrodynamic model FES98 differences with the equilibrium ocean tide. (a) In-phase component, (b) quadrature component.

solutions has long been an open question. As we will see in this section, the new solutions issued from recent modeling experiments and from the analysis of T/P data clearly demonstrate that the semi-monthly and the monthly components have significant departures from an equilibrium response. However, the full validation of the new proposed solutions still has to be done. From the point of view of general oceanography, the question is also of interest because progress in the understanding of the ocean response to the low-frequency tidal forcing may shed light on the dynamics of the more general motions in the ocean at low frequencies. The problem however is difficult to address because:

(1) The signals are small: the maximum equilibrium amplitudes of the major semi-monthly component M_f (period 13.66 days) are 2.94 cm at the pole and 1.7 cm at the equator; the maximum amplitude is 1.54 cm for the monthly component M_m, and 1.7 cm for the semi-annual component Ssa (182.62 days)

(2) The signal-to-noise ratio in these frequency bands are much lower than for the semidiurnal and diurnal components. The recent analysis by Ponchaut et al. (2000) of the long- and high-quality records collected during the TOGA and WOCE experiments gives an edifying illustration of this contrast. Whereas the noise-to-signal ratio is of the order of 1% for the semidiurnal M_2, it rises to the order of 10% for M_f in the tropics and 50% to higher at mid- and high-latitudes.

Wunsch (1967) carefully analyzed the then-available observations and concluded that there exist significant departures from equilibrium of as much as 30° in phase and 60% in amplitude, for both M_f and M_m. These conclusions agreed with the argument of Proudman (1959) that frictional effects might cause tidal currents at periods longer than 50 days to be damped and then semi-annual and annual tides to be in equilibrium, while fortnightly and monthly tides should have departures from an equilibrium response. On the basis of a quasi-geostrophic flat-bottomed box model at the scale of the Pacific Ocean, Wunsch suggested that the non-equilibrium part of the oceanic response to long-period tide forcing would be dominated by resonant barotropic Rossby waves generated at the lateral boundaries. Kagan et al's numerical solutions of the LTE did not support Wunsch's conclusions. However their 5° coarse grid did not resolve the O (1000 km) Rossby waves involved in the Wunsch's theory. Carton (1983) also shows that the response of long-period tides should approach equilibrium at period longer than the M_m period, and attributed the M_f departure to weak Rossby waves. Schwiderski's fortnightly and monthly solutions showed basin-scale responses, but his "hydrodynamic" interpolation scheme involved many observations of doubtful statistical significance. Ray and Cartwright (1994) extracted from Geosat data meridional distributions of zonally

averaged tidal admittances for M_f and M_m, but they appear smooth and without small-scale features. Miller et al. (1993), using a numerical model, have suggested that gravity waves may also contribute to the departure from equilibrium response, through a Kelvin wave response propagating from a generation region in the Arctic Ocean. More recently, Wunsch et al. (1997), from investigations with a barotropic spectral element model, have effectively concluded that the ocean response to low-frequency forcing has both the character of a large-scale and superposed Rossby wave patterns, thus reconciling the different interpretations. However, from model sensitivity studies, they found that, although the Arctic contributes substantially to the global budget of tidal energy in the M_f band, it does so locally, and thus do not appear to play a predominant remote role in the large scale response of the ocean. It is thus expected that the high quality of the T/P data and the long duration of the mission may help to make further progress on this still open question and to supply solutions very close to reality.

Desai and Wahr (1995) have derived empirical solutions for M_f and M_m using cycles 10–78, and Desai (1996) did the same with 10–110 cycles. Ray (personal communication), Eanes (personal communication), and probably others have also produced similar solutions. There is a rough quantitative agreement between these solutions. The maps of the M_f departure from the self-consistent equilibrium response published by Desai and Wahr (1995) is displayed on Figure 8. The rms of this difference is 5 mm. The major amplitudes, larger than 1 cm, are in the North Pacific Ocean and over the South Indian Ocean, with an opposition in phase in the large scale structure between the Atlantic Ocean and the Pacific Ocean, and between the Pacific and the Southern Ocean. These maps exhibit "trackiness" which are certainly related to the weak signal to noise ratio already noticed in the analysis of the in situ data, and the difficulty to extract the tidal peaks from the continuum. Desai et al. (1997) addressed the convergence of their solutions when extending the data sets from cycles 10–50 to cycles 10–130. They did this evaluation through comparisons with a subset of 14 M_m and 23 M_f tide gauge observations provided by Miller et al. (1993). Note that all these tide gauges are concentrated in Pacific Ocean between 140° and 250°E, and 30°S and 30°N. And they observed that increasing the length of the analyzed data set allows a decrease in the rms differences between the solutions and the observations. With observed variance of the in-phase and quadrature components of 67.2 and 14.7 mm^2 for M_f, the differences dropped to 6 and 4 mm^2, respectively, for analysis of cycles 10–110, with still a gain of 1 mm^2 by including the 10 cycles from 101–110.

Contamination of the solution by the energy in the general ocean circulation is the major problem. Kantha et al. (1998) and Lyard et al. (1999) have explored if a combination of dynamics constraints and data (from altimetry and *in situ*) can improve estimation of these long-period solutions. The modeling and assimilation methods are those applied by Kantha (1995) and Le Provost *et al.* (1998) for modeling the semi-diurnal and diurnal tides presented in the previous sections. In Kantha *et al.* (1998), M_f and M_m solutions are obtained by constraining a $1° \times 1°$ version of their hydrodynamic model with the T/P altimeter-derived solutions of Desai and Wahr (1995), updated to cycle 130 and subsampled to $3°$. Although the final solutions are issued from a very small weighting of the altimeter information (1% for M_f), significant improvements are observed. On the basis of comparison to the data set used by Desai *et al.* (1997), the rms of the free hydrodynamic solution (0.216 cm) is reduced to 0.186 cm, while RMS of the altimetric solution of 0.324 cm.

A quantitative comparison of the different long-period solutions now available, purely empirical from altimetry, or combining hydrodynamic modeling, *in situ*, and altimetric information (see Figure 9), has yet to be made. However, it is now agreed that the semi-monthly tidal waves depart significantly from the equilibrium response, that the monthly components are closer to equilibrium, and that these departures from equilibrium of the long-period tides result from a combination of a large-scale gravity mode response of the ocean and of planetary and topographic Rossby waves.

7. INTERNAL TIDES

Oceanic internal tides are internal waves with periods corresponding to those of the oceanic tides. Internal waves involve vertical oscillations of isotherms in the ocean which can have several tens to hundred meters in amplitude. They induce surface vertical displacements which are reduced by the effective gravity ratio, which is typically of order 10^{-3} (Apel, 1987), i.e., amplitudes at the surface of the order of only a few centimeters. Therefore, they cannot be seen at the surface, except in very extreme nonlinear cases of tidally generated solitons, which lead to patches of enhanced surface roughness in radar images (Osborne and Burch, 1980). Some examples from visible and infrared images have also been discussed recently by Pingree and New (1995). But their observation has mainly relied primarily on subsurface measurements. They are, however, within the resolvability of modern satellite altimetry, and we will report in this section on the very recent results showing the potential of T/P data analysis on that field.

The generation of internal tidal waves in the ocean results from interactions between barotropic tides and bottom topography. Oscillating tidal currents caused by barotropic tides force vertical displacements of isopycnals by their passage over topographic changes and thus generate internal tidal waves which then propagate from their location of generation (Siedler and Paul, 1991; Vlasenko and Morozov,

1993; Dushaw *et al.*, 1995). Mid-ocean ridges, seamounts, and continental shelves are thus thought to be major generation sites of internal waves (Morozof, 1995). However, since internal tides are dominated by the lowest vertical modes (Wunsch, 1975), their typical wavelengths are of the order of 100 km for the semidiurnal, with higher modes having even shorter wavelengths. Moreover, as changes in temperature and salinity modify the density of the ocean at different time scales, and because of the variability of the background oceanic currents, the characteristics of baroclinic ocean waves are subject to large variations. This explains why observations have given us a picture of internal tides highly incoherent in both time and space (Wunsch, 1975). Therefore, it was a surprise when Ray and Mitchum (1996) demonstrated that they have detected internal tide signatures in T/P signals off Hawaii. Since then Ray and Mitchum (1997) and Kantha and Tierney (1997) have conducted a systematic analysis to extract from the full T/P data set the surface signature of internal tides at the world ocean scale.

7.1. Methodology for Internal Tide Extraction from Satellite Data

As we said above, surface signatures of internal tides are at most only a few centimeters. This explains why they have rarely been considered in the past in sea level records. But high-precision satellite altimetry can resolve signals at the centimeter level, especially if the signals are temporally coherent, which allows repeat sampling and averaging to reduce noise. Although internal tides did not remain perfectly coherent with the astronomical potential, because of the natural variability of the ocean parameters which control their characteristics, a significant fraction of them does maintain coherency (or phase-lock) with the astronomical forcing, allowing it to be extracted from a long-time series of high-precision satellite data, and to map it along tracks.

We have presented in Section 4.1 the problem of tidal aliasing in the altimeter data, and reviewed in Section 4.2 the binning procedure applied to overcome the difficulties, which sacrifices the spatial resolution but allows combining of data from neighboring tracks. Unfortunately, the small spatial scales of internal tides (wavelengths of the order of 100 km) are completely smoothed out by following this method. It is only through point-by-point along-track analysis that these scales can be recovered. This was hardly feasible during the first years of T/P. But after several years it became possible to attempt such along-track analysis and get significant new results, at least for the semidiurnal major constituent M_2 thanks to the fact that M_2 to S_2 become separable, from the Rayleigh's criterion point of view. Unfortunately, even after 6 years of T/P data, the K_1 to Ssa coupling remains problematic and prevents along-track analysis in their aliased frequency bands.

Along-track analysis has been specifically addressed by Tierney *et al.* (1998). They have tested both classical methods: harmonic and orthotide. For harmonic analysis, 12 constituents are estimated (M_2, S_2, N_2, K_2, K_1, O_1, P_1, Q_1, M_m, M_f, Sa, Ssa). For orthotide analysis, 22 orthoweights are estimated, 6 for each semi-diurnal and diurnal tidal bands, and 2 each around the 5 long-period bands (M_m, M_f, Mt, Sa, Ssa).

7.2. Recent Results on Internal Tides from Satellite Altimetry

Global maps showing the amplitude of the baroclinic M_2 waves along ascending and descending tracks have been published by Kantha and Tierney (1997). They correspond to the part of internal waves phased-locked with the tidal potential, and averaged in time over the period of analysis. In general, there is an excellent correlation between the amplitude of the internal tides and the topographic features. Strong baroclinic signals can be seen above major topographic features in good qualitative agreement with the primary regions listed and discussed by Morozof (1995): the Mascaren Ridge east of Madagascar, the Maldivian Ridge, the Kusu-Palau Ridge south of Japan, the Hawaii island chain, the Tuamotou archipelago, the Melanesian and Micronesian island chains in the western equatorial Pacific, the Reykjanes Ridge south of Iceland, the Great Meteor Banks in the North Atlantic, the Ridge off Trinidad east of Brazil, the Mendocino Ridge

off the west coast of the United States, the Macquarie Ridge south of New Zealand, the Ninety East Ridge and Carlsberg Ridge in the Northern Indian Ocean, and the Walvis Ridge in the South Atlantic Ocean. Some differences are observable between ascending and descending track estimates: they are caused by the different angles at which the tracks cross the topographic features. Kantha and Tierney also extracted the corresponding signals for S_2 and N_2. The latter was considered mainly to identify the probable regions of high mesoscale noise contamination and thus could be used as a guide for interpretation of the M_2 internal tide distribution. In spite of the excellent correlation with mid-ocean bathymetric features, the authors recommended to consider these results as only suggestive of the distribution of internal tides at the global scale.

Ray and Mitchum (1997) have shown, indeed, that a simplistic interpretation of these maps, without regard to variability, noise, or *in situ* information may be highly misleading. They have concentrated their analysis on a few areas to illustrate the level of significance of the results over these areas. In the extension of their 1996 paper, they have first focused their discussion on the Hawaiian Islands. There, along-track analysis of T/P leads to clear signatures of internal waves spatially coherent and propagating on great distances (larger than 1000 km) before decaying below the background noise. We reproduce on Figure 10 the results for M_2 along an ascending track crossing the Hawaiian Ridge. In this figure, the ragged lines are estimates for amplitudes

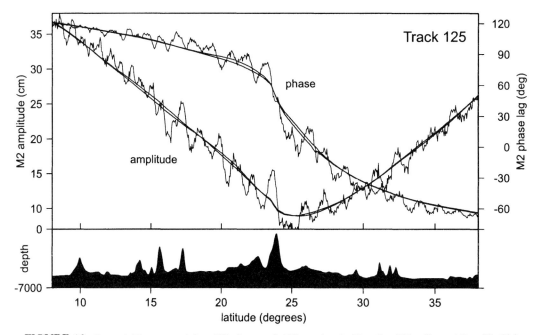

FIGURE 10 Internal tides extracted from T/P, along track 125 crossing the Hawaiian Ridge. Ragged line: M_2 tidal amplitude estimated every 5.75 km, from approximately three years of collinear difference data, using the response method of analysis. Smooth line: M_2 tidal amplitude from the global model SR95.1. The bathymetric profile is extracted from Terrain Base global topography. (Courtesy of Ray and Mitchum (1997).)

and Greenwich phase lags, from approximately 3.5 years of T/P data, computed every 5.75 km. Standard errors for the amplitudes are of the order of 5–10 mm and 3–6° for the phases. The smooth lines correspond to the T/P models of Eanes (CSR3.0) and Schrama and Ray (version 960104). These curves show clear oscillations of several centimeters and wavelengths of the order of 150 km or shorter. These oscillations (present on all the tracks—ascending and descending—crossing the Ridge) have their largest peaks near the main ridge system. Through a careful analysis of the whole T/P-derived picture around the Hawaii Islands, supported also by a parallel study of five Hawaiian long-duration tide gauge observations, it has been demonstrated that these oscillations correspond to internal waves with the right expected wavelength of the first baroclinic mode for M_2, as also previously deduced from CTD data by Chiswell (1994) and Dushaw *et al.* (1995). The phase lags of the "residual tides" obtained after removing the barotropic signal indicate propagating waves, and without exception they propagate away from the ridge.

In their 1997 paper, Ray and Mitchum gave three other examples aiming to highlight the wealth of information in the global maps they have obtained similar to those of Kantha *et al.* (1995) and the difficulties of unambiguous interpretation of these results, as pointed out by Kantha *et al.* The first example is near the Tuamotu Archipelago (near Tahiti and the Society Islands): this is another typical example of well-coherent internal-wave signatures propagating away from the topographic features. The second example is over the Mid-Atlantic Ridge between Brazil and West Africa. Some characteristic wave trains are evident, but of much smaller amplitudes, except off the Amazon shelf. The complex ridge pattern and the shallow slopes in this area must indeed imply very complex internal tide fields more difficult to catch from along-track satellite observations, and also with less coherency. The third example is over the North East Atlantic and the European shelf where it is known that internal tides there are among the largest in the world. Clear but small amplitude wave patterns are observable, also with less spatial coherency. Lack of temporal coherence is the likely explanation, supported by *in situ* observations (New, 1988; Pingree and New, 1995). This example illustrates that any attempt to deduce internal tide properties from these maps will be there of little value.

8. THE TIDES OVER SHALLOW WATERS

In the preceding sections, we focused on the deep ocean. We have simply noted that in shallow waters, much larger discrepancies are evident in the different recent ocean tide models. Shallow water regions are however important for many reasons, both scientific and practical. From the scientific point of view, these areas are important in the global tidal energy budget, because most of the tidal energy is dissipated in shallow water by bottom friction. Also, tidal heights tend to be larger, resulting in an increased loading of the sea bed, with implications in geodesy and gravimetry. From the practical point of view, tides and tidal currents are the major source of variability in shallow waters, and are of considerable interest for shipping, engineering, and for environmental studies.

Tides are much more complex in shallow waters, both in their spatial and temporal characteristics. In space, their wavelengths are 10 times shorter in 40-m coastal waters than in 4000-m oceans. This is because ocean tides manifest themselves as gravity waves with wavelengths of $T(gH)^{1/2}$, with T being their period and H the water depth. In frequency, their spectrum is much more complex because of nonlinear dynamics resulting in deformations of the tidal wave which in terms of harmonic representation correspond to the generation of harmonic and compound new constituents (for reviews, see Parker, 1991 and Le Provost, 1991). The higher complexity of the tides explains why, up to now, shallow water regions have been ignored for the most part in the studies reported above. Indeed, the binning methods are no longer applicable, especially to investigate the harmonic and compound nonlinear constituents which have generally smaller periods, and thus smaller wavelengths. And assimilation methods which generally impose linearized dynamics and linear bottom friction are in some sense compromised. Also, and we have already noted the point in Section 3.2, bottom friction, the sink of energy in shallow waters, is controlled by tidal currents and need to be resolved accurately. Hence the need for increased resolution in numerical hydrodynamic models; hence the interest of the finite element approach developed by Le Provost *et al.* (1994), and of their FES94.1 hydrodynamic solutions. Although containing large-scale errors of the order of 6 cm or even more (see Section 3.2), FES94.1 solutions are considered as valuable a priori in several studies reported in Section 4.2. Recent studies are under development aiming at improving the global tidal solutions over the continental shelves and coastal zones.

8.1. Nonlinear Constituents Solutions from Altimetry

Andersen (1999) has recently addressed this problem in an attempt to infer the major nonlinear constituents from T/P data. The aliasing of the main nonlinear constituents are not problematic. The aliased periods for M_4, the first harmonic of M_2 of the period of 6.21 h is 31.05 days. MN_4 resulting from interactions between M_2 and N_2, is aliased at 244.40 days. M_6, second harmonic of M_2, is aliased at 20.70 days. The major problem is for the main nonlinear interaction between M_2 and S_2. MS_4 is aliased at 1088.60 days and thus almost impossible to separate from the inter-annual variation of the sea level at 3-year periods (1106 days). With their high-spatial sampling, one could have hoped to take advantage of combining T/P with ERS1 and ERS2 data (now available for more than 6 years). But this is not feasible, at first, because of the sun synchronism of ERS, which implies that S_2 is frozen, and consequently some important nonlinear components resulting from interactions between S_2 and M_2, like $2MS_2$ and $2MS_6$, are aliased just like M_4. Others like $2SM_2$, MS_4, M_{Sf}, are aliased like M_2.

Andersen (1999) has shown that reliable empirical estimates of some of these nonlinear constituents can be obtained from T/P, by combining along-track and crossover observations. His domain of investigation has been focused on the European continental shelf where nonlinear constituents are known to reach amplitudes as high as 50 cm for the M_4 constituent. Estimates for M_4, M_6, and MS_4 have been determined and favorably compared to in situ data and existing hydrodynamic models. Tests including more shallow-water constituents like MN_4, $2SM_2$, and M_8 seem to have failed, mainly because of difficulties to recover the phase of these smaller constituents having amplitudes of the order of a few centimeters. The interest of this preliminary study is to pave the way for further studies. Generalizing this empirical approach to other shelves is indeed not straightforward, because of the nonuniform spatial resolution of T/P with latitude. Andersen judiciously noted, in the perspective of the launch of Jason-1, the interest of a tandem mission, T/P and Jason-1, with ground tracks interlaced in such a way to increase the spatial resolution.

8.2. Improving Shallow-water Tide Solution Through Assimilation

Assimilation in high-resolution hydrodynamic models could help face the problem. One major limitation of the hydrodynamic model is the lack of reliable bathymetry over large areas of the world ocean, and especially over shelves, for the present purpose. But assimilation of in situ data and information from altimetric data analysis could help to supply this limitation. Tierney et al. (2000) have followed this approach, with a near-global barotropic tidal model with $1/4°$ resolution (the Arctic Ocean is not included, as in Kantha [1995]). They assimilate data, in the same way as Kantha, through a simple blending technique relying on an empirical weighting of the dynamic constraint and of the data. In order to get tidal estimates reliable for both shallow and deep waters, the T/P data processing has been performed through binning of the altimeter data on the $1/4°$-model grid cells for the shallow waters and on $1°$ cells for the deep ocean. The methodology to process the T/P data followed that of Desai and Wahr (1995) and Tierney et al. (1998). The output of these T/P analyses are blended in the hydrodynamic model, together with harmonic constants from 508

in situ tide gauges. The result is a tidal model that must be useful in both shallow waters and deep areas of the global ocean. Pelagic tide gauge comparisons show that in the deep ocean, this new model is comparable in accuracy with the best models presented above. But the crossover test (cf. Section 5.2.3) suggests markedly improved performance in shallow waters (shallower than 800 m), with crossover residuals reduced from 14 cm for CSR3.0 to 10 cm. Note that Tierney did not consider the question of the nonlinear constituents addressed in the preceding section.

9. TIDAL ENERGETICS AND SATELLITE ALTIMETRY

The subject of tidal energetic has a long history and is still now an open question. Recent reviews have been given by Kagan and Sündermann (1996), Ray (1994), and Munk (1997). Thanks to the application of modern methods of space geodesy and to the major progress in ocean tide modeling induced by high-precision altimetry that we have reviewed above, the total global rate of energy dissipation is now finally well established: 2.5 ± 0.1 TW (1 Terawatt $= 10^{12}$W) for the lunar M_2 component. Note that tides in the solid earth dissipate 0.1 TW, leaving 2.4 TW for ocean dissipation. On one hand, laser ranging to the moon and to an artificial satellite, which has been routine for more than 2 decades, enables the rate of working of the tides over our planet to be deduced from analysis of orbit perturbations (Christodoulidis et al., 1988; Dickey et al., 1994): the more recent estimate is 2.51 TW. On the other hand, the impressive improvements of our knowledge of ocean tides has led to a reliable estimate of global ocean tide dissipation inferred from the spherical harmonic development of the new available cotidal solutions: this estimate varies from 2.36 TW (Egbert, 1997) to 2.47 TW (Kantha et al., 1995), with, among others, 2.42 TW for Le Provost et al. (1998), 2.45 TW for Ray et al. (1994), 2.45 TW for Schrama and Ray (1994), and 2.46 TW for Eanes and Bettadpur (1996). Such a convergence allowed Ray et al. (1996) to infer from the difference between the T/P-derived M_2 solutions and the results of satellite laser ranging reliable information on the anelasticity in the solid earth and the Q of the mantle at the semidiurnal periods.

Note that this amount of energy dissipated by tides is comparable to the global-installed electric capacity in the world but is very small compared to the solar radiation of 2×10^5 TW. These estimates are global, but the determination of how and where the energy is dissipated remains problematic. Hence an intense ongoing debate which has been initiated by the new results presented in the preceding sections (see the review by Munk [1997] and several papers dedicated to the problem published in the *Progress in*

Oceanography, volume 40, edited by Ray and Woodworth [1997]). It is agreed that most of these 2.4 TW are dissipated through bottom friction in shallow seas. But a significant amount of this tidal energy must transit in the deep ocean through conversion of barotropic tidal motions into baroclinic modes, and be dissipated through internal waves.

To investigate the details of tide energetic at regional scales, it is necessary to know not only the tidal characteristics of the sea-surface variations, but also tidal currents. Cartwright and Ray (1989) showed the difficulty of obtaining reliable regional estimates from tidal solutions derived diagnostically from satellite altimetry. The flux divergence methods have been traditionally used to compute dissipation of tides caused by bottom friction (since Taylor, 1919), but they are indeed subject to large errors since they imply to estimate differences between large quantities involving tidal power fluxes at the boundaries of the studied areas. Instead, direct computations of dissipation induced by bottom friction is easier. It requires, however, accurate estimates of tidal currents, since dissipation rates go with the cube of current magnitude. Accurate tidal current solutions are essential: some of the models introduced above are intending to do so, combining high resolution hydrodynamic modeling and data assimilation.

9.1. Dissipation Through Bottom Friction

Tentative estimates of M_2 dissipation rates over the continental shelves of the world oceans have been given by Egbert (1997), Kantha *et al.* (1995) and Le Provost and Lyard (1997), hereafter noted E97, K95, and LPL97. The resolutions are 0.2° for K95, 0.7° for E97, and a variable finite element grid specially designed to increase the resolution down to 10–15 km along the coasts of the world ocean for LPL97 (note that the energetic analysis presented in LPL97 is for the non-T/P assimilated solution FES94.1). Global estimates are consistent between K95 and LPL97, with a total of 2.1 TW and 2 TW, respectively. The results for E97 is much lower, of the order of 1.59 TW: this could possibly be related to the coarser resolution of E97. The major areas of dissipation are known to be located in the Hudson Bay, the European Shelf-English Channel-North Sea system, and the Patagonian shelf, the Yellow-China Sea system, and the Indonesian Through Flow-North Australian plateau system. Figure 11a shows the distribution of the rate of M_2 dissipation per unit of area of tidal bottom friction given by LPL97; a similar map has been produced by E97. Quantitatively, however, the regional estimates can vary substantially from one author to the other. Similar values were obtained by LPL97 and E97 in the Hudson Bay, the European shelf, and the Indonesia-North Australia area. But others are more scattered: if we take as an example the Patagonian shelf, which has been subject to several evaluations, the different estimates are 135 GW for E97, 144 GW for

K95, and 185 GW for LPL97. Previous estimates were from Miller (1966): 130 GW, based on *in situ* observations and the flux method, and more recently Cartwright and Ray (1989): 245 ± 25 GW, computed through the flux method from a tidal solution derived from Geosat altimetric data. A recent regional fully nonlinear hydrodynamic model run by Glorioso and Flather (1997), with a resolution of 1/6° in longitude and 1/8° in latitude, has computed a dissipation rate over the area of 228 GW. This clearly illustrates how difficult it is to still evaluate these dissipation rates, even now with the very sophisticated models and high quality solutions obtained thanks to high-precision satellite altimetry.

9.2. Tidal Fluxes

Although quantitatively uncertain, the energy budgets established by K95 and LPL97 have led to interesting converging conclusions. From the tidal sea level and current solutions, they have been able to compute the different components of the local tidal budgets. Over a given area W, the energy flux through its boundaries corresponds to the balance between the input from the global tidal potential (astronomic + loading + self-attraction) and the dissipation. From the evaluation of these quantities, K95 and LPL97 presented similar comments on how energy is dispatched from areas of net input by the tidal potential to areas where it is dissipated. Figure 11b is from LPL97, but the same picture can be found in K97. The main comments are:

• Over the Pacific Ocean, energy fluxes bring energy from the Central North Pacific (CNP) to the areas of dissipation of the Bering Sea, Okhotsk Sea, and China Sea. Part of the energy dissipated in the China Sea also comes from the Western Tropical Pacific. A large part of the energy input in the CNP propagates south, reinforced by energy from the Central Tropical south Pacific (CTSP), to feed the area around New Zealand, where the tides work against astronomical forcing, and the area of tidal dissipation of the East Australian Coral Reef. A small amount of this energy also propagates along the coast of Antarctica towards the Indian Ocean. Another part goes towards the north, joining the energy flux from the Tropical North East Pacific (TNEP), bringing energy to the areas of dissipation along the western coasts of Canada and Alaska, and the Bering Sea. The energy injected in the TNEP also goes partly south along the west coast of South America, where the ocean tides work against astronomical forcing, and towards Drake Passage.

• Part of this energy enters the South Atlantic Ocean (SAO) through the Drake Passage, and is dissipated over the Patagonian shelf.

• Most of the energy input over the SAO propagates towards the north to feed the North Atlantic Ocean, where almost 40% of the total energy dissipation by bottom friction takes place.

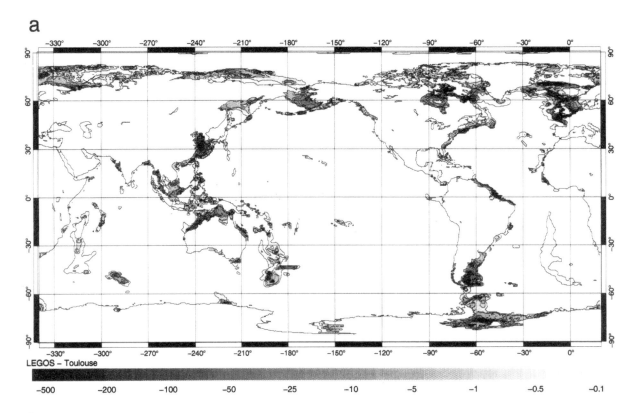

LEGOS – Toulouse

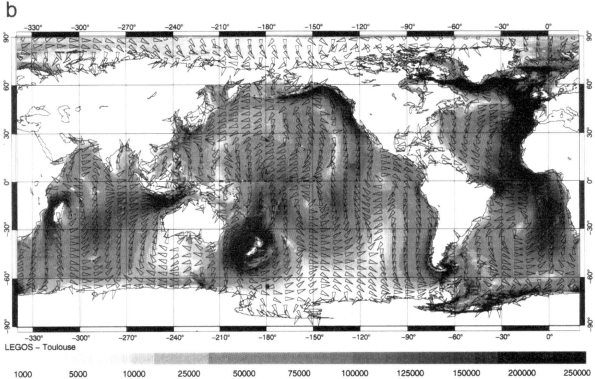

LEGOS – Toulouse

FIGURE 11   (a) rate of dissipation per unit of area by tidal bottom friction for M_2, computed from the hydrodynamic model FES94.1. Units are in KW/km^2. (b) Energy fluxes for the M_2 tide, estimated from FES94.1. Units are in KW/km.

• The energy dissipated in the Weddel Sea comes from both the Atlantic Ocean and the Indian Ocean.

• Part of the energy input in the Indian Ocean is dissipated in the Timor Sea and Arafura Sea.

However, these conclusions need to be taken with care because of the uncertainties in these calculations relying on best fits of imperfect models with nonperfect data. One major question is related to the imbalance in the energy budgets of these different models between the energy input by the tidal potential and dissipation by bottom friction: 0.47 TW for K97, 0.35 TW for LPL97, and 0.5 TW for E97. Part of this difference must correspond to a sink of energy neglected in the hydrodynamic models: the energy conversion from the barotropic tides to internal waves and internal tide dissipation in the deep ocean.

9.3. Internal Tide Dissipation

We have presented in the preceding section the unexpected detection of internal tides by satellite altimetry. These results lead us to believe that internal tides are ubiquitous in the oceans. However, to estimate the energy involved in internal tides from these surface signatures, we need to know the ocean density structure. Ray and Mitchum (1997) have taken advantage of the large amount of data available for the region near Hawaii to compute for the M_2 tide the mean baroclinic energy density at the ridge and the mean energy flux. They got a mean flux of 4000 Wm^{-1}, directed northward and southward from the ridge. This flux is roughly 10% of the barotropic flux there. Over the 2000-km ridge, this corresponds to 15 GW of tidal power converted from barotropic to baroclinic motion: this value is small (less than 1% of the global M_2 dissipation rate) but significant given the small extent of the Hawaiian ridge. However, Ray and Mitchum warned that extending the Hawaiian results for a global internal tide energy estimate is premature. What is extracted from the altimetric signal, up to now, is only the phase-locked part of the surface signatures of internal tides: inferred energy estimates may be biased low, or even totally misleading (remember the examples given by Ray and Mitchum for the European shelf).

Egbert (1997) has suggested that inverse methods are now sufficiently well-advanced to provide a rational framework for investigating this question. Through the assimilation of data (and in particular altimetric data) in a dynamic model, it is possible to rigorously evaluate the consistency of the two. He has presented results of local and global inversions which lead to the conclusion that over elongated bathymetric features oriented perpendicular to tidal flows, energy dissipation in the open ocean must be significantly enhanced to reconcile his model with data. Presumably, this enhancement must correspond to conversion of barotropic tidal motions into internal waves. His preliminary results suggest that

perhaps as much as 0.5 TW of energy could be dissipated through this mechanism. However, Egbert noted that significant uncertainties are associated with these results due to the simplified linear dynamics of his model and its limited spatial resolution.

Kantha and Tierney (1997) have also tried to go further by combining altimetric information and modeling. They used Baines' theory (1982) and the barotropic currents computed from their 1/5° barotropic tidal model. They adopted a constant value for the buoyancy frequency and for the speed of the first baroclinic mode over the global ocean. And they tuned their model for energy density and vertical energy flux (i.e., dissipation rate) to the M_2 energy in the first baroclinic mode estimated from their along-track analysis of the altimetric data. They arrived at an estimate of 0.36 TW dissipated by the M_2 baroclinic tides and 0.52 TW in the whole first mode baroclinic tides (to be compared to the 3.5 TW of tidal power input by the luni-solar forces). However, because of the various uncertainties of their approach, Kantha and Tierney concluded that this global dissipation rate might be anywhere between 0.4 and 0.8 TW.

Note that Morozov (1995), from a compilation of all the major ridge systems, and using both models and observations of internal waves, has estimated the total amount of energy fluxes into internal tides to 1.1 TW. This is a very large value. Ray and Mitchum (1997) have suggested that this estimate might be several times too high. They indeed observed that, while over the Hawaiian ridge their estimate (15 GW) is larger than the one of Morozov (8 GW), the amplitudes of the surface signals extracted from altimetry over the areas of Morozov's largest fluxes (Mascarene Ridge, Mid-Atlantic Ridge) are only comparable or smaller than the ones observed at Hawaii. But they are also reminded that their altimetric results probably provide a lower band which may possibly be very weak.

These wide-range estimates of energy dissipation rates through internal tides clearly show that this is a totally open issue which needs to be clarified over the next number of years. The question is of importance not only for our understanding of the tidal physics, but also because it is thought that pelagic turbulence in the deep ocean is probably partly maintained by topographic scattering of barotropic into baroclinic tidal energy, via internal tides and internal waves (Munk, 1997; Munk and Wunsch, 1997).

10. APPLICATIONS

The aim here is not to give an exhaustive review of all the applications related to the new tidal products presented above. We will only illustrate some of the applications which recently have benefited the progress presented in our understanding of ocean tides.

10.1. Earth Rotation

Due to friction, ocean tides are, in the very long-term, responsible for the spin down of the earth and the slow increase of the length of the day (LOD) which is approximately 2 msec per century. But tides also have periodic effects on the earth's rotation, which are important in the accurate determination of Universal Time (UT1) and in the orientation of the earth rotation axis (polar motion). These effects are caused by two mechanisms: the tidal sea-surface elevation fluctuations modulate the moment of inertia tensor of the earth (mass redistribution), and tidal current fluctuations cause changes of angular momentum. Satellite Doppler Tracking, lunar laser ranging, satellite laser ranging (SLR), Very Long Baseline Interferometry (VLBI), and the Global Positioning System (GPS) have provided earth rotation data that have on daily basis an accuracy of a fraction of a milliarc second. Chao and Ray (1997) have reviewed the oceanic tidal angular momentum predicted from some of the T/P models for the eight major constituents, and their excitations on both the earth rotational speed variation and polar motion. They have compared these results with observations derived from long-term VLBI measurements (from Sovers *et al.* [1993], Herring and Dong [1994], and Gipson [1996]). The agreement is good with discrepancies typically within 1 to 2 msec for UT1 and 10–30 microarc seconds for polar motion. The results clearly show that a substantial part of the signal is caused by oceanic tides. A better understanding of the short-period oceanic tidal angular momentum (OTAM) contributions to earth's rotation fluctuations are thus of major interest to pave the way for non-OTAM contribution studies, such as atmospheric and oceanic angular momentum variations, earthquakes, thermally induced atmospheric tides, and earth librations. Chao and Ray illustrate a breakdown of the tidal height and current contributions for each individual ocean basin, according to one of the T/P models from Egbert *et al.* (1994), which helps us to understand the complex role of the different basins and the large cancellations occurring among the basin contributions. They also noted that significant improvements to OTAM estimates must be expected from improvements in models of deep-ocean tidal currents, the dominant term in most of the OTAM integrals. The on-going developments aiming at further improvements of the modeling-assimilation methods for ocean tide respond to this yearning (Egbert, 1997; Lefevre *et al.*, 2000; Tierney *et al.*, 1999).

10.2. Tidal Loading

Accurate estimates of load tides are important to many applications such as in gravimetry, global positioning systems, and geophysics. They are caused by the deformation of the solid earth, including regions over land, because of the fluctuating load of the oceanic tides above the ocean floor. They can now be computed routinely from the ocean tide models (Ray and Sanchez, 1989; Francis and Mazzega, 1990). In Figure 12, we display as an illustration the ocean-loading radial displacement computed by Llubes and Mazzega (1997). Melchior and Francis (1996) and Llubes and Mazzega (1997) compared the loading computed from most of the new ocean tide models to data from gravimeters. They conclude that significant improvements are noticeable by reference to results issued from Schwiderski's solutions. But discrepancies remain in some areas such as the Indonesian Straits, around New Zealand, and the Patagonian Shelf. Indeed, these are areas where ocean tide models differ the most, especially all the coastal zones. As it is precisely there that tidal-loading estimates need to be accurate, for all the above-noted applications, new improvements of the tidal solutions in coastal areas are clearly required.

10.3. Tidal Currents

Recently, interest is focusing on tidal currents. The observations collected during the World Ocean Circulation Experiment (WOCE) have led to enhance the importance of internal tidal wave breaking as a source of vertical mixing in the deep ocean (Munk and Wunsch, 1997). The development of new technologies for ocean observation like towed instruments, moored ADCP (Firing, 1998), and acoustic tomography (Dushaw *et al.*, 1997) also requires tidal current prediction to eliminate their contribution to the measured signals.

Among the many tidal solutions presented above, only a few of them have produced in parallel sea level and tidal current solutions. Indeed, most of these solutions are derived from the analysis of altimetry and thus restricted to sea surface elevation variations. For deducing tidal current solution from them, it is necessary to take the gradient of the elevation solutions, hence a degraded precision, which has been noticed by Cartwright and Ray (1991) when computing tidal energy fluxes over the Patagonian shelf from their altimetric tidal solutions. The advantage of the methodology presented in Section 4.2.4, combining hydrodynamic modeling and data assimilation, is that it computes simultaneously tidal elevation and tidal currents, with the same level of precision. Up to now, very little has been done to validate and exploit these solutions. It is indeed very difficult to systematically carry such a validation, because of the importance of baroclinicity which prevent one from isolating the barotropic contribution to tidal current in most of the current meter arrays which are also not very numerous, even now after the end of the WOCE program. Few studies pave the way for this very difficult validation. Dushaw *et al.* (1997) have presented such a test on two specific sites in the Western North Atlantic and in the North Pacific ocean. They have compared the tidal flows predicted by TPX0.2 (from Egbert *et al.*, 1994) to the barotropic tidal currents determined

a

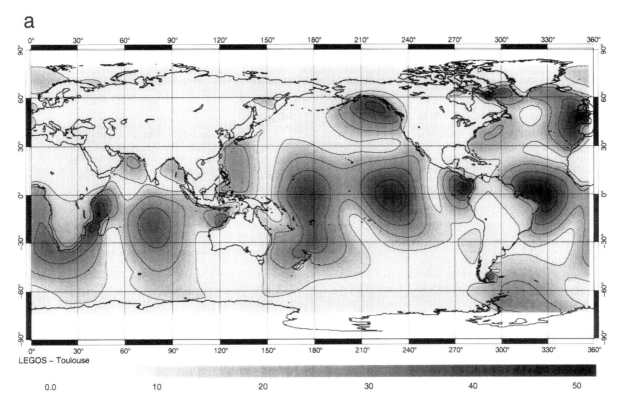

LEGOS – Toulouse

0.0   10   20   30   40   50

b

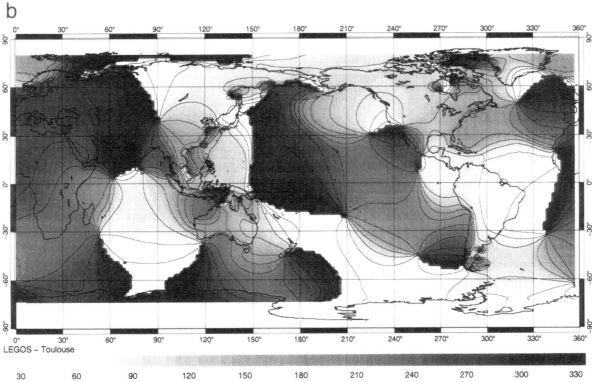

LEGOS – Toulouse

30   60   90   120   150   180   210   240   270   300   330

FIGURE 12   Amplitude of the vertical displacement of the ocean floor under the effect of tidal loading. (Courtesy of M. Llubes and P. Mazzega (1997).)

from long-range acoustic transmissions, and from current meter records (more specifically the M_2 component of these tidal currents) reported by Luyten and Stommel (1991) and Dick and Siedler (1985). Their paper illustrates the difficulty of these comparisons. However, even if only qualitative, the tidal current prediction models now available from Egbert *et al.* (1994), Kantha (1995), Le Provost *et al.* (1998), and others can help a lot to respond to the oceanographic demand above noted.

10.4. Tides and Coastal Engineering

Tidal currents are needed for coastal engineering studies. Observations from current meter moorings are generally difficult to get in shallow waters due to severe environmental conditions. It is then preferred to develop regional models, generally now three-dimensional models with very high horizontal and vertical resolution. But the implementation of these models requires boundary conditions to be prescribed along the open limits of the modeled area. In the past, Schwiderski's solutions have been extensively used. It must now be recommended to use one of these new tidal solutions presented in this chapter, and in the near future the new coming solutions which will be improved over the continental shelves, and will include nonlinear components necessary for prescribing correct boundary conditions near the coasts.

11. CONCLUSIONS

The aim of this chapter was to present the impressive progress observed during this last decade in tidal science. This progress first concerned the determination of the tidal characteristics over the whole ocean. Highly accurate tidal prediction models are required for oceanographic exploitation of altimetric data, but in turn, altimetry itself has brought a revolutionary instrument for observing tides at the global scale. The era of high-precision altimetry will certainly be associated in the future to the period where, after a few years of observation, the scientific community definitively mapped the characteristics of the ocean tides to their best possible accuracy. We have noted that about 10–15 tide solutions are now available, almost all with the same level of accuracy. The rms maximum differences to the standard set of *in situ* tide gauge (ST103) are of the order of 1.6 cm for the major lunar component M_2, and 1 cm for most of the other constituents, leading to RSS estimates of 2.6 cm on the eight major components. These models allow us to predict tides everywhere in the deep ocean, with an accuracy comparable to what was possible before at a given location after several months to years of observations, i.e., typically 2–4 cm precision over periods of several months to years. One might

ask where the major discrepancies are located. In this chapter there are maps of the vector differences between some of these solutions, which can be considered as indicative of the areas where possible problems remain. Very little is expected from the analysis of longer time series in the deep ocean, because of the high level of accuracy already attained. The main effort is now oriented towards the shallow-water areas, where the different models are less accurate, and tidal predictions require more constituents to be considered.

The progress reported here are not only because of altimetry but is also because of the development of new hydrodynamic models and methods for data assimilation, which allow one to optimally combine dynamic constraints and observations. These models also offer the major interest to simultaneously allow the computation of tidal elevations and of the barotropic component of tidal currents.

Beside these impressive improvements in the quality of the description of the major tidal constituents, interesting new results have been noted on the determination and the understanding of long-period tides. It is now clear that the ocean response to semi-monthly and monthly forcing is a mix of gravity mode and planetary-topographically trapped modes. Another significant progress is the unexpected discovery of internal tide signatures in along-track analysis of the residual altimetric signals. This is of interest not only for internal tidal dynamics understanding, but also as a way for further investigating where internal tides are active as a possible source of turbulence for deep-ocean mixing.

All these improvements have contributed to reactivate the old debate on where and how tidal energy is dissipated. The more recent models converge towards the same values for global tidal dissipation, in agreement with the previous estimates from space geodesy. The combined hydrodynamic-assimilation models have allowed us to compute where the tidal forcing is acting, where bottom tidal friction is dissipating part of the energy, and how the energy is propagating from the areas where the tidal forcing is driving the ocean tides to areas of dissipation. And the energy budgets suggest that a significant part of the energy is converted from barotropic to baroclinic modes, and dissipated through internal tide breaking. These questions are still in the stage of scientific discussions and need further investigations.

We have also pointed out some recent applications which are benefitting from these improvements of our knowledge on tides. This is the case in geophysics, for earth-rotation studies, and for earth-deformation investigations. This is also true in ocean technology where new measuring techniques ask for improved tidal current predictions, and in coastal engineering developments.

As a final conclusion, it must be hoped that in the future, new efforts will be put to improve our description and understanding of ocean tides over coastal areas, on one side, and of internal tides on the other side. This requires further analysis of residual altimetric signals, and their combination with hydrodynamic modeling and data assimilation methods, both for theoretical studies on tides and for supplying more valuable information for oceanography, geophysics, and engineering.

ACKNOWLEDGMENTS

The author thanks E.J.O. Schrama for his careful review of the chapter and his detailed comments. He also thanks Fabien Lefevre, Florent Lyard, and Frederique Ponchaut for their help for the creation of the figures, Richard Ray for allowing to use and reproduce Figure 10, and all the authors who have made feely available their tidal solutions.

References

Accad, Y. and Pekeris, C. L. (1978). Solution of the tidal equations for the M2 and S2 tides in the world ocean from a knowledge of tidal potential alone. *Phil. Trans. Roy. Soc. London*, **A290**, 235–266.

Agnew, D. C. and Farell, W. E. (1978). Self-consistent equilibrium ocean tides. *Geophys. J. R. Astr. Soc.*, **55**, 171–181.

Apel, J. R. (1987). *In* "Principles of Ocean Physics." Academic Press, London.

Andersen, O. B. (1995). Global ocean tides from ERS1 and TOPEX/POSEIDON, *J. Geophys. Res.*, **100**, C12, 25249–25259.

Andersen, O. B. and Knudsen, P. (1997). Multi-satellite ocean tide modeling—the K1 constituent. *Prog. In Oceanog.*, **40**, 197–216.

Andersen, O. B. (1999). Shallow water tides in the Northwest European Shelf region from TOPEX/POSEIDON altimetry. *J. Geophys. Res.*, **104**, 64, 7729.

Baines, P. G. (1982). On internal tide generation models. *Deep Sea Res.*, **80**, 320–327.

Bennett, A. F. (1992). Inverse methods in physical oceanography. *In* "Monographs on mechanical and applied mathematics." Cambridge University Press, Cambridge, 346 pp.

Bennett, A. F. and McIntosh, P. C. (1982). Open ocean modeling as an inverse problem: Tidal theory. *J. Phys. Oceanog.*, **12**, 1004–1018.

Bogdanov, K. T. and Magarik, V. A. (1967). A numerical solution of the problem of tidal wave propagation in the world ocean. *Izv. Atmos. Oceanic Phys.*, **5**, 12, 1309–1317.

Carton, J. A. (1983). The variation with frequency of the long-period tides. *J. Geophys. Res.*, **88**, 7563–7571.

Cartwright, D. E. and Tayler, R. J. (1971). New computations of the tide-generating potential, *Geophys. J. Roy. Astron. Soc.*, **23**, 45–74.

Cartwright, D. E., Edden, A. C., Spencer, R., and Vassie, J. (1980). The tides of the Northeast Atlantic Ocean. *Phil. Trans. R. Soc., London*, **A298**, 87–139.

Cartwright, D. E. and Alcock, G. A. (1983). Altimeter measurements of ocean topography, *In* "Satellite microwave remote sensing," (T. Allan, Ed.) 309–319. Ellis Horwood, Chichester, England.

Cartwright, D. E. and Ray, R. D. (1989). New estimates of oceanic tidal energy dissipation from satellite altimetry. *Geophys. Res. Lett.*, **16**, No. 1, p. 73–76.

Cartwright, D. E. and Ray, R. D. (1990). Ocean tides from Geosat altimetry. *J. Geophys. Res.*, **95**, C3, 3069–3090.

Cartwright, D. E. and Ray, R. D. (1991). Energetics of global ocean tides from Geosat altimetry. *J. Geophys. Res.* **96**, C9, 16897–16912.

Cartwright, D. E. and Ray, R. D. (1994). On the radiational anomaly in the global ocean tide, with reference to satellite altimetry. *Oceanol. Acta*, **17**, 453–459.

Chao, B. F. and Ray, R. D. (1997). Oceanic tidal angular variation and Earth's rotation variations. *Prog. in Oceanog.*, **40**, 399–421.

Chiswell, S. M. (1994). Vertical structure of the baroclinic tides in the central north Pacific subtropical gyre. *J. Phys. Oceanog.*, **24**, 2032–2039.

Christodoulidis, D. C., Smith, D. E., Williamson, R. G., and Klosko, S. M. (1988). Observed tidal braking in the Earth/Moon/Sun system. *J. Geophys. Res.*, **93**, 6216–6236.

Darwin, G. H. (1886). On the dynamical theory of the tides of long period. *Proc. Roy. Soc., Ser. A* **41**, 319–336.

Desai, S. D. and Wahr, J. (1995). Empirical ocean tide models estimated from TOPEX/POSEIDON. *J. Geophys. Res.* **100**, C12, 25205–25228.

Desai, S. D. (1996). Ocean tides from TOPEX/POSEIDON altimetry with some geophysical applications. PhD thesis, Univ. Colorado, Boulder.

Desai, S. D., Wahr, J. M., and Chao, Y. (1997). Error analysis of empirical tide models estimated from TOPEX/POSEIDON altimetry. *J. Geophys. Res.*, **102**, 25157–25172.

Dick, G. and Siedler, G. (1985). Barotropic tides in the Northeast Atlantic inferred from moored current meter data. *Deutsche Hydrograph. Zeitschrift*, **38**, 7–22.

Dickey, J. O., Bender, P. L., Faller, J. E., Newhall, X. X., Ries, J. G., Shelus, P. J., Veillet, C., Whipple, A. L., Wiant, J. R., Williams, J. G., and Yoder, C. F. (1994). Lunar laser ranging: a continuing legacy of the Apollo Programme, *Science*, **265**, 482–490.

Doodson, A. T. (1921). The harmonic development of the tide generating potential. *Proc. Roy. Soc. of London*, **100**, 305–328.

Dushaw, B. D., Cornuelle, B. D., Worcestor, P. F., Cornuelle, B. D., Howe, B. M., and Metzger, K. (1997). A TOPEX/POSEIDON global tidal model (TPX0.2) and barotropic tidal currents determined from long-range acoustic transmissions. *Prog. in Oceanog.*, **40**, 337–367.

Dushaw, B. D., Egbert, G. D., Worcestor, P. F., Howe, B. M., and Luther, D. S. (1995). Barotropic and baroclinic tides in the central North Pacific Ocean determined from long-range reciprocal acoustic transmissions. *J. Phys. Oceanog.*, **25**, 631–647.

Eanes, R. J. and Bettadpur, S. (1996). The CSR3.0 global ocean tide model: diurnal and semi-diurnal ocean tides from TOPEX/POSEIDON altimetry. CSR-TM-96-05, The University of Texas Center for Space Research.

Egbert, G. D., Bennett A. F., and Foreman, M. G. (1994). TOPEX/POSEIDON tides estimated using a global inverse model. *J. Geophys. Res.*, **99**, C12, 24821–24852.

Egbert, G. D. and Bennett, A. F. (1996). Data assimilation methods for ocean tides, *In* "Modern Approaches to Data Assimilation in Ocean Modelling," (P. Malanotte-Rizzoli, Ed.) Elsevier Press, Amsterdam.

Egbert, G. D. (1997). Tidal data inversion: interpolation and inference. *Prog. In Oceanog.*, **40**, 53–80.

Eyries, M. (1968). Les marégraphes de grandes profondeurs, *Cahier Océanograph.*, **20**, 355–368.

Firing, E. (1998). Lowered ADCP development and use in WOCE, *Int. WOCE Newslett.*, **30**, 12–14.

Francis, O. and Mazzega, P. (1990). Global charts of ocean tide loading effects. *J. Geophys. Res.*, **95**, 11411–11424.

Fu, L. L., Christansen, E. J., Yamarone, C. A., Lefebvre, M., Ménard, Y., Dorrer, M., and Escudier, P. (1994). TOPEX/POSEIDON mission overview, *J. Geophys. Res.*, **99**, C12, 24369–24381.

Fu, L. L. and Smith, R. D. (1996). Global ocean circulation from satellite altimetry and high resolution computer simulations. *Bull. Am. Meteorol. Soc.*, **77**, 2525–2636.

Gipson, J. M. (1996). VLBI determination of neglected terms in high frequency Earth orientation parameter variation. *J. Geophys. Res.*, **101**, 28051–28064.

Glorioso, P. D. and Flather, R. A. (1997). The Patagonian Shelf tides. *Prog. Oceanog.*, **40**, 263–283.

Groves, G. W. and Reynolds, R. W. (1975). An orthogonalized convolution method of tide prediction. *J. Geophys. Res.*, **80**, 4131–4138.

Harwitz, B. and Cooley, A. (1973). The diurnal and semidiurnal barometric oscillations, global distribution and annual variation. *Pure App. Geophys.*, **102**, 193–222.

Hendershott, M. C. (1977). Numerical models of ocean tides, *In* "The Sea," vol. 6, (E. D. Goldberg, I. N. McCave, J. J. O'Brien and J. H. Steele, Eds.) 47–95, John Wiley & Sons, New York.

Herring, T. A. and Dong, D. (1994). Measurement of diurnal and semi-diurnal rotational variations and tidal parameters of the Earth. *J. Geophys. Res.*, **99**, 18051–18071.

Hough, S. S. (1897). On the application of harmonic analysis to the dynamical theory of the tide, 1. *Phil. Trans. R. Soc. London*, **A189**, 201–257.

Kagan, B. A., Rivkind, V. Y., and Chernayev, P. K. (1976). The fortnightly lunar tides in the global ocean. Izv. Acad. Sci. USSR, *Atmos. Oceanic Phys.*, **12**, 274–276.

Kagan, B. A. and Sundermann, J. (1996). Dissipation of tidal energy, paleotides, and the evolution of the Eart–Moon system. *Adv. Geophys.*, **38**, 179–266.

Kantha, L. H. (1995). Barotropic tides in the global oceans from a non linear tidal model assimilating altimetric tides, I, Model description and results, *J. Geophys. Res.*, **100**, 25283–25308.

Kantha, L. H., Tierney, C., Lopez, J. W., Desai, S. D., Parke, M. E., and Dexler, L. (1995). Barotropic tides in the global oceans from a nonlinear tidal model assimilating altimetric tides 2. Altimetric and geophysical implications, *J. Geophys. Res.*, **100**, C12, 25,309–25,317.

Kantha, L. H. and Tierney, C. C. (1997). Global baroclinic tides. *Prog. Oceanog.*, **40**, 163–178.

Kantha, L. H., Stewart, J. S., and Desai, S. D. (1998). Long period lunar fortnightly and monthly tides. *J. Geophys. Res.*, **103**, 12639–12647.

King, C., Stammer, D., and Wunsch, C. (1995). Tide model comparison at CPMO/MIT. Working paper for the T/P Science working Team Tide Model Study Group, MIT Department of Earth, Atmospheric and Planetary Sciences report, May 5.

Kowalik, Z. (1994). Modelling of topographically amplified diurnal tides in the Nordic Sea. *J. Phys. Oceanog.*, **24**, 1717–1731.

Krohn, J. (1984). A global ocean tide model with resolution in shelf areas. *Marine Geophys. Res.*, **7**, 231.

Kuo, J. T. (1991). Synoptic prediction of tides and currents everywhere in the ocean waters, *In* "Tidal hydrodynamics," (B. Parker Ed.) John Wiley & Sons, New York, 61–75.

Laplace, P. S. (1776). Recherches sur plusieurs points du system du monde, *Mémoires de l'Académie Royale des Sciences de Paris*, Reprinted in *Œuvres complètes de Laplace*, Gauthier Villard, Paris, **9**, 1893.

Li, X., Shum, C. K., and Tapley, B. D. (1996). Accuracy evaluation of global ocean tide models. Center for Space Research, Austin, Texas. CSR-96-03, 80 p.

Lefevre, F., Le Provost, C., and Lyard, F. (2000). How can we improve a global ocean tide model at a regional scale? A test on the Yellow Sea and the East China Sea. *J. Geophys. Res.* **105**, 64, 8707–8725.

Le Provost, C. (1983). An Analysis of Seasat altimeter measurements over a coastal area: the English Channel. *J. Geophys. Res.*, **88**, C3, 1647–1654.

Le Provost, C. and Vincent, P., (1986). Extensive tests of precision for a finite element model of Ocean Tides. *J. Comput. Phys.*, **65**, 273–291.

Le Provost, C. (1991). Generation of overtides and compound tides (review). *In* "Tidal hydrodynamics," (B. Parker Ed.) John Wiley & Sons, New York, 269–296.

Le Provost, C., Lyard, F., and Molines, J. M. (1991). Improving ocean tide predictions by using additional semidiurnal constituents from spline interpolation in the frequency domain, *Geophys. Res. Let.*, **18**, 845–848.

Le Provost, C. (1994). A new *in situ* reference data set for ocean tides, *AVISO Altimetry Newslett.*, **3**, Toulouse.

Le Provost, C., Genco, M. L., Lyard, F., Vincent, P., and Canceil, P. (1994). Tidal spectroscopy of the world ocean tides from a finite element hydrodynamic model. *J. Geophys. Res.*, **99**, C12, 24777–24798.

Le Provost, C., Bennett, A. F., and Cartwright, D. E. (1995). Ocean tides for and from TOPEX/POSEIDON, Science, **267**, 639–642.

Le Provost, C., Lyard, F., Molines, J. M., Genco, M. L., and Rabilloud, F. (1998). A Hydrodynamic Ocean Tide Model Improved by Assimilating

a Satellite Altimeter-derived Data Set. *J. Geophys. Res.*, **103**, C3, 5513–5529.

Le Provost, C. and Lyard, F. (1997). Energetics of the M2 barotropic ocean tides: an estimate of bottom friction dissipation from a hydrodynamic model. *Prog. Oceanog.*, **40**, 37–52.

Llubes, M. and Mazzega, P. (1997). Testing recent global ocean tide models with loading gravimetric data. *Prog. Oceanog.*, **40**, 369–383.

Luther, D. S. (1980). Observations of long period waves in the tropical oceans and atmosphere. Phd. dissertation, MIT-WHOI Joint Programme in Oceanography, 210 p.

Luyten, J. R. and Stommel, H. M. (1991). Comparison of M_2 tidal currents observed by some deep moored current meters with those of the Schwiderski and Laplace models. *Deep Sea Res.*, **38**, 573–589.

Lyard, F. (1997). The tides in the Arctic from a finite element model. *J. Geophys. Res.*, **102**, C7, 15611–15638.

Lyard, F. (1998). Data assimilation in a wave equation: a variational representer approach for the Grenoble tidal model, *J. Comput. Phys.*, **143**, 1–31.

Lyard, F., Ponchaut, F., and Le Provost, C. (1999). Long period tides in the global ocean from a high resolution hydrodynamic model and tide gauge data assimilation. *J. A. O. T.*, submitted.

Ma X. C., Shum, C. K., Eanes, R. J., and Tapley, B. D. (1994). Determination of ocean tides from the first year of TOPEX/POSEIDON altimeter measurements. *J. Geophys. Res.*, **99**, C12, 24809–24820.

Matsumoto, K., Ooe, M., Sato, T., and Segawa, J. (1995). Ocean tide model obtained from TOPEX/POSEIDON altimetry data. *J. Geophys. Res.*, **100**, 25319–25330.

Mazzega, P. (1985). M2 model of global ocean tide derived from Seasat altimetry. *Marine Geodesy*, **9**, 3, 335–363.

Mazzega, P. and F. Jourdin, (1991). Inverting SEASAT Altimetry for Tides in the Northeast Atlantic: Preliminary results, *In* "Advances in Tidal Hydrodynamics," (B. Parker, Ed.) John Wiley & Sons, New York, 569–592.

McIntosh, P. C. and Bennett, A. F. (1984). Open ocean modeling as an inverse problem: M2 tides in the Bass Strait. *J. Phys. Oceanogr.* **14**, 601–614.

Melchior, P. (1994). Checking and correcting the tidal gravity parameters on the ICET Data Bank, *Bulletin d'Information des marees Terrestres*, AIG, **119**, 8899–8936.

Melchior, P. and Francis, O. (1996). Comparison of recent ocean tide models using ground based tidal gravity measurements. *Mar. Geod.*, **19**, 291–330.

Miller, G. R. (1966). The flux of tidal energy out of the deep oceans. *J. Geophys. Res.*, **71**, 2485–2489.

Miller, G. R., Luther, D. S., and Hendershott, M. (1993). The forthnightly and monthly tides: Resonant Rossby waves or nearly equilibrium gravity waves. *J. Phys. Oceanog.*, **23**, 879–899.

Molines, J. M., Le Provost, C., Lyard, F., Ray, R. D., Shum, C. K., and Eanes, R. J. (1994). Tidal corrections in the TOPEX/POSEIDON GDR's, *J. Geophys. Res.*, **75**, 24749–24760.

Morozov, E. G. (1995). Semidiurnal internal wave global field, *Deep Sea Res.*, **42**, 135–148.

Munk, W. and Cartwright, D. E. (1966). Tidal spectroscopy and prediction. *Phil. Trans. Roy. Soc. London*, **259**, 533–581.

Munk, W. (1997). Once again: once again tidal friction. *Prog. Oceanog.*, **40**, 7–35.

Munk, W. and Wunsch, C. (1997). The moon, of course . . . *Oceanography*, **10**, 3, 132–134.

New, A. L. (1988). Internal tidal mixing in the Bay of Biscay. *Deep Sea Res.*, **35**, 691–709.

Newton, I. (1687). Philosophiae Naturalis Principia Mathematica, See Newton's Principia *Cjori's 1946, Revision of Motte's 1729 translation*, University of California Press, Berkeley, 680 p.

Osborne, A. R. and Burch, T. L. (1980). Internal solitons in the Andaman Sea, *Science*, **208**, 451–457.

Parke, M. E. and Hendershott, M. C. (1980). M2, S2, K2 models of global tide on an elastic earth. *Mar. Geod.*, **3**, 379–408.

Parke, M. E., Stewart, R. H., Farless, D. L., and Cartwright, D. E. (1987). On the choice of orbits for an altimetric satellite to study ocean circulation and tides. *J. Geophys. Res.*, **92**, 11693–11707.

Parker, B. B. (1991). The relative importance of the various non-linear mechanisms in a wide range of tidal interactions (review), *In* "Tidal hydrodynamics," (B. Parker, Ed.) John Wiley & Sons, New York, 237–268.

Pekeris, C. L. and Accad, A. (1969). Solution of Laplace's equation for the M2 tide in the world oceans. *Phil. Trans. Roy. Soc. London*, **A265**, 413–436.

Pingree, R. D. and Griffiths, D. K. (1984). Trapped diurnal waves on Porcupine and Rockall banks, J. Mar. Biol. Ass. U.K., **64**, 889–897.

Pingree, R. D. and New, A. L. (1995). Structure, seasonal development and spatial coherence of the internal tide on the Celtic and Armorican shelves and in the Gulf of Biscay. *Deep Sea Res.*, **42**, 245–284.

Platzman, G. W. (1981). Normal modes of the world ocean, II, Description of modes in the period range 8 to 80 hours, *J. Phys. Oceanog.*, **11**, 579–603.

Platzman, G. W. (1984). Planetary energy balance for tidal dissipation, *Rev. Geophys.*, **22**, 73–84.

Platzman, G. W. (1984). Normal modes of the world ocean, III, a procedure for tidal synthesis, *J. Phys. Oceanog.*, **14**, 1521–1531.

Platzman, G. W. (1984). Normal modes of the world ocean, III, Synthesis of diurnal and semidiurnal tides, *J. Phys. Oceanog.*, **14**, 1532–1550.

Poincaré, H. (1910). Leçons de mécanique céleste. **III**, *Gauthier Villard*, Paris.

Ponchaut, F., Lyard, F., and Le Provost, C. (2000). A comprehensive analysis of the tidal and oceanic signals in the WOCE Sea Level data and its applications. *J.A.O.T.*, in press.

Proudman, J. (1913). Limiting forms of long period tides, *Proc. London Math. Soc.*, **13**, 2, 273–306.

Proudman, J. (1959). The condition that long period tide shall follow equilibrium law. *Geophys. J.*, **2**, 244–249.

Ray, R. D. and Sanchez, B. V. (1989). Radial deformation of the earth by oceanic loading. *NASA Techn. Memo. 100743*, Goddard Space Flight Center.

Ray, R. D. (1994). Tidal energy dissipation: observations from astronomy, geodesy, and oceanography, *In* "The oceans: physical–chemical dynamics and human impact," (S. K. Majumdar, Ed.), Peam. Acad. Science, 171–185.

Ray, R. D., Sanchez, B.V., and Cartwright, D. E. (1994). Some extensions to the response method of tidal analysis applied to TOPEX/POSEIDON altimetry (abstract). *EOS*, **75** (16), 108.

Ray, R. D. and Cartwright, D. E. (1994). Satellite observations of the Mf and Mm ocean tides, with simultaneous orbit corrections, IUGGI, Vienna, Austria.

Ray, R. D., Eanes, R. J., and Chao, B. J. (1996). Detection of tidal dissipation in the solid earth by satellite tracking and altimetry, *Nature*, **381**, 595–597.

Ray, R. D. and Mitchum, G. T. (1996). Surface manifestation of internal tides generated near Hawaii. *Geophys. Res. Lett.*, **23**, 21101–2104.

Ray, R. D. and Mitchum, G. T. (1997). Surface manifestation of internal tides in the deep ocean. *Prog. Oceanog.*, **40**, 135–162.

Ray, R. D. and Woodworth, P. L. (1997). Special issue on tidal science in honour of D. E. Cartwright, *Prog. Oceanog.*, **40**, 1–4, 437 pp.

Sanchez, B. V. and Pavlis, N. K. (1995). The estimation of main tidal constituents from TOPEX/POSEIDON altimetry using a Proudman function expansion. *J. Geophys. Res.*, **100**, C12, 25229–25248.

Schrama, E. J. O. and Ray, R. D. (1994). A preliminary tidal analysis of TOPEX/POSEIDON altimetry. *J. Geophys. Res.*, **99**, C12, 24799–24808.

Schureman, P. (1958). Manual of harmonic analysis and prediction of Tides, *Special pub N°98*, U.S. Department of Commerce, Coastal and Geodetic Survey, Washington, 204 pp.

Schwiderski, E. W. (1980). Ocean tides, I, Global tidal equations, *Mar. Geod.*, **3**, 161–217.

Schwiderski, E. W. (1980). Ocean tides, II, A hydrodynamic interpolation model, *Mar. Geod.*, **3**, 219–255.

Schwiderski, E. W. (1982). Global ocean tides, 10, the Fortnightly lunar tide (Mf). *Atlas of tidal charts and maps*, Rep. TR 82-151, Naval Surface Weapons Center, Dahlgren, Virginia.

Schwiderski, E. W. (1983). Atlas of ocean tidal charts and maps, I, The semidiurnal principal lunar tide M2, *Mar. Geod.*, **6**, 219–265.

Seidler, G. and Paul, U. (1991). Barotropic and baroclinic tidal currents in the eastern basins of the North Atlantic. *J. Geophys. Res.*, **96**, 22259–22271.

Shum, C. K., Woodworth, P. L., Andersen, O. B., Egbert, G., Francis, O., King, C., Klosko, S., Le Provost, C., Li, X., Molines, J. M., Parke, P., Ray, R., Schlax, M., Stammer, D., Tierney, C., Vincent, P., and Wunsch, C. (1997). Accuracy Assessment of recent Ocean Tide Models. *J. Geophys. Res.*, **102**, 25173–25194.

Smith, R. D., Dukowicz, J. K., and Malone, R. C. (1982). Parallel ocean general circulation modeling, *Physica D*, **60**, 36–61.

Smithson, M. J. (1992). Pelagic tidal constants–3 *IAPSO Publication Scientifique*, N **35**, 191 pp.

Snodgrass, F. E. (1968). Deep sea instrument capsule, *Science*, **162**, 78–87.

Sovers, O. J., Jacobs, C. S., and Gross R. S. (1993). Measuring rapid ocean Earth orientation variations with VLBI. *J. Geophys. Res*, **98**, 19959–19971.

Taylor, G. I. (1919). Tidal friction in the Irish Sea, *Phil. Trans. Roy. Soc. London*, **A220**, 1–93.

Tierney, C. C., Parke, M. E., and Born, G. H. (1998). Ocean tides from along track altimetry. *J. Geophys. Res.*, **103**, 10273–10287.

Tierney, C. C., Kantha, L. H., and Born, G. H. (2000). Shallow and deep water global ocean tides from satellite altimetry and numerical modeling. *J. Geophys. Res.*, submitted.

Vlasenko, V. I. and Morozov, Y. G. (1993). Generation of semidiurnal internal waves near submarine ridge, *Oceanology*, **33**, 282–286.

Woodworth, P. L. (1985). Accuracy of existing ocean tide models. *Proc. Conf. Satellite Data in Climate Models*, Alpach, Autriche, 10–12, ESA SP–244.

Wunsch, C. (1967). The long-period tides. *Rev. Geophys. Space Phys.*, **5**, 447–475.

Wunsch C. (1975). Internal tides in the ocean. *Rev. Geophys. Space Phys.*, **13**, 1, 167–182.

Wunsch, C. and Gaposchkin, E. M. (1980). On using satellite altimetry to determine the general circulation of the ocean with application to geoid improvement. *Rev. Geophys.*, **18**, 725–745.

Wunsch, C. and Stammer, D. (1995). The global frequency wave number spectrum of oceanic variability estimated from TOPEX/POSEIDON altimetric measurements. *J. Geophys. Res.*, **100**, 24895–24910.

Wunsch, C., Haidvogel, D. B., Iskandarani, M., and Hugues, R. (1997). Dynamics of long period tides, in *Prog. Oceanog.*, **40**, 81–108.

Zahel, W. (1977). A global hydrodynamic degree model of the ocean tides: oscillation system for M2 tide and its distribution of energy dissipation. *Ann. Geophys.*, **33**, 31–40.

Zahel, W. (1991). Modelling ocean tides with and without assimilating data. *J. Geophys. Res.*, **96**, 20379–20391.

7

Ocean Surface Waves

J.-M. LEFÈVRE* and P. D. COTTON[†]

*Météo-France, 42 Av G
Coriolis, Toulouse 31057
France
[†]Satellite Observing Systems
15 Church Street
Godalming, Surrey GU7 1EL
U.K.

1. INTRODUCTION

Ocean surface waves provide perhaps the most spectacular manifestation of the sea-state. They reflect the local weather conditions under which they are developing, and also the conditions from other locations at an earlier time. Energy from the wind is propagated as waves and dispersed over the ocean surface until the waves finally meet an obstacle which causes them to break. The energy per unit time transported by the waves is considerable, hence the numerous incidences of damage caused to ships and fixed structures. The monitoring and prediction of sea-state conditions is therefore of primary importance for all activities which are exposed to waves. Knowledge of usual and extreme conditions for a given location, with seasonal information, is also very important to enable optimal design for a range of offshore activities.

Waves are often used to symbolize the dynamics of the oceans, even though they do not themselves actually represent a significant movement of water mass. This paradox precisely illustrates one of the main features of waves: a wave transports a disturbance from one place to another, but without a significant displacement of the medium itself.

It is possible to distinguish two major families of waves according to the type of differential equation which describe them: hyperbolic waves and dispersive waves. Sound waves and electromagnetic waves belong to the first family, but the majority of waves, including ocean swell, belong to the second family. A detailed description of wave theory can be found in Whitham (1974).

Ocean waves can be classified according to scale, in terms of their characteristic wavelengths and periods. At the smallest periods or wavelengths are capillary waves which are generated by the wind and maintained by surface tension at the air-sea interface. The characteristic wavelength of these capillary waves is about 1 cm. Next in scale are capillary-gravity wind waves, with wave lengths ranging typically between 1 and 10 cm. These waves play a fundamental role in the backscattering of radar signals. Capillary-gravity waves can be modulated by longer waves, which are also generated by the wind with gravity providing the restoring force. Gravity waves are the most familiar of the oceanic surface waves, and have wavelengths ranging between 10 cm and 1 km. On the very largest scales are seiches, tsunami, storm surges, and astronomical tides.

The total wave field is a quasi-linear superposition of many single waves, and so the distribution of surface elevations is almost Gaussian. This property is important and is used to model the response of a radar signal to the sea surface.

When the generating force for wind waves disappears, wave development ceases, but a swell wave remains. Such swell waves are then maintained purely by the gravitational restoring force. Now, smaller wavelength waves dissipate more quickly than the longer waves, and deep water waves are dispersive (waves with different wavelengths travel at different velocities). Thus waves of a certain wavelength and which have been generated by a common source will separate out as they travel (the longest waves traveling the fastest), with the consequence that swell often appears more regular than wind sea. Such a swell can propagate over very long distances, being able to cross several oceans, until it comes across an obstacle such as a coastline, shallow-water areas, or ice. This propagation occurs along great circle tracks, the shortest paths between two points on a sphere.

This capability of long waves to propagate over great distances from their point of origin is particularly helpful when attempting to predict swell. Moreover, the rapid improvement in computing facilities during the last 2 decades has made possible the development of global operational weather and wave prediction models. Suitable validation of such models requires observations with global coverage, now possible with over 12 years of global satellite altimeter wave and wind observations. Moreover the satellite altimeter data, which are more accurate than visual observations currently provided by selected ships, can be assimilated into numerical wave prediction models in order to improve the description of the sea-state and the wave forecast.

Model output can in turn be very useful for the rapid validation of altimeter wind wave products shortly after the satellite launch, since they are very dense in space and in time.

Another important application of altimeter wave measurements follows from the global coverage and continuity that these data provide, permitting the generation of global wave climatologies. These climatologies have enabled us for the first time to look in detail at large-scale patterns of variability in the ocean wave climate, and consider the connections between climate in different oceanic regions. Such studies have not been previously possible, due to the sparse and uneven distribution of surface data and the limitations of global scale wave models.

In the following sections of this chapter we first consider the use of altimeter data in conjunction with wind/wave models (Section 2), before moving on to their use in large-scale studies of wave climate (Section 3). A glossary of special terminology is provided at the end of the text.

2. WAVE MODELING AND ALTIMETRY

2.1. Numerical Wave Prediction

2.1.1. Basic Physics

When the wind blows over the sea, waves are generated in a range of directions and wavelengths, but the most energetic waves propagate in the wind direction. Initially generated by the turbulence existing near the water surface, waves then propagate under the action of gravity, developing further under the action of the wind by disturbing the air flow over the ocean surface. The distribution of the forces exerted by the wind, primarily the pressure forces, then lead to an exponential growth of the energy of each wave. A good description of wave physics can be found in Phillips (1977). Through

nonlinear energy-transfer mechanisms between waves, the wavelengths of the largest waves lengthen progressively as they develop, resulting in a decrease in the frequency of the energy peak of the wave spectrum. Meanwhile, the smallest waves continue to grow in height until they reach a maximum steepness beyond which they break. A balance between the energy dissipation and the wind input energy intervenes quickly for the shortest waves. A consequence of this balance is that the shape of the energy spectrum for these short-wavelength waves is simply a function of the wave number itself.

For longer waves, the spectrum has a form which depends only on the ratio between the wind speed and the phase velocity of the waves which contain most of the energy. The wind provides energy to those waves whose velocity is lower than the wind speed at the air-sea interface. This ratio characterizes the degree of development of the waves and completely determines the wave energy spectrum for a given wind. The resulting spectrum is characterized by large directional and frequency spreading, resulting in a sea-state which has a rather chaotic aspect, known as "wind sea."

When waves escape from their generation zone they become swell. Because small waves are quickly dissipated, and because gravity waves are dispersive in deep water (long waves are faster than short waves) swell has an energy spectrum narrower than wind sea, hence it is much more regular appearance. This regularity is increased when the swell approaches the shore. This is because the swell then moves through shallow water where its propagation velocity depends only on depth. In these situations, all the waves at a given location will move at the same speed.

2.1.2. Basic Equation and Validity Domain

Modern numerical model predictions of sea-state are based on the solution of the energy balance equation. This equation assumes that, in the absence of source/sink terms, the mean spectral energy density is conserved when waves are propagating in a medium of variable depth and in the presence of a uniform ambient current. It is valid when the amplitudes, wavelengths and periods of waves are slowly varying with their own space and time scales (the geometric optics approximation).

The energy balance equation therefore does not apply in the immediate vicinity of the shore, in the surf zone, or in zones which have very strong gradients in the bottom topography or very strong currents. The first numerical wave-prediction models (which simulated the evolution of the energy spectrum) date from the end of the 1950s, when the theories of Phillips (1957) and Miles (1957) on wave generation were not yet published. This period marked the beginning of the development of "first-generation" models, when nonlinear interactions were either not taken into account or only allowed for in a very partial way, so that the "overshoot" effect could not be reproduced. The "overshoot" phenomenon is a decrease of the energy peak which occurs just before the wind-sea energy spectrum becomes saturated, following a phase when the energy has been constantly increasing.

The importance of the nonlinear interactions in the wave growth stage was recognized after the JONSWAP Experiment (Hasselmann et al., 1973). Towards the end of the 1970s, a new family of "second-generation" wave models was then designed, and this type of model is still widely used by many National Weather Services. In these models, the nonlinear interactions are implicitly parameterized. Also, the wind-sea part of the energy spectrum is isolated from the remainder of the spectrum after the growth and dissipation terms are taken into account. The energy of the wind sea is then redistributed according to a spectrum determined by a few parameters. The JONSWAP spectrum is a famous example of wind sea spectrum.

Once well-calibrated, this type of model gives very satisfactory results, though it does retain some weaknesses, for instance when one is interested in a high temporal resolution of the evolution of the wave-energy spectrum and there is fast wind rotation. An explicit calculation of the non linear wave-wave interactions would remove the problem of separation between the wind sea and the swell, and hence avoid the need to use a prescribed spectrum. Nevertheless, for operational needs, the complete calculation of these interactions remains beyond the capabilities of existing computers. "Third-generation" wave models were only developed once it was found that only a limited number of the quadruplets had a significant contribution to the energy transfer.

Most recently in third-generation models, for instance, the WAM model (WAMDI, 1988), which was developed in the end of the 1980s, the nonlinear interactions have been explicitly parameterized, although still in an incomplete way. Such models are now running in several major weather centers.

Examples of spectra obtained from WAM and VAG (an operational French numerical wave model) are shown in Figure 1. They are compared to a wave spectrum obtained from an airborne scanning-beam radar (Jackson et al., 1985) during the SEMAPHORE experiment (Fradon et al., 2000). The radar spectrum has a 180° ambiguity, and therefore produces a symmetrical spectrum. The main components of the radar spectra are represented in both VAG and WAM. The wind sea (propagating to North West) is symmetrical by construction in VAG whereas it is not fully symmetrical in WAM.

For more details about the state of the art regarding the dynamics and modeling of ocean waves, see Komen et al. (1994).

2.1.3. Numerics

Numerical calculations in a wave model consist of the temporal integration of the balance equation at each point of a geographical grid, the resolution of which varies according to computer resources. Nowadays, for a global wave model (e.g., the WAM model operated at ECMWF, the European Centre for Medium-range Weather Forecasting), the typical spatial resolution is 0.5° while the spectral resolution provides 24 directions and 25 frequencies. This spatial resolution corresponds to that of the global weather atmospheric models which provide the winds needed by the numerical wave prediction models. For regional models, like the VAGMED model (Lefèvre, 1990) operated by the French Weather service, Météo-France, for the Western Mediterranean Sea, or the WAMED model operated by ECMWF, the spatial resolution is typically 0.25°.

In global Numerical Wave Prediction models, the spatial resolution improved from 3–0.5° within 6 years, thanks to the rapid increase in the capacities of super-computers. However, this progression is likely to slow down in the next few years because the model resolution is now close to the limits of validity of the equations governing the evolution of the wave spectra. Moreover, meso-scale models (mesh lower than typically 5–10 km) will run up against the problem of finding suitably high-resolution data for initialization.

2.1.4. Synthetic Parameters

For many applications, a simple knowledge of the main features of a wave spectrum is adequate. For instance, marine broadcast bulletins often provide the total significant wave height (SWH—defined as four times the square root of the variance of the sea surface elevation), but also provide an estimate of the height, period, and direction of the primary swell component. In order to produce such information from numerical wave prediction models, a wave spectrum is broken up into separate wave fields representing the wind sea, the dominant swell, the secondary swell, and so on. For each of these wave fields, one separately calculates the significant wave height, the average wave direction and the average wave period. The partitioning of a wave spectrum in this way was introduced by Gerling (1992).

The altimeter wave parameter most relevant, and most commonly used for the monitoring and validation of wave models, is the total SWH, although attempts have also been made to extract swell information from the altimeter measurement (e.g., Mognard, 1984). Further studies have also tried to characterize the sea-state in particular situations (Mognard et al., 1986, Mognard et al., 1991). In theory, when the energy in the altimeter SWH measurement is greater than that for the fully developed sea which would be generated by the local wind speed (as measured by the altimeter), the difference must be due to the presence of swell, thus a minimum swell energy can be calculated. The energy difference between that calculated for the measured significant wave height and that due to wind sea is exactly equal to the swell energy only when the wind sea is fully developed. If the wind sea is not fully developed, then the actual swell

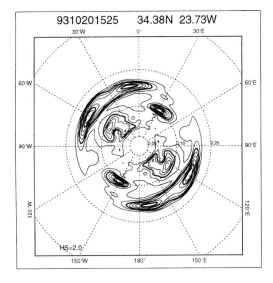

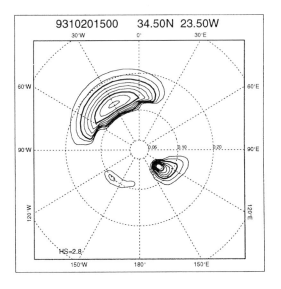

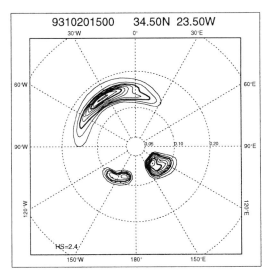

FIGURE 1   An example of a wave spectrum, obtained from an airborne scanning-beam radar (top left), with the wave model VAG (top right) and with the wave model WAM (bottom right), at the same location and time.

will contain more energy than this "minumum" swell. If the energy in the SWH measurement is lower than the energy of the fully developed sea, then no minimum significant swell height can be derived.

Recent research has also investigated the possibility of deriving an altimeter wave period parameter. Two approaches have been taken. Hwang *et al.* (1998) take the altimeter measurements of significant wave height and wind speed, and derive a function relating wave height, wind speed and wave period based on the assumptions of a "saturated" sea condition and negligible swell. The theory of Davies *et al.* (1998) notes that the zeroth- and fourth-order moments of the sea-surface spectrum are functions of SWH and radar backscatter, respectively. They then combine these

functions to derive an algorithm for wave period which they test, and empirically fit, against *in situ* data. However, both these studies are in their early stages, and large-scale verification is necessary before an altimeter wave-period parameter reaches a suitable stage of maturity to enable its acceptance by the research community.

2.2. Altimetric Applications

2.2.1. Altimeter Wind/wave Measurements

Satellite radar altimeters provide an estimate of SWH by measuring the slope of the return pulse leading edge, which is stretched out in time because of the delay between reflections from the wave crests and the wave troughs (Brown,

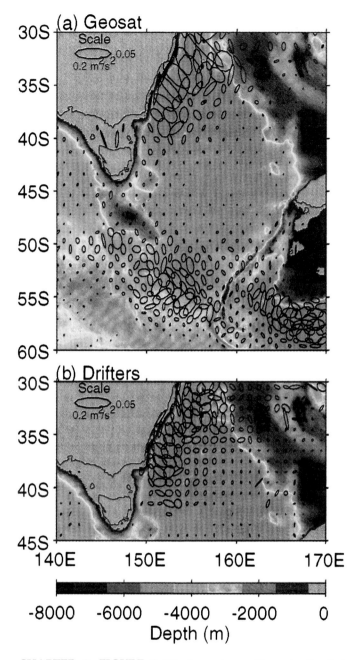

CHAPTER 3, FIGURE 9 Velocity variance ellipses in the East Australian Current region from (a) Geosat observations at crossover points and (b) long-term surface drifter data plotted over bathymetry. (From Wilkin, J., and Morrow, R.A., 1994. With permission.)

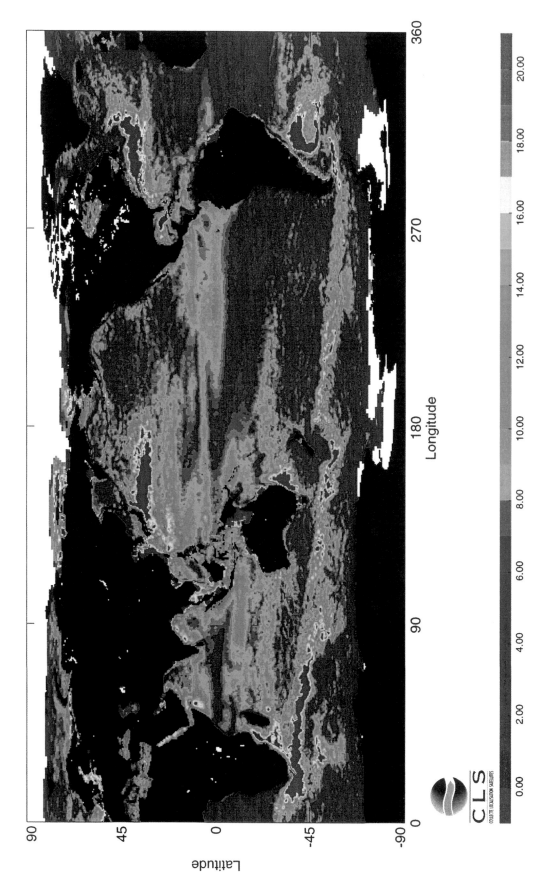

CHAPTER 3, FIGURE 17 The rms sea level anomaly from 5 years of TOPEX/POSEIDON and ERS-1/2 combined maps. Units are in centimeters. (From Ducet, N., and Le Traon, P.Y., 1999. With permission.)

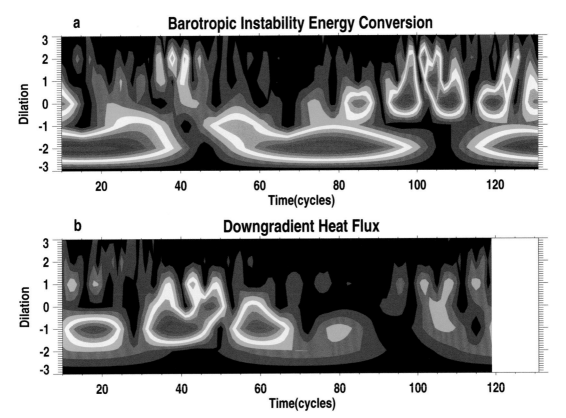

CHAPTER 3, FIGURE 24 Wavelet transform of the energy conversion for 33°N to 35°N due to (a) barotropic instability and (b) downgradient heat flux. The abscissa is the time in 10-day TOPEX/POSEIDON cycle numbers: Cycle 11 corresponds to January 1993, cycle 120 to December 1995, and the change in response occurs around cycle 70 (August 1994). The ordinate ("dilation") is a function of frequency: frequency doubles for each integer value of dilation, and zero dilation corresponds to mesoscale frequencies. The warmer shades indicate larger energy values within a frequency (dilation). (From Adamec, D., 1998. With permission.)

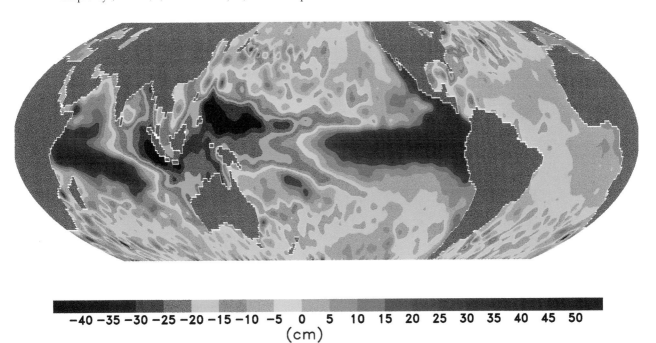

CHAPTER 4, FIGURE 1 Global sea-level transition from the height of El Niño to La Niña as illustrated by TOPEX/POSEIDON anomalies from December 1997 minus December 1998.

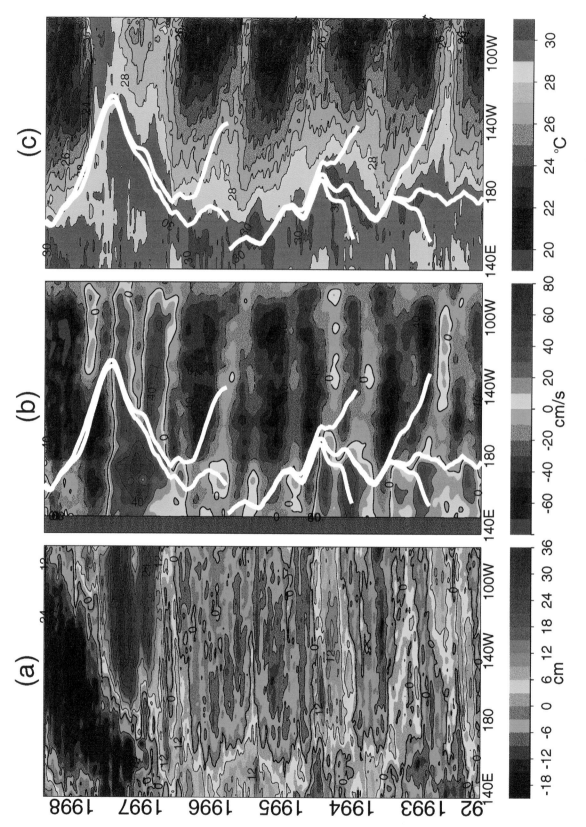

CHAPTER 4, FIGURE 7 Longitude-time distribution of the following parameters averaged within 2°N to 2°S: (a) TOPEX/POSEIDON anomalous sea level. (b) Sum of the TOPEX/POSEIDON derived zonal surface current anomalies and the climatological zonal currents along the equatorial band estimated from hundreds of near-surface drifter observations. Superimposed as thick white lines are the trajectories of hypothetical drifters moved by the zonal currents averaged from 2°N to 3°S. (c) Sea surface temperature; thick lines as in panel (b); the convergence of additional hypothetical drifters into a single trajectory emphasizes the existence of an oceanic zone of convergence at the eastern edge of the warm pool.

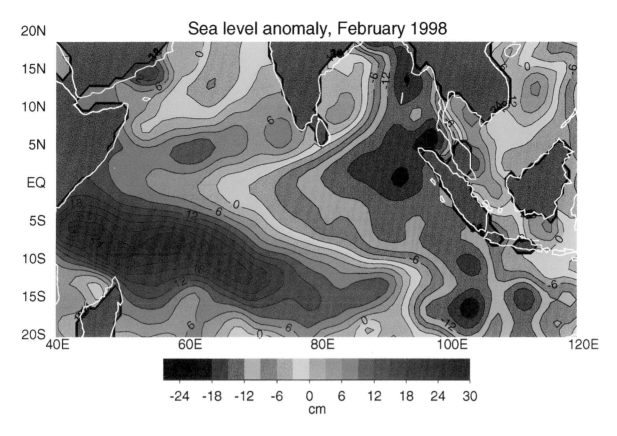

Sea level anomaly, February 1998

-24 -18 -12 -6 0 6 12 18 24 30
cm

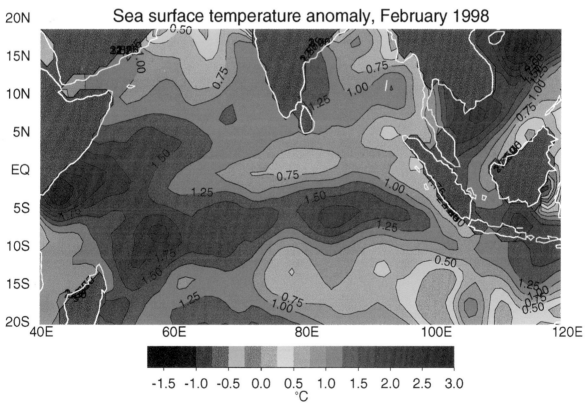

Sea surface temperature anomaly, February 1998

-1.5 -1.0 -0.5 0.0 0.5 1.0 1.5 2.0 2.5 3.0
°C

CHAPTER 4, FIGURE 8 Monthly mean fields in the Indian Ocean in February 1998. (a) TOPEX/POSEIDON sea level anomaly, (b) NCEP sea-surface temperature anomaly.

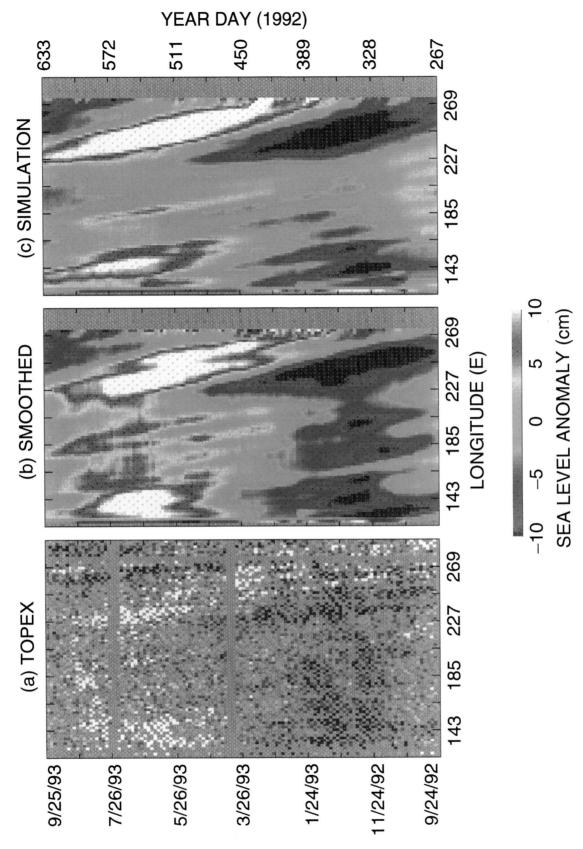

CHAPTER 5, FIGURE 1 Longitude vs. time plot of sea level anomalies along 12.5°N measured by TOPEX (a), assimilation (b), and model simulation (c). Color scale is between ±10 cm. Gray denotes missing values either because of land or missing data (in case of TOPEX). The resolution of the plots are 2° by 1° horizontally and 3-days in time. The model is a wind-driven, reduced gravity shallow water model, and the assimilation is performed by an approximate Kalman filter/smoother. (Adapted from Fukumori, I., 1995, Plate 2, p. 25030.)

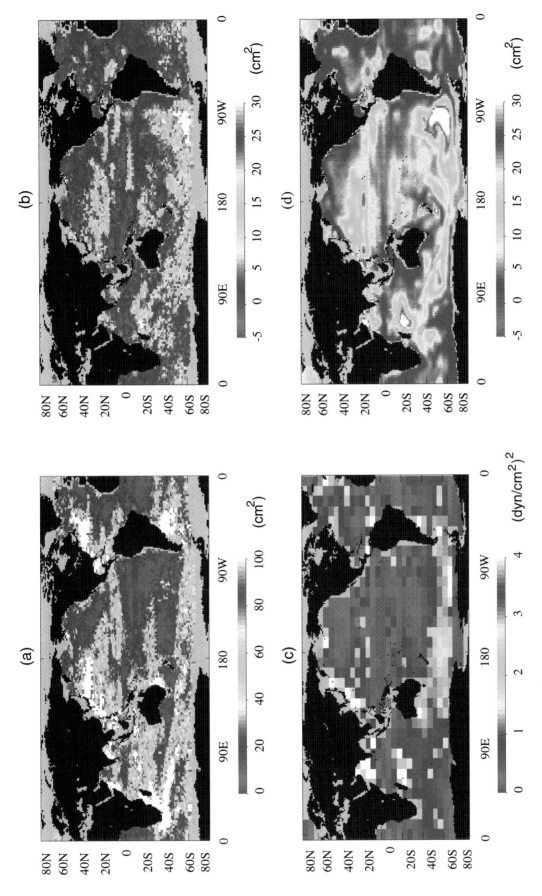

CHAPTER 5, FIGURE 9 An example of Error Calibration; (a) errors of altimetric sea level constraint (Eq. [23]), (b) errors of model simulated sea level (Eq. [24]), (c) calibrated (zonal) wind stress error (units in (dyn/cm²)²), (d) model simulated error based on (c). All units are sea level variance (cm²) except (c). The model is based on the GFDL MOM with TOPEX/POSEIDON data assimilated using an approximate Kalman filter and smoother. (Adapted from Fukumori, I., *et al.*,1999, Plate 1.)

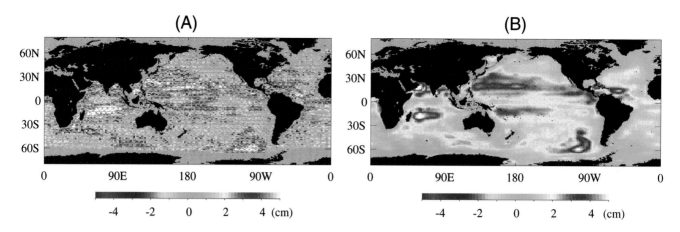

CHAPTER 5, FIGURE 12 Comparison of an altimetric assimilation with expectations. The figures show reductions in root-mean-square model-data sea level differences by assimilation of TOPEX/POSEIDON data into a global OGCM; (A) actual improvement, (B) expected value. Units in centimeters. Positive numbers indicate smaller residuals after assimilation. Model and assimilation are based on those described in Fig. 9. The residuals of the assimilation are for the innovation vector (Section V.A.2). (Adapted from Fukumori, I., *et al.*, 1999, Plate 3.)

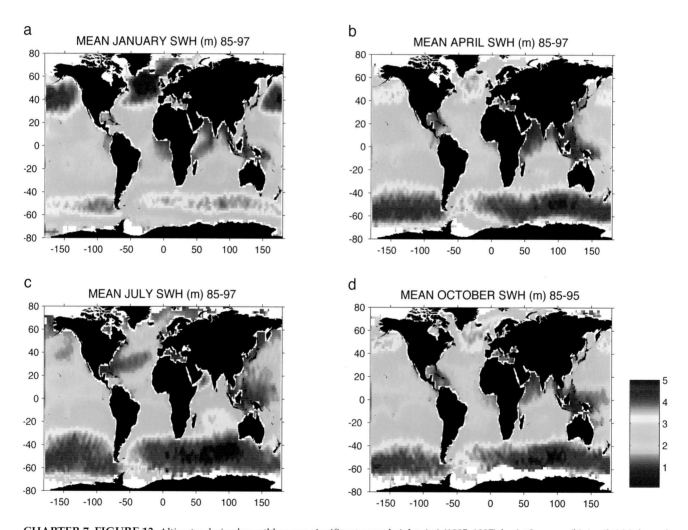

CHAPTER 7, FIGURE 13 Altimeter derived monthly mean significant wave heights (m) (1985–1997) for (a) January, (b) April, (c) July, and (d) October.

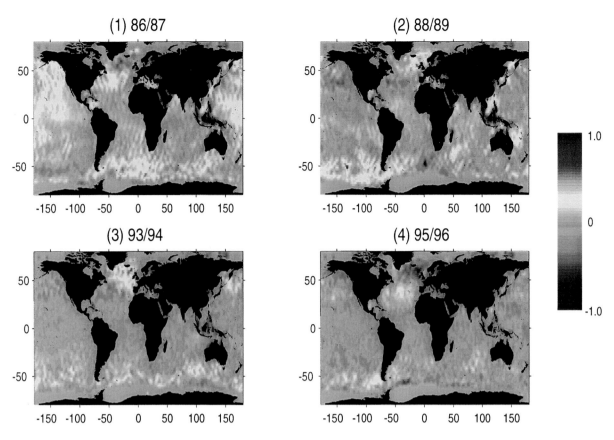

CHAPTER 7, FIGURE 22 Significant wave height anomalies from the long–term winter mean for December–January averages of: (1), 1986/87; (2), 1988/89; (3), 1993/94, and (4), 1995/96.

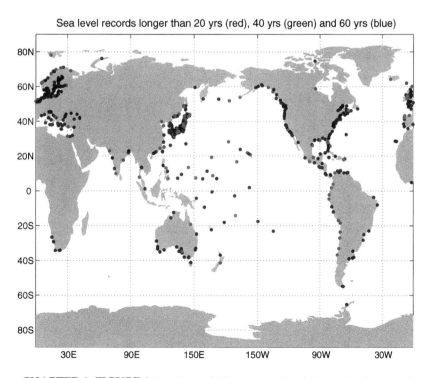

CHAPTER 8, FIGURE 1 Locations of tide gauge sea level time series that exceed 20 years (red), 40 years (green), and 60 years (blue) in length.

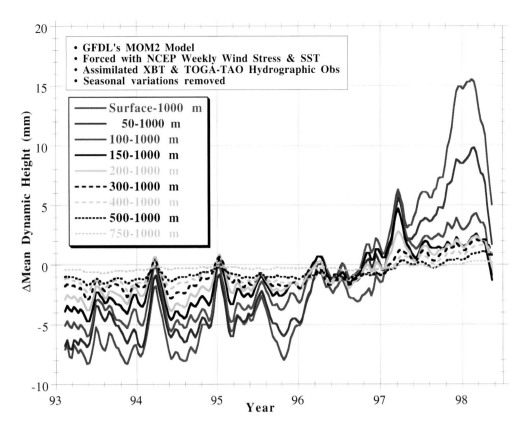

CHAPTER 8, FIGURE 14 Global mean dynamic height variations, at several depths, using the MOM2 model and a level of no motion of 1000 m.

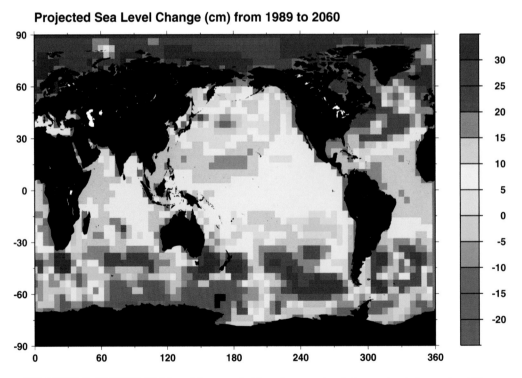

CHAPTER 8, FIGURE 15 Map of the expected long-term sea level change due to increasing CO_2 as predicted by a global climate model. (From Russell, G.L., *et al.*, 1999. With permission.)

Local trends of sea level

Local trends, first 4 EOFs removed

cm/yr

CHAPTER 8, FIGURE 16 Sea level trends during the T/P mission (1993–1998) (top map), and the same trends after removing the 4 leading EOF modes (bottom panel). For these maps, an IB correction was applied to the data.

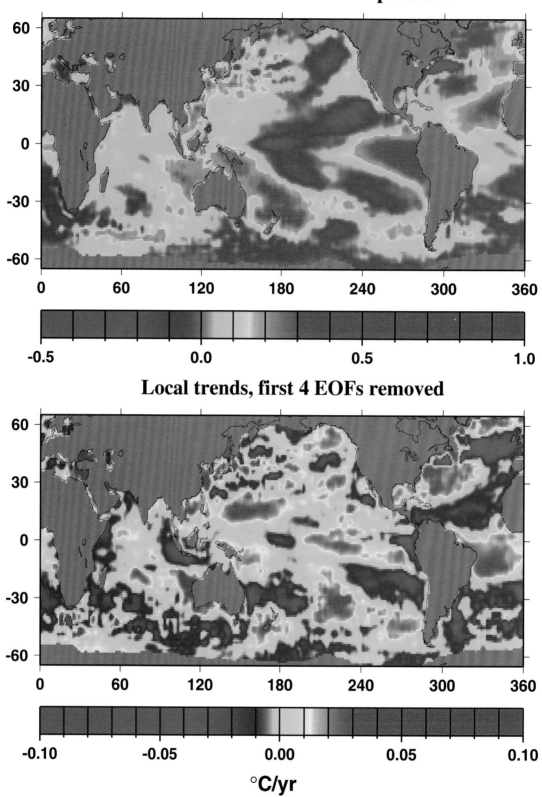

CHAPTER 8, FIGURE 17 Same as Fig. 16, but for changes in sea surface temperature during the T/P mission.

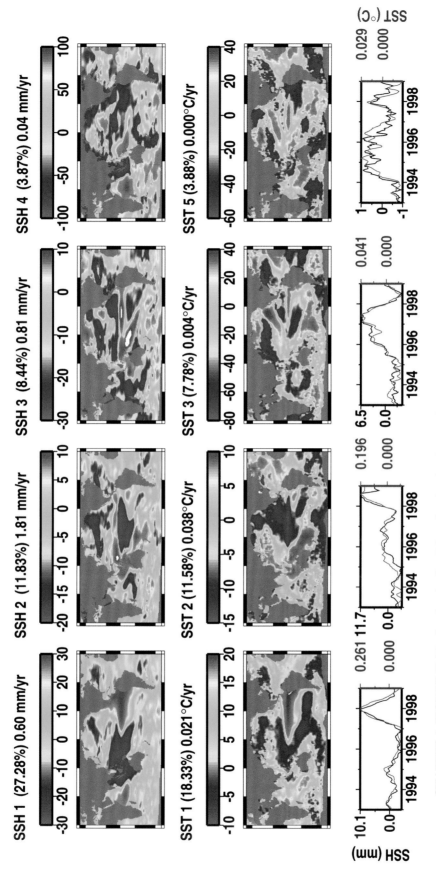

CHAPTER 8, FIGURE 18 The first four leading EOFs of sea level and sea-surface temperature. The temporal modes have been scaled to represent their contribution to the global mean variations in these quantities as shown in Fig. 8. To prevent contamination of the EOF modes, an IB correction was applied to the data.

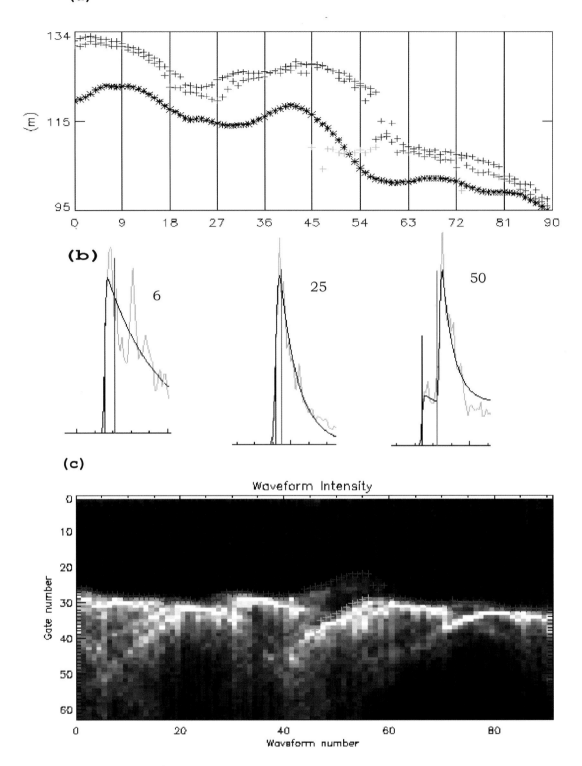

CHAPTER 9, FIGURE 7 (a) Height profile over a portion of the Antarctic Ice Sheet (corresponds with Fig. 7 waveforms). Black—unretracked, Blue—20% threshold retracker, Red—GSFC V4 retracker, Green—GSFC V4 retracker 2nd peak. (b) ERS-1 Ice Mode Altimeter Return Waveforms over the Antarctic Ice Sheet with the GSFC V4 retracker function. (c) Waveform intensity plot corresponding with data in 7a. The red line shows the elevation from the first ramp from the GSFC V4 retracked surface and the green show the elevation from the second ramp from the GSFC V4 retracker.

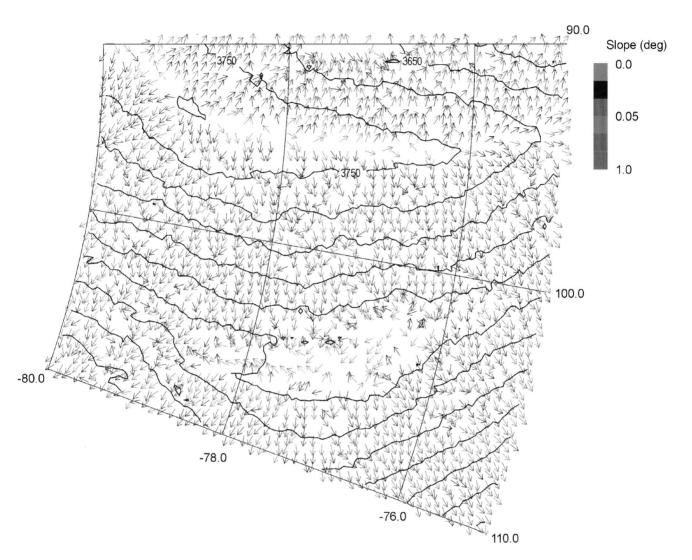

CHAPTER 9, FIGURE 9 Sample area of East Antarctica showing surface topography in 100-m contours and slope-gradient vectors for squares of 5 km on the side. The features illustrated are described in the text.

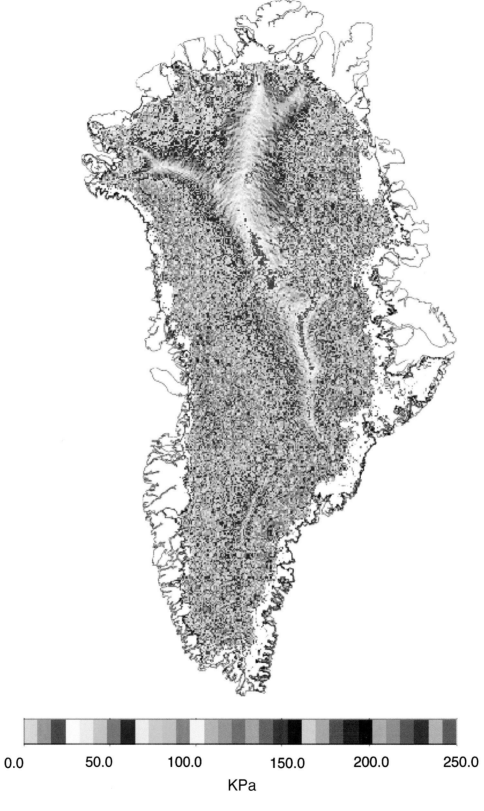

CHAPTER 9, FIGURE 10 Basal shear stresses for Greenland derived from surface typography from satellite altimetry and bottom topography from airborne radar sounding. Minimum stresses occur at ice divides.

1977). Altimeters provide an estimate of the ocean surface wind speed through the measurement of the radar cross section, which is a function of the small scale roughness of the sea surface. The first algorithms to convert radar cross section to surface wind speed were developed for Geos 3 altimeter data (Mognard and Lago, 1988; Brown, 1979). Subsequently, further wind algorithms were developed and tuned using comparisons with buoy data (Brown *et al.*, 1981, Goldhirsh and Dobson, 1985), with satellite scatterometer data (Chelton and McCabe, 1985; Chelton and Wentz, 1986; Witter and Chelton, 1991) or with output from Numerical Weather Prediction models (NWP) analyses (Freilich and Dunbar, 1993; Lefèvre *et al.*, 1994). A review of most of these algorithms, together with an assessment of their accuracy, can be found in Lefèvre *et al.*, (1994).

Since the radar footprint has a typical size of about 7–9 km, depending on the sea-state, the satellite altimeter measurement represents a spatial average of wind speed and significant wave height over an area of the ocean. When wind and wave altimeter measurements are compared to buoy measurements or NWP analyses, the standard deviation of their difference is typically about 0.5 m (or 10%) for the SWH and 1.5–2 m sec^{-1} for the wind speed (depending on the algorithm).

2.2.2. Sea-state Monitoring

The ability of a satellite altimeter to measure ocean wave heights was first demonstrated following the launch of GEOS3 (Rufenach and Alpers, 1978, Mognard and Lago 1979) followed by Seasat in 1978 (Fedor and Brown, 1982), but it has only been in the last few years since the launch of ERS-1 in July 1991, that near real-time data has been distributed to the national weather services. The ability of the satellite altimeter to measure wave heights and wind speed accurately has been demonstrated by several studies (Dobson *et al.*, 1987; Glazman and Pilorz 1990; Carter *et al.*, 1992, Cotton and Carter, 1994; Queffeulou *et al.*, 1994). Data from Geosat (1985–1990), TOPEX/POSEIDON (1992–present) ERS-1 (1991–1996), and ERS-2 (1995–present) are now available and the combination of these data enables the generation of multi-year climatologies.

Surface observations are much less dense over sea than they are over land, due to the significantly higher cost of making sea based measurements. As an illustration of this, we can consider the Western Mediterranean Sea, which is a region with a high level of traffic, but which only received on average approximately 100 observations per day in 1996 (from the IGOOS, International Global Ocean Observing System). For an equivalent terrestrial area such as France, the number of daily observations from synoptic stations put on the GTS (Global transmitting System) is 25 times greater, at about 2500. Moreover, the number of observations at sea falls dramatically during storm events. Satellite altimeters generate observations in all conditions (except during very heavy rainfall) and significantly increases the number of observations at a cost bearable to the NWSs.

Satellite wave data, in particular altimeter data, are a significant additional source of information for experiments at sea. Thus, satellite wave data were widely used for studies (Fradon *et al.*, 1996, 1999) which followed the SEMAPHORE experiment (Eymard *et al.*, 1996) held in the vicinity of the Azores in 1993. Satellite data will be also used with the data collected during the FETCH experiment (Hauser *et al.*, 2000), which took place in spring 1998 in the Gulf of Lions (Mediterranean Sea). One important goal of this experiment is the study of wave generation at short fetches and under strong wind conditions (in particular during Mistral winds events).

2.2.3. Numerical Model Validation

While the spatial coverage from a satellite is much higher than that given by *in situ* measurements, the temporal coverage for any given location is, in general, poorer. It should also be noted that measurements from altimeters (and scatterometers) are unreliable in the coastal zones. Thus conventional and remote-sensed data are in many ways complementary. The importance of altimetric data with regard to the validation of NWP models has been demonstrated in a number of papers, for instance by Guillaume *et al.* (1992). In particular, altimeter data from Geosat and ERS have made it possible to validate numerical wind and wave models in areas where the wind and wave fields showed a strong spatial variability (Guillaume *et al.*, 1992). The ability of an altimeter to provide estimates of the wind velocity is very useful when analyzing wave model errors. In particular, it allows one to distinguish between errors in the input wind field from those caused by the wave model itself. In certain areas, it is also possible to use the wind and wave model outputs together to validate the winds from altimeters (Guillaume and Mognard 1992), or even to derive a geophysical model function for the wind speed (Lefèvre *et al.*, 1994).

2.2.4. Monitoring Tropical Cyclones

Another valuable application for altimeter data follows from their ability to measure the sea-state and the surface wind speed generated by tropical cyclones (hurricanes). Although it is difficult to estimate wind speeds higher than 25 m sec^{-1} with a reliable accuracy, altimeter data can provide useful information on the structure of a cyclone. The degree of asymmetry can be estimated, as can the extent of the regions (along the satellite track) where the wind speed is higher than about 17 m sec^{-1} or 34 knots (gale force 8). In fact, altimeter wind-speed algorithms have been proposed for winds higher than 25 m sec^{-1} (Young 1993).

Because the fetch is much larger in the areas of a cyclone where the wind speed is less than 50 knots than in the regions of higher wind speed, the structure of these lower wind speed areas of a cyclone is of great importance with

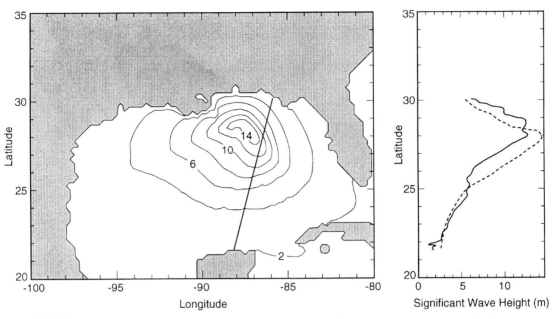

FIGURE 2   (Left panel) contour map of the significant wave height obtained from a numerical wave model driven by winds from a parametric and analytic model during hurricane conditions in the Gulf of Mexico. (Right panel) numerical model SWH (dashed) compared to TOPEX/POSEIDON altimeter SWH (solid). Contour intervals are 1 m. (From Lefèvre *et al.*, 1998. With permission.)

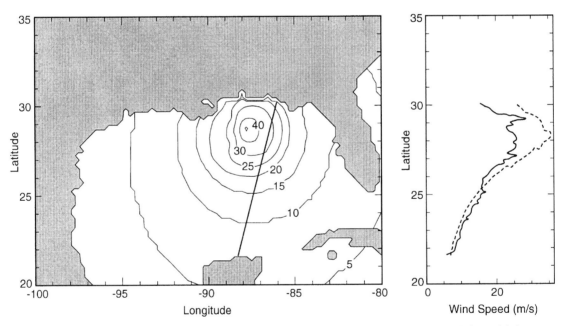

FIGURE 3   Left panel, contour map of the wind velocity obtained from a parametric and analytic model during hurricane conditions in the Gulf of Mexico. Right panel, numerical model wind speed (dashed) compared to TOPEX/POSEIDON altimeter wind speed (solid). Contour intervals are 10 m sec^{-1}. (From Guesquin and Lefèvre, 1998. With permission.)

regard to the prediction of hurricane generated waves. Thus when an altimeter flies over a hurricane, it allows a partial validation of the warning messages produced and transmitted by the Tropical Prediction Center based at the National Hurricane Center (NHC) in Miami. It also enables correc-

tions to be appended to these warning messages. An example simulation of hurricane winds and waves is given in Figures 2 and 3. Using information from NHC warning bulletins, a numerical wave model is driven by winds generated from a parametric hurricane model. The wave model is derived from the VAG model operated at the French Weather Service Météo-France. The significant wave height (Figure 2, left panel) and the wind speed (Figure 3, left panel) from the numerical models are compared to values from the TOPEX/POSEIDON altimeter (right hand panels) at positions along the satellite ground track. Despite the fact that the altimeter wind speed data in this study were limited to values below 25 m sec^{-1}, it is possible to partially assess the validity of the bulletins from NHC. One can see in Figure 3 that for wind speeds less than 15 m sec^{-1} the wind speed deduced from the NHC bulletins are in good agreement with altimeter in the southern part of the hurricane. However, for speeds greater than 15 m sec^{-1}, the winds from NHC seems too high in the southern part and too low in the north. Since the significant wave height is not fully correlated with the wind speed, it is difficult to draw any definite conclusion when comparing Figure 2 with Figure 3, although it does seem that the NHC overestimation of higher winds in the southern section is well correlated with an overestimation in the model significant wave height. Unfortunately, however, a similar correlation is not seen in the northern sector of the hurricane.

Clearly, if one is to develop a full understanding of the wave field generated by a hurricane it is important to take into account the history of the hurricane development. Unfortunately, in practice it is often difficult to get altimeter information at the location of a hurricane at a time close to the hurricane's passage. Even so, for much of the time the altimeter provides the only information that can be used to validate a wave prediction system in hurricane conditions. Thus the altimeter allows an assessment of the NHC bulletins, based on independent data, and enables corrections to be made to the forecast. The possibility in the near future of several altimeters delivering real-time sea-state information will further improve our ability to monitor and forecast waves generated by hurricanes.

2.3. Operational Data Assimilation

2.3.1. Introduction

For many years, operational wave models did not routinely assimilate wave data to correct the initial model state. Part of the reason for this lack of action lay in two perceptions about the influence such assimilation was likely to have. First, it was believed that the initial state of the wave field had a limited long-term influence on subsequent wave fields, the state of which was dominated by other external parameters. While for weather forecasts the evolution of atmospheric conditions is mainly controlled by the atmospheric initial state, in wave models the initial wave field loses its influence after a time which depends primarily on the basin size and on the atmospheric dynamic time scale (from a few hours to a few days). Second, in contrast to atmospheric

models, the wave forecast is strongly constrained by an external data source, the wind field. In theory, a perfect wave model driven by perfect winds should, after a certain time, produce a perfect wave field, whatever the initial state. However, the same cannot be said for models of a real atmosphere whose unstable behavior makes it very sensitive to its initial conditions.

In earlier years, before the advent of the satellites, little reliable wave data were available for assimilation, and so it is not surprising that a positive contribution has been seen to follow from the assimilation of altimeter wave data in an open ocean NWP model. Only local applications could be expected to significantly benefit from the assimilation of *in situ* data, in areas where instrumental measurements were available on a useful scale (for example in the North Sea, where platforms provide regular sea-state measurements). In fact, most *in situ* data are still derived from visual observations, when the sea-state is necessarily estimated in a rather subjective way. This can lead to significant uncertainties in the data rendering them unsuitable for numerical forecasts, although the data are of unquestionable value in the control and monitoring of NWP models.

The advent of satellite data encouraged NWSs to investigate the possibility of including data assimilation schemes in their operational wave forecast suite. As the winds provided by the analyses are from atmospheric models, and the numerical wave models are themselves not perfect, it seems clear that the assimilation of satellite data must improve the initial sea-state analyses. This assimilation must then also improve the wave forecasts, because the corrected initial wave field is then propagated by the NWP model. Moreover, the assimilation of wave data also allows the wind speeds from the atmospheric model analyses to be adjusted so that they are consistent with the assimilated wave fields.

The level of improvement in wave prediction following the assimilation of wave data depends on the assimilation procedure used. The methods which were first developed were obviously the simplest and the least expensive in terms of computer resources. A numerical wave model requires a computer memory size similar to that necessary for an atmospheric model of the same horizontal resolution, and thus there are significant constraints to be considered when adding extra processing stages. The most frequently used procedures for operational applications are based on instantaneous sequential methods such as Optimum Interpolation (O-I) and successive corrections.

More advanced methods, based on variational schemes (De la Heras *et al.*, 1994), have now been developed. Such methods often use the adjoint technique which, under the constraint of the wave model dynamics, take into account the history of the observations. However, in the case of operational forecasting, the high cost of these methods has slowed down their development. Significant research efforts are still necessary in order to develop them for operational use.

An alternative to the adjoint technique, based on Green's functions, has been proposed by Bauer *et al.* (1996). Although this new technique is much less computationally expensive, it does rely on certain strong assumptions. In this technique, Green's functions are used to approximate the response of the wave spectrum to a disturbance in the forcing wind field. Then, in contrast to four-dimensional (4-D) variational methods, a cost function which takes into account observations at a single time only is minimized. However, this technique does rely on the assumption that the model perfectly represents the wave physics. Moreover the Green's function technique cannot give an accurate estimate of the error on the restored wave field nor on the model forcing variables (e.g., the wind vector).

In contrast, Kalman Filter (KF) techniques can provide estimates of error, both in the forecast wave field and the forcing wind field. However, the basic KF techniques are very computationally expensive, and so must be simplified before they can be used effectively in operational models. Chapter 5 provides a more detailed discussion of assimilation techniques.

2.3.2. Examples of Operational Altimeter Data Assimilation

One of the first operational satellite data assimilation procedures was based on the O-I method. Such a procedure, based on the work of Lionello *et al.* (1992), was implemented on WAM at ECMWF in 1994. The main advantage of this method was its low computer cost, though some limitations should be recognized. The originality of the Lionello method, but also the limitations, stem from the way the altimeter information (SWH only) is distributed into the energy spectrum. The model variables are components of the discretized wave spectrum. The SWH is calculated as the integral of the wave spectrum over the frequency domain and the directional domain, and so strong assumptions are required to link the altimeter information to the model wave spectrum. The total wave energy, obtained by integrating the energy spectrum, is modified through an O-I process with the altimeter data. It is then assumed that the ratio between wind sea and swell energies is constant, and the total energy obtained after O-I of the SWH is redistributed in the wind sea and swell parts of the energy spectrum, based on certain properties of the wind sea and of the swell.

Consider first the wind sea. The JONSWAP spectrum used in the model depends only on the wave age. One can therefore rebuild a new wind-sea spectrum, based on the new total energy, assuming that the degree of development of the waves was correctly estimated in the model first guess. This procedure also assumes that the wind duration was correctly estimated in the model first guess. A further advantage of this method is that it also provides a new wind-speed estimate, which will be consistent with that of the wind sea. Second, we consider the swell. The redistribution of the

swell energy is based on the property (identified from analyses of the WAM model) that the average mean steepness of the swell is the same for any given decay time. If it is assumed that the decay time of the swell was correctly estimated by the model, one can thus compute the swell mean wavelength, and consequently its corresponding mean frequency. The method used in WAM does not allow a modification of the directional energy distribution or of the ratio between wind sea and swell energy.

Some of the consequences of this method, in particular the limitations on the wave prediction, were discussed by Lefèvre (1992). Nevertheless, the method does improve the significant wave-height forecast. This improvement is significant for short-range forecasts (up to 2 days) in windy areas and for medium-range forecasts (3–10 days typically) in areas with low wind where swell is dominant. We can assess the impact of assimilation in terms of the dispersion index, which is defined as the ratio between the standard deviation error and the mean value of the reference. By this measure, after 1 day the impact of assimilation is reduced by a factor of about three in the tropics and by four on the whole globe (Komen *et al.*, 1994). The impact is reduced by a factor of 10 after 5 days in the tropics, and after 2 days globally.

This significant short-term reduction in the impact is probably partially caused by the assumptions used in the assimilation. In particular, a better estimate of the swell should increase the duration of the impact. A method to improve the swell estimate, based upon modifications to the WAM method, was established and tested in the VAG model operated at Météo-France. The most significant of the modifications was the use of altimeter wind information, allowing the wind sea to be distinguished from the swell. This, in turn, meant that the previously held assumption of a conservation of the ratio between the swell and the wind-sea energy was dropped.

For a given degree of development, a wind-sea spectrum can be entirely determined either by the wind speed, the peak frequency, or the energy. In the WAM assimilation method, the wind-sea energy is obtained by assuming that the wind sea and swell energies maintain a constant ratio with respect to each other, whereas an option within the VAG method allows the altimeter wind data to be used to estimate a new wind sea energy. Note that this option can be used only for the winds lower than 25 m sec^{-1}, because of the possible inaccuracies in higher values of altimeter wind speed (Lefèvre *et al.*, 1994). The impact of assimilating the altimeter winds was tested during the SEMAPHORE experiment in the North Atlantic (Eymard *et al.*, 1996). Unfortunately, there were no clear improvements in the predictions of significant wave height. If repeated, the experiment could be improved by including an analysis of the spectra, the use of a longer testing period (the SEMAPHORE test lasted 2 months), and the careful examination of individual case studies.

nodata waves 93.10.04, 14 UT   TOPEX SWH

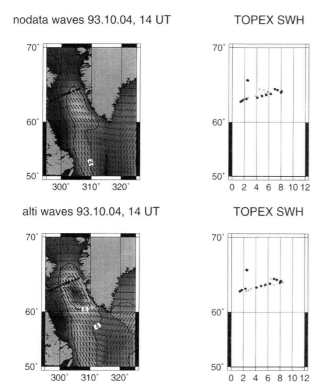

alti waves 93.10.04, 14 UT   TOPEX SWH

An example of the impact of assimilating ERS-2 altimeter wave heights is shown in Figure 4. ERS-2 data have been assimilated into a wave model, whose output is then compared to independent TOPEX/POSEIDON measurements. The left-hand panels of Figure 4 show the contours of the model outputs; the right-hand panels show observations (squares) and model outputs along the T/P satellite track. In the top panels, no ERS data were assimilated, whereas in the bottom panel, ERS data have been assimilated (at a different place and different time to the T/P measurements). In this case, there remains some wave-height underestimation in the model output (compared with TOPEX/POSEIDON data), because of a quasi-local model wind underestimation (not shown). The modified assimilation (Le Meur *et al.*, 1995) method was then implemented in a global version of the VAG model operated by the French Weather Service, Météo-France (Lefèvre *et al.*, 1996, 1998). This version, with a spatial resolution of $1° \times 1°$, has been running on a quasi-operational basis to produce daily analyses. An example of the significant wave height field from a global wave analysis is shown on Figure 5. This figure provides a typical overview of the number of stormy areas and their size. ERS tracks during a typical wave-correlation period of 6 hr are given in Figure 6, and demonstrate the coverage a single satellite can provide. The effect of the assimilation on the analysis can be seen in the analysis increment field, which gives the difference between the first guess fields and the modified fields (Figure 7).

FIGURE 4 (Top-left panel) contour map of SWH from the VAG wave model driven by the wind analyses from the ARPEGE atmospheric model (both models are operated by Météo-France), without assimilation (NODATA). (Top-right panel) model SWH compared to the TOPEX/POSEIDON SWH. Contour intervals are 1 m. Bottom panels, as for top but after assimilation of ERS2 altimeter wind/wave data.

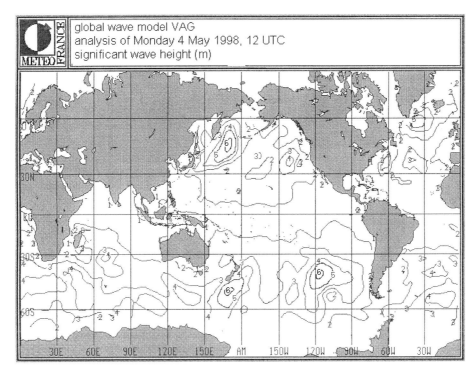

FIGURE 5 Analysis of the SWH from the global version of the Météo-France VAG wave model, in which ERS2 altimeter wind/wave data have been assimilated.

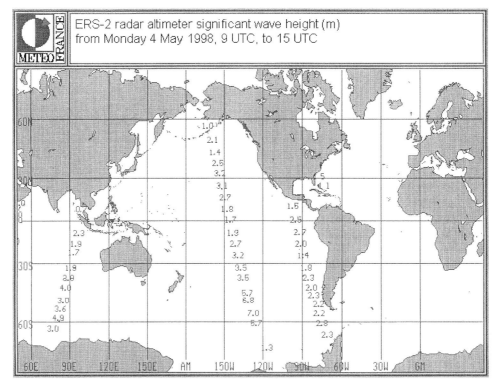

FIGURE 6   Typical ERS2 satellite coverage obtained during a time window of 6 hr, the SWH correlation time scale.

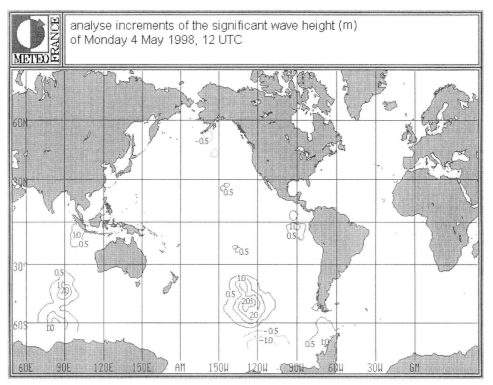

FIGURE 7   Example of analysis increments, giving the difference between the analysis and the model first guess of the SWH from the Météo-France global VAG model. Contour intervals are 0.5 m.

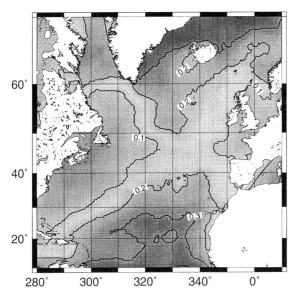

FIGURE 8   Mean difference between the SWH fields obtained with and without assimilation of ERS2 altimeter data during a 2-month period on the North Atlantic. (From Le Meur *et al.*, 1995. With permission.)

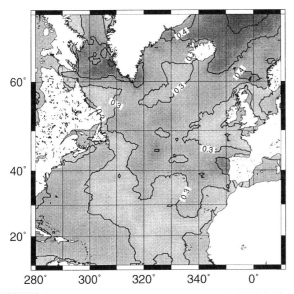

FIGURE 9   Root mean square difference between the SWH fields obtained with and without assimilation of ERS2 altimeter data during a 2-month period on the North Atlantic. (From Le Meur *et al.*, 1995. With permission.)

The impact of the ERS altimeter data assimilation during the SEMAPHORE experiment was evaluated by comparing the significant wave-height fields (with and without assimilation of ERS data) against the altimeter SWH measurements from the TOPEX/POSEIDON satellite. About 40,000 data points were available for analysis after quality control and averaging to the wave-model analysis resolution. The mean differences between the significant wave height obtained with and without the assimilation of ERS data over the whole period is shown on Figure 8. The positive signifi-

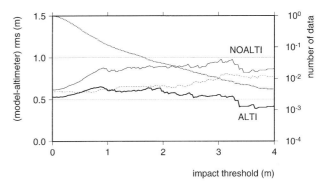

FIGURE 10   Root mean square error of the SWH fields obtained with assimilation of ERS2 altimeter data (ALTI) and without any assimilation (NODATA). Error calculated with respect to TOPEX/POSEIDON measurements as a function of the impact of the altimeter data on the mean sea-state. The dotted line gives the ratio of data points exceeding a given impact threshold, the dashed line indicates the proportion of favorable ALTI cases between 0 and 1 in the same scale as the error in meters. (From Le Meur *et al.*, 1995. With permission.)

cant bias found at low latitudes suggests an underestimation (in the unassimilated model output) of the tradewinds, a previously recognized shortcoming of weather model forecasts. The standard deviations of the differences (Figure 9) exhibits a main impact in terms of variability at mid- and high latitudes with maxima for the SWH of about 50 cm. When the model output is evaluated against TOPEX/POSEIDON data, the errors curves obtained (Figure 10) denote a clear improvement in the presence of ERS altimetric data.

The improvement is significant everywhere, with a reduction of the root mean-square error by more than 10% (64–57 cm), and is significant at a 95% confidence level for all waves up to a threshold in SWH of 2.5 m (the number of data points exceeding this threshold, indicated by the dotted line, is about 400 or 1% of the total number of data). The ratio of favorable cases, when the ERS altimeter has improved the analysis, is always above 55% (dashed line on Figure 10).

2.4. Use of Model Data to Improve Altimetry

As noted earlier, model output can, in its own right, be very useful in the rapid validation of altimeter wind-wave products in the important period shortly after the satellite launch. Numerical weather and wave prediction models now provide dat, at a global scale with an horizontal resolution of up to 0.5°, typically every 6 hr, and the good quality of the analyses makes them suitable for use in the validation of altimeter data. The main advantage of model output over buoy data is the much larger number of satellite and model collocations that can be generated within a short period after the launch of the satellite. Thus, model data from Météo-France and ECMWF have been used to validate wind and wave products from ERS-1 (Queffeulou and Lefèvre, 1992; Hanssen and Guenther, 1992) and

TOPEX/POSEIDON (Lefèvre *et al.*, 1994). In particular, the comparison of altimeter wind speed and significant wave height with the corresponding model output from ECMWF helped identify a number of problems in the ERS-1 retrieval algorithms for wind speed and wave height. Since altimeter data are now assimilated in some wave models, one must be careful when considering using such data to validate other altimeter data. Of course, the use of *in situ* data to provide accurate validation is still necessary.

Wave models also provide wave spectra, which are unavailable from altimeters. In the future, this information should help develop a better understanding of the radar signal as returned from the sea surface. The current forms of algorithms used to compute the "EM Bias" range correction (Electro Magnetic Bias—a bias in the altimeter range measurement resulting from the irregular nature of the sea surface) are functions of altimeter wave height and wind speed, based on the fact that a correlation has been found between the EM Bias and the sea maturity (Fu and Glazman, 1991). Since the wave spectrum contains a great deal more information about the nature of the sea surface than the single parameter represented by the maturity, it is very likely that the model wave spectra should be useful in providing an improved estimate for the EM Bias.

3. WAVE CLIMATE STUDIES WITH ALTIMETER DATA

3.1. Introduction

A knowledge of the large-scale climate of the ocean surface, in terms of the seasonal patterns and natural variability, is of central importance to climate studies, and is also of great value in a wide range of practical applications, including transport, offshore exploration, fisheries, insurance, and marine design technology. Clearly, previous assumptions of a largely stationary climate, with predictable seasonal behavior, are not true, and a large inter-annual variability exists. The need to understand, and perhaps eventually predict, this inter-annual variability has provided the motivation behind the studies presented below.

Satellite data are the only data which permit such studies. *In situ* wind and wave measurements are only sparsely and unevenly distributed around the world's oceans, and certainly cannot give a true global representation of the state of the ocean surface. Estimates from visual observations are more widely distributed, but are of (at best) variable quality. Output from global wave model simulations can give this coverage, but ultimately rely on how well the model represents the true ocean physics, the accuracy of the input wind fields to force the model, and on *in situ* data for providing sea-truth.

3.2. The Data

The altimeter-derived wave height climate data set now available covers the period 1985–1999, with a gap from late 1989 to mid 1991, and is generated from five satellite altimeters: Geosat, ERS-1, ERS-2, TOPEX, and POSEIDON (Table 1). The TOPEX and POSEIDON altimeters were both mounted upon a single satellite, such that only one could operate at any given time.

3.2.1. Accuracy and Calibrations

Extensive comparisons with co-located *in situ* data have shown that the altimeter measurement of significant wave height (SWH) is remarkably accurate, these comparisons giving rms values as low as 0.3 m (Cotton and Carter, 1994; Cotton *et al.*, 1997; Gower, 1996). However, in order to generate a long-term data set it has been necessary to combine measurements from different satellites, and because of the different characteristics of each altimeter, both in the engineering and data processing, wave measurements from different altimeters are not consistent. Comparisons with wave buoy data has confirmed that each altimeter wave data set requires the application of individual calibration corrections.

Cotton *et al.* (1997) extracted a large set of co-located and near simultaneous altimeter and buoy wind and wave data, using maximum separation criteria of 50 km and 30 min between the individual altimeter and buoy measurements. Buoy data were taken from 24 open ocean data buoys operated by the NOAA Data Buoy Center (selected because of their reliable records and their open ocean locations), altimeter data were taken from the regular offline Geophysical Data Records. A principle components regression, a procedure which accounts for the presence of variability in both data sets being compared, was carried out on the co-located data sets for each satellite altimeter, and linear calibration corrections derived. These calibration corrections are given in Eqs. (1) to (5).

Geosat GDR (Geophysical Data Records)
$$\text{SWH(cor)} = 0.089 + 1.114\,\text{SWH (Geosat)}$$
$$\text{rrms} = 0.3779\,\text{m} \tag{1}$$

ERS-1 OPR (Ocean Product)
$$\text{SWH(cor)} = 0.336 + 1.109\,\text{SWH (ERS-1)}$$
$$\text{rrms} = 0.4531\,\text{m} \tag{2}$$

TOPEX GDR
$$\text{SWH(cor)} = -0.094 + 1.052\,\text{SWH (TOPEX)}$$
$$\text{rrms} = 0.2619\,\text{m} \tag{3}$$

POSEIDON GDR
$$\text{SWH(cor)} = 0.033 + 0.979\,\text{SWH (POSEIDON)}$$
$$\text{rrms} = 0.2811\,\text{m} \tag{4}$$

TABLE 1. Satellite Altimeters Providing Wave Height Data Incorporated in the Global Wave Climatology

Altimeter	Data coverage	Max. latitude	Repeat cycle
Geosat	04/85–09/89	72°	17/176 day
ERS-1	08/91–05/96	81.5°	3/35/168 day
ERS-2	05/95–12/97	81.5°	35 day
TOPEX	10/92–12/97 (90%)	66°	10 day
POSEIDON	10/92–12/97 (10%)	66°	10 day

ERS-2 OPR

$$\text{SWH(cor)} = 0.035 + 1.061 \text{ SWH (ERS-2)}$$

$$\text{rrms} = 0.3089 \text{ m.} \tag{5}$$

N.B. Recent analyses (Challenor and Cotton, personal communication, Queffeulou, 1999) have indicated that the TOPEX calibration may have started to drift around October 1996. The evidence to date is that this drift is small, but significant. It is not taken into account here. (As a consequence the "B" side TOPEX altimeter was switched into operation in February 1999.)

The rrms (residual root mean square) values indicated in Eqs. (1) to (5) provide an indication of the accuracy of individual significant wave-height measurements from the five data sets. Thus all altimeters have provided significant wave-height data to an accuracy of better than 0.5 m, with TOPEX providing the most accurate measurements with an rrms of close to 0.25 m. It is important here to emphasize the importance of networks of reliable *in situ* data for calibrating satellite measurements. Without ground truth, satellite data cannot have a genuine reference point. Ongoing campaigns of calibration, continuing throughout a satellite's lifetime are necessary to check for possible drift in the satellite measurements. There have been a number of recent cases in the field of environmental physics when invalid conclusions drawn from inadequately validated satellite data sets were corrected only after reference to long-term data sets (e.g., stratospheric ozone depletion, sea-level rise).

Once calibrated and quality controlled to exclude any non-ocean data (i.e., those contaminated by rain, sea ice, or other effects) the altimeter data can be combined to form a consistent data set, now covering the period 1985–1999.

3.2.2. Altimeter Sampling Issues

Altimeter satellites have had a variety of orbital patterns to suit their particular mission objectives. Typically, their tracks repeat at intervals of 3–35 days, and so they will revisit an individual location at these intervals. Figure 11 illustrates the ground tracks, to the west of the British Isles, of altimeters on 3, 10, 17, and 35 day repeat orbits. This figure clearly shows how a gain in spatial resolution is obtained at the expense of a loss in temporal resolution. The

different inclination angles of the Geosat, TOPEX and ERS orbits can also be seen (108°, 66°, and 98.5°, respectively). However, although there are several days between successive transects, individual measurements are provided once every second along an orbit track, i.e., once every 6–7 km on the ground (these data are subsequently referred to as 1 Hz data records). Unfortunately, SWH values along such a transect are highly correlated. Thus, if one wishes to generate wave statistics, it is generally preferable to generate uncorrelated samples through the mean or median value from each transect of a selected region, rather than directly using the 1 Hz data.

When generating climatological data sets, it is important to consider the sampling characteristics of the satellite altimeter in combination with the variability of oceanic wave fields, so that one can ensure the altimeter data are capable of providing a genuinely representative climatology. However, prior to the availability of satellite altimeter data, not enough was known about the characteristics of spatial variability of the (open ocean) oceanic wave field to enable an investigation of suitable sampling schemes and grid sizes.

It is simple to establish (e.g., see Figure 11) that the ground track of an individual altimeter in a 10 to 30 day repeat orbit will pass through a 2° latitude by 2° longitude square on 5–10 separate occasions every month, providing 5–10 independent samples (one sample per satellite transect). Note, however, that the 3-day orbit misses completely a large number of 2° grid squares in Figure 11 and is thus not suited to the generation of wave climate data. Cotton and Carter (1994) compared the average altimeter significant wave height taken from the medians from all satellite transects of a 2° × 2° square within a month against the monthly mean measured by a buoy located within the same square (Figure 12). They showed that the correlation between altimeter- and buoy-derived means did not significantly improve with an increase above five in the number of satellite transects through the grid square. Thus, provided that data are available from at least five satellite transects, it can be concluded that the altimeter provides a reliable representation of the monthly mean wave climate within a 2° × 2° square.

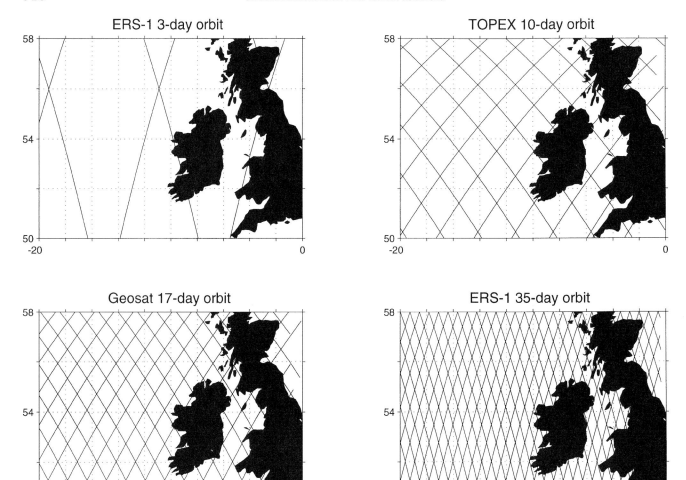

FIGURE 11   Altimeter ground tracks for (a) ERS-1 3-day repeat orbit, (b) TOPEX 10-day repeat orbit, (c) Geosat 17-day repeat orbit, and (d) ERS-1 35-day repeat orbit.

3.2.3. An Altimeter-derived Wave Climate Sata Set

The analysis presented in this chapter is based on a wave climate data set comprising monthly means on a global 2° latitude by 2° longitude grid. Table 1 lists the five altimeters whose data were included. The data from each satellite were quality controlled, calibrated according to Eqs. (1) to (5) and monthly means generated. Where data from more than one satellite were available, separate gridded files were generated, and these files merged, with appropriate weighting. This weighting was calculated individually for each grid square in each month and was proportional to the number of transects of the grid square by each satellite in the month. A global significant wave height data set, comprising 130 sets of monthly means and spanning the period 1985–1997 (though missing the last three months of 1989, all of 1990, and most of 1991) was thus generated. This data set enabled the first truly global-scale studies of oceanic wave climate.

3.3. Global Ocean Wave Field

3.3.1. Seasonal Variability

3.3.1.1. Large-Scale Spatial Features   The altimeter wave data set can be used to describe the mean seasonal cycle, at least for the years over which altimeter data are available. Figure 13 (see color insert) illustrates the mean wave climate (1985–1997) for January, April, July, and October. The expected characteristics can be clearly seen. Panel 1 represents a Northern Hemisphere winter. Highest waves are found in the mid-latitudes in the central North Atlantic and North Pacific oceans, while a lower wave climate predominates in the tropics. In the Southern Oceans a broad band of high waves, which remains throughout the year, and has a peak centered on 50° S, circles all longitudes. In July, the Northern Hemisphere summer, the high waves of winter almost completely die away in the North, whereas in the Southern Ocean the band of high waves increases in mag-

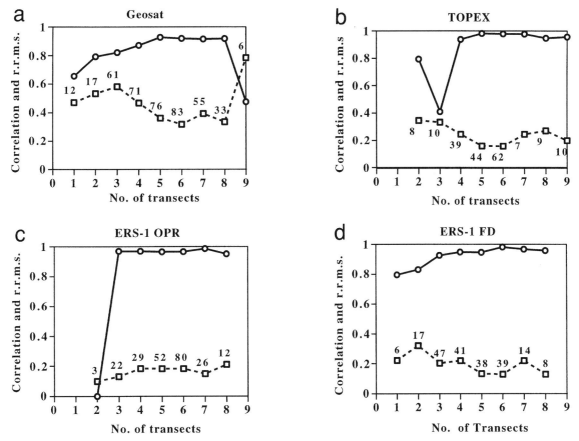

FIGURE 12   Results from regression of altimeter against buoy monthly mean significant wave heights: Correlation coefficient (solid line) and residual RMS in meters (dashed line) against number of altimeter transects in grid square: (a) Geosat, (b) TOPEX, (c) ERS-1 offline data, (d) ERS-1 Fast Delivery data. The number of data points used in each regression are indicated. (From Cotton and Carter, 1994.)

nitude and extent. The effect of the South-West monsoon, increasing wave heights in the Arabian Sea, can also be seen in this panel. The spring and autumn maps (April and October) are similar, and represent the transition between the extremes of summer and winter.

The seasonal cycle is the largest signal of variability in most climatological data sets. For the altimeter wave data we have chosen to represent this cycle by a simple cosine model, Eq. (6). SWH(m) is the mean wave height in a given month, m. In Eq. (6), B1 is the annual mean, B2 the amplitude of the cosine curve, ϕ the phase, and ε_m represents the residual.

$$\text{SWH}(m) = \text{B1} + \text{B2}\ \cos(m/12 + \phi) + \varepsilon_m. \qquad (6)$$

This model was fitted individually to each grid square of the altimeter-derived wave climate data set. This simple annual model explains much more of the variability in the data in the Northern Hemisphere than it does in the South. For instance, for much of the Northern Hemisphere between 80–90% of the variance in the monthly means is explained by the model, whereas in the tropics, less than 10% reduction in variance is obtained. Even across mid and high latitude

regions in the Southern Oceans only 40–50% reduction in variance is achieved. Taken as an average over the whole global ocean, the simple model for the annual cycle characterized by Eq. (6) explains 64% of the total variance.

Figure 14 shows parameters for the mean annual cycle fitted to the altimeter data between 1985 and 1996. We can see from panel (a) that the annual mean wave height is at a maximum in mid-ocean at mid-latitudes, and is a minimum in the tropics and in coastal regions sheltered from swell. The mean wave height is greatest in the Southern Oceans, being higher than 4 m across most longitudes at approximately 50° S, whereas it is between 3 and 4 m in the central North Atlantic and the North Pacific Oceans.

The biggest range in the annual cycle (calculated as the difference between the maximum and minimum monthly mean significant wave heights in the fitted cycle) occurs in the Northern Hemisphere, middle panel (b). The annual range in the central North Atlantic and North Pacific reaches 3 m, illustrating the great difference between the winter and summer months in these regions. In contrast, across most of the Southern Ocean the annual range is 1–2 m, demon-

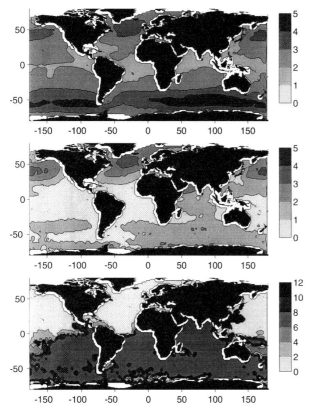

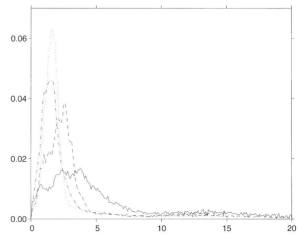

FIGURE 14 Parameters calculated from the annual cycle fitted to monthly mean altimeter data from 1985–1996. Top panel, annual mean significant wave height. Middle panel, range of annual cycle. Bottom panel, month of maximum wave height. (From Cotton *et al.*, 1997b. With permission.)

FIGURE 15 Significant wave height probability distribution functions for winter mid-latitude (solid line), winter low latitude (dashed line), summer mid latitude (dash-dot line), and summer low latitude (dotted line).

strating that higher waves (and winds) remain throughout the year, because of the lack of large continental land masses and to the persistent year-round high winds in this region.

Semi-annual cycles can similarly be fitted to the data. A significant semi-annual cycle with a range of 1–3 m is found in the Northern Indian Ocean, and is a consequence of the two monsoon seasons in this region. A significant semi-annual cycle is also found the Caribbean.

The phase of the annual cycle can be illustrated by identifying the month with the maximum mean wave height, Figure 14 panel (c). As would be expected, the maximum wave height consistently occurs in January/February throughout the Northern Hemisphere, apart from localized regions on the eastern boundaries of the Pacific and Atlantic, and in the Arabian Sea. The Southern hemisphere oceans show more variability, with the maximum wave height usually occurring between June and August. In areas where the annual cycle is not well defined the displayed month of maximum wave height is not particularly meaningful (primarily in the tropics, but also some regions in the southern oceans, e.g., to the east of Australia).

3.3.1.2. Wave Height Distribution Functions
Figure 15 shows how the shapes of wave-height distribution functions vary with latitude and season. During the winter at mid- to high latitudes (solid line), the distribution is broad, demonstrating a large range of recorded wave heights, and has a mode at about 3 m. In the same area in summer (dash-dot line), the mode of the distribution reduces to less than 2 m, the distribution becomes narrower, and the characteristic skewed shape of a wave height distribution is more evident. At low latitudes in winter (dashed line), the distribution is narrower than it was for the equivalent mid-latitude data, and has a mode between 2–3 m. The summer low-latitude distribution has a mode close to that of the mid-latitude data, but is even more highly peaked, indicating a very limited range of wave height data. The key features of these distribution functions are well known from *in situ* data for certain regions, but altimeter data allow distribution functions to be generated for any oceanic region, without the need for buoy deployment. These data can then be fitted to suitable forms of probability distribution functions [e.g., Fisher Tippett Type 1 for mid-latitudes, Tucker, (1991)]. Important statistical parameters, such as the 50-year return value (the wave height that can be expected to be exceeded once every 50 years), can then be estimated.

3.4. Climate Variability

3.4.1. Introduction

Analysis of historical *in situ* data has shown that the wave climate is not stationary, and that significant changes have occurred in the past. For instance, Bacon and Carter (1991) showed that average wave conditions had been increasing in the North Atlantic Ocean by more than 1% per annum for 25 years since the mid 1960s. More recent studies (WASA,

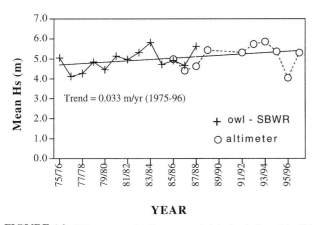

FIGURE 16 Winter mean significant wave height (m; indicated by Hs) from OWS Lima (+), and altimeter data (○). (From Cotton *et al.*, 1997b. With permission.)

1998) using data from other sources (including model hindcasts and ship routing charts) have confirmed this increase, and also shown a statistical connection to sea level pressure fields (Bacon and Carter, 1993; Kushnir *et al.*, 1997).

Global wave-height climatologies derived from satellite altimeter measurements provide full spatial coverage, and so allow us to address some of the important questions that arise from such studies. For instance, what is the spatial extent of the region affected by such increases in significant wave height? Do all ocean regions exhibit significant variability? On what frequency and spatial scale is inter-annual variability observed in different ocean regions? Is the variability in different regions connected in any way? Is it possible to develop an understanding of the causal physical links between variability in wave climate and associated changes in forcing fields and atmospheric climate indices?

Having established that a simple annual cycle can be used describe a significant proportion of the observed variability of wave climate, the remaining or residual variability can be characterized on two-time scales, long-term decadal scale variability (sometimes referred to as "climate trends"), and short-term inter-annual, even intra-annual variability (year to year, and month to month).

3.4.2. Long Term Trends

Long term trends are most likely the consequence of long term changes in atmospheric circulation. Indeed, Bacon and Carter (1993) found that the increase in wave heights in the North-East Atlantic was correlated with an increase in the sea-level pressure gradient between the Azores and Iceland (related to the North Atlantic Oscillation Index, the NAO). To confirm the continuance of the trend observed by Bacon and Carter (and others) mean wintertime (December, January, February, March) significant wave heights were generated from the altimeter data set at the location of Ocean Weather Station Lima (57°N, 20°W) and then appended to

Ship Borne Wave Recorder data taken at this location, Figure 16.

A linear trend has been fitted to these data, and gives an increase of 0.033 m per year between 1975 and 1996. It is clear that the winter of 1995–1996 was unusually calm (in the context of recent years), and that up until that year a steeper trend of about 0.075 m year^{-1} was in evidence (achieved from fitting a trend to 1975–1994 data). It is too early to say whether the winter of 1995/1996 represents a turning point in the long-term trend, or is merely a short-term anomaly.

The *in situ* data are, however, unable to give an indication of the spatial structure of the region of the North Atlantic affected by any such increase in wave heights. To investigate the spatial nature of this trend, the altimeter data were divided into two sets, 1985–1989 and 1991–1996, and annual cycles fitted separately to the data from these periods. The cycle maxima (absolute monthly maxima) and cycle means were then compared (Figure 17). The mean data (bottom panel) do not show any significantly interesting features, however the differenced maxima data (top panel) are more revealing. We see that the biggest increase between the years 1985–1989 and 1991–1996 is concentrated in the north-eastern corner of the North Atlantic, with the magnitude increasing into the far north-eastern corner. Contours of equal increase appear to run roughly north-west to southeast. When we consider the global picture we see that only one other oceanic region of any size, a region to the south of Australia, shows a change in wave climate of a similar magnitude. However, the major consequences of the El Niño phenomena in the Southern Hemisphere, and the fact that the two selected periods contain different numbers of El Niño cycles means that in its current form this analysis is not appropriate for the identification of trends in the Southern Hemisphere.

Nonetheless the altimeter data indicate that long term variability has been limited to specific ocean regions, and is not apparently a global phenomenon.

3.4.3. Inter-Annual Variability

3.4.3.1. Time Series It is a complex task to characterize variability occurring over a range of frequencies (months to decades) and spatial scales (within individual seas to global). One option is to consider specific locations of particular interest. We have already focused on the area around the site of Ocean Weather Station Lima.

Figure 18(a) presents a time series of altimeter-derived monthly mean-significant wave-height data for the 2° grid square containing Lima, with the seasonal cycle removed (solid line). This is here referred to as the residual significant wave height. The dashed line gives the index of the North Atlantic Oscillation (NAO) for the same period. The NAO is a measure of the anomaly in the sea-level pressure gradient across the North Atlantic between the Azores and Iceland.

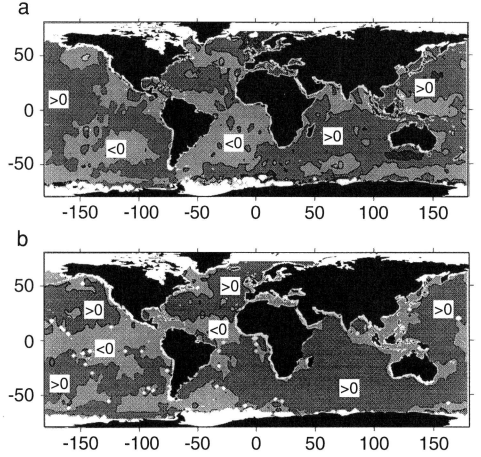

FIGURE 17   Difference in significant wave height annual cycle maximum (top) and mean (bottom) between the periods 1985–1989 and 1991–1995. Contours are at 0.5 m intervals, with the lightest gray areas showing a *decrease* (from 1985–1989 to 1991–1995) of between −0.5 to 0 m, the next lightest an *increase* of 0–0.5 m, and the darkest gray an *increase* of more than 0.5 m.

When the pressure is abnormally low over Iceland and high over the Azores, the pressure gradient is higher than normal, and so the NAO is positive. When the pressure is unusually high over Iceland and low over the Azores, then the pressure gradient is lower than usual and the NAO is negative. There is a clear correlation between the NAO index and the residual significant wave height at both long and short time scales (the correlation coefficient is 0.78). This suggests that the wave climate in this region of the North Atlantic is responding to changes in the gradient of the sea level pressure across the region at both time scales. This is perhaps an expected result, as this gradient is related to the strength of the westerly winds in the North Eastern Atlantic. Nonetheless the high correlation provides important confirmation of a connection between the sea level pressure field and the wave climate. An upward trend is also evident in both time series between the years between 1985 and 1995, but it is difficult to see any cyclical patterns in this presentation of a relatively short time series of data.

Figure 18(b) demonstrates an equivalent comparison between the Southern Oscillation Index (the normalized anomaly in the sea-level pressure difference between Darwin, Australia, and Tahiti—dashed line) and the scaled residual wave height for a region in the North Eastern Pacific around 47°N, 165°E (solid line). In this case the residual wave heights appear to show an anti-correlation with the climate index (correlation coefficient of −0.48).

Although the time series of residual wave heights presented in Figure 18 do not appear to exhibit any coherent cyclical behavior, further analysis reveals that some years are anomalously calm (low wave heights), whereas others are anomalously rough. These swings are significant in size, and are often 0.5 m or more in a signal of 4–5 m (i.e., > 10%). Further analysis revealed an unexpected correlation between these anomalously calm and rough years in the North Atlantic with similar anomalies (though of an opposite sense) in the North Eastern Pacific. Figure 19 overlays the time series of residual wave heights for the North East Atlantic (Ocean Weather Station Lima) and for the location in the North Eastern Pacific identified above. An anti-correlation between the residual wave heights of the two regions, particularly at longer time scales, is evident in this figure. In

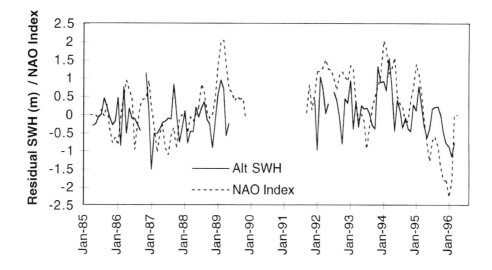

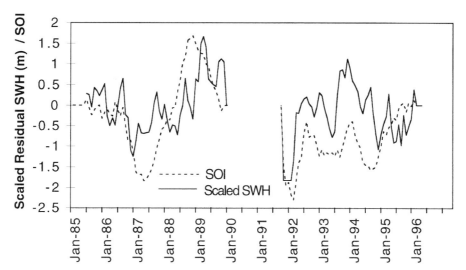

FIGURE 18   Residual significant wave height (monthly mean wave height with annual cycle removed) compared to climate indices. (Top) the monthly North Atlantic Oscillation index (dashed line) and wave-height residual (solid line) at OWS Lima (59°N, 20°W). (Bottom) the monthly Southern Oscillation index (dashed line) and the scaled (multiplied by −3) wave height residual (solid line) at a location in the NE Pacific (47°N 165°E). (From Cotton *et al.*, 1997b. With permission.)

the years when the residual wave height is anomalously low in the North Eastern Atlantic (1987, 1988), it is high in the North-East Pacific, and vice versa (1989, 1994). This correlation seems strongest on longer time scales (for all the data the correlation between these two time series is quite low, at −0.21). This anti-correlation between physically separated ocean basins suggests an atmospheric link, possibly through a large-scale atmospheric mode of circulation.

3.4.3.2. Spatial Patterns   To generate a picture of the spatial structure of these inter-annual variations, orthogonal modes of variability were identified in the residual wave height climate data set (i.e., with the annual cycle re-

moved) using the technique of Empirical Orthogonal Functions (Preisendorfer, 1988). The altimeter wave data were first smoothed in time (5 month running mean) and in space (9 point, nearest neighbor, Gaussian filter).

i. North Atlantic

First, modes for the North Atlantic were generated. Figure 20 illustrates the spatial structure and time series of the most significant eigen mode, which accounted for over 42% of the variance in the residual wave-height data. The North Atlantic Oscillation Index (smoothed with a 5-month running mean) is shown as a dashed line. The figure clearly shows a bipolar structure in which the south-western North

Atlantic is anti-correlated with the north-eastern North Atlantic, with the dividing line running South East from the southern tip of Greenland toward the west coast of the Iberian peninsula. This pattern matches well with a pattern

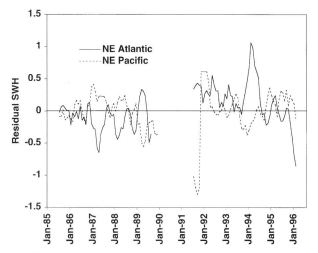

FIGURE 19   Residual monthly mean significant wave height for two regions, OWS Lima in the northeast North Atlantic (59°N, 20°W, solid line), and a region in the NE Pacific (47°N 165°E, dashed line)). (From Cotton *et al.*, 1997b. With permission.)

identified by Kushnir *et al.* (1997) from a model wave-height climatology. Through a canonical correlation analysis, coupled with sea level pressure fields, they connected this pattern to the two main phases of the NAO. When the NAO is in its negative phase (i.e., the pressure gradient across the North Atlantic is lower than normal) westerly winds over the Atlantic are weaker than usual and wave heights are lower than normal in the North Eastern Atlantic. In the converse case (the positive NAO phase, more common in recent years) westerly winds are stronger and hence wave heights greater in the North Eastern Atlantic. According to the altimeter derived wave climate data presented here (see also Figure 23), the southern extent of the affected region may continue into the sub-tropical North Atlantic. The time series of the first eigen mode of the altimeter data (bottom panel of Figure 20) clearly shows that the pattern was negative (i.e., lower than average wave heights in the North East Atlantic) in the winters of 1986/1987, 1987/1988, and 1995/1996, but positive (higher than average waves) in the winters of 1988/1989, 1993/1994, and 1994/1995. The correlation between this time series of the first eigen and the smoothed NAO (as displayed) is 0.78, confirming the strength of the connection.

The next three modes of variability (not illustrated) together explain a further 29% of the variability in the data, thus the first four modes account for over 70%. Further stud-

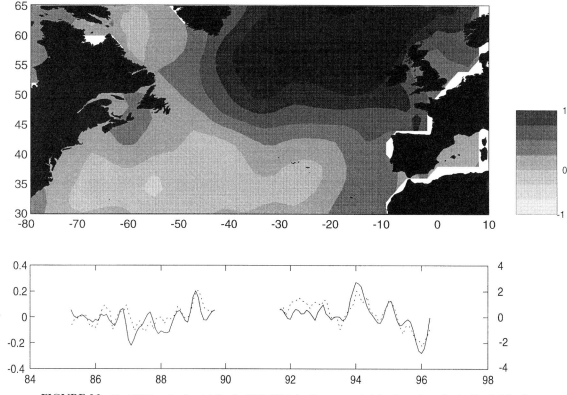

FIGURE 20   First EOF mode of variability in 1985–1996 significant wave height climatology for the North Atlantic. The lower panel gives the time series for the spatial pattern in the upper plot (solid line), and the North Atlantic Oscillation index smoothed by a 5-month running mean (dashed line). (From Cotton and Challenor, 1999. With permission.)

ies will aim to investigate links between these higher order modes and other atmospheric circulation patterns.

ii. North Pacific

When the wave climate of the North Pacific is analysed in a similar fashion, the first eigen mode (explaining 28% of data variability) is characterized as a pattern having a single centre of activity, with the region of greatest magnitude centered along a latitude of 40°N running from 180°E to 220°E (Figure 21). This feature appeared to occur most strongly (with lower than normal wave heights) during the winter 1988/1989, with expressions of the opposite sense in 1986/1987 and 1993/1994. The second mode (not illustrated) shows a bipolar structure with the anti-correlated centers of activity to the South-West and North-East of the North Pacific. The first four modes explain 63% of the variability in this region.

iii. Connected Pacific-Atlantic Variability

Clearly the observations of connected variability noted in section 3.3.1 merit further investigation. A simple illustration of the spatial nature of this anti-correlation is given in Figure 22 (see color insert), where the anomalies in mean wave heights for four winters are shown (the anomaly is calculated by subtracting the long term winter mean). It is clear that the north-eastern sector of the North Atlantic had anomalously high wave heights in the winters of 1988/1989

and 1993/1994, corresponding with seasons of unusually low waves for a significant region in the North Pacific. Anti-correlated anomalies of the opposite sense occurred in these regions in 1987/1988 and 1995/1996. The affected regions appear to be the area of the north-eastern North Atlantic identified by the EOF analysis in Figure 20 and a slightly less well-defined region in the North Pacific, which also resembles the most significant mode of variability picked out in Figure 21. To confirm these observations, and to get a measure of significance, the EOF variability analysis was applied to whole of the Northern Hemisphere (Figure 23). The first mode, explaining 23% of the non-seasonal variability in the Northern Hemisphere, confirms a structure whereby the wave height climate of the north-eastern Atlantic Ocean is anti-correlated with that of the eastern North Pacific Ocean. This EOF analysis therefore serves to confirm the evidence of Figures 19 and 22, and provides a spatial structure and time series for the Atlantic-Pacific anti-correlation. The time series (bottom panel) shows that the winters 1986–1987 and 1987–1988 represent positive phases of the pattern illustrated in the upper panel (low in the North Atlantic and high in the North Pacific), and the winters 1988–1989 and 1993/1994 represent negative phases.

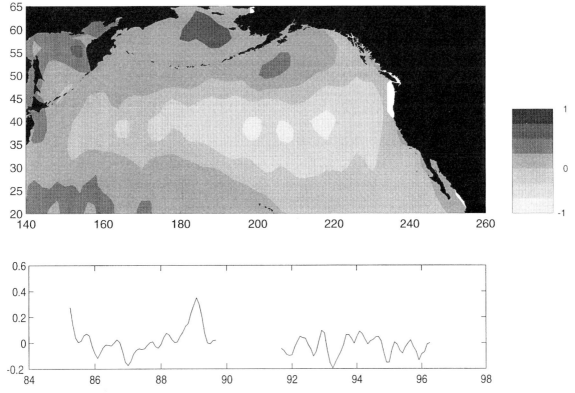

FIGURE 21 First EOF mode of variability in 1985–1996 significant wave height climatology for the North Pacific. The lower panel gives the time series for the spatial pattern in the upper plot.

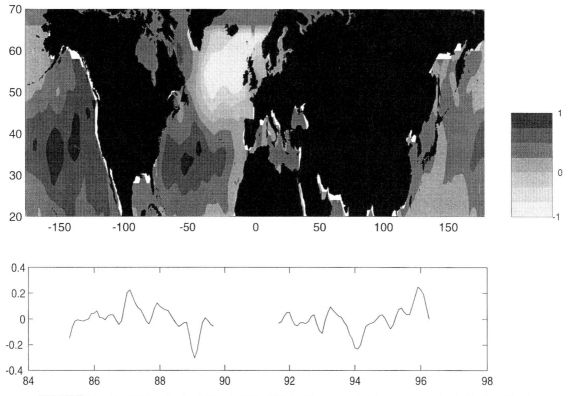

FIGURE 23 First EOF mode of variability in 1985–1996 significant wave-height climatology for the Northern Hemisphere. The lower panel gives the time series for the spatial pattern in the upper plot.

3.5. Summary

The altimeter wave data have allowed us for the first time to look in detail at large-scale patterns of variability in the global ocean wave climate. We have found confirmation of increasing winter wave heights in the North-East Atlantic previously seen in analyses of *in situ* data, and have also been able to generate a spatial picture of the regions affected. Strong links with the North Atlantic Oscillation have been demonstrated, and further work will be directed to further characterizing and understanding such links, with an eventual aim of developing some prediction capability. A recent study based on output from global scale model simulations, Sterl *et al.* (1998), shows trends which are only marginally significant and which vary from month to month. As a whole, the model simulations showed smaller scale variability than the satellite data. This perhaps reflects an inability of wave models, at the current stage of development, to fully represent the variability present in the oceans.

Perhaps most importantly, the altimeter data have provided the first indications that the wave climates of the North Atlantic and North Pacific Oceans are connected. This connection is such that when one of these regions is experiencing a winter of higher than normal wave heights, the other is experiencing calmer conditions than usual. It seems most likely that the physical connection linking the wave climate of these two separated ocean regions must apply through the atmosphere, and further work will aim to develop an understanding of this connection.

4. CONCLUSIONS

We conclude that the outlook for applications of wind and wave measurements from altimeters, and from instruments based on the altimeter principle, is very promising. In terms of operational modeling, these measurements will play a significant role in improving the monitoring and forecasting of the ocean surface state. A better estimation of the sea-state will in turn contribute significantly to reducing the altimeter range error budget arising from the EM bias correction.

The altimeter wave data have also allowed the construction of the first global wave climatologies based on measurements rather than models. Analyses of these data have provided new insights into the nature of wave climate, demonstrating convincingly that large scale wave climate is highly variable on inter- and intra-annual time scales, and providing a tool whereby this variability can be characterized and understood. The data have also provided the very first evidence of connections between the surface wave climates of physically separated ocean regions.

Together, these studies provide a foretaste of the exciting new discoveries that will surely continue to flow as a consequence of the availability of wave and wind data from satellites.

ACKNOWLEDGMENTS

The authors acknowledge the support of Didier Le Meur and Joël Geusquin from Météo-France, and David Carter from Satellite Observing Systems United Kingdom.

5. GLOSSARY

EM Bias	Electro Magnetic Bias—A bias in the altimeter range measurement, resulting from the irregular nature of the sea surface
ECMWF	European Center for Medium-range Weather Forecasting
EOF	Empirical Orthogonal Function
GDR	Geophysical Data Record
GTS	Global Transmitting System
SWH	Significant Wave Height
KF	Kalman Filter
IGOOS	International Global Ocean Observing System
JONSWAP	JOint North Sea WAve Project
Météo-France	The French Weather Service
NAO	North Atlantic Oscillation
NHC	(US) National Hurricane Center
NWP	Numerical Wave Prediction
NWS	National Weather Service
OPR	Ocean Product
O-I	Optimum Interpolation
OWL	Ocean Weather Station Lima
rms	root mean square
rrms	residual root mean square
SOI	Southern Oscillation Index
VAG	French operational numerical wave model

References

Bacon, S., and Carter, D. J. T. (1991). Wave climate changes in the North Atlantic and North Sea. *Int J. Climatol.*, **11**, 545–558.

Bacon, S., and Carter, D. J. T. (1993). A connection between mean wave height and atmospheric pressure gradient in the North Atlantic. *Int J. Climatol.*, **13**, 423–436.

Bauer, E., Hasselmann, K., Young, I. R., and Hasselmann, S. (1996). Assimilation of wave data into the wave model WAM using an impulse response function method. *J. Geophys. Res.*, **16**, 1087–1102.

Brown, G. S. (1977). The average impulse response of a rough surface and its applications. *IEEE Trans. Antennas Propag.*, **AP-25(1)**, 67–74.

Brown, G. S. (1979). Estimation of surface winds using satellite-borne radar measurements at normal incidence. *J. Geophys. Res.*, **84**, 3974–3978.

Brown, G. S., Stanley, H. R., and Roy, N. A. (1981). The wind speed measurement capability of space borne radar altimeters. *IEEE J. Oceanic. Eng.*, **OE-6(2)**, 59–63.

Carter, D. J. T., Challenor, P. G., and Srokosz, M. A. (1992). An assessment of Geosat wave height and wind speed measurements. *J. Geophys. Res.*, **97**, 11383–11392.

Chelton, D. B., and McCabe, P. J. (1985). A review of satellite altimeter measurement of sea surface wind speed: With a proposed new algorithm. *J. Geophys. Res.*, **90**, 4707–4720.

Chelton, D. B., and Wentz, F. J. (1986). Further development of an improved altimeter wind speed algorithm. *J. Geophys. Res.*, **91**, 14250–14260.

Cotton, P. D., and Carter, D. J. T. (1994). Cross calibration of TOPEX, ERS-1 and Geosat wave heights. *J. Geophys Res.*, **99**, 25025–25033.

Cotton, P. D., Challenor, P. G., and Carter, D. J. T. (1997a). An assessment of the accuracy and reliability of Geosat, ERS-1, ERS-2 and TOPEX altimeter measurements of significant wave height and wind speed. *In* "Proceedings CEOS Wind and wave validation workshop," pp. 81–93, 3–5 June, 1997b, ESTEC, Noordwijk, The Netherlands.

Cotton, P. D., Challenor, P. G., and Carter, D. J. T. (1997b). Variability in altimeter global wave climate data. *In* "Ocean Wave Measurement and Analysis: Proceedings of the third International Symposium, WAVES '97," pp. 809–818, Nov. 3–7, Virginia Beach, VA.

Cotton, P. D., and Challenor, P. G. (1999). North Atlantic wave climate variability and the North Atlantic Oscillation Index. *In* "Proc. 9th International Offshore and Polar Engineering Conference," Brest, France, **III**, 153–157.

Davies, C. G., Challenor, P. G., and Cotton, P. D. (1998). Measurements of wave period from radar altimeter. *In* "Ocean Wave Measurement and Analysis: proceedings of the third international symposium WAVES '97," pp. 819–826, Nov. 3–7, Virginia Beach, VA.

De la Heras, M. M., Burgers, G., and Janssen, P. A. E. M. (1994). Variational wave data assimilation in a third-generation wave model. *J. Atmos. Ocean. Tech.*, **11**, 1350–1369.

Dobson, E. B., Monaldo, F. M., Goldhirsh, J., and Wilkerson, J. (1987). Validation of Geosat altimeter-derived wind speeds and significant wave heights using buoy data. *J. Geophys. Res.*, **92**, 10719–10731.

Eymard, L., Planton, S., Durand, P., Le Visage, C., Le Traon, P. Y., Prieur, L., Weill, A., Hauser, D., Rolland, J., Pelon, J., Baudin, F., Bénech, B., Brenguier, J. L., Caniaux, G., De Mey, P., Dombrowski, E., Druilhet, A., Dupuis, H., Ferret, B., Flamant, C., Hernandez, F., Jourdan, D., Katsaros, K., Lambert, D., Lefèvre, J.-M., Le Borgne, P., Le Squere, B., Marsoin, A., Roquet, H., Tournadre, J., Trouillet, V., Tychensky, A., and Zakardjian, B. (1996). Study of the air-sea interactions at the mesoscale: the SEMAPHORE experiment. *Annal. Geophys.*, **14**, 986–1015.

Fedor, L. S., and Brown, G. S. (1982). Wave height and wind speed measurements from the Seasat altimeter. *J. Geophys. Res.*, **87**, 3254–3260.

Fradon, B., Hauser, D., and Lefèvre, J-M. (1996). Performance des modèles de prévision de l'état de la mer VAG et WAM pendant l'expérience SEMAPHORE. *In* "Actes de l'Atelier de Modélisation de l'Atmosphère," 3–4 Décembre 1996, Toulouse, France.

Fradon, B., Hauser, D., and Lefèvre, J.-M. (2000). Comparison study of a second-generation and of a third-generation wave prediction model in the context of the SEMAPHORE experiment, *J. Atmos. Ocean. Tech.*, **17**, 191–214.

Freilich, M. H., and Dunbar, R. S. (1993). Derivation of satellite wind model functions using operational surface wind analyses: An altimeter example. *J. Geophys. Res.*, **98**, 14633–14649.

Fu, L.-L., and Glazman, R. (1991). The effect of the degree of wave development on the Sea State Bias in the Radar Altimetry Measurement. *J. Geophys. Res.*, **96**, 829–834.

Gerling, T. W. (1992). Partitioning sequences and arrays of directional ocean wave spectra into component wave systems. *J. Atmos. Ocean. Tech.*, **9**, 444–458.

Glazman, R. A., and Pilorz, S. H. (1990). Effect of sea maturity on satellite altimeter measurement. *J. Geophys Res.*, **95**, 2857–2870.

Goldhirsh, R. E., and Dobson, E. B. (1985). A recommended algorithm for the determination of ocean surface wind speed using a satellite-borne radar altimeter, *Rep. JHU/APLSIR85U-005*, Appl. Phys. Lab., Johns Hopkins University, Laurel, MD.

Gower, J. F. R. (1996). Intercalibration of wave and wind data from Topex/POSEIDON and moored buoys off the west coast of Canada. *J. Geophys Res.*, **101**, 3817–3829.

Guillaume, A., and Mognard, N. M. (1992). A new method for the validation of the altimeter-derived sea-state parameters with results from wind and wave models. *J. Geophys. Res.*, **97**, 9705–9717.

Guillaume, A., Lefèvre, J-M., and Mognard, N. M. (1992). The use of altimeter data to study wind wave variability in the western Mediterranean sea. *Oceanol. Acta*, **15**, 5, 555–561.

Hanssen, B., and Guenther, H. (1992). ERS-1 Radar Altimeter Validation with the WAM model. *In* "Proceedings of the ERS-1 Geophysical Validation Workshop," Penhors, Bretagne, France, April 1992, European Space Agency, ESA wpp-36, pp. 157–161.

Hasselmann, K., Barnett, T. P., Bouws, E., Carlson, H., Cartwright, D. E., Enke, K., Ewing, J. A., Gienapp, H., Hasselmann, D. E., Kruseman, P., Meerburg, A., Müller, P., Olbers, D. J., Richter, K., Sell, W., and Walden, H. (1973). Measurements of wind-wave growth and swell decay during the Joint North Sea Wave Project (JONSWAP). *Deut. Hydrogr. Z. Suppl. A8* (12).

Hauser, D., Dupuis, H., Durrieu de Madron, X., Estournel, C., Flamant, C., Pelon, J., Queffeulou, P., and Lefèvre, J.-M. (2000). La Campagne FETCH: étude des échanges océan/atmosphère dans le Golfe du Lion, *La Météorologie*, 8è série, **29**, 14–31.

Hwang, P. A., Teague, W. J., Jacobs, G. A., and Wang, D. W. (1998). A statistical comparison of wind speed, wave height, and wave period derived from satellite altimeters and ocean buoys in the Gulf of Mexico region. *J. Geophys. Res.*, **103**, 10451–10468.

Jackson, F. C., Walton, W. T., and Baker, P. L. (1985). Aircraft and satellite measurements of ocean wave directional spectra using scanning-beam microwave radars. *J. Geophys. Res.*, **90**, 987–1004.

Komen, G. J., Cavaleri, L., Donelan, M., Hasselmann, K., and Janssen, P. A. E. M. (1994). *In* "Dynamics and Modelling of Oceans Waves." Cambridge University Press, 532 pp.

Kushnir, Y., Cardone, V. J., Greenwood, J. G. and Cane, M. A. (1997). The recent increase in North Atlantic wave heights. *J. Climate*, **10**, 2107–2113.

Lefèvre, J-M. (1990). Vagmed devient operationnel. *Metmar* n° 149, Météorologie Maritime, Météo-France, 4 ème trimestre 1990.

Lefèvre, J-M. (1992). The impact of altimeter data assimilation for wave forecasting in the Mediterranean sea. *In* "Proceedings of the third international workshop on wave hindcasting and forecasting," May 19–22, 1992 Montreal.

Lefèvre, J-M., Barckicke, J., and Ménard, Y. (1994). A Significant Wave Height-Dependent model function for Topex/POSEIDON wind speed retrieval. *J. Geophys. Res.*, **99**, 25,025–25,049.

Lefèvre, J-M., Roquet, H., and Le Meur, D. (1996). Towards operational altimeter data assimilation into a numerical wave model at Météo-France. *In* "Proceedings of the AGU, 15–19 December 1996," San-Francisco, CA.

Lefèvre, J-M., Le Meur, D., Fradon, B., Guesquin, J., Duret, F., and Potevin, J. (1998). Altimetry and sea-sate forecasting at Météo-France, 1998. *AVISO Altimetry Newsletter* N°6, CNES, France, NASA, US, April 1998.

Le Meur, D., Lefèvre, J-M., and Roquet, H. (1995). Apport des capteurs actifs micro-onde d'ERS-1 et de Topex/POSEIDON à la modelisation numérique des vagues. *Actes de l'Atelier de Modélisation de l'Atmosphère*, 26–28 Novembre 1995, Toulouse, France.

Lionello, P., and Janssen, P. A. E. M. (1992). Assimilation of altimeter wave data in a global third generation wave model. *J. Geophys. Res.*, **97**, 14453–14474.

Miles, J. W. (1957). On the generation of surface waves by shear flows. *J. Fluid Mech.*, **3**, 185–204.

Mognard, N. M. (1984.) Swell in the Pacific Ocean observed by Seasat Radar Altimeter. *Mar. Geodesy*, **8**, 183–209.

Mognard, N. M., and Lago, B. (1979). The computation of wind speed and wave heights from Geos 3 Data. *J. Geophys. Res.*, **93** *(C3)*, 2285–2302.

Mognard, N. M., Campbell, W. J., Cheney, R. E., Marsh, J. G., and Ross, D. B. (1986). Southern Ocean Waves and Winds Derived from Seasat Altimeter measurements. *In* "Wave Dynamics and Radio Probing of the Ocean Surface," eds. O. Phillips and K. Hasselmann, Plenum Press, 479–489.

Mognard, N. M., Johannessen, J. A., Livingstone, C. E., Lyzenga, D., Shuchman, R., and Russel, C. (1991). Simultaneous observations of ocean surface winds and waves by Geosat radar altimeter and airborne Synthetic Aperture Radar during the 1988 Norwegian Continental Shelf Experiment. *J. Geophys. Res.*, **96**, C6, 10467–10486.

Phillips, O. M. (1957). On the generation of waves by turbulent wind. *J. Fluid Mech.*, **2**, 417–445.

Phillips, O. M. (1977). *In* "The dynamics of the upper ocean," Cambridge University Press, Cambridge, 336 pp.

Preisendorfer, R. (1988). *In* "Principal component analysis in meteorology and oceanography," Elsevier Pub. Co., N.Y.

Queffeulou, P., and Lefèvre, J-M. (1992). Validation of altimeter wave and wind fast delivery product. *In* "ERS-1 Geophysical Validation Proceedings," 27–30 April 1992, Eur. Space Agency, Penhors, Brittany, France.

Queffeulou, P., Bentamy, A., Quilfen, Y., and Tournadre, J. (1994). Validation of ERS-1 and TOPEX-POSEIDON altimeter wind and wave measurements. *Document de travail*, DRO-OS 94–08, December 1994, Ifremer, Plouzane, France.

Queffeulou, P. (1999). Long term comparison of ERS, TOPEX and POSEIDON altimeter wind and wave measurements. *In* "Proc. 9th International Offshore and Polar Engineering Conference," Brest, France, **III**, 114–120.

Rufenach, C. L., and Alpers, W. R. (1978). Measurement of ocean wave heights using the Geos 3 altimeter. *J. Geophys. Res.*, **83**, 5001–5018.

Sterl, A., Komen, G. J., and Cotton, P. D. (1998). Fifteen years of global wave hindcasts winds from the European Centre for Medium Range Weather Forecasts reanalysis: Validating the reanalyzed winds and assessing the wave climate. *J. Geophys Res.*, **103**, 5477–5492.

Tucker, M. J. (1991). *In* "Waves in ocean engineering: measurement, analysis, interpretation." Ellis Horwood, Chichester, UK.

WAMDI Group. (1988). The WAM Model—A third generation ocean wave prediction model. *J. Phys. Oceanogr.*, **18**, 1775–1810.

WASA group (1998). Changing storm and wave climate in the North-East Atlantic? *Bull. Am. Meteor. Soc.*, **79**, No. 5, 741–760.

Whitham, G. B. (1974). *In* "Linear and nonlinear waves." John Wiley & Sons, New York, 636 p.

Witter, D. L., and Chelton, D. B. (1991). A Geosat altimeter wind speed algorithm and a method for altimeter wind speed algorithm development. *J. Geophys Res.*, **96**, 8853–8860.

Young, I. R. (1993). An estimate of the Geosat altimeter wind speed algorithm at high wind speeds. *J. Geophys. Res.*, **98**, 20,275–20,285.

CHAPTER

8

Sea Level Change

R. S. NEREM[*] and G. T. MITCHUM[†]

[*]Center for Space Research
The University of Texas at Austin
Austin, Texas
[†]Department of Marine Science
University of South Florida
Tampa, Florida

1. INTRODUCTION

Long-term sea-level change is a topic of considerable interest to many diverse groups in our society. A large segment of the world's population lives in the coastal zone, thus they are concerned about the socioeconomic consequences of sea-level change. Stewards of the world's wetlands, one of our most sensitive environmental zones, are concerned that a rapid change in sea level might dramatically reduce the size of our arable wetlands. In addition, the rate of sea-level rise is expected to increase over the next century in response to increasing "greenhouse gases," and thus sea-level change can be used as a barometer to validate these predictions from climate models (Houghton *et al.*, 1996). This response is mainly caused by thermal expansion of the oceans as well as the melting of polar ice in Greenland and Antarctica, although the relative contributions of these effects is highly uncertain (Houghton *et al.*, 1996). Thus, measuring the temporal and spatial characteristics of sea-level change has long been a goal of our society so that we may perhaps understand the sources of the change, and then mitigate these sources, or plan for the inevitable socioeconomic changes that would occur.

Over the last century, long-term sea-level change has been estimated from tide gauge measurements. However, two fundamental problems are encountered when using tide gauge measurements for this purpose. First, tide gauges only measure sea level change *relative* to a crustal reference point, which may move vertically at rates comparable to the true sea-level signals (Douglas, 1995). Second, tide gauges

have limited spatial distribution and suboptimal coastal locations (Barnett, 1984; Groger and Plag, 1993), and thus they provide poor spatial sampling of the open ocean. Douglas (1991, 1992) has argued that by selecting tide gauge records of at least 50 years in length and away from tectonically active areas, even a limited set of poorly distributed tide gauges can give a useful estimate of global sea-level rise. However, averaging over such a long time period makes investigating shorter term changes difficult.

Clearly, an independent global measurement technique is needed to investigate the important issues associated with sea-level change. In principle, satellite altimeters should provide improved measurements of global sea-level change over shorter averaging periods because of their truly global coverage and direct tie to the earth's center-of-mass. Satellite altimeters provide a measure of *absolute* sea level relative to a precise reference frame realized through the satellite tracking stations whose origin coincides with the earth's center-of-mass [Nerem *et al.* (1998) among others]. However, for altimeter missions such as Seasat, Geosat, and ERS-1/2, errors in the satellite altitude and measurement corrections obscured the sea-level change signal (Wagner and Cheney, 1992), although recent reprocessing of ERS-1 data show encouraging progress (Anzenhofer and Gruber, 1998; Cazenave *et al.*, 1998). Many of the limitations of previous altimeter missions have been corrected or improved with the TOPEX/POSEIDON (T/P) mission (Fu *et al.*, 1994). This chapter summarizes the current ability of satellite altimetry for precisely measuring long-term sea-level variations, identifies limitations, and suggests future improvements. There is less emphasis on the current interpretation of satellite-

Satellite Altimetry and Earth Sciences

Copyright © 2001 by Academic Press
All rights of reproduction in any form reserved.

observed sea-level change, as the current record from satellite altimetry is too short to permit definitive conclusions.

2. THE TIDE GAUGE RECORD AND ITS LIMITATIONS

Sea-level data from tide gauges have long been used to provide useful, high-quality information about oceanic variations, and these records comprise a sizable fraction of the multidecadal time series available to oceanographers. It is not our purpose here to review the history of such applications of tide gauge data, but the interested reader is referred to Chelton and Enfield (1986), who discuss in some detail the variety of ocean signals that are captured by sea-level time series and the methods for extracting useful information from these data, and to Mitchum and Wyrtki (1988), who provide an extensive review of the application of tide gauge data to studies of oceanic variability in the Pacific Ocean. To date, sea-level data have also been the major source of information about low-frequency changes in the total ocean sea-level, as measured by the globally averaged sea-surface height. This application is the focus of this chapter, and it seems obvious that the global tide gauge network should be useful for this purpose. But despite the extensive use made of sea-level data in ocean circulation studies, these data have serious limitations for determining very-low-frequency sea-level change, and before proceeding we will more critically examine the limitations of the global tide gauge network for this application.

Probably the major limitation of the tide gauge data is the spatial distribution, which is of course restricted to coastlines and open ocean islands, and therefore does not adequately represent the global oceans. This problem is illustrated in Figure 1 (see color insert), which shows the locations of sea level time series that exceed 20, 40, and 60 years in length. Note that although the length of time series available is one of the strengths of the tide gauge network, the longest series are almost all along continental margins, are primarily found in Europe, and are nearly completely lacking in the Southern Hemisphere. Although there are a reasonable number of open ocean sites in the tropical Pacific, most of these stations only pass the 20-year hurdle, and the Atlantic, Indian, and high-latitude ocean areas are very sparsely covered.

This limitation in spatial coverage is a serious one for studies of low-frequency sea-level change. Consider first the change in sea level caused by ocean volume change. In this case, the spatial coverage is problematic because of the necessity of eliminating ocean signals that correspond to redistributions of ocean mass without any change in the total ocean volume. This problem is easily recognized if one considers trying to analyze data from a single tide gauge. If a long-term increase in sea level is noted, it is at best

difficult to determine from this sea-level time series alone if the rise is caused by a global volume change, or by a low-frequency shift in ocean circulation patterns that leads to sea-level changes through the geostrophic relationship. For example, there is nearly a meter of ocean topography across the Gulf Stream and low-frequency changes in this current could certainly affect tide gauge records along the east coast of the United States. Even with a sparse global network of tide gauges, the problem remains, because even though some cancellation is expected when the data from the different gauges are averaged, the uncertainty caused by these ocean signals is still large. And we must remember that this distributional problem is most severe for the longest sea-level records that are the best source of information about possible volume changes. Of course, there are also interesting sea-level changes that do not correspond to ocean-volume changes. Even in this case, however, the tide gauges do not adequately resolve the spatial patterns associated with such changes, and diagnosing the causes of such changes is difficult. That is, the altimetric analyses are still an important complement to tide gauge analyses because of the relatively dense and nearly global coverage. Only ocean models have better spatial coverage and resolution, and then one has to contend with uncertainties caused by incomplete model physics and inadequate knowledge of the forcings.

There are additional problems with the tide gauge sea-level data for determining sea-level changes that go beyond the basic spatial sampling inadequacies. The two that we will discuss here are the uncertainties caused by land motion and the inadequate quality control of the historical records. In the case of land motion, it is important to realize that sea level, strictly defined, is not the level of the sea surface, but is the distance of the sea surface to a fixed point on the adjacent land. Consequently what we measure is actually the differential vertical motion of the sea and the land. Land motion caused by post-glacial rebound (PGR) [see Tushingham and Peltier (1991) and Douglas (1991)] is relatively well-known, but there are other sources of land motion as well. An example is found in the Hawaiian Islands (Figure 2) when we plot the time series at Honolulu, which is about midway up the main Hawaiian archipelago, and at Hilo, which is at the extreme southeastern end of the chain near the hot spot that forms the islands. Honolulu shows a relatively consistent long-term rise rate of 1–2 mm/year, while Hilo, over a somewhat shorter time period, shows a trend almost twice as large. Note that the trend estimates given on the figure (1.6 and 3.5 mm/year) were computed over the common time period when both stations have data. The additional rise at Hilo is actually caused by land motion as the Big Island moves away from the hot spot. It is tempting to interpret the Honolulu record as representative of the global rate, but it is not at all clear that the island of Oahu where Honolulu is located can be considered to have completely reached steady state. This example is only intended as an illustration of non-

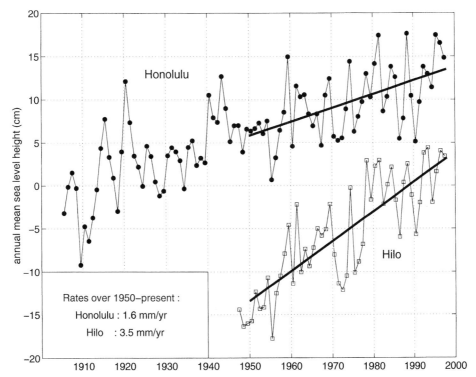

FIGURE 2 Annual mean sea level time series measured by the tide gauges at Honolulu and Hilo in Hawaii over, 1950 to the present.

PGR land motions that complicate the interpretation of sea-level changes from tide gauges in terms of global sea-level change, and there are certainly other significant causes for local land motion at individual tide gauges. Some additional examples could be given (Pugh, 1987).

In addition to the land motion signals, which from the point of view of sea-level measurements is a true signal rather than an error in the observations, there are also occasional problems with individual gauges because of inadequate maintenance of the tide gauges or the siting of the gauge, and to less-than-desirable quality control of the time series. One rather extreme example is shown on Figure 3, which shows the long-term sea-level series at Manila in the Philippines and at Bermuda in the open Atlantic Ocean. The sea-level series from Bermuda is of very high quality, and these variations have been extensively compared with nearby hydrographic data and interpreted in terms of ocean dynamics [e.g., Sturges and Hong (1995)]. On the other hand, the curve at Manila is reasonable until about 1960, but then the rise rate accelerates tremendously. If one were to interpret this as an acceleration in the global sea-level rise rate, however, it would be seriously in error, as this change actually corresponds to major harbor construction work at Manila, which most likely resulted in long-term subsidence of the land around the tide gauge.

The sort of problem that occurred at Manila could have been avoided with careful attention to the siting of the gauge and to the leveling to the geodetic benchmarks on land, but careful quality control of the tide gauge time series for the purpose of insuring scientific utility of the data is a relatively recent development. On a positive note, however, finding and correcting such errors can now be done quite easily, and level shifts of only a few millimeters can be detected even in daily averages of sea level (Mitchum *et al.*, 1994). The precision on longer time scales is much better than a few millimeters. But uncertainties remain in the historical records. And, again, the land motion is a true signal in the tide gauge records, so if we desire to eliminate it entirely, for the purpose of estimating sea-level change for example, then independent estimates of the land motion at the tide gauge sites must be included.

Despite these limitations in the tide gauge records, useful estimates of the historical rate of sea level change have been made, and these will be reviewed in a later section. The purpose here is not to condemn the tide gauge estimates, but rather to understand clearly the limitations of these data, and therefore to ask if there is a sensible way to improve upon them. In particular, using sea surface heights from altimetric satellites, such as T/P, is very promising. Why should the satellites help? First, the obvious advantage is that the satellites provide us with the capability of computing nearly global averages at intervals of tens of days. This allows a direct elimination of the ocean signals that correspond to redistributions of mass, which are a major noise source for the

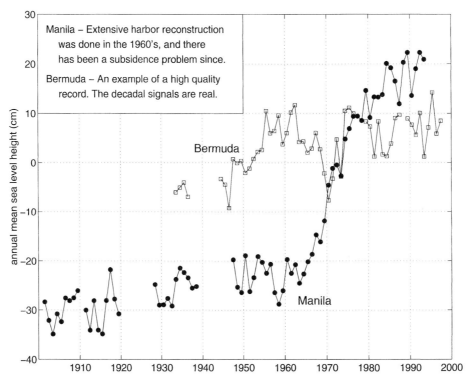

FIGURE 3   Annual mean sea-level time series measured by the tide gauges at Manila and Bermuda.

estimates derived from tide gauges. Consequently, we should be able to obtain estimates of the rate of sea-level change using much shorter time series than would be required using the sea-level network. This advantage is particularly important if one notes that the most important parameter is not necessarily the rate of sea-level change, but the acceleration; i.e., we want to know whether the rate of sea-level change is increasing or decreasing because of anthropogenic influences. With estimates available from recent altimetric time series, one can imagine differencing from the background rate computed from the long sea-level series in order to address this important question. It should be noted that the estimation of this acceleration directly from the tide gauge data has been attempted, but without success (Douglas, 1992; Woodworth, 1990).

There is an obvious problem, however, with using the altimetric sea surface heights in this fashion. In order to make very-low frequency estimates of globally averaged sea surface height, it is necessary to assume that the time series available from the altimeters are not subject to slow temporal drifts. This assumption is problematic with any measurement device and is particularly so with a system as complex as an altimetric satellite. Further, since individual satellites have a limited lifetime, it is necessary to combine multiple missions, possibly with different drift characteristics, in order to make estimates of long-term sea level change. Note that even tide gauges, which are relatively simple instruments, are not assumed to be free of temporal drift.

The stability of these measurements is continuously monitored against the observations made from an even simpler and more direct system, which are the tide staff, or tide pole, measurements [see Pugh (1987), for a description]. Although the tide staff observations are temporally sparse and noisy, the measurements are very direct and can be used to monitor and correct instrumental drift in the tide gauge system.

Over the past decade, a strategy for using the tide gauges to provide an analogous stability check for the satellite altimeters has been developed (Mitchum, 1994; Mitchum, 1998; Wyrtki and Mitchum, 1990), and this technique and recent improvements to it will be described in Section 4. Before doing that, however, we will turn to a more complete description of the measurements of sea-surface height from satellite altimeters, and the methods used to make estimates of sea-level change from these data.

3. SATELLITE ALTIMETER MEASUREMENTS OF SEA-LEVEL CHANGE

In principle, satellite altimetry provides a means of overcoming the limitations of tide gauge measurements because the measurements are truly global in distribution (within the latitude bands covered by the satellite) and tied to the earth's center-of-mass in a well-defined reference frame defined by the satellite tracking stations, in combination with other precise geodetic techniques [e.g., Nerem *et al.* (1998)]. However, satellite altimetry is subject to its own unique set of errors, such as satellite orbit errors, errors in computing the atmospheric delay (ionosphere, troposphere) and sea-state corrections, and instrument errors. For many years, these errors severely limited the analysis of long-term changes in sea level using satellite altimetry, and T/P was the first mission with sufficient accuracy to allow real changes in mean sea level to be detected.

The first attempts to measure long-term sea-level variations using satellite altimetry were made using data from the Seasat (1978) and Geosat (1985–1989) missions. Seasat's 3-day repeat orbit showed 7 cm variations for estimates of global mean sea level over a month (Born *et al.*, 1986). Tapley *et al.* (1992) used 2 years of Geosat altimeter data to determine 17-day values of variations in mean sea level with an rms of 2 cm and a rate of 0 ± 5 mm/year. The largest errors were attributed to the orbit determination, the ionosphere and wet troposphere delay corrections, and unknown drift in the altimeter bias, which was not independently calibrated for Geosat. Wagner and Cheney (1992) used a collinear differencing scheme and 2.5 years of Geosat altimeter data to determine a rate of global sea level change of -12 ± 3 mm/year. When compared to a 17-day Seasat data set, a value of $+10$ mm/year was found. The rms of the Geosat-derived mean sea level variations was still a few cm, even after the application of several improved measurement corrections. The ionosphere path-delay correction was identified as the single largest error source, but there were many other contributions including errors in the orbit, wet troposphere correction, ocean tide models, altimeter clock drift, and drift in the altimeter electronic calibration. Since the Geosat study of Wagner and Cheney (1992), several improvements have been made to the Geosat altimeter measurement corrections (ionosphere, tides) and the orbit determination. However, Nerem (1995b) and Guman (1997) still find the Geosat mean sea-level measurements are not of sufficient quality to allow a determination of the rate of mean sea-level change accurate to a millimeter/year, although the latter study succeeded in developing a tie between Geosat and T/P that allowed a reasonable estimate of sea-level change ($+1.0 \pm 2.1$ mm/year) over the decade spanning the two missions. Some investigators have examined the use of ERS-1 data for long-term sea level change (Anzenhofer and Gruber, 1998; Cazenave *et al.*, 1998; Guman, 1997), but in general these results are less accurate because the ERS-1 altimeter is single frequency, and the orbits are less precise than T/P. Because of instrument problems, ERS-2 will be of limited use for studies of mean sea-level change (Moore *et al.*, 1999).

With the launch of T/P in August, 1992, many of the error sources which limited the measurement of long-term sea-level variations by previous missions have been eliminated or significantly reduced. The precision orbits have been improved to 2–3 cm rms radially (Marshall *et al.*, 1995; Nouel *et al.*, 1994; Tapley *et al.*, 1994b); an ionosphere correction is produced directly from the dual-frequency altimeter measurements (Imel, 1994); a wet troposphere correction is provided by a three frequency microwave radiometer measurements of the integrated water column (Ruf *et al.*, 1994); and the altimeter system calibration is monitored at several verification sites (Christensen *et al.*, 1994). T/P flys in a circular orbit at an altitude of 1336 km and an inclination of $66°$ with respect to the equator, which is optimal with respect to tidal aliasing. The orbit selection, in combination with the improved measurement accuracy, has led to vastly improved models of the ocean tides (Shum *et al.*, 1997).

Numerous papers have described the global mean sea-level variations observed by T/P (Cazenave *et al.*, 1998; Minster *et al.*, 1995; Minster *et al.*, 1999; Nerem, 1995a; Nerem, 1995b; Nerem *et al.*, 1999; Nerem *et al.*, 1997a) and their spatial variation (Hendricks *et al.*, 1996; Nerem *et al.*, 1997b). Altimeter measurements have also been used to monitor mean water level in semi-enclosed or enclosed seas [(Cazenave *et al.*, 1997; Larnicol *et al.*, 1995; Le Traon and Gauzelin, 1997), among others], lakes [e.g., Morris and Gill (1994); Birkett (1995)], and rivers [e.g., Birkett (1998)], although these applications are outside the scope of this chapter.

For the purpose of completeness, and to update the analysis with the most recent data, the latest results from T/P will be reviewed here. The data processing employed for the results presented here is essentially identical to that used in Nerem (1995a, 1995b) and Nerem (1997a) and thus will not be reproduced in detail. To summarize, mean sea-level variations are computed every 10 days by using equi-area weighted averages (Nerem, 1995b) of the deviation of sea level from a 6-year along-track mean (1993-1998) computed exclusively from T/P data (as opposed to using a more general multi-mission mean sea surface). Note that T/P cannot measure "global" mean sea level because it covers only $\pm66°$ latitude, however tests have indicated the mean sea-level estimates are very insensitive to this gap in latitudinal coverage (Minster *et al.*, 1995; Nerem, 1995b). All of the usual altimeter corrections (ionosphere, wet/dry troposphere, ocean tides, sea state, etc.) have been applied to the data. No inverted barometer (IB) correction was applied to these data (Nerem, 1995a, 1995b), as we were concerned about errors in the IB correction (Fu and Pihos, 1994; Raofi, 1998). Although improvements in the IB correction have been developed (Dorandeu and Le Traon, 1999; Raofi, 1998), we argue it is the total sea-level change signal that is of interest, and not its IB-corrected equivalent (e.g., if there was a secular change in mean atmospheric pressure, resulting in a secular change in mean sea level, we do not want to remove this signal from the results). Nevertheless, the IB contribution to secular changes in mean sea level over the T/P mission is less than 1 mm/year (Dorandeu and Le Traon, 1999). Since the CSR 3.0 ocean tide model was used (which was developed with IB-corrected T/P data), this introduced a slight error with a 58.7-day period associated with the S_2 atmospheric pressure tide normally modeled in the IB correction. Modified Geophysical Data Records (MGDRs) covering Cycles 10–233 [Cycles 1–9 were omitted because they are suspect (Nerem, 1995b)] from both the TOPEX and POSEIDON altimeters have been used in this study [the single-frequency POSEIDON altimeter does not provide an ionosphere correction directly, but since this altimeter is only used about 10% of the mission, the somewhat lower accuracy of the DORIS-derived ionosphere correction (Minster et al., 1995) does not significantly affect the results presented here]. The MGDRs differ from the GDRs (used in earlier studies) in that an improved EM bias algorithm is employed (Gaspar et al., 1994), and updated estimates of the σ_0 calibration were used (Callahan et al., 1994). The onboard TOPEX altimeter internal calibration estimates have also been applied (Hayne et al., 1994), which is designed to measure changes in the instrument calibration using measurements from a calibration loop in the instrument electronics. In addition, a correction of 1.2 mm/year has been applied for an apparent drift in the TOPEX Microwave Radiometer (TMR), which provides the wet troposphere correction, has been applied (Keihm et al., 2000). The latest improved orbits using the improved JGM-3 gravity model (Tapley et al., 1996) have been employed. While some data editing is performed (shallow water, outliers, high-mesoscale variability, etc.), Nerem (1995b) and Minster et al. (1995) have shown the mean sea-level estimates to be very insensitive to this editing.

Figure 4 shows the cycle-by-cycle (10 days) estimates of global mean sea level for Cycles 10–233 computed using the techniques described in Nerem (1995b). The RMS of the mean sea-level variations is roughly 7 mm after removing a trend. A spectral analysis reveals much of the variability lies near periods of 511, 365, 182, and 59 days, which all have amplitudes of 2 mm and greater. The variability at a period of 59 days is near the fundamental period that the T/P orbit samples semi-diurnal varying phenomena, such as atmospheric and ocean tides (such as the aforementioned S_2 error), the ionosphere, etc. This variability can be significantly reduced by smoothing the mean sea level values using a 60-day boxcar filter. As shown in Figure 4, this reduces the rms of the time series to less than 4 mm after removing a trend. Also note the large 15–20 mm increase in mean sea level at the end of the time series, which we will show later is related to the 1997–1998 ENSO event (Nerem et al., 1999). The power of satellite altimetry to spatially map the sea-level change signal will also be reviewed later.

The analysis of the altimeter data to this point has been reasonably straightforward. The difficulty arises in determining the errors in the mean sea level estimates, which are directly related to the long-term performance of the satellite instruments.

4. CALIBRATION OF SATELLITE ALTIMETER MEASUREMENTS USING TIDE GAUGE DATA

As discussed earlier, to fully exploit the altimetric sea-surface heights for the purpose of evaluating possible acceleration in the rate of global sea level change, or even simply to best estimate the character of the sea level changes, it is necessary to address the issue of the stability of the altimetric series. That is, it is not considered reasonable to assume that these time series are free of low-frequency drifts. In this section, we will summarize ongoing efforts to make estimates of the altimetric drift errors through combination of the altimetric sea surface heights with sea levels measured by tide gauges. The basic rationale for this approach has been developed and discussed at length by Wyrkti and Mitchum (1990) and Mitchum (1994; 1998), and only a short review is given here. Since the last description of the method (Mitchum, 1998), significant improvements have been made, and these will be briefly reviewed. An error analysis for the technique will be presented in Section 5 along with a similar analysis for the altimetric calculation of the sea-level change curve.

In order to describe the basic idea behind using the tide gauge sea levels to monitor drift in altimetric satellites, imagine the following highly idealized situation. A more realistic scenario will be presented shortly. First, imagine having tide gauges that perfectly measure the sea-surface height signals in the vicinity of the gauge. That is, there are no instrumental errors, no errors caused by small-scale deformations of the sea surface because of the topography on which the tide gauge sits, and the land is perfectly stable in the vertical. Second, imagine that the altimetric satellite is similarly ideal, except that there may be a low-frequency temporal drift caused by instrument errors, environmental corrections, or some other cause. Even in this idealized case, analyzing the altimetric data alone would not allow an unambiguous determination of the drift error. Partly the problem is statistical, since the ocean signals would mask the trend caused by the drift, leading to an uncomfortably small sensitivity to modest, but significant, drift rates. And the problem is partly one of interpretation, because any low-frequency trend detected in the heights could not be unambiquously separated from real ocean trends, because of the sea-level change that we want to detect, for example.

Continuing with this idealized situation, consider what happens when the altimetric heights and the sea levels are differenced. Under the assumptions of this (admittedly unrealistic) model, the ocean signals cancel exactly, leaving only the altimeter drift signal. This is the crux of the idea, that differencing two independent sources of the same information

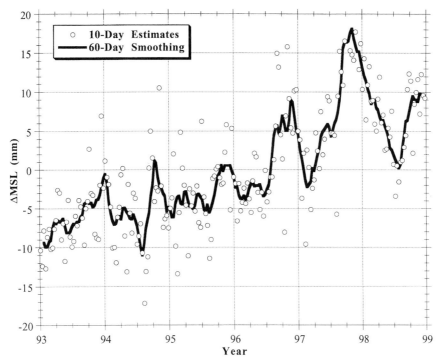

FIGURE 4 10-day estimates of global mean sea level from the TOPEX/POSEIDON mission and after smoothing using a 60-day boxcar filter.

will greatly improve the signal-to-noise ratio for the determination of errors in one or both of the measurement systems. The idealized model, though, is unrealistic in several ways. First, although we are confident that the tide gauge sea levels can be accurately maintained relative to the adjacent land (as discussed in Section 2), which provides the long-term stability that we want to exploit, we must still consider errors caused by low-frequency vertical land motions at the gauges. And, second, both the tide gauges and the altimeters have random errors that must be accounted for. We write a more realistic, but still schematic, model as:

$$\eta_n(t) = s_n(t) + \lambda_n(t) + \varepsilon'_n(t) \tag{1}$$

$$h_n(t) = s_n(t) + \Delta_n(t) + \varepsilon''_n(t) \tag{2}$$

$$\delta_n(t) = h_n(t) - \eta_n(t) = \Delta_n(t) - \lambda_n(t) + \varepsilon_n(t) \tag{3}$$

where Eqs. (1) and (2) define the tide gauge sea level at one site $[\eta_n(t)]$ and the analogous altimetric heights $[h_n(t)]$ as the sum of ocean signals $[s_n(t)]$ that are observed by both systems, noise in the sea-level measurements $[\varepsilon'_n(t)]$, noise in the altimetric observations $[\varepsilon''_n(t)]$, vertical land motion at the gauge site $[\lambda_n(t)]$, and the altimeter drift $[\Delta_n(t)]$ that we wish to estimate. Eq. (3) defines the difference series at an individual station (n) that we use to attack the drift problem. Note that the ocean signal contributions have cancelled, and that the random errors have been combined into a single term $[\varepsilon_n(t)]$.

In order for this schematic model to be useful, the random-error component in the differences $[\varepsilon_n(t)]$ must be significantly smaller than the ocean signal contribution $[s_n(t)]$, otherwise the signal-to-noise ratio will not be much improved. The test for this is simply that the difference series have a variance that is much smaller than the individual sea-level and sea-surface height series from the tide gauges and the altimeter, respectively. Several representative series are shown (Figure 5) that demonstrate that this is, in fact, a valid assumption. The difference series plotted in the right-hand panels have a much smaller range than the individual sea-level and sea-surface height series that are plotted together in the left-hand panels. Looking at the left-hand panels reveals an obvious high correlation between these series at any particular station that is because of the fact that both systems are primarily measuring the same ocean signals. The high degree of coherence also implies that the random errors and the land motion and drift errors are relatively small compared to the ocean signals. So the computation of the differences accomplished exactly what we intend, which is to significantly increase the signal-to-noise ratio for determining the altimeter drift term by eliminating the large ocean signal contribution.

After computing difference series at each station, the method proceeds by spatially averaging these series from many stations at each point in time in order to reduce the magnitude of the random error, which is assumed to be uncorrelated from one station to another, while retaining the

Representative TOPEX (circles), Tide Gauge (line) and Difference Series (all in mm)

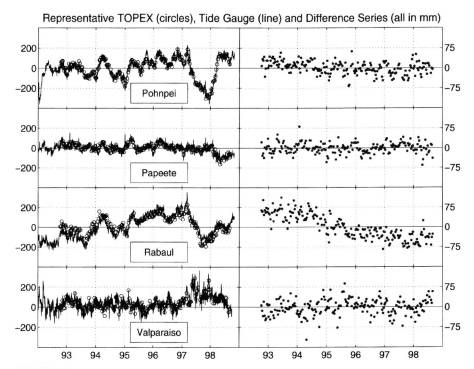

FIGURE 5 Comparison of TOPEX/POSEIDON and tide gauge sea level for selected sites. The left panels show the time series from the altimeter (dots) and daily sea levels (lines). The daily sea levels are non-tidal residuals that have been low-pass filtered to eliminate periods shorter than 2–3 days in order to avoid aliasing. The right panels show the differences between the altimetric heights and the sea levels. Note the change in the height scale between the left and right panels.

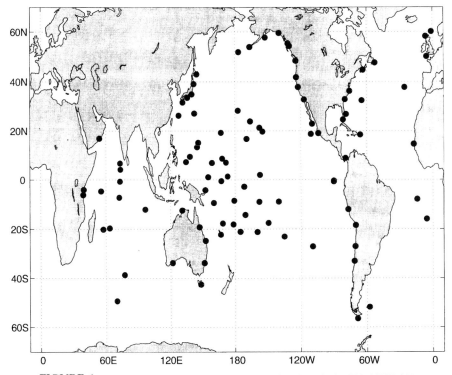

FIGURE 6 Map of tide gauge locations used in the calibration of TOPEX/POSEIDON.

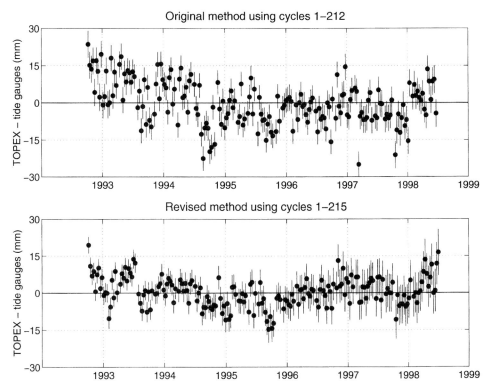

FIGURE 7 Time series of TOPEX-tide gauge sea level differences using the original method (upper panel) (Mitchum, 1998) and the improved method (lower panel).

altimeter drift-error term, which is assumed to be slowly varying enough that the drift term can be considered constant over the averaging period of about 10 days. Naturally, in order for the random error to be reduced these errors must be statistically independent from one station to another. This assumption was checked in some detail by Mitchum (1998) and was found to be appropriate. Therefore, at the end of this spatial averaging we have a single, globally averaged, time series of altimeter minus gauge differences that can be used to estimate the drift error, assuming that the land motion has also cancelled or is much smaller to begin with. Setting aside the land motion issue for a moment, we present the results from the technique, which are summarized in Figure 6, which shows the stations used in the analysis, and Figure 7, which shows the globally averaged difference series. The upper panel is from using the technique described by Mitchum (1998), while the lower panel also incorporates a number of improvements that have been made since that paper was prepared. Looking at either figure we can see that the linear trend is relatively small, indicating that the TOPEX altimeter is stable to at least the order of a few mm/year. We also see a suggestion of a quadratic trend, but the reason for this is currently not understood. The main difference between the upper and lower curves is that the point-to-point noise, which is caused by the random errors, has been somewhat reduced by the improvements made to the method (the trend is also

smaller because of the TMR correction). This should allow a more accurate determination of linear drift errors, as well as a more sensitive determination of drift errors that are temporally more complex than simple linear trends. Before returning to the land motion question, we will briefly describe the improvements made to the technique since the Mitchum (1998) description.

Most of the changes made to the method are aimed at reducing the random-error contribution. This is accomplished by doing a more thorough job of matching up the altimeter and tide gauge series in order to allow the best cancellation of the ocean signal contribution. As a first step, more altimeter data were incorporated into the analysis at each tide gauge. Previously, only the nearest four TOPEX passes were used in the analysis, but it was found that useful correlations existed at larger separations that could be exploited to better describe the ocean signals. At present the nearest eight passes are used, albeit with a weighting that favors the TOPEX data that is nearest the tide gauge. Second, for each TOPEX pass, a more careful matching of the altimetric and tide gauge data was done. In the original method the TOPEX data at the point of nearest approach was used with zero temporal lag. It was known, however (Mitchum, 1994), that there were a number of stations where significant temporal lags existed because of the spatial separation between the tide gauge and the altimetric data, which is important because at some locations a large portion of the ocean signal is in propagating signals, mostly in the form of westward travelling Rossby waves. In the improved technique, therefore, spatial and temporal lags were allowed when matching up the time series before differencing. Finally, partly as a result of the earlier versions of the tide gauge estimates of the TOPEX drift given by Mitchum (1998), the wet tropospheric range correction derived from the microwave radiometer was reexamined (Keihm *et al.*, 2000) and was found to contain an error. This error term has been accounted for in the new technique as well. Several other less-significant changes were also made, but these three changes account for the majority of the changes seen from the upper to the lower panels in Figure 7.

Returning to Figure 5, we can see in the difference series in the right-hand panel that at least one of the stations, Rabaul, has a large temporal drift that is not seen at the other stations. This is in fact why the Rabaul record was included in this plot. This trend is even obvious in the left hand panel upon close examination. This trend is almost certainly due to vertical crustal motion at Rabaul, which is in a volcanically active area. The question remains, then, as to how much bias remains in the globally averaged difference series due to signals such as this. That is, if the land motion signals, the $\lambda_n(t)$ in Eq. (3), do not average to zero, then it is not possible to interpret the trend in the record as being caused by the altimeter drift. The only true solution to this problem is to obtain independent estimates of the land motion rate at each gauge, by collocating geodetic instruments (e.g., GPS, DORIS, etc.) with the gauges, for example. In the earlier version of this work, Mitchum (1998) estimated that the uncertainty associated with determining the net effect of the land motion was of order 1 mm/year (1σ), as opposed to 0.6 mm/year from the random errors. Clearly, reducing the land motion errors using independent geodetic information will be necessary to improve the technique further, and plans have been made to obtain these data, as will be discussed in the next section.

5. DETECTING CHANGES IN THE RATE OF SEA-LEVEL RISE

As mentioned in the first section, we are primarily interested in determining whether the rate of mean sea-level change is accelerating, in response to global climate change, for example, as suggested by the Intergovernmental Panel on Climate Change (IPCC) in their recent report (Houghton *et al.*, 1996). In order to do this, it is necessary either to determine the acceleration directly, which has not been possible using the tide gauges alone (Douglas, 1992; Woodworth, 1990), or to make an estimate of the sea-level change rate from only very recent data and difference this from an estimate determined over a longer period of time. One strategy is to use the tide gauge estimate of the historical rate of 1.8 (± 0.1) mm/year (Douglas, 1991) as the background rate, to then determine independently a rate from the more recent altimetric data, and then to ask whether the altimetric rate is significantly different than the background rate. Although we will describe the present altimetric estimate of the recent sea-level changes, we are mainly concerned with a careful error analysis of this approach in order to know what is required to make this calculation as sensitive as possible, and that is the main point of this section. In the near future, however, we believe that it may be possible to quantitatively test the IPCC projections using this method. It will be much better when the acceleration of the rate can be measured directly using satellite altimetry, but this will require several decades of measurements.

Before discussing errors and the improvements that will be necessary to reach this goal, we should discuss the reasons for our adopting the Douglas (1991) rate as the background rate. In his paper, Douglas reviews a number of earlier estimates of the background global sea-level change rate, and several issues are addressed that are especially noteworthy. First, he correctly notes that it is essential to deal with land motion at the tide gauges in some fashion. Douglas explicitly corrects for post-glacial rebound using the results of Tushingham and Peltier (1991), and he further examines all of the individual sea-level records for evidence of serious land motion problems, such as that shown earlier (Figure 3) at Manila. Second, in addition to identifying problems such as that seen at Manila, Douglas has made a careful gauge by gauge effort at quality control, explicitly attempting to use the long records best suited for the estimation of long-term trends. Finally, as pointed out earlier by Barnett (1983), it is not possible to consider closely spaced tide gauges as being statistically independent for the purposes of this calculation. Real ocean signals make up the majority of the "noise" in which the sea-level change signal is embedded, and such signals can have long spatial and temporal decorrelation scales. Thus, it is necessary to combine gauges into regional averages and to only consider this smaller number of time series to be independent; i.e., there are fewer degrees of freedom spatially than might be expected. Douglas also incorporates this consideration into his analysis.

Most of the earlier studies cited by Douglas incorporated one of more of these considerations into their analyses, but the Douglas study stands out in being particularly thorough in incorporating all of these improvements into his calculation of the sea level change estimate. We therefore adopt his result, which is 1.8 (± 0.1) mm/year and uses data over approximately the past 60 years. We interpret this rate to be an average of a possibly changing rate over that time period, and the challenge now is to obtain a more recent estimate with a comparable uncertainty that can be used to test the hypothesis that the rate is increasing.

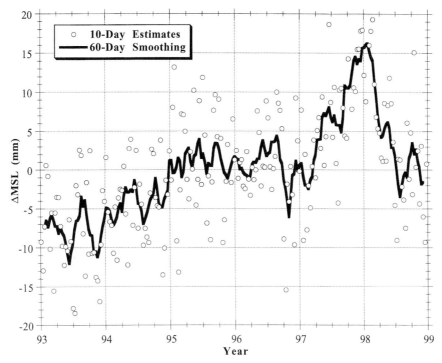

FIGURE 8   Same as Figure 4, but after correcting for instrument effects using the tide gauge calibration time series (lower panel, Figure 7) and removing annual and semi-annual variations.

We now want to correct the T/P mean sea level time series (Figure 4) using the tide gauge instrument calibration time series (Figure 7) to get our final estimate of the rate of sea-level change. Note that in general we prefer to identify the altimeter measurement error being detected by the tide gauges and directly correct the altimeter data, such has been done in the past for the oscillator correction (Nerem, 1997) and the wet troposphere correction (Keihm *et al.*, 1998). However, not all of the altimeter measurement errors can currently be identified and corrected, thus using the tide gauge calibration as a correction is the only way to proceed. For this discussion, we also remove annual and semi-annual variations from the time series because there is some evidence that the tide gauge calibration might have small errors at this frequency [the tide gauges are more sensitive to localized seasonal variations in heating, freshwater flux, and coastally trapped waves (Mitchum, 1998)]. Figure 8 shows the final time series of global mean sea level change from T/P. A simple least squares linear fit to these data give a rate of +2.5 mm/year with a formal error ±0.25 mm/year (based on the scatter of the fit). We use formal error here because we cannot assemble a realistic error budget for global mean sea level otherwise, since we need to know the globally integrated errors caused by mismodeling of the orbit, ionosphere, troposphere, tides, etc. The formal error assumes the point-to-point measurement errors (one 10-day estimate of mean sea level to the next) are uncorrelated. If we account for the autocorrelation of the fit residuals [e.g.,

Maul and Martin (1993)], this increases the formal error to 0.4 mm/year. However, we also corrected the T/P data using the tide gauge calibration values (to account for measurement errors that are otherwise inaccessible to us), and the estimated error in these values needs to be added to the formal error. Our estimate of the error in the tide gauge determination of the instrument drift is ±1.2 mm/year, which is dominated by the land motion errors. If we root-sum-square (rss) the formal error and the instrument drift error, we get a total error of 1.3 mm/year. Our final estimate of +2.5 ± 1.3 mm/year is valid only over the 6 years covering, 1993–1998; we really can't say if it is representative of the rate of long-term sea-level change (as we will show later, the effects of ENSO variability significantly affect the rate estimate). Note that in any case, this estimate of sea-level rise over, 1993–1998 is statistically indistinguishable from the long-term rate determined from the tide gauge data.

In the above determination of the present rate of sea-level change, two distinct calculations are required in order to arrive at the final rate. First, the altimetric estimate of the sea-level change rate must be estimated, and second, this estimate must be corrected for altimeter drift using the results of the tide gauge analysis described in Section 4 above. We will first discuss the errors caused by the tide gauge analysis, turn then to the errors in the altimetric analysis, and will conclude this section by combining these results in order to project the future prospects for this method.

Starting with the uncertainty caused by applying the application of the tide gauge estimate of the altimeter drift, we recall that in the previous section it was noted that Mitchum (1998) has already estimated uncertainties for the technique. The errors were of two types, those caused by random scatter in the time series of the drift estimate and those caused by the uncertainty in estimating the error associated with the net land motion at the gauges after averaging all the gauges together. Since the TOPEX time series were approximately 3 years in length at that time, the uncertainties were quoted as appropriate for that record length after considering the effect of serial correlation in the time series. Mitchum found that the random errors contributed an error to the trend estimate of 0.6 mm/year (1 σ) and also estimated that the uncertainty caused by residual land motion was of order 1 mm/year (also 1 σ). Combining these errors resulted in a final uncertainty estimate of 1.2 mm/year. In order to extrapolate these uncertainties to the longer records available now, one can use the fact that errors in trend estimates decrease as the record length to the −1.5 power. For example, with 10 years of data we expect that the same analysis would yield a final uncertainty of 0.2 mm/year.

Clearly the major area of improvement is in determining the land motion at the gauges, although some reduction in the random-error component has also been accomplished, as described above in Section 4 (see also Figure 7). Mitchum (1997) considered the land motion problem, and evaluated the requirements for independent land motion estimates, from GPS or DORIS geodetic measurements for example, that would be necessary in order to significantly reduce the error budget. The most significant result from this analysis is that if the land motion can be estimated at a single tide gauge site to a precision of 10 mm/year using 1 year of observations, and if the land motion errors are independent at the different tide gauge sites, then the land motion error will not contribute at lowest order to the error budget for the drift estimate. That is, the error will be completely dominated by the random errors. For example, with these assumptions for a 3-year record length at a single tide gauge site, the error caused by random errors is about 5 mm/year, while the land motion contribution is reduced to 2 mm/year, or to about 10% of the total variance associated with both errors. This is encouraging since a precision of 10 mm/year for vertical rates should be easily within reach of the GPS and DORIS technologies.

In the same report, Mitchum (1997) further estimated that having 30 independent tide gauge sites that were equipped with independent land motion estimates would result in an error for the drift estimate of order 1 mm/year for a 3-year record, which is comparable to the more extensive analysis using nearly 100 stations without land motion information, and the error drops to the order of 0.2 mm/year once a 10-year record has been obtained. A careful selection of tide gauge sites was made and a strawman network was proposed (Figure 9) as a backbone network for this application. In the time since that proposal, excellent progress has been

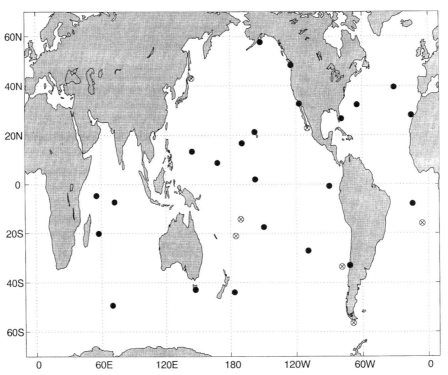

FIGURE 9   Strawman tide gauge network proposed for calibrating satellite altimeter measurements.

made in instrumenting this backbone network. Of course, it is expected that more than 30 stations will be available for making the drift estimate, so the error of ±0.2 mm/year after 10 years is probably a conservative estimate of what is possible. If the estimates of sea-level change from the altimeters can be made equally precise, then we could expect to be at the point of being able to quantitatively test the IPCC projections (Houghton *et al.*, 1996) at the end of the 10-year period, or very early in the next millenium. We still need to consider, however, the errors associated with estimating the sea level change from the altimetric data, which we will turn to now.

We will show later that the 20 mm rise and fall of global mean sea level during, 1997–1998 is almost certainly related to the ENSO phenomena. The presence of ENSO-scale variability in global mean sea level means that a longer time series will be needed to reduce this variability through averaging to reveal the smaller climate signals, unless reliable techniques can be developed to remove the ENSO signal from the data. To assess the impact of such variability, Nerem *et al.* (1999) developed two simulated long-time series of global mean sea-level variations, both containing ENSO variability. The first was a ~ 100-year-long (1882–1998) time series of the Southern Oscillation Index (SOI) which was linearly regressed against T/P mean sea level over, 1993–1998 to determine the proper regression coef-

ficients. The regression coefficients were then used to scale the SOI and simulate a ~ 100-year-long time series of global mean sea level. A similar technique was applied to a set of reconstructed global mean SST anomalies (Smith *et al.*, 1996) covering 1950–1998. In addition, the rate and acceleration of each of these time series was set to zero. Note that these estimates do not include errors in the tide gauge calibrations, but as mentioned previously, these errors should be significantly smaller than those caused by ENSO variability if geodetic monitoring is performed at each gauge.

Each of these time series was then used to perform two simulations. In the first simulation, a 2 mm/year secular sea-level change was added to the time series, and a series of Monte Carlo solutions were performed to determine the error in the estimated sea level rate by a least squares solution using varying data spans with a random midpoint time. The result of these simulations is a plot of the accuracy of the sea level rise estimate versus the length of the data span of altimeter data employed (Figure 10). Both simulations (using SOI or SST simulated sea level) suggest 10 years of T/P class altimetry will be required to determine the rate of mean sea level change to an accuracy of 0.5 mm/year. These simulations were repeated by adding an acceleration (using a $0.5at^2$ convention) of mean sea level of 0.06 mm/year^2, and estimating both the rate and acceleration of sea level change. Figure 11 shows the error in the estimated acceler-

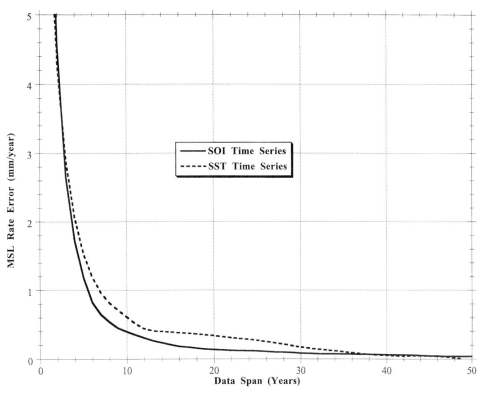

FIGURE 10   Plot of expect accuracy of the T/P determination of the rate of sea level rise versus record length, using the simulated mean sea level time series derived from long SST and SOI records.

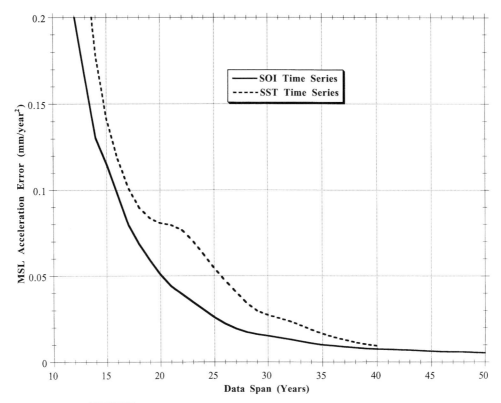

FIGURE 11   Same as Figure 10, but for the acceleration of mean sea level.

ation as a function of data span. The SST results are invalid after about 20 years because the time series is of insufficient length to test longer data spans. However, the SOI results suggest ∼ 30 years of T/P class altimetry will be needed to detect an acceleration of mean sea level to an accuracy of 0.02 mm/year^2.

In summary, we should be able to detect a difference from the Douglas (1991) tide gauge-determined rate with a decade of T/P-class altimeter data, which will provide a limited test of the IPCC predictions. A more meaningful test, where the acceleration of mean sea level change is measured, will require at least 2–3 decades of similar data, unless decadal variability in global mean sea level is a significant contaminant to the computation. There is currently no way to assess the level of decadal variability in time series of global mean sea level computed using satellite altimetry.

6. GLOBAL MEAN CHANGES IN SEA LEVEL, SEA-SURFACE TEMPERATURE, AND PRECIPITABLE WATER

When studying measurements of sea-level change, measurements of ocean temperature are often examined to determine if the observed changes in sea level are caused by ther-

mal expansion. Unfortunately, measurements of temperature over the water column can only be made with in situ instrumentation (hydrographic measurements from ships, buoys, sondes, etc.), which have limited spatial extent. Sea-surface temperature (SST) can be observed from space, using either infrared or microwave techniques, thus making SST a natural variable for comparison to the global measurements of sea level collected by T/P. Unfortunately, SST measurements reveal nothing about the temperature of the entire water column, and thus some assumptions (mixed-layer thickness and temperature, thermal expansion coefficient, etc.) must be invoked to directly compare these measurements to sea level. For this discussion, we will qualitatively compare the observed spatial and temporal variations of sea level and SST, but we need to remember that they need not be in agreement if the SST is not representative of the temperature of the mixed layer, or if other phenomena besides thermal expansion of the mixed layer are driving the sea level change.

We will use a global 1° × 1° SST dataset covering, 1982 to the present compiled at weekly intervals by Reynolds and Smith (1994), which is based on thermal infrared images collected by the Advanced Very High Resolution Radiometer (AVHRR) onboard the NOAA satellites. Using equi-area weighting, we can compute variations in global mean SST since, 1982. This is dominated by a large annual change in global mean SST as described by Chen et al. (1998), which

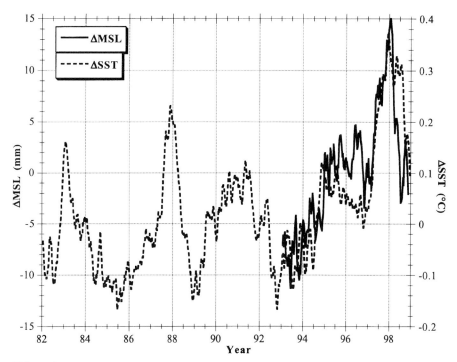

FIGURE 12 Comparison of global variations of mean sea level (Figure 8) and SST, after removing annual and semi-annual variations.

is initially surprising since the annual change in global mean sea level is only 2–3 mm. Chen *et al.* (1998) and Minster *et al.* (1999) reconciled this difference by observing that annual changes in continental water storage effectively cancel out the annual sea level change caused by changes in SST.

For the present discussion, we will simply remove the annual and semi-annual SST variations from the original data set, and compute the global mean SST variations, as shown in Figure 12. In addition, we apply 60-day smoothing to be consistent with the smoothing of the sea level time series. A large increase in global mean SST is observed for each major ENSO event since, 1982, with the largest change (0.35°C) occurring during the, 1997–1998 event. Also shown in Figure 12 is the variation in global mean sea level from T/P, after correcting for instrument effects using the tide gauge calibration (Figure 7), removing seasonal variations, and applying 60-day smoothing. Qualitatively, the comparison is quite compelling. Note that the increase in sea level and SST at the beginning of, 1997 occurs nearly simultaneously, while the decrease of SST in 1998 lags that of sea level by several months, suggesting temperatures in the subsurface returned to normal before the surface temperatures. These results suggest that the rise and fall in sea level during, 1997–1998 was directly related to the ENSO event, and we will further demonstrate this by examining the spatial variation of these changes in Section 7.

These results show that, on average globally, the ocean retains heat during an ENSO event. How does this occur?

For the ocean to retain heat, there could be an associated decrease in the Outgoing Longwave Radiation (OLR). OLR is quite sensitive to changes in precipitable water vapor (PWV) in the troposphere, since water vapor effectively insulates the sea surface. Fortunately, the TOPEX Microwave Radiometer (TMR) provides along-track estimates of columnar water vapor content in order to correct the altimeter measurements of sea level for the delay this causes (Ruf *et al.*, 1994). We have computed global mean variations in PWV in a manner analogous to the sea-level computations.

Nerem *et al.* (1999) computed global mean anomalies of precipitable water vapor (PWV) using measurements from the TOPEX Microwave Radiometer (TMR). These measurements have been calibrated using the aforementioned drift estimate (Keihm *et al.*, 2000). As shown in Figure 13, the correlation of global mean variations of sea level, SST, and PWV is quite high, although the spatial pattern of the PWV maps are noticeably different from those sea level and SST (next section), and thus are not shown here. The correlation of the smoothed variations of SST and PWV peaks at 0.91 with zero lag. Coupled with evidence that global mean Outgoing Longwave Radiation (OLR) decreased in a similar manner during the ENSO event (Jackson and Stephens, 1995; Wong *et al.*, 1998), one mechanism for the observed sea level change is an increase in water vapor associated with ENSO, which in turn decreased the OLR causing the ocean to retain more heat during ENSO. However, this is difficult to verify, as a globally averaged change in heat flux of 7 W/m^2

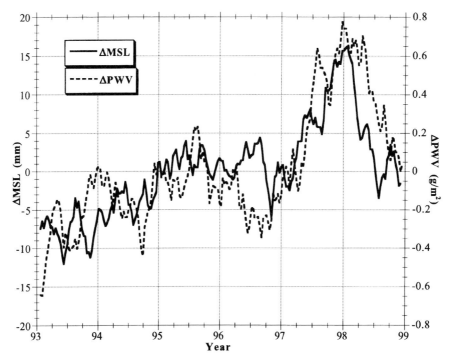

FIGURE 13 Global mean variations of sea level (Figure 8) and precipitable water vapor (from the TOPEX Microwave Radiometer).

is needed to change sea level by 20 mm at annual and inter-annual timescales, but the observed fluxes are only accuracy to 20 W/m^2.

Nerem *et al.* (1999) also showed that the observed change in global mean sea level can be reproduced using global mean dynamic height from a numerical ocean model. The model incorporated an optimal interpolation scheme which assimilates expendable bathythermograph (XBT) and mechanical bathythermograph (MBT) temperature data from the NODC (Levitus and Boyer, 1994) data set and thermister data from the TOGA-TAO moorings. The model was forced with weekly NCEP winds (Kalnay *et al.*, 1996). Sea surface temperature was damped to NCEP weekly values and the sea surface salinity is damped to monthly mean climatology from the comprehensive ocean atmosphere data set (COADS) (daSilva *et al.*, 1994). The model was spun up, from 1985–1991, using data assimilation of XBT, MBT, and TOGA-TAO temperature data.

The model temperature and salinity fields were used to calculate dynamic height at several depths assuming a level of no motion of 1000 m, as shown in Figure 14 (see color insert) (1000 m was chosen because of a lack of confidence in the model representation of the deep currents, although using a deeper level of motion of 2000 and 4000 m had little effect on the results). Dynamic height at the surface should most closely resemble sea level height as observed by T/P, and in fact the two curves do closely match each other. Since T/P observations are not used in the assimilation, this con-firms that the model and satellite are independently detecting a real oceanic signal. By comparing the mean dynamic heights for different reference depths, we found that most of the dynamic height variability associated with the ENSO event is confined to the upper few hundred meters. The total contribution to dynamic height from below 200 m is characterized by a linear trend from the beginning of the record, and does not show a distinct ENSO structure. The confinement of dynamic height anomalies to the surface suggests a surface forcing instead of a dynamic forcing that would presumably be located near the thermocline, and therefore be deeper in the water column.

7. SPATIAL VARIATIONS OF SEA-LEVEL CHANGE AND SEA-SURFACE TEMPERATURE

While variations in global mean sea level have great scientific and public interest as a barometer of environmental change, the true scientific and social implications will be determined by the spatial patterns of sea-level change. As an example, anthropogenic-induced climate change, while likely to cause a rise in globally-averaged sea level, will actually cause sea level to decline at some locations and rise in others. This type of information will be critical for planning mitigation efforts related to sea level change. In addition, the spatial pattern of sea-level change, if it could be mapped, provides another powerful constraint on climate change models by mapping the geographic "fingerprint" of the change. An example of sea-level change predicted from a representative climate model is shown in Figure 15 (see color insert) (Russell *et al.*, 1999), which shows the sea-level difference between a control run of the model, and a model that has increasing CO_2. If a sea-level change signal could be detected that is in good agreement with Figure 15, then this would provide one source of corroboration of the climate models.

The true power of satellite altimetry lies in its ability to map the geographic variation of sea-level change. There are many different spatio-temporal analysis techniques that may be employed to accomplish this. A relatively simple technique for mapping the geographic variation of long-term sea level change is to compute the linear trend of sea level at each geographic location, as shown in Figure 16 (see color insert) for the T/P mission. These trends were determined via a least squares fit of secular, annual, and semi-annual terms at each location along the T/P groundtrack. The problem with this technique is that sea level change at a single location often has considerable deviations from a linear trend, and these deviations tend to alias the trend unless a long time series is available. Currently, the trends during the T/P mission are dominated by variability from recent ENSO events, as they are also clearly manifested in the satellite observed sea-surface temperature results (Reynolds and Smith, 1994) of the same time period (Figure 17) (see color insert), as well as in numerical ocean models (Stammer *et al.*, 1996). However, as the sea-level record from satellite altimetry lengthens, the ENSO variations will gradually average out, hopefully allowing the detection of signals such as shown in Figure 15. However, decadal variability may limit this approach until many decades of data are available.

More quantitative results can be obtained using statistically-based analysis methods. The method of Empirical Orthogonal Functions (EOFs, also called Principal Component Analysis) identifies linear transformations of the dataset that concentrates as much of the variance as possible into a small number of variables (Preisendorfer, 1988). This method has been used in a variety of oceanographic and meteorological analyses to identify the principal modes of variability. If the spatial and temporal characteristics of a mode can be used to identify a physical cause, then the method can be a very powerful data analysis tool. Several investigators have used EOF techniques to help isolate the cause of the sea-level change signals observed by T/P (Hendricks *et al.*, 1996; Nerem *et al.*, 1997b). Unfortunately, EOFs have so far been unable to separate ENSO-related variations from long-term sea-level change, possibly because these processes may be interrelated (Cane *et al.*, 1996; Trenberth and Hoar, 1996). We will nevertheless review the EOF results here because

they provide considerable insight into the cause of the mean sea-level variations observed by T/P.

We start by converting the raw T/P sea level measurements into $1° \times 1°$ maps of sea level at 10-day intervals as described by Tapley *et al.* (1994a). For these computations, we apply the IB correction so that these signals do not contaminate the EOF modes. If sea level at a given latitude (ϕ) and longitude (λ) map location is represented by $h(\phi, \lambda, t)$ where t ranges over each T/P cycle, then the EOF representation of sea level can be written as:

$$h(\phi, \lambda, t) = \sum_{i=1}^{N} s_i(\phi, \lambda)a_i(t)$$

where $s_i(\phi, \lambda)$ is the ith spatial EOF, and $a_i(t)$ is its temporal history. These are determined from the eigenmodes of the spatial-temporal covariance matrix of the T/P sea level grids, thus each mode attempts to describe as much of the sea-level variance as possible. Figure 18 (see color insert) shows the four leading (highest-variance) EOF modes for sea level and SST (Reynolds and Smith, 1994) over the timeframe of the T/P mission, where the temporal history of these modes has been scaled to represent their contribution to global mean sea level and smoothed using a 60-day boxcar filter. Note that the first two "ENSO modes" describe most of the, 1997–1998 ENSO event in global mean sea level. Mode 1 began rising at the beginning of 1997, peaked in early 1998 at 10 mm, and then fell back to zero by the end of 1998. Mode 2 began rising at the beginning of 1998, peaking at 11 mm near the end of 1998. Both of these modes show large signals in the tropics, as expected, but also large extratropical signals, especially in the Southern Ocean. Peterson and White (1998) have seen similar ENSO-related extratropical signals in sea surface temperature, which they attribute to atmospheric coupling between the two regions. Also note the large signal in the southwestern Indian Ocean between 5–10°S, which has been described by Chambers *et al.* (1999) and is correlated with ENSO in the Pacific. The EOF analysis establishes that most of the observed variations in global mean sea level are related to the ENSO phenomena.

Leuliette and Wahr (1999) have taken this idea one step further by conducting a coupled pattern analysis (Bretherton *et al.*, 1992; Wallace *et al.*, 1992) of the variations in mean sea level and SST (basically an EOF analysis of the cross covariance matrix of sea level and SST). This technique allows a more rigorous method of identifying the common EOF mode causing the changes in mean sea level, although the conclusions are basically the same as computing EOFs of the individual fields. Thus, they have also concluded that most of the long-term sea level change signal observed by T/P is being caused by changes in sea surface temperature related to the ENSO phenomena.

Clearly it would be nice to remove the ENSO signals from the T/P sea-level record so that we could begin to search for smaller signals, such as those related to climate change. It would be tempting to use the leading EOF modes (the ENSO modes) of sea level and SST in such a computation, as shown in Figures 16 and 17. However, these results must be interpreted carefully, as there is evidence that ENSO and climate change are not unrelated processes (Trenberth and Hoar, 1996), and thus removing the leading EOFs might also remove climate signal. In addition, the EOF technique by no means guarantees that each mode will correspond to only a single process. In fact, processes that have spatial simularities will tend to coalesce into a single mode. There are many other statistical techniques that can be explored, so this should be considered an area of active research as of this writing.

8. LINKING TOGETHER DIFFERENT SATELLITE ALTIMETER MISSIONS

As shown in Section 5, a sea-level time series of several decades in length derived from satellite altimeter measurements will be needed in order to test different sea-level change predictions from climate change models. Such a time series cannot be provided by a single satellite mission such as T/P, thus measurements from multiple missions will be needed to assemble a time series of sufficient length. A single 10-day averaged global mean sea level measurement from T/P has a precision of roughly 3–5 mm, thus we would like to link measurements from future missions to T/P with roughly the same accuracy.

Historically, this has proven to be a difficult task, as each mission has a unique set of time-varying measurement errors. Guman (1997) presented an analysis combining measurement from Geosat, ERS-1, and T/P. Because the ERS-1 and T/P missions significantly overlapped in time, establishing the relative bias between the two sets of measurements can be obtained through a straightforward differencing of their measurements at crossover points. However, the multi-year gap between the end of the Geosat mission and the beginning of the ERS-1 and T/P missions is more difficult to overcome. Tide gauge measurements spanning these missions provides the best technique for linking the sea level time series provided by these missions. However, the errors in this approach will be significant because of the unknown vertical movement of the tide gauges, and because of their poor geographic sampling of the altimeter measurement errors. This is particularly true for Geosat, which had much larger measurement and orbit errors than T/P. Guman (1997) implement a novel method for correcting the Geosat measurements using tide gauge comparisons, however the combined Geosat/ERS-1/T/P sea level time series only loosely constrained the sea level change over the intervening decade.

The first missions likely to be used to extend the T/P measurement record are Envisat and Jason-1, both scheduled for launch in the year 2000. However, Envisat is in a sun-synchronous orbit, and thus tidal aliasing (Parke *et al.*, 1987) may be a significant error source since errors in the model of the solar tides will alias to zero frequency, potentially contaminating measurements of global mean sea level. If T/P is still operating when these spacecraft are launched, then computing the relative biases between these missions at a given time epoch will be relatively straightforward, and changes in these biases can be monitored using the tide gauges as discussed in Section 4. Indeed, it is currently planned to fly T/P and Jason-1 in tandem (in the same orbit separated in time by only a few minutes) for a short time in order to conduct a detailed study of the measurement differences.

However, if the instruments onboard T/P fail before the launch of these missions (there is plenty of extra propellant to maintain the orbit, but some of the satellite instruments have already exceeded their designed lifetime), it may be necessary to use tide gauges to bridge the measurement gap (Geosat Follow-On and ERS-2 may also be useful for bridging the gap, despite their larger errors relative to T/P). Envisat and Jason-1 will provide measurements of comparable accuracy to T/P, so that task of linking these measurements together will be much easier than encountered by Guman (1997) with Geosat. In this case, the largest error source is likely to be the vertical movement of the tide gauges. For a one year measurement gap, each tide gauge could be expected to move between 1–5 mm if gauges in areas of known tectonic activity or subsidence are disregarded. For this reason, efforts to monitor the vertical motion of several tide gauges in the Pacific using the Global Positioning System have recently been initiated [Merrifield and Bevis, personal communication]. A GPS-derived height time series of several years in length can determine the rate of vertical land motion to better than 1 mm/year.

Another issue that is currently being studied is if the accuracy of the reference frame to which the altimeter measurements are referred can be maintained over several decades (Nerem *et al.*, 1998). The International Terrestrial Reference Frame (ITRF) is defined by the positions and velocities of several hundred points on the earth's surface using a myriad of geodetic techniques such as Very Long Baseline Interferometry (VLBI), Satellite Laser Ranging (SLR), the Global Positioning System (GPS), DORIS, etc. As new positioning techniques are developed (such as GPS and DORIS), older techniques (such as SLR and VLBI) take on a smaller role in defining the reference frame, thus the maintenance of the reference frame definition over time is a matter of concern. As an example, SLR is currently an important technique for determining the centering of the reference frame (the tie to the earth's center-of-mass), thus if SLR were to be phased out over the next decade (as has been proposed by some), what would the impact be on the reference frame definition, and how would this affect a multi-decadal time series of altimeter-derived sea level measurements? These issues are only likely to become significant after several decades, but nevertheless remain important since this is roughly the time frame within which we expect to begin observing climate change signals.

Thus, while the problem will require careful study, it is likely that the T/P sea level time series can be extended in a seamless fashion using Envisat and/or Jason-1 measurements (the latter being preferable because they will fly in identical orbits).

9. CONCLUSIONS

The first 6 years of T/P data have demonstrated that very precise measurements of global mean sea level can be made using satellite altimeters. In addition, the T/P results have shown that global mean sea level contains significant ENSO variability, which can be regarded as an error source for determining the long-term rate of sea level change, but is also of significant scientific interest in itself.

We have shown that a series of T/P-Jason class satellite altimeter missions should be able to measure the long-term rate of sea-level rise with about a decade of measurements, provided the instruments are precisely monitored using GPS-positioned tide gauges. Measuring the acceleration of sea-level rise, a more important quantity for corroborating the climate models, will require 2–3 decades of measurements. We believe these objectives can be achieved given the current performance of T/P and the tide gauge calibration technique. Uncertainties in this assessment include (1) the unknown effects of decadal variability in global mean sea level, (2) the possibility of gaps in the time series caused by satellite failure, and (3) the successful implementation of GPS monitoring at each of the tide gauges. Despite these uncertainties, satellite altimetry is already defining a new paradigm in studies of sea-level change.

ACKNOWLEDGMENTS

Thanks to Eric Leuliette for producing the GMT plots used in this paper. The MOM2 simulations of global mean dynamic height were computed by Ben Giese. Gary Russell provided Figure 15. Also, thanks to Don Chambers for participating in many fruitful discussions regarding these results. The altimeter database used in this study was developed by researchers at the Center for Space Research under the direction of Byron Tapley. This work was supported by a NASA TOPEX Project Science Investigation.

References

Anzenhofer, M., and Gruber, T. (1998). Fully reprocessed ERS-1 altimeter data from 1992 to 1995: Feasibility of the detection of long-term sea level change. *J. Geophys. Res.*, **103(CC4)**, 8089–8112.

Barnett, T. (1983). Recent changes in sea level and their possible causes. *Climate Change*, **5**, 15–38.

Barnett, T. P. (1984). The estimation of "global" sea level change: a problem of uniqueness. *J. Geophys. Res.*, **89(C5)**, 7980–7988.

Birkett, C. M. (1995). The contribution of TOPEX/POSEIDON to the global monitoring of climatically sensitive lakes. *J. Geophys. Res.*, **100(C12)**, 25,179–25,204.

Birkett, C. M. (1998). Contribution of the TOPEX NASA radar altimeter to the global monitoring of large rivers and wetlands. *Water Resour. Res.*, **34(5)**, 1223–1240.

Born, G. H., Tapley, B. D., Ries, J. C., and Stewart, R. H. (1986). Accurate measurement of mean sea level changes by altimetric satellites. *J. Geophys. Res.*, **91(C10)**, 11775–11782.

Bretherton, C. S., Smith, C., and Wallace, J. M. (1992). An intercomparison of methods for finding coupled patterns in climate data. *J. Climate*, **5**, 541–560.

Callahan, P. S., Hancock, D. W., and Hayne, G. S. (1994). New sigma calibration for the TOPEX altimeter. *TOPEX/POSEIDON Res. News*, **3**, 28–32.

Cane, M. A., Kaplan, A., Miller, R. N., Tang, B., Hackert, E. C., and Busalacchi, A. J. (1996). Mapping tropical Pacific sea level: Data assimilation via a reduced state space Kalman filter. *J. Geophys. Res.*, **101(C10)**, 22,599–22,618.

Cazenave, A., Bonnefond, P., Dominh, K., and Schaeffer, P. (1997). Caspian sea level from TOPEX-POSEIDON altimetry: Level now falling. *Geophys. Res. Lett.*, **24(8)**, 881–884.

Cazenave, A., Dominh, K., Gennero, M. C., and Ferret, B. (1998). Global mean sea level changes observed by Topex-POSEIDON and ERS-1. *Phys. Chem. Earth*, **23(9–10)**, 1069–1075.

Chambers, D. P., Tapley, B. D., and Stewart, R. H. (1999). Anomalous warming in the Indian Ocean coincident with El Nino. *J. Geophy. Res.*, **104(C2)**, 3035–3047.

Chelton, D., and Enfield, D. (1986). Ocean signals in tide gauge records. *J. Geophys. Res.*, **91**, 9081–9098.

Chen, J. L., Wilson, C. R., Chambers, D. P., Nerem, R. S., and Tapley, B. D. (1998). Seasonal global water mass budget and mean sea level variations. *Geophys. Res. Lett.*, **25(19)**, 3555–3558.

Christensen, E. J., Haines, B. J., Keihm, S. J., Morris, C. S., Norman, R. A., Purcell, G. H., Williams, B. G., Wilson, B. D., Born, G. H., Parke, M. E., Gill, S. K., Shum, C. K., Tapley, B. D., Kolenkiewicz, R., and Nerem, R. S. (1994). Calibration of TOPEX/POSEIDON at platform harvest. *J. Geophys. Res.*, **99(C12)**, 24,465–24,486.

daSilva, A. M., Young, C. C., and Levitus, S. (1994). *In* "Atlas of surface marine data, 1994," Vol. 1. *NOAA Atlas NESDIS*, 6.

Dorandeu, J., and Le Traon, P. Y. (1999). Effects of global mean atmospheric pressure variations on mean sea level changes from TOPEX/POSEIDON. *J. Atmos. Oceanic Tech.*, **16(9)**, 1279, 1999.

Douglas, B. C. (1991). Global sea level rise. *J. Geophys. Res.*, **96(C4)**, 6981–6992.

Douglas, B. C. (1992). Global sea level acceleration. *J. Geophys. Res.*, **97(C8)**, 12,699–12,706.

Douglas, B. C. (1995). Global sea level change: Determination and interpretation. *Rev. Geophys.*, **33**, 1425–1432.

Fu, L.-L., Christensen, E. J., Yamarone, C. A., Jr., Lefebvre, M., Menard, Y. M. E. N., Dorrer, M., and Escudier, P. (1994). TOPEX/POSEIDON mission overview. *J. Geophys. Res.*, **99(C12)**, 24,369–24,382.

Fu, L.-L., and Pihos, G. (1994). Determining the response of sea level to atmospheric pressure forcing using TOPEX/POSEIDON data. *J. Geophys. Res.*, **99(C12)**, 24,633–24,642.

Gaspar, P., Ogor, F., Le Traon, P.-Y., and Zanife, O.-Z. (1994). Estimating the sea state bias of the TOPEX and POSEIDON altimeters from crossover differences. *J. Geophys. Res.*, **99(C12)**, 24,981–24,994.

Groger, M., and Plag, H.-P. (1993). Estimations of a global sea level trend: Limitations from the structure of the PSMSL global sea level data set. *Global Planet. Change*, **8**, 161–179.

Guman, M. D. (1997). Determination of global mean sea level variations using multi-satellite altimetry, Ph.D. Dissertation, The University of Texas at Austin, Austin.

Hayne, G. S., Hancock, D. W., and Purdy, C. L. (1994). TOPEX altimeter range stability estimates from calibration mode data. *TOPEX/POSEIDON Res. News*, **3**, 18–22.

Hendricks, J. R., Leben, R. R., Born, G. H., and Koblinsky, C. J. (1996). Empirical orthogonal function analysis of global TOPEX/POSEIDON altimeter data and implications for detection of global sea level rise. *J. Geophys. Res.*, **101(C6)**, 14,131–14,146.

Houghton, J. T., Meira Filho, L. G., Callander, B. A., Harris, N., Kattenberg, A., and Maskell, K. (1996). *Climate Change, 1995*, Cambridge University Press, Cambridge, England.

Imel, D. A. (1994). Evaluation of the TOPEX/POSEIDON dual-frequency ionosphere correction. *J. Geophys. Res.*, **99(C12)**, 24,895–24,906.

Jackson, D. L., and Stephens, G. L. (1995). A Study of SSM/I-Derived Columnar Water Vapor over the Global Oceans. *J. Climate*, **8**, 2025–2038.

Kalnay, E., Kanamitus, M., Kistler, R., Collins, W., Deaven, L., Gandin, L., Iredell, M., Saha, S., White, G., Woollen, J., Zhu, Y., Chelliah, M., Ebisuzaki, W., Higgins, W., Janowiak, J., Mo, C., Ropelewski, C., Wang, J., Leetmaa, A., Reynolds, R., Jenne, R., and Joseph, D. (1996). The NCEP/NCAR 40-year reanalysis project. *Bull. Am. Met. Soc.*, **77**, 437–471.

Keihm, S., Zlotnicki, V., and Ruf, C. (2000). TOPEX/microwave radiometer performance evaluation. *IEEE Trans. Geosci. Remote Sens.*, **38(3)**, 1379–1386.

Larnicol, G., Le Traon, P.-Y., Ayoub, N., and De Mey, P. (1995). Mean sea level and surface circulation variability of the Mediterranean Sea from 2 years of TOPEX/POSEIDON altimetry. *J. Geophys. Res.*, **100(C12)**, 25,163–25,178.

Le Traon, P. Y., and Gauzelin, P. (1997). Response of the Mediterranean mean sea level to atmospheric pressure forcing. *J. Geophys. Res.*, **102(C1)**, 973–984.

Leuliette, E. W., and Wahr, J. M. (1999). Coupled pattern analysis of sea surface temperature and TOPEX/POSEIDON sea surface height. *J. Phys. Oceanography*, **29(April)**, 599–611.

Levitus, S., and Boyer, T. (1994). *In* "World Ocean Atlas, 1994, Volume 4." U.S. Dept. of Commerce.

Marshall, J. A., Zelensky, N. P., Klosko, S. M., Chinn, D. S., Luthcke, S. B., Rachlin, K. E., and Williamson, R. G. (1995). The temporal and spatial characteristics of TOPEX/POSEIDON radial orbit error. *J. Geophys. Res.*, **100(C12)**, 25,331–25,352.

Maul, G. A., and Martin, D. M. (1993). Sea level rise at Key West, Florida, 1846–1992: America's longest instrument record? *Geophys. Res. Lett.*, **20(18)**, 1955–1958.

Minster, J.-F., Brossier, C., and Rogel, P. (1995). Variation of the mean sea level from TOPEX/POSEIDON data. *J. Geophys. Res.*, **100(C12)**, 25,153–25,162.

Minster, J. F., Cazenave, A., Serafini, Y. V., Mercier, F., Gennero, M. C., and Rogel, P. (1999). Annual cycle in mean sea level from Topex-POSEIDON and ERS-1: Inference on the global hydrological cycle. *Global Planet. Change*, **20**, 57–66.

Mitchum, G. A tide gauge network for altimeter calibration. *IGS/PSMSL Sea Level Workshop*, Pasadena, California.

Mitchum, G., Kilonsky, B., and Miyamoto, B. Methods for maintaining a stable datum in a sea level monitoring system. *OCEANS 94 OSATES*, Brest, France.

Mitchum, G., and Wyrtki, K. (1988). Overview of Pacific sea level variability. *Mar. Geodesy*, **12**, 235–245.

Mitchum, G. T. (1994). Comparison of TOPEX sea surface heights and tide gauge sea levels. *J. Geophys. Res.*, **99(C12)**, 24,541–24,554.

Mitchum, G. T. (1998). Monitoring the stability of satellite altimeters with tide gauges. *J. Atmos. Oceanic Tech.*, **15(June)**, 721–730.

Moore, P., Carnochan, S., and Walmsley, R. J. (1999). Stability of ERS altimetry during the tandem mission. *Geophys. Res. Lett.*, **26(3)**, 373–376.

Morris, C. S., and Gill, S. K. (1994). Evaluation of the TOPEX/POSEIDON altimeter system over the Great Lakes. *J. Geophys. Res.*, **99(C12)**, 24,527–24,540.

Nerem, R. S. (1995a). Global mean sea level variations from TOPEX/POSEIDON altimeter data. *Science*, **268**, 708–710.

Nerem, R. S. (1995b). Measuring global mean sea level variations using TOPEX/POSEIDON altimeter data. *J. Geophys. Res.*, **100(C12)**, 25,135–25,152.

Nerem, R. S. (1997). Global mean sea level change: Correction. *Science*, **275(February 21)**, 1053.

Nerem, R. S., Chambers, D. P., Leuliette, E. W., Mitchum, G. T., and Giese, B. S. (1999). Variations in global mean sea level associated with the, 1997–1998 ENSO event: Implications for measuring long term sea level changes. *Geophys. Res. Lett.*, **26(19)**, 3005–3008.

Nerem, R. S., Eanes, R. J., Ries, J. C., and Mitchum, G. T. (1998). The use of a precise reference frame for sea level change studies, *In* "Integeated Global Geodetic Observing System," International Association of Geodesy, Munich.

Nerem, R. S., Haines, B. J., Hendricks, J., Minster, J. F., Mitchum, G. T., and White, W. B. (1997a). Improved determination of global mean sea level variations using TOPEX/POSEIDON altimeter data. *Geophys. Res. Lett.*, **24(11)**, 1331–1334.

Nerem, R. S., Rachlin, K. E., and Beckley, B. D. (1997b). Characterization of global mean sea level variations observed by TOPEX/POSEIDON using empirical orthogonal functions. *Surv. Geophys.*, **18**, 293–302.

Nouel, F. N. E. L., Berthias, J. P., Deleuze, M., Guitart, A., Laudet, P., Piuzzi, A., Pradines, D., Valorge, C., Dejoie, C., Susini, M. F., and Taburiau, D. (1994). Precise Centre National d'Etudes Spatiales orbits for TOPEX/POSEIDON: Is reaching 2 cm still a challenge? *J. Geophys. Res.*, **99(C12)**, 24,405–24,420.

Parke, M. E., Stewart, R. H., Farless, D. L., and Cartwright, D. E. (1987). On the choice of orbits for an altimetric satellite to study ocean circulation and tides. *J. Geophys. Res.*, **92**, 11693–11707.

Peterson, R. G., and White, W. B. (1998). Slow oceanic teleconnections linking the Anarctic circumpolar wave with the tropical El Nino-Southern oscillation. *J. Geophy. Res.*, **103(C11)**, 24573–24583.

Preisendorfer, R. W. (1988). *In* "Principal Component Analysis in Meteorology and Oceanography," Elsevier.

Pugh, D. (1987). *In* "Tides, surges, and mean sea level," John Wiley & Sons, New York.

Raofi, B. (1998). "Ocean response to atmospheric pressure loading: The inverted barometer correction for altimetric measurements," Ph.D. Dissertation. The University of Texas at Austin, Austin.

Reynolds, R. W., and Smith, T. S. (1994). Improved global sea surface temperature analysis. *J. Climate*, **7**, 929–948.

Ruf, C. S., Keihm, S. J., Subramanya, B., and Janssen, M. A. (1994). TOPEX/POSEIDON microwave radiometer performance and in-flight calibration. *J. Geophys. Res.*, **99(C12)**, 24,915–24,926.

Russell, G. L., Miller, J. R., Rind, D., Ruedy, R. A., Schmidt, G. A., and Sheth, S. (1999). Climate simulations by the GISS atmosphere-ocean model:, 1950 to 2099. *Geophys. Res. Lett.*, in review.

Shum, C. K., Woodworth, P. L., Andersen, O. B., Egbert, G. D., Francis, O., King, C., Klosko, S. M., Le Provost, C., Li, X., Molines, J.-M., Parke, M. E., Ray, R. D., Schlax, M. G., Stammer, D., Tierney, C. C., Vincent, P., and Wunsch, C. I. (1997). Accuracy assessment of recent tide models. *J. Geophys. Res.*, **102(C11)**, 25173–25194.

Smith, T. M., Reynolds, R. W., Livezey, R. E., and Stokes, D. C. (1996). Reconstruction of historical sea surface temperatures using empirical orthogonal functions. *J. Climate*, **9(June)**, 1403–1420.

Stammer, D., Tokmakian, R., Semtner, A., and Wunsch, C. (1996). How well does a 1/4deg. global circulation model simulate large-scale oceanic observations? *J. Geophys. Res.*, **101(C11)**, 25,779–25,812.

Sturges, W., and Hong, B. (1995). Wind forcing of the Atlantic thermo-cline along 32°N at low frequencies. *J. Phys. Oceanogr.*, **25**, 1706–1715.

Tapley, B. D., Chambers, D. P., Shum, C. K., Eanes, R. J., Ries, J. C., and Stewart, R. H. (1994a). Accuracy assessment of the large-scale dy-namic ocean topography from TOPEX/POSEIDON altimetry. *J. Geo-phys. Res.*, **99(C12)**, 24,605–24,618.

Tapley, B. D., Ries, J. C., Davis, G. W., Eanes, R. J., Schutz, B. E., Shum, C. K., Watkins, M. M., Marshall, J. A., Nerem, R. S., Put-ney, B. H., Klosko, S. M., Luthcke, S. B., Pavlis, D., Williamson, R. G., and Zelensky, N. P. (1994b). Precision orbit determination for TOPEX/POSEIDON. *J. Geophys. Res.*, **99(C12)**, 24,383–24,404.

Tapley, B. D., Shum, C. K., Ries, J. C., Suter, R., and Schutz, B. E. (1992). *In* "Global mean sea level using the geosat altimeter." Sea Level Changes: Determination and Effects, American Geophysical Union, IUGG.

Tapley, B. D., Watkins, M. M., Ries, J. C., Davis, G. W., Eanes, R. J., Poole, S. R., Rim, H. J., Schutz, B. E., Shum, C. K., Nerem, R. S., Lerch, F. J., Marshall, J. A., Klosko, S. M., Pavlis, N. K., and Williamson, R. G. (1996). The joint gravity model 3. *J. Geophys. Res.*, **101(B12)**, 28,029–28,050.

Trenberth, K. E., and Hoar, T. J. (1996). The 1990–1995 El Nino-Southern oscillation event: Longest on record. *Geophys. Res. Lett.*, **23(1)**, 57–60.

Tushingham, A. M., and Peltier, W. R. (1991). Ice-3G: A new global model of late Pleistocene deglaciation based upon geophysical predictions of post-glacial relative sea level change. *J. Geophys. Res.*, **96(B3)**, 4497–4523.

Wagner, C. A., and Cheney, R. E. (1992). Global sea level change from satellite altimetry. *J. Geophys. Res.*, **97(C10)**, 15,607–15,615.

Wallace, J. M., Smith, C., and Bretherton, C. S. (1992). Singular value de-composition of wintertime sea surface temperature and 500-mb height anomalies. *J. Climate*, **5**, 561–576.

Wong, T., Barkstrom, B. R., Gibson, G. G., and Weckmann, S. (1998). Ob-servations of the radiative impacts of ENSO events using data from ERBE and CERES. *Eos Trans.*, **79(45)**, F147.

Woodworth, P. L. (1990). A search for acceleration in records of European mean sea level. *Int. J. Climatol.*, **10**, 129–143.

Wyrtki, K., and Mitchum, G. (1990). Interannual differences of Geosat al-timeter heights and sea level: The importance of a datum. *J. Geophys. Res.*, **95(C3)**, 2969–2975.

9

Ice Sheet Dynamics and Mass Balance

H. JAY ZWALLY*
ANITA C. BRENNER[†]

*NASA/Goddard Space Flight Center
Oceans and Ice Branch, Code 971
Greenbelt, MD 20771
[†]Raytheon Technical Services Company
at NASA/Goddard Space Flight Center
Oceans and Ice Branch, Code 971
Greenbelt, MD 20771

1. INTRODUCTION

The Greenland and Antarctic Ice Sheets contain 77% of the earth's freshwater, 99% of all the glacier ice, and cover 10% of the earth's land area. The average thickness of the Greenland Ice Sheet is 1800 m, and its ice volume is approximately 3×10^6 km^3. The Antarctic Ice Sheet is about 10 times larger with an average thickness of 2400 m and ice volume of 29×10^6 km^3. Although the weight of the ice depresses the earth's crust by hundreds of meters, the bases of the ice sheets are, on average, close to sea level. However, the 12% portion of the Antarctic Ice Sheet that lies mostly in the Western Hemisphere, called the West Antarctic Ice Sheet, is grounded as much as 2500 m below sea level. If the ice sheets completely melted, sea level would rise about 80 m.

Most of the surface of the ice sheets is composed of compacted snow, called firn. New snow is compacted by wind drifting and other processes to a specific density of about 0.3 near the surface. The firn is further compacted by the weight of overlying layers to solid ice with a specific density of 0.92 at about 50–100 m depth. Where the total ice sheet thickness exceeds a few hundred meters with surface slopes of 1/100, for example, the ice deforms under the force of gravity and flows in the downslope direction. At

the margins, the ice melts or flows into the ocean in icebergs.

Ice sheet mass balance is the difference between the mass input and the mass output. The total balance includes both the surface mass balance processes and the ice flow components. Mass is added to the surface from snowfall, condensation, and occasional rainfall. Mass is removed by evaporation (sublimation), surface and bottom melting, water runoff, and iceberg discharge. Snow drifting transfers mass across the surface and accounts for some removal from the ice sheet.

The boundary between the higher surface elevations, which have positive ice accumulation, and the lower elevations, which have a net ablation, is called the equilibrium line. In southern Greenland, the equilibrium line is at approximately 1800 m elevation, decreasing to 300 m in the north. The average rate of ice accumulation is 273 mm year^{-1} of water equivalent in the accumulation zone, which is about 86% of the total ice sheet area of 1.9×10^6 km^2. About 1/3 of the surface of the accumulation zone remains dry and frozen all year, whereas intermittent melting occurs on the other 2/3. Above the equilibrium line, most water from surface melting refreezes in the firn. Below the equilibrium line, water runoff from melting accounts for about half of the mass loss from the ice sheet. The other half is from direct ice discharge into the ocean.

a

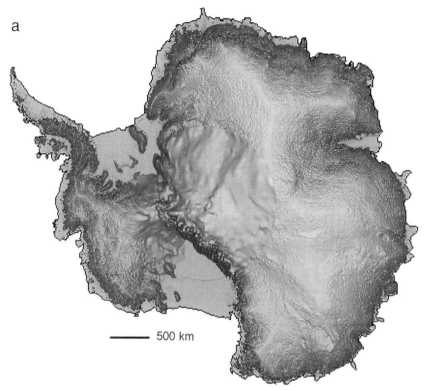

─── 500 km

FIGURE 1a   Antarctic shaded topography from ERS-1 and Geosat Altimetric DEM (airborne data set used south of 81.5°).

In Antarctica, the equilibrium line is close to sea level, so less than 1% of the total ice sheet area of 13.5×10^6 km^2 has a net ablation loss. The average accumulation rate is 162 mm year^{-1} of water equivalent. At some locations around Antarctica, ice flowing from the grounded portion of the ice sheet forms large floating ice shelves, which account for about 10% of the total continental ice area. The three largest ice shelves are the Ross and the Ronne-Filchner Ice Shelves in west Antarctica, and the Amery Ice Shelf in east Antarctica.

The shape of the ice sheet is defined by the thickness and surface slope, which together determine the driving stresses in the ice and the velocity of ice-flow. The shape is in turn modified by the ice flow, which usually tends to adjust to reach a balance between the mass output and input. However, the time scales for the principal processes affecting the surface mass balance are much shorter than the time-scales for changes in the ice flow. The mass and energy exchange processes at the surface change on seasonal to interannual time scales, as well as on longer time scales due to long-term changes in the ice elevations and climate. In contrast, the driving stresses in the ice and the ice velocities change more slowly as the ice sheet shape changes in response to a mass imbalance. Furthermore, the ice flow relationship between the ice strain rates and the driving stresses is a function of ice

temperature, ice fabric, and other slowly changing factors. In particular, the internal ice temperature changes slowly in response to changes in surface heat flux, the downward advection of colder surface ice, and changes in the basal heat flux to the ice.

At present, the mass balance of the ice sheets is not known to within about ±25% (Warrick et al., 1996). Each year, the equivalent of about 8 mm of water from the entire surface of the earth's oceans accumulates as snow on the Greenland and Antarctic Ice Sheets. Therefore, the overall uncertainty in the ice sheet mass balance corresponds to an uncertainty in the ice sheet contribution to sea level rise or fall of about ±2 mm year^{-1}. About half of the current sea-level rise of 2 mm year^{-1} has been attributed to the recent melting of small glaciers and to thermal expansion of the ocean with climate warming. Whether the ice sheets are currently contributing to sea-level rise, or perhaps removing water, is still unknown and could increase the unexplained part of the rise due to other sources.

Robin (1966) first proposed mapping of ice sheet topography by satellite radar altimetry. The first ocean radar altimetry satellite, GEOS-3 in 1975–1978, mapped the southern tip of the Greenland Ice Sheet below 65°N (Brooks et al., 1978). GEOS-3 measurements were followed by the Seasat mapping of Greenland to 72.2°N and part of the East Antarc-

b

250 km

FIGURE 1b Greenland shaded topography from ERS-1 and
Geosat Altimetric DEM.

tic Ice Sheet to 72.2°S, (Zwally *et al.*, 1983), and later by
Geosat mapping to the same latitudes (Zwally *et al.*, 1987a).
ERS-1 and ERS-2 significantly extended the measurements
to 81.5°N and 81.5°S (Bamber, 1994b; Ekholm, 1996;
Bamber and Bindschadler, 1997; Phillips, 1998). Therefore,
satellite radar altimeters have given the most accurate near-
global coverage of ice sheet surface elevations to date, as
shown in the topographic maps of Antarctica and Greenland
(Figures 1a and b). The detail of the Antarctic Ice Sheet
North of 81.5°S is obtained from satellite radar altimetry
measurements.

The performance of the radar altimeters over the sloping
and undulating ice sheet surfaces is significantly different
than over the relatively flat ocean surfaces. The characteristic design of a small pulse-limited footprint, located within
a broader beam-limited footprint, becomes limited in range

accuracy and range-tracking capability as the slope increases
and fluctuates because of surface undulations. In addition,
the location of the pulse-limited footprint on the surface is
not well-defined and more than one footprint from nearly
equidistant surfaces may be detected. Absolute accuracies of
the elevations depend primarily on the magnitudes of the surface slope and undulations. Comparisons with aircraft laser
altimetry over Greenland (Bamber *et al.*, 1998) showed the
satellite radar accuracy to range from 84 ± 79 cm for slopes
less than 0.1° to 10.3 m ± 8.4 m for slopes of 0.7°. Over the
Amery Ice Shelf in east Antarctica, Phillips *et al.*, (1998)
found a mean difference of 0.0 m ± 0.1 m and an rms of
1.7 m between ERS radar altimetry and surface GPS observations. Better accuracy is achieved over the relatively flat
ice shelves, because the small surface slopes do not cause a
significant slope induced error, as described in Section 2.

A significant use for satellite altimetry is to measure surface-elevation changes to use in studying ice sheet mass balance, which was proposed by Zwally (1975) before the limitations of radar altimeters over non-ocean surfaces were fully realized. Fortunately, the principal effect of surface slope on the absolute accuracy tends to be the same for repeat measurements from the same satellite location on successive orbits. Therefore, the relevant accuracy for elevation change measurement is the relative accuracy (or precision), which is about 5–10 times better than the absolute accuracy. Even so, many elevation differences at orbital crossovers must be averaged, and systematic errors in orbits, timing, and other factors can also affect results.

Increases in the elevation of the Greenland Ice Sheet south of 72°N first reported by Zwally et al., (1989) and Zwally (1989) were recently shown to be significantly smaller (Davis et al., 1998a, 1998b; Zwally et al., 1998) because of correction of systematic errors in orbits and timing. Other studies of ice shelf and ice sheet elevation change include Lingle et al., (1994) and Lingle and Covey (1998). Recently, Wingham et al., (1998) reported that the elevation of the Antarctic Ice Sheet interior fell by 0.9 $\pm\, 0.5$ cm year^{-1} between 1992–1996. In 2001, the narrow beam laser altimeter on NASA's ICESat (Ice, Cloud and land Elevation Satellite) will make elevation measurements with an absolute accuracy of 15–20 cm for laser footprints that are 65 m in diameter and located to a horizontal accuracy of about 5 m. ICESat coverage will extend to 86° latitude and cover the steeper margins of the ice sheets where the radar altimeter tracking fails.

In addition to surface topography, detailed maps of surface slope derived from the topographic maps provide details of the ice flow directions [Phillips (1998)], locations of ice divides and drainage basins, and the direction and magnitude of the driving stress acting on the ice (Section 4). The location of ice shelf fronts and ice margins has been mapped (Zwally et al., 1987b and Section 5), using a technique described by Thomas et al. (1983). The position of the grounding line between the floating ice shelves and ice sheets has also been mapped (Herzfeld et al., 1994 and Phillips, 1998). Information on the characteristics and changes of the surface firn, including surface roughness and volume scattering (Partington et al., 1989; Davis and Zwally, 1993; Legresy and Remy, 1998), and information on surface melt streams (Phillips, 1998) is also derived from radar altimetry.

2. RADAR ALTIMETER MEASUREMENT OF ICE SHEET SURFACE ELEVATIONS

The altimeter measurement of ice sheet elevations is significantly different from measurement of ocean surface because of the sloping nature of the ice sheets, variations in the surface reflection, penetration of the radar signal in the snow surface, and the generally irregular surface geometry. Ice sheet surfaces have small-scale wind-driven snow drifts called *sastrugi*, which are analogous to ocean waves. Larger-scale undulations, with amplitudes of tens of meters and wavelengths of kilometers to tens of kilometers, are caused mainly by ice flow over the irregular bedrock topography and are modified by preferential deposition of drifting snow.

Along-track variations in the ice surface elevation, and the consequent variations in the altimeter range to the surface between pulses, have several effects. First, the geodetic location of the altimeter footprint on the surface is ambiguous, and usually not at the satellite nadir location as it is over the oceans. Second, range variations during the pulse-averaging period may broaden the waveform depending on the ability of the tracking circuit to estimate the pulse-to-pulse range variations and properly stack (align) the pulses in, for example, 10 or 20/sec waveform averages. Errors in stacking cause a range error, which can not be corrected. The 10 or 20/sec averages, which correspond to about 660 or 330 m along the ground track depending on the altimeter design, must be used, and not the typical 1- or 10-sec averages used over the oceans to reduce the random errors. Third, surface penetration of the radar signal and internal volume back-scattering causes irregularities in the altimeter waveform shape, which makes it harder to define the midpoint of the leading waveform edge. Fourth, the elevation variations cause migration of the midpoint of the leading waveform edge with respect to the central altimeter tracking gate. The retracking range correction, as described below, varies from a few cm to tens of meters. If the range variations exceed the capability of the altimeter to keep the waveform within the window of range gates, the altimeter loses track and no measurement is obtained.

Figure 2, adapted from Ridley and Partington (1988), depicts the development of the pulse over a diffuse, horizontal, and planar surface. The transmitted pulse first intercepts the closest surface within the beam-limited footprint, forming the center of the pulse-limited footprint. The beam-limited footprint, which is defined by the 3-db attenuation points of the antenna-gain function and the satellite altitude, varies from 16–38 km diameter depending on the satellite. The pulse-limited footprint is defined as the maximum circular area from which radar back-scattering is simultaneously received (Brooks et al., 1978), and corresponds in time to the peak of the return waveform. After the back of the pulse shell intersects the surface, the annulus expands and the return becomes increasingly attenuated by the antenna beam pattern, as depicted by the drop-off of the trailing edge. On the sloping surfaces of the ice sheets, the pulse-limited footprint can be located anywhere within the beam-limited footprint, and is strongly attenuated when it is outside the center of the beam. The waveform shape is also affected by surface roughness, which increases the size of the pulse-limited footprint

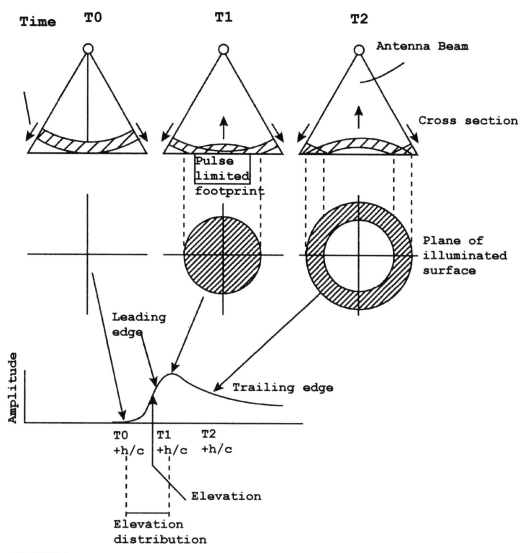

FIGURE 2 The development of the radar pulse over a diffuse, horizontal and planar surface from its initial interaction (time t_0) through the intersection of the back of the pulse shell with the surface (t_1), to a later stage where the pulse begins to be attenualted by the antenna beam (t_2). The return is from the surface only.

and broadens the leading ramp. The mid-point of the leading ramp corresponds to the range to the mean surface within the pulse-limited footprint.

2.1. The Effect of Surface Slope and Undulations

The uncertainty in the geodetic location of the measurement on the ice sheet surface causes the largest elevation error. As depicted in Figure 3, over a smooth surface of constant slope, α, the vertical displacement (ΔH) with respect to nadir is

$$\Delta H = H(1 - \cos\alpha) \approx (H\alpha^2)/2 \qquad (1)$$

where H is the spacecraft altitude and the approximation is for small slopes typical of the ice sheets. The horizontal displacement is

$$\delta = H\cos\alpha\sin\alpha. \qquad (2)$$

For a 0.5° slope and typical satellite altitudes, ΔH is 30 m and increases to more than 100 m near the ice sheet margins.

Figure 4 from Gundestrup et al., (1986) shows how the location of the altimeter reflection varies along a track on the Greenland Ice Sheet, as the altimeter tracks the closest points up-slope from the satellite ground track and does not sense the valleys or troughs. However, without prior detailed knowledge of the surface topography, one cannot determine the surface location from which the reflection originates. One difficulty in correcting for the slope-induced error

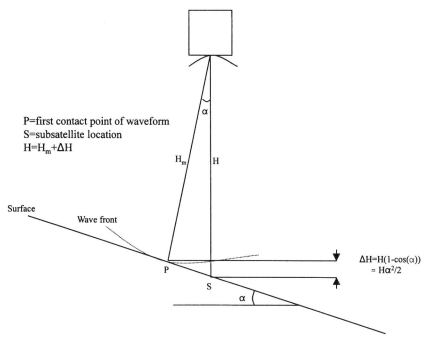

P=first contact point of waveform
S=subsatellite location
H=H$_m$+ΔH

Surface

Wave front

ΔH=H(1-cos(α))
≈ Hα^2/2

FIGURE 3   Schematic description of slope-induced error, ΔH, over planar surface of slope α.

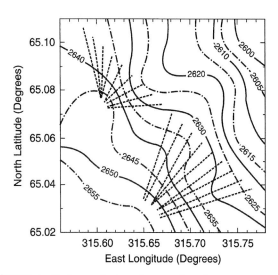

FIGURE 4   Surface projection of vectors from Seasat to nearest surface location for orbit 1236 in the vicinity of the Dye-3 station in southern Greenland. (From Gundestrup *et al.*, 1986. With permission.)

is the variation of α across the beam-limited footprint, which makes the selection of an appropriate α difficult. Another difficulty is that the slope varies both along and across the satellite ground track. Any one profile gives an estimate of α along the track, but only where the tracks are very dense is accurate cross track knowledge available.

Several different methodologies have been developed for correcting for the slope-induced error. These include: the relocation method of migrating the measurement horizontally

based on known topography (Brenner *et al.*, 1983; Bamber, 1994a; Stenoien and Bentley, 1997), the direct method of using the α at the subsatellite location (Brenner *et al.*, 1983), or an intermediate method (Remy *et al.*, 1989). In all methods, the topography obtained before slope corrections are applied is used in the correction procedure and its accuracy is a limiting factor.

2.2. The Effect of Penetration and Sub-surface Volume Scattering

The waveform models described in Figure 2 assume that the return is formed only by surface scattering of the observed radar reflection. However, Davis and Poznyak (1993), for example, showed the penetration depth in the cold, dry regions for microwave altimetry was at least 4.7 m. Therefore, as shown by Ridley and Partington (1988), the ice sheet returns consist of a combination of surface return and subsurface volume scattering. Penetration of part of the radar signal through the surface is followed by volume scattering, as shown in Figure 5. At the end of the rise from the surface-scattering portion, there is a small point of inflection. Again, the mean surface elevation corresponds to the mid-point on the surface return ramp. The effect of penetration and volume scattering on the accuracy of the surface elevation calculated from the altimetry is heavily dependent on the retracking method in Section 2.3.

The strength of the volume scattering depends on the size of the snow grains in the firn, which are largest in regions of surface melting and smallest in cold regions with larger ac-

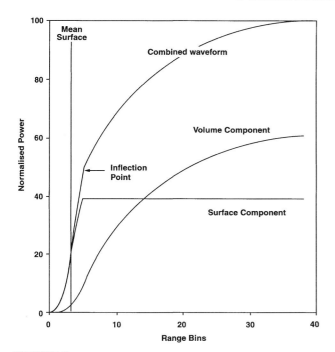

FIGURE 5 Main components of an ideal modeled ice sheet return from Ridley and Partington 1988. The surface component is similar to an ocean return, and is added to a volume component which produces a return with an "inflection point," where surface scattering becomes constant. The mean surface is shown as the half power position on the leading edge of the surface component. The mean surface can be calculated, if the inflection point is located.

cumulation rates (Zwally, 1977). Where grain sizes are large and volume scattering is strong, the penetration is small and the volume-scattering component follows closely behind the surface return and may be inseparable. Paradoxically, where volume scattering is weak and the penetration extends to greater depths, the effect on the waveform shape is largest. Legresy and Remy (1998) calculated a map of penetration depth over Antarctica based on analysis of temporal variations of the shapes of ERS-1 waveforms.

2.3. Waveform Shape Fitting and Retracking

A software simulator that generates return waveforms for the Seasat altimeter over a specified surface illustrates how the ice surface slope and undulations affect the altimeter measurements. The simulator includes the effects of lags in the α-β tracking circuit that are specific to the Seasat altimeter, but other altimeters have analogous lags. Figure 6 from Bindschadler *et al.* (1989) shows the character of the return pulses and the tracking error over a three-dimensional surface modeled from a surface undulation function for a typical ice-sheet surface (Zwally *et al.*, 1981). The satellite groundtrack is perpendicular to the large-scale surface undulations and the waveforms change continually as different portions of the surface come into view. The altimeter mea-

sures the range to the nearest surface, ranging forward toward the elevation peaks and continuing to range backward after it has passed over the peaks. When a forward and backward surface are at nearly equal ranges, a double-ramped waveform is observed, as for example for waveforms 33–41. The simulated waveforms look very similar to those from the altimeter. Retracking, which corrects for the drift of the waveform away from the center of the waveform window as described below, significantly improves the elevation accuracy. For double-ramped waveforms, two retracked ranges are obtained. The retracked elevations match the surface at the topographic highs, but the residual differences elsewhere are because of the slope-induced error. In general, the closest surface to the altimeter may occur to the sides of the groundtrack.

When the range to the surface changes faster than the altimeter tracking circuitry can accommodate, the waveform drifts outside the range window, which is described as loss-of-track. The altimeter trackers for Seasat and Geosat were designed to maintain track over range changes up to 50 m sec^{-1} or relative slopes of $0.43°$ between the satellite orbit and the surface, which is sufficient over the oceans. Although the slopes of the ice sheets relative to the satellite orbit over most areas are sufficiently small, the instantaneous slopes over undulations and near the ice margins are large enough to cause loss-of-track. For example, a sharp upturn in slope followed by a sharp downturn can cause the ramp to drift out of the center of the window to the right until there is no signal in the range window.

Several methods have been used for retracking ice-sheet altimeter data (Martin *et al.*, 1983; Wingham, 1986; Zwally *et al.*, 1994; Bamber, 1994a; and Davis, 1996) with various advantages and disadvantages. Retracking algorithms can be mostly divided into two categories: threshold or functional fit. Threshold retrackers, which define the retracking point where the waveform exceeds a certain percentage of the maximum power relative to the noise level, are the most simple. However, the relationship of a given threshold track point to the mean surface elevation within the altimeter footprint varies with the characteristics of the surface and subsurface scattering. Although, the threshold can be chosen to reduce the effect of subsurface scattering, as explained by Davis (1997), it is not practical to automatically select a threshold value as a function of the surface characteristics. If a waveform is dominated by surface scattering, the 50% threshold best represents the mean surface elevation. As the volume-scattered signal increases relative to the surface return, a 10 or 20% threshold better represents the mean surface elevation.

Functional-fit retracking algorithms range from pure surface-scattering models (Martin *et al.*, 1983) to models that encompass both surface and volume scattering (Davis, 1993; Newkirk and Brown, 1996; Yi and Bentley, 1994; and Yi and Bentley, 1996). The volume scattering models work

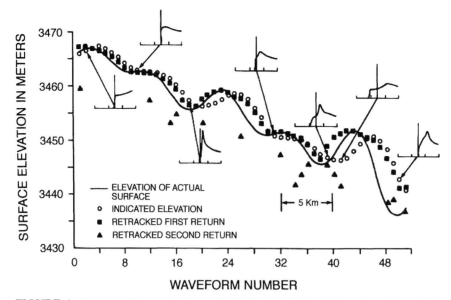

FIGURE 6 Simulated altimeter elevation profile (indicated surface) and selected waveforms. (Figure 3 from Bindschadler *et al.*, 1989.) Actual surface is realistic ice surface model (Zwally *et al.*, 1981). Elevation differences are sum of slope-induced error and tracking error.

well for surfaces that produce a single-ramp return, however they are computationally difficult and do not account for double-ramped waveforms that are quite common. The model used by Martin *et al.* (1983) has been modified and referred to in literature as the NASA GSFC V4 retracker. The model function represents the signal as reflections from a gaussian surface height distribution, as is used for ocean returns (Parsons, 1979). Two functions are used, one with a single ramp and one with a double ramp.

The trailing edge of the return from the top of the last ramp to the end of the return is modeled as an exponential to account for the antenna-beam attenuation of the waveform, which is enhanced in the coarser ice mode of the ERS altimeters, and for the rapid decline of off-nadir reflections over the more specular sea ice. In ice mode, the range window is increased by a factor of almost 4 from the ocean mode, so the ice mode waveforms have sharper ramps and exponential tails. The GSFC V4 retracker fits both the Seasat and Geosat waveforms, which are similar to ERS-1 ocean waveforms, and the ERS-1 ice mode waveforms.

The form of the 5-parameter single-ramp function used for V4 is

$$y = \beta_1 + \beta_2 e^{-\beta_5 Q_1} P\left\{\frac{(t - \beta_3)}{\beta_4}\right\}. \qquad (3)$$

The 9-parameter V4 double-ramp function is

$$y = \beta_1 + \beta_2(1 + \beta_9 Q_1) P\left\{\frac{(t - \beta_3)}{\beta_4}\right\}$$
$$+ \beta_5 e^{-\beta_8 Q_2} P\left\{\frac{(t - \beta_6)}{\beta_7}\right\} \qquad (4)$$

where $Q1 = 0$ for $t < \beta_3 - 0.5\beta_4$
       $= t - \beta_3 - 0.5\beta_4$ for $t \geq \beta_3 - 0.5\beta_4$
   $Q2 = 0$ for $t < \beta_6 - 0.5\beta_7$
       $= t - \beta_6 - 0.5\beta_7$ for $t \geq \beta_6 - 0.5\beta_7$
   $t =$ gate number

$$P(W) = \int_{-\infty}^{w} \frac{1}{\sqrt{2\pi}} \exp\left(\frac{-q^2}{2}\right).$$

Bayesian least squares is used to fit these functions to the waveforms. Fine-tuning of the a priori values and the weighting of the individual gates is carried out to obtain optimal results that best portray the ice sheet topography, and reduce "retracking jitter" caused by the retracker switching between the single and double ramp functions.

Figure 7 (see color insert) shows typical results from ERS-1 in ice mode over Antarctica, highlighting the differences between the threshold and functional-fit retracking range corrections. Figure 7a shows the elevation profile without retracking in black, with a 20% threshold retracked elevation in blue, and the GSFC V4 retracked elevation corresponding to the first ramp in red, and the occasional second ramp in green. Figure 7b shows representative waveforms with single and double ramps. The smooth line in Figure 7b is the functional fit to the return. The retracking corrections are calculated as the difference between the midpoint of the ramps and the tracking gate. Figure 7c shows the GSFC V4 retracking points on a waveform intensity plot, for which the amplitude

TABLE 1. Differences in the Functional Fit and Threshold Retracking Corrections over Greenland

| Greenland elevation band (m) | Average slope (deg) | Retracking correction differences between the GSFC functional fit Version 4 retracking and the 3 threshold retrackers (m) | | | | | |
| | | GSFC V4–10% Threshold | | GSFC V4–20% Threshold | | GSFC V4–50% Threshold | |
		Mean (m)	SD (m)	Mean (m)	SD (m)	Mean (m)	SD (m)
<100 sea ice		1.00	1.60	0.32	1.30	−0.90	1.92
700–1200	0.64	2.87	2.84	0.88	2.64	−2.51	3.08
1200–1700	0.54	2.56	2.63	0.53	2.32	−3.52	2.79
1700–2200	0.36	1.82	2.22	0.33	1.85	−3.01	2.63
2200–2700	0.25	1.10	1.47	0.19	1.22	−2.41	2.27
>2700	0.15	0.62	0.69	0.19	0.60	−1.62	1.57

of each successive waveform is displayed from top to bottom in a grey scale. The second reflecting surface is clearly indicated by the pulse-to-pulse continuity of the grey scale intensity, even when the algorithm does not always detect a second ramp in the individual waveforms. The waveform-to-waveform continuity of the second ramp and its parabolic shape also demonstrates that the second peak is caused by double reflections from the undulating surface.

During the initial portion of the profile, the bias between the two retrackers is fairly consistent at about 1 m. The waveforms show distinct single ramps, as exemplified in Figure 7b waveform 6, which illustrates the variability of the elevation derived from the threshold tracker as may be caused by variations in surface roughness within the footprint. Near waveform 25, where the surface becomes concave, the waveform becomes sharper with a large post-peak attenuation. At waveform 27, where the slope of the ramp is largest, the relative bias increases to about 5 m. Double waveforms starting at waveform 45 indicate ranging to two separate surfaces, as shown by waveform 50. Near waveform 56, there is downward stepping in the profile as the signal from the closer surface weakens and the tracking switches to the surface previously represented by the second ramp. For the threshold retracker, the transition to the second surface occurs abruptly at waveform 56.

Studies comparing elevations calculated using the threshold and functional-fit algorithms to aircraft laser data over Greenland, Ferraro and Swift (1995) showed that the GSFC retracker results in a lower bias than the threshold or volume and surface scattering models, but produced more noise in the percolation zone. Ridley and Partington (1988) found the GSFC retracker to be more insensitive to the effects of penetration and subsurface volume scattering and therefore more accurate than the threshold algorithms when volume scatter-

ing occurs. The insensitivity occurs because the double-ramp function fits many of the returns with large volume scattering by starting a second ramp near the point of inflection (cf. Figure 5) between the surface return and the volume-scattering component, so the mid-point of the first ramp correctly represents the range to the surface.

The differences between the different threshold retracking values and the GSFC V4 are a function of the threshold value used and the type of surface, as indicated by different elevation bands in Table 1 for the Geosat GM Greenland data. The 20% threshold gives the closest approximation to the functional fit, but there are significant biases ranging from 19 cm over the higher plateau regions, to 88 cm in the ablation zone. The 10% threshold differences are much larger, varying from 62 cm over the plateau regions to 2.87 m over the ablation zone. Both the 10% and the 20% thresholds give elevations higher than the functional fit, while the 50% threshold gives significantly lower elevations.

Davis (1997) showed that the 10% threshold retracker gives 30–35% lower standard deviations for crossover differences than using the GSFC V4 retracker. At crossovers using the functional fit, some of the measurements in the crossover comparison are retracked using a double-ramp and some using a single-ramp function, based on slight differences in the waveforms. The threshold tracker could be preferred for elevation change studies, because it gives slightly more repeatable elevations at crossovers, even though it has a variable relation to the mean surface elevation.

2.4. Summary of Radar Altimetry Missions Used for Ice Sheet Elevation Studies

A summary of satellite missions over the polar ice sheets, with time durations, ground coverage, and precisions is given in Table 2. GEOS-C (1975–1978) covered the south-

TABLE 2. Altimetry Missions Over the Polar Ice Sheets

Satellite/mission	Dates of operation	Ground spatial coverage (deg latitude)	Ground track spacing at equator (km)	Measured[b] (precision over ice sheets) (cm)	Beam-limited footprint diameter (km)
GEOS-C	1975–78	±65	Variable	>200	38.1
Seasat	July–Oct 1978	±72	163	>40	22.3
Geosat/GM	Apr '85–Sep '86	±72	13.6	>40	29.3
Geosat/ERM	Nov '86–Dec '89	±72	163	>40	29.3
ERS-1[a]/GM	May '94–Apr '95	±81.5	8.5	>73 (ice mode)	16.2
ERS-1[a]/ERM	Aug '91–May '94	±81.5	79	>40 (ocean mode)	16.2
	Apr '95–Jul '96			>73 (ice mode)	
TOPEX/POSEIDON	Aug '92–present	±66	275	N/A	25.6
ERS-2[a]/ERM	Apr '95–present	±81.5	79	>40 (ocean mode)	16.2
				>73 (ice mode)	
ENVISAT	2000 launch	±81.5	79	Not yet known	16.2
ICESat (laser)	2001 launch	±86	14.5	15	0.07

[a]ERS-1,2 altimeters operated in two modes; ocean mode with a resolution similar to Seasat and Geosat and in a coarser ice mode that allowed it to maintain track over larger surface slopes at a reduced range resolution.

[b]The precisions are calculated by taking the standard deviation of all measurement crossover residuals within a 30-day time period, which biases the results to smaller values because more measurements are made over the flatter parts of the ice sheets.

ern tip of Greenland to 65°N, but the range precision was only about 2 m. The first really usable elevation data set over the ice sheets came from the 90 days of Seasat data during the summer of 1978, followed by the similar Geosat altimeter (1985–1989). The ERS-1 (1991–1996) and ERS-2 (1996 to the present) radar altimeters with their dual-tracking mode extended coverage to 81.5° and maintained track better. TOPEX (1992 to the present) only covered the southern portion of the Greenland Ice Sheet to 66°N, but had a dual-frequency altimeter, which enabled study of frequency-dependent penetration and volume-scattering effects, (Remy et al., 1996). ENVISAT is scheduled to launch in 2001 and is an ESA follow-on to the ERS missions with a similar altimeter. It will operate in the two modes of the ERS altimeters, but will also have a coarser mode to enable it to track better over the steeper regions. The ICESat laser altimeter, beginning in July 2001, will extend the coverage to 86° with improved vertical accuracy.

3. GREENLAND AND ANTARCTICA ICE SHEET TOPOGRAPHY

Several methods have been used to create DEMs (Digital Elevation Models) of ice sheet topography from satellite altimetry. All methods involve mapping the data onto evenly spaced grids and correcting for the slope error either before or after the mapping. Full DEMs of Greenland and Antarctica have been produced using triangularization (Bamber, 1994b; Ekholm, 1996) or gridding procedures sim-

ilar to that described in this section. A three-step inversion technique, which accommodates a priori information on the expected topography and propagates the data errors, has also been demonstrated in small regions (Remy et al., 1989). Remy first estimates a large-scale reference surface, then maps the residuals related to the undulations, and finally iteratively corrects for the slope error. Geostatistical variogram and kriging methods have also been used for characterizing the noise levels in the altimeter data and gridding the elevations (Herzfeld et al., 1993).

DEMs of Antarctica and Greenland, created using methods described below, are displayed in shaded relief maps in Figures 1a and b. The center of Antarctica south of 81.5°, the farthest extent of the altimetry, has been filled in using data from airborne radar altimeters (Drewry, 1983). The DEM grids with interactive extraction software is available on CD-ROM from the National Snow and Ice Data Center, NSIDC. Also on the CD-ROM are grids of the ice sheet slopes calculated from the DEM, the relative precision of each grid value, and the results of the comparison with aircraft laser data over Greenland (see next section).

3.1. Data Preparation

The data consists of the 18-month Geosat GM data set, the 11-month ERS-1 GM data set, and two 35-day repeat cycles of ERS-1 altimetry surrounding the ERS-1 GM data. All of the ERS-1 data is in ice mode. The Geosat data fills in details between the ERS-1 ground tracks below 72°. All data is processed using a consistent set of corrections, models, and orbits which include: the GSFC V4 range retracking correction, the IRI 95 ionosphere model for the ionosphere delay, the ECMWF meteorological data for the dry and wet troposphere delay, JGM3 orbits for Geosat and DGM04 orbits for ERS-1, and the University of Texas Center for Space Research tide models.

3.2. Gridding Procedure

The data is gridded at 5-km spacing on a polar stereographic projection taking into account the different precision between the missions and the variable data density with latitude and ice surface slope (because of loss-of-track). A variable cap size (area surrounding the grid point from which to use data) optimizes the accuracy of the grid where the density of data is greatest and still obtains valid grid values at the locations where density is low. The gridding procedure is explained in detail in Zwally *et al.* (1990). In summary, for each grid location, all elevations within a given cap size are fitted to a bi-quadratic surface using singular value decomposition, SVD. In the fitting process, a weight (Wt) is assigned to each elevation with

$$Wt = 1/\left(D^* \sigma_{dat}^2\right) \qquad (5)$$

where D is the distance from the data location to the grid node and the values for σ_{dat} are set to 0.73 and 0.40 m for ERS-1 and Geosat, respectively, based on the standard deviations of the crossover residuals for each satellite. When output from the SVD signifies insufficient information for a valid fit to a bi-quadratic function, the cap size is increased in steps up to the maximum value. Three cap size radii are used: 7, 10, and 30 km. At the smallest cap size where the bi-quadratic fit is successful, the standard deviation, σ_{grid}, of the differences in the data elevations to the functional fit is calculated. Data is then edited if it disagrees with the function by more than $3\sigma_{grid}$, and a new value for the grid elevation and σ_{grid} is calculated. If σ_{grid} is greater than 100 m, the cap size is increased and the procedure is repeated. If a satisfactory solution is not found at any of the cap sizes, a bilinear functional fit is then tried.

A preliminary 10-km resolution grid is first created using the above methodology from the non-slope-corrected data. This grid is then used to evaluate the surface slope at every data point location. The data is then corrected for the slope-induced error using Eq. (1), and a 5-km resolution grid is created from these corrected data. The 5-km grid is referenced to a geoid derived from the EGM 96 gravity model (Lemoine *et al.*, 1998).

3.3. Accuracy Analysis of Digital Elevation Models

The accuracy of the Greenland DEMs is evaluated by comparison with aircraft laser data, which is reported to have

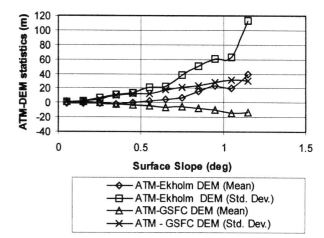

Comparison of Laser ATM data to DEMs created from Satellite Radar Altimetry

FIGURE 8 Comparison of aircraft laser measurements with Altimetric DEMs over Greenland. Comparison with Ekholm DEM comes from Ekholm 1996.

a 10 cm accuracy (Krabill *et al.*, 1999). All aircraft data within 600 m of a 5 km grid point are used. The overall mean of the differences is 6 cm, and the standard deviation is 6.71 m. The mean and standard deviations of the differences as a function of surface slope are shown in Figure 8. This confirms that the radar altimeter measurements are most precise in the flatter regions. At slopes less than 0.2°, the mean difference is near zero and the standard deviation is 1–3 m. Over larger sloping surfaces, the error increases and tends to give a higher elevation than the true value, which indicates that we have under-corrected for the slope-induced error. At about 0.4°, the mean difference for the GSFC DEM becomes negative and decreases to −10 m at 1° slope while the standard deviation increases to 30 m.

Ekholm (1996) created a Greenland DEM of 2-km resolution using satellite and aircraft altimetry over the ice sheets and photogrammetry and map scannings over the coastal regions. The ice sheet data used included: all the Geosat GM data, seven 3-day and four 35-day repeat cycles of ERS-1 data in ocean mode, and airborne radar and laser altimetry. Figure 8 also shows a comparison of the Ekholm 1996 DEM to the aircraft laser data as a function of slope. At the lower slopes, the Ekholm and GSFC DEMs have the same agreement with the laser data. At slopes between 0.5° and 0.7° the Ekholm DEM yields elevations lower than the GSFC DEM, but the absolute value of the mean differences are the same. Above 0.7°, the GSFC DEM agrees better with the laser data, which may be because of the better coverage of the ERS-1 GM data used in the GSFC DEM. In addition to smaller ground-track spacing, the ERS-1 GM data is in ice mode, which had better tracking over the higher sloping surfaces than the ocean mode data used by Ekholm (1996).

4. ICE SURFACE SLOPES AND DRIVING STRESSES

Ice flows downhill under the force of gravity in the direction of the surface slope, and the driving force depends on the magnitude of the surface slope. The shear stress, τ_b, acting on the ice is

$$\tau_b = \rho g h \sin(\alpha) \qquad (6)$$

where ρ is the density (920 kg/m^3), g is the acceleration of gravity (9.81 m/s^2), h is the ice thickness (m), and α is the surface slope (e.g., Paterson, 1994). For the small basal slopes typical of ice sheets, the additional effect of a sloping base is negligible. The shear stress, usually called the driving stress, is balanced at the bottom by the basal drag. Because very small scale undulations in the surface and the underlying bedrock base do not significantly affect the ice flow, the derivation of Eq. (6) implicitly assumes that the surface slope is averaged over several ice thicknesses.

The surface-slope vector maps derived from the surface topography provide details of the ice flow patterns, as shown in Figure 9 (see color insert), for a portion of the East Antarctic Plateau. The slope vectors, which are determined for squares 5 km on a side, enhance the information obtained from the elevation contours that are drawn at 100-m increments. The distribution of vectors may be used to delineate drainage divides, as well as to detect and/or determine approximate limits for areas of sub-glacial melting. The western (upper) half of the figure illustrates the resolution that may be achieved in delineating a segment of the divide for the Amery Ice Shelf drainage system along the ridge that extends to the NNE of Dome Argus (off the figure and centered at approximately 81°S, 78°E). The eastern (lower) half illustrates the randomness in vector orientation over an area of low surface slope (around 77°S and 105°E near Vostok Station), which was identified by Oswald and Robin (1973) with airborne radio-echo sounding as having a substantial subglacial lake. The ice divide and the location of the Vostok station are also evident in Figure 1a. Remy and Minster (1997) also used Antarctic Ice Sheet topography created from ERS-1data to analyze the relationship between curvature of the ice surface in the across-slope direction and ice flow. Their maps of ice sheet curvature showed coherent patterns, which they compared with flowlines deduced from surface slopes. Near the coast, they found ice flow anomalies that appear to be correlated with bedrock features.

For Greenland, ice thicknesses, h, are obtained from the basal elevations determined from airborne radar sounding measurements compiled by S. Ekholm (private communication). Multiplying the slope by $\rho g h$ gives the corresponding map of the ice sheet's driving stresses τ_b (Figure 10, see color insert). Since the grid size of the slope map is 5 km, the averaging distance for the surface slope is greater than

1.5 times the ice thickness. However, the ice thicknesses are smoothed over 10 times longer horizontal distances than the surface topography, because the grid size of the bedrock topography data is on a larger scale (100 + km), which affects only the magnitude and not direction of the driving stress. Therefore, the directions of the driving stresses and the corresponding flow directions are mapped on a scale of 5 km, retaining features of the flow deviations resulting from bedrock undulations on these scales. Although the surface-slope component of the magnitude of τ_b is smoothed on a 5 km scale and the thickness component is smoothed on longer scales, the fractional changes in the surface slope in the vicinity of undulations is generally greater than the fractional changes in ice thickness on this scale. Therefore, both the magnitude and the direction of the driving stresses show small scale variations related to the surface undulations.

Ice flow velocities can be derived from the driving stresses using a stress-strain deformation relation for ice. For example, the Glen flow law gives the ice strain rate as

$$\dot{\varepsilon} = A\tau^n \qquad (7)$$

where n is a constant and A depends on the ice temperature, fabric, impurities and other factors (e.g., Paterson, 1994). In general, the deformation properties of ice are intermediate between viscous-flow properties, for which the strain rate is proportional to the shear stress ($n = 1$), and perfect plastic properties, for which $n = \infty$. Various studies of ice sheet flow have used values of n from about 1.5–4. For example, assuming perfect plasticity and that the ice thickness everywhere adjusts so that the basal sheet stress is equal to the yield stress τ_b, the height profile of the ice sheet is described by a parabola $h = \{(2\tau_b/\rho g)(L-x)\}^{1/2}$, where L is the distance from the margin to the center and x is the distance from the center. Using $\tau_b = 100$ Kpa (1 bar) and $L = 450$ km, gives a maximum height of 3160 m, which is a good approximation for the central part of Greenland (e.g., Paterson, 1994). Reeh (1982) calculated the three-dimensional surface topography of the Greenland Ice Sheet using analytical solutions of the three-dimensional equations with plastic ice rheology, and obtained good definition of the flow patterns, ice divides, and major ice streams.

The driving stresses τ_b mapped in Figure 10 for Greenland have a distribution peak of 55 kPa, which is only about 1/2 of the 100 kPa value commonly used for the yield stress of ice. The tail of the distribution has only about 15% of the values above 100 kPa. Reeh (1982) noted that the mean value of basal shear stress along a flow line varies in the range of 50–150 kPa, depending on the accumulation rate and basal ice temperature. However, the calculated values appear to be somewhat lower, ranging from about 20 kPa near the ice divides to 100–150 kPa toward the margins in most regions.

5. MEASUREMENT OF ICE MARGINS

The Antarctic coastline is mapped using the slant range technique described by Thomas *et al.* (1983). The technique relies on the tendency of the altimeter to continue to measure slant ranges to the highly reflective sea ice after it passes over an ice shelf front. The measured ranges are oblique distances to the nearest portion of the sea ice along the ice front. When a slant range is measured, there is an apparent decrease in elevation after the satellite passes the ice boundary (Figure 11). This decrease in elevation occurs because the slant range backward to the sea ice is longer than the height of the satellite above the geoid. The slant range is measured because the altimeter can not detect the abrupt increase in elevation of about 40 m at the ice front. The sea ice signal is also stronger than the diffuse return from the firn on the ice shelf, which is outside the range-gate window.

Figure 11 shows the altimeter-measured change in waveform signal intensity and surface elevation as the satellite passes over the edge of the ice shelf. Data points marked 1 through 11 appear to have surface elevations lower by Δ than the geoid. They represent oblique-range measurements to the nearby sea-ice, with range increasing and apparent elevation decreasing as the satellite moves inland. The distance X to the ice front from each sub-satellite position corresponding to these 11 data points is calculated from the following relationship:

$$E^2 + X^2 = R^2 \qquad (8)$$

where E is the satellite elevation above the geoid ($\sim$800 km) and R is the measured oblique range to the sea ice. The decrease in apparent surface elevation is defined as $\Delta = R - E$. Because $E \gg \Delta$, Eq. (8) simplifies to

$$X \approx (2E\Delta)^{1/2}. \qquad (9)$$

Solving Eq. (9) for each of the first 11 data points gives a set of circles of radius X, centered on each sub-satellite location as shown in Figure 12. As discussed in Thomas *et al.* (1983) the position of the ice front nearest to a given subsatellite location lies somewhere on its circle. The midpoints between intersections of the circles are chosen as the average reflecting points at the boundary. A left-right ambiguity occurs at each crossing, with the mirror images of the two possible ice boundaries forming two arms of a "V." In most cases, the left-right ambiguity is resolved when the results from several tracks are plotted and two arms of partially overlapping "V"s line up with one another. The alignment of these "V"s is used to map the ice front position (Zwally *et al.*, 1987b). Figure 13 shows the advance of the front of the Amery Ice Shelf over two decades using data from Seasat (1978), Geosat(1985–1986), and ERS-1(1994–1995). Periodic calving of large tabular icebergs from the Amery and other Antarctic ice shelves tends to reestablish the front at a position defined by the grounded boundaries of the embayment.

6. SURFACE ELEVATION CHANGES AND MASS BALANCE

Conventional methods of studying ice sheet mass balance examine the difference between the mass input and output quantities, but significant errors in these quantities have limited the determination to $\pm 25\%$ (Warrick *et al.*, 1996 and Giovinetto and Zwally, 1995). Measurements of surface elevation changes can provide a determination of the overall ice sheet mass balance (e.g., Zwally, 1989; Davis *et al.*, 1998a; and Wingham *et al.*, 1998), because of their relationship to

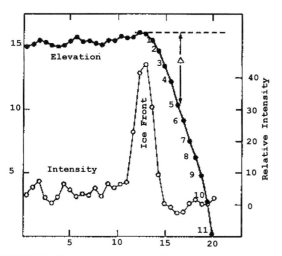

FIGURE 11 Waveform intensity (unfilled circles) and elevation at the edge of an ice shelf.

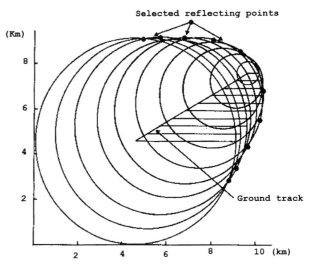

FIGURE 12 Each circle shows loci of possible points from which individual reflections could have come from. The intersection of the circles shows selected possible reflecting points.

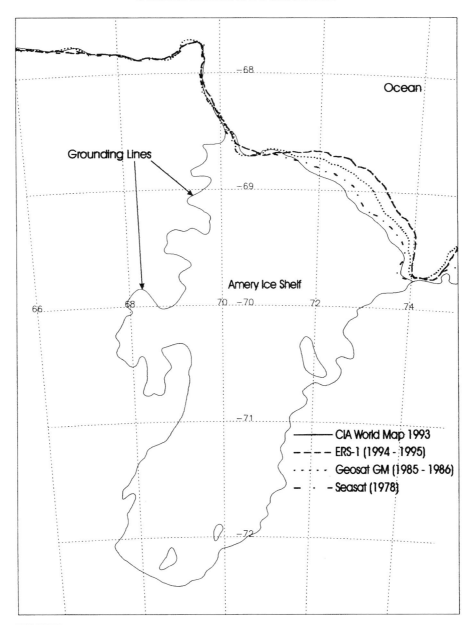

FIGURE 13   Changes in the Amery ice shelf, measured from Seasat 1978, Geosat 1985, and ERS-1 1994.

changes in ice thickness and therefore ice mass. Surface elevation changes are equivalent to ice thickness changes minus the vertical motion of the bedrock, which is primarily caused by isostatic adjustments caused by long-term changes in the ice mass loading of the earth's crust.

Surface elevation changes are derived from elevation differences, $dH_{12} = H_2 - H_1$, measured at crossover locations where sub-satellite paths intersect at successive times t_1 and t_2 (Zwally *et al.*, 1989). Since the measurement error for each dH_{12} is usually larger than the actual elevation change, a set of N values of $(dH_{12})_i$ must be averaged to reduce the error of the mean by $\sqrt{N}$. The altimeter measurement precision, estimated from crossover analysis using a

convergent $2 - \sigma$ edit, ranges from 0.3 m to several meters depending on the surface slope, which generally varies with elevation (Table 1).

Several methods have been used for analyzing a set of $(dH_{12})_k$ to obtain an average vertical velocity. The long-interval method is appropriate for measurements made during two distinct time periods, separated by a relatively long time interval (dt). Then, $dH/dt = [\sum (dH_{12})_k]/Ndt$. This method was used by Zwally *et al.* (1989), and Davis *et al.* (1998) for analysis of Seasat-Geosat data separated by 7–11 years. Possible biases between height measurements A_2 made on ascending passes versus heights D_1 on descending passes were accounted by Zwally *et al.* (1989) by averaging ascending-descending average differences and descending-ascending average differences

$$dH_{12} = [1/N_{12} \sum (A_2 - D_1)_k$$
$$+ 1/M_{12} \sum (D_2 - A_1)_k]/2 \quad (10)$$

where dH_{12} is the average elevation change between time t_1 and a later time t_2, N_{12} is the number of crossovers between ascending passes during t_2 and descending passes during t_1, and similarly for the second term. Assuming each of the $(A_i)_k$ have a bias a_k and each $(D_i)_k$ have a bias d_k, then their average values, a and d, tend to cancel in Eg. (10). Furthermore, the ascending-descending (A-D) bias difference is

$$B_{12} = [1/N_{12} \sum (A_2 - D_1)_k - 1/M_{12} \sum (D_2 - A_1)_k]/2$$
$$= a - d. \quad (11)$$

For inter-satellite comparisons, for which t_2 are for one satellite with a (A-D) bias of b_G and t_1 for another satellite with an A-D bias of b_S, the relative A-D bias of the two satellites (e.g., G and S) is

$$2B_{12} = b_G + b_S. \quad (12)$$

A-D biases can be caused by errors in orbital calculations, along-track timing errors interacting with the slope of the orbit with respect to the surface slope, or perhaps differences in the directional measurement properties of the radar. These equations were used to analyze biases in the early Seasat and Geosat data that were believed to be orbit or timing errors.

Another method, dH/dt, is appropriate when the time intervals, $(dt_{ij})_k = (t_j - t_i)_k$, are widely distributed and the rate of change, dH/dt, can be assumed to be constant over the period of measurement. In the dH/dt method, dH/dt is derived from a linear fit to the $(dH_{ij})_k$ versus $(dt_{ij})_k$ values. In contrast, simple averaging of $(dH/dt)_{ij}$ values gives a poor result, because the measurements with small time intervals contribute large errors to the average dH/dt. The dH/dt method was used for Geosat/Geosat analysis in Zwally et al. (1989). Plotting $(dH)_k = (H_a - H_d)_k$, regardless of whether $t_a > t_d$ or $t_a < t_d$, gave the A-D bias as the dH value at the $dt = 0$ intercept of the linear fit.

A time series of elevation changes can be created from a continued period of measurements using crossovers between successive time intervals t_i and t_j, where for example t_i and t_j are 30 days. For example, using Eq. (10)

$$dH_{ij} = [1/N_{ij} \sum (A_j - D_i)_k + 1/M_{ij} \sum (D_j - A_i)_k]/2 \quad (13)$$

gives the set dH_{ij} of average elevation changes between time interval t_i and later time intervals t_j. The ascending-descending (A-D) bias difference

$$B_{ij} = [1/N_{ij} \sum (A_j - D_i)_k - 1/M_{ij} \sum (D_j - A_i)_k]/2 \quad (14)$$

may also be a function of time. The series of elevation changes with respect to t_1 is

$$DS_{1N}(t) =$$
$$D_{11}(t_1), D_{12}(t_2), D_{13}(t_3) D_{14}(t_4), \ldots D_{1N}(t_N) \quad (15)$$

where

$$D_{ij} \equiv dH_{ij} = [1/N_{ij} \sum (A_j - D_i)_k + 1/$$
$$M_{ij} \sum (D_j - A_i)_k]/2 \quad (16)$$

is the elevation change between time i and time j. The first term $D_{11} = 0$. Wingham et al. (1998) used Eg. (16) with ERS crossovers between successive 35-day repeat-orbit periods, and then applied a linear-fit to the $dH(t)$ to estimate a linear rate of elevation change. In the Thwaites Glacier drainage basin around 105°W in West Antarctica, their analysis showed a 40 cm surface lowering between 1994 and 1996.

However, the series $DS_{1N}(t)$ of Eq. (16) is only the first of a series of similar series of elevation changes, which can be created with respect to successive time periods $t_1, t_2, \ldots t_N$, and therefore utilizes only a fraction of the available crossover data. A more comprehensive $H(t)$ time-series method combines successive time-series $DS_{iN}(t)$ and uses all crossover elevation differences in selected areas. The $DS_{iN}(t)$ series are combined using $D_{11} = 0$ as the reference level for all series. The combined series is:

$$H(t) = H_1(t_1), H_2(t_2), H_3(t_3), H_4(t_4), \ldots H_N(t_N) \quad (17)$$

where $[H_1 = D_{11} = 0, H_2 = D_{12}, H_3 = 1/2[(D_{13} + (H_2 + D_{23})]$,

$$H_4 = 1/3[(D_{14} + (H_2 + D_{24}) + (H_3 + D_{34})], \ldots$$

$$H_N = 1/(N-1)[D_{1N} + (H_2 + D_{2N}) + \cdots$$
$$+ (H_{N-1} + D_{N-1,N})]. \quad (18)$$

Each series $DS_{jN}(t)$ has its own reference level $D_{jj}(t_j) = 0$. The reference level $D_{22}(t_2) = 0$ of the second series is tied to the second point $H_2 = D_{12}$ of the first series. The third point H_3 then becomes the average of the change D_{13} from 1–3 and a second value, which is the change H_2 from 1 to 2 plus the change D_{23} from 2–3 from the second series, and so forth for all H_N. Standard errors are also calculated for each D_{ij} and used to weight the averages in calculating H_N.

The $H(t)$ time series is then analyzed to infer the character of the changes, for example, a seasonal cycle imposed on a linear trend. Such analysis is very important, because seasonal variations in snowfall and surface melting, as well as inter-annual variations in these parameters, are significant and will affect the analysis of longer-term trends in the surface elevations if not properly accounted for. Appropriate characterization of the seasonal cycle will enable more accurate statistical evaluation of trends in the data. In addition,

TABLE 3.   Elevation Change Greenland South of 72°N

Elevation band (m)	Fractional area	Seasat 1978–Geosat 1985–1989 (late summer–late summer) (cm year^{-1})[a]	Geosat 1985–1989 (cm year^{-1})[b]	Seasonal amplitude (peak to peak) (cm)	Date of minimum amplitude
2700–3300	0.23	+0.6 ± 0.1	+3.1 ± 1.1	10 ± 3	July 4
2200–2700	0.31	+3.7 ± 1.1	+8.2 ± 1.6	13 ± 4	June 9
1700–2200	0.20	+7.6 ± 1.5	+7.3 ± 1.9	10 ± 5	April 21
1200–1700	0.15	−4.9 ± 0.7	−6.9 ± 9.2	198 ± 24	Oct. 14
700–1200	0.12	+15.6 ± 1.9	+21.1 ± 19.4	396 ± 48	Sept. 26
Area Weighted Average		+5.4 ± 1.6	+6.2 ± 2.8		

[a]Using long-interval method Eq. (10).
[b]Using time-series $H(t)$ analysis Eq. (18) and multivariate regression with linear and sine-cosine functions with fitted phase and amplitude.

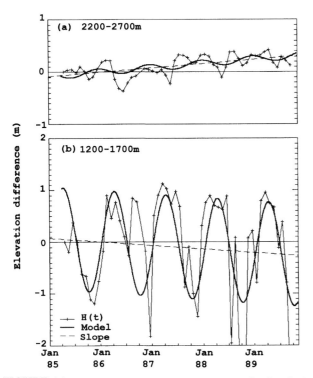

FIGURE 14 $H(t)$s calculated from Geosat crossovers for Greenland. Model of form $A + B^{*}t + C^{*}\sin\omega t) + D^{*}\cos(\omega t)$ fit to function. (a) Elevation band 2200–2700 m; slope = 8.20 cm year^{-1}, peak-to-peak amplitude = 0.13 m. (b) Elevation band 1200–1700 m; slope = −6.93 cm year^{-1}, peak-to-peak amplitude = 1.97 m.

measurement of the seasonal and inter-annual variations in the surface elevations will provide unique information on the variations in snowfall and surface melting that drive the climate-induced changes in the surface mass balance.

Figure 14 shows the $H(t)$ series constructed from 4 years of Geosat data for two elevation bands in Greenland. Super-imposed on the $H(t)$ series are multi-variate regression fits to a linear function and a sine-cosine function with the phase and amplitude as fitted variables. In the elevation band of 2200–2700 m (Figure 14a), the surface elevation increased at a rate of 8.2 ± 1.6 cm year^{-1}. The seasonal cycle of 13 ± 4 cm amplitude with a maximum in early June is does not match the data as well in this elevation band as it does at lower elevations, but may be indicative of seasonal variations in the snowfall and firn compaction. The elevation band of 1200–1700 m (Figure 14b) in Southern Greenland is mostly just above the equilibrium line where extensive surface melting occurs in summer. Because the eastern side of Southern Greenland is steeper than the Western side, most of the data included for this elevation band is from the Western side. The inferred amplitude of the seasonal cycle is 197 ± 24 cm, with a maximum in mid-April just before the onset of melting for much of the region and a minimum in mid-October after the melt season. Because the melt season is only shorter than 1/2 year, a better fit may be obtained by using an appropriately asymmetric function. The linear part of the fit is -6.9 ± 9.2 cm year^{-1}, which is affected by the summer of 1989 that may have had more intense melting. Since most of the melt water above the equilibrium line refreezes in the firn, decreases at these elevations do not necessarily represent ice volume changes.

Based on an analysis of Seasat and Geosat ERM data on Greenland south of 72°N, Davis *et al.* (1998a) reported that surface elevations above 2000 m increased at an average rate of 2.0 ± 0.5 cm year^{-1}. After including Geosat GM data, they obtained 2.2 ± 0.9 cm year^{-1} for their spatial average elevation change (Davis *et al.*, 1998b). Their estimate of the corrections for isostatic adjustment would reduce these values by about 0.5 cm year^{-1}. A major limitation in assessing the state of ice sheet balance with radar altimetry data, is the sparsity of data coverage at lower elevations.

Analysis by elevation bands (Table 3), which uses somewhat more data at lower elevations than Davis *et al.* (1998b), gives area-weighted averages of $+5.4 \pm 1.6$ cm year^{-1} for Geosat-Seasat and $+6.2 \pm 2.8$ cm year^{-1} for Geosat-Geosat. Below 1700 m the increases are less significant because of fewer crossovers and larger altimeter errors, and most of the crossovers are on the western side of the ice sheet. For elevations above 1700 m, the average thickening rates are 3.8 ± 0.2 cm year^{-1} for Geosat-Seasat and 6.4 ± 0.9 cm year^{-1} Geosat-Geosat.

Recently, the results of airborne-laser-altimeter surveys of the Greenland Ice Sheet south of 72°N in 1993 and 1998 showed thinning rates that exceeded 1 m year^{-1} on East coast glaciers (Krabill *et al.*, 1999). They found three large areas of thickening by more than 10 cm year^{-1}, and generally a mixed pattern of thickening and thinning. They attributed the high thinning rates at lower elevations to increased rates of ice creep, rather than excessive melting. However, the results are also consistent with expectations that the Greenland Ice Sheet will grow in the interior regions because of higher precipitation in a warmer climate and decrease at the lower elevations because of increased melting. As a percentage of mass balance, a thickening rate of 5 cm year^{-1} is about 10% of the average rate of accumulation for Southern Greenland (Chen *et al.*, 1997), which is a significant growth rate. Relative to climate change, 10% is within the range of predicted changes for a 1 K climate warming (note 26 in Zwally, 1989 and Warrick *et al.*, 1996).

In East Antarctica from 68° to 72°S and 80° to 150°E, Remy and Legresy (1999) found a positive mass imbalance of 20% in the western part at high elevations and a negative imbalance in some of the lower elevation areas. Lingle and Covey (1998) assessed the relative errors of Seasat, Geosat, and ERS-1 in East Antarctica North of 72°S, the use of sea ice as a reference surface, and evaluated their derived inter-satellite elevation changes. Yi *et al.* (1997) and Zwally (1994) also examined seasonal variations of about 0.2–0.6 m in the apparent height of East Antarctica Ice Sheet covered by Geosat ERM data, but were unable to conclude if they were due to real seasonal elevation changes or residual errors. While Seasat and Geosat had similar altimeter designs and orbit characteristics, their differences compared to ERS-1 and -2 introduced significant uncertainties in the indicated elevation changes. As previously mentioned the Wingham *et al.* analysis of ERS data from 1992–1996 gave -0.9 ± 0.5 cm year^{-1} for the interior of the ice sheet, a result which may be more reliable because it used data from only the ice mode of a single satellite. For long-term studies of ice sheet elevation changes, the use of nearly-identical instrumentation and inter-calibration to a relative accuracy better than 1 cm is essential.

In general, the surface mass budget of the ice sheets (snow accumulation, sublimation and/or melting) may vary with seasonal and interannual changes in polar climate, while the ice dynamics (flow and iceberg discharge) varies on much longer time scales. Therefore, measurement of some ice sheet parameters such as ice velocity need not be repeated more frequently than a decade or longer. However, determination of ice sheet mass balance requires observations over at least a 3- to 5-year interval to provide sufficient averaging over interannual variations in the surface accumulation and melting. Even so, short-term climate changes, such as changes in atmospheric moisture fluxes and possibly El Niño related effects (Figure 2 in Robasky and Bromwich, 1994) or Pinatubo volcanic effects on temperature and surface melting over Greenland (Abdalati and Steffen, 1997), can significantly affect short-term time series of elevation change. For the purpose, of predicting future ice sheet behavior with climate change, long-term time-series of elevation change are especially important to improve our understanding of the sensitivity of the mass balance to climate (e.g., temperature, precipitation, radiation, cloudiness, etc). ICESat-1, which is to be launched in 2001 as the first is a series for long-term systematic measurements, will initiate the capability of monitoring the seasonal and interannual changes in the surface mass budget, as well as the decadal scale overall ice sheet mass balance.

ACKNOWLEDGMENTS

The authors wish to acknowledge John DiMarzio, Matthew Beckley, Suneel Bhardwaj, Helen Cornejo, and Jack Saba for their help in processing and analysis of the data used to create the results presented in this chapter. We also thank two reviewers and Dr. Waleed Abdalati for their comments and suggestions.

References

Abdalati, W., and K. Steffen (1997). The apparent effects of the Mt. Pinatubo eruption on the Greenland Ice Sheet. *Geophys. Res. Lett.*, **24**, 14, 1795–1797.

Bamber, J. L. (1994a). Ice sheet altimeter processing scheme. *Int. J. Remote Sensing*, **15**, 4, 925–938.

Bamber. J. L. (1994b). A digital elevation model of the Antarctic Ice Sheet derived from ERS-1 altimeter data and comparison with terrestrial measurement. *Ann. Glaciol.*, **20**, 48–54.

Bamber, J. L., and R. A. Bindschadler (1997). An improved elevation dataset for climate and ice-sheet modelling: validation with satellite imagery. *Ann. Glaciol.*, **25**, 439–444.

Bamber, J. L., S. Ekholm, and W. Krabill (1998). The accuracy of satellite radar altimeter data over the Greenland Ice Sheet determined from airborne laser data. *Geophys. Res. Lett.*, **25**, 16, 3177–3180.

Bilitza, D. (Ed.) (1990). *International Reference Ionosphere 1990*, Rep NSSDC 90-22, National Space Science Data Center, Greenbelt, MD.

Bindshadler, R. A., Zwally, H. J., Major, J. A., and Brenner, A. C. (1989). *Surface Topography of the Greenland Ice Sheet from Satellite Radar Altimetry*, NASA SP503.

Brenner, A. C., R. A. Bindschadler, R. H. Thomas, and H. J. Zwally (1983). Slope-induced errors in radar altimetry over continental ice sheets. *J. Geophys. Res.*, **88**, 1617–1623.

Brooks, R. L., W. J. Campbell, R. O. Ramseier, H. R. Stanley, and H. J. Zwally (1978). Ice sheet topography by satellite altimetry. *Nature*, **274**, 539–543.

Chen, Q.-S., D. H. Bromwich, and L. Bai (1997). Precipitation over Greenland retrieved by a dynamic method and its relation to cyclonic activity. *J. Climate*, **10(5)**, 839.

Davis, C. H., and H. J. Zwally (1993). Geographic and seasonal variations in the surface properties of the ice sheets from satellite radar altimetry. *J. Glaciol.*, **39**, 687–697.

Davis, C. H. (1993). A surface and volume scattering retracking algorithm for ice sheet altimetry. *IEEE Trans. Geosci. Remote Sensing*, **31(4)**, 811–818.

Davis, C. H., and V. I. Poznyak (1993). The depth of penetration in Antarctic Firn at 10 Ghz. *IEEE Trans. Geosci. Remote Sensing*, **31(5)**, 1107–1111.

Davis, C. H. (1996). Comparison of ice-sheet satellite altimeter retracking algorithms. *IEEE Trans. Geosci. and Remote Sensing*, **34(1)**, 229–236.

Davis, C. H. (1997). A robust threshold retracking algorithm for measureing ice sheet surface elevation change from satellite radar altimeters. *IEEE Trans. Geosci. and Remote Sensing*, **35(4)**, 974–979.

Davis, C. H., C. A. Cluever, and B. J. Haines (1998a). Elevation change of the Southern Greenland Ice Sheet. *Science*, **279**, 2086–2088.

Davis, C. H., C. A. Cluever, and B. J. Haines (1998b). Growth of the Southern Greenland Ice Sheet letter. *Science*, **281(5381)**, 1251.

Drewry, D. J. (1983). Surface of Antarctic Ice Sheet, *In* "*Antarctica: Glaciological and Geophysical Folio*, (D. J. Drewry, ed.), Scott Polar Research Institute, Cambridge.

Ekholm, S. (1996). A full coverage, high-resolution, topographic model of Greenland computed from a variety of digital elevation data. *J. Geophys. Res.*, **101(B10)**, 21,961–21,972.

Ferraro, E. J. F., and C. T. Swift (1995). Comparison of retracking algorithms using airborne radar and laser altimeter measurements of the Greenland Ice Sheet, *IEEE Trans. Geosci. Remote Sensing*, **33(3)**, 700–707.

Giovinetto, M. B., and H. J. Zwally (1995). An assessment of the mass budgets of Antarctica and Greenland using accumulation derived from remotely sensed data in areas of dried snow. *Zeitschrift für Gleitscherkunde und Glazialgeologie*, **31**, 25–37.

Gundestrup, N. S., R. A. Bindschadler, and H. J. Zwally (1986). Seasat range measurements verified on a 3-d ice sheet. *Ann. Glaciol.*, **8**, 69–72.

Herzfeld, U. C., Lingle, C. S., and Lee, L.-H. (1993). Geostatistical evaluation of satellite radar altimetry for high-resolution mapping of Lambert Glacier, Antarctica, *Ann. Galciol.*, **17**, 77–85.

Herzfeld, U. C., Lingle, C. S., and Lee, L.-H. (1994). Recent advance of the grounding line of Lambert Glacier, Antarctica, deduced from satellite altimetry, *Ann. Galciol.*, **20**, 43–47.

Krabill, W., E. Frederick, S. Manizade, C. Martin, J. Sonntag, R. Swift, R. Thomas, W. Wright, and J. Yungel (1999). Rapid thinning of prts of the Southern Greenland ice sheet, *Science*, **283**, 1522–1524.

Legresy, B., and F. Remy (1998). Using the temporal variability of satellite radar altimetric observations to map surface properties of the Antarctic ice sheet. *J. Glaciol.*, **44(147)**, 197–206.

Lemoine, F. G., S. C. Kenyon, J. K. Factor, R. G. Trimmer, N. K. Pavlis, D. S. Chinn, C. M. Cox, S. M. Klosko, S. B. Luthcke, M. H. Torrence, Y. M. Wang, R. G. Williamson, E. C. Pavlis, R. H. Rapp, and T. R. Olson (1998). *The Development of the Joint NASA GSFC and the National Imagery and Mapping Agency (NIMA) Geopoetential Model EGM96.* NASA/TP-1998-206861.

Lingle, C. S., L.-H. Lee, H. J. Zwally and T. C. Seiss (1994). Recent elevation increase on Lambert Glacier, Antarctica, from orbit crossover analysis of satellite radar altimetry. *Ann. Glaciol.*, **20**, 26–32.

Lingle, C. S., and D. N. Covey (1998). Elevation changes on the East Antarctic ice sheet, 1978–93, from satellite radar altimetry: A preliminary assessment. *Ann. Glaciol.*, **27**, 7–18.

Martin, T. V., H. J. Zwally, A. C. Brenner, and R. A. Bindschadler (1983). Analysis and retracking of continental ice sheet radar altimeter waveforms, *J. Geophys. Res.*, **88(C3)**, 1608–1616.

Newkirk, M. H. and G. S. Brown (1996). A waveform model for surface and volume scattering from ice and snow. *IEEE Trans. Geosci. Remote Sensing*, **34(2)**, 444–454.

Oswald, G. K. A. and G. DE. Q. Robin (1973). Lakes beneath the Antarctic ice sheet. *Nature*, **245**, 251–254.

Parsons, C. L. (1979). *In* "An assessment of geos-3 waveheight measurements, ocean wave climate." (M. D. Earle and A. Malahoff, eds.). Plenum, New York.

Partington, K. C., J. K. Ridley, C. G. Rapley, and H. J. Zwally (1989). Observations of the surface properties of the ice sheets by satellite radar altimetry. *J. Glaciol.* **35(120)**, 267–275.

Paterson, W. S. B. (1994). *In* "The physics of glaciers," Pergamon, Oxford.

Phillips, H. A. (1998). *In* "' Applications of ERS satellite radar altimetry in the Lanbert Glacier-America ice shelf system, East Antarctica, Antarctic CRC and Institude of Antarctic and Southern Ocean Studies." University of Tasmania, Hobart, Australia, Doctoral Thesis.

Phillips, H. A., I. Allison, R. Coleman, G. Hyland, P. Morgan, and N. W. Young (1998). Comparison of ERS satellite radar altimeter heights with GPS-derived heights on the Amery ice shelf, East Antarctica. *Ann. Glaciol.*, **27**, 19–24.

Reeh, N. (1982). A plasticity theory approach to the steady-state shape of a three-dimensional ice sheet. *J. Glaciol.*, **28**, 431–455.

Remy, F., P. Mazzega, S. Houry, C. Brossier, and J. F. Minster (1989). Mapping of the topography of continental ice by inversion of satellite-altimeter data. *J. Glaciol.*, **35(119)**, 98–107.

Remy, R. F., B. Legresy, S. Bleuzen, F. Vincent, and J. F. Minster (1996). Dual-frequency Topex altimeter observations of Greenland. *J. Electromagnetic Waves Applications*, **10**, 1507–1525.

Remy, F., and J. F. Minster (1997). Antarctica ice sheet curvature and its relation with ice flow, *Geophys. Res. Lett.*, **24(9)**, 1039–1042.

Remy, F. and B. Legresy (1999). Antarctic non-stationary signals derived from Seasat-ERS-1 altimetry comparison, *Ann. Glaciol.*, **27**, 81–85.

Ridley, J. K., and K. C. Partington (1988). A model of satellite radar altimeter return from ice sheets. *Int. J. Remote Sensing*, **9(4)**, 601–624.

Robasky, F. M. and D. H. Bromwich (1994). Greenland precipitation estimates form the atmospheric moisture budget. *Geophys. Res. Lett.*, **21(23)**, 2495–2498.

Robin G. DE Q. (1966). Mapping the Antarctic ice sheet by satellite altimetry. *Can. J. Earth Sci.*, **3**, 893–901.

Stenoien, M., and C. R. Bentley (1997). "Topography Estimation in W. Antarctica Directly from Level-2 Radar Altimeter Data." Third ERS Symposium on Space at the Service of our Environment, Florence, Italy, ESA SP-414, Vol 11, 837–842.

Thomas, R. H., T. V. Martin, and H. J. Zwally (1983). Mapping ice-sheet margins from radar altimetry data. *Ann. Glaciol.*, **4**, 283–288.

Warrick, R. A., C. Le Provost, M. F. Meier, J. Oerlemans, and P. L. Woodward (1996). *In* "Climate Change 1995: The Science of Climate Change," IPCC Report, (J. T. Houghton *et al.*, eds.), Ch. 7, Cambridge Press, Cambridge.

Wingham, D. J., C. G. Rapley, and H. D. Griffiths (1986). New techniques in satellite altimetry tracking systems. *Proceedings of the IGARSS Symposium, Zurich, 8–11 Sept. 1986*, edited by T. D. Guyenne and J. J. Hunt (ESA) SP-254, 1339–1344.

Wingham, D. J., A. J. Ridout, R. Scharroo, R. Arthern, and C. K. Shum (1998). Antarctic elevation change from 1992 to 1996. *Science*, **282(5388)**, 369–580.

Yi, D., and C. R. Bentley (1994). Analysis of satellite radar altimeter return waveforms over the East Antarctic ice sheet. *Ann. Glaciol.*, **20**, 137–142.

Yi, D. and C. R. Bentley (1996). A retracking algorithm for satellite radar altimetry over an ice sheet and its applications. *US. Army Corps of Engineer, CRREL, Special Report 96-27*, 112–120.

Yi, D., C. R. Bentley, and M. D. Stenoien (1997). Seasonal variation in the apparent height of the East Antarctic ice sheet. *Ann. Glaciol.*, **24**, 191–198.

Zwally, H. J. (1975). Untitled discussion point. *J. Glaciol.*, **15(73)**, 444.

Zwally, H. J. (1977). Microwave emissivity and accumulation rate of polar firn. *J. Glaciol.*, **18(79)**, 195–215.

Zwally, H. J., R. H. Thomas, and R. A. Bindschadler (1981). Ice-sheet dynamics by satellite laser altimetry. *Proc. IEEE Int. Geosci. Remote Sensing Symp.*, **2**, 1012–1022.

Zwally, H. J., R. A. Bindschadler, A. C. Brenner, T. V. Martin, and R. H. Thomas (1983). Surface elevation contours of Greenland and Antarctic ice sheets. *J. Geophys. Res.*, **88(3)**, 1589–1596.

Zwally, H. J., J. A. Major, A. C. Brenner, and R. A. Bindschadler (1987a). Ice measurements by Geosat radar altimetry. *Johns Hopkins APL Tech. Dig.*, **8(2)**, 251–254.

Zwally, H. J., S. N. Stephenson, R. A. Bindschadler, and R. H. Thomas, (1987b). Antarctic ice-shelf boundaries and elevations from satellite radar altimetry. *Ann. Glaciol.*, **9**, 229–235.

Zwally, H. J., A. C. Brenner, J. A. Major, R. A. Bindschadler, and J. G. Marsh (1989a). Growth of Greenland ice sheet: Measurement. *Science*, **246**, 1587–1589.

Zwally, H. J. (1989b). Growth of Greenland ice sheet: Interpretation. *Science*, **246**, 1589–1591.

Zwally, H. J., A. C. Brenner, J. A. Major, and R. A. Bindschadler (1990). Satellite radar altimetry over ice, Vol 1, 2, and 4. *NASA Ref. Pub. 1233*.

Zwally, H. J., A. C. Brenner, J. P. DiMarzio, and T. Seiss (1994). Ice sheet topography from retracked ERS-1 altimetry. *Proc. Sec. ERS-1 Symp., 11–14 Oct 1993*, ESA SP-361, 159–164.

Zwally, H. J. (1994). Detection of change in antarctica, *In* "Antarctic science," (G. Hempel, ed.), Springer-Verlag, Berlin, Heidelberg.

Zwally, H. J., A. C. Brenner, and J. P. DiMarzio (1998). Growth of the Southern Greenland ice sheet. *Science*, **281(5381)**, 1251.

10

Applications to Geodesy

BYRON D. TAPLEY and MYUNG-CHAN KIM

Center for Space Research
The University of Texas at Austin
Austin, TX 78759-5321

1. INTRODUCTION

Geodesy is the science devoted to the study of the figure and the external gravity field of the earth. Aspects of this science include surveying, positioning, navigation, topography and bathymetry mapping, and the study of diverse terrestrial, atmospheric, marine, submarine, and even extraterrestrial gravity phenomena from a variety of stationary and mobile instrumentation. The instruments have been installed on field-surveying, land-vehicular, coastal, offshore, shipboard, airborne, and spaceborne platforms. In this chapter, we focus our main interest on the problems of mapping the physical shape of the global mean sea surface and recovering the associated gravity field by use of satellite altimetry. Some aspects of time-varying gravity phenomena and their influence on the ocean surfaces are discussed in conjunction with the coming space-geodetic missions dedicated to gravity recovery.

As a science, there are essentially two approaches to geodesy. The deductive one starts from theoretical studies to model the geopotential and the inductive one starts from analysis of real data collected from passive observations and active experiments. Determination of the deduced geopotential model parameters using the induced geodetic measurements is the primary focus of contemporary geodetic interests. The requirement is satisfied by fitting data to the geopotential model parameters and by analyzing their error characteristics. (The most important two operational tools in geodesy come from works of Carl Friedrich Gauss, whose life spanned the period from 1777–1855. He provided the foundations for the potential theory, an essential tool in the deductive approach to geodesy, and he proposed the method of least squares, a fundamental tool in the inductive approach to geodesy.)

While the deductive approach has advanced in parallel with the development of the classical potential theory, there have been inadequate global data to fully implement the inductive approach. This is especially true in the marine environment. Before the space age, accessibility of the oceans for the study of marine geodesy relied on sparse measurements made from surveying ships, buoys, drifting floats, and tide gauge stations located along irregular local coastline segments. With the advent of various space-geodetic techniques and other diverse remote sensing capabilities, the situation is changing rapidly. The application of satellite radar altimetry, supported by the ever-improved precision orbit determination techniques (Tapley *et al.*, 1994), has spurred the study of geodesy as a global experimental science after an extend history of development, which was mostly confined to geopolitically accessible terrestrial regions, and of theoretically inclined global studies. With the extensive set of sea-surface height measurements collected from satellite altimetry, the days of marine geodesy with regionally sparse and frequently unreliable data are over. We are in an era where questions can be posed in the context of a global geodetic database.

One continuing quest of geodesy is the establishment of an accurate global vertical datum. The definition of a geodetic datum entails the classical concept of the mean sea level, which is assumed to be an equipotential surface of the earth's gravity field. As a long-term averaged value of the sea-surface heights observed at a tide gauge, the mean sea level serves as a local datum for geodetic benchmarks. And a traditional procedure for establishing regional height networks is based on leveling of several tide gauge stations. However, comparing the vertical datum referred to remotely located networks is inconsistent because the averaged sea-surface heights do not coincide exactly with an equipotential surface. The discrepancy is partly caused by local coastal phenomena associated with currents and winds, and partly by vertical motions of tide gauge stations caused by post-glacial rebound and by other tectonic plate variations over geological time scales. However another major effect is the steady-state response of the air-sea interface to the general ocean circulation and atmospheric pressure variations. Accounting for this discrepancy, here we use the terminology "mean sea surface" rather than the more conventionally used "mean sea level."

If the problem of geodesy is to determine the figure and the external gravity field of the earth, we can say that the precise knowledge of the physical size and shape of the sea surface, which covers seven-tenths of our globe, is a significant component of the overall problem. If the problem of physical oceanography is to monitor the temporal variations of the sea surface which occur globally, regionally, and locally, with secular, periodic, sporadic, and impulsive natures, a global high-accuracy high-resolution mean sea-surface model can play the role of an absolute oceanic datum for this purpose. Satellite altimetry measures the real shape of the oceans almost directly. The precision of altimeter-measured sea-surface heights has improved by well over a hundred-fold from the experimental SKYLAB era (McGoogan *et al.*, 1974) to the TOPEX/POSEIDON era (Fu *et al.*, 1994). With the increased precision and spatial coverage of the altimeter measurements, the accuracy and resolution of mean sea-surface models have been improved across all spatial and temporal scales. In the second section of this chapter, we review the evolution of the global mean sea-surface modeling and describe some data analysis techniques associated with the construction of a mean sea-surface model from satellite altimeter-derived sea-surface height measurements.

The equipotential surface of the earth's gravity field that most closely coincides with the global mean sea surface is termed the geoid. The marine geoid is the physical figure of equilibrium that the ocean would have if there were no motion of the water relative to the earth and no atmospheric effect (winds and pressure variations). While the geoid extends through the land surfaces as a mathematical figure derived from the earth's global gravity field (Figure 1), the significance of the geoid determination is most prominent over the oceans. The mean sea surface is static vertically but dynamic horizontally. There is water flow along the mean sea surface. The mean sea surface measured by satellite altimetry with a 1-cm precision is a composite of the marine geoid, with variations as large as 100 m, the mean ocean dynamic topography that reflects the steady-state ocean circulation, with 1-m order variations, and the time-averaged response of the ocean surface to the atmospheric pressure changes and winds, with 10-cm order variations. The departure of the gra-

dient of the mean sea surface from that of the geoid is related to the geostrophic velocity of the ocean currents (Wunsch, 1993). The mean ocean dynamic topography, which occurs because of the horizontal motion of the mean sea surface, is intertwined with the marine geoid, so physical oceanography is an essential partner of marine geodesy.

To provide meaningful answers to important problems related to a broad range of disciplines in earth sciences, it is necessary to model the earth's gravity field with sufficient accuracy and resolution. Analysis of satellite tracking data has provided the best means to determine the long-wavelength features of the gravity field. Although shipboard gravimetry remains the most precise direct means to measure the short-wavelength features of the marine gravity field, satellite altimeter data, with near-global coverage and almost homogeneous error characteristics, provide the best overall approach for determining the minute spatial variations of the marine gravity field. With the advent of satellite altimetry, the earth's gravity field is now known more completely over the oceans than over the continents, except for a few limited areas. Although the problem of separating the mean ocean dynamic topography remains, the mean sea surface is a good first approximation of the marine geoid. For geophysical applications of the altimetric mean sea-surface models to form a global gravity data bank, one must transform the geoid heights over the oceans to gravity anomalies. In the third section of this chapter, we describe the use of satellite altimeter measurements in the gravity-recovery issues that include the problem of obtaining short-wavelength geotectonic marine gravity anomalies, and the global long-wavelength geopotential modeling.

Since the first use of satellites to sense the earth's gravity field, we have tremendous improvement in the geoid accuracy. Contemporary geodesists discuss the need for a geoid with 1-cm accuracy (Sanso and Rummel, 1997) to support the full application of satellite altimetry. In most current studies, the geoid has been treated as though it were static. Reflecting a continuously changing mass distribution within the dynamically evolving earth system, the geoid actually undergoes variations on a variety of temporal scales. The age of the geoid, defined as the time in the past before it was not correlated with the present geoid, is about 50 million years (Ricard *et al.*, 1993). Besides such a very-low-frequency geomorphologic deformation, and apart from hourly and daily tides, the geoid also exhibits other diverse fluctuations. For instance, representing the longest wavelength geoid components, the earth's center of mass and inertia tensor exhibit the signature of many subtle variations. As such, the next-generation geoid will be inevitably a space-time model. In fact, subcentimeter accuracy global monitoring of the time-varying gravity field will soon become available from dedicated satellite gravity missions (e.g., Tapley and Reigber, 1998). It is also likely that a series of satellite altimeter missions will orbit during the next decade. Along with intro-

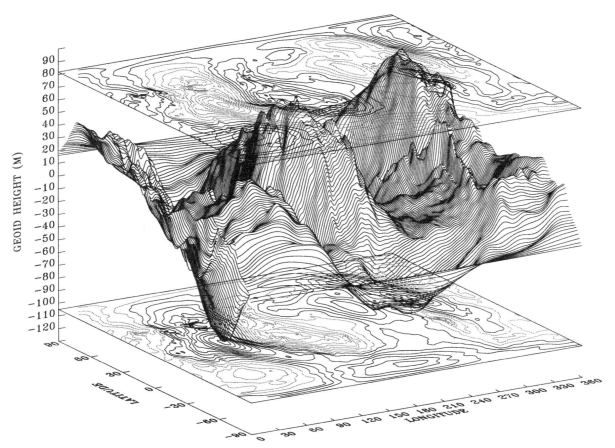

FIGURE 1 Three-dimensional projection of the global geoid computed using the TEG 3 geopotential model (Tapley *et al.*, 1997), which is complete to spherical harmonic degree 70. This long-wavelength static geoid model has an undulation range that varies from −105 to 82 m about the earth ellipsoid that best fits the geoid. Parallel geoid height profiles are shown at 1° latitudinal intervals. A geoid undulation contour map with 10-m intervals is overlaid at both of the extremal geoid heights. A spherical harmonic geopotential model complete to degree L can resolve the geoidal features down to the half-wavelength of $\pi a/\sqrt{L(L+1)}$, where $a = 6378.1363$ km is the semi-major axis of the earth ellipsoid. With $L = 70$, the resolution of the TEG 3 geoid will be approximately 300 km. The cumulated long-wavelength geoid undulation errors committed up to this resolution are a few tens of centimeters for the current best geoid models, and the accuracy and error characteristics vary from area to area.

duction of some recent applications of satellite altimetry to resolve the geocenter and angular momentum variations of the earth, the fourth section of the chapter overviews the impact of these new-generation satellite altimetry and satellite gravity data sets.

2. MEAN SEA SURFACE MAPPING

Mean sea surface is a particular time-averaged geographic configuration of the continuously fluctuating boundary between the oceans and the atmosphere. The physical definition is simple, however, it is by no means trivial to obtain a global homogeneous determination since the available measurements are scattered both in space and time. An instantaneous sea-surface height measurement is composed of the mean sea-surface height and the sea-surface height vari-

ability (Figure 2). For a complete separation of the mean and variability components, uninterrupted records of measurements at the same geographic location are required. Time-wise continuous samplings can be achieved at space-wise discrete geographic locations by conventional tide gauges for a specified time span, but such records can never be achieved globally. In fact, the only means of determining the mean sea surface on a global basis is satellite altimetry.

After applying a host of instrumental and environmental corrections to the altimeter-measured instantaneous sea-surface height, the variability is mostly composed of ocean tides having a wide spectrum of harmonic constituents. Nontidal variability exhibits a stochastic nature in addition to regular patterns of annual, semi-annual, seasonal, and other periodic components. For many of oceanographic applications, under the assumption that the ocean acts as an inverted barometer (IB), the ocean's static reaction to the atmospheric

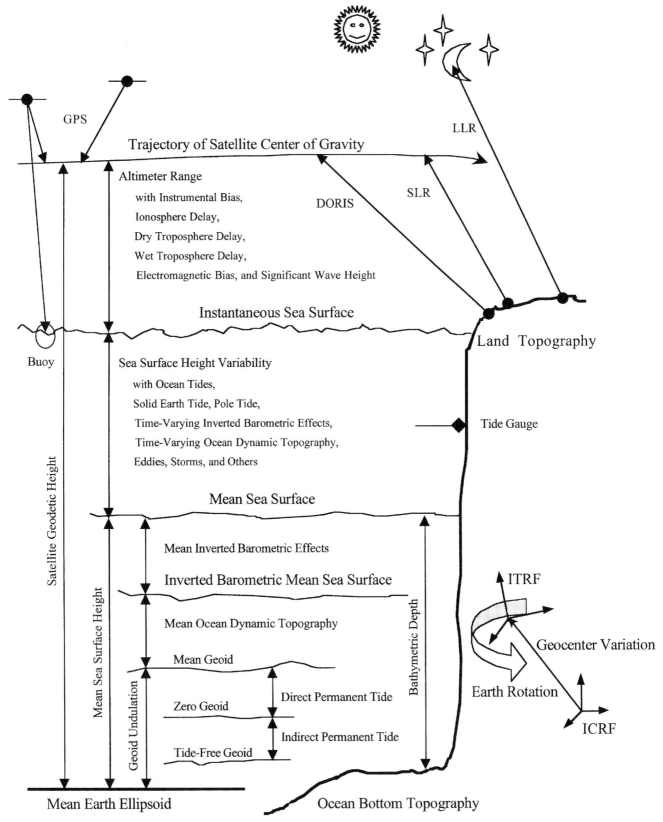

FIGURE 2 Schematic of geodesy from satellite altimetry.

pressure variations is removed from the physical mean sea surface. We refer to the resulting somewhat hypothetical surface as the inverted barometric mean sea surface (IBMSS). One may refer to the true physical mean sea surface as the non-inverted barometric mean sea surface (NIBMSS).

Although the magnitude of ocean signals in the tide and IB corrected sea-surface height variability can reach several meters at some geographic locations, the global root mean square (rms) value is less than a decimeter including the residual correction errors. Mean sea-surface mapping from satellite altimetry has, as a central challenge, the problem of filtering the temporal variability to retrieve the spatial variations of the sea-surface heights with as high a resolution as possible.

2.1. Historical Review

Early determinations of the mean sea surface were derived from the GEOS 3 and Seasat measurements (e.g., Marsh and Martin, 1982; Rapp, 1986; Marsh et al., 1992). In spite of the high noise level and regional sparseness of the GEOS 3 measurements and the short temporal coverage of the Seasat measurements, these studies demonstrated the potential of satellite altimetry in the global mean sea-surface mapping from space. Improved determinations were obtained by including measurements collected from the Geosat altimeter during the exact repeat mission (ERM) which began in October 1986 after completing the primary geodetic mission (GM). Geosat-era mean sea-surface models were developed by Basic and Rapp (1992), in a combination with bathymetric data, and by Kim (1993) and Anzenhofer and Gruber (1995), by combining the Geosat ERM data with some early fast delivery data from the ERS 1 mission.

However the accuracy and resolution of these surfaces were limited across all spatial scales. At long-wavelength scales, the limitation was primarily caused by uncertainties in the radial component of the spacecraft orbit. At short-wavelength scales, the limitation was due to the cross-track gaps of observations inherent in missions with short ground-track repeat periods. In fact, no significant direct usage of these mean sea surfaces has been reported, except for some use in differentiated forms to correct the so-called geoid gradient effects, or to compute short-wavelength marine gravity anomalies. A better surface was required for applications in almost all fields of marine geodesy and physical oceanography.

A culmination of more than 2 decades of efforts by the satellite altimetry community leads to the unique accuracy of the TOPEX/POSEIDON (T/P) mission. Along with improved understanding and modeling of nonconservative surface forces (Ries et al., 1993), the serial developments of the T/P-prelaunch JGM 1, T/P-postlaunch JGM 2 (Nerem et al.,

1994), and JGM 3 (Tapley et al., 1996) geopotential models have played a major role in satisfying the stringent orbit accuracy requirements for the T/P mission. Tests based on tracking data fits, the geopotential error covariance analysis, and various orbit comparisons indicate that the accuracy of the radial orbit of the T/P orbital ephemerides based on the JGM 3 model is in the range of 3–4 cm (Tapley et al., 1994; Smith et al., 1996). This accuracy is nearly an order of magnitude better than the orbit accuracy achieved for any altimeter mission prior to T/P. The achievement is dependent on the rapidly evolving satellite tracking technologies, e.g., the Satellite Laser Ranging (SLR), the Doppler Orbitography and Radiopositioning Integrated by Satellite (DORIS), and the Global Positioning System (GPS). Such orbit precision, together with the highly precise dual-frequency altimeter measurements produced during the mission, has provided a means to monitor the sea surface at an unprecedented accuracy level.

If the satellite orbit is known, a satellite altimeter is essentially a profiler that measures the sea-surface height along the subsatellite ground-tracks. Profiling from a single altimeter platform entails a compromise between temporal and spatial sampling of the sea-surface heights. A mission with a short ground-track repeat period has advantages for constructing highly precise along-track averaged sea-surface profiles, but leaves significant areas between the ground-tracks without coverage. A mission with a long repeat period is better suited to map the sea-surface topography with higher spatial resolution, but at any given geographic location, the repeat profiling occurs only after a long duration. To obtain improved coverage and resolution, it is desirable to combine measurements collected from several missions with distinct ground-track coverage patterns.

In 1995 alone, utilizing repeat-track averaged sea-surface height profiles constructed from the first 2-year T/P data as a calibration datum for combining multi-satellite altimeter measurements, higher resolution global mean sea-surface models have been developed by several groups of investigators. These include the Center for Space Research (CSR) at the University of Texas at Austin, The Ohio State University (OSU), and the Goddard Space Flight Center (GSFC) in the United States, GeoForschungsZentum (GFZ) in Germany, and Centre National d'Etudes Spatiales (CNES) in France. From this time, the significance of the ocean's static response to the atmospheric loading has been recognized in the mean sea-surface modeling. (The CSR and OSU models are given in both the IBMSS and NIBMSS forms.) The OSU IBMSS model (Yi, 1995) was adopted in the update of T/P-merged geophysical data records. A decimeter-level sea-surface height-variability signal might be extracted using any of these T/P-era mean sea-surface models at a resolution of around 10 km. However the time-series analysis of the variability signal will be inevitably limited at temporal scales longer than 2 years.

As of 2000, the satellite altimetry data bank is continuously expanding. The TOPEX and POSEIDON altimeters are still providing quality measurements after the successful completion of the almost uninterrupted six-year primary/extended mission period (August 10, 1992 to August 10, 1998). The high-inclination ERS 1 and 2 missions have provided more than 8 years of measurements. In the period from April 1995 to June 1996, when the two ERS satellites were in a tandem orbit, three altimeter platforms were simultaneously operated. The current T/P, ERS 2, and Geosat Follow-On (GFO) also provide a distinct triple platform configuration. Such a simultaneous operation of multiple satellite altimeter platforms will be more common, as we have the planned JASON, ENVISAT, and other mission series. Furthermore the high-density Geosat GM measurements are now completely declassified. The full Geosat GM and ERM measurements were reprocessed and released with the more precise JGM 3 orbits and improved corrections (Lillibridge and Cheney, 1997). With these data, investigators will seek continuous improvement in the mean sea-surface models for the indefinite future (e.g., Cullen et al., 1997; Schaeffer et al., 1998; Wang, 1998).

Figure 3 illustrates the improvement in resolving geotectonic features in an area (340°E to 355°E, 10°S to 50°S) of the southern Mid-Atlantic ridge system. One can more clearly identify the upwelling features of spreading-center segments and the fractured transform faults with the right panel (a T/P-era mean sea-surface model) than with the reduced resolution left panel (a Seasat-era mean sea-surface model). While larger-scale geoidal features in the mean sea surface are caused by heterogeneity in deep-earth mass density structure and mantle wide mass convection that drives plate tectonics, shorter-scale geoidal features reflect mass density anomalies in the oceanic lithosphere. The mean sea surface mimics the ocean bottom topography at the short-wavelength scales. For geophysical and tectonic interpretations of the geoidal features in the mean sea surface, we refer to Vanicek and Christou (1994), Cazenave et al. (1996), and Chapter 12 in this book.

2.2. Repeat-Track Averaging

One of the initial techniques for analyzing satellite altimeter data is based on repeat-track analysis. An altimeter samples discrete sea-surface heights along the satellite ground-tracks, generally in 1-s condensed forms giving a data point approximately every 6–7 km. The spatial and temporal characteristics of data coverage and distribution are strictly dependent on the satellite's motion relative to the earth's surface. If an altimeter satellite has an orbit with an exactly repeating ground-track, the separation of the mean and variability components can be accomplished by analyzing the repeated sea-surface height profiles. This assertion is based on the assumption that the frequencies of the sea surface height variability components are not correlated with the ground-track repeat frequency and that the data sampling is sufficient to perform along-track interpolations of the discrete measurements on each of the profiles.

The ground-track repeatability is a central consideration in planning satellite altimeter missions. Many of the historic, current, and future missions, including the planned deployment of the ICESAT Observatory (Schutz, 1998), are designed to closely maintain the frozen repeat orbit (Born et al., 1987). The term "frozen" implies that the orbit is chosen so that the secular and long-period perturbations in its perigee motion, induced by the even and odd zonal harmonics of the geopotential, respectively, closely cancel each other resulting in an almost stationary mean perigee. By "repeat," we imply that the mean semi-major axis and the mean inclination of the orbit are chosen so that the satellite's in-plane orbital motion and the orbit plane's motion relative to the earth's rotation are in deep resonance and thus its ground-track retraces itself during every repeat period.

However, along with other time-dependent gravitational perturbations, the stochastic nature of nonconservative forces acting on the spacecraft, such as the atmospheric drag, and solar and terrestrial radiation pressures, can induce significant variations of the true ground-track from the designed ground-track. Therefore, the successive altimeter-measured sea-surface heights are not referred to exactly the same locations. The sea-surface height profiles can also contain irregular data gaps, caused by systematic excursions in spacecraft attitude, which adversely impact the altimeter's ability to track the reflected radar pulse, and caused by rapid changes in land-sea topography. These data gaps affect the repeat-track analysis. For example, averaging a single profile having an unknown bias and a data gap together with other profiles having no data gap may produce an artificial dip at the location of the data gap. For high latitude regions where data density is variable because of seasonal ice cap change, data gaps can significantly affect a mean profile.

The satellites are maneuvered to compensate for the ground-track variations as much as possible. Nevertheless, a satellite cannot be maintained in a perfectly repeating ground-track. On different cycles the satellite will sample a shifted sea-surface height profile. This introduces the cross-track geoid gradient problem (Brenner et al., 1990). In extreme cases of ground-track mismatch, the magnitude of the slope-induced height error may reach tens of centimeters where the geoid gradient is high. A more typical value might be a few centimeters per kilometer, which is still significant for high-precision sea-surface height-variability studies.

An extensive list of oceanographic applications of satellite altimetry has been based on the repeat-track analysis. Several treatments are employed to alleviate the geoid gradient effects. In earlier analysis, to extract mesoscale variability signals from Seasat profiles, Thompson et al. (1983)

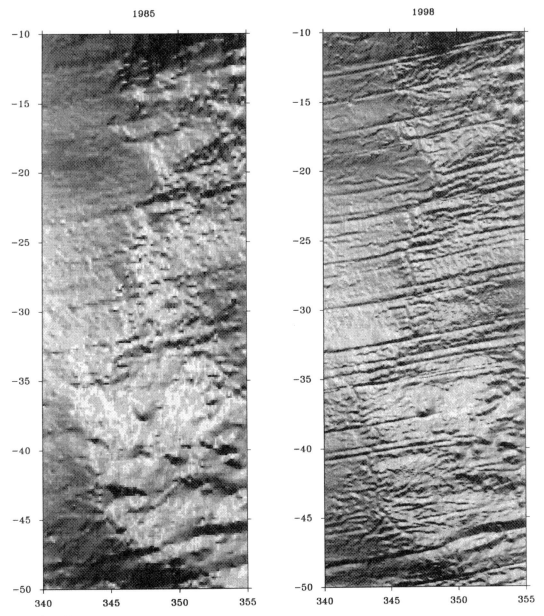

1985

1998

FIGURE 3 Advances in mean sea-surface modeling. The left panel is a Goddard Space Flight Center model developed in 1985 based on GEOS 3 and Seasat data. The right panel is a University of Texas Center for Space Research model developed in 1998 based on Geosat, ERS 1, ERS 2, and TOPEX/POSEIDON data. Both images show sea-surface relief features illuminated from the north.

ignore the effects assuming that the selected sea-surface profiles are geographically collinear. For the construction of a 1-year average Geosat sea-surface height profile, Wang and Rapp (1992) apply external corrections to individual profiles using a high-resolution geoid model. For T/P profiles, Chambers *et al.* (1998) develop internal corrections by fitting cycle-by-cycle sampled sea-surface heights to bin-by-bin local plane models. When one investigates the variability in sea-surface height profiles, such approaches can be applied under the assumption that the external or internal correction models are reliable. On the other hand, when one's concern is the mean rather than the variability, a fruitful way of avoiding the cross-track geoid gradient problem is to generate a mean of the ground-tracks. One can simply average the ground-track variations as well as the sea-surface heights so that the averaged sea-surface height profile is defined along the mean ground-track rather than a pre-defined reference ground-track. The starting point of this repeat-track averaging process is to adopt a simple analytical nominal ground-track.

A widely used nominal ground-track model is based on the geocentric projection of a uniformly rotating circular orbit of a constant inclination onto the surface of a uniformly rotating spherical earth model. Taking into account the secular perturbation effects of the earth's oblateness, such an idealized analytical ground-track can be defined by a few constant parameters. Kim (1997) analyzes the fundamental geometry of a family of three-parameter nominal ground-tracks. The three parameters are the orbital inclination (i), the angular velocity ratio (q) between the satellite's in-plane orbital plane motion and the orbit plane's motion relative to the earth rotation, and the equator crossing longitude (λ_e) of an ascending ground-track pass.

Referenced to the nominal ground-track, the geocentric unit vector $\hat{r}$, with its latitude ϕ and longitude λ, can be represented by

$$\hat{r}(\phi, \lambda) = R_3(-\vartheta)R_1(-i)R_3(-\tau)R_2(-\varepsilon)\hat{e}_x \qquad (1)$$

with $\vartheta = \lambda_e - q\tau$. Here $R_k(k = 1, 2, 3)$ denote Eulerian rotations, τ is the argument of latitude that represents the satellite's in-plane orbital motion, ϑ is the longitude of an ascending orbital node, ε is the out-of-plane angle of $\hat{r}$ with respect to the nominal ground-track, and $\hat{e}_x$ is the unit vector pointing along the x-axis of a terrestrial reference frame. The nominal ground-track, with $\varepsilon \equiv 0$, is a one-variable curve of τ. Referenced to the nominal ground-track, a geographic location on the true ground-track can be identified with the two variables, τ and ε, which will be also referred to as the along-track location and the cross-track variation, respectively (Figure 4).

A global repeat-track analysis of T/P with respect to its nominal ground-track indicates that the instantaneous sea-surface heights, collected from more than 183 cycles during a complete 5-year span of 1993–1997, are scattered within a ±1 km band. Figure 5 effectively describes the repeat-track averaging procedure at a 0.1° by 0.1° geographic bin located near the Philippine islands. The mean ground-track location has a global rms cross-track variation from the nominal ground-track of 155 m. The global rms cross-track deviation of the individual ground-track locations with respect to the mean ground-track is around 338 m. The global rms variation of the instantaneous sea-surface heights is less than 9 cm. This number reflects the radial orbit error components, media correction errors, ocean tides correction errors, cross-track geoid gradient effects, and interpolation errors, as well as the actual signals of sea-surface height variability. The global uncertainty of the averaged 1-s knot sea-surface heights is well below 1 cm, indicating the ultra-high precision of the 5-year repeat-track averaged T/P sea-surface height profile. The precision improves as the square root of the number of profiles used in the repeat-track averaging.

2.3. Crossover Adjustment

The locations where the ground-track of a satellite intersects itself on the earth's surface are called (single-satellite) ground-track crossover points. When the intersections are from ground-tracks of two distinct satellites, those are called dual-satellite ground-track crossover points. Satellite altimeter missions are actually designed to interweave nets of sea-surface height profiles (Figure 6). Analysis of the spatial distribution of crossovers and the number of their occurrences is an important element of satellite altimeter mission design in satisfying some scientific requirements. For example, the ground-tracks must cross at sufficiently large angles to deter-

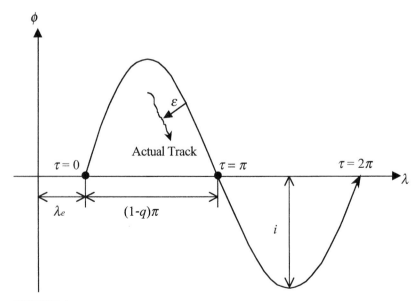

FIGURE 4   Schematic geometry of a single revolution of an eastward nominal ground-track.

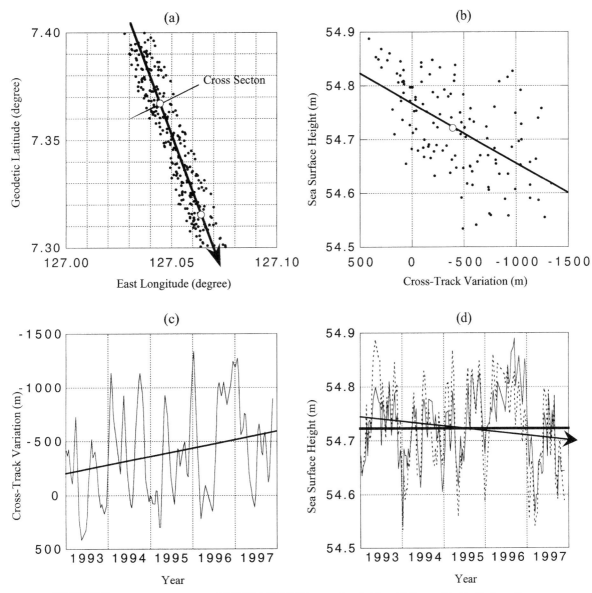

FIGURE 5 Repeat-track analysis of TOPEX/POSEIDON sea-surface heights. (a) Geographic distribution of instantaneous sea-surface heights. The grid interval is around 1 km. The linear fit corresponds to a descending segment of the mean ground-track. The two open circles are the 1-sec knot locations on the mean ground-track. (b) A cross section of the sea-surface height profiles viewed from the south on the mean ground-track. The open circle denotes the point that the mean profile passes through. The linear fit indicates an eastward downhill slope of 11 cm/km. Thus a decimeter level height error can be easily introduced from the cross-track geoid gradients. The repeat-track averaging used is to find a simple two-dimensional arithmetic mean of the fiber bundle of actual sea-surface height profiles in the cross section of the bundle. (c) The time series of cross-track variations shown in (b). The linear fit indicates eastward progression of the ground-tracks at the rate of 80 m/yr during the five-year time span at this specific location. The cross-track variations introduce the geoid gradient problem in comparing sea-surface heights. In the panel (d), the dashed line is the time series of along-track interpolated sea-surface heights without applying the cross-track corrections. Its linear-fit is the straight line with an arrowhead. The sea-surface height time series after applying the cross-track corrections is shown by the solid line. Its linear-fit is drawn as the bold horizontal line. The cross-track effects introduce a secular trend of almost -1 cm/year, which can disguise a huge signal associated with local sea-level change. After along-track interpolations and cross-track corrections, the horizontal linear fit actually indicates no secular trend.

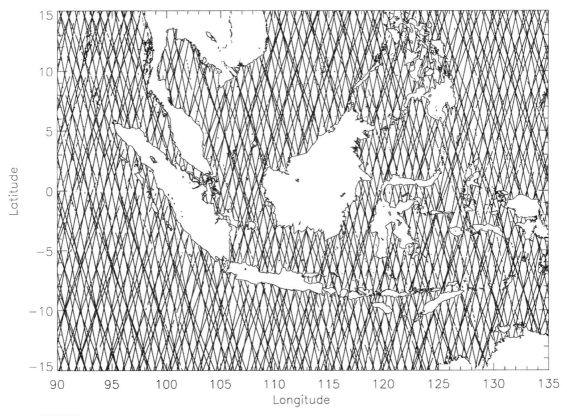

FIGURE 6 Geographic distribution of three sets of repeat-track averaged sea-surface heights over an equatorial region. These are obtained from the 10-day repeat TOPEX/POSEIDON mission, the 17-day repeat Geosat ERM, and the 35-day repeat phases in ERS 1 and 2 missions, respectively. We effectively see three sets of mean ground-tracks that produce three sets of dual-satellite crossover points as well as three sets of single-satellite crossover points.

mine the two orthogonal components of the surface gradient (i.e., deflections of vertical) to which the geostrophic oceanic current velocity components are related.

Techniques to determine crossover locations have been introduced in several studies as an important procedure for the diverse applications of satellite altimetry. The conventional procedure is based on the search of intersections between piecewise ground-track passes, which are segmented by ascending and descending traces. Given the orbital states of actual satellites and a proper definition of the nadir mapping of the satellite trajectories onto the earth's surface to draw the ground-tracks, the crossover locating problem is reduced to the determination of the ground-track crossing time-tag pairs. The problem is to solve $\hat{r}_1(t_1) \cdot \hat{r}_2(t_2) = 1$ for the time tags t_1 and t_2, where the unit vectors $\hat{r}_1$ and $\hat{r}_2$ denote ground-track time traces, and the timetags are the independent variable of the two traces. Considering the problem as a two-dimensional root finding problem, Kim (1997) presents numerical methodologies for the crossover locating from satellite ephemeredes tables, and discusses the issue of predicting the number of crossovers occurring in the general dual-satellite nominal ground-track configurations.

At a crossover point $\hat{r}^p = \hat{r}_1(t_1^p) = \hat{r}_2(t_2^p)$, where the index p identifies a crossover time-tag pair $\{t_1^p, t_2^p\}$, one obtains two sea-surface heights, $h(t_1^p, \hat{r}^p)$ and $h(t_2^p, \hat{r}^p)$, derived by interpolating discrete sea-surface height measurements sampled along their individual ground-track passes. The difference

$$X\left(t_1^p, t_2^p, \hat{r}^p\right) = h\left(t_1^p, \hat{r}^p\right) - h\left(t_2^p, \hat{r}^p\right) \qquad (2)$$

is called (the sea-surface height) crossover difference, which has been widely used as a means of comparing sea-surface heights at the same geographic location, but separated in time. In addition to numerous sea-surface height-variability studies, crossover differences are also used in the modeling of ocean tides (Egbert *et al.*, 1994), and in the accuracy assessment of gravity-field models (Moore *et al.*, 1998; Klokocnik *et al.*, 1999), etc.

As a dynamically and kinematically integrated nonlinear statistical estimation process, the precision orbit determination (POD) techniques are subject to several error sources. The major part of the orbit error is caused by the stochastic nature of surface forces acting on the satellite and inaccuracies in the geopotential model used to determine the spacecraft trajectories. Since inaccurate orbits will corrupt

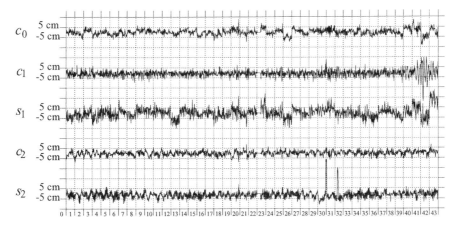

GEOSAT-ERM cycle number

FIGURE 7 Variations of orbit error correction parameters derived from a global simultaneous crossover adjustment of the first 43 Geosat ERM cycles. Altimeter corrections are from the earlier geophysical data records (Cheney *et al.*, 1987), and orbits are based on the TEG 2B geopotential model (Tapley *et al.*, 1991).

the sea-surface height measurements, variability studies can be performed only if the orbit error is properly treated. In addition, the crossover difference can be affected significantly by errors in the corrections for the medium and instrument delays in the altimeter measurement of the round-trip travel time. Errors in the altimeter time tags also induce crossover differences, coupled with the mismatch of sea-surface height locations along the ground-track.

A number of studies describe the periodic nature of the orbit error induced by the limited geopotential accuracy (e.g., Tapley and Rosborough, 1985; Schrama, 1992). This error has geographically correlated and anti-correlated parts (frequently called mean and variability, respectively), and manifests large-scale low-frequency characteristics with most of its power modulated at the 1-cpr (cycle per orbital revolution) spectral line. The anti-correlated part can be effectively detected by fitting the crossover differences to some smoothly parameterized radial orbit error model. This leads to the sometimes called "geometric" or "nondynamic" orbit adjustment, which has been used as one of the more popular ways to refine the radial orbit components of altimeter satellites. Jolly and Moore (1994) have discussed a novel for using double-differenced crossovers as a technique for assessing some orbit errors.

The simplest orbit-error parameterization assigns a bias to each of the sea-surface height profiles passing through a given local area. The global oceans can be divided into subareas to use the simplicity of the bias model. One may add more parameters for a better fit, e.g., the bias-and-tilt model. However such approaches produce at least one arbitrary bias at each of the subareas and the adjustment problem becomes physically singular. Even though the singularity can be mathematically avoided by adopting a proper

constraint (e.g., by fixing one track), or taking a generalized inverse (van Gysen and Coleman, 1999), the multiple local crossover adjustment may degrade the global consistency of the corrected orbits and associated sea-surface heights. Since the crossover differences are relative quantities, any of the geometric crossover adjustment schemes is inherently singular. Thus, after Shum (1983), the crossover techniques have been often dynamically integrated into POD as an additional data type. Use of dual satellite crossovers connecting separate altimeter mission phases now plays an increasingly important role in POD (e.g., Kozel *et al.*, 1994), as well as in oceanographic and geodetic applications (e.g., Wagner *et al.*, 1997).

One other effective parameterization is the piecewise sinusoidal model (Tai, 1988) that expresses an approximation of the radial orbit error in the 1-cpr form, $\delta r = c_1 \cos \tau + s_1 \sin \tau$, for each or some members of the orbital revolutions. This model, which also effectively detects orbit errors induced by uncertainties in surface force models and satellite epoch conditions, has been extensively used with a number of refinements (e.g., Chelton and Schlax, 1993). Figure 7 illustrates some earlier result (Kim, 1993) of a geometric crossover adjustment applied to the first 2-year Geosat ERM data, employing a five-parameter error model with a bias (c_0), a 1-cpr pair (c_1 and s_1), and a 2-cpr pair (c_2 and s_2) for each of half-day orbit segments. The large variations of c_1 in Cycle 42 reflect the influences of strong solar activity during October 1988 on the atmospheric density and solar radiation pressure. The larger variations of s_1 compared to other four time-series reflect the unconstrained nature of the Geosat orbit tracking in the spin-axis direction. The peaks of s_2 in Cycles 30, 31, and 32 reflect altimeter time-tag errors.

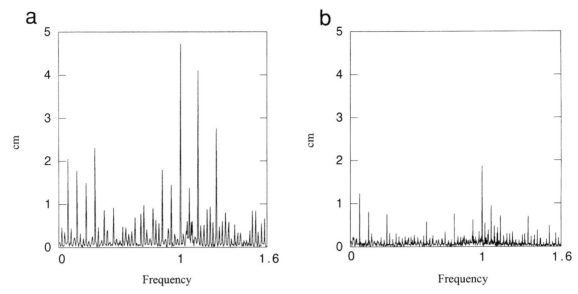

FIGURE 8 Amplitude spectra of the periodic orbit error in the Geosat ERM and ERS 1 and 2 mean sea-surface height profiles. Note the dominance of 1-cpr spectral lines.

Before the orbit improvement achieved for the T/P mission, the crossover adjustment to improve orbit accuracy was a routine part of the satellite altimeter data processing. One particular concern was the indications that the geometric adjustment can degrade the actual ocean variability signals subject to characteristics of the empirically adopted orbit error model (Wagner and Tai, 1994). The study by van Gysen and Coleman (1997) indicates that the T/P orbit corrections estimated by the crossover adjustment also contain long-wavelength small amplitude signals in the sea-surface height variability.

Orbit errors have characteristic scales both in time and space. For the purpose of mean sea-surface modeling, both the orbit errors and the ocean variability need to be suppressed. By using a number of repeat cycles, many of time-dependent effects can be suppressed by repeat-track averaging. The repeat-track averaged sea-surface height profiles will approximate the mean sea surface along the mean ground-track, but are contaminated by the gravity-induced geographically correlated and anti-correlated orbit error components. Douglas *et al.* (1984) employ a Fourier series to model such geographically periodic orbit errors.

Figure 8 shows the amplitude spectra of the Fourier series error models derived from a global crossover adjustment of three along-track averaged sea-surface height profiles obtained from 183 T/P cycles, 43 Geosat ERM cycles, and 45 cycles of the 35-day repeat phases in the ERS 1 and 2 missions, respectively. The T/P profile was fixed, and the other two profiles were adjusted in terms of Fourier series models complete to 1.6 cpr. Before adjustments, the rms crossover differences within or between the three profiles were in the range of 7–12 cm, except for the number for

intra-T/P crossovers which is around 2 cm. While one may obtain different statistics depending on different approaches to the crossover adjustments, these numbers indicate the order of some pessimistic magnitude of the periodic orbit errors in the three mean profiles. After removing the Fourier series errors from the Geosat and ERS profiles, the numbers fall in the range of 3–6 cm, except for the intra-T/P crossover number, which is fixed.

2.4. Weighted Least-Squares Objective Analysis

A common practice in geodesy is to draw contour maps from irregularly scattered data. The first step in drawing a contour map is to obtain the data values on a regular geographic grid, e.g., a rectangular grid defined by equally spaced meridians and parallels. This procedure, which is called gridding, is a key component in data reduction, archiving, and retrieving. Once the gridding step is achieved, data display, map projections to other grid systems (Mercartor, Miller, polar stereographic, and orthographic projections, etc.), and a number of mathematical treatments (Fourier transformations, wavelet transformations, and collocations, etc.) can be performed in more straightforward ways. Most of the software and hardware for scientific image processing and visualization also rely on inputs in a regular form.

Our concern here is the mean sea-surface gridding. In practice, sea-surface height measurements derived from a satellite altimeter mission phase have a somewhat regular pattern of geographic distribution and coverage governed by the satellite's orbital motion relative to the earth's surface. Jolly and Moore (1996) employ such a pattern in analyzing

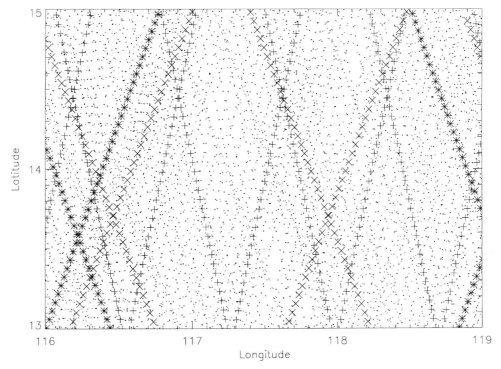

FIGURE 9 Overlay of five data sets. This is actually an enlarged view of Figure 6, after adding dots for two sets of 1-sec data obtained from the Geosat GM and 168-day repeat ERS 1 geodetic phases. The asterisks (*), cross signs (×), and plus signs (+) represent the three repeat-phase data sets of TOPEX/POSEIDON, Geosat ERM, and ERS 1 and 2, respectively.

the 35-day repeat-phase data of ERS 1. However data collected from several distinct missions reveal a more complex pattern (Figure 9).

Gridding is basically an interpolation process. The extent of sophistication in the process depends on the qualitative and quantitative characteristics of data. It depends on the investigation objective as well. A simple bin-by-bin averaging or fitting works well for a quick look result. But it is by no means acceptable when one requires a reasonable error estimate of the gridded mean sea-surface height values. The gridded values are also expected to be more accurate for regions surrounded by dense data points than at locations effectively extrapolated from distant data points. To account for the distance of data to grid points, one may employ a weighted-averaging scheme. A distance weighting function, such as the Gaussian function, usually smoothes the data. To control the extent of the smoothness, the issue of optimal smoothing arises. The question of optimality is even more complicated when one needs to combine hybrid data sets with heterogeneous accuracy estimates, e.g., sea-surface heights and slopes collected from several distinct missions.

The spline is an interpolation tool used in the nongeodetic environment. As a classical draftsman's technique, the cubic spline is used to find the smoothest curve that passes through a set of irregularly spaced data points. The term "smoothest" requires that the curve satisfy the mini-

mum curvature property. The one-dimensional spline is generalized to higher dimensions, e.g., a wide class of the spherical surface splines (e.g., Freeden *et al.*, 1994). In applying the spline interpolation technique to the analysis of altimeter data, we consider a set of n sea-surface height data $\{\hat{r}_k, h_k\}_{k=1}^{n}$ scattered on the unit sphere. Among all smooth functions passing through these measurements, the minimum curvature spherical surface spline is represented by

$$h(\hat{r}) = \alpha_0 + \sum_{k=1}^{n} \alpha_k K(\psi_k), \quad K(\psi_k) = \sum_{l=1}^{\infty} d_l P_l(\cos \psi_k),$$
$$d_l = (2l + 1)l^{-2}(l + 1)^{-2}. \tag{3}$$

Here the spline kernel $K(\psi_k)$ is a function of the spherical angular distance $\psi_k = \cos^{-1}(\hat{r} \cdot \hat{r}_k)$, P_l is the Legendre polynomial of degree l, the degree variance d_l represents the power spectra of the spline kernel, and α_0 forms the spline's null space to which the curvature is insensitive. Subject to the natural boundary condition of spline, $\sum_{k=1}^{n} \alpha_k = 0$, the $n + 1$ model parameters (α_k and α_0) are obtained by a pure-interpolation that exactly passes the spline through all of the n data points.

In addition to passing through the height data, the spline can be forced to osculate a set of sea-surface slope data. This leads from interpolation, in which the data set is homogeneous, to collocation, in which different data types can

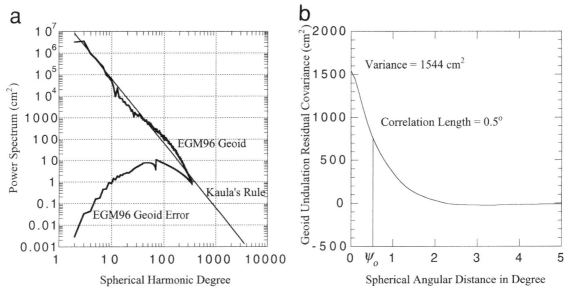

FIGURE 10 Covariance information represented in spectral and spatial domains. (a) The power spectra of geoid undulation computed from the EGM96 geopotential model (Lemonie *et al.*, 1998), the power spectra of the associated geoid error, and the Kaula's rule. The apparent deviation of the EGM96 geoid from the (straight-line) Kaula's rule indicates that some other more complex scale-variant relations are present in the earth's gravity field, along with amplification of errors in the EGM96 solution. (b) An isotropic covariance function for the residual geoid undulation field, estimated by use of the EGM96 geoid error power spectra (degrees 2–360) and Kaula's rule (degrees 361–3600), which account for the commission and omission parts of the residual marine geoid field, respectively. The covariance value at the origin is called the variance, and the distance for which the covariance value has decreased to one-half of the variance is called the correlation length (ψ_o).

be combined. The pure-interpolation or pure-collocation is not desirable for noisy data collected from different sources. When the standard deviations of the data sets are available, one may perform the surface fitting in a more rigorous statistical sense. One can also use data more than the number of assigned spline kernels. Furthermore there is no reason to work with a predefined kernel. The kernel can be adjusted in its shape by changing its power spectra to reflect the actual spectral characteristics of the interpolation field. These refinements lead the interpolation to the weighted least-squares (WLS) objective analysis, which is a statistical technique of fitting hybrid data with heterogeneous accuracy and scattered distribution to a realistic mathematical model. This is closely allied to the Wiener filtering (Press *et al.*, 1992), the Gauss-Markov process (Gelb, 1974), and the universal kriging (Olea, 1974), in different disciplines. The so-called least-squares collocation (LSC) (Moritz, 1980), in which the number of data is equal to the number of unknown parameters, is a special limit of the WLS objective analysis.

With the adjustment of the power spectra, the spline kernel becomes equivalent to the so-called covariance function, which represents an average product of a geodetic field value at a pair of points. It is usually defined to be dependent only on the distance between the two points, describing the statistical behavior of a homogeneous isotropic field. By homo-

geneous, the field would look the same to all observers, no matter where they are on the surface. By isotropic, the field looks the same in all directions from the point of view of a given observer.

As an example for the power spectra of a homogeneous isotropic field, we recall the Kaula's rule. This rule of thumb states that the geoid undulation power spectrum is proportional to the reciprocal cube of the spherical harmonic degree (Kaula, 1966). If this approximation is true, the geoid is a fractal surface with scale invariant self-similarity (Turcotte, 1989). As pointed out by van Gysen and Merry (1989), the power spectrum of the minimum curvature spherical surface spline kernel falls off eventually at the same rate as the Kaula's rule. One may say that the Kaula's rule closely states that the geoid is a minimum curvature surface. As such, splines have been used for the gravity field modeling, and the Kaula type power rules are widely used as a constraint in terrestrial, planetary, and lunar gravity field recoveries (e.g., Konopliv *et al.*, 1998).

In reality, some of the long-wavelength geoid components are already well known. The basic principle in applying the objective analysis is to use the already known "maximum likelihood" information as much as possible. A discussion of this nature in the mean sea-surface modeling can be found in several studies (e.g., Wunsch and Zlotnicki, 1984).

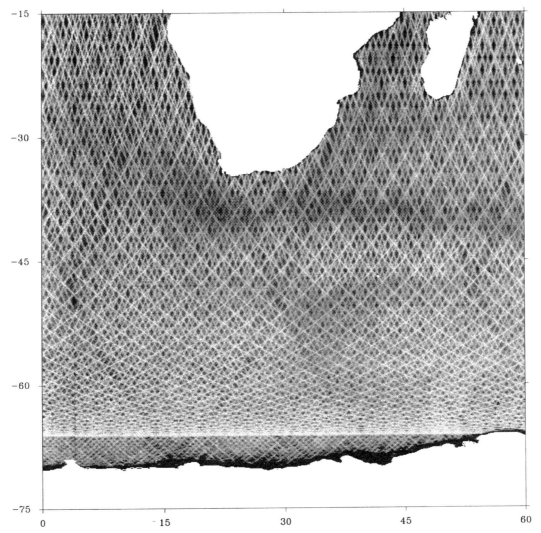

FIGURE 12 A standard deviation map (0°E to 75°E, 15°S to 75°S) of the CSR98 mean sea-surface, with a subcentimeter precision along the TOPEX/POSEIDON mean ground-track, and 1–3 cm precision along the Geosat and ERS 1 and 2 mean ground-tracks. White areas are lands. The predicted standard deviations seldom exceed 5 cm, except near coastline regions.

Bhaskaran and Rosborough (1993) discuss incorporation of the orbit-error covariance information. Assume that a reference geoid model and an error-free mean ocean dynamic topography model have been removed from the mean sea-surface to yield a synthesized residual marine geoid field. Then, as shown in Figure 10, the wavelength characteristics of the residual field and its covariance function can be described by using the power spectra of the reference geoid error and the Kaula's rule.

Let column vectors $\mathbf{x}$ and $\mathbf{y}$ represent a set of unknown parameters and a set of measurements with random noises, respectively. As the Gauss-Markov estimation process, the weighted least-squares objective analysis predicts the solution vector accompanied with its error covariance matrix,

i.e.,

$$\mathbf{x} = \mathbf{P}\left[\mathbf{H}^T\mathbf{R}^{-1}(\mathbf{y} - \bar{\mathbf{y}}) + \bar{\mathbf{P}}^{-1}\bar{\mathbf{x}}\right] \quad \text{with}$$

$$\mathbf{P} = \left(\mathbf{H}^T\mathbf{R}^{-1}\mathbf{H} + \bar{\mathbf{P}}^{-1}\right)^{-1}. \tag{4}$$

Here $\mathbf{H}$ is the mapping matrix between the observation $\mathbf{y}$ and the prediction $\mathbf{x}$, $\mathbf{H}^T$ is its transpose, and $\mathbf{R}$ is the covariance matrix of the errors in $\mathbf{y}$. Both the observation and the prediction can consist of hybrid geodetic quantities, e.g., geoid heights, geoid gradients, and gravity anomalies. Note that by use of the a priori information $\bar{\mathbf{x}}$ and $\bar{\mathbf{P}}$, the a priori signal $\bar{\mathbf{y}}$ is removed from $\mathbf{y}$ to apply the objective analysis to the residual observation $\mathbf{y} - \bar{\mathbf{y}}$, and $\bar{\mathbf{x}}$ is effectively restored in generating the prediction $\mathbf{x}$. This corresponds to the remove-restore procedure (Forsberg and Tscherning, 1981).

Employing a long-wavelength reference field, we can consider the residual geoid field to be defined on local flat-earth domains. The synthesized covariance function shown in Figure 10(b) has a correlation length of 0.5°, and indicates that the geoid residuals at distances larger than about 2° are hardly correlated. In contrast, the minimum curvature spherical surface spline has a correlation length of 48°. Working on the residuals, one may perform the objective analysis on small portions of the oceans separately. In fact, most of the current global high-resolution mean sea-surface models are produced by some smooth patching of overlapped local least-squares fittings.

Figure 11 (see color insert) shows a color-coded image of a mean sea-surface model produced by the objective analysis of fitting hybrid data sets to the patched surface model with the covariance function shown in Figure 10(b). This mean sea-surface model is obtained by use of three repeat-track averaged sea-surface height data sets (T/P, Geosat ERM, and 35-day repeat phases of ERS 1 and 2) and two sea-surface slope data sets (Geosat GM, and ERS 1 geodetic phases). The remove-restore procedure is employed; by subtracting the EGM96 geoid heights and slopes from the sea-surface heights and slopes and adding them back after the fitting and patching operations. Figure 12 shows a regional map of the predicted standard deviations of the resulting mean sea-surface. The standard deviations largely reflect the density of data used, as a direct result of objective analysis.

3. GRAVITY RECOVERY

Gravity recovery is the geodetic operation of fitting data to a gravity field model so that one can retrieve the gravity field information at any location, algebraically or statistically, in original or transformed forms. For instance, one can recover a spherical harmonic series expansion of the geopotential through geophysical inversion of pure surface gravimetry, and compute the gravity force acting on a satellite by upward continuation of the expansion. Conversely, one can recover a satellite-only geopotential model through space-geodetic inversion of spacecraft orbit perturbations, and generate a set of area-mean free-air surface gravity anomalies by downward continuation of the model. The general solution of modern gravity recovery involves both the geophysical and space-geodetic approaches.

While the spherical harmonic series expansion of the geopotential is the standard representation for the earth's gravitational field, the gravity recovery from surface geodetic data can be performed more appropriately using the ellipsoidal harmonic series expansion, which is referring to the earth (reference) ellipsoid. These harmonic series expansions are "global representations" of the gravity field. Once a set of the harmonic coefficients is obtained, the gravity

information can be computed at any location at any wavelength scale. If reliable error covariance estimates of the coefficients are available, one may also provide some answers to such a question as the resolutions at which one of the geophysical and space-geodetic analyses has more strength than the other one. On the other hand, in physical geodesy, we frequently employ "geographic representations" of the gravity field, such as the geoid undulation, gravity anomaly, gravity disturbance, vertical deflections, and gravity gradients. Alternatively, "local representations" have been also proposed, e.g., point masses, mass layers, polyhedra, finite elements, splines, and wavelets, etc. Each of these diverse representations has its own advantages and forms some content of the gravity recovery problem. In the followings, after presenting some background materials, we discuss the gravity recovery from satellite altimetry, including the recovery of the short-wavelength marine geo-tectonic gravity field, and the long-wavelength spherical harmonic geopotential modeling.

3.1. Geoid Undulation and Gravity Anomaly

The gravity g, at the spatial point $\vec{r}$, referring to a terrestrial reference frame is the magnitude of the gravity force $\vec{g}$, which is the gradient of the gravity potential W that combines the gravitational potential $V = G \iiint \rho(\vec{r}') |\vec{r} - \vec{r}'|^{-1} d^3\vec{r}'$ and the centrifugal potential $R = 1/2 |\vec{\omega} \times \vec{r}|^2$. Here G is the universal gravitational constant, $\rho(\vec{r}')$ denotes the earth's mass density at the material point $\vec{r}'$, $d^3\vec{r}'$ is an infinitesimal volume, and $\vec{\omega}$ is the angular velocity vector of the earth rotation in the inertial space. The gravity potential is decomposed into harmonic and inharmonic parts by satisfying the generalized Poisson's differential equation (Heiskanen and Moritz, 1967), $\nabla^2 W = -4\pi G\rho + 2\omega^2$. While the inharmonic part needs a more general treatment, the harmonic part is represented by a combination of outer and inner spherical harmonic series expansions, which, respectively, reflect the mass distribution interior and exterior of a spherical earth bound.

We consider the outer harmonic part of the static gravitational potential represented by the solid spherical harmonic series expansion complete to degree L,

$$V(\vec{r}) = \frac{GM}{r} \sum_{l=0}^{L} \left(\frac{a}{r}\right)^l \left[P_l(\sin\phi)C_{l0} \right.$$

$$\left. + \sum_{m=1}^{l} P_{lm}(\sin\phi)(C_{lm}\cos m\lambda + S_{lm}\sin m\lambda) \right]. \quad (5)$$

Here M is the earth's total mass, and P_{lm} is the associated Legendre function of the first kind of degree l and order m (Heiskanen and Moritz, 1967). The zonal coefficients, $C_{l0} = -J_l$, and tesseral coefficients, C_{lm} and S_{lm}, are convolution integrals of the earth's mass distribution interior of

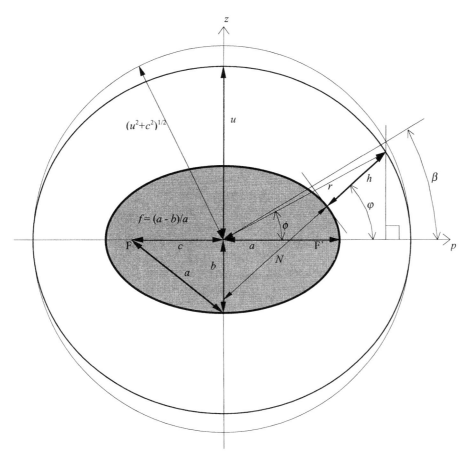

FIGURE 13 Geometry and coordinates associated with the earth reference ellipoid in the meridian plane (p, z). The geodetic coordinate system has the geodetic height h, which is the normal distance from the surface of the ellipsoid, and the geodetic latitude φ, which is the angle between the ellipsoidal normal and the equatorial plane. The ellipsoidal coordinate system has the semi-minor axis u and the reduced latitude β for a family of ellipsoids with the same foci (F and F') and linear eccentricity $c = \sqrt{a^2 - b^2}$ as the reference ellipsoid.

a selected spherical earth bound (e.g., the Bjerhammar or Brillouin spheres). The particular tesseral coefficients with $l = m$ are called the sectorial coefficients. These "conventional" spherical harmonic geopotential model parameters are referred to as the Stokes coefficients. The "fully normalized" spherical harmonic geopotential model parameters ($\bar{C}_{l0}$, $\bar{C}_{lm}$ and $\bar{S}_{lm}$) are also widely used. We imply the relationship between the two sets of parameters by writing the spherical harmonic power spectrum (degree variance)

$$d_l = \bar{C}_{l0}^2 + \sum_{m=1}^{l} \left(\bar{C}_{lm}^2 + \bar{S}_{lm}^2 \right)$$

$$= \frac{C_{l0}^2}{2l+1} + \sum_{m=1}^{l} \frac{(C_{lm}^2 + S_{lm}^2)}{2(2l+1)} \frac{(l+m)!}{(l-m)!} \qquad (6)$$

which is a basis for quantifying the power spectrum of geodetic measurements on a spherical earth.

Surface geodetic measurements are actually referred to the earth ellipsoid, whose shape is given by the semi-major

axis $a = 6378136.3$ m, and the geometric flattening, $f = 1/298.257$ (Figure 13). It is a hypothetical homogeneous self-gravitating body having the geocentric gravitational parameter $\mu = GM = 398600.4415$ km^3/sec^2, and the earth's mean angular velocity $\omega = 7.292115 \times 10^{-5}$ rad/sec. The earth ellipsoid can be defined in accordance with the situation of the tide-free, mean, and zero geoid definitions (Rapp *et al.*, 1991a). The values of the four parameters given above are for the zero-definition ellipsoid adopted in the analysis of T/P altimetry (Tapley *et al.*, 1994). One may employ a mean-definition ellipsoid by fully including the permanent response of the earth's oblateness to the luni-solar gravitational tides.

The gravity potential due to the earth ellipsoid is called the normal gravity potential U. This is used to form the anomalous potential $T = W - U$, during linearized analysis of surface-geodetic data. The most frequently used two linearized surface-geodetic representations of the earth's gravity field are the geoid undulation Δh, and the gravity anomaly Δg. These have global rms variations around 31 m and 31 mGal, respectively (up to the resolution of 55 km). The unit Gal is named after Galileo Galilei. One mGal equals 10^{-5} m/sec^2 and is approximately 10^{-6} of the absolute gravity value on the earth's surface.

The geoid undulation, which is also called geoid height or geoid anomaly, is the straight normal height of the geoid above or below the earth ellipsoid. To rigorously compute the geoid undulation at a given geographic point (φ, λ), we need to solve the nonlinear algebraic level equation $W(\Delta h, \varphi, \lambda) = U_o$, where the constant geoidal potential U_o is chosen as the normal potential value at the surface of the ellipsoid to satisfy a minimum condition for the deviation of the geoid from the ellipsoid.

The classical definition of the gravity anomaly is the difference of the gravity on the geoid g_G, and the normal gravity on the ellipsoid γ_o, which is given by the Somigliana's, Clairaut's, and Pizetti's closed formulas (Hotine, 1969, p. 202–203). After the formulation of Molodensky's problem (Moritz, 1980, p. 330–336), the modern definition is the difference of the gravity g_P on the actual earth's surface, and the normal gravity γ_Q on the telluroid (which is a hypothetical surface on which the normal potential U_Q is equal to the geopotential W_P on the earth's surface). Over the oceans, the difference between the classical and modern definitions is insignificant. (The marine gravity anomaly $\Delta g = g_G - \gamma_o$ differs from the magnitude of the vector difference $\vec{g}_G - \vec{\gamma}_o$ unless $\vec{g}_G$ and $\vec{\gamma}_o$ are aligned. The difference in their directions is represented by the vertical deflections ξ and η, which are the northward and eastward surface gradient components of the geoid with respect to the earth ellipsoid.)

Based on the Brun's formula and the fundamental equation of physical geodesy (Heiskanen and Moritz, 1967), the geoid undulation and the gravity anomaly can be represented as

$$\Delta h \approx \frac{\mu}{\gamma_o a} \sum_{l=0}^{\infty} \sum_{m=-l}^{l} \bar{Y}_m^l(\hat{v}) \bar{e}_m^l,$$

$$\Delta g \approx \frac{\mu}{Na} \sum_{l=0}^{\infty} \sum_{m=-l}^{l} (l-1) \bar{Y}_m^l(\hat{v}) \bar{e}_m^l. \qquad (7)$$

Here notational conventions are due to Kim and Tapley (2000), such that N is the meridian radius of curvature on the ellipsoid, $\bar{Y}_m^l(\hat{v})$ represents the real-valued orthonomal surface spherical harmonic functions on the unit sphere $\hat{v} = (\beta, \lambda)$ associated with the reduced latitude β, and $\bar{e}_m^l$ represents the ellipsoidal harmonic coefficients of the anomalous geopotential.

Given a global set of geoid undulations or gravity anomalies on the ellipsoid, by use of the orthonomality of the surface spherical harmonic functions, the ellipsoidal harmonic coefficients $\{\bar{e}_m^l\}$ can be obtained by the surface convolution of $\{\bar{Y}_m^l\}$ over $\gamma_o \Delta h$ or $N \Delta g$ on the unit sphere $\hat{v}$. The grav-

ity anomalies or geoid undulations can be then obtained by de-convolutions. These convolution and de-convolution operations are also referred to as the analysis and synthesis of surface spherical harmonics. Colombo (1979) notes that the analysis and synthesis, to a very high degree, need insignificant computations, especially for a global data set defined on a geographically regular grid.

In physical geodesy, the Stokes' integral formula has been widely used to create geoid undulations from gravity anomalies measured by terrestrial surface gravimetry. With the advent of satellite altimetry, the inverse Stokes' formula, which is also known as the Molodensky's formula, was used to create the marine gravity anomalies from satellite altimeter-derived mean sea-surface heights (e.g., Balmino et al., 1987). The global evaluations of these geodetic surface integral formulas are allied to the surface spherical harmonic analysis and synthesis processes. Indeed, global cross-convolutions between the two equations in Eq. (7) followed by the use of the spherical harmonic decomposition formula will lead to

$$\Delta h = \frac{1}{4\pi\gamma_o} \oiint N \Delta g S(\psi) \mathrm{d}^2 \hat{v},$$

$$\Delta g = \frac{1}{4\pi N} \oiint \gamma_o \Delta h I(\psi) \mathrm{d}^2 \hat{v}. \qquad (8)$$

Here $S(\psi)$ is the Stokes' kernel, $I(\psi)$ is the Molodensky's (inverse Stokes') kernel, and ψ is the angular distance between the geoid undulation and the gravity anomaly on the unit sphere $\hat{v}$. Instead of the direct and inverse Stokes' formulas, one can also consider the direct and inverse Hotine's formulas between the geoid undulation and the gravity disturbance (Zhang and Sideris, 1996), or the direct and inverse Vening Meinesz's formulas between the vertical deflections and the gravity anomaly (Hwang, 1998). As approximations, evaluations of these geodetic surface integrals are frequently limited to a spherical cap centered at the evaluation point.

We apply the global surface spherical harmonic analysis to the mean sea-surface model shown in Figure 11. We actually obtain a synthetic geoid model by combining (1) the CSR98 mean sea-surface heights corrected for the ocean dynamic topography inferred by Semtner and Chervin's POCM model (Stammer, 1996), and (2) the EGM96 geoid over the land areas. A set of the ellipsoidal harmonic coefficients $\{\bar{e}_m^l\}$ complete to degree 3600 is obtained by convolution. These are then converted to the equivalent set of the fully normalized spherical harmonic coefficients by use of the two-way transformations between the spherical and ellipsoidal harmonics (Jekeli, 1988).

Figure 14 (see color insert) shows the spectral power ratio $(d_l/\bar{d}_l)$ of the synthetic geoid model with respect to Kaula's rule $\bar{d}_l = 1.6 \times 10^{-10} \times l^{-3}$. Spectral power ratios of some other geoid models complete to spherical harmonic degree 360 are also shown. Note that the spectral power of the geoid models has been continuously increased. This reflects the enhancement in the density of altimeter data sets incorporated in deriving the geopotential models. The synthetic geoid model has power up to the spherical harmonic degree 2000, which corresponds to a half-wavelength resolution of 10 km. The curves indicate the actual wavelength characteristics of the geoid, e.g., Kaula's rule is too large at long-wavelength scales (degrees between 10 and 30), and too small at intermediate wavelength scales (degrees between 50 and 300).

In Figure 15 (see color insert) we show a global map of the gravity anomaly field computed by de-convolution of the decomposed ellipsoidal harmonic coefficients. Although many of the spectral characteristics in the non-Euclidean spheroidal surfaces and proper treatments of data discontinuities need further investigation, the spherical harmonic analysis and synthesis operations are one of warranted techniques for efficient recovery of gravity signals. As such, Wenzel (1998) obtains a degree 1800 geopotential model tailored to Europe by using the analysis operation with iterative enhancements.

Geoid undulations obtained from satellite altimetry and gravity anomalies obtained from surface gravimetry describe the earth's gravitational field with resolutions well beyond that of satellite tracking data. After Shum (1983), satellite altimeter range measurements have been directly used in the long-wavelength spherical harmonic geopotential modeling under the dynamic tracking concept. This brute-force dynamic approach is currently limited at shorter-wavelength geopotential modeling. For the high-degree spherical harmonic geopotential modeling, the indirectly derived altimetric marine gravity anomalies have been widely used under the geophysical boundary-value problem concept. Rather than the derived gravity anomalies, one can also consider the more direct use of the altimeter-derived marine geoid undulations (after correcting for the ocean dynamic topography) in a combination of the terrestrial surface gravity anomalies. Formulating and solving this problem in terms of a surface integral over the composite land-sea domain is referred to as the altimetry-gravimetry problem (Mainville, 1986; Brovelli and Migliaccio, 1993). The existence, uniqueness, and stability of the solution of this geophysical mixed-boundary value problem have been extensively studied with some hypothetical land-sea distribution. Keller (1996) reports that the solution, if the problem is sufficiently regular, exists and is unique for an arbitrary land-sea distribution on a fixed earth-bounding sphere. Klees *et al.* (1997) present a good classification of the problem.

3.2. Short-Wavelength Marine Gravity Field

Marine gravity can be precisely measured using modern shipboard instruments and the towed deep-ocean gravimeter traveling close to the seafloor gravitational mass sources. The sea-surface gravity measurements suffer exponential

falloff because of the distance between the point of measurements and the seafloor mass. As such, the more complicated seafloor gravity survey is also used to acquire the unattenuated gravity signal. However, due to the sparsity of shipboard and seafloor surveys, satellite altimetry provides the most valuable data sets for the recovery of the marine gravity field.

Since the advent of satellite altimetry, investigators created numerous local and global marine gravity field models using a variety of successful techniques. The first regional (Haxby *et al.*, 1983), and global (Haxby, 1987) color portrayals, created from the 1978 Seasat data, demonstrated the promising potential of satellite altimetry for the global recovery of the marine gravity field. These results are based on the planar spectral method of using two-dimensional fast Fourier transform (FFT) to convert the altimeter-derived sea-surface slopes to gravity anomalies on flat-earth domains. In an alternate study, a global simultaneous recovery of the sea-surface height and the marine gravity anomaly field was developed from the Seasat data (Rapp, 1983) using the least-squares collocation technique.

The planar spectral method and the least-squares collocation are the most widely used tools in the short-wavelength marine gravity recovery. The major advantage, claimed for the least-squares collocation, is that randomly spaced hybrid type data can be incorporated using statistical information about errors in the data, while providing corresponding information about the quality of the output gravity anomalies at arbitrary locations. As such, the accuracy of the result is dependent on the validity of the statistical information used. On the other hand, the spectral method has great simplicity and computational efficiency when compared with any least-squares techniques. Olgiati *et al.* (1995) report that the spectral method applied to sea-surface slopes yields better recovery of relative gravity variations than the use of sea-surface heights.

The effectiveness of using slopes as a data type for marine gravity recovery was noted at the very beginning of marine geotectonic applications of satellite altimeter measurements. In a flat-earth local Cartesian space with (x, y, z) as the eastward, northward, and outward coordinates, Haxby *et al.* (1983) differentiate a gridded sea-surface height field to obtain the vertical deflection fields $\eta(x, y)$ and $\xi(x, y)$. Then, they employ the Fourier-transformed locally approximated Laplace's differential equation

$$F(\Delta g) = j\bar{g}\frac{k_x F(\eta) + k_y F(\xi)}{\sqrt{k_x^2 + k_y^2}} \tag{9}$$

to obtain the gravity anomaly field $\Delta g(x, y)$ by forward and backward FFT techniques. Here F is the planar Fourier operator, $\bar{g}$ stands for a local mean gravity value, k_x and k_y denote horizontal wavenumbers, and j is the unit imaginary number.

Rather than differentiating a pre-gridded sea-surface height field to obtain the vertical deflections, investigators have also used the sea-surface slopes directly derived along satellite ground-tracks. In particular, use of along-track averaged slopes greatly enhanced the signal-to-noise ratio of marine gravity recovery along the repeat mission ground-tracks. After a number of refinements in the data processing algorithms, the full potential of this approach is realized by Sandwell and Smith (1997) using the high-density data collected from the geodetic phases of the Geosat and ERS 1 missions. This study, in which measurements collected from several missions with distinct ground-track coverage patterns and different noise characteristics have been combined efficiently in a bin-by-bin weighted least-squares fitting, provided the first detailed view of the geotectonic features of the ocean basins at some 10-km resolution. They report that the accuracy of the derived gravity field along some selected ship tracks is about 4–7 mGal.

Remarkably, the marine gravity recovery has been performed over areas covered by permanent sea ice. Using ERS 1 slope data with full retracking analysis of altimeter waveforms, Laxon and McAdoo (1994), McAdoo and Laxon (1997), and Sarrailh et al. (1998) demonstrate that satellite altimetry provides new geophysical information about the structure and development of the Arctic and Antarctic seafloors. Before these investigations, much of the polar areas were devoid of gravity information since geophysical and oceanographic surveying with traditional shipboard techniques was difficult or impossible.

As the slope of the mean sea surface, after correcting for non-gravitational effects, the geoid gradients are essentially free of many systematic long-wavelength errors inherent in satellite orbits, ocean currents, and ocean tides. The largest error source can be poor identification of coastlines, sea-ice, and mid-ocean islands along the altimeter footprint, which are unrelated to the elevation changes of the ocean surfaces. A good example of the slope error budget is given in Sandwell (1991, Table 1). He lists "Land and Ice, Dynamic Topography, Altimeter Noise, Wet Troposphere, E/M Bias, Height Acceleration, GDR Orbit, Ocean Tide, Ionosphere, and GEM T1 Orbit," in diminishing order of significance as error sources in some earlier slope data obtained from the Geosat ERM. With improvements in most of the global environmental correction models, Sandwell and Smith (1997) state that the most important correction to the slopes is now the ocean tide, especially over the shallow continental margins. Sea-surface slope variability essentially represents eddy kinetic energy, and varies geographically. In the quietest regions, its magnitude is around 1 μrad. Yale (1997) reports that slopes of the ocean tide correction exceed 2 μrad on most of the continental shelves, whereas other environmental corrections are less than 0.5 μrad almost everywhere.

Resolution of along-track averaged slope profiles has been continuously improved. Yale et al. (1995) find the Geosat ERM and TOPEX resolutions of 24 and 22 km, respectively, both with a 31-cycle stack of sea-surface slope profiles. The resolution criterion they adopted is the wavelength at which the mean square coherence between two collinear profiles becomes 0.5. By using this coherence analysis, Maus et al. (1998) show that the along-track resolution of the ERS 1 ocean product records (OPRs) can be significantly improved when a realistic spectral model is employed for analysis of the retracked sequence of altimeter waveforms.

Parallel to the spectral techniques, the least-squares collocation methods have been also used to recover the marine gravity field. In addition to the GEOS 3, Seasat, and Geosat ERM altimeter measurements, Rapp and Basic (1992) incorporate the ETOPO5U bathymetric data to take into account the effects of the seafloor topographic irregularities. One of the most recent recoveries of the global marine gravity field is achieved by Hwang et al. (1998) using the Seasat, Geosat, ERS 1, and T/P data. They use the least-squares collocation to obtain the vertical deflection fields on a regular grid, and the planar spectral method to convert the vertical deflections to the gravity anomalies. This technique, referring to the inverse Vening Meinesz formula and deflection-geoid formula, is summarized in Hwang (1998). They report that comparison of ship-measured gravity and altimeter-derived gravity measurements yields rms agreements from 5 up to 14 mGal at tectonically active areas.

Closely related to the concept of multi-resolution analysis in wavelet theories (Liu et al., 1998; Freeden and Schneider, 1998), localization in classical gravimetry is essentially a tailored potential theory for measurements available only on relatively small areas. As Sanso (1993, p. 322) states, understanding the local behavior of the geopotential will be much more difficult than the global problem from the point of view of potential theory. Nevertheless, to obtain high-resolution local gravity anomaly maps accompanied with some proper error estimates derived by linear covariance propagation techniques, the least-squares collocation method has been widely used for the altimetric recovery of the marine gravity field. We see that the computationally efficient planar spectral method is also based on a localization concept, i.e., patching of flat-earth domains.

Accompanied with the remove-restore of the long-wavelength reference gravity information, both the planar spectral method and the least-squares collocation method have provided many high-resolution marine gravity field models (e.g., Andersen and Knudsen, 1998), which have the capability to resolve many of the geophysically significant short-wavelength seafloor tectonic features. Accurate short-wavelength marine gravity information is critical for many applications, such as oil exploration, marine and submarine navigation and positioning. However one cannot use the patched fields for the study of large-scale oceanographic phenomena that requires information about the correlation of errors between distant points. To satisfy this requirement, we must utilize a global potential theory approach, which, in general, imposes significant computational demands to achieve a rigorous error analysis.

3.3. Global Gravity Recovery

The need to improve our knowledge of the earth's global gravity field is required to support interdisciplinary studies in geodesy, geophysics, oceanography, and climate-change phenomena. In the field of geodesy, accurate geopotential models are required to support the precision orbit determination of diverse spacecraft. These spacecraft include the altimeter satellites used to infer accurate sea-surface height measurements. In areas of geophysics, we need gravity information to improve our knowledge of the three-dimensional structure and long-term variations of the mass and energy transport within and between the gaseous, watery, liquid, and solid earth components. In oceanography and climate change studies, orders of improvement in the accuracy of the marine geoid are required to determine oceanic current velocities and to better understand the ocean's role in heat and nutrient transport.

The use of artificial satellites as sensors of the global gravitational field was anticipated before the Sputnik 1 was launched in 1957. In the late 1950s, orbit perturbation analysis of early geodetic satellites provided a few low degree zonal coefficients (C_{20}, then C_{40}, followed by C_{30}). In the early 1960s, some sectorial (C_{22}, S_{22}), and tesseral (C_{41}, S_{41}, . . .) coefficients were estimated. Starting in 1966, the Smithsonian Astrophysical Observatory published a series of the Standard Earth models, i.e., SE 1, 2, and 3, which are complete to spherical harmonic degree 8, 16, and 18, respectively (Gaposchkin, 1974). Since then, the size of the global geopotential model has been continuously increased; the maximum degree has been approximately doubled during each 10-year interval. For a review of global gravity determinations up to the early 1990s, we refer Yuan (1991), Exertier (1993), and Nerem (1995). Contemporary global models have been developed from research institutions in Europe (CNES/GRGS, GFZ) and United States (NASA/GSFC, The Ohio State University, and the University of Texas at Austin). Today we use global models complete to degree 70 with full error covariance estimates, complete to degree 360 with limited error covariance estimates, and higher without error covariance estimates.

The classical approach to geopotential modeling relies on the analysis of surface gravity anomaly measurements. If global coverage is provided, the simple convolution technique will provide an explicit spherical harmonic series expansion of the geopotential as the linearized solution of a geophysical boundary-value problem. However surface gravimetry involves time-consuming, expensive surveys. Despite recent acquisition of new data sets from Eastern Europe, the former Soviet Union, and other areas, the surface gravity data archive, which includes terrestrial, shipboard, and airborne measurements, is still far from forming a uniform dense global coverage.

The most productive program contributing to the improvement of global gravity recovery has been satellite altimetry, which improved the lack of information over the oceans. By merging the surface gravimetry and the satellite altimetry, one can obtain almost global coverage of measurements. Indeed, altimeter-derived marine gravity anomalies are one of the dominant ingredients of high-degree geopotential models. The OSU91A model incorporates collocated gravity anomalies, and the EGM96 model incorporates spectral anomalies as well as the collocated anomalies. These models are complete to spherical harmonic degree 360, and reveal relatively broad features, with half-wavelength scale of 55 km, as compared to the incorporated pure-altimeter marine gravity anomaly fields, which have spatial resolution of around 10 km.

Modern solutions of the geopotential coefficients are a result of combining measurements collected from satellite altimetry, surface gravimetry, and a variety of dynamic satellite tracking. The incorporation of all three kinds of data sets into a single solution is referred to as the comprehensive combination solution. Rigorously the problem has to be formulated and solved in an optimally weighted nonlinear least-squares sense that involves iterative refinements of parameters in a comprehensively large-scale inverse problem. In principle, one may minimize the weighted least-squares joint objective index

$$J(\mathbf{x}) = f_s \frac{(\mathbf{s} - \mathbf{x})^T \mathbf{P}_s^{-1}(\mathbf{s} - \mathbf{x})}{n} + f_h \left\langle \left(\frac{\Delta h - H(\mathbf{x})}{\sigma_h} \right)^2 \Theta \right\rangle_e$$
$$+ f_g \left\langle \left(\frac{\Delta g - G(\mathbf{x})}{\sigma_g} \right)^2 (1 - \Theta) \right\rangle_e . \quad (10)$$

Here $\mathbf{x}$ is the unknown geopotential coefficients complete to degree L, $n = (L + 1)^2$ is the number of the coefficients, $\mathbf{s}$ and $\mathbf{P}_s$ are an a priori solution and the associated error covariance matrix obtained from satellite tracking measurements (which may include some dynamically processed satellite altimeter ranges and their differences at ground-track crossovers). The differences $\Delta h - H(\mathbf{x})$ and $\Delta g - G(\mathbf{x})$ denote residuals of the geoid undulation and gravity anomaly with standard deviations σ_h and σ_g, respectively, Θ is the ocean function defined as unity over oceans and zero over lands, $\langle \cdot \rangle_e$ is the area-mean operator on the earth reference ellipsoid, and f_s, f_h, and f_g are the weighting factors for the three information sets.

One important aspect of combining the hybrid information equations is in selecting the weighting factors. Extensive efforts are devoted to optimally assign them and to calibrate the resulting error covariance matrix associated with the comprehensive gravity field solution. Lerch *et al.* (1988) adopt an iterative approach by use of dependent and independent subset data solutions. Yuan (1991) presents another iterative numerical algorithm to simultaneously estimate the weighting factors and the solution parameters.

The rigorous procedure forms the information equations from all available data to estimate the geopotential coefficients to the maximum degree consistent with the data resolution. Exploiting the wealth of information contained in the altimeter measurements with wavelengths down to some tens of km in terms of the full-covariance geopotential modeling is a challenging task. The main obstacle is the heavy computational resource required to process the large amount of data required to estimate the large number of unknown geopotential coefficients, together with a host of other parameters required to model the ocean dynamic topography, ocean tides, earth orientation, satellite tracking station coordinates, satellite epoch conditions, and other empirically known phenomena. Therefore, alternative techniques have been used to obtain high-degree models by exploiting sparseness of matrix elements under some restrictions in data characteristics. For instance, the block-diagonal normal matrix scheme used in the high-degree extensions of the GFZ96 (Gruber et al., 1997) and EGM96 (Lemoine *et al.*, 1998) geopotential models is based on longitudinal independence of data weights, and obtains the coefficients beyond degree 72 and 70, respectively, without a full-error covariance information.

A geodetic operation of fitting data to models without estimating the related error budget has an inherent limitation in interpreting the results. Errors arise both from the data used in the computations and from a truncated solution. These are called commission errors and omission errors, respectively. Errors can be stochastic, but more likely are systematic. In fact, today's geodetic measurements are highly precise but relatively inaccurate. The least-squares techniques yield biases for most solutions, since the omission errors cannot be random noises in the real world. Nevertheless, a number of geodetic diagnoses are based on the use of geopotential error covariance matrices produced by least-squares techniques.

The covariance matrix contains information relevant to a number of important questions. What is the geoid error at a geographic location? What is the correlation of the geoid errors at two locations at a certain wavelength? To what accuracy and resolution are the marine geoid and the ocean dynamic topography separately identifiable by use of satellite altimetry? What are the orbit error characteristics of a future satellite altimeter mission orbiting with a certain altitude and a certain inclination? To answer these questions, one needs a well-calibrated and complete error covariance matrix. With the lack of higher degree error covariance in-

formation, the current models also introduce difficulties in rigorous error analysis at the current and near-future level of accuracy and resolution requirements. The more rigorous treatment of satellite altimeter data for the comprehensive full-covariance high-degree geopotential modeling will be an active and rewarding research area (e.g., Kim and Tapley, 2000).

3.4. Marine Geoid Error

In many of sophisticated ocean modeling studies to assimilate satellite altimetry (e.g., Tsaoussi and Koblinsky, 1994), there has been a growing need to accurately quantify the geographic characteristics of the marine geoid error in various forms. Given the error covariance matrix of a global geopotential model, the auto and cross error covariance estimates of the geoid undulations, gravity anomalies, vertical deflections, and their effects on the geostrophic current velocities, can be obtained by linear covariance propagation. Consider two geoid undulation errors δh and $\delta h'$ at different locations on the earth ellipsoid. We denote the geoid undulation error covariance by $[\delta h \delta h']$ where the bracket is the statistical expectation symbol. The error covariance with δh and $\delta h'$ at the same location forms the error variance $[\delta h^2]$. Figure 16 shows geoid undulation error variance and covariance maps computed using the JGM 3 geopotential error covariance matrix. Note that the marine geoid is better known than the geoid over the land areas. Also note that the spatial structure of the geoid error is far from that of a homogeneous and isotropic field.

A single index that quantifies the global marine geoid accuracy might be its area-mean variance computed over the ocean area. The square root of this number is currently a few tens of centimeters and would be at some 10-m level before the advent of satellite altimetry. Let δh_l be the l-th degree portion of the geoid error, such that $\delta h = \sum_{l=0}^{L} \delta h_l$. We write the area-mean variance of the marine geoid error committed to the maximum degree L as

$$E(L) = \sum_{l=0}^{L} \sum_{l'=0}^{L} E_{ll'} \quad \text{with}$$

$$E_{ll'} = \frac{\langle [\delta h_l \delta h_{l'}] \Theta \rangle_e}{\langle \Theta \rangle_e}. \tag{11}$$

Here $E_{ll'}$ represents the degree covariances of the marine geoid error.

For a global spherical domain, with the orthonormality of spherical harmonics, one needs to account for the degree variances, E_{ll}, only. In general such a decorrelated error rep-

Variance

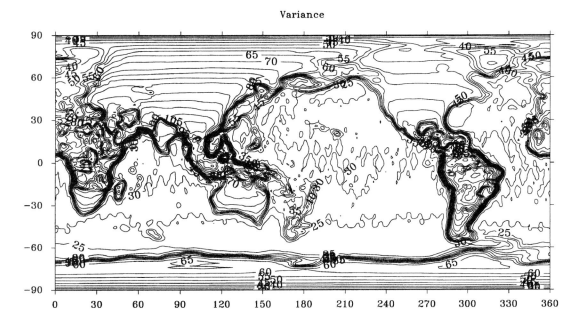

Covariance

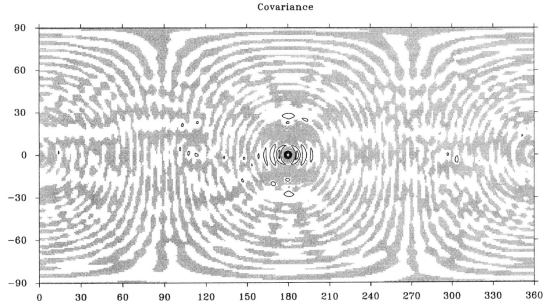

FIGURE 16   Geoid undulation error variance and covariance maps computed using the JGM 3 geopotential error covariance matrix complete to spherical harmonic degree 70. The top map shows the square-root geoid undulation error variance, with its range from 20.2–117.3 cm, and rms values of 53.5, 79.9, and 37.4 cm over global, land, and ocean domains, respectively. The bottom map shows the geoid undulation error covariance with respect to the central point where the square-root variance is 28.1 cm. Contour interval is 5 cm. Negative covariance areas, which reach down to −12.3 cm, are shaded.

resentation is not sufficient for a non-global elliptic domain of satellite altimetry, where the spherical harmonics are not an orthogonal basis. We have to keep the full degree covariances to imply the wavelength characteristics of the marine geoid error to the maximum commission degree. Figure 17 shows the $(L + 1) \times (L + 1)$ degree covariance matrix of

the marine geoid error inferred from the JGM 3 ($L = 70$) model. While the degree error variances in the main diagonal are dominant, the magnitudes of degree covariances are significant and reflect the extent of correlation between different spherical harmonic degrees over the oceanic areas. Thus, special care must be taken to quantify the magnitude of the

```
                                   DEGREE L2

              37 38 39 40 41 42 43 44 45 46 47 48 49 50 51 52 53 54 55 56 57 58 59 60 61 62 63 64 65 66 67 68 69 70

   0  0      +70-20-14-16-12-11 -6 -5 -5 -4 -5 -3 -4 -3 -4 -3 -3 -3 -3 -2 -2 -2 -2 -2 -2 -2 -2 -2 -2 -1 -2 -1 -1 -1   37
   1  0  0      +71-21-15-16-12-11 -6 -5 -5 -4 -5 -3 -4 -3 -4 -3 -3 -3 -3 -2 -2 -2 -2 -2 -2 -2 -2 -2 -2 -1 -1 -1 -1   38
   2  0  0 +0      +74-22-15-17-13-12 -7 -5 -5 -5 -5 -4 -4 -3 -4 -3 -3 -3 -3 -2 -3 -2 -2 -2 -2 -2 -2 -2 -2 -1 -1 -1   39
   3  0  0 -0 +1      +75-23-16-17-13-12 -7 -6 -5 -5 -5 -4 -4 -3 -4 -3 -3 -3 -3 -2 -3 -2 -2 -2 -2 -2 -2 -2 -2 -1 -1   40
   4  0  0 +0 -0 +2      +78-23-16-17-14-12 -7 -6 -6 -5 -5 -4 -4 -4 -4 -3 -3 -3 -3 -2 -3 -2 -2 -2 -2 -2 -2 -2 -2 -1   41
   5  0  0 +0 +0 -0 +4      +78-23-16-17-13-12 -7 -6 -6 -5 -5 -4 -4 -4 -4 -3 -3 -3 -3 -2 -3 -2 -2 -2 -2 -2 -2 -2 -2   42
   6  0  0 -0 -0 +1 -0 +5      +80-24-17-17-14-12 -7 -6 -6 -5 -5 -4 -4 -4 -4 -4 -3 -3 -3 -3 -3 -2 -2 -2 -2 -2 -2 -2   43
   7  0  0 -0 -0 -0 +1 -1 +7      +80-24-17-17-14-12 -7 -6 -6 -5 -5 -4 -4 -4 -4 -3 -3 -3 -3 -2 -3 -2 -2 -2 -2 -2 -2   44
   8  0  0 +0 +0 -1 -0 +1 -1+10      +82-24-17-17-14-12 -7 -6 -6 -5 -5 -4 -4 -4 -4 -4 -3 -3 -3 -3 -3 -2 -2 -2 -2 -2   45
   9  0  0 +0 +0 +0 -1 -1 +2 -2+13      +81-25-17-17-14-12 -7 -6 -6 -5 -5 -4 -4 -4 -4 -3 -3 -3 -3 -2 -3 -2 -2 -2 -2   46
  10  0  0 +0 +0 +0 +0 -1 -1 +3 -2+15      +83-25-17-17-14-12 -7 -6 -6 -5 -5 -4 -4 -4 -4 -3 -3 -3 -3 -2 -3 -2 -2 -2   47
  11  0  0 -0 +0 -0 +1 +1 -2 -2 +3 -3+20      +83-25-17-17-15-12 -7 -6 -6 -5 -5 -4 -4 -4 -4 -3 -3 -3 -3 -2 -2 -2 -2   48
  12  0  0 -0 -0 +0 -0 +1 +1 -2 -2 +4 -3+22      +83-25-17-17-15-12 -7 -6 -6 -5 -5 -4 -4 -4 -4 -3 -3 -3 -3 -2 -2 -2   49  D
  13  0  0 +0 -0 +0 +0 -1 +1 +2 -3 -3 +4 -5+26      +84-25-17-17-15-12 -7 -6 -6 -5 -5 -4 -4 -4 -4 -3 -3 -3 -3 -2 -2   50  E
  14  0  0 +0 +0 -0 +0 +0 -1 +1 +1 -3 -3 +4 -5+27      +84-25-17-17-15-12 -7 -6 -6 -5 -5 -4 -4 -4 -4 -3 -3 -3 -3 -2   51  G
  15  0  0 +0 +0 -0 -0 +0 -0 -1 +1 +2 -4 -4 +4 -7+31      +84-25-17-17-15-12 -7 -6 -6 -5 -5 -4 -4 -4 -4 -3 -3 -3 -2   52  R
  16  0  0 +0 +0 +0 -0 -0 -0 +0 -1 +1 +2 -4 -5 +3 -8+30      +84-25-17-17-15-12 -7 -6 -6 -5 -5 -4 -4 -3 -4 -3 -3 -3   53  E
  17  0  0 -0 +0 +0 +0 -0 -0 +0 -0 -2 -1 +3 -5 -6 +3 -9+33      +84-24-17-17-15-12 -7 -6 -6 -5 -5 -4 -4 -3 -4 -3 -3   54  E
  18  0  0 -0 -0 -0 +0 +0 -0 -1 -1 +1 -2 -1 +3 -5 -7 -1-10+34      +84-24-17-17-15-12 -7 -6 -6 -5 -5 -4 -4 -3 -3 -3   55
  19  0  0 +0 -0 -0 -0 +0 +0 -1 -1 -1 -1 -3 -2 +3 -6 -8 -3-11+36      +84-24-17-17-15-12 -7 -6 -6 -5 -5 -4 -4 -3 -3   56  L
  20  0  0 +0 +0 -0 -0 -1 +1 +1 -1 -1 -1 -0 -3 -2 +3 -7 -8 -4-11+37      +84-24-17-17-14-12 -7 -6 -6 -5 -5 -4 -4 -3   57  1
  21  0  0 -0 -0 +0 +0 -0 -1 +1 +1 -1 -2 -2 -1 -3 -3 +2 -7 -9 -4-12+40      +84-24-17-16-15-12 -7 -6 -6 -5 -5 -3 -3   58
  22  0  0 +0 +0 +0 +0 -0 -0 -1 +1 +1 -1 -1 -2 -1 -3 -3 +2 -7 -9 -5-13+42      +83-24-17-16-14-12 -7 -6 -6 -5 -5 -3   59
  23  0  0 -0 +0 -0 -0 +0 -0 -1 -1 +1 +1 -1 -2 -2 -1 -3 -3 +1 -8-10 -5-13+44      +83-24-17-16-14-12 -7 -6 -5 -4 -4   60
  24  0  0 -0 -0 +0 -0 +0 -0 -1 -1 -1 +0 +1 -1 -2 -3 -1 -4 -3 -2 -8-11 -6-14+46      +82-23-17-16-14-12 -6 -5 -5 -4   61
  25  0  0 -0 +0 -0 +0 -0 -0 -0 -1 -1 -2 -0 -0 -2 -3 -3 -1 -4 -4 -2 -8-11 -7-14+48      +82-23-17-16-14-11 -6 -5 -5   62
  26  0  0 -0 -0 -0 -0 -0 -0 -0 -0 -1 -1 -2 -1 -1 -2 -3 -3 -1 -4 -4 -3 -9-12 -8-15+51      +81-23-16-16-13-11 -6 -5   63
  27  0  0 -0 +0 -0 -0 -0 +0 -0 -0 -1 -1 -1 -2 -1 -1 -2 -3 -3 -1 -4 -4 -4 -9-12 -8-15+53      +81-23-16-15-13-10 -6   64
  28  0  0 +0 -0 -0 -0 +0 -0 -0 -0 -1 -1 -1 -1 -2 -1 -1 -2 -3 -4 -2 -4 -4 -5 -9-13 -9-16+56      +80-22-16-15-12-10   65
  29  0  0 -0 -0 -0 +0 -0 +0 -0 -0 -1 -0 -1 -1 -1 -3 -2 -2 -2 -4 -4 -2 -4 -5 -5-10-13-10-17+58      +78-22-15-14-11   66
  30  0  0 -0 +0 +0 -0 -0 +0 -0 -0 -0 -1 -1 -1 -2 -1 -3 -2 -2 -2 -4 -4 -2 -4 -5 -6-10-14-11-17+60      +76-21-14-13   67
D 31  0  0 -0 +0 -0 -0 -0 +0 +0 -0 -1 -1 -1 -1 -1 -2 -1 -3 -2 -2 -2 -4 -4 -2 -4 -5 -7-10-14-11-18+62      +74-20-13   68
E 32  0  0 +0 -0 +0 +0 -0 -0 -0 -0 -0 -1 -1 -1 -1 -2 -2 -1 -3 -2 -2 -2 -4 -4 -3 -4 -5 -8-11-14-12-18+64      +71-18   69
G 33  0  0 -0 -0 +0 +0 -0 -0 -0 -0 +0 -0 -1 -1 -1 -1 -2 -2 -1 -3 -2 -2 -2 -4 -4 -3 -4 -5 -8-11-14-12-18+65      +67   70
R 34  0  0 +0 +0 -0 +0 +0 -0 -0 -0 -0 -0 -0 -1 -1 -1 -1 -2 -2 -1 -3 -2 -2 -2 -4 -4 -3 -4 -6 -9-11-15-13-19+66
E 35  0  0 -0 +0 -0 -0 -0 +0 -0 -0 -1 -1 -0 -0 -1 -1 -1 -1 -2 -2 -1 -3 -2 -2 -2 -4 -4 -3 -4 -6 -9-11-15-13-19+68
E 36  0  0 +0 -0 -0 -0 -0 -0 +0 -0 -0 -1 -1 -0 -0 -1 -1 -1 -1 -2 -2 -1 -3 -2 -2 -2 -4 -4 -3 -4 -6-10-11-15-13-20+69
  37  0  0 -0 +0 -0 -0 -0 -0 -0 -0 -0 -0 -1 -1 -0 -0 -1 -1 -1 -1 -2 -2 -1 -3 -2 -2 -2 -4 -4 -4 -5 -6-10-11-15-13-20
L 38  0  0 +0 +0 -0 -0 +0 -0 -0 -0 -0 -0 -0 -1 -1 -1 -1 -1 -1 -1 -1 -2 -2 -1 -3 -2 -2 -3 -4 -4 -4 -5 -6-10-12-15-14
1 39  0  0 +0 +0 -0 -0 -0 +0 -0 -0 -0 -1 -0 -1 -1 -1 -1 -1 -1 -1 -1 -1 -2 -2 -2 -3 -2 -2 -3 -4 -4 -4 -5 -6-10-12-16
  40  0  0 -0 +0 -0 -0 -0 -0 +0 -0 -0 -1 -1 -0 -1 -1 -1 -1 -1 -1 -1 -1 -2 -2 -2 -2 -3 -3 -3 -3 -4 -4 -4 -5 -6-11-12
  41  0  0 -0 +0 -0 +0 -0 -0 -0 -0 -0 -0 -1 -1 -0 -1 -1 -1 -1 -1 -1 -1 -1 -2 -2 -2 -2 -3 -3 -3 -3 -5 -4 -4 -5 -6-11
  42  0  0 -0 +0 -0 -0 +0 -0 -0 -0 -0 -0 -1 -1 -1 -0 -1 -1 -1 -1 -1 -1 -2 -2 -2 -2 -3 -2 -3 -3 -3 -3 -5 -4 -4 -5 -6
  43  0  0 +0 +0 -0 -0 -0 -0 -0 -0 -0 -0 -1 -1 -1 -1 -1 -1 -1 -1 -1 -1 -1 -2 -2 -2 -2 -3 -2 -4 -3 -3 -3 -5 -4 -5 -5
  44  0  0 -0 -0 +0 -0 -0 -0 +0 -0 -0 -0 -0 -1 -1 -1 -1 -1 -1 -1 -1 -1 -1 -1 -2 -2 -2 -2 -3 -2 -4 -3 -3 -3 -5 -4 -5
  45  0  0 -0 -0 +0 +0 -0 -0 -0 -0 -0 -0 -1 -1 -1 -1 -1 -1 -1 -1 -1 -1 -1 -1 -2 -2 -2 -2 -3 -3 -3 -4 -3 -3 -3 -5 -4
  46  0  0 +0 -0 -0 -0 +0 -0 -0 -0 -0 -0 -1 -1 -1 -1 -1 -1 -1 -1 -1 -1 -1 -1 -1 -2 -2 -2 -2 -3 -3 -3 -4 -3 -3 -3 -5
  47  0  0 +0 +0 -0 -0 -0 -0 -0 -0 -0 -0 -0 -1 -1 -1 -1 -1 -1 -1 -1 -1 -1 -1 -1 -1 -2 -2 -2 -2 -3 -3 -3 -4 -3 -3 -3
  48  0  0 -0 +0 -0 -0 -0 -0 -0 -0 -0 -0 -0 -0 -1 -1 -1 -1 -1 -1 -1 -1 -1 -1 -2 -1 -1 -2 -2 -2 -2 -3 -3 -3 -4 -3 -3
  49  0  0 -0 -0 -0 -0 -0 -0 -0 -0 -0 -0 -0 -0 -0 -1 -1 -1 -1 -1 -1 -1 -1 -1 -1 -2 -1 -1 -2 -2 -2 -2 -3 -3 -3 -4 -3
  50  0  0 -0 +0 -0 -0 -0 -0 -0 -0 -0 -0 -0 -0 -0 -0 -1 -1 -1 -1 -1 -1 -1 -1 -1 -2 -2 -1 -2 -2 -2 -2 -2 -3 -3 -3 -4
  51  0  0 +0 +0 +0 -0 -0 -0 -0 -0 -0 -0 -0 -0 -0 -1 -1 -1 -1 -1 -1 -1 -1 -1 -1 -1 -2 -2 -2 -2 -2 -2 -2 -2 -3 -3 -3
  52  0  0 +0 +0 -0 -0 -0 -0 -0 -0 -0 -0 -0 -1 -0 -1 -1 -1 -1 -1 -1 -1 -1 -1 -1 -1 -1 -2 -2 -2 -2 -2 -2 -2 -2 -3 -3
  53  0  0 +0 -0 +0 -0 -0 -0 -0 -0 -0 -0 -0 -0 -1 -1 -1 -1 -1 -1 -1 -1 -1 -1 -1 -1 -1 -2 -2 -2 -2 -2 -2 -2 -2 -2 -3
  54  0  0 -0 -0 -0 -0 -0 -0 -0 -0 -0 -0 -0 -0 -1 -1 -1 -1 -1 -1 -1 -1 -1 -1 -1 -1 -1 -1 -2 -2 -2 -2 -2 -2 -2 -2 -3
  55  0  0 -0 -0 -0 -0 -0 -0 -0 -0 -0 -0 -0 -0 -0 -1 -1 -1 -1 -1 -1 -1 -1 -1 -1 -1 -1 -2 -1 -2 -2 -2 -2 -2 -2 -2 -2
  56  0  0 -0 -0 -0 -0 -0 -0 -0 -0 -0 -0 -0 -0 -0 -1 -1 -1 -1 -1 -1 -1 -1 -1 -1 -1 -1 -1 -2 -1 -2 -2 -2 -2 -2 -2 -2
  57  0  0 +0 -0 -0 +0 -0 -0 -0 -0 -0 -0 -0 -0 -0 -0 -1 -1 -1 -1 -1 -1 -1 -1 -1 -1 -1 -1 -1 -2 -1 -2 -2 -2 -2 -2 -2
  58  0  0 +0 +0 -0 -0 -0 -0 -0 -0 -0 -0 -0 -0 -0 -0 -1 -1 -1 -1 -1 -1 -1 -1 -1 -1 -1 -1 -1 -1 -2 -1 -2 -2 -2 -2 -2
  59  0  0 -0 -0 -0 -0 -0 +0 -0 -0 -0 -0 -0 -0 -0 -0 -1 -1 -1 -1 -1 -1 -1 -1 -1 -1 -1 -1 -1 -1 -1 -2 -1 -2 -2 -2 -2
  60  0  0 +0 -0 -0 -0 -0 -0 -0 -0 -0 -0 -0 -0 -0 -0 -1 -1 -1 -1 -1 -1 -1 -1 -1 -1 -1 -1 -1 -1 -1 -1 -2 -1 -2 -2 -2
  61  0  0 -0 +0 -0 -0 -0 -0 -0 -0 -0 -0 -0 -0 -0 -0 -1 -1 -1 -1 -1 -1 -1 -1 -1 -1 -1 -1 -1 -1 -1 -1 -1 -2 -2 -2 -2
  62  0  0 +0 +0 -0 -0 -0 -0 -0 -0 -0 -0 -0 -0 -0 -0 -1 -1 -1 -1 -1 -1 -1 -1 -1 -1 -1 -1 -1 -1 -1 -1 -1 -1 -2 -2 -2
  63  0  0 +0 +0 -0 -0 -0 -0 -0 -0 -0 -0 -0 -0 -0 -0 -1 -1 -1 -1 -1 -1 -1 -1 -1 -1 -1 -1 -1 -1 -1 -1 -1 -1 -1 -2 -2
  64  0  0 +0 +0 +0 -0 -0 -0 -0 -0 -0 -0 -0 -0 -0 -0 -1 -1 -1 -1 -1 -1 -1 -1 -1 -1 -1 -1 -1 -1 -1 -1 -1 -1 -2 -1 -2
  65  0  0 +0 -0 -0 -0 -0 -0 -0 -0 -0 -0 -0 -0 -0 -0 -0 -1 -1 -1 -1 -1 -1 -1 -1 -1 -1 -1 -1 -1 -1 -1 -1 -1 -1 -1 -1
  66  0  0 +0 -0 -0 -0 -0 -0 -0 -0 -0 -0 -0 -0 -0 -0 -0 -0 -1 -1 -1 -1 -1 -1 -1 -1 -1 -1 -1 -1 -1 -1 -1 -1 -1 -1 -2
  67  0  0 -0 +0 -0 -0 -0 -0 -0 -0 -0 -0 -0 -0 -0 -0 -0 -0 -0 -1 -1 -1 -1 -1 -1 -1 -1 -1 -1 -1 -1 -1 -1 -1 -1 -1 -1
  68  0  0 -0 -0 +0 -0 -0 -0 -0 -0 -0 -0 -0 -0 -0 -0 -0 -0 -0 -0 -1 -1 -1 -1 -1 -1 -1 -1 -1 -1 -1 -1 -1 -1 -1 -1 -1
  69  0  0 +0 +0 -0 -0 -0 -0 -0 -0 -0 -0 -0 -0 -0 -0 -0 -0 -0 -0 -0 -1 -1 -1 -1 -1 -1 -1 -1 -1 -1 -1 -1 -1 -1 -1 -1
  70  0  0 +0 +0 -0 -0 -0 -0 -0 -0 -0 -0 -0 -0 -0 -0 -0 -0 -0 -0 -0 -0 -1 -1 -1 -1 -1 -1 -1 -1 -1 -1 -1 -1 -1 -1 -1

      0  1  2  3  4  5  6  7  8  9 10 11 12 13 14 15 16 17 18 19 20 21 22 23 24 25 26 27 28 29 30 31 32 33 34 35 36

                                   DEGREE L2
```

FIGURE 17 Marine geoid error degree covariances computed using the JGM 3 error covariance matrix complete to spherical harmonic degree 70. Magnitudes of the displayed matrix elements are the square root absolute degree covariance values (identified by L1 and L2). Units are millimeters. Nearest integer values are displayed. Signs indicate positive, zero, and negative covariances. Degree zero and degree one correlated parts are zero by the definition of the analomous geopotential.

marine geoid error from geopotential error covariance matrices.

Figure 18 illustrates the improvement in the knowledge of the long-wavelength portion of the marine geoid in terms of the cumulated degree error amplitude $\sqrt{E(L)}$. Although the knowledge of the global geoid has been significantly improved with the progresses in terrestrial gravimetry, together with the increasing number of diverse satellite tracking databases available, it is satellite altimetry that provides the current marine geoid information across wide

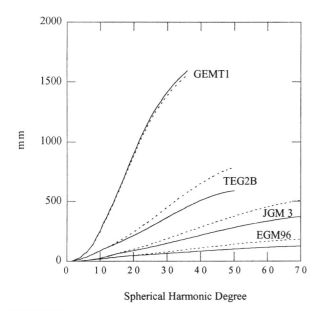

FIGURE 18 Advances of marine geoid accuracy during the last decade. Displayed by solid lines are the amplitudes of the cumulated marine geoid error computed using the geopotential error covariance matrices of the GEM T1 (Marsh *et al.*, 1988), TEG 2B (Tapley *et al.*, 1991), JGM 3 (Tapley *et al.*, 1996), and EGM96 (Lemoine *et al.*, 1998) geopotential models. (Dashed lines are corresponding amplitudes for the global geoid accuracy, computed using the usual degree variances only.) Note the 20-fold improvement of the marine geoid accuracy from GEM T1 (1593 mm) to EGM96 (75 mm) committed to the spherical harmonic degree 36, which correspond to half-wavelength resolution of 550 km.

scales of spatial wavelengths. Nevertheless, as investigated by Ganachaud *et al.* (1997), the current accuracy of the marine geoid is not adequate to provide geodetic constraints for many global oceanographic studies.

4. NEW FRONTIERS

Satellite altimetry has provided a revolution in mapping the physical shape of the oceans and the static marine gravity field. Averaged over several years, one can now determine the mean sea surface at a 1-cm level precision along the T/P ground-tracks. Merged over several distinct missions, one can also produce quality-assured detailed short-wavelength features of the marine gravity field at better than a 10-km level spatial resolution. This is a major step forward in the general area of the earth and space science disciplines. This advance holds not just for marine geodesy, but leads to enhanced knowledge in lithosphere structure, mantle composition, and rheology. Indeed, the spectacular improvement in the knowledge of the global mean sea surface and associated marine gravity field accomplished by many investigators must be counted among the most important early scientific achievements of satellite altimetry.

While satellite altimetry has freed the traditional physical geodesy from constraint to the two-dimensional terrestrial surface domain, we now see the emergence of a new dimension in the field of space geodesy. The new dimension is, of course, time. The importance of four-dimensional geodesy is not a new issue. In the first volume of *Marine Geodesy*, Ronald S. Mather (1978) provides extensive discussions of the role of the geoid in four-dimensional geodesy. One may also refer to the selected papers in the monograph entitled "Developments in Four-Dimensional Geodesy" (Brunner and Rizos, 1990), and others. However many of the relevant notions have relied on the deductive approach, due to the limited accessibility of the dimension-raised domain to most of the current space-geodetic techniques. When viewed in the space-time domain, a dense spatial coverage of measurements may no longer be dense, and a detailed marine gravity field model may no longer be detailed. With its ever-enlarging sphere of influence, satellite altimetry begins to reveal the complex curvature structure of marine geodesy in the space-time domain where the time-varying gravity signals are recorded.

4.1. Time-Varying Gravity Field

The earth constitutes a dynamically coupled complex evolutionary system that deforms over a wide range of spatial and temporal scales, from plate tectonics to regional weather perturbations, and from geomorphology to seismology. The earth's gravity field is viable as a passive target of the celestial mechanics, and as an active self-gravitating source of geology, geophysics, and of their evolutions. Many of the outer gravitational forces are well known, i.e., the tidal forces occurring with astronomically regulated frequencies and intensities. Many of the inner gravitational forces are poorly known, e.g., the driving forces of plate tectonics and the three-dimensional density structure of the planetary interiors that essentially remain vitual terra incognita and only the seismic waves and the media-free gravity waves can penetrate this structure.

The time-varying part of the earth's gravity field constitutes less than 1% of the departure of its total from a uniformly rotating homogeneous fluid figure of the earth in its gyrostatic equilibrium (e.g., the Jacobi ellipsoid). As noted, this departure reflects the shifting of mass sources in the terrestrial earth, oceans, ice sheets, glaciers, snow covers, ground and underground water storage, atmosphere, and liquid and solid earth cores as an ensemble. The mechanisms that cause this mass redistribution include solid-earth and ocean tides, non-tidal mass transport within and between the atmosphere and oceans, atmospheric and oceanic pressure field variations, sea-level rise caused by polar ice and groundwater storage changes, earthquakes, postglacial rebound, and some possible anthropogenic greenhouse effects, etc. For lists of the interacting volume and

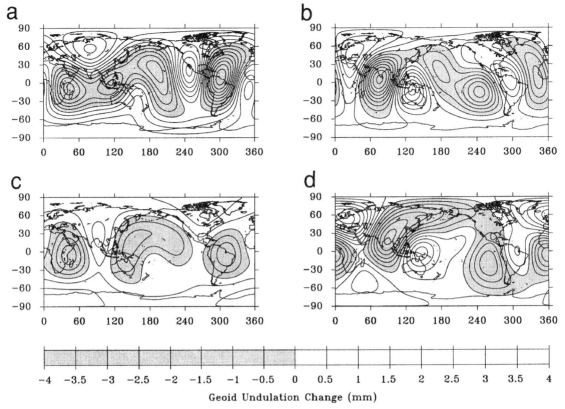

FIGURE 19   Annual variation of long-wavelength geoid undulation determined from space-geodetic techniques, and from geophysical modeling. The variation is effectively implied by $\bar{h}(\phi, \lambda, t) = c(\phi, \lambda)\cos(2\pi t) + s(\phi, \lambda)\sin(2\pi t)$, where the geographic functions $c(\phi, \lambda)$ and $s(\phi, \lambda)$ represent surface spherical harmonic expansions complete to degree 4, and t is the time (in years) after January 1, 1994. The top panels, (a) and (b), respectively, display $c(\phi, \lambda)$ and $s(\phi, \lambda)$ obtained from SLR data analysis. In contrast, the corresponding bottom panels, (c) and (d), are obtained from geophysical data analysis, which accounts for the European Center for Medium-range Weather Forecasts (ECMWF) atmospheric pressure variations, hydrological and oceanic effects given in Wahr *et al.* (1998b). One notes considerable correlation between the two sets of variations. These four panels are adapted from Cheng and Tapley (1999b).

surface forces that cause the time-varying gravity potential, one may refer to references cited in Dickman (1998), and Grafarend *et al.* (1997), where a detailed space-time spherical harmonic gravitational potential theory is presented.

The dominant component of the temporal variations in the gravity field is tidal accelerations caused by the lunar and solar gravitational forces acting on the rotating earth system under the earth's complex precession, nutation, polar and spin-axis motions. Other temporal variations are orders of magnitude smaller than the tidal effects. While, considerable understanding of the wide class of rhythmic ocean tide constituents is in hand by analysis of satellite altimeter-derived sea-surface height measurements (e.g., Eanes and Bettadpur, 1996), the current knowledge of the temporal fluctuations of the non-tidal gravity phenomena is less certain.

However, there has been some significant progress. For instances, by analyzing SLR data collected during the previous two decades from a number of geodetic satellites (Star-

lette, Lageos 1 and 2, Ajisai, Etalon 1 and 2, Stella, and BE-C), Cheng *et al.* (1997), and Cheng and Tapley (1999a, 1999b) obtain some secular and seasonal variations of low degree (2 to 6, and 8) zonal components, the 18.6-year periodic tidal components that provide information on the earth's mantle an-elasticity, and a synoptic view of some dominant long-wavelength annual variations (Figure 19). Although these studies suggest that temporal extension of SLR data sets will enhance the secular solution for zonal coefficients beyond degree 6, the current satellite-derived time-varying gravity signal is limited at resolutions in excess of 5000 km with poor sensitivity to individual harmonic constituents, especially at the longitudinal variation scales.

Mass redistribution in the various earth components is mostly caused by water movement. Driven by climate change, the exchange between oceanic water, atmospheric water vapor, continental soil moisture, snow cover, and polar ice sheets are constrained by the global water mass conservation. Since the sea-surface height variability involves sufficient mass variations, satellite altimetry can monitor a significant part of the non-tidal variations of the marine gravity field. A set of the global space-time frequency-wavenumber spectra of the sea-surface variability obtained from satellite altimetry (e.g., Wunsch and Stammer, 1995) may be converted to some equivalent time-varying gravity spectra. This is true if complementary oceanographic and climatology databases, such as the three-dimensional salinity and temperature fields (Levitus *et al.*, 1994), are accurate enough to account for the (thermally variant) steric sea-surface height changes.

Geoid height variation is strongly correlated with long-wavelength changes in the sea-surface height. Based on a thin-layer assumption for the oceans (Wahr *et al.*, 1998b), the contribution of sea-surface height changes to the geoid height variations can be represented approximately as

$$\Delta \tilde{h}_l(\varphi, \lambda, t) = \frac{3\rho_w}{\rho_e} \frac{1 + k_l}{2l + 1} \Delta \hat{h}_l(\varphi, \lambda, t). \quad (12)$$

Here $\Delta \tilde{h}_l$ is the variation in the l-th spherical harmonic degree portion of the geoid height, and $\Delta \hat{h}_l$ is the corresponding variation in the sea-surface height, $\rho_w = 1035$ kg/m^3 and $\rho_e = 5517$ kg/m^3 are the mean densities of the sea water and the total earth, respectively, and k_l represents the load Love numbers that quantify the wavelength-dependent elastic deformation of the solid earth under the surface loading.

The ocean dynamic topography is related directly to the potential and kinetic energy of the ocean circulation and its net weight is sufficient to depress the seafloor. This non-tidal ocean loading induces changes in the pressure on the ocean bottom, and deforms the marine geoid and the nearby crust. Indeed, van Dam *et al.* (1997) use T/P sea-surface heights to infer changes in the bottom pressure, and estimate the possible time-dependent fluctuations of the marine geoid caused by oceanic mass redistribution. They found that ocean variability might cause geoid undulation variations with amplitudes as large as 10 mm, and surface gravity changes up to 5 µGal. Such a use of time series of satellite altimeter data to infer the non-tidal time-varying marine gravity field will command more and more attention in the coming years.

4.2. Variations in the Geocenter and Earth Rotation Parameters

The currently accepted geocentric gravitational parameter value has a two part-per-billion uncertainty, i.e., 0.0008 km^3/sec^2 (Ries *et al.*, 1992), which corresponds to a 13-mm uncertainty in the global geoid scale. Bounded by this global mass conservation, we consider the time-variations of the earth's gravity field at the longest wavelength scales. These are variations in the earth's translation and rotation motions,

which are governed by fundamental physical laws in inertial space. Motion of the earth's mass elements as well as motion of satellites orbiting in the vicinity of the earth are usually referenced to a body-fixed Cartesian coordinate system, where its origin is placed at the earth's mass center at a specific epoch. Because of the mass redistribution as time evolves, the origin of the reference frame is displaced from the actual mass center and the displacement will vary with time as the mass distribution changes. In addition, the mass redistribution will change the earth's moments and products of inertia and will lead to a change in the earth's rotation. The translation motion is represented by variations in the geocenter coordinates (x_g, y_g, z_g), i.e., the earth's center of mass, and the rotation motion is represented by variations in the earth rotation parameters, i.e., the length-of-day (LOD) and polar motion coordinates (X_p, Y_p). Monitoring of subtle variations in these six global scale geodetic parameters involves an interdisciplinary research and observational efforts among astronomy, geodesy, geophysics, oceanography, climatology, and navigation. Satellite altimetry is beginning to play an important role in accounting for the ocean's contributions to those variations.

The geocenter, the origin of the International Terrestrial Reference Frame (ITRF), moves relative to the origin of the International Celestial Reference Frame (ICRF), over a variety of time scales (Sillard *et al.*, 1998). One can identify the geocenter variation as time series of the first-degree spherical harmonic coefficients of a geopotential model, i.e.,

$$\left\{ \begin{array}{c} \bar{C}_{11}(t) \\ \bar{S}_{11}(t) \\ \bar{C}_{10}(t) \end{array} \right\} = \frac{-1}{a\sqrt{3}} \left\{ \begin{array}{c} x_g(t) \\ y_g(t) \\ z_g(t) \end{array} \right\}. \quad (13)$$

Excited at a given temporal frequency Ω, the equatorial components of the geocenter variation are also expressed as $x_g + j y_g = A^+ \exp j(\theta^+ + \Omega t) + A^- \exp j(\theta^- - \Omega t)$, with the amplitudes (A^+, A^-) and phases (θ^+, θ^-) of the prograde $(+)$ and retrograde $(-)$ motions.

As early as the 1970 decade, before the establishment of suitable reference frames, and with limited surface observations, geophysicists estimated the magnitude of seasonal geocenter variations due to composite shifting of air mass, ocean and groundwater to be less than 1 cm (Stolz and Larden, 1979). Contemporary geophysicists produce the geocenter time-series by using much better quality geophysical surface observations. These include atmospheric surface pressure data, general ocean circulation models, global surface water/snow/ice information, mass balance data of glaciers, and tide gauge data. Recent studies (Dong *et al.*, 1997; Chen *et al.*, 1999) report that the lumped sum geocenter variations due to the surface mass load is indeed at the 1-cm level. They also indicate that the geocenter variations are predominantly at the annual and semiannual periods, and the variation due to the oceans is less than that of ground water but comparable to that of atmosphere.

Linking the satellite dynamics governed by celestial inertial laws and the crust-fixed satellite tracking stations influenced by the plate tectonics and other surface deformation, the ever-refined SLR data has provided one of the most precise methods for monitoring the geocenter variations. Kar (1997) presents several multi-satellite combination SLR solutions, and reports that the geocenter time-series are dominated by annual and semiannual variations. One of his solutions uses data from Lageos 1, Lageos 2, and T/P, and shows rms variations of 1–2, 2–5, and 7–9 mm in the prograde, retrograde, and axial components, i.e., A^+, A^-, and z_g, respectively. Cazenave *et al.* (1999) present a multi-satellite SLR/DORIS combination solution (SLR to Lageos 1 and Lageos 2, and DORIS from T/P), and report the annual excitation amplitudes of 2 mm for x_g and y_g, and 3.5 mm for z_g. They also report that the annual geocenter variation appears significantly larger than the annual polar motion, and the dynamic flattening variations, which are implied by changes in the pair of $\bar{C}_{21}$ and $\bar{S}_{21}$, and $\bar{C}_{20}$, respectively (Gegout and Cazenave, 1993).

The two entirely different approaches, i.e. the space-geodetic analysis and the geophysical surface data budget analysis, yield comparable result with the same level of uncertainties. The mature satellite altimetry now provides a unique way to investigate the role of oceans in the geocenter variations. The top panel in Figure 20 shows a 5-year geocenter variation history due to T/P-derived sea-surface height variability after removing thermal effects, compared with an SLR solution (Eanes *et al.*, 1997). We see correlation between the two completely independent solutions, especially in the equatorial geocenter components. However, the current geocenter uncertainty estimates are larger than the variations themselves, indicating that further work and longer-term data sets are needed to obtain more reliable geocenter time series, which is also an important element in the precise centering of satellite altimeter platforms and must be accounted for in the determination of multi-year analysis of altimeter measurements.

The earth's rotation has important dynamical effects on the fluid environment of the planetary system. As a deformable body subject to external torque, internal mass redistribution, and geophysical fluid dynamics, the earth's rotation is bound up with the angular momentum. As well as the free rotational Chandler wobble modes, the earth rotation manifests its irregularities in its rate and direction consistent with the exchange of angular momentum among the various material portions of the earth (Hide and Dickey, 1991). The earth rotation is governed by the Liouville equations of motion (Moritz and Mueller, 1987) that contain the earth's time-varying inertia tensor represented by the second-degree

spherical harmonic coefficients of the geopotential, i.e.,

$$
\left\{ \begin{array}{c} \bar{S}_{22}(t) \\ \bar{S}_{21}(t) \\ \bar{C}_{20}(t) \\ \bar{C}_{21}(t) \\ \bar{C}_{22}(t) \end{array} \right\} =
$$

$$
\frac{1}{a^2 M} \sqrt{\frac{3}{5}} \left\{ \begin{array}{c} -I_{xy}(t) \\ -I_{yz}(t) \\ \sqrt{\frac{1}{3}} \left[\frac{1}{2} I_{xx}(t) + \frac{1}{2} I_{yy}(t) - I_{zz}(t) \right] \\ -I_{xz}(t) \\ \frac{1}{2} I_{yy}(t) - \frac{1}{2} I_{xx}(t) \end{array} \right\}. \quad (14)
$$

Here the inertia tensor components are given as $I_{xy}(t) = \int (-xy) dM(t)$, $I_{xx}(t) = \int (y^2 + z^2) dM(t)$, etc., where $dM(t)$ implies the mass element at the material point (x, y, z) referring to the ITRF. In principle, one can recover the inertia tensor from the earth rotation parameters, and vice versa.

As the mission of the International Earth Rotation Service (IERS), the coordinated use of the Very Long Baseline Interferometry (VLBI), Lunar Laser Ranging (LLR), and satellite techniques (GPS, SLR, and DORIS) has been providing the reference data for the determination of the earth rotation. The data consist of time series of polar motion coordinates and the universal time (UT1). While variations of atmospheric angular momentum (de Viron *et al.*, 1999) are mirrored clearly in the time series, ocean dynamics and core-mantle dynamics are other important geophysical considerations as well. The core-mantle dynamics are poorly known, but manifest lower frequency characteristics related to long-term climate change via geomagnetic field variations (Jochmann and Greiner-Mai, 1996). Chao *et al.* (1995, 1996) report that the agreement between the IERS time series and the T/P altimetry-predicted diurnal and semidiurnal oceanic tidal angular momentum is as good as 10–30 micro-arc-second for polar motion and 2–3 μsec for UT1.

Johnson (1998) uses 9-year data product from the Parallel Ocean Climate Model (Semtner and Chervin, 1992) to investigate the oceanic role in the planetary angular momentum budget. This study reports that the non-tidal ocean variability, driven by wind fields and heat fluxes, can explain a significant amount of the previously unaccounted for variations in the earth rotation. The use of satellite altimeter data to infer the non-tidal variations in the earth rotation parameters is almost a brand new research activity. The bottom panel in Figure 20 shows T/P-derived polar motion and LOD variations (UT1–UTC) after removing seasonal variations, compared with space-geodetic time series (Gross,

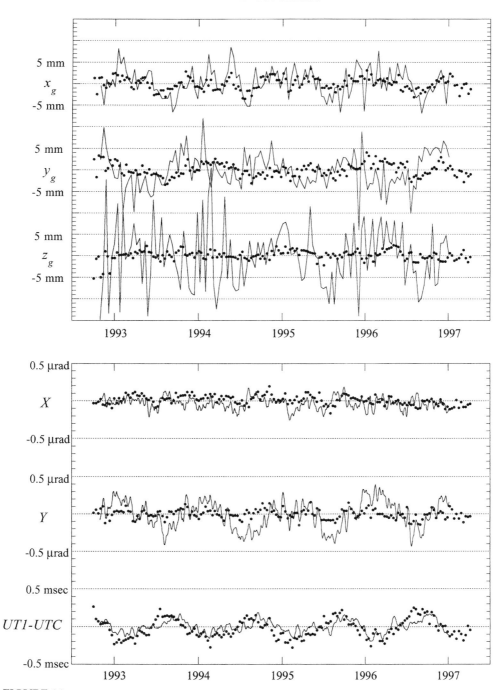

FIGURE 20 Geocenter time series (the top panel) and earth rotation parameter time series (the bottom panel) due to oceanic mass variations derived from TOPEX/POSEIDON measured sea-surface height variability (dots), compared to the Lageos 1 and 2 combination solutions (solid lines). Data used to generate these two panels are from Chen *et al.* (1999).

1996) after removing atmospheric and continental hydrological excitations. With ever-refined space-geodetic technologies, which include the GPS data for high-frequency analysis (Nam, 1999), the role of satellite altimetry in monitoring the earth rotation will manifest its full strength across a wide range of spectrum.

4.3. Roles of Satellite Gravity Data

After an extensive period of diverse concept studies, several dedicated space-geodetic gravity recovery missions are scheduled for the next decade. The DLR/GFZ CHAMP (A Challenging Micro-Satellite Payload for Geophysical Research and Applications) is planned for launch in 2000. The NASA/DLR GRACE (Gravity Recovery and Climate Experiment) mission is planned for launch in 2001. Other missions are also actively pursued, such as the ESA GOCE (Gravity Field and Steady-State Ocean Circulation Explorer) mission.

Opening a new frontier in studies of the solid earth and its fluid envelopes, these satellite gravity missions will address a wide range of issues in earth system science modeling. For instance, during the planned 5-year lifetime in a near-polar near-circular tandem orbit, the GPS-tracked satellite-to-satellite microwave tracking GRACE mission will sense the spatial and temporal variations of surface water storages, averaged over a few hundred kilometers and over a few weeks, to a sub-centimeter geoid undulation equivalent accuracy (Tapley and Reigber, 1998). For extensive discussions on the impact of the satellite gravity missions, one may refer to numerous presentations, articles, and reports (e.g., European Space Agency, 1996; National Research Council, 1997). Some prospective roles of such a mission warrant discussions here in connection with applications of satellite altimeter measurements to the mapping and monitoring of the static and time-varying gravity field.

An important use of the satellite gravity data will be the calibration of surface gravity anomaly data. Along with the notorious datum inconsistency in the terrestrial gravity data, both shipboard and altimeter-derived marine gravity databases have been also suffered from very long wavelength scale errors. Such errors in the shipboard gravity data can be attributed to the quality of calibration at navigation ports (Wessel and Watts, 1988), and to navigation errors especially for measurements collected prior to the GPS era. Errors in the altimeter-derived gravity data are mainly due to the lack of knowledge of the ocean dynamic topography. These errors are detectable by comparing the two independent data sets, but are not correctable in an absolute sense. Only a satellite gravity mission can provide absolute long-wavelength gravity information. A multi-year mean gravity field model derived from a satellite gravity mission will calibrate and validate the altimeter-derived marine gravity maps to make an unbiased gravimetric geoid that can be used to precisely separate the real mean ocean dynamic topography from the altimeter-derived mean sea-surface models.

Figure 21 illustrates some qualitative and quantitative characteristics of two geopotential models in terms of the gravity anomaly errors predicted from their error covariance matrices. The top panel is from one of the state-of-art models, EGM96, which is derived from diverse satellite tracking data sets as well as terrestrial, shipboard, and altimeter-derived surface gravity anomaly data sets (Lemoine et al., 1998). Note that many of the contour lines are along the coastlines, reflecting the hybrid characteristics of the data sets used to derive the field. The map unit is milli-Gal. In contrast, the bottom panel is obtained from a hypothetical gravity field model generated from a realistic simulation of the GRACE mission gravity recovery with its 200-km inter-satellite spacing and 400-km altitude. The unit of the gravity anomaly error map is now micro-Gal, implying orders of accuracy improvement in the recovery of the long-wavelength gravity field. One notes the dominance of zonal contour lines and some satellite ground-track patterns in equatorial areas, as typical error characteristics of a geopotential model derived from a single circumpolar orbiting gravity recovery mission. The satellite gravity missions, which manifest recovery of the highest accuracy gravity information at the longest wavelength scales, will certainly establish a global absolute datum of surface gravity data by improving the continuity of databases across all geopolitical boundaries and shorelines.

Along with the CHAMP and GRACE missions, the extension of T/P by the joint NASA/CNES JASON mission, and the extension of ERS 1 and 2 by the ESA ENVISAT mission, the year 2001 will see another very important satellite in geodesy. It is the ICESAT (Ice, Cloud and land Elevation Satellite) that will carry the GLAS (Geoscience Laser Altimeter System) and extend the domain of altimetric marine geodesy north and south to 86° latitudes. As an integral part of the NASA Earth Science Enterprise, GLAS has the objectives of measuring ice-sheet topography, and monitoring mass change in polar ice caps. In addition, operation of GLAS over land and water will provide topography profiles (with a 70-m surface spot size along the planned 183-day repeat 94° inclination frozen orbit ground-track). The GRACE and GLAS combination can separate the effects of ongoing changes in polar ice volume from those of visco-elastic crustal motion (Wahr et al., 1998a). Monitoring the volumetric changes of the Arctic and Antarctic polar ice sheets is of significant importance in interpreting the signals present in sea level rise. Despite recent impressive progresses achieved by using the current radar altimeter data (Davis et al., 1998; Wingham et al., 1998), the study of global scale glaciology is in its infancy. The GLAS will deliver many fruitful answers over the ice-covered regions, which still contain the most poorly understood areas in geodesy.

It is now almost two decades since Wunsch and Gaposchkin (1980) described the problem of estimating the general ocean circulation through the combination of satellite altimetry and hydrography, stating that "The critical problem is to understand the error budgets of four fields: orbit, height measurement, geoid, and ocean water density." Orbits and (altimeter) height measurements are now approaching 1-cm precision levels. However, despite the significant improvements made in estimates of the earth's static gravity field during that time frame, the marine geoid, along with the inadequate three-dimensional density distribution mapping of the oceans, remains as major unknowns in physical oceanography.

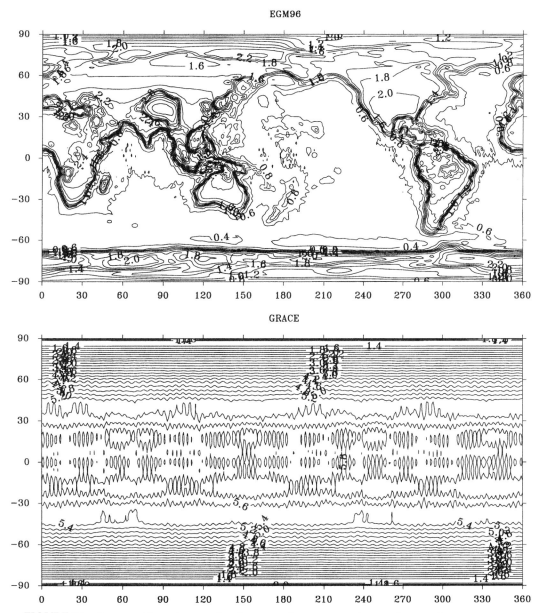

FIGURE 21 Current and future long-wavelength mean gravity error comparison: gravity anomaly error maps computed using error covariance matrices (complete to degree 70) of the EGM96 field (contour interval of 0.2 mGal), and a simulated GRACE field (contour interval of 0.2 μGal). The magnitude of the EGM96 gravity anomaly error has a minimum of 0.3 mGal at mid-oceans, a maximum of 3.9 mGal at the Tibetian Plateau, and global rms of 1.4 mGal (0.9 mGal over oceans, and 2.1 mGal over lands). The magnitude of the GRACE gravity anomaly error has a minimum of 1.2 μGal, a maximum of 6.1 μGal, and an rms value of 5.3 μGal.

Satellite altimetry is a global tool for determining the ocean circulation, only if the marine geoid is precisely known, and vice versa. As noted above, separate determination of the marine geoid and the mean ocean dynamic topography has been actually one of the most demanding research tasks in marine geodesy and physical oceanography. As knowledge of the marine geoid improves, so will our understanding of the ocean dynamics. In the intermediary period, oceanographers claim that considerable a priori knowledge of the ocean circulation already exists. Rather than the conventional gravimetric approach, Wunsch (1996) proposes an oceanographic approach that an accurate marine geoid can be constructed by using state-of-art ocean dynamic topography and altimetric mean sea-surface models. Ganachaud et al. (1997) explore estimates of global oceanic general circulation obtained from a geostrophic inversion of in-situ hydrographic data, T/P altimetry with the JGM 3 geoid, and a combination of the two. This study reports that

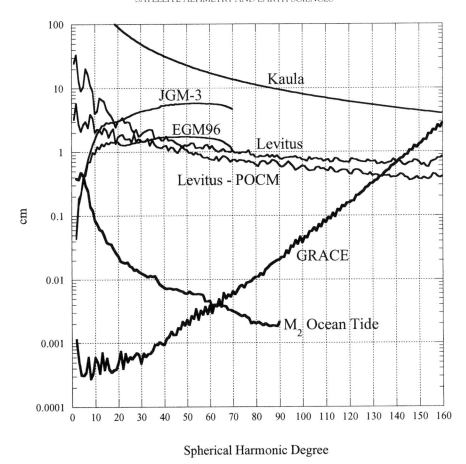

Spherical Harmonic Degree

FIGURE 22 Current and future marine geoid accuracy estimates, and the magnitude of ocean dynamic topography. A comparison of this nature can be found in several studies (Rapp *et al.*, 1996; Tapley *et al.*, 1996; National Research Council, 1997) to estimate the highest spherical harmonic degree for which the determination of ocean dynamic topography appears reasonable. The magnitude of M_2 ocean tide signal (Eanes and Bettadpur, 1996) is also shown. The EGM96 and JGM 3 marine geoid error curves meet the Levitus dynamic topography curve at degrees lower than 36 and 18, respectively, indicating that the current geoid accuracy is limited at resolutions shorter than wavelengths of 1000–2000 km. Note that the simulated future GRACE geoid will bring several factors of improvement in resolving the time-varying tide signal as well as the mean ocean dynamic topography, since the GRACE curve crosses the M_2 ocean tide and Levitus curves at degrees 60 and 140, respectively.

the addition of the T/P altimetry reduces the uncertainty of the ocean circulation results by only a small percent because of the significant JGM 3 geoid error magnitude relative to the purely oceanographic inferences.

The comparatively large geoid errors present in the current gravity field models limit the ability of satellite altimetry to improve the general ocean circulation models derived directly from in situ measurements (see Figure 22). Before the arrival of data from dedicated gravity missions, the best marine geoid model might be formulated by combining the gravimetric and oceanographic approaches. A satellite gravity mission, with capabilities of determining a 1-cm accuracy long-wavelength global geoid for well above the spherical harmonic degree 100, will clearly explain the origin of the significant differences between the current outputs

of numerical ocean models and satellite altimeter-derived ocean topography models. Identifications of the static and time-varying ocean dynamic topography, the static and time-varying marine geoid, and the steric and non-steric effects in the local and global sea level variations will allow the determination of seafloor pressure variations over the global oceans, thereby variable deep ocean currents could be inferred.

Satellite gravity data will substantially reduce the marine geoid as an error source in analyzing a broad spectrum of hydrological and oceanographic phenomena (Wahr *et al.*, 1998b). The mature satellite altimetry and long-term global monitoring of minute temporal variations in the gravity field from the gravity missions will assure the full opening of the new dimension in geodesy and provide a truly integrated view of the globe and its climate. Synthesis of satellite altimetry and satellite gravity data, and other complementary geophysical, oceanography, and climatology data will eventually demonstrate a global absolute monitoring of the time-evolving large-scale ocean circulation system—a primary objective of the Earth Observing System (Rothrock *et al.*, 1999).

5. CONCLUDING REMARKS

The coming years promise to be an exciting era in the field of geodesy. The reason is that the study of geodesy will be heralded by the synergetic advent of new-generation satellite altimeter missions and dedicated space-geodetic gravity recovery missions. As an experimental science, geodesy has continuously extended its domain. During the previous decades, satellite altimetry established the study of marine geodesy as an essential realm of earth sciences. On the other hand, today's geodesy needs a global multi-disciplinary approach for studying the minutely varying diverse space-time geodetic phenomena occurring in the total earth system. In the near future one could hardly approach geodesy without encountering globally integrated four-dimensional concepts.

It is true that many of the advances in geodesy have relied on the rapid improvements in geodetic instrumentation and the power of electronic computers for data acquisition and processing. In fact, we must elaborate many of our current geodetic theories as well, so the revolution in the ultra high precision new-generation technologies will not be limited by lack of analysis tools. The oceans are vast, deep, and opaque. We have to refine our tools to explore the vastness, to understand the deep structure, and to penetrate the opaqueness. In particular, the oceans are vividly active in the new dimension-raised domain. The ocean surface is the face of marine geodesy. The triumph of satellite altimetry to monitor and understand this face will depend not only on healthy satellite missions, but also on enthusiastic activities of dedicated scientists to refine theories, to advance models, and to convene better methods for deciphering the ever-increasing data in the virtually open space-time domain of integrated geodesy.

ACKNOWLEDGMENTS

Thanks are due to some help from Drs. J. L. Chen, M. K. Cheng, and R. J. Eanes. This work is supported from NASA under the Earth Observing System project (NAG5-6309) and the TOPEX/POSEIDON Precise Orbit Determination and Verification project (JPL 956689).

References

Andersen, O. B., and Knudsen, P. (1998). Global marine gravity field from the ERS 1 and Geosat geodetic mission altimetry. *J. Geophys. Res.*, **103**, 8129–8137.

Anzenhofer, M., and Gruber, T. (1995). MSS93A: A new stationary sea surface combining one year upgraded ERS 1 fast delivery data and 1987 Geosat altimeter data. *Bull. Geod.*, **69**, 157–163.

Balmino, G., Moynot, B., Sarrailh, M., and Vales, N. (1987). Free air gravity anomalies over the oceans from Seasat and GEOS 3 altimeter data. *EOS Trans.*, **68**, 17–19.

Basic, T., and Rapp, R. H. (1992). "Oceanwide prediction of gravity anomalies and sea surface heights using GEOS 3, Seasat and Geosat altimeter data and ETOPO5U bathymetric data." *DGSS Rep. 416*, Ohio State University, Columbus, OH.

Bhaskaran, S., and Rosborough, G. W. (1993). Computation of regional mean sea surfaces from altimetry data. *Manuscripta Geodaetica*, **18**, 147–157.

Born, G. H., Mitchell, J. L., and Heyler, G. A. (1987). Design of the Geosat exact repeat mission. *Johns Hopkins APL Tech. Dig.*, **8**, 260–266.

Brenner, A. C., Koblinsky, C. J., and Beckley B. D. (1990). A preliminary estimate of geoid-induced variations in repeat orbit satellite altimeter observations. *J. Geophys. Res.*, **95**, 3033–3040.

Brovelli, M. A., and Migliaccio, F. (1993). The direct estimation of the potential coefficients by biorthogonal sequences. *In* "Satellite Altimetry in Geodesy and Oceanography," 421–441, (Rummel, R. and Sanso, F. eds.), Springer-Verlag, Berlin, Heidelberg, New York.

Brunner, F. K., and Rizos, C., Eds., (1990). *In* "Developments in Four-Dimensional Geodesy" Springer-Verlag, Berlin, Heidelberg, New York.

Cazenave, A., Schaeffer, P., Berge, M., Brossier, C., Dominh, K., and Gennero, M. C. (1996). High-resolution mean sea surface computed with altimeter data of ERS 1 (geodetic mission) and TOPEX/POSEIDON. *Geophys. J. Int.*, **125**, 696–704.

Cazenave, A., Mercier, F., Bouille, F., and Lemonie, J. M. (1999). Global-scale interactions between the solid earth and its fluid envelopes at the seasonal time scale. *Earth Planet. Sci. Lett.*, **171**, 549–559.

Chambers, D. P., Tapley, B. D., and Stewart, R. H. (1998). Reduction of geoid gradient error in ocean variability from satellite altimetry. *Mar. Geod.*, **21**, 25–39.

Chao, B. F., Ray, R. D., and Egbert, G. D. (1995). Diurnal/semidiurnal oceanic tidal angular momentum: TOPEX/POSEIDON models in comparison with earth's rotation rate. *Geophys. Res. Lett.*, **22**, 1993–1996.

Chao, B. F., Ray, R. D., Gipson, J. M., Egbert, G. D., and Ma, C. (1996). Diurnal/semidiurnal polar motion excited by oceanic tidal angular momentum. *J. Geophys. Res.*, **101**, 20151–20163.

Chelton, D. B., and Schlax, M. G. (1993). Spectral characteristics of time-dependent orbit errors in altimeter height measurements. *J. Geophys. Res.*, **98**, 12579–12600.

Chen, J. L., Wilson, C. R., Eanes, R. J., and Nerem, R. S. (1999). Geophysical interpretation of observed geocenter variations. *J. Geophys. Res.*, **104**, 2683–2690.

Cheney, R. E., Douglas, B. C., Agreen, R. W., Miller, L., Porter, D. L., and Doyle, N. S. (1987). *In* "Geosat Altimeter Geophysical Data Record User Handbook," NOAA, Rockville, MD.

Cheng, M. K., Shum, C. K., and Tapley, B. D. (1997). Determination of long-term changes in the earth's gravity field from satellite laser ranging observations. *J. Geophys. Res.*, **102**, 22377–22390.

Cheng, M. K., and Tapley, B. D. (1999a). Seasonal variations in low degree zonal harmonics of the earth's gravity field from satellite laser ranging observations. *J. Geophys. Res.*, **104**, 2667–2681.

Cheng, M. K., and Tapley, B. D. (1999b). Observing the time-varying earth's gravity field from satellite tracking data. XXII General Assembly of IUGG, Birmingham, England.

Colombo, O. L. (1979). "Optimal estimation from data regularly sampled on a sphere with applications in geodesy," *DGSS Rep. 291*, Ohio State University, Columbus, OH.

Cullen, R., Moore, P., and Reynolds, M. (1997). Global altimetric mean sea surface derived from the geodetic phase of the ESA ERS 1 mission utilizing a spectral least-squares collocation technique. The 3rd ERS Symposium, Florence, Italy.

Davis, C. H., Kluever, C. A., and Haines, B. J. (1998). Elevation change of the southern Greenland ice sheet. *Science*, **279**, 2086–2088.

de Viron, O., Bizouard, C., Salstein, D., and Dehant V. (1999). Atmospheric torque on the earth and comparison with atmospheric angular momentum variations. *J. Geophys. Res.*, **104**, 4861–4875.

Dickman, S. R. (1998). Determination of oceanic dynamic barometer corrections to atmospheric excitation of earth rotation. *J. Geophys. Res.*, **103**, 15127–15143.

Dong, D., Dickey, J. O., Chao, Y., and Cheng, M. K. (1997). Geocenter variations caused by atmosphere, ocean and surface ground water. *Geophys. Res. Lett.*, **24**, 1867–1870.

Douglas, B. C., Agreen, R. W., and Sandwell, D. T. (1984). Observing global ocean circulation with Seasat altimeter data. *Mar. Geod.*, **8**, 67–83.

Eanes, R. J., Kar, S., Bettadpur, S. V., and Watkins, M. M. (1997). Low-frequency geocenter motion determined from SLR tracking, AGU Fall Meeting, San Francisco, CA.

Eanes, R. J., and Bettadpur, S. V. (1996). "The CSR 3.0 global ocean tide model" CSR-TM-96-05, The University of Texas at Austin.

Egbert, G. D., Bennett, A. F., and Foreman, M. G. G. (1994). TOPEX/POSEIDON tides estimated using a global inverse model. *J. Geophys. Res.*, **99**, 24821–24852.

European Space Agency (1996). "Gravity field and steady-state ocean circulation mission," ESA SP-1196(1), Noordwijk, The Netherlands.

Exertier, P. (1993). Geopotential from space techniques. *Cel. Mech. Dyn. Astron.*, **57**, 137–153.

Forsberg, R., and Tscherning, C. C. (1981). The use of height data in gravity field approximation by collocation. *J. Geophys. Res.*, **86**, 7843–7854.

Freeden, W., Gervens, T., and Schreiner, M. (1994). Tensor spherical harmonics and tensor spherical splines. *Manuscripta Geodaetica*, **19**, 70–100.

Freeden, W., and Schneider, F. (1998). An integrated wavelet concept of physical geodesy. *J. Geodesy*, **72**, 259–281.

Fu, L. L., Christensen, E. J., Yamarone, C. A., Lefebvre, M., Menard, Y., Dorrer, M., and Escudier, P. (1994). TOPEX/POSEIDON mission overview. *J. Geophys. Res.*, **99**, 24369–24381.

Ganachaud, A., Wunsch, C., Kim, M. C., and Tapley, B. D. (1997). Combination of TOPEX/POSEIDON data with a hydrographic inversion for determination of the oceanic general circulation and its relation to geoid accuracy. *Geophys. J. Int.*, **128**, 708–722.

Gaposchkin, E. M. (1974). Earth's gravity field to the eighteenth degree and geocentric coordinates for 104 stations from satellite and terrestrial data. *J. Geophys. Res.*, **79**, 5377–5411.

Gegout, P., and Cazenave, A. (1993). Temporal variations of the earth gravity field for 1985–1989 derived from Lageos. *Geophys. J. Int.*, **114**, 347–359.

Gelb, A., ed., (1974). *In* "Applied optimal estimation," MIT Press, Cambridge, Massachusetts.

Grafarend, E. W., Engels, J., and Varga, P. (1997). The spacetime gravitational field of a deformable body. *J. Geodesy*, **72**, 11–30.

Gross, R. S. (1996). Combinations of earth orientation measurements: SPACE94, COMB94, and POLE94. *J. Geophys. Res.*, **101**, 8729–8740.

Gruber, T., Anzenhofer, M., Rentsch, M., and Schwintzer, P. (1997). Improvements in high resolution gravity field modeling in GFZ, *In* "Gravity, geoid and marine geodesy" (Segawa, J., Fujimoto, H., and Okubo, S., eds.), *IAG Symposia*, **117**, 445–452, Springer-Verlag, Berlin, Heidelberg, New York.

Haxby, W. F., Karner, G. D., La Brecque, J. L., and Weissel, J. K. (1983). Digital images of combined oceanic and continental data sets and their use in tectonic studies. *EOS Trans.*, **64**, 995–1004.

Haxby, W. F. (1987). *In* "Gravity field of the world's oceans. A portrayal of gridded geophysical data derived from Seasat radar altimeter measurements of the shape of the ocean surface," National Geophysical Data Center, Boulder, CO.

Heiskanen, W. A., and Moritz, H. (1967). *In* "Physical geodesy," W. H. Freeman and Company, San Francisco and London.

Hide, R., and Dickey J. O. (1991). Earth's variable rotation. *Science*, **253**, 629–637.

Hotine, M. (1969). "Mathematical geodesy," *ESSA Monograph 2*, U. S. Dept. of Commerce, Washington, D.C.

Hwang, C. (1998). Inverse Vening Meinesz formula and deflection-geoid formula: applications to the predictions of gravity and geoid over the South China Sea. *J. Geodesy*, **72**, 304–312.

Hwang, C., Kao, E. C., and Parsons, B. (1998). Global derivation of marine gravity anomalies from Seasat, Geosat, ERS 1 and TOPEX/POSEIDON altimeter data. *Geophys. J. Int.*, **134**, 449–459.

Jekeli, C. (1988). The exact transformation between ellipsoidal and spherical harmonic expansions. *Manuscripta Geodaetica*, **13**, 106–113.

Jekeli, C. (1999). An analysis of vertical deflections derived from high-degree spherical harmonic models. *J. Geodesy*, **73**, 10–22.

Jochmann, H., and Greiner-Mai, H. (1996). Climate variations and the earth's rotation. *J. Geodynamics*, **21**, 161–176.

Johnson, T. J. (1998). *In* "The role of the ocean in the planetary angular momentum budget," CSR-98-2, The University of Texas at Austin.

Jolly, G. W., and Moore, P. (1994). Validation of empirical orbit error corrections using crossover difference differences. *J. Geophys. Res.*, **99**, 5237–5248.

Jolly, G. W., and Moore, P. (1996). Analysis of global ERS 1 altimetry by optimal Fourier transform interpolation. *Mar. Geod.*, **19**, 331–357.

Kar, S. (1997). *In* "Long-period variations in the geocenter observed from laser tracking of multiple satellites," CSR-97-2, The University of Texas at Austin.

Kaula, W. M. (1966). *In* "Theory of Satellite Geodesy" Blaisdell Publishing Company, Waltham, Massachusetts.

Keller, W. (1996). On a scalar fixed altimetry-gravimetry boundary value problem. *J. Geodesy*, **70**, 459–469.

Kim, M. C. (1993). *In* "Determination of high resolution mean sea surface and marine gravity field using satellite altimetry," CSR-93-2, The University of Texas at Austin.

Kim, M. C. (1997). Theory of satellite ground-track crossovers. *J. Geodesy*, **71**, 749–767.

Kim, M. C., and Tapley, B. D. (2000). Formation of surface spherical harmonic normal matrices and application to high degree geopotential modeling. *J. Geodesy*, **74**, 359–375.

Klees, R., Ritter, S., Lehmann, R. (1997). Integral equation formulations for geodetic mixed boundary value problems. *DEOS Progr. Lett.*, **97.1**, 1–8, Delft University of Technology, Delf, The Netherlands.

Klokocnik, J., Wagner, C. A., and Kostelecky, J. (1999). Spectral accuracy of JGM 3 from satellite crossover altimetry. *J. Geodesy*, **73**, 138–146.

Konopliv, A. S., Binder, A. B., Hood, L. L., Kucinskas, A. B., Sjogren, W. L., and Williams, J. G. (1998). Improved gravity filed of the moon from lunar prospector. *Science*, **281**, 1476–1480.

Kozel, B. J., Shum, C. K., Ries, J. C., and Tapley, B. D. (1994). Precision orbit determination using dual satellite crossover measurements. AAS/AIAA Astrodynamics Conf., Paper 94–181.

Laxon, S., and McAdoo, D. (1994). Arctic ocean gravity field derived from ERS 1 satellite altimetry. *Science*, **265**, 621–624.

Lemoine, F. G., *et al.*, (1998). "The development of the Joint NASA GSFC and the National Imagery and Mapping Agency geopotential model EGM96" NASA TP 206861, NASA GSFC, Greenbelt, MD.

Lerch, F. J., Marsh, J. G., Klosko, S. M., Pavlis, E. C., Patel, G. B., Chinn, D. S., and Wagner, C. A. (1988). *In* "An improved error assessment for the GEM T1 gravitational model," NASA TM 100713, NASA GSFC, Greenbelt, MD.

Levitus, S., Burgett, R., and Boyer, T. (1994). *In* "World Ocean Atlas 1994," NOAA and U. S. Department of Commerce, Washington, D.C.

Lillibridge, J., and Cheney, B. (1997). "The Geosat Altimeter JGM 3 GDRs on CD-ROM" NOAA and U. S. Department of Commerce, Washington, D.C.

Liu, L. T., Hsu, H. T., and Gao, B. X. (1998). A new family of orthonomal wavelet bases. *J. Geodesy*, **72**, 294–303.

Mainville, A. (1986). The altimetry-gravimetry problem using orthonomal base functions, *DGSS Rep. 373*, Ohio State University, Columbus, OH.

Marsh, J. G., and Martin, T. V. (1982). The Seasat altimeter mean sea surface model. *J. Geophys. Res.*, **87**, 3269–3280.

Marsh J. G., *et al.*, (1988). A new gravitational model for the earth from satellite tracking data: GEM T1. *J. Geophys. Res.*, **93**, 6169–6215.

Marsh, J. G., Koblinsky, C. J., Zwally, H. J., Brenner, A. C., and Beckley, B. D. (1992). A global mean sea surface based upon GEOS 3 and Seasat altimeter data. *J. Geophys. Res.*, **97**, 4915–4921.

Mather, R. S. (1978). The role of the geoid in four-dimensional geodesy. *Mar. Geod.*, **1**, 217–252.

Maus, S., Green, C. M., and Fairhead, J. D. (1998). Improved ocean-geoid resolution from retracked ERS 1 satellite altimeter waveforms. *Geophys. J. Int.*, **134**, 243–253.

McAdoo, D., and Laxon, S. (1997). Antarctic Tectonics: constraints from an ERS 1 satellite marine gravity field. *Science*, **276**, 556–560.

McGoogan, J. T., Miller, L. S., Brown, G. S., and Hayne, G. S. (1974). The S-193 radar altimeter experiment. *Proc. IEEE*, **62**, 793–803.

Moore, P., Ehlers, S., and Carnochan, S. (1998). Accuracy assessment and refinement of the JGM 2 and JGM 3 gravity fields for radial positioning of ERS 1. *J. Geodesy*, **72**, 373–384.

Moritz, H. (1980). *In* "Advanced physical geodesy." Helbert Wichmann Verlag, Karlsruhe, Germany.

Moritz, H., and Mueller, I. I. (1987). *In* "Earth rotation, theory and observation." The Ungar Publishing Company, Ungar, New York.

Nam, Y. S. (1999). *In* "GPS determination of diurnal and semidiurnal variations in earth rotation parameters and the geocenter," CSR-99-2, The University of Texas at Austin.

National Research Council (1997). *In* "Satellite Gravity and the Geosphere." National Academy Press, Washington, D. C.

Nerem, R. S., *et al.*, (1994). Gravity model development for TOPEX/POSEIDON: Joint Gravity Models 1 and 2. *J. Geophys. Res.*, **99**, 24421–24447.

Nerem, R. S. (1995). Terrestrial and planetary gravity fields, *In* "Reviews of geophysics, Supplememt, U. S. National Report to IUGG 1991–1994." 469–476.

Olea, R. A. (1974). Optimal contour mapping using universal kriging. *J. Geophys. Res.*, **79**, 695–702.

Olgiati, A., Balmino, G., Sarrailh, M., and Green, C. M. (1995). Gravity anomalies from satellite altimetry: comparison between computation via geoid heights and via deflections of the vertical. *Bull. Geod.*, **69**, 252–260.

Press, W. H., Teukolsky, S. A., Vetterling, W. T., and Flannery, B. P. (1992). *In* "Numerical recipes in FORTRAN," 2nd ed. Cambridge University Press, Cambridge.

Rapp, R. H. (1983). The determination of geoid undulations and gravity anomalies from Seasat altimeter data. *J. Geophys. Res.*, **88**, 1552–1562.

Rapp, R. H. (1986). Gravity anomalies and sea surface heights derived from a combined GEOS 3/Seasat altimeter data set. *J. Geophys. Res.*, **91**, 4867–4876.

Rapp, R. H., and Basic, T. (1992). Oceanic gravity anomalies from GEOS 3, Seasat, and Geosat altimeter data. *Geophys. Res. Lett.*, **19**, 1979–1982.

Rapp, R. H., and Cruz, J. Y. (1986). *In* "Spherical harmonic expansions of the earth's gravitational potential to degree 360 using 30' mean anomalies," *DGSS Rep. 376*, Ohio State University, Columbus, OH.

Rapp, R. H., Nerem, R. S., Shum, C. K., Klosko, S. M., and Williamson, R. G. (1991a). *In* "Consideration of permanent tidal deformation in the orbit determination and data analysis for the TOPEX/POSEIDON mission," NASA TM 100775, NASA GSFC, Greenbelt, MD.

Rapp, R. H., Wang, Y. M., and Pavlis, N. K. (1991b). *In* "The Ohio State 1991 geopotential and sea surface topography harmonic coefficient models," *DGSS Rep. 410*, Ohio State University, Columbus, OH.

Rapp, R. H., Zhang, C., and Yi, Y. (1996). Analysis of dynamic ocean topography using TOPEX data and orthonomal functions. *J. Geophys. Res.*, **101**, 22583–22598.

Ricard, Y., Richards, M., Lithgow-Bertelloni, C., and Stunff, Y. L. (1993). A geodynamic model of mantle density heterogeneity. *J. Geophys. Res.*, **98**, 21895–21909.

Ries, J. C., Eanes, R. J., Shum, C. K., Watkins, M. M. (1992). Progress in the determination of the gravitational coefficient of the earth, *Geophys Res. Lett.*, **19**, 529–531.

Ries, J. C., Shum, C. K., and Tapley, B. D. (1993). Surface force modeling for precision orbit determination, *In* "Environmental effects on spacecraft positioning and trajectories," *Geophys. Monogr.*, 73, IUGG V13, 111–124.

Rothrock, D. A., *et al.*, (1999). Ocean circulation, productivity, and exchange with the atmosphere, *In* "EOS Science Plan" (King, M. D., ed.), 115–160, NASA GSFC, Greenbelt, MD.

Sandwell, D. T. (1991). Geophysical applications of satellite altimetry. *In* "Reviews of geophysics. Supplement, U. S. National Report to IUGG 1987–1990," 132–137.

Sandwell, D. T., and Smith, W. H. F. (1997). Marine gravity anomaly from Geosat and ERS 1 satellite altimetry. *J. Geophys. Res.*, **102**, 10039–10054.

Sanso, F. (1993). Theory of geodetic B.V.P.s applied to the analysis of altimeter data. *In* "Satellite Altimetry in Geodesy and Oceanography," (Rummel, R. and Sanso, F., eds.), 317–371, Springer-Verlag, Berlin, Heidelberg, New York.

Sanso, F., and Rummel, R., eds., (1997). *In* "Geodetic Boundary Value Problems in View of the One Centimeter Geoid," Springer-Verlag, Berlin, Heidelberg, New York.

Sarrailh, M., Balmino, G., and Doublet, D. (1998). The Arctic and Antarctic oceans gravity field from ERS 1 altimeter data. *In* "Gravity, Geoid and Marine Geodesy" (Segawa, J., Fujimoto, H., and Okubo, S., eds.). *IAG Symposia*, **117**, 437–444, Springer-Verlag, Berlin, Heidelberg, New York.

Schaeffer, P., Hernandez, F., Le Traon, P.-Y., Mertz, F., and Bahurel, P. (1998). *In* "A mean sea surface dedicated to ocean studies: global estimation." EGS, Nice, France.

Schrama, E. J. O. (1992). Some remarks on several definitions of geographically correlated orbit errors: consequences for satellite altimetry. *Manuscripta Geodaetica*, **17**, 282–294.

Schutz, B. E. (1998). Spaceborne laser altimetry: 2001 and beyond. *In* "WEGENER 98" (Plag, H. P., ed.), 7–10, Statens Kartverk, Honefoss, Norway.

Semtner, A. J., and Chervin R. M. (1992). Ocean general circulation from a global eddy-resolving model, *J. Geophys.*, **97**, 5493–5550.

Shum, C. K. (1983). *In* "Altimeter methods for satellite geodesy," CSR-83-2, The University of Texas at Austin.

Sillard, P., Altamimi, Z., and Boucher, C. (1998). The ITRF96 realization and its associated velocity field. *Geophys. Res. Lett.*, **25**, 3223–3226.

Smith, A. J. E., Hesper, E. T., Kuijper, D. C., Mets, G. J., Visser, P. N. A. M., Ambrosius, B. A. C., and Wakker, K. F. (1996). TOPEX/POSEIDON orbit error assessment. *J. Geodesy.*, **70**, 546–553.

Stammer, D., Tokmakian, R., Semtner, A., and Wunsch, C. (1996). How well does a 1/4° global circulation model simulate large-scale oceanic observations? *J. Geophys. Res.*, **101**, 25779–25811.

Stolz, A., and Larden, D. R. (1979). Seasonal displacement and deformation of the earth by the atmosphere. *J. Geophys. Res.*, **84**, 6185–6194.

Tai, C. K. (1988). Geosat crossover analysis in the tropical pacific, 1. Constrained sinusoidal crossover adjustment. *J. Geophys. Res.*, **93**, 10621–10629.

Tapley, B. D., and Reigber, C. (1998). GRACE: A satellite-to-satellite tracking geopotential mapping mission, AGU Fall Meeting, San Francisco, CA.

Tapley, B. D., and Rosborough, G. W. (1985). Geographically correlated orbit error and its effect on satellite altimetry missions. *J. Geophys. Res.*, **90**, 11817–11831.

Tapley, B. D., Shum, C. K., Yuan, D. N., Ries, J. C., Watkins, M. M., and Schutz, B. E (1991). The University of Texas earth gravity field model. XX General Assembly of IUGG, Vienna, Autria.

Tapley, B. D., *et al.*, (1994). Precision orbit determination for TOPEX/POSEIDON. *J. Geophys. Res.*, **99**, 24383–24404.

Tapley, B. D., *et al.*, (1996). The Joint Gravity Model 3. *J. Geophys. Res.*, **101**, 28029–28049.

Tapley, B. D., Shum, C. K., Ries, J. C., Poole, S. R., Abusali, P. A. M., Bettadpur, S. V., Eanes, R. J., Kim, M. C., Rim, H. J., and Schutz, B. E. (1997). The TEG 3 geopotential model. *In* "Gravity, Geoid and Marine Geodesy." (Segawa, J., Fujimoto, H., and Okubo, S., eds.) *IAG Symposia*, **117**, 453–460, Springer-Verlag, Berlin, Heidelberg, New York.

Thompson, J. D., Born, G. H., and Maul, G. A. (1983). Collinear-track altimetry in the Gulf of Mexico from Seasat: measurements, models, and surface truth. *J. Geophys. Res.*, **88**, 1625–1636.

Tsaoussi, L. S., and Koblinsky, C. J. (1994). An error covariance model for sea surface topography and velocity derived from TOPEX/POSEIDON altimetry. *J. Geophys. Res.*, **99**, 24669–24683.

Turcotte, D. L. (1989). Fractals in geology and geophysics. *In* "Fractals in Geophysics." (Scholz, C. H., and Mandelbrot, B. B., eds.), 171–196, Birkhauser Verlag, Basel.

van Dam, T. M., Wahr, J., Chao, Y., and Leuliette, E. (1997). Predictions of crustal deformation and of geoid and sea-level variability caused by oceanic and atmospheric loading. *Geophys. J. Int.*, **129**, 507–517.

van Gysen, H., and Coleman, R. (1997). On the satellite altimeter crossover problem. *J. Geodesy*, **71**, 83–96.

van Gysen, H., and Coleman, R. (1999). On the analysis of repeated geodetic experiments. *J. Geodesy*, **73**, 237–245.

van Gysen, H., and Merry, C. L. (1989). Towards a cross-validated spherical spline geoid for the south-western cape, South Africa. *In* "Sea Surface Topography and the Geoid." (Sunkel, H., and Baker, T., eds.) *IAG Symposia*, **104**, 53–60, Springer-Verlag, Berlin, Heidelberg, New York.

Vanicek, P., and Christou, N. T., eds. (1994). *In* "Geoid and Its Geophysical Interpretations." CRC Press Inc., Boca Raton.

Wagner, C. A., Klokocnik, J., and Kostelecky, J. (1997). Dual-satellite crossover latitude-lumped coefficients, their use in geodesy and oceanography. *J. Geodesy*, **71**, 603–616.

Wagner, C. A., and Tai, C. K. (1994). Degradation of ocean signals in satellite altimetry due to orbit error removal processes. *J. Geophys. Res.*, **99**, 16255–16267.

Wahr, J. M., Bentley, C., and Wingham, D. (1998a). What can be learned about Antarctica by combining GRACE and ICESAT (GLAS) satellite data? AGU Fall Meeting, San Francisco, CA.

Wahr, J. M., Molenaar, M., and Bryan, F. (1998b). Time variability of the earth's gravity field: hydrological and oceanic effects and their possible detection using GRACE. *J. Geophys. Res.*, **103**, 30205–30229.

Wang, Y. M. (1998). A high-resolution mean sea surface from satellite altimeter data and its accuracy assessment. AGU Fall Meeting, San Francisco, CA.

Wang, Y. M., and Rapp, R. H. (1992). The determination of a one-year mean sea surface height track from Geosat altimeter data and ocean variability implications. *Bull. Geod.*, **66**, 336–345.

Wenzel, H. G. (1998). Ultra high degree geopotential model GPM3E97A to degree and order 1800 tailored to Europe. The 2nd Continental Workshop on the Geoid in Europe, Budapest.

Wessel, P., Watts, A. B. (1988). On the accuracy of marine gravity measurements. *J. Geophys. Res.*, **93**, 393–413.

Wingham, D. J., Ridout, A. J., Scharroo, R., Arthern, R. J., and Shum, C. K. (1998). Antarctic elevation change from 1992 to 1996. *Science*, **282**, 456–458.

Wunsch, C. (1993). Physics of the ocean circulation. *In* "Satellite Altimetry in Geodesy and Oceanography," (Rummel, R., and Sanso, F., eds.), 9–98, Springer-Verlag, Berlin, Heidelberg, New York.

Wunsch, C. (1996). Requirements on marine geoid accuracy for significant improvement in knowledge of the ocean circulation. AGU Spring Meeting, Baltimore, MD.

Wunsch, C., and Gaposchkin, E. M. (1980). On using satellite altimetry to determine the general circulation of the oceans with application to geoid improvement. *Rev. of Geophys. Space Phys.*, **18**, 725–745.

Wunsch, C., and Stammer, D. (1995). The global frequency-wavenumber spectrum of oceanic variability estimated from TOPEX/POSEIDON altimetric measurements. *J. Geophys. Res.*, **100**, 24895–24910.

Wunsch, C., and Zlotnicki, V. (1984). The accuracy of altimetric surfaces. *Geophys. J. R. Astr. Soc.*, **78**, 795–808.

Yale, M. M., Sandwell, D. T., and Smith, W. H. F. (1995). Comparison of alongtrack resolution of stacked Geosat, ERS 1, and TOPEX satellite altimeters. *J. Geophys. Res.*, **100**, 15117–15127.

Yale, M. M. (1997). *In* "Modelling upper mantle rheology with numerical experiments and mapping marine gravity with satellite altimetry," Ph.D. Thesis, University of California at San Diego, Scripps Inst. of Oceanography.

Yi, Y. (1995). "Determination of gridded mean sea surface from TOPEX, ERS 1 and Geosat altimeter data," *DGSS Rep. 434*, Ohio State University, Columbus, OH.

Yuan, D. N. (1991). "The determination and error assessment of the earth's gravity field model," CSR-91–1, The University of Texas at Austin.

Zhang, C., Sideris, M. G. (1996). Oceanic gravity by analytical inversion of Hotine's formula. *Mar. Geod.*, **19**, 115–136.

11

Applications to Marine Geophysics

ANNY CAZENAVE* and JEAN YVES ROYER†

*Laboratoire d'Etudes en Geophysique et Oceanographie Spatiales
Centre National d'Etudes Spatiales
18, Av. Edouard Belin
31401 Toulouse Cedex 4
France
†UBO-IUEM Domaines Océaniques
Place Nicolas Copernic—29280 Plouzane
France

1. INTRODUCTION

In the mid 1970s, the Geos3 and Seasat missions successfully demonstrated the interest of radar altimeter measurements for mapping undulations of the sea surface. Recovery of the time variable ocean surface signal of interest for oceanography, began only a decade later with the Geosat mission and especially since the early 1990s with the ERS and TOPEX-POSEIDON missions. Thus during nearly 15 years altimetry missions have mostly served marine geophysicists who abundantly used altimetry-derived geoid observations over the oceanic domain to study the oceanic lithosphere and the mantle-lithosphere interactions. The high-density data set acquired by the ERS-1 satellite altimeter during its geodetic mission (April 1994 to March 1995) let to a new step in the use of geoid data for marine geophysics. Indeed together with the dense Geosat altimeter data collected ten years earlier but classified for a decade, this new data set provided a detailed view of the marine gravity field with a resolution better than 5 km everywhere, opening new perspectives for global studies in marine geophysics. Figure 1 (see color insert) presents a map of the marine geoid based on the dense Geosat and ERS-1 altimeter data. This map shows in great details the gravitational signature of submarine tectonic features and their isostatic compensation. In addition to the well-known geoid anomalies associated with seamounts, volcanic chains, fracture zones and deep-sea trenches, the map reveals details of the mid-ocean ridge segmentation, off-axis V-shaped structures and many other features such as microplates, lineated patterns, in particular in the Pacific ocean, or fossil plate boundaries (extinct spreading ridges, extinct triple junctions, etc.).

The altimetry-derived geoid data have been widely used since 2 decades to model the thermal and mechanical structure of the oceanic lithosphere, to study the interaction between the convecting upper mantle with the lithosphere, spreading ridges, hotspots and in some off-ridge regions to study plate kinematics and reconstruct past motions of tectonic plates. These observations have no doubt greatly improved our knowledge of the dynamics of the lithosphere and upper mantle.

In this chapter, we first discuss classical compensation models usually considered in order to interpret geoid anomalies over marine features and then we review the various areas which have benefited over the past 2 decades of high-accuracy and high-resolution altimeter geoid data. Topics discussed here concern: seamount loading and the mechanical structure of oceanic plates, thermal evolution of the oceanic lithosphere, hotspot swells, geoid lineation patterns, plate kinematics, seamount production, and mid-ocean ridge segmentation.

2. FILTERING THE LONG-WAVELENGTH GEOID SIGNAL

Geoid anomalies associated with most submarine tectonic features have wavelengths typically shorter than $\approx$ 3000 km. To extract the geoid signal from altimetry data, it is first necessary to remove the long-wavelength geoid associated with density variations occurring in the deep mantle (Hager and Richards, 1989; Hager and Clayton, 1989). A widely applied approach consists of subtracting from the altimeter geoid, a reference long-wavelength geoid N based on geopotential solutions given by:

$$N(R, \varphi, \lambda) = R\left[\sum_{l=2}^{\infty}\sum_{m=0}^{l}\{[C_{lm}\cos m\lambda + S_{lm}\sin m\lambda]\right.$$
$$\left.\times P_{lm}(\sin\varphi)\}\right] \qquad (1)$$

ϕ, λ are latitude and longitude, R is earth's mean radius. C_{lm}, S_{lm} are the Stokes' coefficients related to the integrated mass distribution inside the earth (Heiskanen and Moritz, 1967).

Current geopotential models give sets of C_{lm}, S_{lm} coefficients up to a maximum degree and order. Recent models such as JGM-3 or GRIM4-C4, are complete to degree 70 (e.g., Tapley $et\ al.$, 1996; Schwintzer $et\ al.$, 1997). The associated geographical wavelength Λ of a geoid undulation is related to the maximum degree 1 through $\Lambda = 2\pi R/(l(l+1))^{1/2}$.

Many studies have provided medium-wavelength geoid maps based on satellite altimetry after removing the long wavelength signal using Eq. (1) developed up to a given degree expansion. However, abrupt truncation of Eq. (1) introduces Gibbs oscillations in the residual geoid [observed geoid minus truncated geoid given by Eq. (1)]. Sandwell and Renkin (1988) proposed minimizing these artifacts by tapering the higher degrees of the spherical harmonics in Eq. (1), prior to summation. The C_{lm}, S_{lm} coefficients are the multiplied by a weighting function such as a Gaussian or a cosine function. Applied to harmonics between degrees l_1 and l_2, this approach leaves in principle unattenuated geoid anomalies of wavelength shorter than $\Lambda = 2\pi R/(l_2(l_2+1))^{1/2}$.

Another approach consists of applying a 2-D high-pass filter to gridded geoid data. Several kinds of 2-D filters may be implemented. A 2-D filter based on inverse methods (Tarantola, 1987) in which the long-wavelength geoid is modelled as a sum of Gaussian functions centered at the data points proved to be a useful approach (e.g., Cazenave $et\ al.$, 1992, 1996). 2-D filtering is by far a better method than subtracting a reference geopotential model. This is so because except for the very low degree harmonics which are accurately determined, large errors still affect geopotential solutions beyond degrees 10–20. For example, at degree 10

(equivalent wavelength of $\sim$4000 km), the cumulative error on the geoid is $\sim$10 cm. The higher the degree, the larger the error.

Besides geoid anomalies, other gravitational informations may be used for geophysical interpretation such as altimetry-derived gravity anomalies (e.g., Sandwell and Smith, 1997; Hwang $et\ al.$, 1998) or deflection of the vertical. These observations enhance short wavelength gravitational signal, hence may be preferred in a number of studies. However, since they poorly reproduce the medium-wavelength (1000–2000 km) gravity signal, they are not very useful for studies related to hotspot swells or upper mantle convection. In the latter case, geoid data may be preferred.

3. GEOID ANOMALIES AND ISOSTATIC COMPENSATION

The geoid represents the integrated mass distribution over the volume of the earth. For this reason, geoid data alone cannot inform on the lateral density structure. In most instances, however, geoid anomalies are associated with other geophysical anomalies, in particular topography. Used together, these observations are able to constrain plausible models of earth's internal structure. In the framework of the classical concept of isostasy, geophysical applications of satellite altimetry simultaneously use geoid and topography information. Isostasy assumes that loads on (or inside) the earth are compensated by internal density variations such that, at depth, pressure is hydrostatic. The depth above which density variations are confined is sometimes called the compensation depth although its definition is not unique. Isostasy may be understood in terms of mass conservation, minimization of strain energy, and mechanical equilibrium (Dahlen, 1982). Various mechanisms are able to insure isostatic equilibrium: crustal thickening, thermal expansion or contraction of mantle rocks, thermal thinning, plate flexure, etc. Dynamic compensation is often opposed to static compensation and assigned to convection. Convective stresses produce deformation of mantle interfaces, in particular of the earth surface, giving rise to geoid anomalies. Internal density variations associated with thermal anomalies produce geoid anomalies of opposite sign and of magnitude which depends in a complex manner on the mantle stratification and viscosity structure. As for static compensation, observed geoid is the net effect between these opposing effects.

Geoid anomalies over marine tectonic structures have been classically interpreted through simple isostatic models. Most topographic loads of wavelengths <50 km are supported by the strength of the lithosphere and are uncompensated. At wavelengths longer than $\sim$500 km, topography is in general locally compensated. At intermediate wavelengths (50–500 km) most loads are compensated by elastic flexure of the upper lithosphere.

3.1. Local Compensation in the Long-Wavelength Approximation

Ockendon and Turcotte (1977), and Haxby and Turcotte (1978) derived useful expressions for geoid anomalies caused by two-dimensional density variations within a layer of horizontal scale large compared to its thickness. Under the assumption of isostatic compensation (i.e., local density variations are in hydrostatic equilibrium), the geoid anomaly is given by:

$$N = -\frac{2\pi G}{g} \int_0^H z \Delta\rho(x, z)\, dz \qquad (2)$$

where N is geoid height, z is depth positive downward, $\Delta\rho(x, z)$ is the 2-D density variation occurring between the surface $z = 0$ and the base of the layer at $z = H$. G is the gravitational constant, and g is the mean surface gravity. Eq. (2) assumes that $\Delta\rho(x, z)$, integrated between $z = 0$ and $z = H$, is equal to zero.

The above relation may be applied to the classical Airy and Pratt isostasy models. For an Airy compensation, the surface topography is compensated by a density contrast at the base of the crust (thickened crust). For a Pratt compensation, the surface topography overlays a layer of constant thickness but of variable density. Compensation occurs at the Moho in the Airy model and at the base of the lithosphere in the Pratt model (Figures 2a and 2b). Application of Eq. (2) in these two situations gives:

Airy model:

$$N = -\frac{\pi G}{g}(\rho_c - \rho_w)\left[2h(z_c - z_w) + \frac{(\rho_m - \rho_w)}{(\rho_m - \rho_c)}h^2\right] \qquad (3)$$

Pratt model:

$$N = -\frac{\pi G}{g}(\rho_m - \rho_w)Hh \qquad (4)$$

where ρ_w, ρ_c, ρ_m are, respectively, seawater, crust, and mantle densities, h is the height of the topography (above the surrounding seafloor), z_w and z_c are the seafloor and Moho

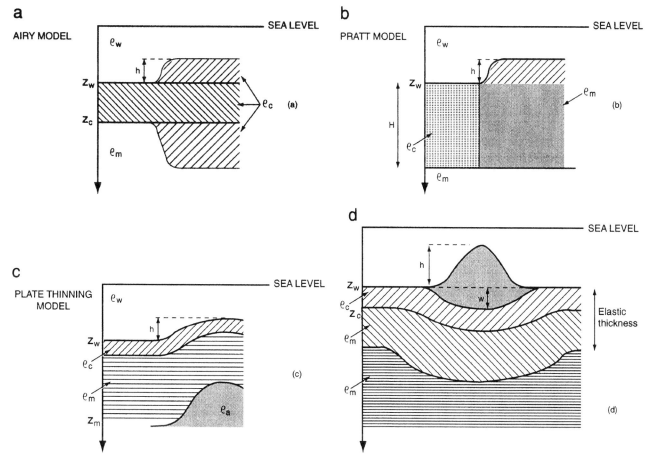

FIGURE 2 Schematic diagrams of isostatic compensation models: (a) Airy compensation; (b) Pratt compensation; (c) Lithospheric thinning compensation; (d) Regional compensation by plate flexure.

depth below sea level, and H is the layer (lithosphere) thickness.

The Airy model has been extended to a two-layer model (Marks and Sandwell, 1991). In the latter model, the base of the crust is underlain by a thin layer of depleted mantle (of density ρ_e, slightly lower than the normal mantle density ρ_m). This model was developed to explain the topography and geoid anomalies over oceanic plateaus not associated with hotspots. The corresponding geoid anomaly is:

$$N = \frac{2\pi G}{g}(\rho_c - \rho_w)h\left[z_c - z_w \right.$$
$$+ \frac{h}{2}[1 + 2\beta - \beta^2]\frac{(\rho_c - \rho_w)}{(\rho_m - \rho_c)}$$
$$\left. + (1 - \beta^2)\frac{(\rho_c - \rho_w)}{(\rho_m - \rho_c)} \right] \quad (5)$$

β is a parameter ranging from 0–1 expressing the fraction of topography compensated by the one layer Airy expression. Eq. (5) reduces to Eq. (2) for $\beta = 1$.

Eq. (2) can also be applied for modeling cooling of the oceanic lithosphere.

The temperature structure $T(x, z)$ inside the oceanic lithosphere is well described by conductive heat transfer models (see Section V). The inferred density structure $\rho(x, z)$ is related to $T(x, z)$ through the equation of state:

$$\rho(x, z) = \rho_m[1 + \alpha(T_m - T(x, z))] \quad (6)$$

where T_m is the average temperature of the underlying mantle and α is the volume coefficient of thermal expansion. The topography of the seafloor is directly related to Eq. (6) if isostatic equilibrium (mass conservation) is assumed. The conductive half-space model (Turcotte and Oxburgh, 1967) predicts a subsidence of the seafloor, proportional to the square root of the distance to mid ocean ridge (or to the square root of seafloor age if the plate is assumed to move at a constant velocity) whereas in the plate model (McKenzie, 1967), the seafloor depth asymptotically approaches a constant value at large age. The latter model assumes that the lithosphere has a constant thickness H unlike the half space model which predicts that the lithosphere continues to thicken at all ages.

The geoid anomaly [Eq. (2)] caused by the temperature distribution inside the cooling plate (assuming the ridge crest as a reference and d being the seafloor depth, positive downwards) is:

$$N(t) = -\frac{2\pi G}{g}\left[\frac{(\rho_m - \rho_w)}{2}d^2(t) \right.$$
$$\left. - \alpha\rho_m \int_0^\infty (d(t) + z)(T_m - T(t, z))dz \right]. \quad (7)$$

For the half-space model, it becomes:

$$N(t) = -\frac{2\pi G}{g}\alpha\rho_m\kappa T_m\left[1 + 2\alpha\rho_m T_m(\rho_m - \rho_w)^{-1}(\pi)^{-1} \right]t \quad (8)$$

where κ is the thermal diffusivity and t is plate age. For the plate model, the expression for $T(t, z)$ is more complicated (Parsons and Sclater, 1977). The corresponding geoid anomaly is given by (e.g., Cazenave, 1984):

$$N(t) = -\frac{2\pi G}{g}H^2\left[\frac{1}{2}(\rho_m - \rho_w)d^2(t) \right.$$
$$\left. + \alpha\rho_m T_m\left(\frac{1}{6} + \frac{2}{\pi^2}\sum_{n=1}^\infty Q_n \right) \right] \quad (9)$$

with:

$$d(t) = -\frac{1}{2}\alpha\rho_m T_m(\rho_m - \rho_w)^{-1}$$
$$\times \left[1 - \frac{8}{\pi^2}\sum_{n=1}^\infty \exp\left(-\beta_{2n-1}\frac{ut}{H} \right)(2n - 1)^{-2} \right]$$
$$Q_n = \frac{(-1)^n}{n^2}\exp\left(-\beta_n\frac{ut}{h} \right) \quad (10)$$
$$\beta_n = (-R^2 + n^2\pi^2)^{1/2} - R$$
$$R = \frac{uH}{(2\kappa)}$$

u is the half-spreading velocity assumed constant. According to Eq. (9), $N(t)$ tends asymptotically toward a constant value at large age.

Another isostatic model has been considered in the literature: the thinning lithosphere model (Figure 2c). This model was first proposed by Crough (1978) to explain the broad topography and geoid swell observed over mid-plate hotspots. In the thinning model, the swell topography is assumed to be compensated by hot, low density asthenospheric material placed in the lower part of the lithosphere: the heat associated with mantle plumes drives the lithospheric isotherms upwards, thins the lithosphere, and produces an uplift of the surface. The geoid anomaly (Sandwell and McKenzie, 1989) is:

$$N = -\frac{2\pi G}{g}h\left[Z_w(\rho_c - \rho_w) + z_c(\rho_m - \rho_c) - z_m(\rho_m - \rho_w) \right] \quad (11)$$

where h is the height of the swell above the surrounding seafloor of depth z_w below sea level, z_c is the average Moho depth, and z_m the average depth of the low density anomaly; z_m is sometimes referred to as the compensation depth.

3.2. Regional Compensation

Expressions given above are valid in the long wavelength approximation of isostatically compensated topography, i.e., if the wavelength of the topography is much larger than the compensation depth. At wavelengths ≤ 500 km, compensation of the topography is better explained by regional rather than local isostasy, assuming elastic flexure of the lithosphere under the surface load (Veining Meinenz, 1964). This model successfully explains medium wavelength (≤ 500 km) geoid anomalies associated with intraplate volcanoes (Watts, 1978). A seamount load causes the upper elastic layer of the lithosphere to flex (see Figure 2d). The downward deflection $w(x, y)$ is given by the solution of the equilibrium equation of a continuous elastic plate overlying a fluid medium subject to normal load $P(x, y)$:

$$D\nabla^4 w + (\rho_m - \rho_c)gw = P \tag{12}$$

where D is the flexural rigidity related to the elastic thickness T_e of the plate through $D = ET_e^3/12(1 - \nu^2)$, with E and ν being the Young modulus and the Poisson ratio.

The load P is related to the topography height h above the unloaded lithosphere through $P \approx (\rho_t - \rho_w)gh$ (ρ_t is the load density). For simple axisymmetric loads analytical solutions of Eq. (12) exist. However, for 3-D seamounts of arbitrary shape, it is necessary to integrate Eq. (12) numerically over the volume of the load, knowing the topography h. The geoid anomaly observed at the surface results from several contribution: (1) the positive density contrast of the topographic load, (2) the negative density contrast due to deflection of the Moho and fill-in of the deflection, and (3) the positive density contrast due to the deflection of the base of the elastic plate. The latter effect is generally considered negligible and omitted. As for the deflection $w(x, y)$, the total geoid anomaly has to be computed numerically. The theoretical geoid anomaly is derived for various values of the flexural rigidity D. Comparison with the observed anomaly allows to estimate D, hence the elastic thickness T_e.

The schematic diagrams in Figure 2 illustrate the above discussed isostatic compensation models.

3.3. Admittance Approach

The expressions of geoid anomalies described in the previous sections are used in analyses conducted in the spatial domain. However, compensation mechanisms eventually produce responses at different wavelengths that may not be easily separated from observed geoid anomalies. Hence, a number of investigators have preferred to work in the spectral domain to avoid assumptions on the characteristic wavelengths of signals. This approach initially developed by Dorman and Lewis (1970) assumes that geoid and topography are linearly related in the spectral domain:

$$G(k_n) = Z(k_n)H(k_n) \tag{13}$$

where k_n is wavenumber, $G(k_n)$ and $H(k_n)$ are Fourier transforms of geoid and topography. $Z(k_n)$ is the transfer function or admittance.

Eq. (13) is valid in the absence of noise (noise in the data or geological noise due to unrelated features). A better estimate of $Z(k_n)$ is given by McKenzie and Bowin (1976) in the presence of noise:

$$Z(k_n) = [G(k_n)\bar{H}(k_n)]/[H(k_n)\bar{H}(k_n)] \tag{14}$$

where bars indicate complex conjugate.

Observed admittance computed as a function of wavenumber with Eq. (14) can be compared to various theoretical admittances derived from models of isostatic compensation. We list below the most widely used theoretical admittances.

The Airy, Pratt, and flexure admittance can be written as:

$$Z(k_n) = Z_{nc}(k_n)[1 - \Psi(k_n)] \tag{15}$$

where $Z_{nc}(k_n)$ is admittance for uncompensated topography equal to:

$$2\pi G(gk_n)^{-1}(\rho_c - \rho_w)\exp(-k_n z_w)$$

and $\psi(k_n)$ is a function ranging from 0 to 1 representing the part of the geoid anomaly due to the topography compensated at depth. It comes (under the approximation that the wavelength is long compared to the layer thickness):
Airy model:

$$\Psi(k_n) = \exp[-k_n(z_c - z_w)]. \tag{16}$$

Pratt model:

$$\Psi(k_n) = \frac{1 - \exp(-k_n H)}{K_n H}. \tag{17}$$

Flexure model:

$$\Psi(k_n) = \left[1 + \frac{Dk_n^4}{g(\rho_m - \rho_c)}\right]^{-1}\exp[-k_n(z_c - z_w)]. \tag{18}$$

Notations are the same as above.

A number of other theoretical admittances have been derived, for example for subsurface loading models of an elastic plate (Forsyth, 1985) and for thermal cooling models (Black and McAdoo, 1988). Admittance models have been also computed for convection in the upper mantle (e.g., Parsons and Daly, 1983; McKenzie, 1977). In these models, analytical expressions have been derived for simple temperature structure using a Green's function approach to solve motion equations. Other approaches solve numerically full convection equations, derive surface topography and geoid due to the convective flow, and deduce the admittance.

4. MECHANICAL BEHAVIOR OF OCEANIC PLATES: FLEXURE UNDER SEAMOUNT LOADING

Altimetry-derived geoid anomalies constitute a basic data set for flexure studies. Indeed 2-D high-resolution geoid (or gravity) grids allow precise computation of the flexural rigidity of the lithosphere under topographic loads of arbitrary shapes. Because of the characteristic wavelengths of flexured geoid anomalies (300–500 km), it is clear that geopotential models are inappropriate, as are ship-board gravity surveys which give only 1-D gravitational information. The flexure model has been widely applied to seamount loading and trench bending to determine the elastic thickness of the lithosphere using geoid and topography anomalies. This approach not only informs on the present mechanical behavior of oceanic plates but also on its evolution.

Since the early work of Watts (1978), it is widely accepted that the upper layer of the lithosphere is well described by an elastic rheology over geological time scales and that the elastic thickness increases with the square root of plate age at the time of loading. The base of the elastic layer coincides with an isotherm of the conductive lithosphere (400–700°C). This temperature range corresponds to the transition between the rigid-elastic to ductile behavior of mantle rocks and is in general agreement with predictions based on rock mechanics experiments (Goetze and Evans, 1979). Flexure studies suggest that the measured elastic thickness is that acquired at the time of loading, an indication that isostatic compensation of the load is not significantly altered by continuing cooling of the plate and that viscous relaxation is either unimportant or occurs on a time scale much shorter than the cooling process.

The relationship between the elastic thickness T_e and the age of a plate at the time of loading was first derived by Watts (1978) over the Hawaiian volcanoes and proved to be valid on a worldwide scale by numerous subsequent studies. It has been very useful to get information on the tectonic setting of volcanoes at the time of formation (Watts *et al.*, 1980; Smith *et al.*, 1989): low T_e indicates that seamounts formed on very young lithosphere, while large T_e indicates that they formed on old lithosphere. Detailed elastic thickness estimates at different locations of individual features have also revealed local variation, hence a composite origin for these features. The T_e vs. (age)$^{1/2}$ relationship has also been used to infer the age of the plate knowing the age of the volcanoes and inversely (Calmant *et al.*, 1990).

Figures 3a and 3b present a summary of the available estimates of T_e as a function of age of plate (at the time of loading in the case of seamounts). Depth of isotherms of the half-space conductive cooling model are superimposed. The T_e values at seamounts presented in Figure 3a are based on the compilation of Wessel (1992) to which we have added or replaced T_e values based on subsequent studies (Goodwillie and Watts, 1995 for some of the Polynesian volcanoes; Wessel and Keating, 1994 for the Hawaiian volcanoes; Kruse *et al.*, 1997 for the Easter seamount chain; Watts *et al.*, 1997, and Canales and Dañobetia, 1998 for the Canary Islands; Goodwillie, 1993 for the Puka Puka ridge in the central Pacific). T_e values at trenches presented in Figure 3b are from

Levitt and Sandwell (1995). Although there is much scatter, it is clear from Figures 3a and 3b that the oceanic lithosphere becomes more rigid as age increases. Moreover, comparing Figure 3a and Figure 3b indicates that elastic thickness deduced from studies at seamounts is significantly lower that estimated at trenches. The base of the elastic layer beneath seamounts follows roughly the 400°C isotherm while in the case of trenches, it follows the 500–700°C isotherm (e.g., Judge and McNutt, 1991). This difference was first interpreted by McNutt (1984) as the result of an upward migration of the plate isotherms due to reheating associated with the emplacement of seamounts. Subsequent observations however made this explanation difficult to support. Indeed, heat flow measurements along the Hawaiian volcanoes (Von Herzen *et al.*, 1989) as well as in the Polynesian region (Stein and Abott, 1991) do not show the extra heat predicted by the reheating hypothesis and rather indicate a normal seafloor heat flow. Wessel (1992) proposed that while the lithospheric reheating during a seamount emplacement may contribute to an apparent low T_e at seamounts, other phenomena generally neglected in flexure studies, such as thermal stresses, may play a significant role. Neglect of thermal stress systematically underestimates the strength of the plate, hence the elastic thickness. Account of it would reconcile T_e estimates at seamounts and at trenches and lead to a more likely temperature range of 500–700°C for the elastic-ductile transition inside oceanic plates, whatever the tectoning setting.

We note that the T_e estimates reported in Figures 3a and 3b fall within three distinct populations. From high to low T_e values, these populations correspond to (1) trenches, (2) world seamounts, (3) some seamounts of the central Pacific and western Pacific. A number of past studies have reported that the elastic thickness under some volcanoes of the Polynesian province is thinner than elsewhere in the world (e.g., Calmant, 1987; Calmant and Cazenave, 1987). A similar observation has been made for some seamounts in the western Pacific, a region characterized by a high seamount concentration and known as the Darwin rise (Smith *et al.*, 1989; Wolfe and McNutt, 1991). The Polynesian province is known as the South Pacific Superswell and is characterized by anomalously shallow seafloor, higher than normal volcanoes concentration, low subsidence rate, enriched volcanism and negative geoid anomaly (McNutt and Fischer, 1987; McNutt and Judge, 1990; McNutt, 1998). Early explanation for the apparent low elastic thickness in the south central Pacific suggested a thinned thermal lithosphere, a result of hotter than normal upper mantle under the South Pacific Superswell. This explanation appears however in conflict with the absence of regional heatflow anomaly over the superswell. Indeed, the thermal hypothesis should be accompanied by elevated temperature in the upper layers of the lithosphere, which is not observed.

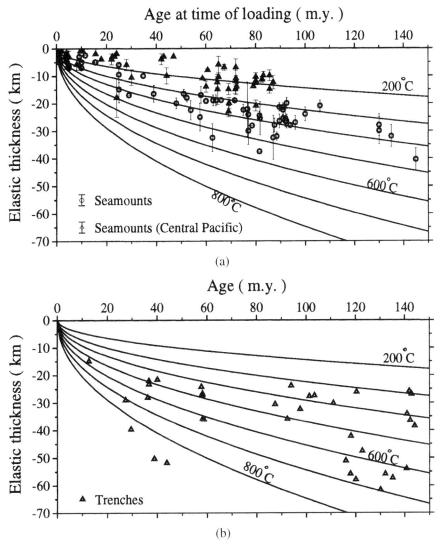

FIGURE 3   (a) Elastic thickness at seamounts versus age of plate at the time of loading (triangles indicate seamounts from the central Pacific while open circles correspond to other seamounts). (b) Elastic thickness versus age of plate at trenches.

Revisited estimates of T_e under the Superswell volcanoes (e.g., Marquesas and Society Islands; Filmer *et al.*, 1993) with more precise bathymetry and geoid data led to normal T_e values in a number of cases. It remains however that even with refined geophysical data, a number of T_e values at seamounts in the central and northwestern Pacific are lower than normal at the time of volcanism. Recent investigations of the origin of the Superswell (McNutt and Judge, 1990; Cazenave and Thoraval, 1994; McNutt, 1998) conclude to a dynamic support through upwelling in a convective mantle. These models suggest that Superswell hotspots may not have a deep, lower mantle origin but rather may originate at the base of the upper mantle. The volcanic chains associated with these hotspots do not show the clear linear age progression seen elsewhere in the world, an indication of

intermittent activity possibly biasing elastic thickness estimates (McNutt *et al.*, 1997; McNutt, 1998). A similar explanation may hold for T_e values at seamounts of the Darwin Rise region. The Darwin Rise presents many characteristics in common to the Polynesian province and may have been an active superswell in Cretaceous time, as is the south Pacific Superswell today (McNutt, 1998).

5. THERMAL EVOLUTION OF THE OCEANIC LITHOSPHERE

Cooling and contraction of oceanic plates has long been considered as a reasonably well understood phenomenon. The problem is classically treated as a purely conductive process and described by simple models (half-space and plate models) giving the temperature distribution inside the plate as solution of the heat transport equation. In the half-space model, the lithosphere is the boundary layer of a large convection cell through which cooling is purely conductive (Turcotte and Oxburgh, 1967). The plate thickens with $(age)^{1/2}$ and its base coincides with an isotherm. A variant of the half-space model is the plate model which assumes a slab of constant thickness with a fixed basal temperature. This model had been originally proposed by McKenzie (1967) to account for nearly constant heatflow observed in old basins. Both models predict that seafloor depth increases with $(age)^{1/2}$ up to $\sim$80 Ma. Beyond, depth tends toward a constant value in the plate model. The plate model has been often preferred to the half-space model because in some regions, in particular in the western Pacific and north western Atlantic, depth data are suggestive of seafloor flattening (Parsons and Sclater, 1977; Stein and Stein, 1992). The plate model assumes a constant temperature at a fixed depth which corresponds to the bottom of the plate. The GDH1 cooling plate model proposed by Stein and Stein (1992) is based on simultaneous inversion of depth and heat flow data. It predicts a depth-age curve which flattens beyond 70 Ma and a mean plate thickness of 95 km. Unlike the halfspace model for which the lithosphere coincides with the thermal boundary of large-scale convective cells, the plate model has no physical meaning but provides a mathematical basis for a system in which additional bottom heatflux balances heat lost by conduction.

Several mechanisms have been invoked to explain seafloor flattening observed in some regions, such as reheating (and subsequent uplift of the lithosphere) by hotspots (Heestand and Crough, 1981), radiogenic heating (Jarvis and Peltier, 1980), asthenospheric return flow (Phipps Morgan and Smith, 1992), supply of heat by small-scale convection (e.g., Parsons and McKenzie, 1978; Buck, 1987), convective destabilization at the base of the cooling plate (Yuen and Fleitout, 1985; Davaille and Jaupart, 1994; Eberle and Forsyth, 1995). In the latter studies, small-scale convection occurs through instabilities growing at the base of the cooling plate and become effective below old plates. An alternative model has been proposed by Doin and Fleitout (1996), considering that convection provides heat at the base of the lithosphere whatever plate age.

It should be noted however that seafloor flattening is principally observed in the western Pacific and north western Atlantic. Maps of the so-called "dynamic topography," (i.e., the observed topography corrected for shallow density contrasts due to crustal thickening and seafloor subsidence) which should reflect dynamic deformation of the Earth surface by large scale convection, presents two antipodal maximas which, in fact, coincide with the regions of observed elevated seafloor (e.g., Cazenave et al., 1989). Such an observation still holds when subsidence effects are removed using GDH1 plate model which empirically fits observations of elevated seafloor topography in the western Pacific. Thus it cannot be ruled out that seafloor flattening, or at least part of it, is not caused by additional heat at the base of the cooling plate but results from dynamic uplift of the earth's surface by large scale convective stresses.

The above discussion concerns mean subsidence estimates. Several observations however suggest significant deviations with respect to the mean, in particular areas still subsiding with $(age)^{1/2}$ at very old ages (Marty and Cazenave, 1989; Calcagno and Cazenave, 1994; Hohertz and Carlson, 1998). This is to be compared with the analysis of surface wave data (Tanimoto and Zhang, 1990) which reports continuous thickening (up to 150 Ma) of oceanic plates. Besides, the seismic data show significant differences from one ocean to another which are not predicted by the standard cooling models. Depth data also show important regional variations in lithospheric subsidence (Hayes, 1988; Kane and Hayes, 1994; Calcagno and Cazenave, 1994; Perrot et al., 1998). These studies indicate seafloor subsidence variations up to 100%, not accounted for by the cooling models unless implausible variations in asthenospheric temperature as high as several hundreds degrees are invoked. In some of these studies, other features such as asymmetrical subsidence or sudden deepening of the topography before flattening are also reported. It is thus quite clear that neither half-space nor plate models are able to account for the complexity of the observations.

With the availability of geoid data from satellite altimetry, geoid anomalies at fracture zones have been much used in the past 15 years for constraining the thermal models. Cooling of the lithosphere causes the geoid height to decrease regularly from mid-ocean ridges with increasing plate age [see Eqs. (8), (9), and (10)]. This behavior is mostly at long-wavelength and is almost impossible to isolate from other long-wavelength geoid components due to mantle convection. The variation in geoid height with age is in turn responsible for the geoid offset observed across fracture zones (FZ), a result of the difference in plate thermal structure and of the isostatically compensated seafloor depth step (see Figure 4). From Eqs. (7) and (8), the geoid step ΔN divided by the age offset Δt across the FZ approximates the first order derivative of $N(t)$. In the half-space model, $\Delta N/\Delta t$ is a constant which depends on the thermal parameters of the model (thermal diffusivity, mantle temperature, thermal expansion), whereas in the plate model, $\Delta N/\Delta t$ decreases with age as shown in Figure 5. Availability of geoid profiles along altimeter tracks roughly perpendicular to the FZ trend has given rise to numerous determinations of $\Delta N/\Delta t$ variations with age (Detrick, 1981; Sandwell and Schubert, 1982; Cazenave, 1984; Marty and Cazenave, 1988; Driscoll and Parsons, 1988; Gibert et al., 1987; Freedman and Parsons, 1990). A difficulty in determining the geoid step ΔN arises because other factors acting at the FZ obscure the

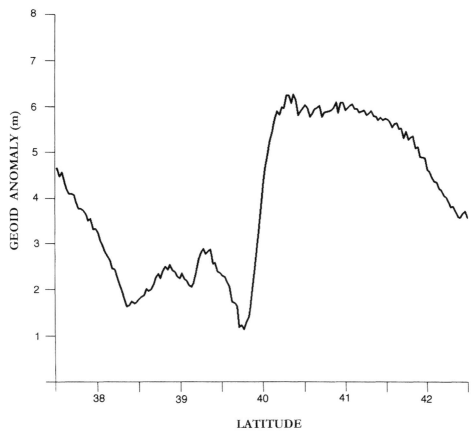

FIGURE 4 Geoid anomaly across the Mendocino fracture zone (Northeast Pacific).

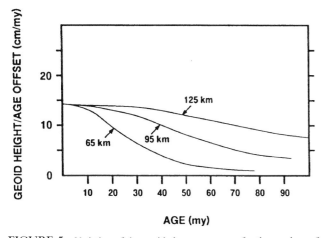

FIGURE 5 Variation of the geoid slope versus age for three values of the plate thickness (65, 95, and 125 km).

pure thermal effect: lateral heat conduction, (Louden and Forsyth, 1976), plate flexure due to differential subsidence and thermal bending stresses (Sandwell, 1984a; Parmentier and Haxby, 1986), and small scale convection developing beneath the FZ (Craig and McKenzie, 1986; Robinson *et al.*, 1988).

There is a considerable disagreement among the various results so far obtained. Early studies based on geoid data at the Mendocino FZ concluded in favor of the plate model, but subsequent studies found that neither plate nor half-space model could fit the data and some of them concluded that the behavior of the geoid step with age is consistent with small scale convection occurring beneath the FZ as predicted by Craig and McKenzie (1986) and Robinson and Parsons (1988). This view was further questioned by Sandwell (1984a) and Wessel and Haxby (1990) who reanalyzed geoid profiles along 4 Pacific FZ (Clarion, Clipperton, Murray and Udintsev), with modeling of lateral heat flux, and flexural response due to differential subsidence and thermal stresses. Putting all together $\Delta N/\Delta t$ results obtained for all four FZ, they conclude that data are compatible with simple conductive models and that previous results were systematically biased. Richardson *et al.* (1995) compiled a large number of published geoid slopes estimated from satellite altimetry. Their study concludes that the data can fit plate model predictions as well, such as those of the GDH1 model. In Figure 6, we have reported geoid slopes with age from 4 Pacific FZ (data are from Marty and Cazenave, 1988). Inspection of this figure which contains essentially the same data set as in Richardson *et al.* (1995) suggests indeed that on the average the observed trend can be explained by purely conductive models (but cannot univoquely discriminate between them) but that additional physical phenomena such as those previously mentionned to explain regional subsidence variations, need to be invoked to explain the scatter of the results.

6. OCEANIC HOTSPOT SWELLS

Most oceanic hotspots are associated with broad topographic and geoid anomalies (swells) of ~1000–2000 km width and amplitude in the range 0.5–1.5 km and ≈1–10 m respectively. Hotspot swells represent the dominant signal

of the medium wavelength geoid as revealed by satellite altimetry maps. Figure 7 shows, as an example, the geoid and topography across the Bermuda swell. It is an axisymmetric feature of 1000-km radius showing a perfect correlation between geoid and topography. In general, other swells do not display such a radial symmetry and are either elongated (Hawaiian or Polynesian swells) or completely irregular (Cape Verde swell).

Over the past two decades, many studies have been devoted to measure the geophysical characteristics of oceanic swells, in particular depth and geoid anomalies. The general correlation observed between depth and geoid anomalies over swells has often been used to infer an apparent compensation depth of the topography through simple isostatic models (Crough, 1983; McNutt and Shure, 1986; Fischer *et al.*,

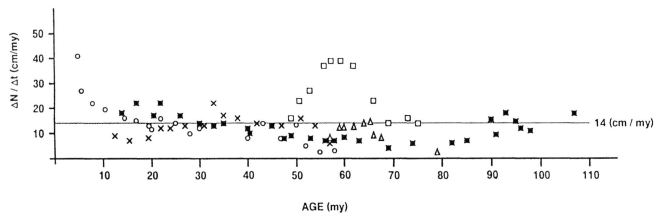

FIGURE 6   Ratios of geoid anomaly to age offset at fracture zones as a function of mean plate age. The data are from four fracture zones (FZ) of the Pacific ocean: Mendocino FZ (dark squares), Clarion FZ (open squares), Murray FZ (triangles), West Udintsev FZ (crosses) and East Udintsev FZ (open circles). The horizontal line corresponds to a constant ratio of 14 cm/my predicted by the half-space cooling model.

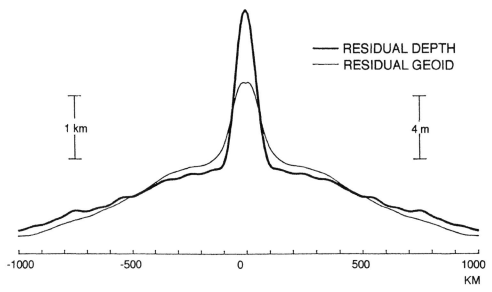

FIGURE 7   Geoid and topography anomalies across the Bermuda swell.

1986; McNutt, 1988; Monnereau and Cazenave, 1988, 1990; Sheehan and McNutt, 1989; Wessel and Keating, 1994; Cañales and Dañobetia, 1998; Moore *et al.*, 1998). If it is assumed that the topography is locally compensated at long wavelength (>500 km), then to a first approximation, geoid and depth anomalies are linearly related [see Eq. (3)] and their ratio depends on the average depth of the compensating density contrast, a parameter which may eventually inform on the mode of compensation of the swell. Although estimates of the depth and geoid anomalies over swells are subject to some uncertainties due to subsidence effect correction for the topography and removal of the long wavelength geoid components, there is a certain agreement between the various results which show a rough linear relationship between geoid and topography and give geoid to depth anomaly ratios in general less than 5–6 m km^{-1} (see for example Monnereau and Cazenave, 1990). Interpreted in terms of compensation depth [Eqs. (2), (3), and (10)], such values indicate that the compensating density contrast is located inside the lithosphere at depths ranging between 50 and 100 km. Such apparent shallow depths led Crough (1983) and Menard and McNutt (1982) to propose that the lithosphere is reheated as it passes over hotspots. In this model, referred as to thermal rejuvenation model, elevated mantle temperature is accompanied by thinning of the rigid plate up to a given depth and the thinned lithosphere is replaced by hot low-density asthenospheric material causing uplift of the surface and explaining the resulting geoid anomaly.

Several observations argue however against a pure lithospheric cause of swells. One of them is the short uplift time (<10 myr) of the Hawaiian swell that heat conduction alone is unable to produce unless the asthenospheric heat flux is excessively large. Another observation concerns surface heat flow around the Hawaiian swell which shows no heat flow anomaly, a result in contradiction with a significantly thinned lithosphere by reheating processes (Von Herzen *et al.*, 1989). Other results indicate that the lithosphere may not be thinned at hotspot swells. Woods *et al.* (1991) and Woods and Okal (1996) measured plate thickness beneath the Hawaiian swell using dispersion of Rayleigh waves between Midway and Oahu and found that the lithosphere has a normal thickness of ~100 km for its age. A similar conclusion was derived by Tarits (1986) from conductivity measurements at several sites in the Pacific. Discontinuities of the mantle conductivity with depth are indeed indicative of partial melting. Such discontinuities are found at depths of 90–100 km around the Hawaiian and Society swells, i.e., at depths expected for the base of a normal lithosphere. Thus, heat flow, seismic, and electric measurements over hotspot swells strongly argue against extensive thinning of the lithosphere.

A recent tomographic investigation of the seismic structure of the mantle along the Hawaiian swell by Katzman *et al.* (1998) has reported a fast region beneath the entire Hawaiian swell located in the uppermost 200–300 km of the mantle which is obviously in conflict with the hypothesis that the uppermost mantle beneath the Hawaiian swell is anomalously hot. Although this result remains controversial, the reheating model leading to a substantially thinned lithosphere beneath hotspot swells is not widely accepted.

A number of authors have proposed that topographic swells result from surface uplift by convective stresses (e.g., Fischer *et al.*, 1986; McNutt, 1988). However, numerical calculations based on constant viscosity convection in the upper mantle give a geoid to depth anomaly ratio in the range 6–10 m km^{-1} (Parsons and Daly, 1983), i.e., a factor ≈ 2 larger than observed. On the other hand, convective calculations by Robinson *et al.* (1987) assuming a low viscosity zone (LVZ) beneath the lithosphere, showed that the geoid to depth anomaly ratio can be much lower than in the constant viscosity convection case. To infer the apparent compensation depth d_c corresponding to their convective calculations, Robinson *et al.* used the Pratt model [see Eq. (3)] to relate the computed geoid to the computed topography and found that d_c can vary from $\approx$20–100 km depending upon the viscosity contrast between the LVZ and underlying upper mantle. Their analysis showed that the LVZ damps the surface uplift while it causes the geoid response to surface and deep density contrasts to change sign inside the LVZ in such a way that the net response is dominated by the shallow contribution. These results may explain why surface observables, when interpreted in terms of simple compensation mechanisms (Airy, Pratt, plate-thinning models), suggest shallow compensation depths whereas the real compensating density distribution may depend in a complex manner on the viscosity structure of the mantle.

Dynamic support of hotspot swells in the presence of a LVZ is also able to explain the linear increase of the geoid to depth anomaly ratio with (age)$^{1/2}$ observed over oceanic swells (Monnereau and Cazenave, 1990). According to Ceuleneer *et al.* (1988), the presence of a LVZ extending to $\approx$200 km depth, of thickness decreasing with plate age at the expense of the growing lithosphere, is able to explain the observed trend providing that the viscosity drops in the LVZ by a factor $\approx$50. Convective calculations by Moriceau *et al.* (1991), Monnereau *et al.* (1993) and Ribe and Christensen (1994) with pressure and temperature dependent rheology have indicated that, unlike in stratified-viscosity models of convection, the deep contribution to the geoid response is negligible, compensating masses being concentrated at a depth corresponding to the base of the thermal plate. These numerical calculations of mantle flow do not produce any lithospheric thinning below hotspot swells, the plume-lithosphere interaction being limited to the base of the lithosphere. Mechanical erosion at the base of the plate may also explain the rapid uplift of swells, as shown by other models of dynamical plume-lithosphere interactions (Sleep,

1990, 1992; Moore *et al.*, 1998). While dynamical support of swells are certainly indicated, other effects may also contribute, such as chemical-differentiation mechanism (Phipps Morgan *et al.*, 1995; Katzman *et al.*, 1998). According to these authors, the swell may be partly supported by chemical buoyancy caused by depletion of the source region from its incompatible elements and extraction of volatiles, as a result of basaltic-differentiation mechanism.

7. SHORT AND MEDIUM WAVELENGTH LINEATIONS IN THE MARINE GEOID

One of the most exciting results based on satellite altimetry was the discovery by Haxby and Weissel (1986) of short-wavelength (200–250 km) geoid undulations of 10–20 cm amplitude elongated in the NW-SE direction over the central Pacific. Other short-wavelength lineations have also been detected in the Indian Ocean and in the south Atlantic (Haxby and Weissel, 1986; Cazenave *et al.*, 1987; Fleitout *et al.*, 1989; Gibert *et al.*, 1989), but it is in the Pacific that the geoid lineations are the most visible, west of the East Pacific Rise between 10°N and 30°S (Figure 8 [see color insert]). They appear oblique with respect to the large Pacific FZ and their orientation coincides with the present motion direction of the Pacific plate in the hotspot reference frame. Some lineations are remarkably continuous over several thousands kilometers.

The geoid lineations coincide with topography lineations of ~200 m amplitude. Closely associated with one of the lineation of the south central Pacific is an elongated topographic ridge, called the Puka Puka ridge, discovered by Sandwell *et al.* (1995) during a research cruise. This volcanic ridge extends over 3000 km and consists of en-echelon individual ridge segments of ~300 km long. A remarkable feature of the Puka Puka ridge is that it is located in a trough of a gravity lineation, while adjacent lineation troughs may be devoid of such volcanic ridges.

Although the short-wavelength lineations have been the object of many studies, their origin remains controversial. A variety of mechanisms has been invoked:

1. Small-scale convection developing in a low viscosity layer below the lithosphere (Haxby and Weissel, 1986; Buck and Parmentier, 1986).
2. Magmatism resulting from lithospheric fracturing under the effects of tensile stresses (Winterer and Sandwell, 1987; Sandwell *et al.*, 1995).
3. Off-ridge origin: magmatic traces left into the plate as it moves over fixed convective plumes more numerous than the classical hotspots (Moriceau and Fleitout, 1989).
4. On-ridge origin: magmatic traces or variations in crustal thickness caused by long-lived, along-strike variations

in melt production between transform faults and overlapping spreading centers (Macdonald *et al.*, 1986; Shen *et al.*, 1993, 1995).

In the small-scale convection hypothesis, the direction of the lineations should coincide with the direction of present absolute motion of the plate. This is obviously the case in the south central Pacific where the lineations are parallel to the present motion of the Pacific plate and are oblique to the direction of fossil fracture zones. To explain the non-observed increase in wavelength with plate age predicted by small-scale convection models, it has been proposed that the lineations are produced by seafloor topography undulations frozen into the plate of young age (when the elastic layer is thin enough) and transported by the motion of the plate (Buck and Parmentier, 1986).

The second model put forward by Winterer and Sandwell (1987) and Sandwell *et al.* (1995) hypotheses that the lineations are surface expressions of tensional cracks filled by magmatic intrusions. In this model, the cause of lithospheric stretching are plate boundary stresses. According to Sandwell *et al.* (1995), three observations argue in favor of the lithospheric stretching model: (1) the very low admittance (geoid to depth ratio) of ~1 m km^{-1} associated with the lineations and (2) the presence of a volcanic ridge, the Puka Puka ridge, located in the trough of a geoid lineation and (3) the absence of clear age progression of the Puka Puka ridge volcanism, inconsistent with a hotspot origin.

The very low admittance of the geoid lineations is suggestive of a compensation at Moho depths. In the stretching model, tensional stress field produces deformations of the lithosphere in form of *boudins* with strain being maximum in the topographic troughs. It is thus in lineation troughs that magmatic intrusion may preferentially occur. This is unlike the small-scale convection model, where volcanism is expected at the crest of the geoid lineations, above the upwelling convective instabilities (see Figure 9).

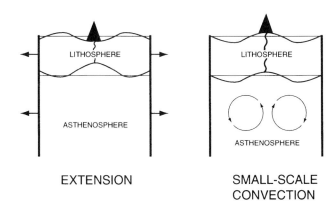

FIGURE 9 Schematic representation of lithospheric stretching (left) and small-scale convection (right). (Redrawn from Sandwell *et al.*, 1995.)

The third and fourth hypotheses are related to a hotspot-type origin for the lineations. The directional analysis performed by Moriceau and Fleitout over the Pacific lineations showed that these tend to be aligned with present and fossil directions of the absolute motion of the Pacific plate. They suggested that the lineations reveal the presence of mini hotspot plumes much more numerous than the classical hotspots, leaving magmatic traces in the lithosphere elongated in the direction of motion.

Another suggestion has been proposed which relates the wavelength of the lineations with that of typical axial depth undulations of the East Pacific Rise (Macdonald *et al.*, 1986; Shen *et al.*, 1993, 1995). The lineations could represent topographic traces left by near-axis seamount volcanism in response to long-lived changes in along-strike melt production. If the magmatic source is fixed with respect to the mantle, the traces will be aligned with the absolute plate motion direction. If the source is fixed with respect to the spreading axis, the traces will coincide with the spreading direction. Numerous linear chains of small seamounts originating at mid-ocean ridges appear parallel to either absolute or relative plate motions. Some of them (those aligning with absolute plate motion) may result from mini hotspots trapped by the spreading center. The hotspot hypothesis appears however inconsistent with the radiogenic dating of the Puka Puka ridge. Over a distance of 1800 km, the Puka Puka volcanoes erupted within less than 5 myr, thus cannot have been formed by a single or even two or three hotspots.

While an hotspot origin should most likely be discarded, it is not yet possible to discriminate between lithospheric stretching or small scale convection as the source of the geoid lineations. It is worth mentioning that the two mechanisms are not mutually exclusive and perhaps both may be invoked to explain the observations.

Longer wavelength geoid lineations have been also identified in the Pacific (Figure 8). Maia and Diament (1991) as well as Baudry and Kroenke (1991) found evidence of 400–600 km wavelength geoid lineations in the south central Pacific. In addition, Cazenave *et al.* (1992, 1995), reported longer wavelength (1000–1200 km) lineations parallel to the short-wavelength lineations discovered by Haxby and Weissel, hence trending in the direction of absolute plate motion. Inspection of Figure 8 shows clearly that the geoid in the south central Pacific exhibits a superposition of parallel lineations of preferential wavebands (150–250 km), (400–600 km), (1000–1200 km). Although not strictly continuous, the 1000–1200 km wavelength lineations extend over several thousands of kilometers. The Polynesian volcanic chains appear trapped in these 1000 km-wavelength lineations. However, the latter extend farther eastward and clearly preceed the active hotspots. On the other hand, north of the Polynesiam province, well-developed geoid lineations do not coincide with known topographic features. The analysis by Cazenave *et al.* (1995) showed that the geoid lin-

eation amplitude, as well as geoid to depth ratio, increases from young to old plate, from 15 to ~45 cm, and from 1.5 to 3 m km^{-1}, respectively, over a plate age range of 10–60 Ma. Wessel *et al.* (1994, 1996) quantitatively demonstrated the three types (200 km, 400–600 km and 1000 km wavelength) of geoid lineations are oriented in the direction of absolute Pacific plate motion in a hotspot reference frame, and that the highest correlations in direction as a function of wavelength are seen for the 200–250 km and 1000–1200 km wavebands, which corresponds to the most visible geoid lineations (see Figure 8). Wessel *et al.* (1994, 1996) incidently noted that the discrete wavebands of maximum correlation have values coinciding with the depth of major mantle discontinuities and inferred that the observed lineations are influenced by mantle dynamics. In fact, a number of phenomena may be able to produce the geoid lineations: crustal thickness variations, lithospheric thickness and density variations, and upper mantle convection. Although the observed low admittance (geoid to depth ratio) is suggestive of a shallow origin, it is not possible to discriminate between a lithospheric or sublithospheric origin with this parameter alone. For instance Robinson *et al.* (1987) showed that convection in the upper mantle within a low viscosity asthenosphere would produce very low admittance values of ≤2 m km^{-1}. The orientation of the geoid lineations is of greater help for discriminating between lithospheric or convective processes. Indeed, if the source of the lineations is inside the crust or the lithosphere, their orientation should coincide with the direction of spreading, which is obviously not the case. The orientation and average wavelength (1000–1200 km) of the geoid lineations is suggestive of the roll-like upper mantle convection pattern observed in experiments conducted by Richter and Parsons (1975). These authors predicted that beneath fast moving plates (as it is the case for the Pacific plate), convective instabilities will align along rolls oriented in the absolute plate motion direction, although this alignement should occur above some critical plate velocity (Rabinowicz *et al.*, 1990). Richter (1973) first noticed that in the Pacific ocean, hotspot chains are ~1000 km apart and proposed that the associated swells are the surface expression of convective rolls. This was further confirmed by the statistical analysis of Yamaji (1992) which showed that the hotspot distribution is periodic, with a typical spacing of 1000 km. The Polynesian hotspot swells are indeed clearly part of the medium-wavelength lineations and very likely share with them a common origin. The experiments of Richter and Parsons (1975) assumed constant upper mantle viscosity and low Rayleigh number. In this case, the depth of the convective layer should equal the half-wavelength of the lineations. In the mantle, where the viscosity varies with depth and the appropriate Rayleigh number is probably much higher, the length-scale of boundary layer instabilities may become much smaller than the depth. If the medium-wavelength geoid lineations are indeed related to mantle convection, we can just say that the observed horizontal length-scale appears to be compatible with a depth of convection of ~600–700 km. Whatever, it is very likely that the ~1000 km geoid lineations are related to mantle convection. The 200 km undulations, on the other hand, may result either from small-scale convection or lithospheric streching, or a combination of both mechanisms. More theoretical and experimental work is indicated to clarify the origin of these curious features.

8. MAPPING THE SEAFLOOR TECTONIC FABRIC

Because of the close correlation between the 20–200 km wavelengths of the geoid height or gravity and the uncompensated seafloor topography (McKenzie and Bowin, 1976), the dense and uniform coverage of the satellite-altimeter measurements reveal in detail the seafloor tectonic fabric in most of the world's ocean, and particularly in the poorly charted southern oceans (Figure 1). Tectonic fabric charts of the ocean floor were initially derived from interpreting altimeter data plotted along satellite tracks (e.g., Vogt et al., 1984; Cande et al., 1988; Gahagan et al., 1988; Royer et al., 1989; Royer et al., 1990; Müller and Roest, 1992). Global geoid or gravity images constructed from gridding satellite-altimeter data progressively unveiled with finer details the topography of the seafloor. The "blurred" images based on Geos-3 and Seasat data (Haxby et al., 1983; Sandwell, 1984b; Haxby, 1987) focused as the dense Geosat Geodetic Mission (GM) data were released (McAdoo and Marks, 1992) and TOPEX and ERS-1 data became available (Laxon and McAdoo, 1994; Cazenave et al., 1996; McAdoo and Laxon, 1997; Sandwell and Smith, 1997). These data now permit an accurate mapping of almost any relief of the seafloor wider than 10 km and higher than about 1 km, such as fracture zones, seamounts and mid-oceanic spreading ridges. This detailed information leads to many improvements in determining the relative and absolute plate motions of the lithospheric plates, and in characterizing in a global perspective the processes that shape the oceanic crust (spreading ridge morphology, ridge segmentation, intraplate deformation, hotspots . . .).

8.1. Fracture Zones

Oceanic fracture zones are perhaps the most striking features visible in a set of parallel satellite-altimeter profiles or in satellite-derived gravity grids (see Figure 1). Fracture zones form long, continuous and linear bathymetric structures reflecting the past relative motions between plates (Wilson, 1965a; Morgan, 1968). Mapping the full extent of fracture zones over 90% of the world's ocean basins proved

very useful for testing and refining plate reconstruction models. Satellite-derived gravity also helped identifying other linear conjugate structures of the ocean floor that rifted and spread apart, adding new constraints on plate reconstructions.

8.1.1. Geoid/gravity Signatures of Fracture Zones

The topography of fracture zones is characterized by very long ridges, troughs or escarpments that delineate corridors of ocean crust of different age. These general characteristics are also observed in the satellite-altimeter data and vary with the spreading rates. In a fast-spreading regime, such as in the Pacific ocean (Figures 10a and 10b [see color insert]), the fracture zone morphology is dominated by a step reflecting the age difference between adjacent corridors, resulting in a geoid step in the order of 1–2 m (see Figure 4) or 20–60 mGal gravity anomalies, for a topographic step of about 1 km (~10 Ma offset; e.g., Sandwell, 1984a). In a slow spreading regime, such as in the Atlantic and western Indian oceans (Figures 10c and 10d), this age step is dominated by a complex topography expressed by deep valleys bounded by elevated and steep walls (e.g., Colette, 1986; Fox and Gallo, 1986), producing up to 1.5 m peak-to-trough geoid anomalies for a ~1.4 km ridge-to-trough height difference (5–10 Ma offset; e.g., Müller et al., 1991). Hence, global tectonic fabric charts of the ocean crust were easily produced by correlating the geoid height anomalies (Vogt et al., 1984; Cande et al., 1988), deflection of the vertical anomalies (Gahagan et al., 1988; Royer et al., 1989; Royer et al., 1990) or vertical gravity anomalies (Müller and Roest, 1992) over a series of satellite passes intersecting fracture zones. Automatic searches either fitting a Gaussian shape to geoid troughs (Shaw and Cande, 1990) or detecting minima in the geoid heights or gravity (Gibert et al., 1989; Royer et al., 1997) were also applied to individual satellite passes in order to accurately map the fracture zone valleys in the South Atlantic and western Indian oceans. The dense spatial resolution of the global gravity images based on the Geosat and ERS-1 geodetic missions would now be suitable for applying terrain analysis or image processing techniques in order to extract all the continuous highs and lows (Figure 11).

Despite the improvements in the spatial resolution of satellite altimeter data, mapping precisely the actual location of fracture zones is generally difficult. Firstly, fracture zones often exhibit a complex topography over a 10–30 km width, particularly along large offset fracture zones (Tasman FZ in Figure 10b) or in areas where a change in the direction of motion occurred (Heezen and Eltanin FZ in Figure 10a). Thus locating precisely the depth/age step between adjacent corridors may be uncertain. Note that the same limitation applies to high-resolution wide-beam bathymetric data (e.g., Caress et al., 1988; Kuykendall et al., 1994). Secondly, since geoid or gravity anomalies are actually caused by density contrasts (mainly between the oceanic crust and sea water),

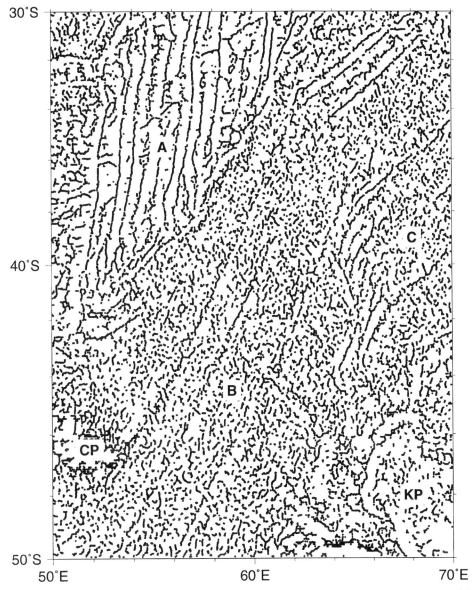

FIGURE 11 Gravity lows extracted in the south-western Indian Ocean from the global gravity anomaly grid of Sandwell and Smith (1997): (A) Ultra-slow (10–20 mm year^{-1}) Southwest Indian Ridge characterized by a series of deep north-south fracture zones and deep east-west spreading ridge axes, (B) SW–NE fracture zones in the Southern Crozet Basin created at ultra-fast to fast spreading rates (100–220 mm year^{-1}), (C) Eastern Crozet Basin created at intermediate spreading rates (60–100 mm year^{-1}). (CP) Crozet Plateau, (KP) Kerguelen Plateau. After Munschy (1997).

they may also reflect density variations within the oceanic crust and upper mantle, which are common in the vicinity of fracture zones (e.g., Detrick *et al.*, 1993). A comparison of the geoid signature and the basement topography along the Kane fracture zone in the Central Atlantic shows that the mismatch in the location of the fracture-zone valley may be as large as 15 km, although the average phase shift between geoid and topography is about 5 km (Müller *et al.*, 1991). Thus satellite-derived gravity can be used to locate a fracture zone axis within ±5 km. These studies also empha-

size that satellite-derived (or shipborne) gravity reveal the actual basement topography, which in some cases is buried by thick sediments. In this respect, the satellite altimeter data helped tracing the fracture zones up to the continental margins (see the Equatorial Atlantic Ocean in Figure 1).

Another difficulty when mapping fracture zones, is that there are other look-alike features which are not representative of relative plate motions. In Figure 10d, the Northern FZ follows a flowline parallel to the Kane FZ and progressively departs from it towards the ridge axis to form a low-angle V-shaped trough. Such V-shaped structure reflects the migration of the mid-Atlantic Ridge relative to an underlying mesospheric framework (hotspot reference frame; Schouten *et al.*, 1987; Müller and Roest, 1992). Some small- to medium-offset fracture zones appear to be meandering on either side of the spreading axis, like the fracture zones immediately south of the Atlantis FZ and of the Kane FZ (Figure 10d); they reflect back and forth migrations of non-transform discontinuities along the spreading ridge and thus record the evolution of the small-scale ridge segmentation through time. Note that the symmetry of these meanders puts tight constraints on plate reconstructions, although they do not follow flowlines. High-angle V-shaped structures are also observed, for instance south of Australia, east of Australian-Antarctic Discordance (Phipps-Morgan and Sandwell, 1994), or along the Pacific-Antarctic Ridge (Figure 1); they result from a fast migration of spreading centers, known as propagating rifts, leaving pseudo-faults in their wake (Hey *et al.*, 1989).

8.1.2. Application to Plate Tectonic Reconstructions

Active transform faults record the instantaneous direction of motion between two plates, and thus follow small circles about the pole of rotation describing these motions (Morgan, 1968). Fracture zones form the fossil extension of these transform faults and hence provide a continuous record of the trajectories or flow-lines of each plate relative to the spreading ridge axis. Several techniques can be applied to take advantage of the satellite-altimeter data in order to better constrain plate motion models.

Some methods use the property that fracture zones follow small circles about instantaneous poles of rotation. Transform azimuths measured from Seasat data in the southern ocean were inverted in order to better constrain the direction of the Pacific-Antarctic instantaneous motions (DeMets *et al.*, 1990, 1994). Satellite-derived trends for all the fracture zones in the South Atlantic, south of the Equator, were simultaneously fitted using a single rotation between Africa and South America for the last 35 Ma (Gibert *et al.*, 1989). A more detailed model was also achieved by adjusting several different small circles instead of a single one to these satellite-derived flowlines (Shaw and Cande, 1990).

Other methods assume that the image of a mid-ocean ridge at a given time is identical on the two plates that shared this plate boundary. This image or isochron, whose shape may vary through time, is defined by a succession of magnetic lineations offset by transform fault segments. Superimposing two isochrons identified in conjugate basins yield the finite rotation describing the past position of one plate relative the other at the corresponding time. Isochrons are mapped from crossings of shipborne or airborne magnetic profiles with the magnetic lineations and from crossings of satellite altimeter profiles across fracture zones. If

the plate motion model is correct, and assuming symmetric seafloor spreading, flowlines derived from the combination of successive finite rotations should match the observed fracture zone trends. This test was for instance applied to compare satellite-derived flowlines with predicted flowlines in the Atlantic (Müller and Roest, 1992). It confirmed that the fracture zones trends in the Equatorial Atlantic, south of the Fifteen Twenty Fracture Zone, are better accounted for by the Africa/South America motion than by the Africa/North America motion. The southwest Indian Ocean is another example where the satellite altimeter data help constraining plate motion (Africa/Antarctica). Seasat data gave a hint that two consecutive major changes of motion occurred along the southwest Indian Ridge. A kinematic model combining satellite altimeter information with magnetic anomalies (Royer *et al.*, 1988) predicted that a series of closely spaced fracture zones developed as the result of the first change of motion and coalesced together, during the second change of direction, into the very large offset fracture zones that now offset the Southwest Indian Ridge. This model was later confirmed when the Geosat GM data unveiled all the details of the Southwest Indian Ridge seafloor fabrics (Marks *et al.*, 1993). Satellite altimeter data helped improve finite reconstruction models in many other areas such as the Atlantic (Cande *et al.*, 1988; Nurnberg and Müller, 1991; Müller and Roest, 1992), the southwest Pacific (Mayes *et al.*, 1990; Cande *et al.*, 1995), the southeast Pacific (Tebbens and Cande, 1997), the Indian Ocean (Royer and Sandwell, 1989; Royer *et al.*, 1997), or the Tasman Sea (Gaina *et al.*, 1998).

For the purpose of plate reconstructions, locating the symmetric traces of a fracture zone is critical, but generally difficult, since the expression of a fracture zone may be different on conjugate flanks. For instance, the fracture immediately west of Tasman FZ (bottom fracture zone in Figure 10b) is better expressed on the Antarctic plate than on the Australian plate; the same can be observed in Figure 10c (bottom fracture zone). Along large offset-fracture zones, any change in the spreading direction will reshape their topographic expression. Depending on the fault geometry and the direction of change, extensional or compressional features will develop as the new direction fabric overprints the former direction fabric, asymmetrically on the conjugate limbs of the fracture zone (Caress *et al.*, 1988; Hey *et al.*, 1988). For this reason, one should avoid using large-offset fracture zones for plate reconstructions. In slow spreading regimes, deep valleys seem representative of the actual fracture zone location; this signature appears to be symmetrical on the two limbs of a fracture zone, as for example along the Kane FZ (Figure 10d). Dispersion analysis of reconstructed fracture zone crossings extracted from satellite passes, along the slow spreading Central Indian and Carlsberg ridges, showed that the standard deviation was in the order of 4 km (Royer *et al.*, 1997), slightly better than the 5 km uncertainty inferred from geoid/topography comparison (Müller *et al.*, 1991). This analysis combined left-lateral and right-lateral transform offsets, hence any systematic bias due to an anomalous density distribution along fracture zones should have canceled out. Along fast to medium spreading ridges, the signature is generally anti-symmetric (e.g., Figure 10a and 10b), with a ridge on the young side of the fracture zone and a trough on the old side; then the inflection points in the geoid or gravity signal would be more representative of the symmetric depth/age steps.

Satellite-derived gravity data proved also very useful to identify and map conjugate features of the ocean floor, such as rifted margins, limits of the continental shelf, edges of submarine plateaus that rifted apart, or troughs left by major ridge jumps (e.g., Henry and Hudson troughs in Figure 10a). In some cases (Figure 12 [see color insert]), the remarkable symmetry of these tectonic scars provide a valuable constraint for paleogeographic reconstructions, particularly in the poorly charted southern oceans or near continental margins where sediments often hide the basement structures. Most recent models unraveling the plate tectonic history around Antarctica have taken advantage of this information (e.g., Royer and Sandwell, 1989; Lawver *et al.*, 1991; Lawver and Gahagan, 1994; Sutherland, 1995; Marks and Stock, 1997; Tikku and Cande, 1998).

8.2. Seamounts

Away from the trenches and mid-oceanic ridges, seamounts, and submarine plateaus are, together with fracture zones, the main type of relief of the ocean floor. Until the advent of satellite altimeters, most seamounts or undersea volcanoes remained uncharted because only a small fraction of the world's ocean has been mapped by surface ships. Since seamounts represent an excess of mass relative to the surrounding abyssal plains, they produce little bumps on the mean sea surface or geoid height. Thus satellite altimetry has proven a very powerful tool not only to systematically detect and locate seamounts, but also to infer, upon certain assumptions, their size and shape, as well as their ages. This information is important for understanding the processes that control their origin and evolution, for determining absolute plate motion, or measuring the mechanical properties of the lithosphere. Other applications include the detection of hazards for submarine navigation, the search for new fishing grounds or modeling the ocean circulation pattern. The main progresses in this field since the launch of GEOS-3 in 1975 come from the increasing coverage and improving quality of the satellite altimeter data.

8.2.1. Seamount Signature

A typical signature of a seamount on the geoid (Figures 13 and 14) consists of a small geoid anomaly superimposed on a broad regional trend [curve (b) in Figure 14].

After removing the long-wavelength component (>200 km), the geoid signal correlates fairly well with the topography of undersea volcanoes [curve (a) in Figure 14]. The geoid anomaly is centered on the seamount and reaches 1–2 m for a 1–3 km high seamount with a typical base-diameter of 10–50 km. The horizontal derivative of this geoid anomaly along the satellite track, known as deflection of the vertical, will form a sharp dipolar anomaly easier to detect; peak-to-trough amplitudes range from 20–200 microradians and the inflection point is at the vertical of the seamount summit [curve (e) in Figure 14]. The geoid signature can also be converted into a sharper gravity anomaly [vertical derivative of the geoid; curve (d) in Figure 14]. Amplitudes of the seamount gravity anomaly typically range from 20–200 mGal. Depending on the age, and therefore on the mechanical strength of the underlying lithosphere at the time when the seamount emplaced, the positive anomaly associated with the seamount will be surrounded by a ring-shaped negative anomaly, marking the deflection of the lithosphere caused by the seamount load (Figure 2d). This depression is unfortunately often indiscernible in the bathymetry due to sediment infill. If the topography of the seamount and associated depression is known, then the thickness of the elastic portion of the lithosphere can be inferred from the amplitude and diameter of this gravity low.

The parameters used to characterize a seamount are generally the amplitude of the geoid or gravity anomaly and the distance between the center of the positive anomaly and the zero-crossing. The amplitude is not a good measure of the height of the seamount because of the flexural response of the lithosphere, which tends to reduce this amplitude. The zero-crossing is also often difficult to locate precisely. Peak-to-trough amplitudes of the deflection of the vertical suffers the same limitation. However the peak-to-trough distance is equal to the characteristic diameter of the seamount and the difference between the deflection diameter (distance between the outer zero crossings) and the seamount diameter is equal to the flexural diameter (Craig and Sandwell, 1988). The amplitude of the vertical gravity gradient, which enhances the short-wavelength relative to the long flexural wavelength, is less sensitive to the loading effect; zero-crossings are also easier to detect as the anomaly is sharper (Wessel and Lyons, 1997).

8.2.2. Seamount Distribution

Techniques for detecting seamounts from the geoid height anomalies measured by satellite altimeters evolved as the satellite data coverage and data quality improved. GEOS-3 data coverage was not uniform, with best coverage occurring near tracking stations. Early failure of Seasat left losange-shaped gaps of about 100 km side between profiles; also, due to sea-ice, recovery of Seasat data south of 60°S was poor. However, these gaps are now filled by the Geosat and ERS-1 geodetic missions.

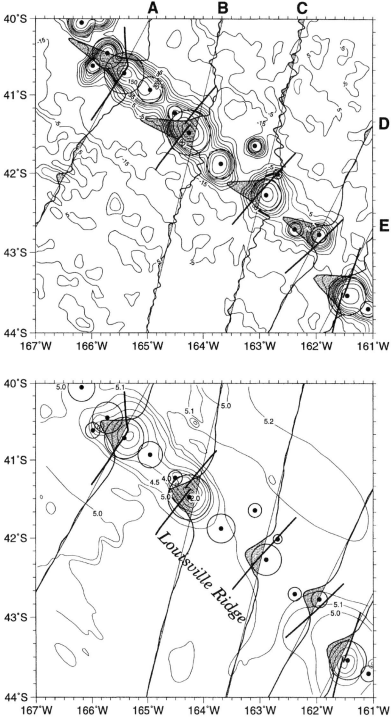

FIGURE 13 Satellite-altimeter profiles along the Louisville Ridge: (A) TOPEX, (B) and (C) ERS-1, (D) Geosat ERM, (E) Seasat. (Bottom) Bathymetric chart based on the GEBCO chart 5–10 (Monahan *et al.*, 1982). Band-path filtered (20–200 km) geoid heights are plotted along satellite tracks, positive to the west (4 m/degree of longitude). (Top) Gravity anomaly contours based on the grid of Sandwell and Smith (1997). Band-path filtered (20–200 km) satellite-derived gravity anomalies are plotted along satellite tracks, positive to the west (120 mgal/degree of longitude). Circle symbols and large circle indicate the location and radius of seamounts as predicted from the global gravity grid (Wessel and Lyons, 1997). Satellite-derived gravity reveals many more seamounts than initially charted. Thick black lines outline the shipboard data shown in Figure 14.

Improvements in the satellite-borne altimeters and storage capabilities reduced the limit of along-track resolution of the altimeter measurements by a factor of three from 75 km for Geos-3, to 50 km for Seasat, to 22–30 km for stacked Geosat Exact Repeat Mission (ERM), TOPEX or ERS-1 profiles (these limits reflect the 50% level of coherence between repeating passes; Sandwell and Smith, 1997). The accuracy in locating a seamount for an individual profile is also a function of the sampling rate along track (e.g., 1 sec or 6.6 km for Seasat). Across track, the resolution limit varies according to the track spacing and thus improves from the Equator towards the latitude of culmination ($\pm 72°$ for Seasat and Geosat, $\pm 81°$ for ERS-1). Merging data from ascending and descending passes or from different satellites partly palliates this limitation.

Different techniques for predicting the presence of seamounts have been developed. Lambeck and Coleman (1982) modeled seamount signatures on the geoid and searched visually Geos-3 and Seasat profiles. Lazarewicz and Schwank (1982) and White et al. (1983) applied a systematic search on Seasat profiles using matched filters. Gairola et al. (1992) applied the same type of filter to Geosat data. Sandwell (1984b) analyzed an image of the sea-surface constructed from deflection of the vertical profiles and detected 72 uncharted seamounts in the southwest Pacific. Baudry et al. (1987) used a least-square fit between modeled geoid anomalies and actual Seasat profiles. Ground-truthing the prediction from satellite altimeter profiles by surface-ship bathymetric survey was initially not very successful (e.g., Keating et al., 1984), casting some doubts about the reliability of this tool; this was mainly due to the across-track uncertainty and in some cases to high noise-to-signal ratios. Detection of seamounts based on at least two adjacent or intersecting profiles proved highly reliable and accurate within 15 km (e.g., Baudry and Diament, 1987). The resolution of the latest gravity grids (e.g., Sandwell and Smith, 1997), in the order of 10 km with an rms accuracy of 3–6 mGal, now enables the detection of all seamounts taller than 1 km, of which 30–50% were not charted previously. Satellite data also demonstrated that some previously charted seamounts were clearly mislocated because of errors in the celestial navigation of the mapping vessels [e.g., Fabert Bank in the Austral-Cook area ($\sim 158.78W$, 24°S as in Mammerickx and Smith, 1982) or the Islas Orcadas Seamounts in the Weddel Sea (as in LaBrecque and Rabinowitz, 1981; see also Figure 13) or even non-existent (e.g., Novarra Knoll southwest of St Paul and Amsterdam islands in the Indian Ocean, as in Fisher et al., 1982)].

Nonetheless, there are only few quantitative studies of seamount global distribution using altimetry data. Craig and Sandwell (1988) were the first to perform a global search for seamounts by analyzing the deflection of the vertical on Seasat profiles and detected as many as 8556 seamounts. Comparison between charted and predicted seamounts in a well-mapped area suggested that less than 25% of the seamount population could be detected by Seasat profiles, because of the track spacing. Another difficulty when analyzing sparse individual profiles was to decipher a seamount from lineated structures associated with fracture zones; both have the same along-track signature in the geoid. Using the gravity grid of Sandwell and Smith (1997), Wessel and Lyons (1997) performed a systematic search for seamounts in the Pacific plate (Figure 15). They actually used the vertical gravity gradient (vertical derivative of gravity) which better reflects the seamount characteristics. On the Pacific plate alone, they isolated and characterized ~ 8882 seamounts out of $\sim 25,000$ volcanic edifices visible in the satellite data, discarding all features detected within 25 km of a known fracture zone and missing all seamounts smaller than ~ 1.5 km, unresolvable by this technique.

These few global studies generalized and confirmed findings from regional analyses, which are mainly available for the Pacific Ocean, based on bathymetric charts or wide-beam soundings (e.g., Menard, 1964; Litvin and Rudenko, 1973; Batiza, 1982; Jordan et al., 1983; Abers et al., 1988; Smith and Jordan, 1988). The seamount population is the largest in the western tropical Pacific, with more than 150 per million square kilometers in the Central Pacific and French Polynesia (Wessel and Lyons, 1997). The northeast Pacific shows only few and generally small seamounts (Figure 15). While there is a general trend showing an increasing seamount population as the crustal ages increase (Wessel and Lyons, 1997), eastern half of the Pacific plate commonly shows high seamount aboundance on the younger sides of large offset fracture zones (Craig and Sandwell, 1988). The largest population of seamounts as well as the largest seamounts are found on the old Pacific crust (mid-Cretaceous age, 90–120 Ma). In the Atlantic Ocean, seamounts cluster around areas showing hotspot activity, whereas seamounts look evenly distributed in the Indian and Southern Oceans (Craig and Sandwell, 1988).

8.2.3. Seamount Characteristics

Deriving the shape and size of a seamount solely from its signature on the geoid is not straightforward, because the solution is non-unique. The signature combines the effects of the topography and of the mode of compensation (local, regional or thermal) of the underlying plate. Another limitation is that satellite altimeters cannot resolve wavelengths shorter than ~ 20 km and hence are unable to detect seamounts whose diameter is smaller than ~ 10 km or whose height is smaller than ~ 1 km.

Seamounts are usually defined by a conical and steep-sided geometry, with sometimes a truncated or flat top. In order to characterize seamount signatures in the geoid, gravity or vertical gravity gradient, this shape is generally approximated by a Gaussian bell (Sandwell, 1984b; Watts and Ribe, 1984; Baudry et al., 1987; Craig and Sandwell, 1988; Wessel

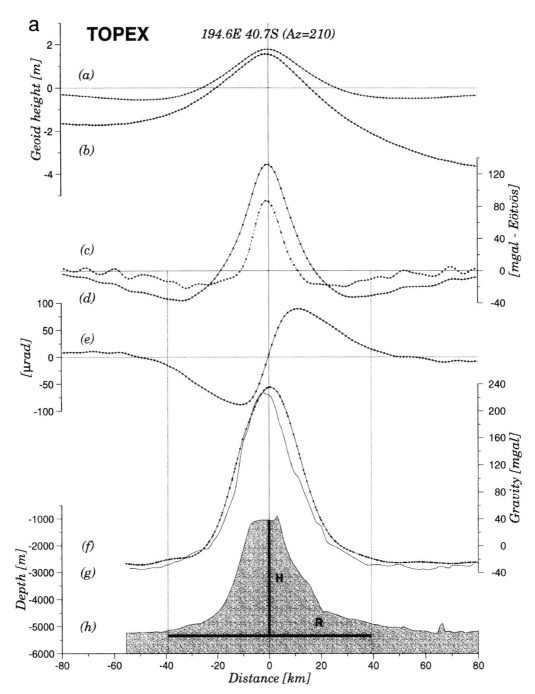

FIGURE 14 Typical satellite-altimeter signatures of seamounts along the Louisville Ridge (see location in Figure 13): (a) band-path filtered (20–200 km) geoid height, (b) unfiltered geoid height, (c) filtered vertical gravity gradient (dashed line), (d) filtered geoid-derived gravity, (e) filtered deflection of the vertical (geoid slope along track), (f) shipborne gravity, (g) gravity predicted along ship-track from the global gravity grid of Sandwell and Smith (1997; line with dots), (h) bathymetric profile. (R) and (H) are the base-radius and height, respectively, predicted by Wessel and Lyons (1997) from inversion of the vertical gravity gradient. Dots along curves give the sampling rate. Shipboard gravity and bathymetric profiles are from the R/V *Vema* cruise 3602 (1979). TOPEX, ERS-1 (35 d. cycle) and Geosat (ERM) profiles are stacks of 38, 9, and 44 repeat passes, respectively. All data are projected along the direction (Az) of the satellite ground tracks. Note the close correspondence between shipboard and satellite-derived gravity (curves f and g). Location mismatches of the seamount tops between the satellite and shipboard observation are due to offsets between the satellite or ship tracks and the projection point. Disagreements between the predicted and observed sizes of seamounts reflect the non-uniqueness of gravity data inversions. Gravity anomalies emphasize a small deflection of the lithosphere for all seamounts.

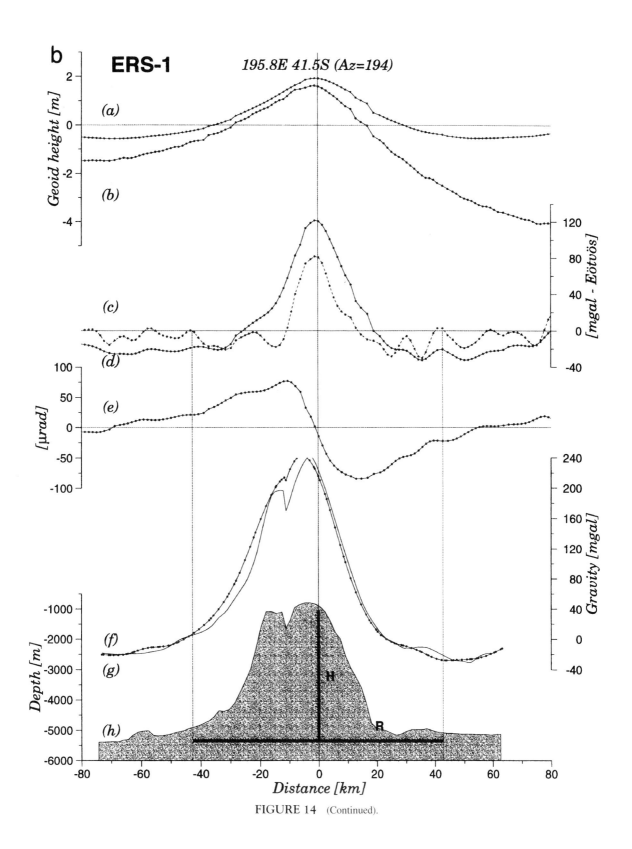

FIGURE 14 (Continued).

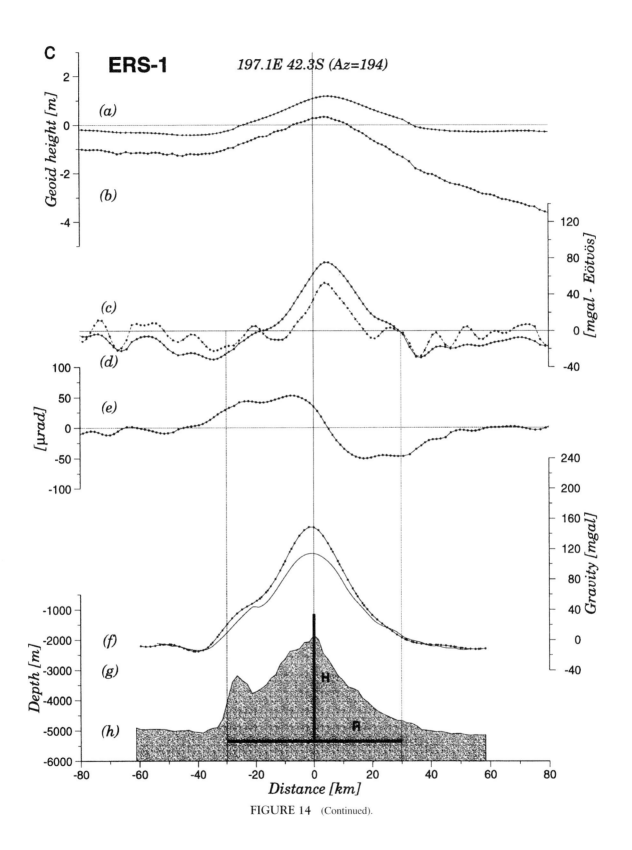

FIGURE 14   (Continued).

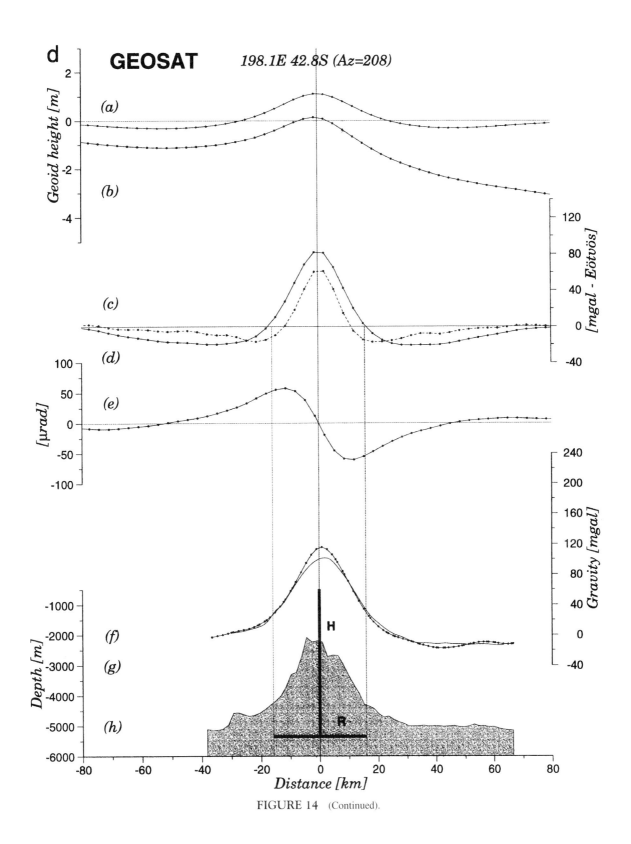

FIGURE 14 (Continued).

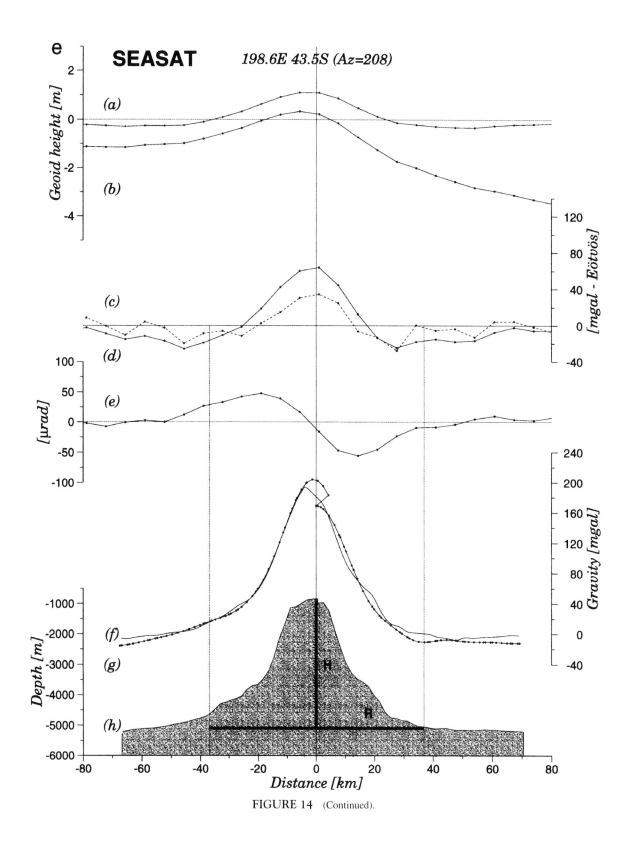

FIGURE 14   (Continued).

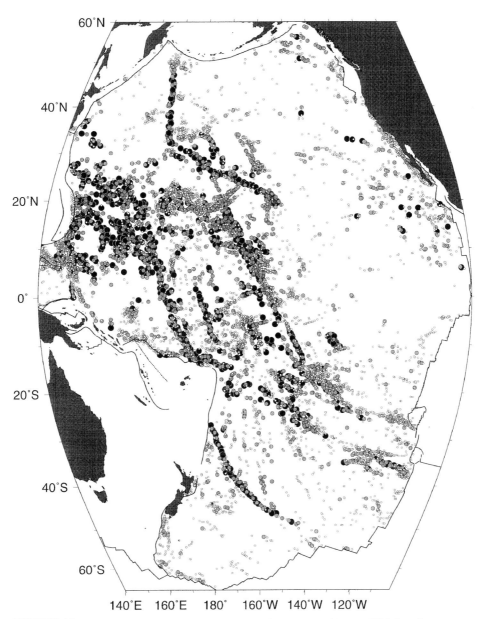

FIGURE 15 Seamount population on the Pacific plate, after Wessel and Lyons (1997). Locations were extracted from the vertical gravity gradient of the global gravity anomaly grid of Sandwell and Smith (1997). The size of circle symbols varies with the amplitude of the vertical gravity gradient: small open-circles 30–60 Eotvos (typically 1–2 km tall), medium grey-circles 60–120 Eotvos (typically 1.5–4 km), large black-circles >120 Eotvos (>2.5 km). The number of seamounts in each category is 4423, 2167, and 895, respectively. Eckert equal-area projection.

and Lyons, 1997). The mean slope is a function of the height to width ratio; hence for a given mean slope, the seamount shape is only controlled by the seamount height. In order to model the flexure of the lithosphere due to the seamount load, the simplest model is to assume that the lithosphere behaves as a continuous and elastic layer, characterized by its elastic thickness T_e (Watts and Ribe, 1984). Comparisons between observed and calculated geoid anomalies, for a known bathymetry, show that a precise knowledge of T_e

is not critical and that it is easy to distinguish a seamount generated near a ridge axis (e.g., $T_e \sim 5$ km) from a seamount emplaced on an old lithosphere ($T_e > 15$ km; Watts and Ribe, 1984; Baudry *et al.*, 1987). Then for a given value or set of elastic thickness, a series of models can be computed with varying heights and slopes, and compared to the observed anomalies (geoid, gravity, or vertical gravity gradient) in order to derive the height and width of a seamount. Predicted and observed seamount sizes for well-charted seamounts may differ by 15–30%; the underestimation of the height is likely due to the Gaussian shape approximation (Baudry and Diament, 1987; Wessel and Lyons, 1997). The satellite-derived volumes are about 10–25% smaller than the volumes determined from bathymetric data sets. The number of seamounts decreases as their diameter increases (Craig and Sandwell, 1988). In the 2–8 km range height, the frequency for seamount height varies as a power law of the seamount predicted heights (Wessel and Lyons, 1997). Perhaps the most promising technique is to invert the satellite-derived gravity grid in conjunction with available shipboard bathymetric data (see Chapter 12; e.g., Baudry and Calmant, 1991; Calmant, 1994; Ramillien and Cazenave, 1997; Smith and Sandwell, 1997). These new predicted topography grids awaits quantitative studies.

The age of the seamount is another parameter that can tentatively be derived from the size of a seamount, its geoid signature and the age of the underlying seafloor. One way would be to compute the elastic thickness in order to infer the age of the lithosphere at the time of loading; knowing the age of the seafloor, an estimate of the seamount age can be deduced (Calmant et al., 1990). However, accurate estimates of T_e require an accurate knowledge of the topography. From a limited number of seamounts, Epp (1984) noted a linear relationship between the seamount heights and the age of the seafloor at the time the seamounts built up. Following this observation and from their collection of seamounts, Wessel and Lyons (1997) determined an empirical relation yielding the pseudo age of a seamount from the age of the seafloor and the vertical gravity gradient amplitude. This empirical law, although highly speculative, suggests that, in the Pacific, intraplate volcanism had a peak activity during the mid-Cretaceous, not in the Paleocene, as found from a previous study (Batiza, 1982).

8.2.4. Hotspot Traces

Perhaps one of the most intriguing question about the numerous observed submarine volcanoes is their origin. Linear chains of seamounts and of elongated narrow ridges such as the Hawaiian-Emperor and Louisville chains are interpreted as the surface expression of stationary deep mantle plumes or hotspots on the overriding plate (Wilson, 1965b; Morgan, 1971). The trajectory of a fixed hotspot on the overriding plate follows a small circle about the absolute pole of motion; if the direction of motion changed through time, the hotspot track will follow a succession of small circles. Thus fitting small circles to at least two congruent hotspot traces yield the location of the stage poles. The length of each small-circle fitted to the seamount locations yields the angle of each stage rotation. Finally, age determinations along the modeled hotspot tracks yields the rate of motion. Thus a correct model of absolute plate motion must both predict the hotspot tracks and the age progression along the tracks (e.g., Clague and Jarrard, 1973; Morgan, 1981).

The identification of many new uncharted seamounts from satellite-altimeter data allows further tests of the hotspot hypothesis. Craig and Sandwell (1988) and Wessel and Lyons (1997) identified many new lineated chains with trends parallel to the Hawaiian-Emperor and Louisville chains, such as the Mariana, Gilbert, Tuamotu and Austral island groups. Some of these chains intersect with one another and are difficult to sort out. In addition, the location of the hotspots from which these chains would have originated is not always known. Wessel and Kroenke (1997) presented a geometric relationship that links hotspots to the seamounts they produced. For a given kinematic model, they constructed the flowlines or trajectories of each seamount relative to the hotspots, instead of reconstructing the hotspot trajectory relative to the overriding plate, the latter requiring to know its present-day location. If the tested model is correct, all the flowlines originating from seamounts related to the same hotspot should intersect at its present-day location, even if extinct. Applied to the Pacific plate, this technique is particularly demonstrative because of the sharp change in the direction of absolute motion which occurred at about 43 Ma, as illustrated by the Hawaiian-Emperor "elbow". However, except for the Hawaiian and Louisville hotspots, which were used to constrain the rotation poles and angles, most flowlines do not converge, or if so, at points that do not coincide with known hotspots such as Macdonald, Society, or Pitcairn. The main implications are either that the kinematic model is incorrect or that hotspots move relative to one another. In addition, as mentioned in Section 7, there are many seamounts, either isolated or in linear chains that do not fit the hotspot hypothesis. Some originated at spreading ridge axes or near transform fracture zones, marking short or long-lived outbursts of magmatism; other may originate from mini-hotspots moving relative to the major ones (Fleitout and Moriceau, 1992) or may result from tensional cracks in the lithosphere (Sandwell et al., 1995). Conversely, some short-wavelength undulations of the geoid, topped by few seamounts or linear ridges, seem to coincide with absolute plate motions (see Section 7), but have not been used to constrain them. Hence, the satellite-altimeter data provide an almost exhaustive set of information to improve the geometry of absolute plate motion model; however, age determinations still remain the limiting factor to fully constrain them and so, to address quantitatively the question of hotspot fixity or of hotspot fluxes trough time.

8.3. Spreading Ridges

Mid-oceanic spreading ridges, the longest mountain chain on Earth, are now almost entirely and accurately mapped from satellite-gravity, except in the Arctic region north of 81.58°N, not covered by satellite altimeters. As discovered since the earliest observation of seafloor morphology

(Heezen, 1960; Menard, 1960; Macdonald, 1982), the topographic expression of spreading ridges varies between two end-members: fast-spreading ridges such as the East Pacific Rise (130–160 mm year^{-1}) are characterized by a small axial rise and smooth flanks whereas slow-spreading ridges such as the mid-Atlantic Ridge (11–25 mm year^{-1}) are characterized by a deep axial valley and rugged flanks. As shown in Figure 10, satellite-derived gravity reflects these characteristics and displays a linear axial positive anomaly and smooth flanks along fast-spreading ridges and a linear negative anomaly flanked by positive anomalies and rough flanks along slow-spreading ridges. Since the intermediate-spreading ridges (Southeast Indian Ridge and Pacific-Antarctic Ridge) are located in the remote southern oceans, the transition between this two end-member configurations used to be poorly ascertained. The uniform coverage of the satellite-derived gravity data now allows one to locate precisely the ridge axes and to compare the gravity signal along most of the world ocean spreading ridge system, and also to characterize the structural gravity grain, also called gravity roughness.

8.3.1. Gravity Signature of Mid-oceanic Ridges

Along slow-spreading ridges, the peak-to-trough amplitude of the gravity anomaly at ridge axes decreases with increasing spreading rates, from −120–0 mGal, whereas the amplitude remains relatively constant on fast spreading ridges, about ~10 mGal (Owens and Parsons, 1994; Small and Sandwell, 1994). The transition from the slow- to fast-spreading structure occurs along the Southeast Indian and Pacific-Antarctic ridges within a narrow range of spreading rates, between 60 and 80 mm year^{-1} (Figure 16; Small and Sandwell, 1989; Small and Sandwell, 1992; Owens and Parsons, 1994; Small and Sandwell, 1994). In some instances, the transition occurs abruptly across a single transform fault, like at 171°W along the Pacific-Antarctic Ridge (60 mm year^{-1}; Sandwell, 1992) or at 78.5°E, 41°S along the Southeast Indian Ridge (68 mm year^{-1}; Owens and Parsons, 1994). In other instances, the transition is more gradual like along the Reykjanes Ridge (19 mm year^{-1}; Hwang et al., 1994) or along the Australian-Antarctic Discordance (74 mm year^{-1}; Palmer et al., 1993). The latter two areas are however notable exceptions in the observed patterns along slow- and fast-spreading ridges, respectively. The range of spreading rates at which the transitions from slow- to fast-spreading structure occur, thus indicates that the spreading rate is not the only controlling factor. The crustal genesis model of Phipps Morgan and Chen (1993a; 1993b) suggests that the axial morphology is ultimately controlled by the thermal structure of the ridge axis, which is a function of the spreading rate and the magma supply. The transition in axial morphology would then occur above a threshold sensitive to small thermal perturbations (Marks and Stock, 1994; Small, 1994; Small and Sandwell, 1994; Sahabi et al., 1996), either

local, or related to a hotspot (Iceland hotspot for the Reykjanes Ridge, Bouvet hotspot for the South Atlantic Ridge; Amsterdam-St. Paul hotspot for the Southeast Indian Ridge) or to asthenospheric flows (Pacific-Antarctic Ridge; Marks and Stock, 1994; Sahabi et al., 1996).

8.3.2. Geoid/gravity Roughness

The previous sections have mainly used the 20–200 km wavelength in the satellite altimeter data for mapping the major structures of the seafloor. This bandwidth is also informative about the structural characteristics of large oceanic areas. Figures 10a and 10d show clearly that the amplitudes of the gravity anomalies are larger and vary over a wider range in the slow-spreading Atlantic Ocean than in the fast-spreading Pacific Ocean. A quantitative comparison between a series of satellite passes in the Atlantic and Pacific oceans show that the gravity field in the Atlantic has more power than in the Pacific for all wavelengths larger than 20 km (Small and Sandwell, 1992).

Variations in the amplitude of the small-wavelength (20–200 km) geoid or gravity field, also called geoid or gravity roughness, can thus be used to characterize on a regional scale these spreading rate dependent processes. Geoid roughness has been estimated along satellite tracks by the envelope of band-path filtered Seasat geoid profiles (Gibert et al., 1989; Goslin and Gibert, 1990) or by a weighted average of least square (rms) of deflection of the vertical in a moving window along band-path filtered and stacked Geosat ERM profiles (Small and Sandwell, 1992). Global roughness map obtained from gridding these data emphasizes the variations of roughness along the spreading ridges. Slow-spreading ridges (Atlantic and western Indian Ocean) display high roughness (>13 cm) on the axis and lower roughness (5–6 cm) on the flanks; fast spreading ridges (East Pacific Rise, northern Pacific-Antarctic Ridge) present a low roughness (<3 cm) both on the axis and the flanks. Gravity roughness at spreading ridge crossings decrease by a factor 10 between 16 and 80 mm year^{-1}, beyond which it remains uniformly low (Small and Sandwell, 1992). The abrupt transition between 60 and 80 mm year^{-1} is consistent with that found in the gravity amplitude or in the topography of spreading axes (Small, 1994, 1998). The roughness of the ridge flanks also increase as the spreading rates decrease, consistent with analyses of the ridge flank topography roughness (Goff, 1991; Hayes and Kane, 1991; Malinverno, 1991). However, except along some portions of the Carslberg and Southeast Indian ridges, the satellite passes are generally not parallel to the fracture pattern (i.e., flowlines), thus making difficult any quantitative comparison between geoid and topography roughness. Geoid roughness also reveals intermediate (90–400 km) wavelength patterns, reflecting the ridge large-scale segmentation (Gibert et al., 1989), small-scale convection pattern (Fleitout and Moriceau, 1992), compensation pro-

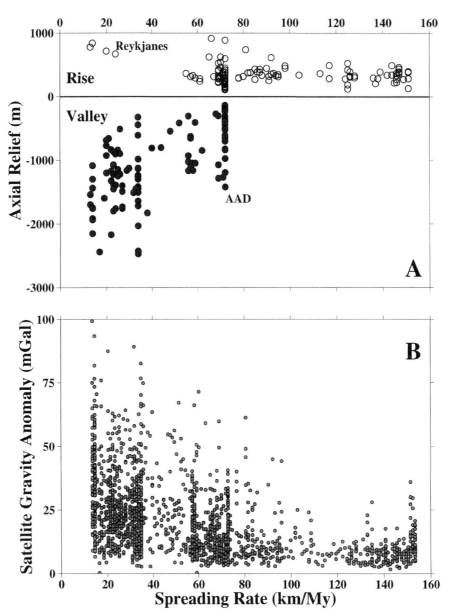

FIGURE 16 Variation of ridge-axis morphology and gravity anomaly with spreading rate, after Small (1998): (a) Amplitude of axial topographic relief measured from bathymetric profiles, (b) Amplitude of axial gravity-anomaly derived from the global gravity grid (Sandwell and Smith, 1997). Note the coincident change in polarity between 60 and 80 mm year^{-1}. The axial rise along the slow-spreading Reykjanes Ridge and the axial valley at the Australian-Antarctic Discordance (AAD) along the intermediate-spreading Southeast Indian Ridge are two notable exceptions in the general pattern.

cesses (Goslin and Gibert, 1990), or lithospheric deformation (McAdoo and Sandwell, 1985; Fleitout *et al.*, 1989) (see also Section 7).

9. CONCLUSION

Satellite altimeters have provided a wealth of information about the structure of the seafloor and about the man-

tle processes that generate them. Perhaps one of the most important result is the confirmation of plate tectonics. The satellite-derived gravity data reveal the whole extension of the active oceanic plate boundaries, transform faults, spreading ridges and trenches, and their triple junctions. These data also outline the complexity of fracture zones which not only reflect the past and present direction of plate motion, but also the large- and fine-scale segmentation of the spreading ridges, and the common occurrence of propagating rifts. The very large number of seamounts that these data unveiled, along with all the small seamounts (<1 km) not resolved by these measurements, indicate that they contribute significantly to the oceanic volcanic layer. The complete mapping of seamount chains confirms and helps improving hotspot models; however, the occurrence of many linear chains that do not fit this model helped formulating alternative explanations, such as small-scale convection or tensional cracks. The satellite-altimeter data also lead to a better knowledge of the mechanical properties of the oceanic lithosphere and its behavior at fracture zones, under vertical stresses such as seamount loads and under horizontal stresses during intraplate deformation. Finally, these data are extremely valuable to plan and design ship cruises for investigating features of the ocean floor at a finer scale.

ACKNOWLEDGMENTS

The authors are grateful to David Sandwell for his review of this chapter.

References

Abers, G. A., Parsons, B., and Weissel, J. K. (1988). Seamount abundances and distributions in the southeast Pacific. *Earth Planet. Sci. Lett.*, **87**, 137–151.

Batiza, R. (1982). Abundances, distribution and sizes of volcanoes in the Pacific Ocean and implications for the origin of non-hotspot volcanoes. *Earth Planet. Sci. Lett.*, **60**, 195–206.

Baudry, N., and Calmant, S. (1991). 3-D Modelling of seamount topography from satellite altimetry. *Geophys. Res. Lett.*, **18**, 1143–1146.

Baudry, N., and Diament, M. (1987). Shipboard confirmation of Seasat bathymetric prediction in the South Central Pacific. *In* "Seamounts, islands and atolls," (Keating, B., Fryer, P., Batiza, R., and Boelhert, G., eds.), Vol. 43, pp. 115–122, American Geophysics Union, Washington D.C.

Baudry, N., and Kroenke, L. (1991). Intermediate-wavelengths (400–600 km) south Pacific undulation: their relationship to linear volcanic chains. *Earth Planet. Sci. Lett.*, **102**, 430–443.

Baudry, N., Diament, M., and Albouy, Y. (1987). Precise location of unsurveyed seamounts in the Austral archipelago area using Seasat data. *Geophys. J. R. Astron. Soc.*, **89**, 869–888.

Black, M. T., and McAdoo, D. C. (1988). Spectral analysis of marine geoid heights and ocean depths: constraints on models of lithospheric and sublithospheric processes. *Mar. Geophys. Res.*, **10**, 157–180.

Buck, W. R. (1987). Analysis of the cooling of a variable-viscosity fluid with applications to the Earth, *Geophys. J.*, **89**, 549–577.

Buck, W. R., and Parmentier, E. M. (1986). Convection beneath young oceanic lithosphere: implications for thermal structure and gravity. *J. Geophys. Res.*, **91**, 1961–1974.

Calcagno, P., and Cazenave, A. (1994). Subsidence of the seafloor in the Atlantic and Pacific oceans: regional and large-scale variations. *Earth Planet. Sci. Lett.*, **126**, 473–492.

Calmant, S. (1987). The elastic thickness of the lithosphere in the Pacific Ocean. *Earth Planet. Sci. Lett.*, **85**, 277–288.

Calmant, S. (1994). Seamount topography from least-squares inversion of altimetric geoid heights and shipborne profiles of bathymetry and/or gravity anomalies. *Geophys. J. Int.*, **119**, 428–452.

Calmant, S., and Cazenave, A. (1987). Anomalous elastic thickness of the oceanic lithosphere in the South-Central Pacific. *Nature*, **328**, 236–236.

Calmant, S., Francheteau, J., and Cazenave, A. (1990). Elastic layer thickening with age of the oceanic lithosphere: a tool for predicting the age of voolcanoes or oceanic crust. *Geophys. J. Int.*, **90**, 100, 59–67.

Cañales, J. P., and Dañobetia, J. J. (1998). The Canary islands swell: a coherence analysis of bathymetry and gravity. *Geophys. J. Int.*, **132**, 479–488.

Cande, S., LaBrecque, J. L., and Haxby, W. B. (1988). Plate kinematics of the South Atlantic: Chron 34 to present. *J. Geophys. Res.*, **93**, 13,479 - 13,492.

Cande, S. C., Raymond, C. A., Stock, J., and Haxby, W. F. (1995). Geophysics of the Pitman Fracture Zone and Pacific-Antarctic plate motions during the Cenozoic. *Science*, **270**, 947–953.

Caress, D. W., Menard, H. W., and Hey, R. N. (1988). Eocene reorganization of the Pacific-Farallon Spreading Center north of the Mendocino Fracture Zone. *J. Geophys. Res.*, **93**, 2813–2838.

Cazenave, A. (1984). Thermal cooling of the oceanic lithosphere: possible evidence of two distinct trends. *Nature*, **310**, 401–403.

Cazenave, A., and Thoraval, C. (1994). Mantle dynamics constrained by degree 6 surface topography, seismic tomography and geoid: inference on the origin of the south Pacific superswell. *Earth Planet. Sci. Lett.*, **122**, 207–219.

Cazenave, A., Monnereau, M., and Gibert, D. (1987). Seasat gravity undulations in the central Indian Ocean. *Phys. Earth Planet. Int.*, **48**, 130–141.

Cazenave, A., Souriau A., and Dominh, K. (1989). Earth surface topography: global coupling with hotspots, geoid and lower mantle heterogeneities. *Nature*, **340**, 54–57.

Cazenave, A., Houry, S., Lago, B., and Dominh, K. (1992). Geosat-derived geoid anomalies at medium-wavelength. *J. Geophys. Res.*, **97**, 7081–7096.

Cazenave, A., Parsons, B., and Calcagno, P. (1995). Geoid lineations of 1000 km wavelength over the central Pacific. *Geophys. Res. Lett.*, **22**, 97–100.

Cazenave, A., Schaeffer, P., Bergé, M., Brossier, C., Dominh, K., and Gennero, M. C. (1996). High resolution mean sea surface computed with altimeter data of ERS-1 (geodetic mission) and T/P. *Geophys. J. Int.*, **125**, 696–704.

Ceuleneer, G., Rabinowicz, M., Monnereau, M., Cazenave, A., and Rosemberg, C. (1988). Viscosity and thickness of the sub-lithospheric low-viscosity zone: constraints from geoid and depth over oceanic swells. *Earth Planet. Sci. Lett.*, **89**, 84–102.

Clague, D. A., and Jarrard, R. D. (1973). Tertiary Pacific plate motion deduced from the Hawaiian-Emperor chain. *Geol. Soc. Am. Bull.*, **84**, 1135–1154.

Craig, C. H., and McKenzie, D. P. (1986). The existence of a thin low-viscosity layer beneath the lithosphere. *Earth Planet. Sci. Lett.*, **78**, 420–426.

Craig, C. H., and Sandwell, D. T. (1988). The global distribution of seamounts from Seasat profiles. *J. Geophys. Res.*, **93**, 10408–10420.

Crough, S. T. (1978). Thermal origin of mid plate hotspot swells. *Geophys. J. R. Astr. Soc.*, **55**, 451–469.

Crough, S. T. (1983). Hotspot swells. *Annu. Rev. Earth Planet. Sci.*, **11**, 165–193.

Dahlen, F. (1982). Isostatic geoid anomalies on a sphere. *J. Geophys. Res.*, **87**, 3943–3947.

Davaille, A., and Jaupart, C. (1994). On set of thermal convection in fluids with temperature-dependant viscosity: application to the oceanic mantle. *J. Geophys. Res.*, **99**, 19853–19866.

DeMets, C., Gordon, R. G., Argus, D. F., and Stein, S. (1990). Current plate motions. *Geophys. J. Int.*, **101**, 425–478.

DeMets, C., Gordon, R. G., Argus, D. F., and Stein, S. (1994). Effect of recent revisions to the geomagnetic reversal timescale on estimates of current plate motion. *Geophys. Res. Lett.*, **21**, 2191–2194.

Detrick, R. S. (1981). An analysis of geoid anomalies across the Mendocino fracture zone: implications for thermal models of the lithosphere. *J. Geophys. Res.*, **86**, 11751–11762.

Detrick, R. S., White, R., and Purdy, G. (1993). Crustal structure of North Atlantic fracture zones. *Rev. Geophys.*, **31**, 439–458.

Doin, M. P., and Fleitout, L. (1996). Thermal evolution of the oceanic lithosphere: an alternative view. *Earth Planet. Sci. Lett.*, **142**, 121–136.

Dorman, L. M., and Lewis, B. T. R. (1970). Experimental isostasy, 1, Theory of the determination of the Earth isostatic response to a concentrated load. *J. Geophys. Res.*, **75**, 3357–3365.

Driscoll, M., and Parsons, B. (1988). Cooling of the oceanic lithosphere—evidence from geoid anomalies across the Udintsev and Eltanin fracture zones. *Earth Planet. Sci. Lett.*, **88**, 289–307.

Eberle, M. A., and Forsyth, D. W. (1995). Regional viscosity variations, small-scale convection and the slope of the depth-age curve. *Geophys. Res. Lett.*, **22**, 473–476.

Epp, D. (1984). Implications of volcano and swell heights for thinning of the lithosphere hotspots. *J. Geophys. Res.*, **89**, 9991–9996.

Filmer, P. E., McNutt, M. K., and Wolfe, C. J. (1993). Elastic thickness of the lithospheric in the Marquesas and society Islands. *J. Geophys. Res.*, **98**, 19,565–19,577.

Fisher, R. L., Jantsch, M. Z., and Comer, R. L. (1982). *In* "General bathymetric chart of the oceans (GEBCO)—Chart 5-09 (scale 1:1.000.000)," Canadian Hydrographic Service, Ottawa, Canada.

Fischer, K. M., McNutt, M. K., and Shure, L. (1986). Thermal and mechanical constraints on the lithosphere beneath the Marquesas swell. *Nature*, **322**, 733–736.

Fleitout, L., and Moriceau, C. (1992). Short-wavelength geoid, bathymetry and the convective pattern beneath the Pacific ocean. *Geophys. J. Int.*, **110**, 6–28.

Fleitout, L., Dalloubeix, C., and Moriceau, C. (1989). Small-wavelength geoid and topography anomalies in the South Atlantic Ocean: a clue to new hot-spot tracks and lithospheric deformation. *Geophys. Res. Lett.*, **16**, 637–640.

Forsyth, D. W. (1985). Subsurface loading and estimates of the flexural rigidity of the continental lithosphere. *J. Geophys. Res.*, **90**, 12632–12632.

Freedman, A. P., and Parsons, B. (1990). Geoid anomalies over two south Atlantic fracture zones. *Earth Planet. Sci. Lett.*, **100**, 18–41.

Gahagan, L. M., Royer J.-Y., Scotese, C. R., Sandwell, D. T., Winn, K., Tomlins, R., Ross, M. I., Newman, J. S., Müller, D., Mayes, C. L., Lawver, L. A., and Huebeck, C. E. (1988). Tectonic fabric map of the ocean basins from satellite altimetry data. *Tectonophysics*, **155**, 1–26.

Gaina, C., Müller, R. D., Royer, J.-Y., Stock, J., Hardebeck, J., and Symonds, P. (1998). The tectonic history of the Tasman Sea: a puzzle with seven pieces. *J. Geophys. Res.*, **103**, 12413–12433.

Gairola, R., Basu, S., and Pandey, P. (1992). A Geosat-derived noise spectrum for detecting seamounts in the Arabian Sea. *Int. J. Remote Sensing*, **13**, 971–979.

Gibert, D., Camerlynck, C., and Courtillot, V. (1987). Geoid anomalies across Ascension fracture zone and the cooling of the lithosphere. *Geophys. Res. Lett.*, **14**, 603–606.

Gibert, D., Courtillot, V., and Olivet, J.-L. (1989). Seasat altimetry and the South Atlantic geoid 2. Short-wavelength undulations. *J. Geophys. Res.*, **94**, 5545–5559.

Goetze, C., and Evans, B. (1979). Stress and temperature in the bending lithosphere as constrained by experimental rock mechanics. *Geophys. J.R. Astr. Soc.*, **59**, 463–478.

Goff, J. A. (1991). A global and regional stochastic analysis of near-ridge abyssal hill morphology. *J. Geophys. Res.*, **96**, 21713–21737.

Goodwillie, A. M. (1993). An altimetric and bathymetric study of elastic thickness in the central Pacific ocean. *Earth Planet. Sci. Lett.*, **118**, 311–326.

Goodwillie, A. M., and Watts, A. B. (1995). Short-wavelength gravity lineations and unusual flexure results at the Puka-Puka volcanic ridge system. *Earth Planet. Sci. Lett.*, **136**, 297–314.

Goslin, J., and Gibert, D. (1990). The geoid roughness: a scanner for isostatic processes in oceanic areas. *Geophys. Res. Lett.*, **17**, 1957–1960.

Hager, B. H., and Clayton, R. W. (1989). Constraints on the structure of mantle convection using seismic observations, flows models and the geoid. *In* "Mantle Convection," (W. R. Peltier, ed.), Gordon and Breach, New York, pp. 657–765.

Hager, B. H., and Richards, M. (1989). Long-wavelength variations in Earth's geoid, physical models and dynamical implications. *Phil. Trans. R. Soc. London*, **7328**, 309–328.

Haxby, W. F. (1987). *In* "Gravity field of the world's oceans," Office of Naval Research, United States Navy, Lamont-Doherty Geological Observatory, Boulder, CO.

Haxby, W. F., and Turcotte, D. L. (1978). On isostatic geoid anomalies. *J. Geophys. Res.*, **83**, 5473–5478.

Haxby, W. F., and Weissel, J. K. (1986). Evidence for small-scale mantle convection from Seasat altimeter data. *J. Geophys. Res.*, **91**, 3507–3520.

Haxby, W. F., Karner, G. D., LaBrecque, J. L., and Weissel, J. K. (1983). Digital images of combined oceanic and continental data sets and their use in tectonic studies. *EOS Trans. Am. Geophys. Un.*, **64**, 995–1004.

Hayes D. (1988). Age-depth relationships and depth anomalies in the southeast Indian Ocean and south Atlantic Ocean. *J. Geophys. Res.*, **93**, 2937–2954.

Hayes, D., and Kane, K. (1991). The dependence of seafloor roughness on spreading rate. *Geophys. Res. Lett.*, **18**, 1425–1428.

Heestand, R. L., and Crough, S. T. (1981). The effect of hotspots on the oceanic age-depth relation. *J. Geophys. Res.*, **86**, 6107–6114.

Heezen, B. (1960). The rift in the ocean floor. *Scientific American*, **203**, 99–110.

Heiskanen, W. A., and Moritz, H. (1967). *In* "Physical geodesy," Freeman, San Francisco.

Hey, R. N., Menard, H. W., Atwater, T. M., and Caress, D. W. (1988). Changes in direction of seafloor spreading revisited. *J. Geophys. Res.*, **93**, 2803–2811.

Hey, R. N., Sinton, J. M., and Duennebier, F. K. (1989). Propagating rifts and spreading centers. *In* "The Eastern Pacific Ocean and Hawaii" (Winterer, E. L., Hussong, D. M., and Decker, R. W., eds.), Vol. N, pp. 161–178, Geological Society of America, Boulder, Colorado.

Hohertz, W. L., and Carlson, R. L. (1998). An independent test of thermal subsidence and asthenospheric flow beneath the Argentine basin. *Earth Planet. Sci. Lett.*, **161**, 73–83.

Hwang, C., Parsons, B., Strange, T., and Bingham, A. (1994). A detailed gravity field over the Reykjanes Ridge from Seasat, Geosat, ERS-1 and TOPEX/POSEIDON altimetry and shipborne gravity. *Geophys. Res. Lett.*, **21**, 2841–2844.

Hwang, C., Kao, E. C., and Parsons, B. (1998). Global derivation of marine gravity anomalies from Seasat, Geosat, ERS-1 and T/P altimeter data. *Geophys. J. Int.*, **134**, 449–459.

Jarvis, G. T., and Peltier, W. R. (1980). Oceanic bathymetry profiles flattened by radiogenic heating in convecting mantle. *Nature*, **285**, 649–651.

Jordan, T., Menard, H., and Smith, D. (1983). Density and size distribution of seamounts in the eastern Pacific inferred from wide-beam sounding data. *J. Geophys. Res.*, **88**, 10508–10518.

Judge, A. V., and McNutt, M. K. (1991). The relationship between plate curvature and elastic plate thickness: a study of the Peru-Chile trench. *J. Geophys. Res.*, **96**, 16625–16640.

Kane, K. A., and Hayes, D. E. (1994). A new relationship between subsidence rate and zero-age depth. *J. Geophys. Res.*, **99**, 21759–21777.

Katzman, R., Zhao, L., and Jordan, T. H. (1998). Seismic structure of the mantle between Ryukyu and Hawaii: implications for the Hawaiian swell. *J. Geophys. Res.*, in press.

Keating, B., Mattey, D., Helsley, C., Naughton, J., and Epp, D. (1984). Evidence for a hot spot origin of the Carolina Islands. *J. Geophys. Res.*, **89**, 9937–9948.

Kruse, S. E., Liu, Z. J., Naar, D. F., and Duncan, R. A. (1997). Effective elastic thickness of the lithosphere along the Easter seamount chain. *J. Geophys. Res.*, **102**, 27305–27317.

Kuykendall, M., Kruse, S., and McNutt, M. (1994). The effects of changes in plate motions on the shape of the Marquesas fracture zone. *Geophys. Res. Lett.*, **21**, 2845–2848.

LaBrecque, J. L., and Rabinowitz, P. D. (1981). *In* "General bathymetric chart of the oceans (GEBCO)—Chart 5–16 (scale 1:1.000.000)," Canadian Hydrographic Service, Ottawa, Canada.

Lambeck, K., and Coleman, R. (1982). A search for seamounts in the southern Cook and Austral islands. *Geophys. Res. Lett.*, **9**, 389–392.

Lawver, L. A., and Gahagan, L. (1994). Constraints on timing of extension in the Ross Sea region. *Terra Antartica*, **1**, 545–552.

Lawver, L. A., Royer, J.-Y., Sandwell, D. T., and Scotese, C. R. (1991). Evolution of the Antarctic continental margins. *In* "Geological Evolution of Antarctica" (Thomson, M. R. A., Crame, J. A., and Thomson, J. D., eds.), pp. 533–539, Cambridge University Press, Cambridge, U.K.

Laxon, S., and McAdoo, D. (1994). Arctic ocean gravity field derived from ERS-1 satellite atlimetry. *Science*, **265**, 621–624.

Lazarewicz, A. R., and Schwank, D. C. (1982). Detection of uncharted seamounts using satellite altimetry. *Geophys. Res. Lett.*, **9**, 385–388.

Levitt, D. A., and Sandwell, D. T. (1995). Lithospheric bending at subduction zones based on depth soundings and satellite gravity. *J. Geophys. Res.*, **100**, 379–400.

Litvin, V., and Rudenko, M. (1973). Distribution of seamounts in the Atlantic. *Dokl. Acad. Sci. USSR, Earth Sci. Sect. (English transl.)*, **213**, 223–225.

Louden, K. E., and Forsyth, D. W. (1976). Thermal conduction across fracture zones and the gravitational edge effects. *J. Geophys. Res.*, **81**, 4869–4874.

Macdonald, K. C., Sempere, J. C., and Fox, P. J. (1986). Reply: the debate concerning overlapping spreading centers and mid-ocean ridges processes. *J. Geophys. Res.*, **91**, 10501–10511.

Maia, M., and Diament, M. (1991). An analysis of the altimetric geoid in various wavebands on the central Pacific Ocean: constraints on the origin of intraplate features. *Tectonophysics*, **190**, 133–135.

Malinverno, A. (1991). Inverse square-root dependence of mid-ocean ridge flank roughness on spreading rate. *Nature*, **352**, 58–60.

Mammerickx, J., and Smith, S. M. (1982). *In* "General bathymetric chart of the oceans (GEBCO)—Chart 5–07 (scale 1:1.000.000)," Canadian Hydrographic Service, Ottawa, Canada.

Marks, K. M., and Sandwell, D. T. (1991). Analysis of geoid height versus topography over oceanic plateaus and swells using non-biased linear regression. *J. Geophys. Res.*, **96**, 8045–8055.

Marks, K. M., and Stock, J. M. (1994). Variations in ridge morphology and depth-age relationships on the Pacific-Antarctic Ridge. *J. Geophys. Res.*, **99**, 531–541.

Marks, K. M., and Stock, J. M. (1997). Early Tertiary gravity field reconstructions of the Southwest Pacific. *Earth Planet. Sci. Lett.*, **152**, 267–274.

Marks, K. M., McAdoo, D. C., and Smith, W. H. F. (1993). Mapping the Southwest Indian Ridge with Geosat. *EOS Trans. Am. Geophys. Un.*, **74**, 81–86.

Marty, J. C., and Cazenave, A. (1988). Thermal evolution of the lithosphere beneath fracture zones inferred from geoid anomalies. *Geophys. Res. Lett.*, **15**, 593–597.

Marty, J. C., and Cazenave, A. (1989). Regional variations in subsidence rate of oceanic plates: a global analysis. *Earth Planet. Sci. Lett.*, **94**, 301–315.

Mayes, C. L., Lawver, L. A., and Sandwell, D. T. (1990). Tectonic history and new isochron chart of the South Pacific. *J. Geophys. Res.*, **95**, 8543–8567.

McAdoo, D. C., and Laxon, S. (1997). Antarctic tectonics: constraints from an ERS-1 satellite marine gravity field. *Science*, **276**, 556–560.

McAdoo, D. C., and Marks, K. M. (1992). Gravity Fields of the Southern Ocean From Geosat Data. *J. Geophys. Res.*, **97**, 3247–3260.

McAdoo, D. C., and Sandwell, D. T. (1985). Folding of Oceanic Lithosphere. *J. Geophys. Res.*, **90**, 8563–8569.

McKenzie, D. P. (1967). Some remarks on heat flow and gravity anomalies. *J. Geophys. Res.*, **72**, 6261–6273.

McKenzie, D. P. (1977). Surface deformation, gravity anomalies and convection. *Geophys. J. R. Astr. Soc.*, **48**, 211–238.

McKenzie, D. P., and Bowin, C. (1976). The relationship between bathymetry and gravity in the Atlantic Ocean. *J. Geophys. Res.*, **81**, 1903–1915.

McNutt, M. K. (1984). Lithospheric flexure and thermal anomalies. *J. Geophys. Res.*, **89**, 11,180–11,194.

McNutt, M. K. (1988). Thermal and mechanical propoerties of the Cape Verde Rise. *J. Geophys. Res.*, **93**, 2784–2794.

McNutt, M. K. (1998). Superswells. *Rev. Geophys.*, **36**, 211–244.

McNutt, M. K., and Fischer, K. M. (1987). The South Pacific Superswell, in Seamounts, Islands and Atolls. *Geophys. Monogr.*, **43**, American Geophysical Union, Washington, D.C., p. 25–34.

McNutt, M. K., and Judge, A. V. (1990). The Superswell and mantle dynamics beneath the south Pacific. *Science*, **248**, 969–975.

McNutt, M. K., and Shure, L. (1986). Estimating the compensation depth of the Hawaiian swell with linear filters. *J. Geophys. Res.*, **91**, 13915–13923.

McNutt, M. K., Caress, D. W., Reynolds, J., Jordahl, K. A., and Duncan, R. A. (1997). Railure of plume theory to explain midplate volcanism in the southern Austral islands. *Nature*, **389**, 479–482.

Menard, H. W. (1960). The East Pacific rise. *Science*, **132**, 1737–1746.

Menard, H. W. (1964). *In* "Marine geology of the Pacific," McGraw-Hill, New York, NY.

Menard, H. W., and McNutt, M. K. (1982). Evidence and consequence of thermal rejuvenation. *J. Geophys. Res.*, **87**, 857–8580.

Monahan, D., Falconer, R. H. K., and Tharp, M. (1982). *In* "General bathymetric chart of the oceans (GEBCO)—Chart 5–10 (scale 1:1.000.000)," Canadian Hydrographic Service, Ottawa, Canada.

Monnereau, M., and Cazenave, A. (1988). Variation of the apparent compensation depth of hotspot swells with age of plate. *Earth Planet. Sci. Lett.*, **91**, 179–197.

Monnereau, M., and Cazenave A., (1990). Depth and geoid anomalies over oceanic hotspot swells: a global survey. *J. Geophys. Res.*, **95**, 15429–15438.

Monnereau, M., Rabinowicz, M., and Arquis, E. (1993). Dynamical and thermal erosion of the lithosphere: a numerical model for hotspot swells. *J. Geophys. Res.*, **98**, 809–823.

Moore, W. B., Schubert, G., and Tackley, P. (1998). Three-dimensional simulations of plume lithosphere interaction at the Hawaiian swell. *Science*, **279**, 1008–1011.

Morgan, W. J. (1968). Rises, trenches, great faults, and crustal blocks. *J. Geophys. Res.*, **73**, 1959–1982.

Morgan, W. J. (1971). Convection plumes in the lower mantle. *Nature*, **230**, 42–43.

Morgan, W. J. (1981). Hotspot tracks and the opening of the Atlantic and Indian Oceans. *In* "The oceanic lithosphere," (Emiliani, C., ed.), Vol. 7, pp. 443–487, John Wiley & Son, New York, N.Y.

Moriceau, C., and Fleitout, L. (1989). A directional analysis of the small wavelength geoid in the Pacific Ocean. *Geophys. Res. Lett.*, **16**, 251–254.

Moriceau, C., Christensen, U., and Fleitout, L., (1991). Geoid and topography associated with sub-lithospheric convection: negligible contribution from deep currents. *Earth Planet. Sci. Lett.*, **103**, 395-408.

Müller, R. D., and Roest, W. R. (1992). Fracture zones in the North Atlantic from combined Geosat and Seasat data. *J. Geophys. Res.*, **97**, 3337–3350.

Müller, R. D., Sandwell, D. T., Tucholke, B. E., Sclater, J. G., and Shaw, P. R. (1991). Depth to basement and geoid expression of the Kane Fracture Zone: a comparison. *Mar. Geophys. Res.*, **13**, 105–129.

Munschy, M. (1997). Dérivée verticale et recherche d'extrema appliquées: la grille d'anomalie gravimétrique des océans. *Colloque Dorsales*, Paris (24–25 Nov. 1997).

Nurnberg, D., and Müller, R. D. (1991). The tectonic evolution of the South Atlantic from Late Jurassic to present. *Tectonophysics*, **191**, 27–53.

Ockendon, J. R., and Turcotte, D. L. (1977). On the gravitational potential and field anomalies due to thin mass layers. *Geophys. J. R.A.S.*, **48**, 479–492.

Owens, R., and Parsons, B. (1994). Gravity field over mid-ocean ridges from Geosat GM data: variations as a function of spreading rate. *Geophys. Res. Lett.*, **21**, 2837–2840.

Palmer, J., Sempéré, J., Christie, D., and Phipps Morgan, J. (1993). Morphology and tectonics of the Australian-Atarctic Discordance between 1238E and 1288E. *Mar. Geophys. Res.*, **15**, 121–152.

Parmentier, E. M., and Haxby, W. F. (1986). Thermal stresses in the oceanic lithosphere: evidence from geoid anomalies at fracture zones. *J. Geophys. Res.*, **91**, 7193–7204.

Parsons, B., and Daly, S. (1983). The relationship between surface topography, gravity anomalies and temperature structure of convection. *J. Geophys. Res.*, **88**, 1129–1144.

Parsons, B., and McKenzie, D. P. (1978). Mantle convection and the thermal structure of the plates. *J. Geophys. Res.*, **83**, 4485–4496.

Parsons, B., and Sclater, J. G. (1977). An analysis of the variation of ocean floor bathymetry and heatflow with age. *J. Geophys. Res.*, **82**, 803–827.

Perrot, K., Francheteau, J., Maia, M., and Tisseau, C. (1998). Spatial and temporal variations of subsidence of the East Pacific rise (0–23°S). *Earth Planet Sci. Lett.*, **160**, 587–592.

Phipps Morgan, J., and Chen, Y. (1993a). Dependence of ridge-axis morphology on magma supply and spreading rate. *Nature*, **364**, 706–708.

Phipps Morgan, J., and Chen, Y. (1993b). The genesis of oceanic crust: magma injection, hydrothermal circulation, and crustal flow. *J. Geophys. Res.*, **98**, 6283–6297.

Phipps Morgan, J., and Smith, W. H. F. (1992). Flattening of the seafloor depth-age curve as a response to asthenospheric flow. *Nature*, **359**, 524–527.

Phipps Morgan, J., and Sandwell, D. (1994). Systematics of ridge propagation south of 300 S. *Earth Planet. Sci. Lett.*, **121**, 245–258.

Phipps Morgan, J., Morgan, W. J., and Price, E. (1995). Hotspot melting generates both hotspot volcanism and a hotspot swell? *J. Geophys. Res.*, **100**, 8045–8062.

Rabinowicz, M., Ceuleneer, G., Monnereau, M., and Rosemberg, C. (1990). Three-dimensional models of mantle flow across a low viscosity zone: implications for hotspot dynamics. *Earth Planet. Sci. Lett.*, **99**, 170–184.

Ramillien, G., and Cazenave, A. (1997). Global bathymetry derived from altimeter data of the ERS-1 geodetic mission. *J. Geodyn.*, **23**, 129–149.

Ribe, N. M., and Christensen, U. R. (1994). Three dimensional modelling of plume lithosphere interaction. *J. Geophys. Res.*, **99**, 669–682.

Richardson, W. P., Stein, S., Stein C. A., and Zuber, M. (1995). Geoid data and thermal structure of the oceanic lithosphere. *Geophys. Res. Lett.*, **22**, 1913–1916.

Richter, F. M., (1973). Convection and the large-scale circulation of the mantle. *J. Geophys. Res.*, **78**, 8735–8745.

Richter, F. M., and Parsons, B. (1975). On the interaction of two scales of convector in the mantle. *J. Geophys. Res.*, **80**, 2529–2541.

Robinson, E. M., and Parsons, B. (1988). Effect of a shallow low-viscosity zone on small scale instabilities under the cooling oceanic plates. *J. Geophys. Res.*, **93**, 3469–3479.

Robinson, E. M., Parsons, B., and Daly, S. F. (1987). The effect of a shallow low visocity zone on the apparent compensation depth of mid plate swells. *Earth Planet. Sci. Lett.*, **82**, 335.

Robinson, E. M., Parsons, B., and Driscoll, M. (1988). The effects of a shallow low-viscosity zone on the mantle flow, the geoid anomalies and depth-age relationship at fracture zones. *Geophys. J. Int.*, **93**, 25–43.

Royer, J.-Y., and Sandwell, D. T. (1989). Evolution of the Eastern Indian Ocean since the Late Cretaceous: Constraints from Geosat altimetry. *J. Geophys. Res.*, **94**, 13755–13782.

Royer, J.-Y., Patriat, P., Bergh, H., and Scotese, C. (1988). Evolution of the southwest Indian Ridge from the Late Cretaceous (anomaly 34) to the Middle Eocene (anomaly 20). *Tectonophysics*, **155**, 235–260.

Royer, J.-Y., Sclater, J. G., and Sandwell, D. T. (1989). A preliminary tectonic chart of the Indian Ocean. *Proc. Indian Acad. Sci. (Earth Planet. Sci.)*, **98**, 7–24.

Royer, J.-Y., Gahagan, L. M., Lawver, L. A., Mayes, C. L., Nurnberg, D., Sandwell, D. T., and Scotese, C. R. (1990). A tectonic chart of the Southern Ocean derived from Geosat altimetry data. *In* "Antarctica as an exploration frontier-hydrocarbon potential, geology, and hazards." (St. John, B., ed.), Vol. 31, pp. 89–99, American Association of Petroleum Geology, Tulsa, OK.

Royer, J.-Y., Gordon, R. G., DeMets, C., and Vogt, P. (1997). New limits on India/Australia motion since Chron 5 (11 Ma) and implications for the lithospheric deformation in the Equatorial Indian Ocean. *Geophys. J. Int.*, **129**, 41–74.

Sahabi, M., Géli, L., Olivet, J.-L., Gilg-Capar, L., Roult, G., Ondréas, H., Beuzart, P., and Aslanian, D. (1996). Morphological reorganization within the Pacific-Antarctic Discordance. *Earth Planet. Sci. Lett.*, **137**, 157–173.

Sandwell, D. T. (1984a). Thermomechanical evolution of oceanic fracture zones. *J. Geophys. Res.*, **89**, 11401–11413.

Sandwell, D. T. (1984b). A detailed view of the South Pacific geoid from satellite altimetry. *J. Geophys. Res.*, **89**, 1089–1104.

Sandwell, D. T. (1992). Antarctic marine gravity field from high-density satellite altimetry. *Geophys. J. Int.*, **109**, 437–448.

Sandwell, D. T., and McKenzie, K. R. (1989). Geoid height versus topography for oceanic plateaus and swells. *J. Geophys. Res.*, **94**, 7403–71418.

Sandwell, D. T., and Renkin, M. L. (1988). Compensation of swells and plateaus in the North Pacific: no diret evidence from mantle convection. *J. Geophys. Res.*, **93**, 2775–2783.

Sandwell, D. T., and Schubert, G. (1982). Geoid-age relation from Seasat altimeter profiles across the Mendocino fracture zone. *J. Geophys. Res.*, **87**, 3949–3958.

Sandwell, D., and Smith, W. H. (1997). Marine gravity anomaly from Geosat and ERS-1 satellite altimetry. *J. Geophys. Res.*, **102**, 10039–10054.

Sandwell, D. T., Winterer, E. L., Mammerichse, R. A., Duncan, R. A., Lynch, M. A., Levitt, D. A., and Johnson, C. L. (1995). Evidence for diffuse extension of the Pacific plate from Puka-Puka ridges and cross grain gravity lineations. *J. Geophys. Res.*, **100**, 15087–15099.

Schouten, H., Dick, H., and Klitgord, K. (1987). Migration of mid-ocean-ridge volcanic segments. *Nature*, **326**, 835–839.

Schwintzer, P. *et al.* (1997). Long-wavelengh global gravity field models: GRIM4-C4, GRIM4-C4. *J. Geodesy*, **71**, 189–208.

Shaw, P. R., and Cande, S. C. (1990). High-resolution inversion for South Atlantic plate kinematics using joint altimeter and magnetic anomaly data. *J. Geophys. Res.*, **95**, 2625–2644.

Sheehan, A. F., and McNutt, M. K. (1989). Constraints on thermal and mechanical structure of the oceanic lithosphere at Bermuda rise from geoid height and depth anomalies. *Earth Planet. Sci. Lett.*, **93**, 377.

Shen, Y., Forsyth, D. W., Scheirer, D. S., and Macdonald, K. C. (1993). Two forms of volcanism: implications for mantle flow and off-axis crustal production on the western flank of the southern East Pacific rise. *J. Geophys. Res.*, **98**, 17875–17889.

Shen, Y., Scheirer, D. S., Forsyth, D. W., and Macdonald, K. C. (1995). Trade-off in production of adjacent seamount chains near the East Pacific rise (17–19°S). *Nature*, **373**, 140–143.

Sleep, N. (1990). Hotspots and mantle plumes: some phenomenology, *J. Geophys. Res.*, **95**, 6715–6736.

Sleep, N. (1992). Hotspot volcanism and mantle plumes, Ann. Rev. *Earth and Planet. Sci.*, **20**, 19–43.

Small, C. (1994). A global analysis of mid-ocean ridge axial topography. *Geophys. J. Int.*, **116**, 64–84.

Small, C. (1998). Global systematics of mid-ocean ridge morphology. *In* "Faulting and magmatism at mid-ocean ridges." (Buck, W., Delaney, P., Karson, J., and Lagabrielle, Y., eds.), *AGU Monograph*, Vol. 106 (in press), American Geophysical Union, Washington, D.C.

Small, C., and Sandwell, D. T. (1989). An abrupt change in ridge-axis gravity with spreading rate. *J. Geophys. Res.*, **94**, 17383–17392.

Small, C., and Sandwell, D. T. (1992). An analysis of ridge axis gravity roughness and spreading rate. *J. Geophys. Res.*, **97**, 3235–3245.

Small, C., and Sandwell, D. T. (1994). Imaging mid-ocean ridge transitions with satellite gravity. *Geology*, **22**, 123–126.

Smith, D. K., and Jordan, T. H. (1988). Seamount statistics in the Pacific Ocean. *J. Geophys. Res.*, **93**, 2899–2918.

Smith, W. H. F., and Sandwell, D. T. (1997). Global seafloor topography from satellite altimetry and ship depth soundings. *Science*, **277**, 1956–1962.

Smith, W. H. F., Staudigel, H., Watts, A. B., and Pringle, M. S. (1989). The Magellan seamounts: early cretaceous record of the south Pacific isotopic and thermal anomaly. *J. Geophys. Res.*, **94**, 10501–10523.

Stein, C., and Abbott, D. H. (1991). Heat low constaints on the south Pacific Superswell. *J. Geophys. Res.*, **96**, 16083–16100.

Stein, C., and Stein, S. (1992). A model for the global variation in oceanic depth and heatflow with lithospheric age. *Nature*, **359**, 123–129.

Sutherland, R. (1995). The Australian-Pacific boundary and Cenozoic plate motions in the SW Pacific: some constraints from Geosat data. *Tectonics*, **14**, 819–831.

Tanimoto, T., and Zhang, Y. (1990). Lithospheric thickness and thermal anomalies in the upper mantle inferred from the Love wave data. *Geophys. Res. Lett.*, **17**, 2405–2408.

Tapley, B. D., *et al.* (1996). The joint gravity model 3. *J. Geophys. Res.*, **101**, 28029–28049.

Tarantola, A. (1987). *In* "Inverse problem theory," Elsevier, Amsterdam.

Tarits, P. (1986). Conductivity and fluid in the oceanic upper mantle. *Physics Earth Planet. Int.*, **42**, 215–226.

Tebbens, S. F., and Cande, S. C. (1997). Southeast Pacific tectonic evolution from early Oligocene to Present. *J. Geophys. Res.*, **102**, 12061–12084.

Tikku, A., and Cande, S. (1999). The oldest magnetic anomalies in the Australian-Antarctic Basin: are they isochrons. *J. Geophys. Res.*, **104**, 661–677.

Turcotte, D. L., and Oxburgh, E. R. (1967). Finite amplitude convective cells and continental drift. *J. Fluid Mech.*, **28**, 29–42.

Veining Meinenz, F. A. (1964). *In* "The earth's Crust and Mantle." Elsevier, Amsterdam.

Vogt, P. R., Zondek, B., Fell, P. W., Cherkis, N. Z., and Perry, R. K. (1984). Seasat altimetry, the North Atlantic geoid, and evaluation by shipborne subsatellite profiles. *J. Geophys. Res.*, **89**, 9885–9903.

Von Herzen, R. P., Cordrey, M. J., Detrick, R. S., and Fang, C. (1989). Heatflow and the thermal origin of hotspot swells: the Hawaiian swell revisited. *J. Geophys. Res.*, **94**, 13783–13799.

Watts, A. B. (1978). An analysis of isostasy in the world's oceans, 1, Hawaiian-Emperor seamount chain. *J. Geophys. Res.*, **83**, 5989–6004.

Watts, A. B., and Ribe, N. M. (1984). On geoid heights and flexure of the lithosphere at seamounts. *J. Geophys. Res.*, **89**, 11152–11170.

Watts, A. B., Bodine, J. H., and Ribe, N. M. (1980). Observations of flexure and the geological evolution of the Pacific ocean basin. *Nature*, **283**, 532–537.

Watts, A. B., Pierce, C., Collier, J., Dalwood, R., Canales, J. P., and Henstock, T. J. (1997). A seismic study of Tenerife, Canary islands: implications of volcanoe growth, lithospheric flexure and magmatic underplating. *Earth Planet. Sci. Lett.*, **146**, 431–447.

Wessel, P. (1992). Thermal stresses and the bimodal distribution of elastic thickness estimates of the oceanic lithosphere. *J. Geophys. Res.*, **97**, 14177–14193.

Wessel, P., and Haxby, W. F. (1990). Thermal stresses, differential subsidence and flexure at oceanic fracture zones. *Geophys. Res. Lett.*, **95**, 375–391.

Wessel, P., and Keating, B. H. (1994). Temporal variations of flexural deformation at Hawaii. *J. Geophys. Res.*, **99**, 2747–2756.

Wessel, P., and Kroenke, L. (1997). A geometric technique for relocating hotspots and refining absolute plate motion. *Nature*, **387**, 365–369.

Wessel, P., and Lyons, S. (1997). Distribution of large Pacific seamounts from Geosat/ERS-1: implications for the history of intraplate volcanism. *J. Geophys. Res.*, **102**, 22459–22475.

Wessel, P., Bercovici, O., and Kroenke, L. W. (1994). The possible reflection of mantle discontinuities in Pacific geoid and bathymetry, *Geophys. Res. Lett.*, **21**, 1943–1946.

Wessel, P., Kroenke, L. W., and Bercovici, D. (1996). Pacific plate motion and undulations in geoid and bathymetry. *Earth Planet. Sci. Lett.*, **140**, 53–66.

White, J., Sailor, R., Lazarewicz, A., and Le Schack, A. (1983). Detection of seamount signature in Seasat altimeter data using matched filters. *J. Geophys. Res.*, **88**, 1541–1551.

Wilson, J. T. (1965a). A new class of faults and their bearing on continental drift. *Nature*, **207**, 343–347.

Wilson, J. T. (1965b). Evidence from ocean islands suggesting movement in the earth. *Phil. Trans. Royal Soc. London*, **A258**, 145–165.

Winterer, E. L., and Sandwell, D. T. (1987). Evidence from an-echelon cross-grain ridges for tensional cracks in the Pacific. *Nature*, **329**, 534–537.

Wolfe, C. J., and McNutt, M. K. (1991). Compensation of cretaceous seamoutns of the Darwin rise, northwest Pacific ocean. *J. Geophys. Res.*, **96**, 2363–2374.

Woods, M. T., and Okal, E. A. (1996). Rayleigh-wave dispersion along the Hawaiian swell: a test of lithospheric thinning by thermal rejuvenation. *Geophys. J. Int.*, **125**, 325–339.

Woods, M. T., Leveque, J. J., Okal, E. A., and Cara, M. (1991). Two-station measurements of Raleigh wave group velocity along the Hawaiian swell. *Geophys. Res. Lett.*, **18**, 105–108.

Yamaji, A. (1992). Periodic hotspot distribution and small-scale convection in the upper mantle. *Earth Planet. Sci. Lett.*, **109**, 107–116.

Yuen, D., and Fleitout, L. (1985). Thinning of the lithosphere by small scale convective destabilization. *Nature*, **313**, 125–128.

12

Bathymetric Estimation

DAVID T. SANDWELL,* and WALTER H. F. SMITH†

*University of California, San Diego
Scripps Institution of Oceanography
La Jolla, CA 92093-0225
†NOAA, Laboratory for Satellite Altimetry
Silver Spring, MD 20910-3282

1. INTRODUCTION

A detailed knowledge of topography is fundamental to the understanding of most earth processes. On the land, weather and climate are controlled by topography on scales ranging from large continental landmasses to small mountain valleys. Since the land is shaped by tectonics, erosion, and sedimentation, detailed topography is essential for any geological investigation. In the oceans, detailed bathymetry is also essential for understanding physical oceanography, biology, and marine geology. Currents and tides are controlled by the overall shapes of the ocean basins as well as by the smaller sharp ocean ridges and seamounts. Sea life is abundant where rapid changes in ocean depth deflect nutrient-rich water toward the surface. Because erosion and sedimentation rates are low in the deep oceans, detailed bathymetry also reveals the mantle convection patterns, the plate boundaries, the cooling/subsidence of the oceanic lithosphere, the oceanic plateaus, and the distribution of off-ridge volcanoes.

Topographic mapping with orbiting laser and radar altimeters has been the focus of current exploration of Venus, the Moon, and Mars and is providing very high resolution topographic maps of the earth's land areas. However, since one cannot directly map the topography of the ocean basins from space, most seafloor mapping is a tedious process that has been carried out over a 30-year period by research vessels equipped with single or multibeam echo sounders (Canadian Hydrographic Service, 1981). For example, a complete mapping of the deep ocean basins at 100-m horizontal resolution would take about 125 ship-years of survey time using the latest multibeam technology. Oddly, some of the most valuable depth measurements of remote ocean areas were collected prior to satellite navigation and shipboard computer facilities and thus the quality of the data is highly non-uniform (Smith, 1993). Moreover, many of the dense surveys of the northern oceans remain classified in military archives (Medea, 1995) or remain proprietary for economic or political reasons. Thus, until recently, our knowledge of the seafloor topography was poor except along the spreading ridge axes where there is almost complete coverage from the RIDGE program.

Two developments have vastly improved our knowledge of seafloor topography. First, the careful efforts by scientists throughout the world (Canadian Hydrographic Service, 1981) to archive the digital sounding data and assemble the data into large databases has provided much improved access to the 30-year mapping effort (Wessel and Watts, 1988; Smith, 1993). Second, radar altimeters aboard the ERS-1 and Geosat spacecraft have surveyed the marine gravity field over nearly all of the world's oceans to a high accuracy and moderate spatial resolution. In March of 1995, ERS-1 completed its dense mapping (~8 km track spacing at the equator) of sea-surface topography between latitudes of ±81.5°. Moreover in July of 1995, all of the high-density radar altimeter data collected by the Geosat spacecraft were declassified. These data have been combined and processed to form a global marine geoid or gravity grids (Cazenave *et al.*, 1996; Sandwell and Smith, 1997; Tapley and Kim, Chapter 10). In the wavelength band 15–200 km, variations in gravity anomaly are highly correlated with seafloor topography and thus, in principal, can be used to recover topography. As described in this chapter there are ongoing efforts to combine ship and satellite data to form a uniform-resolution grid of seafloor topography (Baudry and Calmant, 1991; Jung and Vogt, 1992; Calmant, 1994; Smith and Sandwell, 1994; Sichoix and Bonneville, 1996; Ramillien and Cazenave, 1997; Smith and Sandwell, 1997). The sparse ship soundings constrain the long wavelength variations in seafloor depth and are also used to calibrate the local variations in topography to gravity ratio associated with varying tectonics and sedimentation. The satellite-derived gravity anomaly provides much of the information on the intermediate to short wavelength topographic variations (160–20 km).

The basic theory for predicting seafloor topography from satellite altimeter measurements is nicely summarized in a paper by Dixon *et al.* (1983). Models of flexurally compensated seafloor topography have been used to develop a spectral transfer function for projecting seafloor topography into gravity anomaly (or geoid height) (McKenzie and Bowin, 1976; Banks *et al.*, 1977; McNutt, 1979; Ribe, 1982). The important parameters are the crustal density, the mean ocean depth, and the thickness of the elastic portion of the lithosphere. The inverse of this transfer function provides a theoretical basis for projecting gravity anomaly measurements into seafloor topography but there are a number of complications that require careful treatment.

(1) The gravity to topography transfer function becomes singular at both short wavelengths ($\lambda \ll 2\pi$ times mean ocean depth) and long wavelengths ($\lambda \gg$ depth of compensation or flexural wavelength) due to upward continuation and isostatic compensation, respectively (Figure 1). Thus bathymetric prediction is only possible over a limited band.

(2) While the short-wavelength portion of the gravity to topography transfer function depends on well known parameters (ocean depth, crustal density), the longer wavelength portion is highly dependent on the elastic thickness of the lithosphere and/or crustal thickness.

(3) Sediments raining down onto the seafloor preferentially fill bathymetric lows and can eventually completely bury the preexisting basement topography. This adds a spatially dependent and non-linear aspect to the gravity-to-topography transfer function.

(4) The transfer function is two-dimensional although usually it is assumed to be isotropic. Thus complete ocean surface gravity coverage is required.

(5) Finally, in areas where the amplitude of the topographic relief approaches the mean depth, the transfer function is inherently non-linear (Parker, 1973).

Many previous studies have identified and addressed some of these issues (Dixon *et al.*, 1983; Baudry and Calmant, 1991; Jung and Vogt, 1992; Smith and Sandwell, 1994; Sichoix and Bonneville, 1996).

2. GRAVITY ANOMALY AND SEA SURFACE SLOPES

As we will see in the next section in greater detail, the accurate recovery of seafloor topography at short wavelengths is critically dependent on the accuracy of the short wavelength gravity field derived from satellite altimetry. As described in Chapters 1 and 10, a satellite altimeter measures the topography of the ocean surface which, to a first approximation, is a measure of geoid height or gravitational potential. The oceanographic studies described in earlier chapters use the deviations in sea-surface height above or below the geoid to investigate currents, tides, and waves. These oceanographic signals are our error sources (Rapp and Yi, 1997) and therefore to understand the magnitude of the errors one must review how gravity anomalies are computed from noisy satellite altimeter measurements.

2.1. Geoid Height, Vertical Deflection, Gravity Gradient, and Gravity Anomaly

The geoid height $N(\mathbf{x})$ and other measurable quantities such as gravity anomaly $\Delta g(\mathbf{x})$ are related to the gravitational potential $V(\mathbf{x}, z)$ (Heiskanen and Moritz, 1967). Since we are primarily interested in short wavelength anomalies, we assume that all of these quantities are deviations from a spherical harmonic reference earth model (e.g., EGM96, Lemoine *et al.*, 1998) so a planar approximation can be used for the gravity computation. In the following equations, the bold $\mathbf{x}$ denotes the coordinate (x, y); similarly $\mathbf{k}$ denotes (k_x, k_y) where $k_x = 1/\lambda_x$, where λ_x is wavelength. To a first approximation, the geoid height is related to the gravitational potential by Brun's formula,

$$N(\mathbf{x}) \cong \frac{1}{g_o} V(\mathbf{x}, 0) \qquad (1)$$

where g_o is the latitude-dependent, average acceleration of gravity (~ 9.8 m sec^{-2}). The gravity anomaly is the vertical derivative of the potential,

$$g(\mathbf{x}) = -\frac{\partial V(\mathbf{x}, 0)}{\partial z}. \qquad (2)$$

The east component and north component of vertical deflection are the slope of the geoid in the x- and y-directions, respectively

$$\eta(\mathbf{x}) \cong \frac{-1}{g_o} \frac{\partial V}{\partial x}, \quad \xi(\mathbf{x}) \cong \frac{-1}{g_o} \frac{\partial V}{\partial y}. \qquad (3)$$

These quantities are related to one another through Laplace's equation.

$$\frac{\partial^2 V}{\partial x^2} + \frac{\partial^2 V}{\partial y^2} + \frac{\partial^2 V}{\partial z^2} = 0. \qquad (4)$$

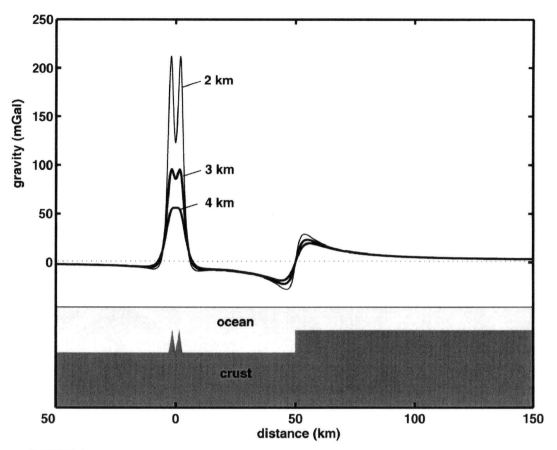

FIGURE 1 Fundamental limitations of topographic recovery from gravity anomaly measurements are illustrated by seamounts (left) and plateau (right). The gravity signatures of the closely-spaced seamounts (4 km apart and 1 km tall) are strong and distinct when the average ocean depth is 2 km or less but their signatures combine and become weak when the ocean depth is 4 km. The isostatically compensated step in depth produces a local gravitational edge effect that is strongly attenuated at a distance of 150 km from the step; thus the gravity far from the step does not provide information on the overall depth offset across the step.

Following Haxby *et al.* (1983) the differential Eq. (4) is reduced to an algebraic equation by fourier transformation

$$\Delta g(\mathbf{k}, 0) = \frac{i g_o}{|\mathbf{k}|} \left[k_x \eta(\mathbf{k}) + k_y \xi(\mathbf{k}) \right]. \tag{5}$$

To compute gravity anomaly from a dense network of satellite altimeter profiles of geoid height, one first constructs grids of east η and north ξ vertical deflection. The grids are then fourier transformed and Eq. (5) is used to compute gravity anomaly (Sandwell, 1992). At this point one could also add the spherical harmonic gravity model back to the gridded gravity values in order to recover the long wavelength gravity field. A more complete description of gravity field recovery from satellite altimetry can be found in (Hwang and Parsons, 1996; Sandwell and Smith, 1997; Rapp and Yi, 1997). The important issue for bathymetric estimation is revealed by a simplified version of Eq. (5). Consider the sea-surface slope and gravity anomaly across a two-dimensional structure. The y-component of slope is zero so conversion from sea-surface slope to gravity anomaly is simply a Hilbert transform.

$$\Delta g(\mathbf{k}, 0) = i g_o \text{sgn}(k_x) \eta(\mathbf{k}). \tag{6}$$

Now it is clear that one μrad of sea-surface slope maps into 0.98 mGal of gravity anomaly and similarly one μrad of slope error will map into 1 mGal of gravity anomaly error. Thus the accuracy of the gravity field recovery and hence the accuracy of the bathymetric prediction is controlled by the accuracy of the sea-surface slope measurement, especially the short-wavelength slope estimates.

3. LIMITATIONS OF RADAR ALITMETRY FOR GRAVITY FIELD RECOVERY

As described in Chapter 1 and in Section 8 (Appendix), a satellite altimeter uses a pulse-limited radar to measure the altitude of the satellite above the closest sea-surface point. Global precise tracking, coupled with orbit dynamic calculations provide an independent measurement of the height of the satellite above the ellipsoid. The difference between these two measurements is equal to the sea-surface height (~geoid height) minus any delays in the propagation of the radar echo due to the ionosphere and troposphere. There are many errors in these measurements but most occur over length scales greater than a few hundred kilometers (Chapter 1 and 10; Sandwell, 1991; Tapley *et al.*, 1994). For gravity field recovery and bathymetric estimation, the major error source is the roughness of the ocean surface due to ocean waves. The radar pulse reflects from an area of ocean surface (footprint) that grows with increasing sea state (see Stewart, 1985 and Section 8). The superposition of the reflections from this larger area stabilizes the shape of the echo but it also smoothes the echo so that the timing of its leading edge less certain. By averaging many echoes (1000 Hz) over multiple repeat cycles one can achieve a 10–20 mm range precision (Noreus, 1995; Yale *et al.*, 1995). Over a distance of 4 km (i.e. 1/4 wavelength) this corresponds to a sea-surface slope error of 4 μrad which maps into a gravity error of about 4 mGal. Thus the only way to improve the resolution is to make many more measurements or stop the ocean waves!

This situation is compounded in the deep oceans (~4000 m) because upward continuation of the gravity field from the ocean floor to the ocean surface provides a strong low-pass filter on the gravity signal (Figure 1) but has no effect on the radar noise. Consider an anomaly on the ocean floor with an 16-km wavelength and a 15-mGal amplitude (i.e., a typical value for oceans). On the ocean surface, this anomaly will be reduced to 3.1 mGal by upward continuation.

Other sources of error include tide model error, ocean variability, mean ocean currents, ionospheric delay, tropospheric delay, and electromagnetic bias. Corrections for many of these errors are supplied with the geophysical data record. However, for gravity field recovery and especially bathymetric prediction not all corrections are relevant or even useful. For example, corrections based on global models (i.e., wet troposphere, dry troposphere, ionosphere, and inverted barometer) typically do not have wavelength components shorter than 1000 km, and their amplitude variations are less than 1 m so they do not contribute more than 1 μrad of error. In addition, the corrections can sometimes have steps associated with geographical boundaries; the numerical derivative of even a 10 mm step will introduce significant noise in the slope estimate. Yale (1997) has examined the slope of the corrections supplied with the TOPEX/POSEIDON GDR and found only the tide correction (Bettadpur and Eanes, 1994) should be applied. The dual frequency altimeter aboard TOPEX/POSEIDON satellite provides an estimate of the ionospheric correction, however, because it is based on the travel-time difference between radar pulses at C-band and Ku-band, the noise in the

difference measurement adds noise to the slope estimate for wavelengths less than about 100 km (Imel, 1994). The most troublesome errors are associated with mesoscale variability and permanent dynamic topography (Rapp and Yi, 1997). The variability signal can be as large as 6 μrad but fortunately it is confined to a few energetic areas of the oceans and given enough redundant slope estimates from nearby tracks (Sandwell and Zhang, 1989), some of this noise can be reduced by averaging. Permanent dynamic topography can have slope up to 6 μrad; this will corrupt both the gravity field recovery and the bathymetric prediction over length scales of 100–200 km.

4. FORWARD MODELS

The forward relationship between topography and gravity anomaly is best described in the two-dimensional fourier transform domain. First consider the case of uncompensated seafloor topography of density ρ_c lying at an average depth of d beneath the ocean surface. Let $h(\mathbf{x})$ be the topography of the seafloor with respect to this mean ocean depth. The exact formula for the gravity anomaly Δg due to the topography is (Parker, 1973)

$$F[\Delta g] = 2\pi \Gamma (\rho_c - \rho_w) \exp[-2\pi |\mathbf{k}| d] \\ \times \sum_{1}^{\infty} \frac{|2\pi \mathbf{k}|^{n-1}}{n!} F[h^n(\mathbf{x})] \qquad (7)$$

where $F[\ ]$ is the two-dimensional Fourier transform operator, Γ is the Newtonian gravitational constant, and ρ_w is the density of seawater. When the topography lies well below the plane of the gravity observations (i.e., sea level), this series converges quite rapidly. Moreover if the maximum amplitude of the topography is much less than the mean ocean depth, then the first term in the series dominates. Indeed it is common to use only the first term because it provides a linear isotropic relationship between the fourier transform of the gravity anomaly $\Delta G(\mathbf{k})$ and the fourier transform of the topography $H(\mathbf{k})$.

$$G(\mathbf{k}) = 2\pi \Gamma (\rho_c - \rho_w) \exp[-2\pi |\mathbf{k}| d] H(\mathbf{k}). \qquad (8)$$

The model is appropriate for all types of topography (e.g., seamounts, fracture zones, trenches).

Next consider a more realistic model where the lithosphere flexes downward in response to the load of the topography. The simplest model consists of a thin elastic plate overlying a fluid asthenosphere (McKenzie and Bowin, 1976; Banks *et al.*, 1977; McNutt, 1979). Note this model also applies to all types of seafloor topography (e.g., seamounts, fracture zones and trenches) as long as the features are not obscured by sediments. As the plate flexes downward (or upward), it displaces the crust-mantle boundary that provides an isostatic restoring force. Assuming thin elastic plate flexure with isotropic and spatially invariant response, the deflection of the crust/mantle interface $M(\mathbf{k})$ is related to the surface topography

$$M(\mathbf{k}) = -\left(\frac{\rho_c - \rho_w}{\rho_m - \rho_c}\right)\Phi(|\mathbf{k}|)H(\mathbf{k}) \qquad (9)$$

where ρ_m is the density of the mantle and the isostatic response function Φ is given by

$$\Phi(|\mathbf{k}|) = [1 + (\lambda|\mathbf{k}|)^4]^{-1}. \qquad (10)$$

The characteristic wavelength of the flexure $\lambda = 2\pi\{D/[g \times (\rho_m - \rho_c)]\}^{1/4}$ depends on the elastic thickness of the lithosphere which in turn depends on the age of the lithosphere at the time the topography formed (Watts, 1979; Wessel, 1992). This flexural wavelength can vary from 100–800 km depending on tectonic conditions. Moreover, since nearby topographic features such as seamounts can have completely different flexural characteristics, the spatially varying flexural wavelength introduces a major complication in bathymetric estimation. Nonetheless the flexure model provides a theoretical framework for the design of downward continuation, low-pass, and high-pass filters used in bathymetric prediction (Dixon et al., 1983; Baudry and Calmant, 1991; Smith and Sandwell, 1994).

The deflection of the crust/mantle interface introduces a second contribution to the gravity anomaly. Again neglecting the nonlinear terms in the Parker expansion [Eq. (7)] there is an isotropic transfer function, or admittance function, Z for mapping seafloor topography into sea-surface gravity anomaly.

$$\begin{aligned} G(\mathbf{k}) &= 2\pi\Gamma(\rho_c - \rho_w)\exp[-2\pi|\mathbf{k}|d] \\ &\quad \times [1 - \Phi(|\mathbf{k}|)\exp(-2\pi|\mathbf{k}|c)]H(\mathbf{k}) \qquad (11) \\ G(\mathbf{k}) &= Z(|\mathbf{k}|)H(\mathbf{k}). \end{aligned}$$

Examples of this transfer function are shown in Figure 2 (upper). Before discussing the inverse of this transfer function and its role in bathymetric prediction, one should be aware of an additional complication due to sediments on the seafloor. Sediments generally accumulate long after the primary basement features are formed so they induce yet another topographic load and flexural response. The overall effects of sedimentation on the transfer function make it nonlinear because sediments preferentially fill bathymetric lows and the sediment density varies with compaction depth. In the extreme case of complete burial, the gravity to basement topography ratio reverses sign so the basement high has a negative gravity anomaly; there is a well-documented case in the Indian Ocean where the 85 Ridge is completely buried by the Bengal Fan (Liu et al., 1982). Of course when the sediments dominate, such as on the shallow continental margins, the topography is flat while the gravity reflects deeper structure; in these areas gravity anomaly is useless for predicting topography.

These limitations and assumptions of the flexural compensation model suggest that the gravity-to-topography transfer function Eq. (11) is only approximate and can only be used in ideal situations. The primary importance of the model is to reveal fundamental limitations in the mapping from gravity anomaly to topography. In the ideal situation one could construct a topographic grid by inverting Eq. (11).

$$G(\mathbf{k}) = Z(|\mathbf{k}|)^{-1}H(\mathbf{k}). \qquad (12)$$

A plot of Z^{-1} for two flexural wavelengths and a typical ocean depth of 4 km, illustrates the three problems (Figure 2—lower):

(1) The inverse transfer function increases exponentially for wavelengths less than 50 km so, for example, at a wavelength of 19 km, 4 mGal of gravity noise will map into 200 m of bathymetric estimation error. Since seafloor has about 200 m of rms relief, the noise begins to dominate the signal at these short wavelengths.

(2) The transfer function grows rapidly at long wavelengths because isostatic compensation causes a near perfect cancellation between the gravity from the topography and the gravity from the relief on the crust/mantle interface; again any noise will be amplified.

(3) The more important issue at long wavelengths is that the topography-to-gravity ratio is very sensitive to the flexural wavelength, and, since this is an unknown parameter that can vary significantly between, for example, a young seamount near an old fracture zone it introduces a major uncertainty in the prediction.

The critical issue is that there is a band of wavelengths where ocean surface gravity can be used to estimate topography but prediction outside this band should be avoided. Other measurements such as ship soundings can be used to recover the long wavelengths. Short-wavelength bathymetry can only be efficiently recovered with multibeam swath surveys.

5. INVERSE APPROACHES

Here we review the various inverse approaches that are used overcome these fundamental limitations. Since there is no one "correct" method and the practical implementation is a matter of taste, one may call these *recipes*. The first part of the recipe, which is common to all approaches, is to use available ship soundings and other depth information to construct a long-wavelength model of the ocean basin and then remove this long-wavelength field from available depth soundings. To be consistent one should remove the same long-wavelength components from the gravity field. The long-wavelength topography is then restored as a final step in the recipe; this is called a *remove/restore* procedure.

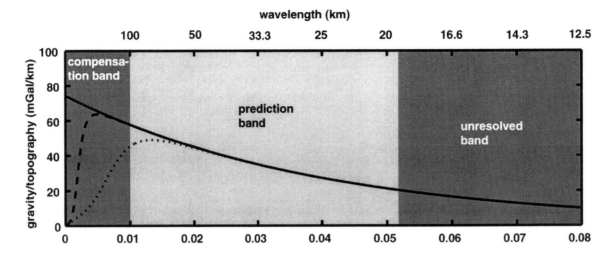

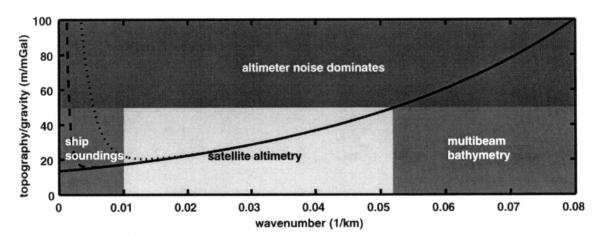

FIGURE 2   (upper) Gravity/topography ratio versus wavelength for uncompensated topography (solid curve) and elastic plate flexure models (dashed—30 km thickness, dotted—5 km thickness). Gravity/topography ratio is highly variable in the *compensation band* but uniform over the *prediction band* and *unresolved band*. (lower) Topography/gravity ratio for same three models illustrates the fundamental limitations of predicting topography from gravity. Errors in the satellite-derived gravity anomaly are typically 4 mGal so when the topography/gravity ratio exceeds 50 m/mGal, the predicted noise of 200 m will exceed the typical seafloor roughness. Sparse ship soundings can be used to constrain longer wavelengths while multibeam bathymetry is needed to constrain shorter wavelengths.

The basic spectral approach [Eq. (12)] overcomes the inherent singularities at long and short wavelengths by *windowing* the inverse transfer function. Window shapes range from a simple boxcar (Dixon *et al.*, 1983) to more complicated functions based on the signal and noise properties of the gravity field and the mean ocean depth (Smith and Sandwell, 1994; Sichoix and Bonneville, 1996). The amplitude and shape of the inverse transfer function depends on the density of the crust and the flexural wavelength, respectively. Some investigators vary these parameters to obtain a good match to known depth soundings (Baudry and Calmant, 1991; Jung and Vogt, 1992; Calmant, 1994; Sichoix and Bonneville, 1996; Ramillien and Cazenave, 1997) while others vary just the crustal density and avoid the com-

pensation issue by eliminating all signals in the compensation band (Smith and Sandwell, 1994; Smith and Sandwell, 1997). Moreover, if the compensation band is eliminated, one can downward-continue the gravity field to the mean ocean depth; this flattens the inverse transfer function so that, in the space domain, a plot of topography versus gravity should lie on a straight line with the slope related to the crustal density. Nettleton (1939) used this approach to estimate density from measurements of gravity and topography so we call this the *inverse Nettleton procedure* (Smith and Sandwell, 1994).

These spectral methods neglect the nonlinear terms in the relationship between topography and gravity (7). These terms are important for tall seamounts, especially when they pierce the ocean surface. Baudry and Calmant (1991) and Calmant (1994) uses an entirely space-domain approach where the non-linear terms are included. Their approach models the complete gravity anomaly given in Eq. (7) and is most appropriate for seamounts more than 500 m tall. Moreover, their approach appears to work best for small regions where a few ship soundings are available for calibration and may also be best for a global solution although it is computationally expensive. Smith and Sandwell (1997) have added an additional step to the spectral/Nettleton approach to force agreement with shipboard depth measurements. In this final step, the predicted depths are subtracted from known depth soundings. The difference is gridded using a minimum curvature approach (Smith and Wessel, 1990) and finally the difference grid is added back to the prediction grid. This processes insures that the final grid matches known soundings exactly and blends smoothly into areas where the gravity field provides the only estimates of topography; we call this *polishing the grid* since it makes things look nice and it can be done over and over again as new depth soundings become available.

A final untested method is the so-called *linear inverse theory* approach (Parker, 1994). Assume that gravity and topography are linearly related such that the topography $h(\mathbf{x})$ is equal to the gravity anomaly $\Delta g(\mathbf{y})$ convolved with an unknown linear, isotropic transfer function $q(|\mathbf{x} - \mathbf{y}|)$.

$$h(\mathbf{x}) = \int \int_A q(|\mathbf{x} - \mathbf{y}|) \Delta g(\mathbf{y}) \, d^2\mathbf{y}$$
$$\text{or} \qquad (13)$$
$$h(\mathbf{x}) = \int \int_A q(|\mathbf{y}|) \Delta g(\mathbf{x} - \mathbf{y}) \, d^2\mathbf{y}.$$

After changing the $\mathbf{y}$-variable of integration from Cartesian coordinates to cylindrical coordinates, the convolution integral becomes:

$$h(x_1, x_2) =$$
$$\int_0^\infty r q(r) \left[\int_0^{2\pi} \Delta g\left(x_1 - r\cos\theta, x_2 - r\sin\theta\right) d\theta \right] dr.$$
$$(14)$$

Given complete 2-dimensional gravity anomaly measurements, the integration in brackets can be performed numerically for any values of x_1, x_2 and r; we'll call this function $G(x_1, x_2, r)$. Finally the convolution integral can be written as:

$$h(x_1, x_2) = \int_0^\infty q(r) G(x_1, x_2, r) r \, dr. \qquad (15)$$

We can view this as a standard linear inverse problem for the function $q(r)$ (Parker, 1994). Bathymetric soundings along isolated ship tracks provide values of $h(\mathbf{x})$ on the left side of the equation and the very dense satellite maps allow us to calculate the kernel function G for every depth sounding. Standard linear inversion techniques can be used to estimate $q(r)$. Finally, once $q(r)$ is known, one can use Eq. (13) to compute the topography from the gravity in the areas between the depth soundings.

The advantage of this method is that in addition to estimating the topography, one can also estimate the uncertainty of the topography in the interpolated areas. Moreover, the transfer function $q(r)$ contains geologic information such as the mean crustal thickness and the flexural wavelength of the lithosphere when the topography formed.

6. DATA AVAILABILITY AND CASE STUDY: BATHYMETRIC ESTIMATION

Ship soundings were assembled from a variety of sources. In all cases, we used only the center beam from multibeam surveys in order to match the single-beam coverage of the older cruise data. We thank many scientists who provided us with soundings from their very recent cruises; these cruises were largely guided by the satellite gravity observations and thus cover some of the more interesting and high relief features in the ocean basins. In particular cruise data from the Pacific Antarctic and Southeast Indian ridges filled some important gaps in ship coverage. Archive data were derived from four overlapping sources. The most important contributions were originally archived at Lamont Doherty Earth Observatory where, over the years, scientists inspected data, corrected blunders, and digitized analog soundings that were acquired prior to digital recordings. Two databases were derived from the original Lamont data. The *Wessel and Smith* (WS) data base was organized, cleaned, and placed in an easily assessable GMT format to investigate the quantity and quality of marine gravity (Wessel and Watts, 1988) and marine bathymetry (Smith, 1993). During his assessment of the data, Smith rescued bad sounding lines that contained obvious but fixable problems such as incorrect units or conversion factors and identified cruises where soundings were hopelessly bad. The *Brown Book* (BB) data base was also converted to a GMT format and contains almost all of the cruises found in the WS database. It is largely optimized for investigations of marine magnetic anomalies but it has also been augmented with many unique or recent cruises to the Southern Ocean. Steven Cande (personal communication, 1995) provided us with complete access to these BB data for our project. The most complete database has been assembled by the *National Geophysical Data Center* (NGDC) in their Geodas-3 CD [1995]. The NGDC database contains many cruises not found in the Lamont-derived databases; these were contributed by other institutions throughout the world. While the NGDC data are more complete, there has been no attempt to clean and rescue bad sounding lines and therefore the original Lamont-derived data provide a more reliable contribution. Finally, the *Scripps Institution of Oceanography* (SIO) database was included to ensure that no important cruises or very recent cruises were omitted. The four data bases were inserted into a common GMT-Plus data format at the SIO Geological Data Center using the data base methods developed at Lamont (Wessel and Watts, 1988; Smith, 1993).

The compilation of the satellite-derived gravity anomaly grid is described in a recent publication (Sandwell and Smith, 1997). Along track sea-surface slopes from 4.5 years of Geosat and 2 years of ERS-1 and were used to construct a global marine gravity grid with cell dimensions of 2 min in longitude and cos(latitude) times 2 min in latitude so that cells are equidimensional but vary in size with latitude. The higher accuracy profiles come from averages of 16 ERS-1 repeat cycles along its 35-day repeat track and up to 66 Geosat repeat cycles along its 17-day repeat track. The higher spatial density profiles come from the 1.5-year Geosat Geodetic Mission and the 1-year ERS-1 Geodetic Mission (Figure 3 [see color insert for c and d]). Away from coastlines, the accuracy of the resulting gravity grid is 3–7 mGal depending on factors such as local sea state and proximity to areas of high mesoscale ocean variability (Marks, 1996; Sandwell and Smith, 1997). As discussed above, the accuracy and resolution of the bathymetric prediction are highly dependent on the accuracy of the gravity anomaly grid.

Our method of forming a grid of seafloor topography from these data follows the approach of Smith and Sandwell (1994; 1997).

(1) Grid available bathymetric soundings on the 2-min Mercator grid between latitudes of $\pm 72°$. Coastline points from GMT (Wessel and Smith, 1996) provide the zero-depth estimates. A finite-difference, minimum-curvature routine is used to interpolate the global grid (Smith and Wessel, 1990).

(2) Separate the bathymetry grid into *low-pass* and *high-pass* components using a Gaussian filter (0.5 gain at 160 km). Filtering and downward continuation are performed with a multiple strip, 2-D FFT that spans 0–360° longitude to avoid Greenwich edge effects.

(3) Form *high-pass filtered gravity* using the same Gaussian filter.

(4) Downward continue the *high-pass filtered gravity* to the *low-pass filtered bathymetry* assuming Laplace's equation is appropriate. A depth-dependent Wiener filter is used to stabilize the downward continuation.

(5) Accumulate *high-pass filtered soundings* and corresponding *high-pass filtered/downward-continued gravity* into small (160 km) overlapping areas and perform a robust regression analysis. In sediment-free areas, where the relief is much less than the mean ocean depth, the topography/gravity transfer function should be flat and equal to $1/2\pi G\Delta\rho$ so in the space domain, a linear regression is appropriate. This works well where sediment cover is thin, such as young seafloor. Where regression shows a poor correlation between topography and gravity, we assume the

seafloor is flat and set the *topography/gravity ratio* to zero. Finally there are intermediate cases where topographic depressions will be sediment filled while the highs protrude above the sediments so the topography/gravity relationship is non-linear. It is these partially sedimented areas that make the bathymetric problem difficult; continental margins and shelves pose similar problems. We believe that the nonlinear effects caused by sediment infill are more serious than the errors caused by neglecting the nonlinear terms in the gravity expansion Eq. (7).

(6) Regional *topography/gravity ratio* estimates are gridded and multiplied by the *high-pass filtered/downward-continued gravity* to form *high-pass filtered predicted bathymetry*.

(7) The *total predicted bathymetry* is equal to the sum of the *high-pass filtered predicted bathymetry* and the *low-pass filtered bathymetry*.

(8) Finally, the pixels constrained by ship soundings or coastline data are reset to the measured values and the finite-difference, minimum curvature routine is used to perturb the predicted values toward the measured values (i.e., *polishing*). This final step (8) dramatically increases the accuracy and resolution of the bathymetric grid in well surveyed areas so it agrees with the best hand-contoured bathymetric charts. Grid cells where depth is constrained buy a sounding or coastline point are set to the nearest odd integer while predicted values are set to the nearest even integer; this encoding can be used to distinguish between measured and estimated depths. The grid is available by anonymous ftp (topex.ucsd.edu); high-resolution images can be copied from our web site (http://topex.ucsd.edu/mar_topo.html).

During the construction of the topography grid, additional bad cruise data became apparent as sharp tears in the imaged grid. These cruises were identified using two methods. First, after step (7) in the prediction sequence, the mean deviation and median absolute deviation between the measured and predicted depths were computed on a cruise-by-cruise basis. Higher than normal differences indicated a problem with the cruise soundings, perhaps associated with a digitizing or scale-factor error. The second approach was to identify the bad cruise sounding data visually. Bad or suspect cruises were eliminated from the "good list" in the ship database and the entire prediction was redone. Currently we are at revision 6.2 of this procedure and we expect to update the topography grid as more ship data become available or when the suspect cruise data can be repaired. Thus the grid is a living document that can improve with time.

6.1. Results, Verification, and Hypsometry

The results are shown in Figure 4 as a color image (see color insert) illuminated from the northwest. All of the familiar features of the ocean basins are apparent. The land to-

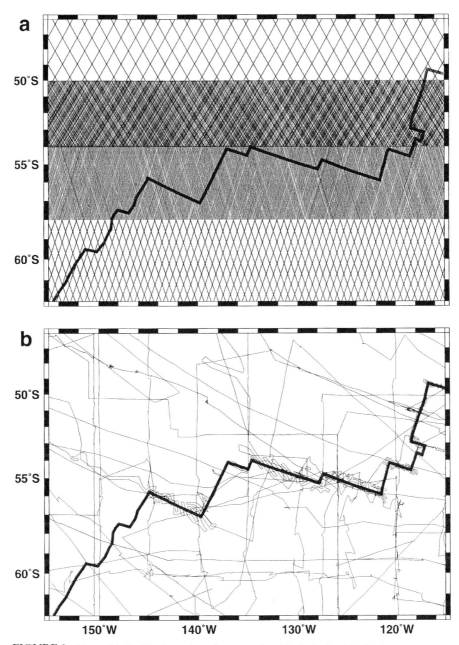

FIGURE 3   Maps of the Pacific-Antarctic seafloor spreading ridge in the South Pacific Ocean. (a) Tracks of stacked Geosat/ERM (17-day repeat cycle), Geosat/GM, ERS-1 Geodetic Phase (168-day repeat cycle) and stacked ERS-1 (35-day repeat). (b) Ship tracks in area of the Eltanin and Udintsev transform faults. Track density is sparse except along the Pacific-Antarctic plate boundary. (See color plates for Chapter 12, Figure 3c and d.)

pography from GTOPO30 is also included to provide a complete view of our planet. In the oceans, the continental margins are displayed as an orange-brown color. The seafloor spreading ridges stand out as broad highs (yellow to green to light blue) with an axial valley along the mid-Atlantic ridge and an axial high along the East Pacific Rise and the Pacific-Antarctic Rise. Fracture zones reflect the opening of the Atlantic basin while in the Pacific they record a more complex history of fast spreading and major reorganizations

of the plates. Numerous seamounts, sometimes occurring in chains, record a complex history of off-axis volcanism. The deepest ocean basins (blue to purple) are also the oldest areas and their depth reflects cooling, thermal contraction, and isostatic subsidence of the oceanic lithosphere (Parsons and Sclater, 1977).

We have assessed the accuracy of the prediction through a comparison with soundings from a recent cruise to the Foundation Seamount Chain in the South Pacific (Figure 5). This

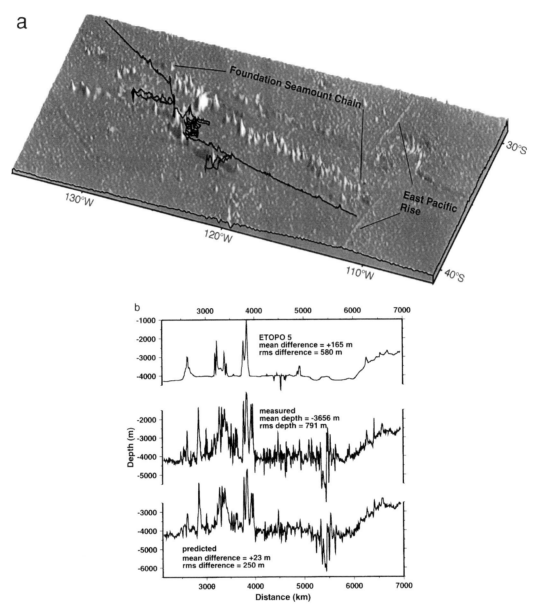

FIGURE 5   (a) 3-D perspective plot of the Foundation seamount chain and ship track used as a blind test of our bathymetric prediction V6.2. (b) Comparison of measured depth along ship track (middle) with ETOPO-5 (upper) and our predicted depth (lower). These ship data were acquired after the V6.2 of the bathymetric grid was completed.

poorly charted area contains a 1600-km long volcanic chain as well as topography associated with microplate tectonics (Mammerickx, 1992). Based on depth predictions from our earlier study (Smith and Sandwell, 1994), an initial mapping and sampling expedition was carried out in 1995 aboard the *R/V Sonne* (Devey *et al.*, 1997) where they charted 44 volcanoes with height ranging from 1500 m to 4000 m; eleven of the uncharted volcanoes come to within 500 m of the ocean surface. These *Sonne-100* sounding data were included in our global seafloor topography map and provide good defini-

tion of the summits of the seamounts (Figure 5a). In January and February of 1997, one of us (Sandwell) participated in a second expedition to the Foundation Seamounts area aboard *R/V L'Atalante* (Maia *et al.*, 1999). The cruise track covers very high relief topography areas that were not surveyed during the *Sonne-100* cruise and thus offers an excellent test of the accuracy of the predicted seafloor depth. The results are shown in Figure 5b where the center beam of the Simrad 12D multibeam echo sounder is plotted versus distance from Tahiti (center profile). A number of large seamounts

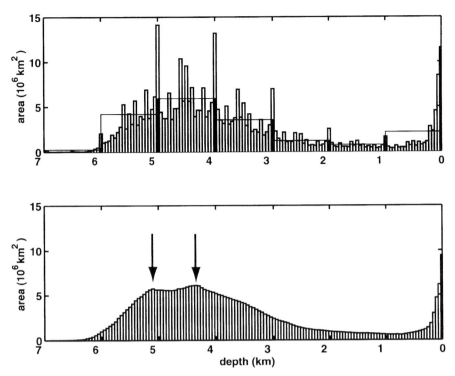

FIGURE 6 Hypsometric curves of the ocean basins based on pre-1977 ship soundings (upper wide bins shading) and ETOPO-5 (upper narrow bins) and our new bathymetric prediction V6.2 (lower). Note that the new solution has two peaks in the hypsometry at 5100 and 4200 m depth.

and ridges were surveyed between distances of 2500 and 4000 km while a 6500-m deep trench was surveyed at a distance of 5800 km. These measurements are compared with depths sampled from the ETOPO-5 grid (upper profile) and our version 6.2 predicted depths (lower profile). The depths from ETOPO-5 show a poor correlation with the measured depths and have an rms misfit of 590 m in an area where the rms signal is large 747 m. The poor fit is due to a lack of ship soundings in the area. The prediction offers a much better match to the observed depths (259 m rms) because the high-resolution gravity field information from the Geosat and ERS-1 satellite altimeters provides most of the depth information. In other well charted areas of the northern oceans, the grid is more tightly constrained by ship soundings. On average, 12% of the grid cells are constrained by ship measurements while the remaining 88% are estimated.

One of the basic measurements of our planet, the hypsometry of the ocean floor (area vs. depth), was poorly constrained prior to this new solution (Figure 6). The earliest curves (e.g., Menard and Smith, 1966), calculated in 1 km intervals, show that most seafloor has a depth of 4–5 km. A more recent solution, ETOPO-5, cannot yield a more detailed curve because of biases toward multiples of 100, 200, and 500 m, the contours that were digitized to produce ETOPO-5. Our solution yields a smooth curve at 50-m intervals. Viewed in 1-km intervals, our solution has more area in the 3–4 km range and less in the 5–6 km range than was seen

previously, reflecting the increased number of seamounts mapped by satellite altimetry. More importantly, the new hypsometry shows two peaks instead of just one. The deeper peak at 5100-m depth is a consequence of the flattening of the depth versus age relation (Parsons and Sclater, 1977; Parsons, 1982). The shallower peak at 4200 m can be understood in terms of a global increase seafloor spreading during the 20–33 Ma period. Seafloor depth increases systematically as it ages (Mueller *et al.*, 1997), cools, and becomes denser.

6.2. Effects on Ocean Currents

Seafloor topography has a significant and perhaps dominant role in ocean circulation, especially at high latitudes. Vertical and horizontal ocean circulation can redistribute water masses, bringing them into contact with the atmosphere where they can buffer or drive regional and global climate change. Much of this water mass formation occurs in the Southern Ocean and especially along the frontal zone of the Antarctic Circumpolar Current (ACC) through vertical and horizontal turbulent mixing processes associated with mesoscale eddy activity. The path of the wind-driven ACC has long been known to be steered by deep seafloor topography (e.g., Gordon and Baker, 1986). More recently, altimetric investigations have shown that eddy activity in the ACC is also modulated by bottom topography (Sandwell and Zhang,

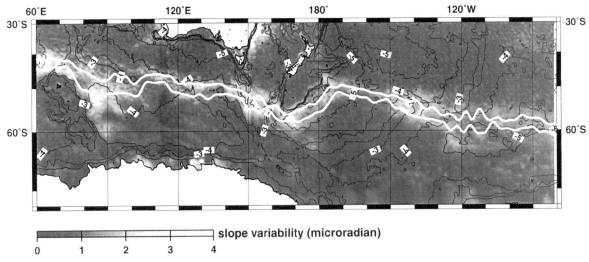

FIGURE 7   rms variability of sea surface slope derived from 116 repeat cycles of TOPEX altimetry (10-day repeat), 16 cycles of ERS-1 altimetry (35-day repeat), and 62 cycles of Geosat altimetry (17-day repeat). Black contours (1-km interval) are based on predicted depth (Smith and Sandwell, 1997). White curves mark the sub-Antarctic front and polar front estimated from Geosat altimetry using a meandering jet model (Gille, 1994). Slope variability is related to both the meandering of the fronts and the variations in ocean depth. The combination of altimeter profiles from T/P, ERS-1, and ERS-2 may provide complete spatial and temporal resolution of high-latitude eddies.

1989; Chelton *et al.*, 1990; Morrow *et al.*, 1992; Gille, 1994); sea-surface variability is higher above the deeper basins (> 3 km) and much lower in the shallower areas (Figure 7). Thus both the path of ACC and the transient eddies associated with it are influenced by ocean depth.

Both observational and numerical studies of the influence of seafloor topography on ocean currents have been severely limited by the lack of accurate bathymetry; this is especially true in the Southern Ocean where areas as large as 2×10^5 km^2 are unsurveyed. A striking example are three major ridges which lie directly in the paths of the ACC between 120°W and 160°W (Figure 3). The shallowest of the three features, first surveyed by a French Group (Geli *et al.*, 1997) in December of 1995, is a 400-km long ridge having a minimum depth of only 135 m. The other two topographic barriers are the 400- and 800-km long transverse ridges of the Udintsev and Eltanin Fracture Zones, respectively; both have a typical crest depth of less than 800 m while the surrounding seafloor is typically 3000-m deep and up to 6000-m deep in the fracture zone valleys.

In addition to the well known interaction of the mean ocean circulation with depth and the less well understood inverse correlation of mesoscale variability with depth, there have been recent suggestions that deep tidal mixing (Polzin *et al.*, 1997) and dissipation of ocean variability (Yale *et al.*, 1998) are enhanced over more rugged seafloor. Ocean floor roughness on scales less than 100 km is comprised of ubiquitous abyssal hill topography and less frequent, but higher amplitude, fracture zone topography. Bathymetric and altimetric studies (e.g., Small, 1994; Smith, 1998) show a pronounced inverse correlation of seafloor rough-

ness with increasing seafloor spreading rate. Seafloor generated at spreading rates lower than ~40 mm/a is characteristically rugged having rms amplitude of 100–200 m while seafloor generated at spreading rates greater than ~80 mm/a is always quite smooth having an rms amplitude of only 20–50 m. Preliminary results (Yale *et al.*, 1998) indicate that eddy kinetic energy (EKE) is greatest in the deeper ocean areas and over smooth seafloor. This anti-correlation between roughness and variability is strongest at higher latitudes suggesting a communication of the surface currents with the deep ocean floor. Rough bathymetry may transfer energy from the 100–300 length scales resolved by altimetry [Chapter 2] to smaller scales resulting in an apparent loss of EKE. Since numerical ocean models do not account for spatial variations in bottom friction and moreover, since they incorporate ad hoc dissipation mechanisms, improvements in seafloor depth and roughness may lead to a better understanding of deep ocean mixing. The link between seafloor roughness and spreading rate provides an interesting possibility that vertical mixing of paleo-oceans depended on the average age of the ocean floor and thus the waxing and waning of the mantle convection patterns.

7. PROSPECTS FOR THE FUTURE

While dense satellite altimeter measurements of the ocean surface have provided a fresh look at the ocean floor in the 20–100 km wavelength band, there are still many important problems in marine geology, physical oceanography, and marine biology where the current bathymetric model is inadequate. The question is, how can we improve our bathymetric models? Do we need to survey all of the oceans or are there more important target areas that could be used as calibration points to address global problems? Would another non-repeat orbit altimeter mission provide a significant improvement?

Of course it is always possible to gather more ship soundings and the new multibeam systems can map swaths of seafloor that are 3–5 times the mean depth at 100 m horizontal resolution and 10 m vertical accuracy. Nevertheless, even with these new systems it will take more than 100 ship-years to completely map the deep ocean floor and the likelihood of obtaining funding for a complete mapping is extremely low. Thanks to the reconnaissance information provided by the ERS and Geosat altimeters during their geodetic missions, is it easier to justify cruises to remote areas and ship time can be spent on the most important targets. The process of complete seafloor mapping will probably not be accomplished in our lifetimes.

An important avenue for improving the bathymetric models, especially in the coastal areas where altimeter-derived bathymetry is unreliable, is to obtain dense surveys of exclusive economic zones that have been compiled by many countries. The difficult task is to assemble a list of available data and convert them to a common format. Similarly, the U.S. Navy and other navies hold a wealth of data in classified archives that have the potential for revolutionizing bathymetric models (Medea Report, 1995). These avenues are being explored by a number of groups, in particular, the engineers who model the potential for tsunami inundation of coastal areas.

Since bathymetric estimation in areas of sparse ship coverage is critically-dependent on the accuracy and resolution of the satellite-derived gravity an important question is: Can gravity field resolution and accuracy can be improved using more measurements or better processing? We feel there are two avenues for improvement. First, in the case of ERS data, better processing of the raw waveform data may lead to significant gains in accuracy but the improvement will be less than a factor of 2 (Maus *et al.*, 1998; Laxon and McAdoo, 1994). Geosat and TOPEX altimeter data do not suffer from the noise problems of ERS onboard tracker and so we don't expect that retracking of the ERS data will provide a major improvement to the global gravity models. Second, in this chapter we discussed how ocean surface waves are the primary factor which limit the accuracy and resolution of the satellite-derived gravity. Since this is a random noise process, a factor of 2 improvement will require four times more data. ERS and Geosat data span 2.5 years in the geodetic mode, so another 5 years of altimetry in a nonrepeat mission would provide dramatic improvement.

8. APPENDIX: INTERACTION OF THE RADAR PULSE WITH THE ROUGH OCEAN SURFACE

8.1. Beam-Limited Footprint

The main advantages of operating an altimeter in the microwave part of the spectrum ($\sim$13 GHz) are that the atmosphere is very transparent and there is little stray radiation coming from the earth. The main limitation is that the illumination pattern on the surface of the ocean is very broad for reasonable sized antennas. Consider an antenna of diameter D_a, at an mean altitude H operating at a wavelength of λ.

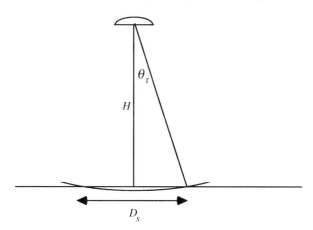

The angular resolution θ_r of a circular aperture having radius D_a is given by $\sin \theta_r = 1.22\lambda/D_a$. Therefore the diameter of the illumination pattern on the ocean surface is

$$D_s = 2H \sin \theta_r = 2.44 H \frac{\lambda}{D_a}. \qquad (16)$$

This illumination diameter is called the *beam width* of the radar. It is typically quite large ($D_s = 43$ km when $D_a = 1$ m, $\lambda = 22$ mm, $H = 800$ km altitude). Using this configuration, it will be impossible to achieve the 16-km wavelength horizontal resolution required for gravity field recovery and bathymetric estimation. However, one benefit of this wide illumination pattern is that small ($\sim$1 degree) pointing errors away from *nadir* have an insignificant effect on the range measurement.

To achieve the 20-mm range precision described in the previous chapters of the book, for example, one must measure the travel time of the radar echo to an accuracy of $\Delta t = 2\Delta h/c = 1.3 \times 10^{-10}$ sec. This can be translated into the bandwidth of the radiation needed to form a sharp pulse $\Delta \nu = 1/\Delta t$. In this case an 8 GHz bandwidth is needed. Note that the carrier frequency of our radar altimeter is 13 GHz so the pulse must span most of the electromagnetic spectrum. Obviously one can't use the entire EM spectrum; in practice one is restricted to a bandwidth of only 0.3 GHz. However, it turns out that ocean waves effectively limit the accuracy of the travel time measurement so a high bandwidth is unnecessary; the desired range resolution can only be achieved by averaging thousands of pulses with the hope that they are randomly distributed around the correct mean value.

8.2. Pulse-limited Footprint

Assume for the moment that the ocean surface is perfectly flat (actually ellipsoidal) but has point scatters to reflect the energy back to the antenna. The radar forms a sharp pulse having length of about 3 nsec corresponding to the 0.3 GHz bandwidth. In practice, to reduce the peak output requirement of the transmitter, the radar emits a long-duration, frequency-modulated chirp. The chirped radar signal reflects from the ocean surface and returns to the antenna where it is convolved with a *matched filter* to regenerate the desired pulse. This is a common signal processing approach used in all radar systems. After the matched filter one can treat the measurement as a pulse. The diagram below illustrates how the pulse interacts with a flat sea surface.

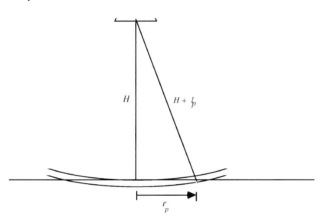

where

H —satellite altitude 800 km

t_p —pulse length 3×10^{-9} sec

c —speed of light 3×10^8 m sec^{-1}

$l_p = ct_p$ —length of pulse 1 m

r_p —radius of pulse on ocean surface

The radius of the leading edge of the pulse is derived as follows.

$$H^2 + r_p^2 = \left(H + l_p\right)^2 = H^2 + l_p^2 + 2Hl_p. \quad (17)$$

The H^2 cancels and we can assume l_p^2 is very small compared with the other terms so the pulse radius is

$$r_p = \sqrt{2Hl_p} = \sqrt{2Hct_p}. \quad (18)$$

For a 3 nsec pulse length the pulse radius is 1.2 km so the diameter or *footprint* of the radar is 2.4 km. Since this footprint

is much less than the beam width, the power that is returned to the radar will be a ramp function.

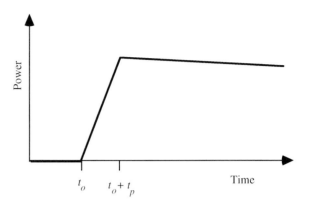

The power begins to ramp-up at time $t_o = 2H/c$ and the ramp extends for the duration of the pulse. At times greater than $t_o + t_p$, the diameter of the radar pulse continues to grow and energy continues to return to the radar. The amplitude of this energy decreases gradually according to the illumination pattern of the radar on the ocean surface. Because of the finite pulse width, the bottom and top of the ramp will be rounded.

8.3. Significant Wave Height

Of course the actual ocean surface has roughness due to ocean waves and swell. This ramp-like return power will be convolved with the height distribution of the waves within the footprint to further smooth the return pulse and make the estimate of the arrival time of the leading edge of the pulse less certain (Walsh *et al.*, 1978). We can investigate the effects of wave height on both return pulse length and footprint diameter using a Gaussian model for the height distribution of ocean waves. This model provides and excellent match to observed wave height (Stewart, 1985).

$$G(h) = \frac{1}{\sqrt{2\pi}\,\Delta h} \exp\left(-\frac{h^2}{2\Delta h^2}\right). \quad (19)$$

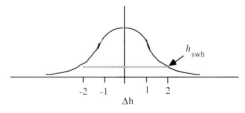

An observer on a ship can accurately report the peak-to-trough amplitude of the highest waves; this is called the significant wave height and it is $h_{swh} = 4\Delta h$. Because the altimeter footprint is broad compared with the wavelength of the swell, this waveheight distribution is convolved with

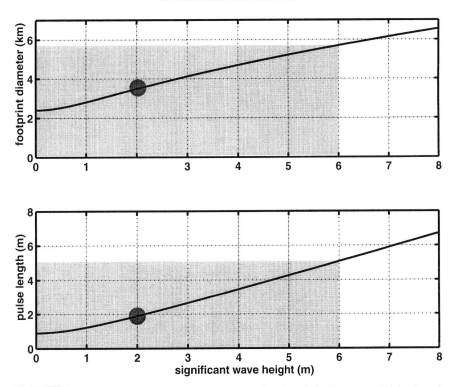

FIGURE 8 (lower) return pulse length in meters as a function of significant wave height. (upper) diameter of the radar pulse on the ocean surface as a function of significant wave height. A typical significant wave height of 2 m is marked by a large gray dot.

the radar pulse. The amplitude of the return pulse A will be a Gaussian function of two-way travel time difference $\tau = 2H/c$.

$$A(\tau) = \exp\left(\frac{-2c^2\tau^2}{h_{swh}^2}\right). \qquad (20)$$

The broadening of the radar return pulse t_w is measured as the full width of the pulse where the power is 1/2.

$$\frac{1}{2} = \exp\left(\frac{-c^2 t_w^2}{h_{swh}^2}\right) \quad \text{so} \quad t_w^2 = \frac{h_{swh}^2}{c^2}\ln 2. \qquad (21)$$

Since we were unable to form a very sharp radar pulse because the radar bandwidth is limited to 0.3 GHz, the total width of the return pulse will be established by convolving the outgoing pulse with the Gaussian wave model. If the outgoing pulse can also be modeled by a Gaussian function having a pulse width of t_p, then the total width of the return pulse is given by

$$t^2 = t_p^2 + \frac{h_{swh}^2}{c^2}\ln 2. \qquad (22)$$

This provides and expression for the pulse width as a function of significant wave height (SWH). Similarly the diameter of the pulse as a function of significant wave height is

$$d = 2\sqrt{(2cHt)}. \qquad (23)$$

Both functions are shown in Figure 8 for SWH ranging from 1 to 8 m. It is clear that the quality of the altimeter measurement will decrease with increasing SWH. In practice we have found that Geosat, ERS, and TOPEX data are unreliable when SWH exceeds about 6 m.

Significant wave height is typically 2 m so the radar footprint is typically 3.5 km and the pulse-length increases from 1–2 m. Now we see that our original plan of having a very narrow pulse of 60 pulses per second to resolve 20 mm height variations was doomed because the ocean surface is usually rough; a 3 nsec (1 m) pulse is all that could be resolved anyway. In addition is clear that the wavelength resolution of the sea-surface height recovery will typically be worse than 7 km. Given this poor inherent range accuracy of 1–2 m, how do we achieve the 20 mm resolution needed for our applications? The way to improve the accuracy by a factor of 10^2 is to average 10^4 measurements and hope the noise is completely random.

The speed of light provides an interesting limitation for space-borne ranging systems. At a typical orbital altitude of 800 km it takes 5.2 msec for the pulse to complete its round trip route. One can have several pulses en-route but because we actually send a long chirp rather than a pulse, the pulse repetition frequency is limited to about 1000 pulses per second; during 1 sec, the altimeter moves about 7000 m along its track. Thus in each second there are 1000 pulses available for averaging; this will reduce the noise from 1.5 m to 0.05 m. Further averaging can be done for many of the oceanographic applications where the horizontal length scale of the feature is > 50 km. Of course one should be careful to remove all of the sharp geoid signals using the full resolution data (Chapter 10, Figure 3) and then smooth the residual data along the profile to achieve higher accuracy. Typically the data are delivered as a geophysical data record (GDR) where 1000 pulses are averaged and then decimated to 1 Hz. This boxcar filter followed by decimation folds the shorter wavelength noise back to longer wavelengths and further decreases the accuracy and horizontal resolution of the data. Data at 10 Hz sampling are also provided with the GDR and we recommend that the user design a convolution filter to achieve the desired spectral output. Our gravity fields and estimated bathymetry are produced with such filters.

8.4. Modeling the Return Waveform

There are several engineering issues related to picking the travel time of the return pulse (Hayne *et al.*, 1994; Rodriguez and Martin, 1994). First, after the return echo is passed through a matched filter to form the pulse, the pulse power is recorded at 64 times in a window that is about 30 nsec (~10 m) long. An adaptive tracker is used to keep the power ramp in the center of the window. The ocean surface is typically smooth at length scales greater than the footprint so keeping the pulse in the window is not a problem. However, over land or ice, it is not usually possible to keep the pulse within the window because 10 m variations in topography over several kilometers of horizontal distance are quite common. The Geosat altimeter loses lock over land and must reacquire the echo soon after moving back over the ocean. The ERS-1/2 altimeters widen the gate spacing over land and ice so they can measure land topography as well as ocean topography.

After recording the waveform of the return pulse, 100 echoes are averaged and an analytic function is fit to each waveform (Maus *et al.*, 1998). The function has three parameters:

(1) the position of the steepest part of the ramp provides the range estimate;
(2) the width of the ramp provides and estimate of SWH;
(3) and the height of the ramp (called sigma-naught σ_o) provides an estimate of surface roughness at the 20–30 mm length scale.

This latter measurement can be related to surface wind speed since wind will roughen the ocean surface. Precise calibration is performed for each of the three measurements. Absolute range calibration is, for example, performed in the open ocean using an oil platform having a GPS receiver and accurate tide gauge (Christensen *et al.*, 1994). Both SWH and wind speed are calibrated using open-ocean shipboard measurements.

References

Banks, R. J., Parker, R. L., and Huestis, S. P. (1977). Isostatic compensation on a continental scale: Local versus regional mechanisms. *Geophys. J. R. Astr. Soc.*, **51**, 431–452.

Baudry, N., and Calmant, S. (1991). 3-D Modeling of seamount topography from satellite altimetry. *Geophys. Res. Lett.*, **18**, 1143–1146.

Bettadpur, S. V., and Eanes, R. J. (1994). Geographical representation of radial orbit perturbations due to ocean tides: Implications for satellite altimetry. *J. Geophys. Res.*, **99**, 24,883–24,898.

Calmant, S. (1994). Seamount topography of least-squares inversion of altimetric geoid heights and shipborne profiles of bathymetry and/or gravity anomalies. *Geophys. J. Int.*, **119**, 428–452.

Canadian Hydrographic Service (1981). *In* "General bathymetric chart of the oceans (GEBCO)." Hydrographic Chart Distribution Office, Ottawa, Canada.

Cazenave, A., Schaeffer, P., Berge, M., and Brossier, C. (1996). High-resolution mean sea-surface computed with altimeter data of ERS-1 (Geodetic Mission) and TOPEX-POSEIDON. *Geophys. J. Int.*, **125**, 696–704.

Chelton, D. B., Schlax, M. G., Witter, D. L., and Richman, J. G. (1990). Geosat altimeter observations of the surface circulation of the Southern Ocean. *J. Geophys. Res.*, **95**, 17,877–17,903.

Christensen, E. J., *et al.* (1994). Calibration of TOPEX/POSEIDON at Platform Harvest. *J. Geophys. Res.*, **99**, 24,465–24,486.

Devey, C., *et al.* (1997). The Foundation Seamount Chain: A first survey and sampling. *Mar. Geol.*, **137**, 191–200.

Dixon, T. H., Naraghi, M., McNutt, M. K., and Smith, S. M. (1983). Bathymetric prediction from Seasat altimeter data. *J. Geophys. Res.*, **88**, 1563–1571.

Geli. L., *et al.* (1997). Evolution of the Pacific-Antarctic Ridge south of the Udintsev Fracture Zone. *Science*, **278**, 1281–1284.

Gille, S. T. (1994). Mean sea surface height of the Antarctic circumpolar current from Geosat data: Method and application. *J. Geophys. Res.*, **99**, 18,255–18,273.

Gordon, A. L., and Baker, T. N. (1986). *In* "Southern ocean atlas." Published for the International Decade of Ocean Exploration, National Science Foundation.

Haxby, W. F., Karner, G. D., LaBrecque, J. L., and Weissel, J. K. (1983). Digital images of combined oceanic and continental data sets and their use in tectonic studies. *EOS Trans. Am. Geophys. Un.*, **64**, 995–1004.

Hayne, G. S., Hancock, D. W., Purdy, C. L., and Callahan, P. S. (1994). The corrections for significant wave height and altitude effects in the TOPEX radar altimeter. *J. Geophys. Res.*, **99**, 24,941–24,955.

Heiskanen, W. A., and Moritz, H. (1967). *In* "Physical Geodesy." W. H. Freeman and Co., San Francisco.

Hwang, C., Kao, E.-C., and Parsons, B. (1998). Global derivation of marine gravity anomalies from Seasat, Geosat, ERS-1 and TOPEX/POSEIDON altimeter data. *Geophys. J. Int.*, **134**, 449–459.

Hwang, C., and Parsons, B. (1996). An optimal procedure for deriving marine gravity from multi-satellite altimetry. *J. Geophys. Int.*, **125**, 705–719.

Imel, D. A. (1994). Evaluation of the TOPEX/POSEIDON dual-frequency ionospher correction. *J. Geophys. Res.*, **99**, 24,895–24,906.

Jung, W. Y., and Vogt, P. R. (1992). Predicting bathymetry from Geosat-ERM and shipborne profiles in the South Atlantic ocean. *Tectonophysics*, **210**, 235–253.

Laxon, S., and McAdoo, D. (1994). Arctic ocean gravity field derived from ERS-1 satellite altimetry. *Science*, **265**, 621–624.

Lemoine, F. G., *et al.* (1998). The development of the joint NASA CSFC and the national Imagery and Mapping Agency (NIMA) geopotential model EGM96. Goddard Space Flight Center, NASA, NASA/TP-1998–206861.

Liu, C.-S., Sandwell, D. T., and Curray, J. R. (1982). The negative gravity field over the 85° Ridge. *J. Geophys. Res.*, **87**, 7673–7686.

Maia, M., *et al.* (1999). The Foundation Seamounts: A ridge-hotspot interaction in the South Pacific. Preprint.

Mammerickx, J. (1992). The Foundation Seamounts: tectonic setting of a newly discovered seamount chain in the South Pacific. *Earth Planet. Sci. Lett.*, **113**, 293–306.

Marks, K. M. (1996). Resolution of the Scripps/NOAA marine gravity field from satellite altimetry. *Geophys. Res. Lett.*, **23**, 2069–2072.

Maus, S., Green, C. M., and Fairhead, J. D. (1998). Improved ocean-geoid resolution from retracked ERS-1 satellite altimeter waveforms. *Geophys. J. Int.*, **134**, 243–253.

McKenzie, D. P. (1976). Some remarks on heat flow and gravity anomalies. *J. Geophys. Res.*, **72**, 6261–6273.

McKenzie, D. P., and Bowin, C. (1976). The relationship between bathymetry and gravity in the Atlantic Ocean. *J. Geophys. Res.*, **81**, 1903–1915.

McNutt, M. (1979). Compensation of oceanic topography: An application of the response function technique to the Surveyor area. *J. Geophys. Res.*, **84**, 7589–7598.

Medea, Scientific Utility of Naval Environmental Data. MEDEA Office, 1995.

Menard, H. W., and Smith, S. M. (1966). Hypsometry of Ocean Basin Provinces. *J. Geophys. Res.*, **71**, 4305–4325.

Morrow, R., Church, J., Coleman, R., Chelton, D., and White, N. (1992). Eddy momentum flux and its contribution to the Southern Ocean momentum balance. *Nature*, **357**, 482–484.

Mueller, R. D., Roest, W. R., Royer, J.-Y., Gahagan, L. M., and Sclater, J. G. (1997). Digital isochrons of the world's ocean floor. *J. Geophys. Res.*, **102**, 3211–3214.

Nettleton, L. L. (1939). Determination of Density for Reduction of Gravity Observations. *Geophysics*, **4**, 176–183.

Noreus, J. P. (1995). Improved resolution of Geosat altimetry using dense sampling and polynomial adjusted averaging. *Int. J. Remote Sensing*, **16**, 2843–2862.

Parker, R. L. (1973). The rapid calculation of potential anomalies. *Geophys. J. R. Astr. Soc.*, **31**, 447–455.

Parker, R. L. (1994). *In* "Geophysical Inverse Theory." Princeton University Press, Princeton.

Parsons, B. (1982). Causes and consequences of the relation between area and age of the ocean floor. *J. Geophys. Res.*, **87**, 289–302.

Parsons, B., and Sclater, J. G. (1977). An analysis of the variation of the ocean floor bathymetry and heat flow with age. *J. Geophys. Res.*, **82**, 803–827.

Polzin, K. L., Toole, J. M., Ledwell, J. R., and Schmitt, R. W. (1997). Spatial variability of turbulent mixing in the abyssal ocean. *Science*, **276**, 93–96.

Ramillien, G., and Cazenave, A. (1997). Global bathymetry derived from altimeter data of the ERS-1 Geodetic Mission. *J. Geodynamics*, **23**, 129–149.

Rapp, R. H., and Yi, Y. (1997). Role of ocean variability and dynamic topography in the recovery of the mean sea surface and gravity anomalies from satellite altimeter data. *J. Geodesy*, **71**, 617–629.

Ribe, N. M. (1982). On the interpretation of frequency response functions for oceanic gravity and bathymetry. *Geophys. J. R. Astron. Soc.*, **70**, 273–294.

Rodriguez, E., and Martin, J. M. (1994). Assessment of the TOPEX altimeter performance using waveform retracking. *J. Geophys. Res.*, **99**, 24,957–24,969.

Sandwell, D. T. (1984a). A detailed view of the South Pacific from satellite altimetry. *J. Geophys. Res.*, **89**, 1089–1104.

Sandwell, D. T. (1991). Geophysical applications of satellite altimetry. *Rev. Geophys. Suppl.*, **29**, 132–137.

Sandwell, D. T. (1992). Antarctic marine gravity field from high-density satellite altimetry. *Geophys. J. Int.*, **109**, 437–448.

Sandwell, D. T., and Smith, W. H. F. (1997). Marine gravity anomaly from Geosat and ERS-1 satellite altimetry. *J. Geophys. Res.*, **102**, 10,039–10,054.

Sandwell, D. T., and Zhang, B. (1989). Global mesoscale variability from the Geosat exact repeat mission: Correlation with ocean depth. *J. Geophys. Res.*, **94**, 17,971–17,984.

Sichoix, L., and Bonneville, A. (1996). Prediction of bathymetry in French Polynesia constrained by shipboard data. *Geophys. Res. Lett.*, **23**, 2469–2472.

Small, C. (1994). A global analysis of mid-ocean ridge axial topography. *Geophys. J. Int.*, **116**, 64–84.

Smith, W. H. F. (1993). On the accuracy of digital bathymetry data. *J. Geophys. Res.*, **98**, 9591–9603.

Smith, W. H. F. (1998). Seafloor tectonic fabric from satellite altimetry. *Ann. Rev. Earth Planet. Sci.*, **26**, 697–738.

Smith, W. H. F., and Sandwell, D. T. (1994). Bathymetric prediction from dense satellite altimetry and sparse shipboard bathymetry. *J. Geophys. Res.*, **99**, 21,803–21,824.

Smith, W. H. F., and Sandwell, D. T. (1997). Global sea floor topography from satellite altimetry and ship depth soundings. *Science*, **277**, 1956–1961.

Smith, W. H. F., and Wessel, P. (1990). Gridding with continuous curvature splines in tension. *Geophysics*, **55**, 293–305.

Stewart, R. H. (1985). *In* "Methods of Satellite Oceanography." University of California Press, Berkeley.

Tapley, B. D., Chambers, D. P., Shum, C. K., Eanes, R. J., Ries, J. C., and Stewart, R. H. (1994). Accuracy assessment of large-scale dynamic ocean topography from TOPEX/POSEIDON altimetry. *J. Geophys. Res.*, **99**, 24,605–24,617.

Walsh, E. J., Uliana, E. A., and Yaplee, B. S. (1978). Ocean wave height measured by a high resolution pulse-limited radar altimeter. *Boundary-Layer Meterology*, **13**, 263–276.

Watts, A. B. (1978). An analysis of isostasy in the world's oceans: 1, Hawaiian-Emperor seamount chain. *J. Geophys. Res.*, **83**, 5989–6004.

Watts, A. B. (1979). On geoid heights derived from Geos-3 altimeter data and flexure of the lithosphere along the Hawaiian-Emperor seamount chain. *J. Geophys. Res.*, **38**, 119–141.

Wessel, P. (1992). Thermal stress and the bimodal distribution of elastic thickness estimates of the oceanic lithosphere, *J. Geophys. Res.*, **97**, 14177–14193.

Wessel, P., and Smith, W. H. F. (1991). Free software helps map and display data. *EOS Trans. AGU*, **72**, 445–446.

Wessel, P., and Smith, W. H. F. (1996). A global, self-consistent, hierarchical, high-resolution shoreline database. *J. Geophys. Res.*, **101**, 8741–8743.

Wessel, P., and Watts, A. B. (1988). On the accuracy of marine gravity measurements. *J. Geophys. Res.*, **93**, 393–413.

Yale, M. M. (1997). Modeling Upper mantle Rheology with Numerical Experiments and Mapping Marine Gravity with Satellite Altimetry. Ph.D. Thesis, University of California, San Diego.

Yale, M. M., Gille, S. T., and Sandwell, D. T. (1998). Ocean mixing-mesoscale EKE, bathymetry, and seafloor roughness seen by ERS-1/2 and Topex. *EOS Trans. AGU*, **79**, F213.

Yale, M. M., Sandwell, D. T., and Smith, W. H. F. (1995). Comparison of along-track resolution of stacked Geosat, ERS-1 and TOPEX satellite altimeters. *J. Geophys. Res.*, **100**, 15,117–15,127.

INDEX